PHYSICAL CONSTANTS

Quantity	Symbol	Value
Universal gravitational constant	G	6.673×10^{-11} m^3/(kg·s^2)
Speed of light in vacuum	c	2.998×10^8 m/s
Elementary charge	e	1.602×10^{-19} C
Planck's constant	h	6.626×10^{-34} J·s
		4.136×10^{-15} eV·s
	$\hbar = h/(2\pi)$	1.055×10^{-34} J·s
		6.582×10^{-16} eV·s
Universal gas constant	R	8.315 J/(mol·K)
Avogadro's number	N_A	6.022×10^{23} mol^{-1}
Boltzmann constant	k_B	1.381×10^{-23} J/K
		8.617×10^{-5} eV/K
Coulomb force constant	$k = 1/(4\pi\epsilon_0)$	8.988×10^9 N·m^2/C^2
Permittivity of free space (electric constant)	ϵ_0	8.854×10^{-12} C^2/(N·m^2)
Permeability of free space (magnetic constant)	μ_0	$4\pi \times 10^{-7}$ T·m/A
Electron mass	m_e	9.109×10^{-31} kg
		$0.000\ 548\ 580$ u
Electron rest energy	$m_e c^2$	0.5110 MeV
Proton mass	m_p	1.673×10^{-27} kg
		$1.007\ 276\ 5$ u
Proton rest energy	$m_p c^2$	938.272 MeV
Neutron mass	m_n	1.675×10^{-27} kg
		$1.008\ 664\ 9$ u
Neutron rest energy	$m_n c^2$	939.565 MeV
Compton wavelength	λ_C	2.426×10^{-12} m
Stefan-Boltzmann constant	σ	5.670×10^{-8} W/(m^2·K^4)
Rydberg constant	R	1.097×10^7 m^{-1}
Bohr radius	a_0	5.292×10^{-11} m

College PHYSICS

Volume One

Alan Giambattista
Cornell University

Betty McCarthy Richardson
Cornell University

Robert C. Richardson
Cornell University

McGraw Hill **Higher Education**

Boston   Burr Ridge, IL   Dubuque, IA   Madison, WI   New York   San Francisco   St. Louis
Bangkok   Bogotá   Caracas   Kuala Lumpur   Lisbon   London   Madrid   Mexico City
Milan   Montreal   New Delhi   Santiago   Seoul   Singapore   Sydney   Taipei   Toronto

COLLEGE PHYSICS, VOLUME ONE

Published by McGraw-Hill, a business unit of The McGraw-Hill Companies, Inc., 1221 Avenue of the Americas, New York, NY 10020. Copyright © 2004 by The McGraw-Hill Companies, Inc. All rights reserved. No part of this publication may be reproduced or distributed in any form or by any means, or stored in a database or retrieval system, without the prior written consent of The McGraw-Hill Companies, Inc., including, but not limited to, in any network or other electronic storage or transmission, or broadcast for distance learning.

Some ancillaries, including electronic and print components, may not be available to customers outside The United States.

This book is printed on acid-free paper.

1 2 3 4 5 6 7 8 9 0 VNH/VNH 0 9 8 7 6 5 4 3 2

ISBN 0–07–253724–8

Publisher: *Kent A. Peterson*
Sponsoring editor: *Daryl Bruflodt*
Developmental editor: *Mary E. Haas*
Marketing manager: *Debra B. Hash*
Lead project manager: *Jill R. Peter*
Senior production supervisor: *Laura Fuller*
Senior media project manager: *Stacy A. Patch*
Media technology producer: *Jeff Huettman*
Design manager: *Stuart Paterson*
Cover/interior designer: *Jamie E. O'Neal*
Cover image: *Stone*
Lead photo research coordinator: *Carrie K. Burger*
Photo research: *Karen Pugliano*
Supplement producer: *Brenda A. Ernzen*
Compositor: *Precision Graphics*
Typeface: *10/12 Times Roman*
Printer: *Von Hoffmann Corporation*

The credits section for this book begins on page 569 and is considered an extension of the copyright page.

Library of Congress has cataloged the main title as follows:

Giambattista, Alan.
   College physics / Alan Giambattista, Betty McCarthy Richardson, Robert C. Richardson. — 1st ed.
      p.   cm.
   Includes index.
   ISBN 0–07–052407–6 (set : acid-free paper) — ISBN 0–07–253724–8 (v. 1 : acid-free paper) — ISBN 0–07–253725–6 (v. 2 : acid-free paper)
   1. Physics. I. Richardson, Betty McCarthy. II. Richardson, Robert C. (Robert Coleman), 1937–. III. Title.

QC21.3 .G53   2004
530–dc21   2002038915
                                                     CIP

ABOUT THE AUTHORS

Alan Giambattista grew up in Nutley, New Jersey. In his junior year at Brigham Young University he decided to pursue a physics major, after having explored math, music, and psychology. He did his graduate studies at Cornell University and has taught introductory college physics in one form or another for the past 20 years. When not found at the computer keyboard working on *College Physics,* he can often be found at the keyboard of a harpsichord or piano. He is a member of the Cayuga Chamber Orchestra and has given performances of the Bach harpsichord concerti at several regional Bach festivals. He met his wife Marion in a singing group and presently they live in an 1824 former parsonage surrounded by a dairy farm. Besides music and taking care of the house, gardens, and fruit trees, they love to travel together.

Betty McCarthy Richardson was born and grew up in Marblehead, Massachusetts, and tried to avoid taking any science classes after eighth grade but managed to avoid only ninth grade science. After discovering that physics tells how things work, she decided to become a physicist. She attended Wellesley College and did graduate work at Duke University. While at Duke, Betty met and married fellow graduate student Bob Richardson and had two daughters, Jennifer and Pamela. Betty began teaching physics at Cornell in 1977. Twenty-five years later, she is still teaching the same course, Physics 101/102, an algebra-based course with all teaching done one-on-one in a Learning Center. From her own early experience of math and science avoidance, Betty has empathy with students who are apprehensive about learning physics. Betty's hobbies include collecting old children's books and vintage toys, reading, enjoying music, travel, and dining with royalty. Betty's highlight during the Nobel Prize festivities in 1996 was being escorted to dinner at the Stockholm Royal Palace on the arm of King Carl XVI Gustav of Sweden and sitting between the king and the prime minister. Currently she is spending spare time pushing wooden trains around wooden tracks with grandson Jasper (the 1-m child in Chapter 1) and caring for the family cats.

Robert C. Richardson was born in Washington, D.C., attended Virginia Polytechnic Institute, spent time in the United States Army, and then returned to graduate school in physics at Duke University where his thesis work involved NMR studies of solid helium-3. In the fall of 1966 Bob began work at Cornell University in the laboratory of David M. Lee. Their research goal was to observe the nuclear magnetic phase transition in solid ^{3}He that could be predicted from Richardson's thesis work with Professor Horst Meyer at Duke. In collaboration with graduate student Douglas D. Osheroff, they worked on cooling techniques and NMR instrumentation for studying low-temperature helium liquids and solids. In the fall of 1971, they made the accidental discovery that liquid ^{3}He undergoes a pairing transition similar to that of superconductors. The three were awarded the Nobel Prize for that work in 1996. Bob is currently the F. R. Newman Professor of Physics and the Vice Provost for Research. He has been active in teaching various introductory physics courses throughout his time at Cornell. In his spare time he enjoys gardening and photography.

For Marion
Alan

In memory of our daughter Pamela,
and for Jasper, Jennifer, and Jim Merlis
Bob and Betty

CONTENTS IN BRIEF

CONTENTS

Part ONE

Mechanics

Part
TWO

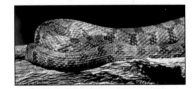

Thermal Physics

LIST OF SELECTED APPLICATIONS

Biology/Life Science

Forces supporting a water strider, Section 2.2
Forces on bird in air, Section 2.4
Function of tendons in the body, Section 2.7
Force exerted by deltoid muscle, Examples 4.3, 8.10
Traction applied to a leg, Example 4.5
Seagull dropping a clam, Section 4.5
Energy conversion in a jumping cricket, Section 6.6
Energy conversion in animal jumping, Sections 6.6, 6.7,
    Example 6.10
Flexor versus extensor muscles, Section 8.5
Forces on the human spine during heavy lifting, Section 8.5
Sphygmomanometer and blood pressure, Section 9.5
Floating and sinking of fish and animals, Example 9.8,
    Section 9.6
Speed of blood flow, Example 9.9
Air circulation in underground burrows, Section 9.8
Arterial blockage, Example 9.12
Surface tension of alveoli in the lungs, Section 9.11,
    Example 9.14
Tension and compression in bone, Section 10.2,
    Example 10.2
Stress on the spine during lifting, Example 10.4
Size limitations on organisms, Section 10.3
Comparison of walking speeds for various creatures,
    Example 10.12
Sound waves from a songbird, Example 12.2
Human ear, Section 12.5
Echolocation of bats and dolphins, Section 12.10
Medical applications of ultrasound, ultrasonic imaging,
    Section 12.10
Evolutionary advantages of warm-blooded versus cold-
    blooded animals, Section 13.7
Diffusion of oxygen into the bloodstream, Example 13.10
Using ice to protect buds from freezing, Section 14.5
Thermal radiation from the human body, Example 14.16
Electrolocation in fish, Section 16.4
Transmission of nerve impulses, Sections 17.2, 18.10,
    Example 17.11
Electrocardiogram (EKG), Section 17.2
Electroencephalogram (EEG), Section 17.2
Defibrillator, Example 17.12, Section 18.11
Magnetotactic bacteria, Section 19.1
Electromagnetic blood flow meter, Section 19.5
Magnetoencephalography, Section 20.3
Thermograms of the human body, Section 22.4
Phosphorescence in fish, coral, and butterflies, Section 22.4
X-rays in medicine and dentistry, CAT scans, Section 22.4,
    Example 27.4
Navigation of bees, Section 22.8

Endoscope, Section 23.4
Human eye, Section 24.3
Correcting myopia/hyperopia, Section 24.3,
    Examples 24.4, 24.5
Iridescent colors in butterfly wings, Section 25.4
Resolution of the human eye, Section 25.8
Images from electron microscopes, Section 28.3
Laser surgery, Section 28.9
Radiocarbon dating, Section 29.4
Biological effects of radioactivity, Section 29.5
Radioactive tracers in medical diagnosis, Section 29.5
Positron emission tomography (PET) scans, Section 29.5
Radiation therapy, Section 29.5
Gamma knife radiosurgery, Section 29.5

Chemistry

Motion of electron in Bohr model of hydrogen,
    Example 5.11
Collision between krypton atom and water molecule,
    Example 7.10
Change of reaction rate with temperature, Section 13.7,
    Example 13.8
Polarization of charge in water, Section 16.1
Current in electrolytes, Section 18.1
Spectroscopic analysis of elements, Sections 27.5, 27.6
Electron orbitals, electron configurations of atoms,
    Section 28.7
Periodic Table, Section 28.7

Geology/Earth Science

Forces on a submersible, Example 2.1
Angular speed of Earth, Example 5.1
Thunderclouds and lightning, Example 17.1, Section 17.6
Earth's magnetic field, Section 19.1, Examples 19.1, 19.2
Intensity of sunlight reaching the Earth, Example 22.6
Colors of the sky during the day and at sunset, Section 22.8
Rainbows, Section 23.3
Cosmic rays, Examples 26.2, 26.4
Radioactive dating of geologic formations, Section 29.4
Neutron activation analysis of geological objects,
    Section 29.5

Astronomy/Space Science

Spacecraft to Mars, Unit Conversions, Section 1.5
Detection of extrasolar planets, Section 2.3
Speed of Hubble Telescope orbiting the Earth, Example 5.8
Kepler's laws of planetary motion, Sections 5.4, 8.8
Orbit of geostationary satellite, Examples 5.9, 6.7

Apparent weightlessness of orbiting astronauts, Section 5.7
Work on an orbiting satellite, Section 6.2
Escape speed from Earth, Example 6.9
Center of mass of a binary star system, Example 7.8
Motion of an exploding model rocket, Example 7.9
Orbital speed of Earth, Example 8.15
Composition of planetary atmospheres, Section 13.6
Global warming and the greenhouse effect,
   Sections 14.7, 14.8
Temperature of the Sun, Example 14.15
Aurorae on Earth, Jupiter, and Saturn, Section 19.4
Cosmic microwave background radiation, Section 22.4
Light from a supernova, Example 22.2
Doppler shift for distant stars and galaxies, Section 22.9
Reflecting telescopes, Section 24.6
Hubble Space Telescope, Section 24.6
Radio telescopes, Section 24.6
Observing active galactic nuclei, Section 26.2
Aging of astronauts during space voyages, Example 26.1
Nuclear fusion in stars, Section 29.8
The Big Bang and the history of the universe, Section 30.3

Architecture

Cantilever building construction, Section 8.4
Strength of building materials, Section 10.3
Vibration of a bridge, Section 10.10
Expansion joints in bridges and buildings, Section 13.3
Heat transfer through window glass,
   Examples 14.11, 14.12
Thermogram of heat leaking from a house, Section 22.4

Technology/Machines

Mechanical advantage of a pulley, Section 2.7, 6.2,
   Example 2.14.
Mechanical advantage of an inclined plane, Section 4.3,
   Examples 4.6, 6.1
Catapults and projectile motion, Example 4.9
Safety features in a modern car, Section 7.3
Recoil of a cannon, Example 7.7
Atwood's machine, Example 8.2
Angular momentum of a gyroscope, Section 8.9
Hydraulic lift, Example 9.2
Mercury manometer, Example 9.5
Hydrometer, Section 9.6
Venturi meter, Example 9.10
Operation of sonar, Section 12.10
Bimetallic strip in a thermostat, Section 13.3
Operation of an internal combustion engine, Section 15.5
Efficiency of a heat engine, Section 15.5, Example 15.7
Efficiency of a refrigerator, Section 15.6
Photocopier, Section 16.2
Cathode ray tube, Examples 16.8, 17.8
Lightning rods, Sections 16.6, 17.6
Electrostatic precipitator, Section 16.6

Battery-powered lantern, Example 17.3
Van de Graaff generator, Section 17.2, Example 17.6
Computer keyboard, Example 17.9
Condenser microphone, Section 17.5
Random-access memory (RAM) chips, Section 17.5
Camera flash attachments, Sections 17.5, 18.10
Electron drift velocity in household wiring, Example 18.3
Resistance thermometer, Section 18.4
Battery connection in a flashlight, Section 18.6
Household wiring, Sections 18.11, 21.2
Magnetic compass, Section 19.1
Bubble chamber, Section 19.3
Mass spectrometer, Section 19.3
Cyclotron, Section 19.3, Example 19.5
Synchrotron, Section 19.3
Electric motor (DC), Section 19.7
Galvanometer, Section 19.7
Audio speakers, Section 19.7
Electromagnets, Section 19.10
Computer hard disks, magnetic tape, Sections 19.10, 20.3
Electric generators, Section 20.2
Ground fault interrupter, Section 20.3
Moving coil microphone, Section 20.3
Back emf in a motor, Section 20.5
Transformers, Section 20.6
Electric power distribution, Section 20.6
Eddy-current braking, Section 20.7
Induction stove, Section 20.7
Betatron, Section 20.8
Radio tuning circuit, Examples 21.3, 21.6
Laptop computer power supply, Example 21.5
Diodes and rectifiers, Section 21.7
Filters for audio tweeters and woofers, Section 21.7
Radio/TV antennas, Section 22.3
Microwave ovens, Section 22.4
Liquid crystal displays, Section 22.8
Radar guns, Example 22.9
Periscope, Section 23.4
Fiber optics, Section 23.4
Zoom lens, Example 23.9
Cameras, Section 24.2
Simple magnifier, Section 24.4
Microscopes, Section 24.5
Telescopes, Section 24.6
Reading a compact disk (CD), Section 25.1
Michelson interferometer, Section 25.2
Interference microscope, Section 25.2
Antireflective coatings, Section 25.4
CD player tracking, Section 25.6
Grating spectroscope, Section 25.6
Photolithography for making integrated circuits,
   Section 25.6
Resolution of a laser printer, Example 25.10
X-ray diffraction, Section 25.9
Photocells used for sound tracks, burglar alarms, garage
   door openers, Section 27.2

*C*ollege Physics is intended for a two-semester college course in introductory physics using algebra and trigonometry. Our main goals in writing this book are

- To present the basic concepts of physics that students need to know for later courses and future careers,
- To emphasize that physics is a tool for understanding the real world, and
- To teach transferable problem-solving skills that students can use throughout their lives.

We have kept these goals in mind while developing the main themes of the book.

COMPREHENSIVE COVERAGE

Students should be able to get the whole story from the book. The manuscript was tested for five semesters in a self-paced course, where students *must* rely on the textbook as their primary learning resource. Nonetheless, completeness and clarity are equally advantageous when the book is used in a more traditional classroom setting. Over a dozen other class tests have shown that *College Physics* frees the instructor from having to try to "cover" everything and from filling in the gaps left by the text. The instructor can then tailor class time to the students' needs, whether it be going over particularly difficult concepts, working through examples, engaging the students in cooperative learning activities, describing applications, or presenting demonstrations.

INTEGRATING CONCEPTUAL PHYSICS INTO A QUANTITATIVE COURSE

Some students approach introductory physics with the idea that physics is just the memorization of a long list of equations and the ability to plug numbers into those equations. We want to help students see that a relatively small number of basic physics concepts are applied to a wide variety of situations. In our presentation, based on years of teaching this course, we blend conceptual understanding with analytical skills, using plain language with simple mathematics throughout the book. In our experience, this approach anticipates many of the conceptual difficulties that students may have.

Physics education research has shown that students do not automatically acquire conceptual understanding; the concepts must be explained and the students given a chance to grapple with them. To that end, we include many **Conceptual Examples** and **Conceptual Practice Problems** in the text and a selection of **Conceptual Questions** and **Multiple Choice Questions** at the end of each chapter.

INTRODUCING CONCEPTS INTUITIVELY

We introduce key concepts and quantities in an informal way by establishing why the quantity is needed, why it is useful, and why it must be defined precisely. Then we make a transition from the informal, intuitive idea to a formal definition and name. We find that concepts motivated in this way are easier for students to grasp and remember than are concepts introduced by seemingly arbitrary, formal definitions.

For example, in Chapter 8, the idea of rotational inertia emerges in a natural way from the concept of rotational kinetic energy. Students can understand that a rotating rigid body has kinetic energy due to the motion of its particles. We discuss why it is useful to be able to write this kinetic energy in terms of a single quantity common to all the particles, rather than as a sum involving particles with many different speeds. Using this approach, the definition of rotational inertia is motivated; students understand why it is

"I think chapter 8 is particularly well-written. Rotational motion, magnetism, and AC circuits spring to mind as the most notoriously difficult subjects to teach in this course. The authors have chosen a number of excellent biomechanical examples in chapter 8 and this chapter's presentation alone might persuade some lecturers to switch texts."—Dr. Nelson E. Bickers, University of Southern California

defined the way it is. Then, equipped with an understanding of rotational inertia, students are better prepared for the concept of torque.

We avoid as much as possible presenting definitions or formulas without any motivation. When an equation cannot be derived using algebra and trigonometry, we at least indicate where the equation comes from or give a plausibility argument. For example, Section 9.9 introduces Poiseuille's law with two identical pipes in series to show why the volume flow rate must be proportional to the pressure drop per unit length. Then we discuss why $\Delta V/\Delta t$ is proportional to the fourth power of the radius (rather than to R^2, as for an ideal fluid).

Similarly, we have found that the definitions of the displacement and velocity vectors seem arbitrary and counterintuitive to students if introduced without any motivation. Therefore, we precede any discussion of kinematic quantities with an introduction to Newton's laws, so students know that forces determine how the state of motion of an object changes. Then, when we define the kinematic quantities to give a precise definition of acceleration, we can apply Newton's second law quantitatively to see how forces affect the motion. We give particular attention to laying the groundwork for a concept when its name is a common English word such as *velocity* or *work*.

HELPING STUDENTS SEE THE RELEVANCE OF PHYSICS IN THEIR LIVES
Students in an introductory college physics course have a wide range of backgrounds and interests. We stimulate interest in physics by relating the principles to applications relevant to students' lives. We appeal to topics familiar to students and in line with their interests.

The text, examples, and end-of-chapter problems draw from the everyday world, from familiar technological applications, and from other fields such as biology, medicine, archaeology, astronomy, sports, environmental science, and geophysics. Applications in

the text are marked with an icon in the margin , **Making The Connection**.

In addition, the **Physics at Home** experiments give the students an opportunity to explore and see physics principles operate in their everyday lives. These activities are chosen for their simplicity and for the effective demonstration of physics principles. Each **Chapter Opener Vignette** is designed to capture student interest and maintain it through the chapter. The question asked in each opener is answered somewhere in the chapter.

WRITTEN IN CLEAR AND FRIENDLY STYLE
We have kept the writing down-to-earth and conversational in tone—the kind of language an experienced teacher uses when sitting at a table working one-on-one with a student. We believe students will find the book pleasant to read, informative, accurate without seeming threatening, and filled with analogies that make abstract concepts easier to grasp. We want students to feel confident that they can learn by studying the textbook.

While learning correct physics terminology is essential, we avoid all *unnecessary* jargon—terminology that just gets in the way of the student's understanding.

PROBLEM-SOLVING APPROACH
Problem-solving skills are central to an introductory physics course. We illustrate these skills in the example problems. Lists of problem-solving strategies are sometimes useful; we provide such strategies when appropriate. However, the most elusive skills—perhaps the most important ones—are subtle points that defy being put into a neat list. To develop real problem-solving expertise, students must learn how to think critically and analytically. Problem solving is a multidimensional, complex process; an algorithmic approach is not adequate to instill real problem-solving skills.

Strategy   We begin each example with a discussion—in language that the students can understand—of the **Strategy** to be used in solving the problem. The strategy illustrates the kind of analytical thinking students must do when attacking a problem: How

"The major strength of this text is its approach, which makes students think out the problems, rather than always relying on a formula to get an answer. The way the authors encourage students to investigate whether the answer makes sense, and compare the magnitude of the answer with common sense is good also."—Dr. Jose D'Arruda, University of North Carolina, Pembroke

do I decide what approach to use? What laws of physics apply to the problem and which of them are *useful* in this solution? What clues are given in the statement of the question? What information is implied rather than stated outright? If there are several valid approaches, how do I determine which is the most efficient? What assumptions can I make? What kind of sketch or graph might help me solve the problem? Is a simplification or approximation called for? If so, how can I tell if the simplification is valid? Can I make a preliminary estimate of the answer? Only after considering these questions can the student effectively solve the problem.

Solution   Next comes the detailed **Solution** to the problem. Explanations are intermingled with equations and step-by-step calculations to help the student understand the approach used to solve the problem. We want the student to be able to follow the mathematics without wondering, "Where did that come from?"

Discussion   The numerical or algebraic answer is not the end of the problem; our examples end with a **Discussion**. Students must learn how to determine whether their answer is consistent and reasonable by checking the order of magnitude of the answer, comparing the answer to a preliminary estimate, verifying the units, and doing an independent calculation when more than one approach is feasible. When there are several different approaches, the Discussion looks at the advantages and disadvantages of each approach. We also discuss the implications of the answer—what can we learn from it? We look at special cases and look at "what if" scenarios. The Discussion sometimes generalizes the problem-solving techniques used in the solution.

Practice Problem   After each Example, a **Practice Problem** gives students a chance to gain experience using the same physics principles and problem-solving tools. By comparing their answers to those provided at the end of each chapter, they can gauge their understanding and decide whether to move on to the next section.

Our many years of experience in teaching college physics in a one-on-one setting has enabled us to anticipate where we can expect students to have difficulty. In addition to the consistent problem-solving approach, we offer several other means of assistance to the student throughout the text. A boxed problem-solving strategy gives detailed information on solving a particular type of problem, while an icon for problem-solving tips draws attention to techniques that can be used in a variety of contexts. A hint in a worked example or end-of-chapter problem provides a clue on what approach to use or what simplification to make. A warning icon emphasizes explanations that clarify possible points of confusion or common student misconceptions.

An important problem-solving skill that many students lack is the ability to extract information from a graph or to sketch a graph without plotting individual data points. Graphs often help students visualize physical relationships more clearly than they can do with algebra alone. We emphasize the use of graphs and sketches in the text, in worked examples, and in the problems.

USING APPROXIMATION, ESTIMATION, AND PROPORTIONAL REASONING

College Physics is up front about the constant use of simplified models and approximations in solving physics problems. One of the most difficult aspects of problem solving that students need to learn is that some kind of simplified model or approximation is usually required. We discuss how to know when it is reasonable to ignore friction or air resistance, treat g as constant, ignore viscosity, treat a charged object as a point charge, or ignore diffraction. A brief discussion of air resistance and terminal velocity in Chapter 3 enables us to discuss when it is reasonable to ignore air resistance—and also to show students that physics can account for these other effects.

"I understood the math, mostly because it was worked out step-by-step, which I like."—student, Bradley University
"The math was really clear. I was impressed with how easy the math and steps involved were to understand."—student, Bradley University

"The 'Strategy & Discussion' in each example were extremely helpful in understanding the ideas."—student, Houston Community College

Some Examples and Problems require the student to make an estimate—a useful skill both in physics problem solving and in many other fields. Similarly, we teach proportional reasoning as not only an elegant shortcut but also as a means to understanding patterns. We frequently use percentages and ratios to give students practice in using and understanding them.

SHOWCASING AN INNOVATIVE ART PROGRAM

Throughout the book we emphasize the value of drawing diagrams to help visualize and understand physical situations. We live up to our own advice by including many drawings, using color to distinguish different kinds of physical quantities so students can better see what concepts are important in each situation.

To help show that physics is more than a collection of principles that explain a set of contrived problems, in every chapter we have developed several innovative **Showcase Illustrations** to bring to life the connections between physics concepts and the complex ways in which they are applied. We believe these illustrations, with subjects ranging from three-dimensional views of electric field lines to the biomechanics of the human body and from representations of waves to the distribution of electricity in the home, will help students see the power and beauty of physics. We also provide many photographs to convey the application of physics to the real world and to help motivate many topics.

INNOVATIVE ORGANIZATION

There are a few places where, for pedagogical reasons, the organization of our text differs from that of most textbooks. The most significant reorganization is in the treatment of forces and motion. In *College Physics,* the central theme of Chapters 2–4 is *force.* Kinematics is woven into the fabric of the three force chapters as a tool to understand how forces affect motion. Overall, we spend less time on kinematics and more time on forces. This approach has these advantages:

"I am thoroughly delighted to see a text which presents Newton's Laws before kinematics. This in my opinion is an intuitively better approach for the students. I believe it is better to deal with why things move before we attempt to quantitatively describe the resulting motion. . . . Compared to the text we have been using, this approach is better."—Dr. Kelly Roos, Bradley University

"It was easier to understand these ideas since we came from reading about Newton's laws."—student, Bradley University

- The first few chapters in any text set up student expectations that are hard to change later. If the course starts with a series of definitions of the kinematic quantities, with no explanation of *why* we are interested in those quantities, students may see physics as a series of equations to memorize and manipulate. Similarly, it is not a good precedent for students to learn all of the vector techniques at the outset when they are not yet needed.

- We explain to students that we develop the kinematic concepts so we can understand the effect of a net force on the motion of an object. Newton's second law is the key reason why we need a precise definition of acceleration; to define acceleration requires that we first define displacement and velocity. If the definitions of these quantities are imprecise, we cannot hope to understand how forces change the state of motion of an object. We have found that laying out this clear program at the start of the course helps students realize that in physics, physical concepts determine understanding, not mathematics.

- Learning kinematics first may suggest to students that physics is not connected to the real world. If they are told that objects all fall with the same acceleration—which they know from experience to be false—they learn not to trust the principles they're learning. With an understanding of forces and Newton's laws, *College Physics* enables the student to learn that constant acceleration is an approximation and to learn how to judge when that approximation is reasonable.

- A gradual introduction to vectors lets students acquire basic vector skills before moving on to more difficult ideas. We spread the vector material over three chapters, introducing vector techniques *as needed,* so students are not overloaded with formalism. We emphasize understanding how the physical vector quantities behave and can be analyzed, so that students are not required to learn the operations of an abstract mathematical entity before they have any inkling of how essential they are to problem solving in physics. It is our experience that by the time students reach Chapter 4, on forces and nonuniform motion in two dimensions, they have already attained a high level of proficiency with vector techniques (and with the conceptual framework of Newton's laws).

- We use correct vector terminology and methods from the very beginning. Even in one dimension, displacements, velocities, and accelerations are treated as vector quantities. Chapter 2 starts with the graphical addition of one-dimensional vectors. Chapter 3 introduces subtraction and component notation for one-dimensional vectors. For example, we write "$v_x = -5$ m/s" rather than "$v = -5$ m/s" when we are talking about a velocity component; we carefully distinguish components from magnitudes. Thus, the students carry over everything they learned about vectors in one dimension to two dimensions, without any change of notation. Some of our class testers report fewer students struggling with the concept of vector components using *College Physics* compared to their previous textbook.

- We begin in Chapter 2 with Newton's laws of motion so the students can build a solid conceptual framework in situations where the mathematics is simple. If forces were not introduced until Chapter 4, the students would have much less time to overcome conceptual difficulties associated with Newton's laws and would have much less practice applying them.

ACCURACY ASSURANCE

The authors and the publisher acknowledge the fact that inaccuracies can be a source of frustration for both the instructor and students. Therefore, throughout the writing and production of this textbook we have worked diligently to eliminate errors. We would like to describe the process used to assure the accuracy of this textbook.

Dr. Larry Rowan, of the University of North Carolina, checked the accuracy of all textual examples, practice problems and solutions, and end-of-chapter questions and problems in the second draft of the manuscript. Corrections were made to the final draft manuscript by the authors.

Dr. Richard Heinz, of Indiana University, and Dr. Marllin Simon, of Auburn University, then independently checked the accuracy of all textual examples and practice problems and solutions and worked all end-of-chapter questions and problems in the final draft of the manuscript.

Bill Fellers of Laurel Technology also conducted an independent accuracy check and worked all end-of-chapter questions and problems in the final draft of the manuscript. He then coordinated his efforts with Dr. Heinz, Dr. Simon, and the authors to resolve any discrepancies to ensure the accuracy of not only the text, but also the end-of-book answer section and the solutions manuals. Corrections were then made to the manuscript before it was typeset.

The page proofs of the text were double-proofread against the manuscript to ensure the correction of any errors introduced when the manuscript was typeset. The textual examples, practice problems and solutions, end-of-chapter questions and problems, and problem answers were accuracy checked by Laurel Technology again at the page proof stage after the manuscript was typeset. This last round of corrections was then cross-checked against the solutions manuals.

TO THE STUDENT: HOW TO USE THIS TEXTBOOK

Welcome! We hope that you enjoy your physics course. While studying physics does require hard work, in writing this book we have tried to remove the obstacles that sometimes make introductory physics unnecessarily difficult. We have also tried to show the beauty inherent in the principles of physics and how these principles are manifest all around you.

In our years of teaching experience, we have found that studying physics is a skill that must be learned. It's much more effective to *study* a physics textbook, which involves *active participation* on your part, rather than to read through passively. Even though active study takes more time initially, in the long run it will *save* you time; you learn more in one active study session than in three or four superficial readings. Here are some suggestions to help you get the most out of your study:

- Study the text *before* attending class on the same topic. Knowing what's in the book helps you take notes selectively rather than scrambling to write everything down. You'll be able to

"I was concerned about having Newton's laws before kinematics until I read the chapters. After reading the chapters I think that this ordering is preferable to the more traditional order. . . . The approach that is taken with Newton's laws, introducing them at the beginning and spreading them out throughout the next three chapters is also very good. Understanding Newton's laws is one of the most difficult things that the students have to do in this class, and teaching them over and over again over a period of time is the best way to do it. I really liked the approach here."—Dr. Grant Hart, Brigham Young University

absorb more of the subtleties and difficult points from class. You'll also have some good questions to ask your instructor.

- Don't try to read an entire chapter in one sitting; study one or two sections at a time. It's difficult to maintain your concentration in a long session with so many new concepts and skills to learn.

- As you study, take particular note of these elements:

Boxed laws, rules, and equations indicate the most important and central concepts.

 indicates a warning about a common student misconception or point of confusion.

indicates a helpful problem-solving tip.

Boxed problem-solving strategies give detailed information on solving a particular type of problem.

- When you come to an Example, pause after you've read the problem. Think about the strategy you would use to solve the problem. See if you can work through the problem on your own. Now study the Strategy, Solution, and Discussion in the textbook. Sometimes you will find that your own solution is right on the mark; if not, you can focus your attention on the place where you got stuck or any mistakes you may have made.

- Work the Practice Problem after each Example to practice applying the physics concepts and problem-solving skills you've just learned. Check your answer with the one given at the end of the chapter. If your answer isn't correct, review the previous section in the textbook to try to find your mistake.

- Take time to study the figures and graphs carefully.

- Find a study partner and get together regularly. Go over the difficulties either of you may be having. Take turns explaining things to each other—you learn a tremendous amount when you teach someone else. Compare your solutions to the Practice Problems. Discuss some of the Conceptual Questions from the end of the chapter.

- Try the Physics at Home experiments. They reinforce key physics concepts and help you see how these concepts operate in the world around you.

- Write your *own* chapter summary or outline and then compare it with the summary provided at the end of the chapter. This will help you identify the most important and fundamental concepts in each chapter.

We hope that these suggestions will help you get the most out of *College Physics*. We have spent many years working with students, both in the classroom and one-on-one in a self-paced course. We wrote this book so you could benefit from our experience as we carefully address the points that have caused difficulties for our students in the past. We also wish to share with you some of the pleasure and excitement we have found in learning about the physical laws that govern our world.

GUIDED TOUR

Forces and Motion Along a Line

Chapter 3

A sailplane (or "glider") is a small, unpowered, high-performance aircraft. A sailplane must be initially towed a few thousand feet into the air by a small airplane, after which it relies on regions of upward-moving air such as thermals and ridge currents to ascend further. Suppose a small plane requires about 120 m of runway to take off by itself. When it is towing a sailplane, how much more runway does it need?

Chapter Opener

Each **Chapter Opener** includes a chapter opening photo and vignette designed to capture student interest and maintain it throughout the chapter. The vignette describes the situation shown in the photo and asks students to consider the relevant physics. This question is then answered within the chapter.

Concepts & Skills to Review

- net force: vector addition (Section 2.4)
- freebody diagrams (Section 2.4)
- gravitational force (Section 2.5)
- internal and external forces (Section 2.4)

Concepts and Skills to Review

Presented on the first page of each chapter, this unique feature lists important material from previous chapters that students should understand before they start reading.

Illustrations and Page Layout to Facilitate and Reinforce Learning

Showcase Illustrations and Diagrammatic Illustrations

Showcase Illustrations and straightforward *Diagrammatic Illustrations* are used in combination to help students grasp the concepts. The Showcase Illustrations are designed to be very appealing in order to capture student interest and help them visualize difficult concepts. The Diagrammatic Illustrations are used to further explain and reinforce concepts.

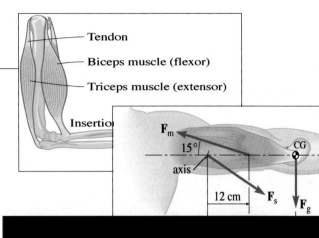

Tendon
Biceps muscle (flexor)
Triceps muscle (extensor)
Insertion

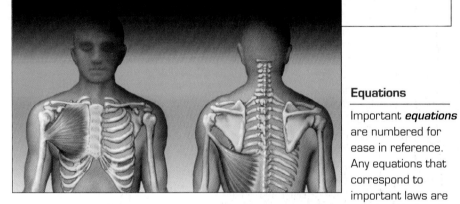

F_m
15°
axis
CG
12 cm
F_s
F_g

Equations

Important *equations* are numbered for ease in reference. Any equations that correspond to important laws are boxed for quick identification.

Reinforcement Notes

Various *Reinforcement Notes* appear in the margin to emphasize the important points in the text.

> The symbol Δ stands for *the change in*. If the initial value of a quantity Q is Q_0 and the final value is Q_f, then $\Delta Q = Q_f - Q_0$. ΔQ is read "delta Q."

Displacement

$$\Delta \vec{r} = \vec{r}_f - \vec{r}_0 \qquad (3\text{-}1)$$

Rules or Laws

Statements of important physics *Rules* or *Laws* are highlighted within the text to help identify the most significant topics.

Newton's Laws of Motion (2-2)

1. If no forces act on an object, then its speed and direction of motion do not change. (If the object is at rest, it remains at rest with a speed of zero; if it is moving, it moves in a straight line with constant speed.)

2. A non-zero net force acting on an object causes its state of motion to change. Quantitatively,

$$F_{net} = ma$$

   where $\mathbf{F}_{net}$ is the net force, m is the mass, and $\mathbf{a}$ is the acceleration (which measures the rate of change of the velocity.)

3. In an interaction between two objects, the forces that each exerts on the other are equal in magnitude and opposite in direction.

Consistent Problem-Solving Approach

Worked Example

A multipart *Quantitative* or *Conceptual Worked Example* appears when an important new concept or skill is introduced. A step-by-step approach is used for every worked example presented in the text.

Four-Step Problem-Solving Approach

1. Strategy
Each example begins with a discussion of the strategy to be used in solving the problem. The strategy illustrates the issues that should be considered when attacking a problem: How do we decide what approach to use? What laws of physics apply to the problem? What clues are given in the statement of the question? What assumptions can be made?

2. Solution
The strategy is followed by a numeric *Solution*. Explanations are intermingled with equations and step-by-step calculations to help the student understand the approach used to solve the problem.

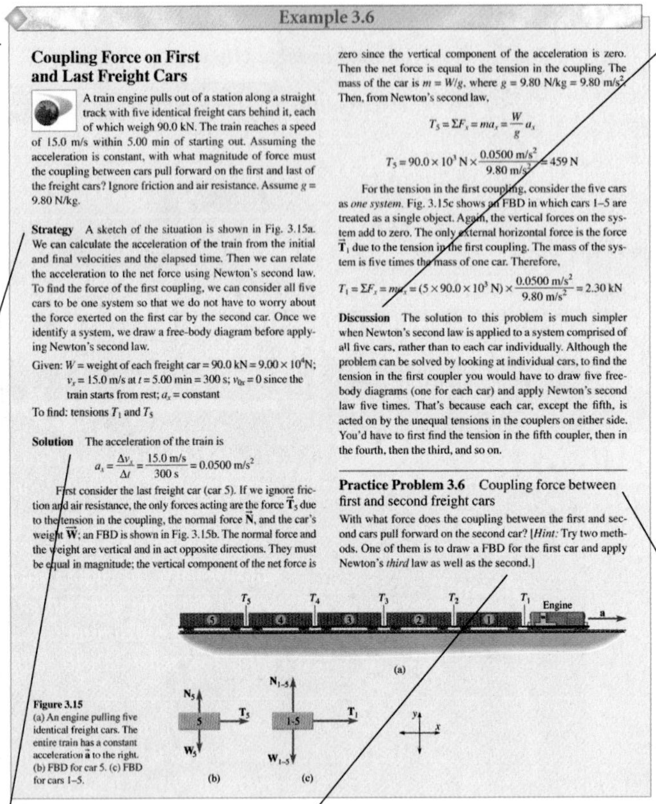

Example 3.6

Coupling Force on First and Last Freight Cars

A train engine pulls out of a station along a straight track with five identical freight cars behind it, each of which weigh 90.0 kN. The train reaches a speed of 15.0 m/s within 5.00 min of starting out. Assuming the acceleration is constant, with what magnitude of force must the coupling between cars pull forward on the first and last of the freight cars? Ignore friction and air resistance. Assume $g = 9.80$ N/kg.

Strategy A sketch of the situation is shown in Fig. 3.15a. We can calculate the acceleration of the train from the initial and final velocities and the elapsed time. Then we can relate the acceleration to the net force using Newton's second law. To find the force of the first coupling, we can consider all five cars to be one system so that we do not have to worry about the force exerted on the first car by the second car. Once we identify a system, we draw a free-body diagram before applying Newton's second law.

Given: $W =$ weight of each freight car $= 90.0$ kN $= 9.00 \times 10^4$N; $v_x = 15.0$ m/s at $t = 5.00$ min $= 300$ s; $v_{0x} = 0$ since the train starts from rest; $a_x =$ constant

To find: tensions T_1 and T_5

Solution The acceleration of the train is

$$a_x = \frac{\Delta v_x}{\Delta t} = \frac{15.0 \text{ m/s}}{300 \text{ s}} = 0.0500 \text{ m/s}^2$$

First consider the last freight car (car 5). If we ignore friction and air resistance, the only forces acting are the force T_5 due to the tension in the coupling, the normal force $\vec{N}$, and the car's weight $\vec{W}$; an FBD is shown in Fig. 3.15b. The normal force and the weight are vertical and in act opposite directions. They must be equal in magnitude; the vertical component of the net force is

zero since the vertical component of the acceleration is zero. Then the net force is equal to the tension in the coupling. The mass of the car is $m = W/g$, where $g = 9.80$ N/kg $= 9.80$ m/s². Then, from Newton's second law,

$$T_5 = \Sigma F_x = ma_x = \frac{W}{g} a_x$$

$$T_5 = 90.0 \times 10^3 \text{ N} \times \frac{0.0500 \text{ m/s}^2}{9.80 \text{ m/s}^2} = 459 \text{ N}$$

For the tension in the first coupling, consider the five cars as *one system*. Fig. 3.15c shows an FBD in which cars 1–5 are treated as a single object. Again, the vertical forces on the system add to zero. The only external horizontal force is the force $\vec{T}_1$ due to the tension in the first coupling. The mass of the system is five times the mass of one car. Therefore,

$$T_1 = \Sigma F_x = ma_x = (5 \times 90.0 \times 10^3 \text{ N}) \times \frac{0.0500 \text{ m/s}^2}{9.80 \text{ m/s}^2} = 2.30 \text{ kN}$$

Discussion The solution to this problem is much simpler when Newton's second law is applied to a system comprised of all five cars, rather than to each car individually. Although the problem can be solved by looking at individual cars, to find the tension in the first coupler you would have to draw five free-body diagrams (one for each car) and apply Newton's second law five times. That's because each car, except the fifth, is acted on by the unequal tensions in the couplers on either side. You'd have to first find the tension in the fifth coupler, then in the fourth, then the third, and so on.

Practice Problem 3.6 Coupling force between first and second freight cars

With what force does the coupling between the first and second cars pull forward on the second car? [*Hint:* Try two methods. One of them is to draw a FBD for the first car and apply Newton's *third* law as well as the second.]

Figure 3.15
(a) An engine pulling five identical freight cars. The entire train has a constant acceleration $\vec{a}$ to the right.
(b) FBD for car 5. (c) FBD for cars 1–5.

---

3. Discussion
The examples do not end with the solution. Students must also learn how to determine whether an answer is consistent and reasonable by checking order-of-magnitude, verifying the units, and doing an independent calculation when more than one approach is feasible. The *Discussion* also draws attention to the problem-solving techniques that were used in the solution.

4. Practice Problems
Every example is immediately followed by a *Practice Problem*. This allows students to further practice the problem-solving skills illustrated in the example and to reinforce what they have learned. The answer for each practice problem appears at the end of the chapter to allow students to check their work.

Hints
Hints appear within the text to help identify important information and to give clues on what approaches to use or laws to apply. These are plentiful in earlier chapters where students are learning to apply problem-solving techniques.

Warning Notes

An icon indicates a *Warning Note* that describes possible points of confusion or any common misconceptions that may apply to a particular concept.

Problem-Solving Tip

An icon points out *Problem-Solving Tips* that guide students in applying problem-solving techniques.

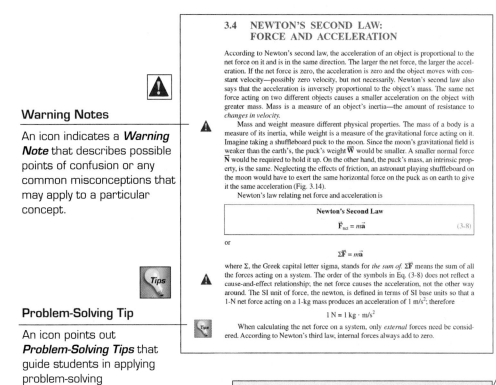

3.4 NEWTON'S SECOND LAW: FORCE AND ACCELERATION

According to Newton's second law, the acceleration of an object is proportional to the net force on it and is in the same direction. The larger the net force, the larger the acceleration. If the net force is zero, the acceleration is zero and the object moves with constant velocity—possibly zero velocity, but not necessarily. Newton's second law also says that the acceleration is inversely proportional to the object's mass. The same net force acting on two different objects causes a smaller acceleration on the object with greater mass. Mass is a measure of an object's inertia—the amount of resistance to *changes in velocity*.

Mass and weight measure different physical properties. The mass of a body is a measure of its inertia, while weight is a measure of the gravitational force acting on it. Imagine taking a shuffleboard puck to the moon. Since the moon's gravitational field is weaker than the earth's, the puck's weight $\vec{W}$ would be smaller. A smaller normal force $\vec{N}$ would be required to hold it up. On the other hand, the puck's mass, an intrinsic property, is the same. Neglecting the effects of friction, an astronaut playing shuffleboard on the moon would have to exert the same horizontal force on the puck as on earth to give it the same acceleration (Fig. 3.14).

Newton's law relating net force and acceleration is

Newton's Second Law

$$\vec{F}_{net} = m\vec{a} \qquad (3-8)$$
---

or

$$\Sigma\vec{F} = m\vec{a}$$

where Σ, the Greek capital letter sigma, stands for *the sum of*. $\Sigma\vec{F}$ means the sum of all the forces acting on a system. The order of the symbols in Eq. (3-8) does not reflect a cause-and-effect relationship; the net force causes the acceleration, not the other way around. The SI unit of force, the newton, is defined in terms of SI base units so that a 1-N net force acting on a 1-kg mass produces an acceleration of 1 m/s²; therefore

$$1\ N = 1\ kg \cdot m/s^2$$

When calculating the net force on a system, only *external* forces need be considered. According to Newton's third law, internal forces always add to zero.

Problem-Solving Strategies for Newton's Second Law

- Decide what objects will have Newton's second law applied to them.
- Identify all the interactions affecting that object.
- Draw a free-body diagram to show all the forces acting on the object.
- Find the net force by adding the forces as vectors.
- Use Newton's second law to relate the net force to the acceleration.

Problem-Solving Strategies

Problem-Solving Strategies give detailed information on solving a particular type of problem. These are supplied for the most fundamental physics rules and laws.

Applying Physics to the Real World

Physics at Home

Go to a balcony or climb up a ladder and drop a basket-style paper coffee filter (or a cup-cake paper) and a penny simultaneously. Air resistance on the penny is negligible unless it is dropped from a very high balcony. At the other extreme, the effect of air resistance on the coffee filter is very noticeable; it reaches its terminal speed almost immediately. Stack several (two to four) coffee filters together and drop them simultaneously with a single coffee filter. Why is the terminal speed higher for the stack? Crumple a coffee filter into a ball and drop it simultaneously with the penny. Air resistance on the coffee filter is now reduced, but still noticeable.

Physics at Home

Throughout the chapters of this text students will find *Physics at Home* activities. These are designed to be performed individually at home to help students further understand and visualize certain physics concepts.

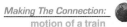
Making The Connection:
motion of a train

problem. The engineer wants to call the railroad office he tell them where to find the train? He might say som of the old trestle bridge." Notice that he uses a point of Then he states how far the train is from that point and of the three pieces (the reference point, the distance, an tion of the train's whereabouts is ambiguous.

The same thing is done in physics. First, we cho **origin.** Then, to describe the location of something, we

Making The Connection

The *Making The Connection* icon indicates areas in the text where physics can be applied to other areas in your students' lives. Familiar topics and interests are discussed in the text and examples, drawing from biology, archaeology, astronomy, sports, and the everyday world.

Master the Concepts

The **Master the Concepts** section gives students a quick review of the chapter. A **Summary** lists important concepts and equations, while the

Highlighted Figures and Tables section identifies important ideas that are graphically represented within the chapter.

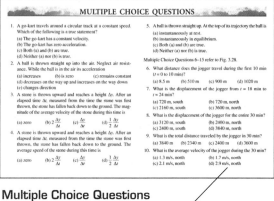

Multiple Choice Questions

The **Multiple Choice Questions** allow students to do a quick self-test and also provide practice for the type of questions found on the MCAT exam. Answers can be found at the end of the book.

Conceptual Questions

The **Conceptual Questions** are designed to test students' qualitative understanding of the key ideas within the chapter.

Problems

A large variety of **Problems** are given by chapter section to check the student's quantitative understanding of the chapter. **Comprehensive Problems** also allow students to test their overall understanding of the chapter.

Paired by Concept
Some problems are paired by concept, and their numbers are connected by a ruled box. Only one problem of the pair has an answer available at the end of the book and a solution available in the Student Solutions Manual.

Color Coded
Blue problem numbers indicate those with solutions available in the College Physics Student Solutions Manual.

Difficulty Level
Difficulty level is designated by ✦ for problems of intermediate difficulty and ✦✦ for more demanding problems.

Conceptual and Quantitative
A **⊙** indicates a problem that combines conceptual and quantitative problem solving.

acceleration when it falls at 75% of its

✦ 52. In free fall, we assume the acceleration
    only is air resistance neglected, but
    strength is assumed to be constant. (a)
    an object fall to the earth's surface suc
    field strength changes less than 1.0%
    most cases, which do we have to wor
    ance becoming significant or g changir

3.7 Apparent Weight

53. Refer to Example 3.17. What is the
    same passenger (weighing 598 N) in th
    In each case, the magnitude of the ele
    0.50 m/s². (a) After having stopped at
    ger pushes 8th floor; the elevator is beg
    ward. (b) Elevator is slowing down as i

⊙ 54. You are standing on a bathroom scale i
    weight is 140 lb, but the reading o
    (a) What is the magnitude and directio

Supplements

Multimedia Supplements

Instructor's Presentation CD

The *Instructor's Presentation CD* is a collection of visual resources that allows instructors to utilize elements from the text such as:
- artwork
- worked examples
- tables
- interactive applets

With the help of these valuable resources, you can create:
- customized classroom presentations
- visually based tests and quizzes
- dynamic course website content
- attractive printed support materials

The digital assets on this cross-platform CD-ROM are grouped by chapter within easy-to-use folders.

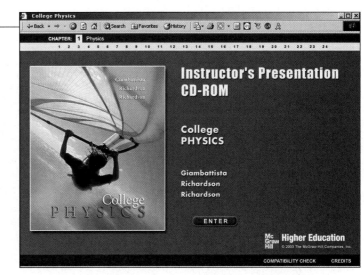

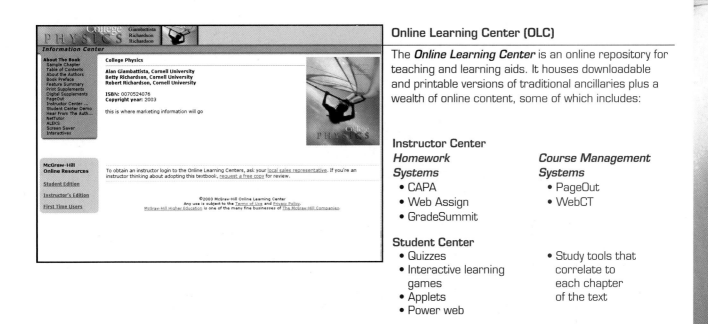

Online Learning Center (OLC)

The *Online Learning Center* is an online repository for teaching and learning aids. It houses downloadable and printable versions of traditional ancillaries plus a wealth of online content, some of which includes:

Instructor Center

Homework Systems
- CAPA
- Web Assign
- GradeSummit

Student Center
- Quizzes
- Interactive learning games
- Applets
- Power web

Course Management Systems
- PageOut
- WebCT

- Study tools that correlate to each chapter of the text

Supplements

Multimedia Supplements

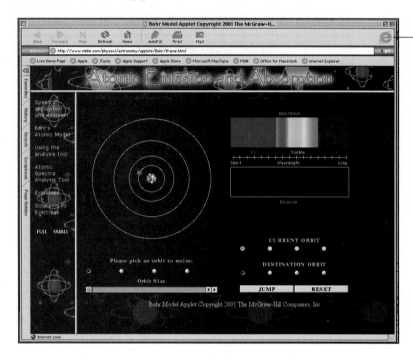

Physics Interactive Applets

McGraw-Hill is proud to bring you an assortment of outstanding *Interactive Applets* like no other. These "Interactives" offer a fresh and dynamic method to teach the physics basics by providing students with applets that are *completely accurate and work with real data*. Interactives allow students to manipulate parameters and gain a better understanding of 20 of the more difficult physics topics by watching the effect of these manipulations. Each Interactive includes:

- Analysis Tool (interactive model)
- Tutorial describing its function
- Content describing its principle themes

Students will easily be able to jump between the Tutorial and the Analysis Tools with just the click of the mouse.

Instructor's Testing and Resource CD

The *Instructor's Testing and Resource CD* contains the *College Physics* Testbank within the Brownstone DIPLOMA© test generator as well as the Word and PDF files of the:

- worked examples and end-of-chapter problems from the text
- multiple-choice version of the end-of-chapter problems
- solutions to the end-of-chapter problems from the Instructor's Resource Guide
- the testbank

The Brownstone software includes a(n):

- test generator
- online testing program
- Internet testing site
- Grade Management System

This user-friendly software's testing capability is consistently ranked number one in evaluations over other products. The Word files for the end-of-chapter problems and the solutions can be used in combination with the Brownstone software or independently.

ALEKS®

Help students master the math skills needed to understand difficult physics problems.

ALEKS (Assessment and LEarning in Knowledge Spaces) is an artificial intelligence-based system for individualized math learning available via the World Wide Web.

ALEKS is . . .

- A robust course management system
  It tells you exactly what your students know and don't know.

- Focused and efficient
  Enables students to quickly master the math needed for college physics.

- Artificial Intelligence
  It totally individualizes assessment and learning.

- Customizable
  Click on or off each course topic.

- Web Based
  Use a standard browser for easy Internet access.

- Inexpensive
  There are no setup fees or site license fees.

ALEKS® is a registered trademark of ALEKS Corporation.

Printed Supplements

Instructor's Resource Guide
The *Instructor's Resource Guide* includes many unique assets for instructors, such as demonstrations, suggested reform ideas from physics education research, and ideas for incorporating Just-In-Time teaching techniques. It also includes answers to the end-of-chapter conceptual questions and complete, worked-out solutions for all the end-of-chapter problems from the text.

Transparencies
This boxed set of 300 full-color *Transparencies* feature images from the text. The images have been modified to ensure maximum readability in both small and large classroom settings.

Testbank
The *College Physics Testbank* is also available in a printed version. The testbank includes over 2000 test questions in multiple-choice format at a variety of difficulty levels.

Student Solutions Manual
The *Student Solutions Manual* contains complete, worked-out solutions to selected end-of-chapter problems from the text.

For more information, contact your McGraw-Hill sales representative at (800) 338-3987, or email www.mhhe.com.
To locate your sales representative, go to www.mhhe.com/catalogs/rep/.

ACKNOWLEDGMENTS

We would like to express our gratitude to the faculty, staff, and students at Cornell University who helped us in a myriad of ways. We especially thank our friend and colleague Bob Lieberman who shepherded us through the process as our literary agent and who inspired us as an exemplary physics teacher. Donald F. Holcomb, Persis Drell, Peter Lepage, and Phil Krasicky read portions of the manuscript and provided us with many helpful suggestions. Raphael Littauer contributed many innovative ideas and served as a model of a highly creative, energetic teacher.

We are indebted to all those who helped us class test the manuscript for five semesters at Cornell: David G. Cassel, Robert M. Cotts, Douglas B. Fitchen, Richard Galik, Robin Hughes, Jeevak Parpia, Joseph Rogers, Janet Scheel, many capable graduate teaching assistants, and, above all, the students in Physics 101 and 103. We also appreciate all the work contributed by our support staff of Bridget Meeds, Karen Fleckenstein, Rosemary French, and Margaret Slaght.

We acknowledge a great debt to David Chelton, who has been aptly called "the best physics developmental editor on the planet," for his expert guidance and sense of humor.

We are grateful for the guidance, patience, and enthusiasm of Daryl Bruflodt and Mary Haas at McGraw-Hill, whose tireless efforts were invaluable in bringing this project to fruition. We would like to thank the entire team of talented and hard-working professionals assembled by McGraw-Hill to publish this book, including Karen Allanson, Carrie Burger, David Dietz, Bill Fellers, Suzanne Guinn, Terri Hamer, Debra Hash, Kevin Kane, Mary Klein, Judith Kromm, Jenni Lang, John Paul Lenney, Richard Mickey, Jay Oberbroeckling, Stuart Paterson, Jill Peter, Kent Peterson, Karen Pugliano, Denise Schank, Pat Steele, and surely many others whose expertise and hard work have contributed to making the book a reality.

We are grateful to Bill Fellers, Richard Heinz, Larry Rowan, and Marllin Simon for accuracy checking the manuscript and for their many helpful suggestions.

Our thanks also go to Janet Scheel, Warren Zipfel, Rebecca Williams, and Mike Nichols for contributing some of the medical and biological applications, to Kyle Dawson and Tom Glickman for their work on the tutorials, and to Nick Taylor and Jason Marshall for writing solutions to the Conceptual Questions.

From Alan: I owe an immense debt of gratitude to Marion, Katie, Charlotte, Julia, and Denisha, without whose love, support, encouragement, and patience this book could never have been written.

From Bob and Betty: We thank our daughter Pamela's classmates and friends at Cornell and in the Vanderbilt Master's in Nursing program who were an early inspiration for the book, and we thank Dr. Philip Massey who was very special to Pamela and is dear to us. We thank our friends at *blur*, Alex, Damon, Dave, and Graham, who love physics and are inspiring young people of Europe to explore the wonders of physics through their work on the European Space Agency's *Beagle2* Mars Express mission. Finally we thank our daughter Jennifer, our grandson Jasper, and son-in-law Jim who endured our protracted hours of distraction while this book was being written.

REVIEWERS, CLASS TESTERS, AND ADVISORS

Valued Input Went into Striving to Meet Your Needs   The content of this first edition text reflects an extensive effort to evaluate the needs of college physics instructors and students, to learn how well we met those needs, and to make improvements where we fell short. The market feedback we received consisted of information gathered from numerous reviews, class tests, focus groups, and a board of advisors.

The primary stage of our market research began with commissioning reviews from instructors across the United States and Canada. We asked them to compare our manuscript to their current text and to submit suggestions for improvement on areas such as content, organization, illustrations, and ancillaries. These over 180 individual reviews constituted the major basis for the revision plan at each stage of the four drafts of manuscript.

The draft manuscript was not only class tested by the authors in their course at Cornell University, but also by thirteen other separate institutions. We solicited feedback from both professors and students, and the authors incorporated this information into their final revision of the manuscript.

We also organized focus groups in six U.S. cities in 2001 and 2002. Participants reviewed draft manuscript in comparison to the books they were using and supplied further input on improvements to the manuscript and ways in which we as publishers could help to improve the content of the college physics course.

Finally, we formed two boards of advisors consisting of professors from regions around the country. We relied on these seasoned instructors to supply opinions during the continued development of the text on discrepancies in reviewer comments, instructional design decisions, the quality of the photo and illustration program, and the content of the print and media ancillary packages.

Considering the total sum of these opinions, this text now embodies the collective knowledge, insight, and experience of more than 200 college physics instructors. Their influence can be seen in everything from the content, accuracy, and organization of the text to the quality of the illustrations.

We are grateful to these instructors for their thoughtful comments and advice:

Reviewers

Wathiq Abdul-Razzaq
West Virginia University

Maris Abolins
Michigan State University

B. N. Narahari Achar
University of Memphis

Brian Adrian
Bethany College

Zaven Altounian
McGill University

Farhang Amiri
Weber State University

Anjum Ansari
University of Illinois at Chicago

Sanjeev Arora
Fort Valley State University

Martina E. Bachlechner
West Virginia University

James R. Benbrook
University of Houston

Cornelius Bennhold
George Washington University

Kyle A. Beran
Saint Mary College

Michael S. Berger
Indiana University

Nelson E. Bickers
University of Southern California

David L. Bixler
Angelo State University

Gary S. Blanpied
University of South Carolina

Donald E. Bowen
Stephen F. Austin
State University

Michael J. Bozack
Auburn University

Matthew Briggs
University of
Wisconsin–Madison

James J. Carroll
Youngstown State University

Gerald B. Cleaver
Baylor University

Sergio Conetti
University of Virginia

Ronald M. Cosby
Ball State University

Stephen R. Cotanch
North Carolina State
University

Michael Crescimanno
Youngstown State University

Pengcheng Dai
University of Tennessee, Knoxville

José D'Arruda
University of North
Carolina–Pembroke

Biman Das
State University of New York College
at Potsdam

Regina Demina
Kansas State University

Peter Douglas
University of Manitoba

Miles J. Dresser
Washington State University

Laurencin Dunbar
St. Louis Community College at
Florissant Valley

John Dykla
Loyola University–Chicago

Donald Elliott
Carroll College

Heidi Fencl
University of Wisconsin–Green Bay

Michael Fisher
Ohio Northern University

James Garner
University of North Florida

Lev Gasparov
University of North Florida

John Gault
Missouri Valley College

Simon George
California State University–Long
Beach

David Gerdes
University of Michigan

Michael Giangrande
Oakland Community College

Martin Guthold
Wake Forest University

Grant W. Hart
Brigham Young University

Richard Holman
    Carnegie Mellon University

Keith R. Honey
    West Virginia University Institute of
    Technology

Joey Huston
    Michigan State University

Michael Jackson
    University of Wisconsin–LaCrosse

Anne Johnson
    Lorain County Community College

J. Bruce Johnson
    Arkansas State University

Brian Jones
    Colorado State University

Sanford Kern
    Colorado State University

Richard Krantz
    Metropolitan State College of Denver

Kwong H. Lau
    University of Houston

Hans Laue
    University of Calgary

David Lederman
    West Virginia University

Paul L. Lee
    California State University,
    Northridge

Bryan H. Long
    Columbia State Community College

Rafael Lopez-Mobilia
    University of Texas at San Antonio

Alfredo A. Louro
    University of Calgary

Donald G. Luttermoser
    East Tennessee State University

Igor Lyuksyutov
    Texas A&M University

Dan MacIsaac
    Northern Arizona University

Phil Matheson
    Salt Lake Community College

Mark E. Mattson
    James Madison University

Bill W. Mayes II
    University of Houston

Marles L. "Mac" McCurdy
    Tarrant County College,
    Northeast

Arthur R. McGurn
    Western Michigan University

Charles R. Meitzler
    Sam Houston State University

John Andrew Milsom
    University of Arizona

Lisa Morris
    Whitworth College

Richard Mowat
    North Carolina State University

O. David Mylander
    California State Polytechnic
    University, Pomona

Blaine E. Norum
    University of Virginia

David P. Norwood
    Southeastern Louisiana University

Ed Oberhofer
    Lake Sumter Community College

Michael J. O'Shea
    Kansas State University

Antonio Pagnamenta
    University of Illinois
    at Chicago

James L. Pazun
    Pfeiffer University

Ralph L. Place
    Ball State University

Steven J. Pollock
    University of Colorado, Boulder

E. W. Prohofsky
    Purdue University

Oren Quist
    South Dakota State University

Syed Rabbani
    Central Texas College

George W. Rainey
    California State Polytechnic
    University, Pomona

Michele Rallis
    Ohio State University

Gordon P. Ramsey
    Loyola University–Chicago

David D. Reid
    Eastern Michigan University

William M. Robertson
    Middle Tennessee State University

Ian Robinson
    University of Illinois at
    Urbana–Champaign

Michael W. Roth
    University of Northern Iowa

Ira Z. Rothstein
    Carnegie Mellon University

John P. Rutherford
    University of Arizona

Earl E. Scime
    West Virginia University

Rob Sells
    Alfred State College

Wesley Shanholtzer
    Marshall University

Bartlett M. Sheinberg
    Houston Community College System

William C. Shockley Jr.
    University of Texas–Pan American

Michael A. Shupe
    University of Arizona

Chandralekha Singh
    University of Pittsburgh

David A. Slimmer
    Lander University

Mark H. Slovak
    Louisiana State University

Xiang-Ning Song
    Richland College

Sudha Srinivas
    Central Michigan University

Noel Stanton
    Kansas State University

Jeffrey Sundquist
    Palm Beach Community College

Wilfred R. Theisen
    St. John's University

J. Marshall Thomsen
    Eastern Michigan University

Lynn P. Thomson
    Brigham Young University–Idaho

Stephen B. Turcotte
    Brigham Young University–Idaho

Ashok Vaseashta
    Marshall University

Michael F. Weber
    Brigham Young University–Hawaii

Martha Riherd Weller
    Middle Tennessee State University

John Wernegreen
    Eastern Kentucky University

C. Steven Whisnant
    James Madison University

Daniel Whitmire
    University of Louisiana–Lafayette

Arthur W. Wiggins
    Oakland Community College

Luc T. Wille
    Florida Atlantic University

Suzanne Willis
    Northern Illinois University

Robert J. Wilson
    Colorado State University

Gary M. Wysin
    Kansas State University

Sanichiro Yoshida
Southeastern Louisiana University

Michael K. Young
Santa Barbara City College

Gary R. Zauft
University of Wisconsin–Barron County

Darin T. Zimmerman
Penn State Altoona

Class Test Schools

Robert W. Arts
Pikeville College

Karla Bier
Stephens College

Regina Demina
Kansas State University

Sharmanthie Fernando
Northern Kentucky University

Esmail Hejazifar
Wilmington College

Susanne M. Lee
State University of New York, Albany

Ntungwa Maasha
Coastal Georgia Community College

Ponn Maheswaranathan
Winthrop University

Kelly R. Roos
Bradley University

Bartlett M. Sheinberg
Houston Community College System

David A. Slimmer
Lander University

Richard Summers
Reinhardt College

Richard J. Zabelka
Lake Superior State University

Focus Group Participants

Ricardo Alarcon
Arizona State University

Wolfgang Bauer
Michigan State University

Cornelius Bennhold
The George Washington University

Nancy Beverly
Mercy College

Donald E. Bowen
Stephen F. Austin State University

Matthew E. Briggs
University of Wisconsin–Madison

Myron Campbell
University of Michigan

John P. Cise
Austin Community College–Rio

Gerald Cleaver
Baylor University

J.M. Collins
Marquette University

Pengcheng Dai
University of Tennessee, Knoxville

Regina Demina
Kansas State University

Renee D. Diehl
Penn State University

Miles J. Dresser
Washington State University

John Dykla
Loyola University–Chicago

Robert Ehrlich
George Mason University

Jerry Feldman
The George Washington University

Nicholas Giordano
Purdue University

Gary Gladding
University of Illinois at Urbana–Champaign

Richard M. Heinz
Indiana University

Paul Holody
Henry Ford Community College

Kwong Lau
University of Houston

Susanne M. Lee
State University of New York, Albany

Rafael Lopez-Mobilia
University of Texas at San Antonio

Igor Lyuksyutov
Texas A&M University

Ponn Maheswaranathan
Winthrop University

David E. Meltzer
Iowa State University

David P. Norwood
Southeastern Louisiana University

Albert Osei
Oakwood College

Antonio Pagnamenta
University of Illinois at Chicago

Mark Plano Clark
Doane College

Gordon P. Ramsey
Loyola University

David Reid
Eastern Michigan University

Carl Rotter
West Virginia University

Larry Rowan
University of North Carolina at Chapel Hill

Arthur G. Schmidt
Northwestern University–Illinois

Bartlett M. Sheinberg
Houston Community College System

Marllin Simon
Auburn University

Mark H. Slovak
Louisiana State University

Sudha Srinivas
Central Michigan University

Richard Summers
Reinhardt College

Linn Van Woerkom
Ohio State University

Gary Westfall
Michigan State University

Suzanne Willis
Northern Illinois University

Board of Advisors

A. Lewis Ford
Texas A&M University

Nicholas Giordano
Purdue University

Richard M. Heinz
Indiana University

Klaus Honscheid
Ohio State University

John Hubisz
North Carolina State University

Susanne M. Lee
State University of New York, Albany

Alan Nathan
University of Illinois at Urbana–Champaign

Richard W. Robinett
Penn State University

Larry Rowan
University of North Carolina at Chapel Hill

Marllin Simon
Auburn University

Media Board of Advisors

Regina Demina
    Kansas State University

Grant Hart
    Brigham Young University

Brian Jones
    Colorado State University

Dale Sayers
    North Carolina State University

CONTRIBUTORS

Of course, we are deeply indebted to:

Professor Richard Heinz of Indiana University, Professor Larry Rowan of the University of North Carolina, and Professor Marllin Simon of Auburn University for their accuracy review of the *College Physics* manuscript;

Professor Suzanne Willis of Northern Illinois University and Professor Susanne M. Lee, Assistant Professor of Physics, University at Albany, SUNY, for their authoring of the instructor resources and demonstrations in the *College Physics* Instructors' Resource Guide;

Professor Larry Rowan of the University of North Carolina and Professor Edward Oberhofer of Lake Sumter Community College for their authoring of the *College Physics* Testbank;

Professor Marllin Simon of Auburn University for his contributions to Share Your Favorite Labs and for authoring the chapter practice quizzes on the *College Physics* Online Learning Center.

Introduction

Science fiction movies, like the *Star Wars* series, are a lot of fun to see. Some of the most exciting scenes involve battles taking place in outer space between the forces for good (Rebel alliance) and the forces for evil (Imperial storm troopers). When the rebels aboard the *Millennium Falcon* blow up a Death Star in *Return of the Jedi*, they see the bright flash and hear the loud roar of the explosion. Is that really what would happen?

Concepts & Skills to Review

• algebra, geometry, and
  trigonometry (Appendix A)

1.1   WHY STUDY PHYSICS?

Physics is the branch of science that describes matter, energy, space, and time at the most fundamental level. Whether you are planning to study biology, architecture, medicine, music, chemistry, or art, there are principles of physics that are relevant to your field.

Physicists look for patterns in the physical phenomena that occur in the universe. They try to explain what is happening and perform experiments to see if the proposed explanation is valid. The goal is to find the most basic laws that govern the universe and to formulate those laws in the most precise and simplified way possible.

The study of physics is valuable for several reasons:

- Since physics describes matter and its basic interactions, all natural sciences are built on a foundation of the laws of physics. A full understanding of chemistry requires a knowledge of the physics of atoms. A full understanding of biological processes in turn is based on the underlying principles of physics and chemistry. Centuries ago, the study of *natural philosophy* encompassed what later became the separate fields of biology, chemistry, geology, astronomy, and physics. Today there are scientists who call themselves biophysicists, chemical physicists, astrophysicists, and geophysicists, demonstrating how thoroughly the sciences are intertwined.

- In today's technological world, many important devices can be understood correctly only with a knowledge of the underlying physics. Just in the medical world, think of laser surgery, magnetic resonance imaging, instant-read thermometers, x-ray imaging, radioactive tracers, heart catheterizations, sonograms, pacemakers, microsurgery guided by optical fibers, ultrasonic dental drills, and radiation therapy.

- By studying physics, you acquire skills that are useful in other disciplines. These include thinking logically and analytically; solving problems; making simplifying assumptions; constructing mathematical models; using valid approximations; and making precise definitions.

- Society's resources are limited, so it is important to use them in beneficial ways and not squander them on scientifically impossible projects. Political leaders and the voting public are too often led astray by a lack of understanding of scientific principles. Can a nuclear power plant supply energy safely to a community? What is the truth about the greenhouse effect, the ozone hole, and the danger of radon in the home? By studying physics, you learn some of the basic scientific principles and acquire some of the intellectual skills necessary to ask probing questions and to formulate informed opinions on these important matters.

- Finally, by studying physics, we hope that you develop a sense of the beauty of the fundamental laws governing the universe. Physicists are driven to find explanations for the phenomena around them. Intellectual satisfaction accompanies a deeper understanding of some of these principles.

We asked if the bright flash and loud roar of the explosion could be seen and heard by the Rebel heroes on board their spaceship. A student of physics watching the film knows that light can easily travel through space—which is why sunlight, moonlight, and starlight can reach Earth. On the other hand, sound can only travel through a physical medium, such as the air in Earth's atmosphere. The near-vacuum of space is not an adequate medium to carry sound waves that can be heard by humans. The Rebel heroes would not hear the explosion.

The study of physics can deepen your understanding of what you observe in the real universe as well as in the imaginary universe of a science-fiction movie.

1.2   TALKING PHYSICS

Some of the words used in physics are familiar from everyday speech. This familiarity can be misleading, since the scientific definition of a word may differ considerably from its common meaning. In physics, words must be precisely defined so that any-one reading a scientific paper or listening to a science lecture understands exactly what is meant. Some of the basic defined quantities, whose names are also words used in everyday speech, include time, length, force, velocity, acceleration, mass, and temperature.

In everyday language, *speed* and *velocity* are synonyms. In physics, there is an important distinction between the two. As defined in Chapter 3, *velocity* includes the *direction* of travel as well as the distance traveled per unit time. *Acceleration* is another word that has a distinct meaning in physics; it can mean to speed up or to slow down or to maintain constant speed while changing direction. The scientific definition of accel-eration includes an associated direction. In everyday use, acceleration means an increase in speed while another word, deceleration, is used for a reduction in speed. This is fine for conversational purposes, but the precise definition of acceleration, including the direction, is required for the term to be useful in physics.

There are two important reasons for the way in which we define physical quanti-ties. First, physics is an experimental science. The results of an experiment must be stated unambiguously so that other scientists can perform similar experiments and com-pare their results. Quantities must be defined precisely to enable experimental measure-ments to be uniform no matter where they are made. Second, physics is a mathematical science. We use mathematics to quantify the relationships among physical quantities. These relationships can be expressed mathematically only if the quantities being inves-tigated have precise definitions.

1.3   THE USE OF MATHEMATICS

A working knowledge of algebra, trigonometry, and geometry is essential to the study of introductory physics. Some of the more important mathematical tools are reviewed in Appendix A. If you know that your mathematics background is shaky, you might want to test your mastery by doing some problems from a math textbook.

Mathematical equations are shortcuts for expressing concisely in symbols relation-ships that are cumbersome to describe in words. Algebraic symbols in the equations stand for quantities that consist of numbers *and units*. The number represents a measurement and the measurement is made in terms of some standard; the unit indicates what standard is used. In physics, a number to specify a quantity is useless unless we know the unit attached to the number. When buying silk to make a sari, do we need a length of 5 mil-limeters, 5 meters, or 5 kilometers? Is the term paper due in 3 minutes, 3 days, or 3 weeks? Systems of units are discussed in Section 1.5.

In the language of physics, the word **factor** is used frequently, often in a rather idio-syncratic way. If the power emitted by a radio transmitter has doubled, we might say that the power has "increased by a factor of two." If the concentration of sodium ions in the bloodstream is half of what it was previously, we might say that the concentration has "decreased by a factor of two," or, in a blatantly inconsistent way, someone else might say that it has "decreased by a factor of one-half." The *factor* is the number by which a quantity is multiplied or divided when it is changed from one value to another. In other words, the factor is really a ratio. In the case of the radio transmitter, if P_0 represents the initial power and P represents the power after new equipment is installed, we write

$$\frac{P}{P_0} = 2$$

It is also common to talk about "increasing 5%" or "decreasing 20%." If a quantity increases $n\%$, that is the same as saying that it is multiplied by a factor of $1 + (n/100)$. If a quantity decreases $n\%$, then it is multiplied by a factor of $1 - (n/100)$. For example, an increase of 5% means something is 1.05 times its original value and a decrease of 4% means it is 0.96 times the original value.

Physicists talk about increasing "by some factor" because it often simplifies a problem to think in terms of **proportions**. When we say that A is proportional to B (written $A \propto B$), we mean that if B increases by some factor, then A must increase by the same factor. For instance, the circumference of a circle equals 2π times the radius: $C = 2\pi r$. Therefore $C \propto r$. If the radius doubles, the circumference also doubles. The area of a circle is proportional to the *square* of the radius ($A = \pi r^2$, so $A \propto r^2$). The area must increase by the same factor as the radius *squared*, so if the radius doubles, the area increases by a factor of $2^2 = 4$.

Example 1.1

Effect of Increasing Radius on the Volume of a Sphere

The volume of a sphere is given by the equation

$$V = \tfrac{4}{3}\pi r^3$$

where V is the volume and r is the radius of the sphere. If the radius of the sphere is increased by a factor of 3, by what factor does the volume of the sphere change?

Strategy   The problem gives us the ratio of the new radius and the old radius:

$$\frac{r_2}{r_1} = 3$$

The subscripts help us keep track of which sphere's volume and radius we mean; the radius of the original sphere is r_1 and the radius of the new sphere is r_2. Since $\tfrac{4}{3}$ and π are constants, we can work in terms of proportions.

Solution   The volume of a sphere is proportional to the cube of its radius:

$$V \propto r^3$$

Since the radius increased by a factor of three, and volume is proportional to the cube of the radius, the new volume should be bigger by a factor of $3^3 = 27$.

Discussion   A slight variation on the solution is to write out the proportionality in terms of ratios of the corresponding sides of the two equations:

$$\frac{V_2}{V_1} = \frac{\tfrac{4}{3}\pi r_2^3}{\tfrac{4}{3}\pi r_1^3} = \left(\frac{r_2}{r_1}\right)^3$$

Substituting the ratio of r_2 to r_1 yields

$$\frac{V_2}{V_1} = 3^3 = 27$$

which says that V_2 is 27 times V_1.

Practice Problem 1.1   Power dissipated by a lightbulb

The electrical power P dissipated by a lightbulb of resistance R is $P = V^2/R$, where V represents the line voltage. During a brownout, the line voltage is 10.0% less than its normal value. How much power is drawn by a lightbulb during the brownout if it normally draws 100.0 watts? Assume that the resistance does not change.

 There are not enough letters in the alphabet to assign a unique letter to each quantity. The same letter V can represent volume in one context and voltage in another. Avoid attempting to solve problems by picking equations that seem to have the correct letters; a skilled problem-solver understands what quantity each symbol in a particular equation represents.

1.4   SCIENTIFIC NOTATION AND SIGNIFICANT FIGURES

In physics, we deal with some numbers that are very small and others that are very large. It can get cumbersome to write numbers in conventional decimal notation. In **scientific notation**, any number is written as the product of a number between 1 and 10 and a whole number power of ten. Thus the radius of Earth, approximately 6,380,000 m at the equator, can be written 6.38×10^6 m; the radius of a hydrogen atom, 0.000 000 000 053 m, can be written 5.3×10^{-11} m. Scientific notation eliminates the need to write zeros to locate the decimal point correctly.

In science, a measurement or the result of a calculation must indicate the **precision** to which the number is known. The precision of a device used to make a measurement is limited by the finest division on the scale. Using a meterstick with millimeter divisions as the smallest separations, we can measure a length to a precise number of millimeters and we can estimate a fraction of a millimeter between two divisions. If the meterstick has centimeter divisions as the smallest separations, we measure a precise number of centimeters and estimate the fraction of a centimeter that remains.

The most basic way to indicate the precision of a quantity is to write it with the correct number of **significant figures**. The significant figures are all the digits that are known accurately plus the one estimated digit. If we say that the distance from here to the state line is 12 km, that does not mean we know the distance to be *exactly* 12 kilometers. Rather, the distance is 12 km *to the nearest kilometer*. If instead we said that the distance is 12.0 km, that would indicate that we know the distance to the nearest *tenth* of a kilometer. More significant figures mean greater precision.

Rules for Identifying Significant Figures

1. Nonzero digits are always significant.
2. Final or ending zeros written to the right of the decimal point are significant.
3. Zeros written on either side of the decimal point for the purpose of spacing the decimal point are not significant.
4. Zeros written between significant figures are significant.

Example 1.2

Identifying the Number of Significant Figures

For each of these values, identify the number of significant figures and rewrite it in standard scientific notation.

(a) 409.8 s   (b) 0.058700 cm

(c) 9500 g   (d) 950.0×10^1 mL

Strategy We follow the rules for identifying significant figures as given. To rewrite a number in scientific notation, we move the decimal point so that the number to the left of the decimal point is between 1 and 10 and compensate by multiplying by the appropriate power of ten.

Solution (a) All four digits are significant. The zero is between two significant figures, so it is significant. To write the number in scientific notation, we move the decimal point two places to the left and compensate by multiplying by 10^2: 4.098×10^2 s.

(b) The first two zeros are not significant; they are used to place the decimal point. The digits 5, 8, and 7 are significant, as are the two final zeros. The answer has five significant figures: 5.8700×10^{-2} cm.

(c) This measurement has two significant figures (9 and 5). The two final zeros are used to space the decimal point: 9.5×10^3 g.

(d) The final zero is significant since it comes after the decimal point. The zero to its left is also significant since it comes between two other significant digits. The result has four significant figures. The number is not in *standard* scientific notation since 950.0 is not between 1 and 10; in scientific notation we write 9.500×10^3 mL.

Discussion Scientific notation clearly indicates the number of significant figures since all zeros are significant; none are used only to place the decimal point. In (c), there would be no way to know if the value had three or four significant figures instead of two. Are those final zeros just place holders, or are they significant? Most likely they are place holders. Writing 9.5×10^3 g makes it clear that the measurement has two significant figures; if there were four, we would write 9.500×10^3 g.

Practice Problem 1.2   Identifying significant figures

State the number of significant figures in each of these measurements and rewrite them in standard scientific notation.

(a) 0.00010544 kg   (b) 0.005800 cm   (c) 602000 s

When two or more quantities are added or subtracted, the result is as precise as the *least precise* of the quantities (Example 1.3). When quantities are multiplied or divided, the result has the same number of significant figures as the quantity with the *smallest number of significant figures* (Example 1.4).

Example 1.3

Significant Figures in Addition

Calculate the sum 44.56005 s + 0.0698 s + 1103.2 s.

Strategy   The sum cannot be more precise than the least precise of the three quantities. The quantity 44.56005 s is known to the nearest 0.00001 s, 0.0698 s is known to the nearest 0.0001 s, and 1103.2 s is known to the nearest 0.1 s. Therefore the least precise is 1103.2 s. The sum has the same precision; it is known to the nearest tenth of a second.

Solution   According to the calculator,

$$44.56005 + 0.0698 + 1103.2 = 1147.82985$$

We do *not* want to write all of those digits in the answer. That would imply greater precision than we actually have. Rounding to the nearest tenth of a second, the sum is written

$$= 1147.8 \text{ s}$$

and there are five significant figures in the result.

Discussion   Note that the least precise measurement is not necessarily the one with the fewest number of significant figures. The least precise is the one whose right-most significant figure represents the largest unit: the "2" in 1103.2 s represents 2 tenths of a second. In addition or subtraction, we are concerned with the precision rather than the number of significant figures. The three quantities to be added have seven, three, and five significant figures, respectively, while the sum has five significant figures.

Practice Problem 1.3   Significant figures in subtraction

Calculate the difference 568.42 m – 3.924 m and write the result in scientific notation. How many significant figures are in the result?

Example 1.4

Significant Figures in Division

Write the solution to 28.84 m divided by 6.2 s with the correct number of significant figures.

Strategy   The quotient should have the same number of significant figures as the factor with the least number of significant figures.

Solution   A calculator gives

$$\frac{28.84}{6.2} = 4.651612903$$

Since the answer should have only two significant figures, we round the answer to

$$\frac{28.84 \text{ m}}{6.2 \text{ s}} = 4.7 \text{ m/s}$$

Discussion   Writing the answer as 4.651612903 m/s would give the false impression that we know the answer to a precision of about 0.000000001 m/s, whereas we actually have a precision of only about 0.1 m/s.

Practice Problem 1.4   Significant figures in multiplication

Find the product of 45.26 m/s and 2.41 s. How many significant figures does the product have?

When an integer, or a fraction of integers, is used in an equation, the precision of the result is not affected by the integer or the fraction; the number of significant figures is limited only by the measured values in the problem. The fraction 1/2 in an equation is *exact*; it does not reduce the number of significant figures to one. In an equation such as $C = 2\pi r$ for the circumference of a circle of radius r, the factors 2 and π are exact. We use as many digits for π as we need to maintain the precision of the other quantities.

Sometimes a problem may be too complicated to solve precisely, or information may be missing that would be necessary for a precise calculation. In such a case, an **order of magnitude** solution is the best we can do. By *order of magnitude,* we mean "roughly what power of ten?" An order of magnitude calculation is done to at most one significant figure. Even when a more precise solution is feasible, it is often a good idea to start with a quick, back-of-the-envelope estimate. Why? Because we can often make a good guess about the correct order of magnitude of the answer to a problem, even before we start solving the problem. If the answer comes out with a different order of magnitude, we go back and search out an error. Suppose a problem concerns a vase that is knocked off a fourth-story window ledge. We can guess by experience the order of magnitude of the time it takes the vase to hit the ground. It might be 1 s, or 2 s, but we are certain that it is *not* 1000 s or 0.00001 s.

1.5   UNITS

A **metric system** of units has been used for many years in scientific work and in European countries. The metric system is based on powers of ten (see Fig. 1.1). In 1960, the General Conference of Weights and Measures, an international authority on units, proposed a revised metric system called the *Système International d'Unités* in French (abbreviated **SI**), which uses the meter (m) for length, the kilogram (kg) for mass, and the second (s) for time. These are the three SI base units used in mechanics; there are four more SI base units (see Table 1.1). **Derived units** are constructed from combinations of the base units. For example, the SI unit of force is $kg \cdot m/s^2$; the combination of $kg \cdot m/s^2$ is given a special name, the newton (N), in honor of Isaac Newton. The newton is a derived unit because it is composed of a combination of base units. When units are named after famous scientists, the name of the unit is written with a lowercase letter, even though it is based on a proper name; the *abbreviation* for the unit is written with an uppercase letter. The inside front cover of the book has a complete listing of the derived SI units used in this book.

As an alternative to explicitly writing powers of ten, SI uses prefixes for units to indicate power of ten factors. Table 1.2 shows some of the powers of ten and the SI prefixes used for them. These are also listed on the inside front cover of the book.

SI units are preferred in physics and are emphasized in this book. Since other units are sometimes used, we must know how to convert units. Various scientific fields, even in physics, do use units other than SI units, whether for historical or practical reasons. For example, in atomic and nuclear physics, the SI unit of energy (the joule) is rarely used, for good reasons; instead the energy unit used is usually the electron-volt. Biologists and chemists use units that are not familiar to physicists. One reason that SI is preferred is that it provides a common denominator—all scientists are familiar with the SI units.

In most of the world, SI units are used in everyday life and in industry. In the United States, the U.S. Customary Units—sometimes called English units—are still used. The base units for this system are the foot (for length), the second (for time), and the pound (for weight—mass is a *derived* unit in this system).

In the autumn of 1999, to the chagrin of NASA, a $125 million spacecraft designed to orbit Mars crashed right into the planet. The company building the booster rocket provided information about the rocket's thrust in U.S. Customary Units, but the NASA scientists who were controlling the rocket thought the figures provided were in metric units. Arthur Stephenson, chairman of the Mars Climate Orbiter Mission Failure Investigation Board, stated that, "The 'root cause' of the loss of the spacecraft was the failed translation of English units into metric units in a segment of ground-based, navigation-related mission software." A journey of 122 million miles ended up a few hundred miles off course thanks to the discrepancy in units, unfortunately resulting in a dramatic failure of the mission.

If the statement of a problem includes a mixture of different units, the units must be converted to a single, consistent set before the problem is solved. Usually the best way is to convert everything to SI units. Unit conversions are carried out by multiplying or dividing units algebraically (Example 1.5). Common conversion factors are listed on the inside front cover of this book.

Some conversions are exact by definition. One meter is defined to be *exactly* equal to 100 centimeters; all SI prefixes are exactly a power of ten. The use of an exact conversion factor such as 1 m = 100 cm, or 1 foot = 12 inches, does not affect the precision of the result; the number of significant figures is limited only by the other quantities in the problem.

Table 1.1

SI Base Units

Base Quantity	Unit Name	Symbol
Length	meter	m
Mass	kilogram	kg
Time	second	s
Electrical current	ampere	A
Temperature	kelvin	K
Amount of substance	mole	mol
Luminous intensity	candela	cd

Table 1.2

SI Prefixes

Prefix (abbreviation)	Power of Ten
peta- (P)	10^{15}
tera- (T)	10^{12}
giga- (G)	10^9
mega- (M)	10^6
kilo- (k)	10^3
deci- (d)	10^{-1}
centi- (c)	10^{-2}
milli- (m)	10^{-3}
micro- (μ)	10^{-6}
nano- (n)	10^{-9}
pico- (p)	10^{-12}
femto- (f)	10^{-15}

Making The Connection:
spacecraft to Mars

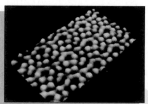

Silicon atoms
(radius 10^{-10} m)

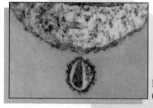

HIV (diameter 10^{-7} m) invading a T-lymphocyte
(a type of white blood cell)

A child
(height 10^0 m)

The Duomo (cathedral) in Florence, Italy
(height 10^2 m)

Earth
(diameter 10^7 m)

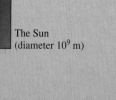

The Sun
(diameter 10^9 m)

A spiral galaxy
(diameter 10^{19} m)

Figure 1.1   Scientific notation uses powers of ten to express quantities that have a wide range of values.

Example 1.5

Buying Clothes in a Foreign Country

 Michel, an exchange student from France, is studying in the United States. He wishes to buy a new pair of jeans, but the sizes are all in *inches*. He does remember that 1 m = 3.28 ft and that 1 ft = 12 in. If his waist size is 82 cm, what is his waist size in inches?

Strategy   Each conversion factor can be written as a fraction. If 1 m = 3.28 ft, then

$$\frac{3.28 \text{ ft}}{1 \text{ m}} = 1$$

We can multiply any quantity by 1 without changing its value. We arrange each conversion factor in a fraction and multiply one at a time to get from cm to in.

Solution   We first convert cm to meters.

$$82 \text{ cm} \times \frac{1 \text{ m}}{100 \text{ cm}}$$

Now, we convert meters to feet.

$$82 \text{ cm} \times \frac{1 \text{ m}}{100 \text{ cm}} \times \frac{3.28 \text{ ft}}{1 \text{ m}}$$

Finally, we convert feet to inches.

$$82 \text{ cm} \times \frac{1 \text{ m}}{100 \text{ cm}} \times \frac{3.28 \text{ ft}}{1 \text{ m}} \times \frac{12 \text{ in}}{1 \text{ ft}} = 32 \text{ in}$$

In each case, the fraction is written so that the unit we are converting *from* cancels out.
   As a check:

$$\text{cm} \times \frac{\text{m}}{\text{cm}} \times \frac{\text{ft}}{\text{m}} \times \frac{\text{in}}{\text{ft}} = \text{in}$$

Discussion   This problem could have been done in one step using a direct conversion factor from inches to cm (1 in. = 2.54 cm). One of the great advantages of SI units is that all the conversion factors are powers of ten (Table 1.2); there is no need to remember that there are 12 inches in a foot, 4 quarts in a gallon, 16 ounces in a pound, 5280 feet in a mile, and so on.

Practice Problem 1.5   Driving on the autobahn
A BMW convertible travels on the German autobahn at a speed of 128 km/h. What is the speed of the car (a) in m/s? (b) in mi/h?

Example 1.6

Conversion of Volume

A beaker of water contains 255 mL of water. (1 mL = 1 milliliter; 1 liter = 1000 cm^3.) What is the volume of the water in (a) cm^3? (b) m^3?

Strategy   First convert milliliters to liters; then convert liters to cm^3. To convert cm^3 to m^3, use 100 cm = 1 m. Since there are *three* factors of cm to convert, we have to multiply by $\left(\frac{1 \text{ m}}{100 \text{ cm}}\right)$ *three times*.

Solution   (a) The prefix milli- means 10^{-3}, so 1 mL = 10^{-3} L. Then

$$255 \text{ mL} \times \frac{10^{-3} \text{ L}}{1 \text{ mL}} \times \frac{1000 \text{ cm}^3}{1 \text{ L}} = 255 \text{ cm}^3$$

(b) 1 m = 100 cm. Since we need to convert *cubic* centimeters to *cubic* meters, we must raise the conversion factor to the third power:

$$255 \text{ cm}^3 \times \left(\frac{1 \text{ m}}{100 \text{ cm}}\right)^3 = 255 \text{ cm}^3 \times \frac{(1 \text{ m})^3}{(100 \text{ cm})^3}$$

$$255 \text{ cm}^3 \times \frac{1 \text{ m}^3}{100^3 \text{ cm}^3} = 2.55 \times 10^{-4} \text{ m}^3$$

Discussion   Be careful when a unit is raised to a power other than one; the conversion factor must be raised to the same power. Writing out the units to make sure they cancel prevents mistakes. When a quantity is raised to a power, both the number and the unit must be raised to the same power. $(100 \text{ cm})^3$ is equal to $100^3 \text{ cm}^3 = 10^6 \text{ cm}^3$; it is *not* equal to 100 cm^3, nor is it equal to 10^6 cm.

Practice Problem 1.6   Surface area of Earth
If the radius of Earth is 6.4×10^3 km, find the surface area of Earth in square meters and in square miles. (Surface area of a sphere = $4\pi r^2$.)

Whenever a calculation is performed, always write out the units with each quantity. Combine the units algebraically to find the units of the result. This small effort has three important benefits:

1. It shows what the units of the result are. A common mistake is to get the correct numerical result of a calculation but to write it with the wrong units, making the answer wrong.

2. It shows where unit conversions must be done. If units that should have canceled do not, we go back and perform the necessary conversion. When a distance is calculated and the result comes out with units of $\frac{m \cdot s}{h}$, we should convert hours (h) to seconds (s).

3. It helps locate mistakes. If a distance is calculated and the units come out as m/s, we know to look for an error.

1.6   DIMENSIONAL ANALYSIS

Dimensions are basic *types* of units, such as time, length, and mass. (Warning: the word *dimension* has several other meanings such as in "three-dimensional space" or "the dimensions of a soccer field.") Many different units of length exist: meters, inches, miles, nautical miles, fathoms, leagues, astronomical units, angstroms, and cubits, just to name a few. All have dimensions of length; each can be converted into any other.

We can add, subtract, or equate quantities only if they have the same dimensions (although they may not necessarily be given in the same units). It is possible to add 3 meters to 2 inches (after converting units), but it is not possible to add 3 meters to 2 kilograms. To analyze dimensions, treat them as algebraic quantities, just as we did with units in Section 1.5. Usually [M], [L], and [T] are used to stand for mass, length, and time dimensions. Equivalently, we can use the SI base units: kg for mass, m for length, and s for time.

Example 1.7

Dimensional Analysis for a Distance Equation

Analyze the dimensions of the equation $d = vt$, where d is distance traveled, v is speed, and t is elapsed time.

Strategy   Replace each quantity with its dimensions. Distance has dimensions [L]. Speed has dimensions of length per unit time [L]/[T]. The equation is dimensionally consistent if the dimensions are the same on both sides.

Solution   The right side has dimensions

$$\frac{[L]}{[T]} \times [T] = [L]$$

Since both sides of the equation have dimensions of length, the equation is dimensionally consistent.

Discussion   If, by mistake, we wrote $d = v/t$ for the relation between distance traveled and elapsed time, we could quickly

catch the mistake by looking at the dimensions. On the right side, v/t would have dimensions $[L]/[T]^2$, which is not the same as the dimensions of d on the left side.

A quick dimensional analysis of this sort is a good way to catch algebraic errors. Whenever we are unsure whether an equation is correct, we can check the dimensions.

Practice Problem 1.7   Testing dimensions of another equation

Test the dimensions of the following equation:

$$x = \tfrac{1}{2}at$$

where x is distance traveled, a is acceleration (which has SI units m/s^2), and t is the elapsed time. If incorrect, can you suggest what might have been omitted?

Dimensional analysis is good for more than just checking equations. In many cases, we can completely solve a problem—up to a dimensionless factor like $1/(2\pi)$ or $\sqrt{3}$—using dimensional analysis. To do this, first list all the relevant quantities upon which the answer might depend. Then determine what combinations of them have the same dimensions as the answer for which we are looking. If only one such combination exists, then we have the answer, except for a possible dimensionless multiplicative constant.

Example 1.8

Violin String Frequency

While it is being played, a violin string produces a tone with frequency f in s^{-1} (Fig. 1.2); the frequency is the number of vibrations *per second* of the string. The string has mass m, length L, and tension T. If the tension is increased 5.0%, how does the frequency change? Tension is a force with SI unit $\frac{kg \cdot m}{s^2}$.

Strategy   We could make a study of violin strings, but let us see what we can find out by dimensional analysis. We want to find out how the frequency f can depend on m, L, and T. We won't know if there is a dimensionless constant involved, but we can work by proportions so any such constant will divide out.

Figure 1.2
A violinist

Solution   The unit of tension T is $\frac{kg \cdot m}{s^2}$. The units of f do not contain kg or m; we can get rid of them from T by dividing the tension by the length and the mass:

$$\frac{T}{mL} \text{ has SI unit } s^{-2}$$

That is almost what we want; all we have to do is take the square root:

$$\sqrt{\frac{T}{mL}} \text{ has SI unit } s^{-1}$$

Therefore,

$$f = C\sqrt{\frac{T}{mL}}$$

where C is some dimensionless constant. To answer the question, let the original frequency and tension be f and T and the new frequency and tension be f' and T', where $T' = 1.050T$. Frequency is proportional to the square root of tension, so

$$\frac{f'}{f} = \sqrt{\frac{T'}{T}} = \sqrt{1.050} = 1.025$$

The frequency increases 2.5%.

Discussion   The value of C is 1/2. That is the *only* thing we cannot get by dimensional analysis. There is *no* other way to combine T, m, and L to come up with a quantity that has the units of frequency.

Practice Problem 1.8   Increase in kinetic energy
When a body of mass m is moving with a speed v, it has kinetic energy associated with its motion. Energy is measured in $kg \cdot m^2 \cdot s^{-2}$. If the velocity of a moving body is increased by 25% while its mass remains constant, by what percentage does the kinetic energy increase?

1.7   PROBLEM-SOLVING TECHNIQUES

No single method exists that can be used to solve every physics problem. We demonstrate useful problem-solving techniques in the examples in every chapter of this text. Even for a particular problem, there may be more than one correct way to approach the solution. Problem-solving techniques are *skills* that must be *practiced* to be learned.

Think of the problem as a puzzle to be solved. Only in the easiest problems is the solution method immediately apparent. When you do not know the entire path to a solution, see where you can get by using the given information—find whatever you can.

Exploration of this sort may lead to a solution by suggesting a path that had not been considered. Be willing to take chances. You may even find the challenge enjoyable!

When having some difficulty, it helps to work with a classmate or two. One way to clarify your thoughts is to put them into words. After you have solved a problem, try to explain it to a friend. If you can explain the problem solution, you really do understand it. Both of you will benefit. But do not rely too much on help from others; the goal is for each of you to develop your own problem-solving skills.

General Guidelines for Problem Solving

1. Read the problem *carefully* and *all the way through.*

2. Reread the problem one sentence at a time and draw a sketch or diagram to help you visualize what is happening.

3. Write down and organize the given information. Some of the information can be written in labels on the diagram. Be sure that the labels are unambiguous. Identify in the diagram the object, the position, the instant of time, or the time interval to which the quantity applies. Sometimes information might be usefully written in a table beside the diagram. Look at the wording of the problem again for information that is implied or stated indirectly.

4. Identify the goal of the problem. What quantities need to be found?

5. If possible, make an estimate to determine the order of magnitude of the answer. This estimate is useful as a check on the final result to see if it is reasonable.

6. Think about how to get from the given information to the final desired information. Do not rush this step. Which principles of physics can be applied to the problem? Which will help get to the solution? How are the known and unknown quantities related? Which equations are relevant and may lead to the solution to the problem? This step requires skills developed only with much practice in problem solving.

7. Frequently, the solution involves more than one step. Intermediate quantities might have to be found first and then used to find the final answer. Try to map out a path from the given information to the solution. Whenever possible, a good strategy is to divide a complex problem into several simpler subproblems.

8. Perform algebraic manipulations with algebraic symbols (letters) as far as possible. Substituting the numbers in too early has a way of hiding mistakes.

9. Finally, if the problem requires a numerical answer, substitute the known numerical quantities, *with their units,* into the appropriate equation. Leaving out the units is a common source of error. Writing the units shows when a unit conversion needs to be done—and also may help identify an algebra mistake.

10. Once the solution is found, don't be in a hurry to move on. Check the answer—is it reasonable? Try to think of other ways to solve the same problem. Many problems can be solved in several different ways. Besides providing a check on the answer, finding more than one method of solution deepens our understanding of the principles of physics and develops problem-solving skills that will help solve other problems.

1.8 APPROXIMATIONS

Physics is about building models and comparing observations of the real world with the model. Simplified models help us to analyze complex situations. In various contexts we assume there is no friction, or no air resistance, no heat loss, or no wind blowing, and so forth. If we tried to take all these things into consideration with every problem, the problems would become vastly more complicated to solve. We never can take account of *every* possible influence. We freely make approximations whenever possible to turn a complex problem into an easier one, as long as the answer will be accurate enough for our purposes.

A valuable skill to develop is the ability to know when an assumption or approximation is reasonable. It might be permissible to neglect air resistance when dropping a stone, but not when dropping a beach ball. Why? We must always be prepared to justify any approximation we make by showing the answer is not changed very much by its use.

As well as making simplifying approximations in models, we also recognize that measurements are approximate. Every measured quantity has some uncertainty; it is impossible for a measurement to be exact to an arbitrarily large number of significant figures. Every measuring device has limits on the precision and accuracy of its measurements.

Sometimes it is difficult or impossible to measure precisely a quantity that is needed for a problem. Then we have to make a reasonable estimate. Suppose we need to know the surface area of a human being to determine the heat loss by radiation in a cold room. We can estimate the height of an average person. We can also estimate the average distance around the waist or hips. Approximating the shape of a human body as a cylinder, we can estimate the surface area by calculating the surface area of a cylinder with the same height and circumference (Fig. 1.3a).

If we need a better estimate, we use a slightly more refined model. For instance, we might approximate the arms, legs, trunk, and head and neck as cylinders of various sizes (Fig. 1.3b). How different is the sum of these areas from the original estimate? That gives an idea of how close the first estimate is.

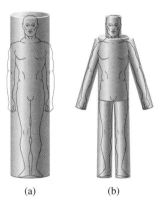

(a)   (b)

Figure 1.3 Approximation of human body by one or more cylinders

---

Example 1.9

Number of Cells in the Human Body

Average-sized animal cells are about 20 μm in length; some human blood cells are shown in Fig. 1.4. How many cells are there in the human body? Make an order-of-magnitude estimate.

Strategy We divide this problem into three subproblems: estimating the volume of a human, estimating the volume of the average cell, and finally estimating the number of cells.

To find the volume of a human body, we approximate the body as a cylinder, as previously discussed. Next we assume the cells are cubical to find the volume of a cell. Third, the ratio of the two volumes (volume of the body to volume of the cell) shows how many cells are in the body.

Solution Model the body as a cylinder. A typical height is about 2 m. A typical *maximum* circumference (think waist or hip

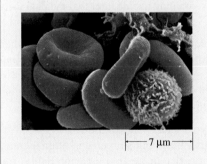

Figure 1.4
Red blood cells are about 7 μm in diameter, which is smaller than the average human cell length.

7 μm

size) is about 1 m. The corresponding radius is $1/(2\pi)$ m or about 1/6 m. The *average* radius is somewhat smaller; say 0.1 m. The volume of a cylinder is the height times the cross-sectional area:

$$V = Ah = \pi r^2 h \approx 3 \times (0.1\ \text{m})^2 \times (2\ \text{m}) = 0.06\ \text{m}^3$$

The volume of a cube is $V = s^3$. Then the volume of an average cell is about

$$V_{\text{cell}} \approx (2 \times 10^{-5}\ \text{m})^3 = 8 \times 10^{-15}\ \text{m}^3$$

The number of cells is the ratio of the two volumes:

$$N = \frac{\text{volume of body}}{\text{average volume of cell}} \approx \frac{6 \times 10^{-2}\ \text{m}^3}{8 \times 10^{-15}\ \text{m}^3} \approx 8 \times 10^{12} \approx 10^{13}$$

Discussion This number is called "10 trillion" (in the United States) or "10 billion" (in Britain). Based on this rough estimate, we cannot rule out the possibility that a better estimate might be 3×10^{12}. On the other hand, we *can* rule out the possibility that the number of cells is, say, 100 million (= 10^8).

Practice Problem 1.9 Drinking water consumed in the United States

How many liters of water are swallowed by the people living in the United States in one year? This is a type of problem made famous by the physicist Enrico Fermi (1901–1954), who was a master at this sort of back-of-the-envelope calculation. Such problems are often called *Fermi problems* in his honor. (1 liter = 10^{-3} m$^3 \approx$ 1 quart.)

Besides estimating quantities that are not well known, another kind of approximation appears frequently in physics. In some cases, we can substitute a mathematical expression with a simpler one that is approximately true. For example, if you were offered a million dollars to calculate the square root of 1.0004 in your head, could you do it? Of course, it would be impossible to give the *exact* answer, which has an infinite number of digits. A calculator shows that the square root of 1.0004 is *very close to* 1.0002. That value can be found from the **binomial approximation** (Appendix A.5), derived in math classes and used in solving physics problems:

$$(1 + x)^n \approx 1 + nx \quad \text{if } |x| \ll 1 \qquad (1\text{-}1)$$

For finding a square root, n is replaced by $\frac{1}{2}$. Then we have

$$(1 + x)^{1/2} \approx 1 + \tfrac{1}{2}x \quad \text{for } |x| \ll 1$$

or $(1 + 0.0004)^{1/2} \approx 1 + \tfrac{1}{2}(0.0004) = 1 + 0.0002 = 1.0002$

Any approximation is of limited validity. This one is valid as long as $|x| \ll 1$. The approximation becomes less useful as $|x|$ gets larger. See Problems 33 and 34 to test the limits of the approximation.

Other important mathematical approximations are found in Appendix A.5. One in particular that bears mentioning here is the **small angle approximation**:

$$\sin \theta \approx \tan \theta \approx \theta \quad \text{for small angles } \theta$$

where θ is measured in radians (rad) (see Appendix A.7). We discuss the radian more fully in Chapter 5. To convert radians to degrees, use 2π rad $= 360°$. As long as θ is small, $\sin \theta$, $\tan \theta$, and θ itself (measured in rad) all have very nearly the same value, so any one can be substituted for any of the others. The approximation is valid for $|\theta| \ll 1$ rad.

The symbol $|x|$ means the absolute value of x.
The symbol $\ll$ means *is much less than*.
The numeral "1" in the binomial approximation is exact.

1.9   GRAPHS

Graphs are used to help us see a pattern in the relationship between two variables. It is much easier to see a pattern on a graph than to see it in a table of numerical values. When we do experiments in physics, we change one quantity (the **independent variable**) and see what happens to another (the **dependent variable**). We want to see how one variable *depends on* another. The value of the independent variable is usually plotted along the horizontal axis of the graph. In a plot of p versus q, normally p is the dependent variable and q is the independent variable.

Some general guidelines for recording data and making graphs are given next.

Recording Data and Making Data Tables

1. Label columns with the names of the data being measured and be sure to include the units for the measurements. Do not erase any data, but just draw a line through data that you think is erroneous. Sometimes you may decide later that the data were correct after all.

2. Try to make a realistic estimate of the precision of the data being taken when recording numbers. For example, if the timer says 2.3673 s, but you know your reaction time can vary by as much as 0.1 s, the time should be recorded as 2.4 s. When doing calculations using measured values, remember to round the final answer to the correct number of significant figures.

3. Do not wait until you have collected all of your data to start a graph. It is much better to graph each data point as it is measured. By doing so, you can often identify equipment malfunction or measurement errors that make your data unreliable. You can also spot where something interesting happens and take data points closer together there. Graphing as you go means that you need to find out the range of values for both the independent and dependent variables.

Graphing Data

1. Make *large, neat* graphs. A tiny graph is not very illuminating. Use at least half a page. A graph made carelessly obscures the pattern between the two variables.

2. Label axes with the name of the quantities graphed and their units. Write a meaningful title.

3. When a linear relation is expected, use a ruler or straightedge to draw the best-fit straight line. Do not *assume* that the line must go through the origin—make a measurement to find out, if possible. Some of the data points will probably fall above the line and some will fall below the line.

4. Determine the slope of a best-fit line by measuring the ratio $\Delta y/\Delta x$ using as large a range of the graph as possible. Do not choose two data points to calculate the slope; instead, read values from two points on the best-fit line. Show the calculations. Do not forget to write the units; slopes of graphs in physics have units, since the quantities graphed have units.

5. When a nonlinear relationship is expected between the two variables, the best way to test that relationship is to manipulate the data algebraically so that a linear graph is expected. The human eye is a good judge of whether a straight line fits a set of data points. It is not so good at deciding whether a curve is parabolic, cubic, or exponential. To test the relationship $x = \frac{1}{2}at^2$, where x and t are the quantities measured, graph x versus t^2 instead of x versus t.

6. If one data point does not lie near the line or smooth curve connecting the other data points, that data point should be investigated to see whether an error was made in the measurement or whether some interesting event is occurring at that point. If something unusual is happening there, obtain additional data points in the vicinity.

7. When the slope of a graph is used to calculate some quantity, pay attention to the equation of the line and the units along the axes. The quantity to be found may be the inverse of the slope or twice the slope or one-half the slope. For example, if you wish to find the value of a in the relationship $x = \frac{1}{2}at^2$, and you graph your measured values of x and t as x versus t^2, then the slope of the line is $\frac{1}{2}a$. The value of a you seek is twice the slope.

The equation of a straight line on a graph of y versus x can be written $y = mx + b$, where m is the slope and b is the y intercept (the value of y corresponding to $x = 0$).

The symbol Δ, the Greek letter delta, stands for the difference between two measurements. The notation Δy is read aloud as "delta y" and represents a change in the value of y.

Example 1.10

Length of a Spring

In an introductory physics laboratory experiment, students are investigating how the length of a spring varies with the weight hanging from it. Various weights (accurately calibrated to 0.01 N) ranging up to 6.00 N can be hung from the spring; then the length of the spring is measured with a meterstick (Fig. 1.5). The goal is to see if the weight F and length L are related by

$$F = kx$$

where $x = (L - L_0)$, L_0 is the length of the spring when no weight is hanging from it, and k is called the *spring constant* of the spring. Graph the data in the table and calculate k for this spring.

F (N):	0	0.50	1.00	2.50	3.00	3.50	4.00	5.00	6.00
L (cm):	9.4	10.2	12.5	17.9	19.7	22.5	23.0	28.8	29.5

Strategy   Weight is the independent variable, so it is plotted on the horizontal axis. After plotting the data points, we draw the best-fit straight line. Then we calculate the slope of the line, using two points on the line that are widely separated and that cross gridlines of the graph (so the values are easy to read). The slope of the graph is not k; we must solve the equation for L, since length is plotted on the vertical axis.

Figure 1.5
A weight causes an extension in the length of a spring.

Solution   Figure 1.6 shows a graph with data points and a best-fit straight line. There is some scatter in the data, but a linear relationship is plausible.

continued on next page

Example 1.10 *continued*

Two points where the line crosses gridlines of the graph are (0.80 N, 12.0 cm) and (4.40 N, 25.0 cm). From these, we calculate the slope:

$$\text{slope} = \frac{\Delta L}{\Delta F} = \frac{25.0 \text{ cm} - 12.0 \text{ cm}}{4.40 \text{ N} - 0.80 \text{ N}} = 3.60 \frac{\text{cm}}{\text{N}}$$

By analyzing the units of the equation $F = k(L - L_0)$, it is clear that the slope cannot be the spring constant; k has the same units as weight divided by length (N/cm). Is the slope equal to $1/k$? The units would be correct for that case. To be sure, we solve the equation of the line for L:

$$L = \frac{F}{k} + L_0$$

We recognize the equation of a line with a slope of $1/k$. Therefore,

$$k = \frac{1}{3.60 \text{ cm/N}} = 0.278 \frac{\text{N}}{\text{cm}}$$

Discussion As discussed in the graphing guidelines, the slope of the straight-line graph is calculated from two widely spaced values *along the best-fit line*. We do not subtract values of actual data points. We are looking for an average value from the data; using two data points to find the slope would defeat the purpose of plotting a graph or of taking more than two data measurements. The values read from the graph, including the units, are indicated in Fig. 1.6. The units for the slope are cm/N since we plotted cm versus N. For this particular problem the *inverse* of the slope is the quantity we seek, the spring constant in N/cm.

Practice Problem 1.10   Another weight on spring

What is the length of the spring of Example 1.10 when a weight of 8.00 N is suspended? Assume that the relationship found in Example 1.10 still holds for this weight.

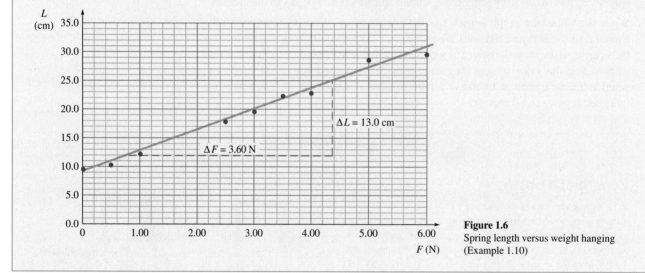

Figure 1.6
Spring length versus weight hanging (Example 1.10)

MASTER THE CONCEPTS

Summary

- Terms used in physics must be precisely defined. The same term may have a different meaning in physics from the meaning of the same word in other contexts.

- A working knowledge of algebra, geometry, and trigonometry is essential in the study of physics.

- The *factor* by which a quantity is increased or decreased is the ratio of the new value to the original value.

- When we say that A is *proportional* to B (written $A \propto B$), we mean that if B increases by some factor, then A must increase by the same factor.

- In *scientific notation,* a number is written as the product of a number between 1 and 10 and a whole number power of ten.

- *Significant figures* are the basic *grammar* of precision. They enable us to communicate quantitative information and indicate the precision to which that information is known.

- When two or more quantities are added or subtracted, the result is as precise as the *least precise* of the quantities. When quantities are multiplied or divided, the result has the same number of significant figures as the quantity with the *smallest number of significant figures.*

- Order of magnitude estimates and calculations are made to be sure that the more precise calculations are realistic.

- The units used for scientific work are those from the *Système International (SI)*. SI uses seven *base units,* which include the meter (m), the kilogram (kg), and the second (s) for length, mass, and time, respectively. Using combinations of the base units, we can construct other *derived units.*

MASTER THE CONCEPTS *continued*

- If the statement of a problem includes a mixture of different units, the units should be converted to a single, consistent set before the problem is solved. Usually the best way is to convert everything to SI units.

- Dimensional analysis is used as a quick check on the validity of equations. Whenever quantities are added, subtracted, or equated, they must have the same dimensions (although they may not necessarily be given in the same units).

- Mathematical approximations aid in simplifying complicated problems.

- Problem-solving techniques are *skills* that must be *practiced* to be learned.

- A graph is plotted to give a picture of the data and to show how one variable changes with respect to another. Graphs are used to help us see a pattern in the relationship between two variables.

- Whenever possible, a careful choice is made of the variables plotted so that the graph displays a linear relationship.

Highlighted Figures and Tables

F1.1 Powers of ten (p. 10)

T1.1 SI base units (p. 9)

T1.2 SI prefixes (p. 9)

CONCEPTUAL QUESTIONS

1. Give a few reasons for studying physics.
2. Why must words be carefully defined for scientific use?
3. Why are models used in scientific study if they do not exactly match real conditions?
4. By what factor does tripling the radius of a circle affect (a) the circumference of the circle? (b) the area of the circle?
5. What are some of the advantages of scientific notation?
6. After which numeral is the decimal point usually placed in scientific notation? What determines the number of numerical digits written in scientific notation?
7. Are all the digits listed as "significant figures" precisely known? Might any of the significant digits be less precisely known than others? Explain.
8. Why is it important to write quantities with the correct number of significant figures?
9. List three of the base units used in SI.
10. What are some of the differences between the SI and the U.S. Customary system of units? Why is SI preferred for scientific work?
11. Sort these units into three groups of dimensions and identify the dimensions: fathoms, grams, years, kilometers, miles, months, kilograms, inches, seconds.
12. What are the first two steps to be followed in solving almost any physics problem?
13. Why do scientists plot graphs of their data instead of just writing them in tables?
14. A student's lab report concludes, "The speed of sound in air is 327." What is wrong with that statement?
15. What are some of the reasons for making order-of-magnitude estimates?
16. Once the solution of a problem has been found, what should be done before moving on to solve another problem?

MULTIPLE CHOICE QUESTIONS

1. A kilometer is approximately
   (a) 2 miles   (b) 1/2 mile   (c) 1/10 mile   (d) 1/4 mile

2. 55 miles/hour is approximately
   (a) 90 km/h   (b) 30 km/h   (c) 10 km/h   (d) 2 km/h

3. By what factor does the volume of a cube increase if the length of the edges are doubled?
   (a) 16   (b) 8   (c) 4   (d) 2   (e) $\sqrt{2}$

4. If the length of a box is reduced to one-third of its original value and the width and height are doubled, by what factor has the volume changed?
   (a) 2/3   (b) 1   (c) 4/3   (d) 3/4
   (e) depends on relative proportion of length to height and width

5. If the area of a circle is found to be half of its original value after the radius is multiplied by a certain factor, what was the factor used?
   (a) $1/(2\pi)$ (b) 1/2   (c) $\sqrt{2}$ (d) $1/\sqrt{2}$ (e) 1/4

6. In terms of the original diameter d, what new diameter will result in a new spherical volume that is a factor of eight times the original volume?
   (a) $8d$ (b) $2d$ (c) $d/2$ (d) $d \times \sqrt[3]{2}$ (e) $d/8$

7. An equation for potential energy states $U = mgh$. If U is in $kg \cdot m^2 \cdot s^{-2}$, m is in kg, and g is in $m \cdot s^{-2}$, what are the units of h?
   (a) s   (b) s^2 (c) m^{-1} (d) m   (e) g^{-1}

8. The equation for the velocity of sound in a gas states that $v = \sqrt{\gamma k_B T/m}$. Velocity v is measured in m/s, γ is a dimensionless constant, T is temperature in kelvin (K), and m is mass in kg. What are the units for the Boltzmann constant, k_B?
   (a) $kg \cdot m^2 \cdot s^2$ (b) $kg \cdot m^2 \cdot s^{-2} \cdot K^{-1}$ (c) N/K
   (d) $kg \cdot m/s$ (e) $kg \cdot m^2 \cdot s^{-2}$

9. How many significant figures are in the sum 4.56 + 9.032 + 580.0078 + 540.439?

   (a) 3   (b) 4   (c) 5   (d) 6   (e) 7

10. How many significant figures are in the product of 0.0078406 × 9.45020?

   (a) 3   (b) 4   (c) 5   (d) 6   (e) 7

PROBLEMS

Note: **C** indicates a combination conceptual/quantitative problem. Gold diamonds ✦, ✦✦ are used to indicate the increasing level of difficulty of each problem. Problem numbers appearing in blue, 9., denote problems that have a detailed solution available in the Student Solutions Manual. Some problems are *paired* by concept; their numbers are connected by a ruled box.

1.3 The Use of Mathematics

1. A spherical balloon expands when it is taken from the cold outdoors to the inside of a warm house. If its surface area increases 16.0%, by what percentage does the radius of the balloon change?

✦ 2. The weight of an object at the surface of a planet is proportional to the planet's mass and inversely proportional to the square of the radius of the planet. Jupiter's radius is 11 times Earth's and its mass is 320 times Earth's. An apple weighs 1.0 N on Earth. How much would it weigh on Jupiter?

3. In cleaning out the artery of a patient, a doctor increases the radius of the opening by a factor of two. By what factor does the cross-sectional area of the artery change?

✦ 4. A scanning electron micrograph of xylem vessels in a corn root shows the vessels magnified by a factor of 600. In the micrograph the xylem vessel is 3.0 cm in diameter. (a) What is the diameter of the vessel itself? (b) By what factor has the cross-sectional area of the vessel been increased in the micrograph?

✦ 5. The average speed of a nitrogen molecule in air is proportional to the square root of the temperature in kelvins. If the average speed is 475 m/s on a warm summer day (temperature = 300.0 kelvins), what is the average speed on a cold winter day (250.0 kelvins)?

✦ 6. The electrical power P drawn from a generator by a lightbulb of resistance R is $P = \dfrac{V^2}{R}$, where V is the line voltage. The resistance of bulb B is 40% greater than the resistance of bulb A. What is the ratio $\dfrac{P_B}{P_A}$ of the power drawn by bulb B to the power drawn by bulb A, if the line voltages are the same?

1.4 Scientific Notation and Significant Figures

7. Given these measurements, identify the number of significant figures and rewrite in scientific notation.

   (a) 0.00574 kg   (b) 2 m   (c) 0.450×10^{-2} m
   (d) 45.0 kg   (e) 10.09×10^4 s   (f) 0.09500×10^5 mL

8. Write these numbers in scientific notation: (a) the U.S. population, 250000000; (b) the diameter of a Helium nucleus, 0.000 000 000 000 003 8 m.

9. Perform these operations with the appropriate number of significant figures.

   (a) $3.783 \times 10^6 + 1.25 \times 10^8$ (b) $(3.783 \times 10^6) \div (3.0 \times 10^{-2})$

10. Write your answers to these problems with the appropriate number of significant figures.

   (a) $6.85 \times 10^{-5} + 2.7 \times 10^{-7}$ (b) 702.35 + 1897.648
   (c) 5.0×4.3 (d) $0.04/\pi$ (e) $0.040/\pi$

11. In these calculations, be sure to use an appropriate number of significant figures.

   (a) $4.759 \times 10^5 - 3.68 \times 10^7$ (b) $\dfrac{(6.497 \times 10^{-31})^5}{5.1037 \times 10^{26}}$

12. How many significant figures are in each of these measurements?

   (a) 7.68 g   (b) 0.420 kg   (c) 0.073 m   (d) 7.68×10^5 g
   (e) 4.20×10^3 kg   (f) 7.3×10^{-2} m   (g) 2.300×10^4 s

1.5 Units

13. Convert 1.00 km/h to m/s using 1 km = 1000 m and 1 h = 3600 s.

14. A sprinter can run at a top speed of 0.32 miles per minute. Express her speed in (a) m/s and (b) mi/h.

15. (a) How many center-stripe road reflectors, separated by 17.6 yards, are required along a 2.20-mile section of curving mountain roadway? (b) Solve the same problem for a road length of 3.54 km with the markers placed every 16.0 m. Would you prefer to be the highway engineer in a country with a metric system or a U.S. Common Unit system?

16. A furlong is 220 yards; a fortnight is 14 days. How fast is 1 furlong per fortnight (a) in μm/s? (b) in km/day?

17. The intensity of the Sun's radiation that reaches Earth's atmosphere is 1.4 kW/m² (kW = kilowatt; W = watt). Convert this to W/cm².

18. You are given these approximate measurements: (a) the radius of Earth is 6×10^6 m, (b) the length of a human body is 6 feet, (c) a cell's diameter is 10^{-6} m, (d) the width of the hemoglobin molecule is 3×10^{-9} m, and (e) the distance between two atoms (carbon and nitrogen) is 3×10^{-10} m. Write these measurements in the simplest possible metric prefix forms (in either nm, Mm, μm, or whatever works best).

19. The smallest "living" thing is probably a type of infectious agent know as a viroid. Viroids are plant pathogens that consist of a circular loop of single-stranded RNA, containing about 300 bases. (Think of the bases as beads strung on a circular RNA string.) The distance from one base to the next (measured along the circumference of the circular loop) is about 0.35 nm. What is the diameter of a viroid in (a) m, (b) μm, and (c) in.?

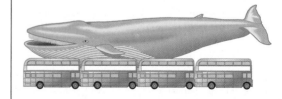

20. The largest living creature on Earth is the blue whale, which has an average length of 70 ft. The largest blue whale on record (and therefore the largest animal ever found) was 1.10×10^2 ft long. (a) Convert this length to meters. (b) If a double-decker London bus is 8.0 m long, how many double-decker-bus lengths is the record whale?

21. Density is the ratio of mass to volume. Mercury has a density of 1.36×10^4 kg/m^3. What is the density of mercury in units of g/cm^3?

22. An air molecule is moving at a speed of 459 m/s. How many meters would the molecule move during 7.00 ms (milliseconds) if it didn't collide with any other molecules?

23. A sheet of paper has length 27.95 cm, width 8.5 in., and thickness 0.10 mm. What is the volume of a sheet of paper in m^3? (Volume = length × width × thickness.)

24. Express this product in units of kg^3 with the appropriate number of significant figures: $(3.2 \text{ kg}) \times (4.0 \text{ g}) \times (13 \times 10^{-3} \text{ mg})$.

◆◆ 25. Three of the fundamental constants of physics are

the universal gravitational constant,

$$G = 6.7 \times 10^{-11} \, \frac{\text{m}^3}{\text{kg} \cdot \text{s}^2}$$

the speed of light, $c = 3.0 \times 10^8$ m/s

and Planck's constant,

$$h = 6.6 \times 10^{-34} \text{ J} \cdot \text{s} \left(1 \text{ J} = \frac{1 \text{ kg} \cdot \text{m}^2}{\text{s}^2} \right)$$

(a) Find a combination of these three constants that has the dimensions of time. This time is called the *Planck time* and represents the age of the universe before which the laws of physics as presently understood cannot be applied.

(b) Using the formula for the Planck time derived in part (a), what is the time in seconds?

26. (a) How many square centimeters are there in 1 square foot? (1 in. = 2.54 cm.) (b) How many square centimeters are there in 1 square meter? (c) Using your answers to parts (a) and (b), but without using your calculator, roughly how many square feet are in one square meter?

27. The largest human chromosome (number 1) consists of a long chain of 2.5×10^8 base pairs of DNA. In the metaphase stage of cell division, the base pairs are all tightly wrapped up into a rod of about 9 μm in length. The distance between base pairs is about 3.5×10^{-10} m. If you could unwrap your chromosome 1, (a) how many meters long would it be? (b) how many inches?

28. The record blue whale in Problem 20 had a mass of 1.9×10^5 kg. Assuming that its average density was 0.85 g/cm^3, as has been measured for other blue whales, what was the volume of the whale in m^3? (Average density is the ratio of mass to volume.)

1.6 Dimensional Analysis

◆ 29. An object moving at speed v around a circle of radius r has an acceleration a directed toward the center of the circle. The SI unit of acceleration is m/s^2. (a) Use dimensional analysis to find a as a function of v and r. (b) If the speed is increased 10.0%, by what percentage does the centripetal acceleration increase?

◆ 30. The speed of ocean waves depends on their wavelength λ (measured in meters) and the gravitational field strength g (measured in m/s^2) in this way:

$$v = K\lambda^p g^q$$

where K is a dimensionless constant. Find the values of the exponents p and q.

◆◆ 31. The Space Shuttle astronauts use a *massing chair* to measure their mass. The chair is attached to a spring and is free to oscillate back and forth. The frequency of the oscillation is measured and that is used to calculate the total mass m attached to the spring. If the spring constant of the spring k is measured in kg/s^2 and the chair's frequency f is 0.50 s^{-1} for a 62-kg astronaut, what is the chair's frequency for a 75-kg astronaut? The chair itself has a mass of 10.0 kg. [*Hint:* use dimensional analysis to find out how f depends on m and k.]

32. Use dimensional analysis to determine how the period T of a pendulum (the elapsed time for a complete cycle of motion) depends on some, or all, of these properties: the length L of the pendulum, the mass m of the pendulum bob, and the gravitational field strength g (in m/s^2). Assume that the amplitude of the swing (the maximum angle that string makes with the vertical) has no effect on the period. (See Fig. 1.7.)

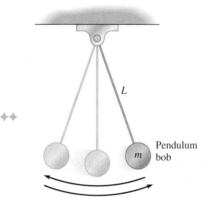

Figure 1.7
Swinging pendulum
(Problem 32)

1.8 Approximations

33. Use the binomial approximation (Eq. 1–1) to find the square root of (a) 1.008; (b) 1.04; (c) 1.4; (d) 1.66. Compare approximations with results from a calculator.

34. Use the binomial approximation (Eq. 1–1) to find the cube root of (a) 0.9994; (b) 0.994; (c) 0.92; (d) 1.3. Compare approximations with results from a calculator.

35. For each of the angles $\theta = 40°$, 30°, 20°, and 10°, use a calculator to find the value of θ in *radians* and the value of tan θ. Organize the values in a table. At which of these angles in degrees does the difference between θ measured in radians and the value of tan θ become approximately: (a) 10% of θ in rad? (b) 1% of θ in rad?

36. For each of the angles $\theta = 35°$, 25°, 18°, and 8°, make a table to compare values of tan θ and sin θ. At which of these angles in degrees does the difference between sin θ and tan θ become approximately: (a) 10% of sin θ? (b) 1% of sin θ?

37. What is the order of magnitude of the height (in meters) of a 40-story building?

38. What is the order of magnitude of the number of seconds in one year?

◆ 39. Without looking up any data, make an order-of-magnitude estimate of the annual consumption of gasoline (in gallons) by passenger cars in the United States. Make reasonable order-of-magnitude estimates for any quantities you need. Think in terms of average quantities. (1 gal ≈ 4 liters.)

◆ 40. How many cups of water are required to fill a bathtub?

1.9 Graphs

41. A nurse recorded the values shown in Table 1.3 for a patient's temperature. Plot a graph of temperature versus elapsed time and from the graph find (a) an estimate of the temperature at noon and (b) the slope of the graph. (c) Would you expect the graph to follow the same trend over the next 12 hours? Explain.

Table 1.3

Temperature Chart for Problem 41

Time	Temp (°F)
10:00 A.M.	100.00
10:30 A.M.	100.45
11:00 A.M.	100.90
11:30 A.M.	101.35
12:45 P.M.	102.48

42. The weight of a baby measured over an 11-month period is given in Table 1.4. (a) Plot the baby's weight versus age over the 11 months. (b) What was the average monthly weight gain for this baby over the period from birth to 5 months? How do you find this value from the graph? (c) What was the average monthly weight gain for the baby over the period from 5 to 10 months? (d) If a baby continued to grow at the same rate as in the first five months of life, what would the child weigh at age 12 years?

43. A physics student plots results of an experiment as v versus t. The equation that describes the line is given by $at = v - v_0$. (a) What is the slope of this line? (b) What is the "y" intercept of this line?

44. A linear plot of speed versus elapsed time has a slope of 6.0 m/s^2 and a "y" intercept of 3.0 m/s. (a) What is the change in speed in the time interval between 4.0 seconds and 6.0 seconds? (b) What is the speed when the elapsed time is equal to 5.0 seconds?

45. The population of a culture of yeast cells is studied in the laboratory to see the effects of limited resources (food, space) on population growth. At 2-hour intervals, the size of the population (measured as total mass of yeast cells) is recorded (Table 1.5). (a) Make a graph of the yeast population as a function of elapsed time. Draw a best-fit smooth curve. (b) Notice from the graph of part (a) that after a long time, the population approaches asymptotically a maximum known as the *carrying capacity*. From the graph, estimate the carrying capacity for this population. (c) When the population is much smaller than the carrying capacity, the growth is expected to be exponential: $m(t) = m_0 e^{rt}$, where m is the population at any time t, m_0 is the initial population, r is the *intrinsic growth rate* (that is, the growth rate in the absence of limits), and e is the base of natural logarithms (see Appendix A.3). To obtain a straight line graph from this exponential relationship, we can plot the natural logarithm of m/m_0:

$$\ln \frac{m}{m_0} = \ln e^{rt} = rt$$

Make a graph of $\ln \dfrac{m}{m_0}$ versus t from $t = 0$ to $t = 6.0$ h, and use it to estimate the intrinsic growth rate r for the yeast population. (The term ln stands for the natural logarithm; see Appendix A.3 if you need help with natural logs.)

Table 1.5

Yeast Population for Problem 45

Time (h)	Mass (g)
0.0	3.2
2.0	5.9
4.0	10.8
6.0	19.1
8.0	31.2
10.0	46.5
12.0	62.0
14.0	74.9
16.0	83.7
18.0	89.3
20.0	92.5
22.0	94.0
24.0	95.1

Table 1.4

Baby's Weight Chart for Problem 42

Weight (lb)	6.6	7.4	9.6	11.2	12.0	13.6	13.8	14.8	15.0	16.6	17.5	18.4
Age (months)	0 (birth)	1.0	2.0	3.0	4.0	5.0	6.0	7.0	8.0	9.0	10.0	11.0

COMPREHENSIVE PROBLEMS

46. A poster advertising a student election candidate is too large according to the election rules. The candidate is told she must reduce the length and width of the poster by 20.0%. By what percentage will the area of the poster be reduced?

47. If the radius of a circular garden plot is increased by 25%, by what percentage does the area of the garden increase?

48. A snail crawls at a pace of 5.0 cm/min. Express the snail's speed in (a) ft/s and (b) mi/h.

49. A marathon race is about 26 miles long. What is the length of the race in kilometers?

C 50. For what radius, *if any,* does the area of a circle equal the circumference of the circle? Explain.

51. The weight W of an object is given by $W = mg$, where m is the object's mass and g is the gravitational field strength. The SI unit of field strength g, expressed in SI base units, is m/s^2. What is the SI unit for weight, expressed in base units?

52. When a force acts over a distance, work is done (Chapter 6). If the force is parallel to the motion, the work W is given by $W = Fd$, where F is the magnitude of the force and d is the distance over which the force acts. (Note that in this context, W stands for work, not for weight.) Find the SI unit of work. Write it two ways: once in terms of base units only and again using the newton $\left(1\ \mathrm{N} = 1\ \mathrm{kg \cdot m/s^2}\right)$.

C 53. One morning you read in the *New York Times* that Bill Gates' net worth is $90,000,000,000. Later that day you see him on the street, and he gives you a $100 bill. What is his net worth now? (Think of significant figures.)

♦♦ 54. Make an order-of-magnitude estimate of the volume of water in the oceans. Do not look up any data in books. (Use your ingenuity to estimate the radius or circumference of Earth.)

♦♦ 55. According to the Stefan-Boltzmann law, an object radiates heat at a rate $\Delta Q/\Delta t = \epsilon \sigma A T^{n}$. In SI units, $\Delta Q/\Delta t$ is measured in watts (W), T is the temperature in kelvins (K), A is the object's surface area, $\sigma = 5.67 \times 10^{-8}$ W·m^{-2}·K^{-4}, and ϵ is a dimensionless number called the emissivity. (a) What is the value of n? (b) If Earth's temperature in kelvins were to increase 5.0%, by how much would Earth's rate of radiation increase?

56. A typical virus is a packet of protein and DNA (or RNA) and can be spherical in shape. The influenza A virus is a spherical virus that has a diameter of 85 nm. If the volume of saliva coughed onto you by your friend with the flu is 0.010 cm^3 and $\dfrac{1}{10^9}$ of that volume consists of viral particles, how many influenza viruses have just landed on you?

57. A patient's temperature was 97.0°F at 8:05 A.M. and 101.0°F at 12:05 P.M. If the temperature change with respect to elapsed time was linear throughout the day, what would the patient's temperature be at 3:35 P.M.?

C ♦ 58. The data in Table 1.6 were obtained in a study of the effects of feeding two equal-sized flocks of hens with special diets. Initially, both flocks were producing an average of 10 eggs per day. The hens in Flock A had a common supplement added to their usual corn diet, while those in Flock B were fed the new NutruCorn. The number of eggs produced per day was then averaged over a week of egg laying for the next few weeks.

Plot the data showing average daily egg production versus time over the first 4 weeks for both groups of hens on one set of axes. Is the relationship linear? If not, can you think of a way to plot the data so that there will be a linear relationship? [*Hint:* You might need to make a new table. Consider the *increase* in average daily egg production per week.] After studying these results from the NutruCorn manufacturers, would you be persuaded to buy their product for your hens?

59. Suppose you have a pair of Seven League Boots. These are magic boots that enable you to stride along a distance of 7.0 leagues with each step. (a) If you march along at a military march pace of 120 paces/minute, what will be your speed in km/h? (b) Assuming you could march on top of the oceans when you step off the continents, how long (in minutes) will it take you to march around the Earth at the equator? (1 league = 3 miles = 4.8 km.)

Table 1.6

Average Daily Egg Production for Problem 58

Data Table	Average Daily # Eggs/Week	
Time (weeks)	Flock A	Flock B
0	10.0	10.0
1	11.2	12.1
2	13.6	16.4
3	17.2	22.8
4	22.0	31.3
5	22.1	31.0
6	22.0	31.2

ANSWERS TO PRACTICE PROBLEMS

1.1 81.0 W   **1.2** (a) five; 1.0544×10^{-4} kg; (b) four; 5.800×10^{-3} cm; (c) three; 6.02×10^5 s   **1.3** The least precise value is to hundredths of a meter; we round the result to the nearest hundredth of a meter: 564.50 m or, in scientific notation, 5.6450×10^2 m; five significant figures.   **1.4** 109 m; three significant figures   **1.5** (a) 35.6 m/s; (b) 79.5 mi/h   **1.6** 5.1×10^{14} m^2; 2.0×10^8 mi^2 **1.7** The equation is dimensionally inconsistent; the right side has dimensions [L]/[T]. To have matching dimensions we must multiply the right side by [T]; the equation must involve time squared: $x = \frac{1}{2}at^2$.   **1.8** kinetic energy = (constant) × mv^2; kinetic energy increases by 56%.   **1.9** 10^{11} liters (Make a rough estimate of the population to be about 2.5×10^8 people; each drinking about 1.5 liters/day.)   **1.10** 37.9 cm

Forces and Introduction to Vectors

How is it possible for a water strider to stand on top of the surface of a pond without sinking into the water? What is holding it up?

Concepts & Skills to Review

- scientific notation and significant figures (Section 1.4)
- units (Section 1.5)
- problem-solving techniques (Section 1.7)

2.1   FORCES

This chapter begins our study of **classical mechanics**. Although it does not apply to the extremely small (atoms, nuclei, elementary particles), the cosmologically large (galaxies), or the incredibly fast (things moving near the speed of light), classical mechanics does apply to just about everything else. Classical mechanics enables us to predict the return of a comet, explain the operation of a centrifuge, control the motion of spacecraft and submarines, construct an artificial knee, design a longer-range golf ball, and understand how a water strider is able to stand on the surface of a pond.

The essence of classical mechanics is stated in three laws of motion, formulated by Isaac Newton (1642–1727), which tell us how the forces acting on an object determine the motion of that object. Together with his law of universal gravitation, Newton's laws showed for the first time that the heavenly bodies (the Sun, the planets and their satellites) obey the same physical laws of motion as do earthly bodies. To pre-Newtonian thinkers, the motion of the heavenly bodies was perfect and eternal, while earthly bodies always come to rest. For example, why does the Moon orbit the Earth month after month, without any engine to drive it, while an airplane that runs out of fuel falls to the ground? Pre-Newtonian thinkers naturally reasoned that there must be two different sets of physical laws.

To understand the causes of motion, we need to know how one body affects the motion of another. In other words, we need to study **interactions** between bodies. There are many sorts of interactions. Some are long-range, such as the gravitational interaction between the Earth and the Moon that keeps the Moon from wandering off. This interaction occurs even though the Earth and the Moon are far apart. Other long-range interactions are electrical or magnetic in nature.

Another type of interaction involves contact, as when a baseball bat contacts the pitcher's fastball and sends it over the fence for a home run. The bat has no noticeable effect on the ball's motion until the two come into contact. This brief interaction between the bat and ball causes a drastic change in the ball's motion. It also affects the bat, which is why the batter can feel the contact with the ball; the bat and ball are actually deformed slightly during the contact. The subsequent trajectory of the ball is affected by interactions with the bat, with the air, and with the Earth (gravity).

Physics at Home

On a dry day, stand in front of a mirror and run a comb vigorously through your hair until you hear some crackling. Now hold the comb a few centimeters from your hair. Observe the long-range electrical interaction between your hair and the comb.

Now take a refrigerator magnet. Hold it near but not touching the refrigerator door. You can feel the effect of a long-range magnetic interaction.

Just as human life would be dull without social interactions, the physical universe would be dull without physical interactions. Social interactions with friends and family change our behavior; physical interactions change the "behavior" (motion, temperature, etc.) of matter. Without interactions, that fastball could not be turned into a home run; there would be no gravity to make it fall to Earth and no air drag to slow it down. What *would* happen to the ball if it did not interact with anything else? According to Newton's first law (presented later in this chapter), the ball would continue on its course in a straight line with an undiminished speed.

Several different models are used in physics to describe the effect of an interaction. Some of the models that we study later in this book include potential energy functions (Chapter 6), momentum exchange (Chapter 7), and the exchange of messenger particles (Chapter 30). In this chapter we use *forces* to describe the interaction between two bodies. A **force** is a push or a pull that one body exerts on another. When you play handball, your hand exerts a force on the ball. The ball also exerts a force on your hand, the effect of

Force: a push or pull that one object exerts on another

which you can feel. A magnetic force pulls a magnet to the refrigerator door. A magnetic force also pulls the refrigerator door toward the magnet; this force is not as easy to observe because the door is so massive and is held in place by other forces. When two bodies interact, they exert forces on each other. We need to be able to analyze the forces acting on a body in order to relate them to changes in the body's state of motion.

Isolated forces do not exist; they always exist in pairs. Nothing can exert a force without having a force exerted on it. When you push open a door, the door pushes you. When two cars collide, each exerts a force on the other. Newton's third law (Section 2.3) specifies how the two forces that make up an interaction are related. Note that the two forces that make up an interaction *act on different objects*—the two objects that are interacting. For any force, we must be able to uniquely name both the object that exerts the force and the object on which the force is exerted. Doing so helps correctly identify the forces in a problem, which is a crucial first step in solving a mechanics problem.

Conceptual Example 2.1

Forces Acting on the *Alvin*

 The submersible *Alvin* was used in 1985 to find the remains of the *Titanic* on the ocean floor. Identify all the forces acting on the submersible while it sits at rest on the deck of the *Titanic*.

Strategy There are two kinds of forces acting on the *Alvin:* long-range forces and contact forces. The possible long-range forces are gravity, electricity, and magnetism. Contact forces must involve contact with some other object, so we consider everything in contact with the *Alvin*. For each force, whether long-range or contact, we should be able to identify what object is exerting the force.

Solution (1) The only long-range force on the *Alvin* is gravity, a downward pull due to the mass of the entire Earth. Long-range electrical and magnetic forces are unlikely, unless the *Alvin* is made of iron and happens to be sitting near some magnetized rocks.

(2) The *Alvin* is sitting on the deck without sinking into it. Therefore, the deck exerts a contact force upward on the *Alvin*.

(3) The only other thing in contact with the *Alvin* is seawater. There are contact forces exerted on the *Alvin* by the water.

Discussion Did we omit the buoyant force—the force that makes some things float and that makes submerged objects *seem* to weigh less under water? The buoyant force is not one of the long-range forces, so it must be a contact force; we have already listed all possible contact forces. The buoyant force is a result of the water squeezing the *Alvin* inward on all sides. Similarly, if water currents exert a sideways force on the *Alvin,* the sideways force arises out of the contact force due to the water. At the microscopic level, the contact forces exerted on the *Alvin* by the water are due to vast numbers of water molecules colliding with the vessel's exterior. It is a great simplification to think of huge numbers of microscopic forces combined into one or two macroscopic forces.

Correctly identifying all of the forces acting on an object is usually the first step to solving a mechanics problem. Overlooking one of the forces, or including a spurious one, makes it impossible to come up with the correct solution to the problem.

Conceptual Practice Problem 2.1 A sailboat on a lake

Identify the major forces acting on a sailboat sailing along the surface of a lake in a gentle breeze (Fig. 2.1).

The submersible *Alvin* on the deck of the *Titanic*.

Figure 2.1
A sailboat is acted upon by several forces.

Measuring Forces

If the concept of force is to be useful in physics, there must be a way to measure forces. Consider a simple spring scale such as those used in a supermarket's produce section. As the scale's pan is pulled down, the spring above is stretched. The harder you pull, the more the spring stretches. As the spring stretches, an attached pointer moves. Then all we have to do to measure forces is to calibrate the scale so that the amount of stretch measures the strength (or *magnitude*) of the force.

As was observed by Robert Hooke (1635–1703), many springs have the property that the extension or compression—the increase or decrease in length from the unstretched length—is roughly proportional to the force exerted on the ends of the spring. **Hooke's law** says the deformation of an object tends to be proportional to the deforming force, as long as the force is not too great:

Hooke's law: the deformation is proportional to the deforming force.

Hooke's law for an ideal spring

$$F = kx \qquad (2\text{-}1)$$

In Eq. (2-1), F is the *magnitude* of the force exerted on each end of the spring, x is the distance that the spring is stretched or compressed from its relaxed length, and k is called the **spring constant** for that particular spring. (Any *real* spring deviates from this simple linear behavior when stretched or compressed too much.) The SI unit for force is called the newton (N), as discussed in Section 1.5, and the SI unit of length is the meter (m), so the SI units for a spring constant are N/m.

The spring constant is a measure of how hard it is to stretch or compress a spring. A stiffer spring has a larger spring constant because larger forces must be exerted on the ends of the spring to stretch or compress it.

A supermarket scale is used to measure the weight of the produce. **Weight** is the force of the Earth's gravity pulling downward. The spring scale has no *direct* way to measure a gravitational force; what it *can* measure is the downward force that the produce exerts on the pan, which is a contact force.

In the United States, supermarket scales are generally calibrated to measure forces in pounds (lb). To convert pounds to newtons, use the approximate conversion factors

The weight of an object near Earth's surface is the force of the Earth's gravity on the object.

$$1.00\ \text{lb} = 4.448\ \text{N} \quad \text{or} \quad 1.00\ \text{N} = 0.2248\ \text{lb}$$

The newton is a derived unit in the SI system. It can be written in terms of three SI base units:

$$1\ \text{N} = 1\ \frac{\text{kg·m}}{\text{s}^2}$$

In most countries other than the United States, produce is sold in mass units (grams or kilograms) rather than in force units (pounds or newtons). The supermarket scale still measures a force, but the scale is calibrated to show the mass of the produce instead of the weight. A 1.0-kg mass has a weight at Earth's surface of approximately 9.8 N or 2.2 lb.

There are more sophisticated means for measuring forces than the supermarket scale. Even so, many operate on the same principle as the supermarket scale: a force is measured by the deformation—change of size or shape—it produces in some object. For the simple spring scale (Fig. 2.2), the larger the force pulling down on the scale, the more the spring stretches. We can measure a force by measuring the extension of the spring. For many springs, the extension is approximately proportional to the force [Eq. (2-1)], which makes calibration easy.

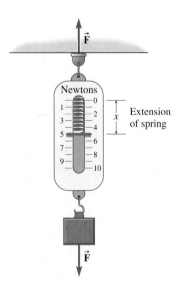

Figure 2.2 A simple spring scale. Note that there is a pull on both ends of the scale; one pull is from the weight hanging below the scale and the other is from the hook attached to the ceiling and supporting the scale from above.

Example 2.2

Getting Down to Nuts and Bolts

 In many hardware stores, bulk nuts and bolts are sold by weight. A spring scale in the store stretches 4.8 cm when 24.0 N of bolts are weighed. On the scale, what is the distance between calibration marks, which are marked in increments of 1 N? Assume an ideal spring.

Strategy   For an ideal spring, Hooke's law holds. Rephrasing the question, we are asked how many centimeters the spring stretches for each newton of force, which is the *reciprocal* of the spring constant: k is a measure of how many newtons of force are required for each centimeter of stretch.

Solution   The reciprocal of the spring constant is

$$\frac{1}{k} = \frac{x}{F} = \frac{4.8 \text{ cm}}{24.0 \text{ N}} = 0.20 \text{ cm/N}$$

The spring stretches an additional 0.20 cm for each additional newton of force. Therefore the marks should be 0.20 cm apart.

Discussion   A variation on the solution is to find the spring constant and then use it to find the distance stretched for 1 N of applied force.

$$k = \frac{F}{x} = \frac{24.0 \text{ N}}{4.8 \text{ cm}} = 5.0 \text{ N/cm}$$

Now let $F = 1.00$ N.

$$x = \frac{F}{k} = \frac{1.00 \text{ N}}{5.0 \text{ N/cm}} = 0.20 \text{ cm}$$

The answer is reasonable: since it takes 5 N to make the spring stretch 1 cm, 1 N makes the spring stretch 1/5 cm.

Practice Problem 2.2   Stretching a spring

16.0 N of nuts are placed in the pan of the scale of Example 2.2. How far does the spring stretch?

Vector Quantities

The magnitude of a force is *not* a complete description of the force. The direction of the force is equally important. The direction of the force exerted by a baseball bat can make the difference between a home run and a foul ball; the direction of force exerted by a club face on a golf ball makes the difference as to whether the golf ball rolls into the hole or misses it.

Force is one of many quantities in physics that are called **vectors**. Because vectors are so important in physics, we introduce the mathematics of vectors gradually in Chapters 2–4 to give you time to learn about them and practice using them. Some vector quantities defined in the next few chapters include force, acceleration, velocity, and momentum. All vectors have a direction as well as a magnitude. Section 1.2 mentioned that velocity, as defined in physics, has magnitude and direction; the magnitude of the velocity vector is the speed and the direction is the direction of motion. The direction of a vector is always a physical direction in space such as up, down, north, or 35° south of west. If a homework or exam question has you calculate a vector quantity such as force or velocity, specify the direction as well as the magnitude in the answer. One without the other is incomplete. Furthermore, all vectors follow the same rules of addition—and these rules take into account the directions of the vectors being added.

Mass is not a vector, it is a **scalar**. A scalar quantity can have magnitude, algebraic sign, and units, but not a direction in space. Scalars add in the usual way: 3 kg of water plus 2 kg of water is always equal to 5 kg of water. Adding vectors is different. A 300-N force added to a 200-N force gives different results, depending on the directions of the two forces. If two friends are trying to push a car out of a snowbank, they help each other by applying forces *in the same direction*. If they push in *opposite* directions, the net effect of the two forces would be *much* smaller. In Section 2.4 we discuss how to add vectors in detail.

In this book, an arrow over a boldface symbol indicates a vector quantity ($\vec{\mathbf{F}}$). Some books use boldface without the arrow. When writing by hand, draw an arrow over a vector symbol to distinguish it from a scalar. When the symbol for a vector is written *without* the arrow and in italics rather than boldface (F), it stands for the *magnitude* of the vector (which is a scalar). Absolute value bars are also used to stand for the magnitude of a vector, so $F = |\vec{\mathbf{F}}|$.

 Tips

Forces have both magnitude and direction.

Vector: a quantity with both magnitude and direction that is added according to special rules

Tips

Conceptual Example 2.3

Vector or Scalar?

Can temperature be a vector quantity?

Strategy   If a quantity is a vector, it must have both a magnitude and a physical direction in space.

Solution and Discussion   Does temperature have a direction? A temperature in Fahrenheit or Celsius can be above or below zero—is that a direction? No. A vector must have a *physical direction* in space. It does not make sense to say that the temperature of your coffee is "85 degrees Celsius in the southwest direction." "The temperature is up 5 degrees today," means that it has increased, not that it is pointing vertically upward. Temperature cannot be a vector.

Conceptual Practice Problem 2.3   Bank balance

Is the balance of your checking account a vector quantity?

2.2   FUNDAMENTAL FORCES

One of the main goals of physics has been to understand the immense variety of forces in the universe in terms of the fewest number of fundamental laws. Physics has made great progress in this quest for *unification;* today all forces are understood in terms of just four fundamental interactions (Fig. 2.3). (At the high temperatures present in the early universe, two of these interactions—the electromagnetic and weak nuclear forces—can be understood as the effects of a single electroweak interaction.) The ultimate goal is to describe all forces in terms of a single interaction.

Gravity

Gravity keeps the planets in their orbits and interferes with any attempt of ours to get away from the Earth's surface, so you may be surprised to learn that gravity is the *weakest* of the fundamental forces. Gravity *seems* strong because in many cases it is the only appreciable force acting between two bodies. Any two objects exert gravitational forces on each other, but the force is tiny unless at least one of the masses is very large. We tend to notice the relatively large gravitational forces exerted by planets and stars, but not the feeble gravitational forces exerted by smaller objects.

Gravity is an unlimited-range force and is always attractive. The force gets weaker as the distance between two objects increases, but it never drops to zero, no matter how far apart the objects get.

Newton's law of gravity is an early example of unification. Before Newton, people did not understand that the same interaction that makes an apple fall from a tree also keeps the Moon in its orbit. Newton unified the Earth's gravitational force on objects with the gravitational force exerted by the Sun on the planets. A single law, Newton's law of universal gravitation, describes both.

The Strong and Weak Nuclear Forces

The strong nuclear force holds protons and neutrons together in the nucleus. The same force binds quarks (a family of elementary particles) in combinations so they can form protons and neutrons and many more exotic subatomic particles. The same force holds protons and neutrons together in the atomic nucleus. The strong nuclear force is by far the strongest of the fundamental forces—hence its name—but its range is short: its effect is negligible at distances much larger than the size of an atomic nucleus (about 10^{-15} m). We study the strong nuclear force in Chapters 29 and 30.

The range of the weak nuclear force is even shorter than that of the strong nuclear force (about 10^{-17} m). It is manifest in certain radioactive decay processes. In the Sun, the weak nuclear interaction allows thermonuclear reactions to occur, without which there would be no sunlight.

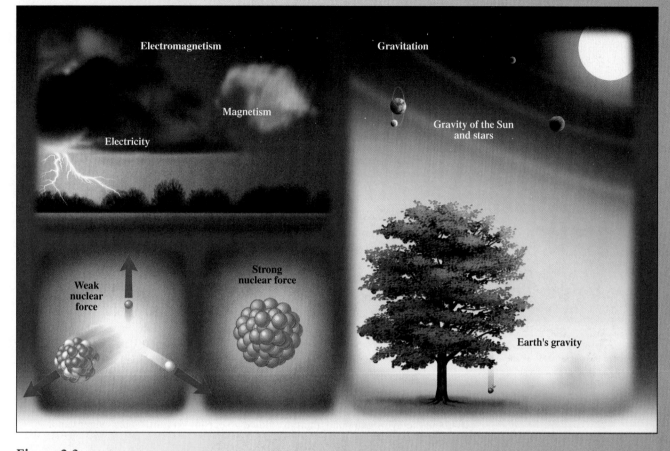

Figure 2.3 Schematic representation of the unification of forces

The Electromagnetic Force

The electromagnetic force is unlimited in range and acts between charged particles. The electrical and magnetic forces themselves were thought of separately until they were unified in the nineteenth century by James Clerk Maxwell (1831–1879). Place two magnets near each other and you will feel the macroscopic effect of the electromagnetic force. By reversing one magnet, you will see that this force can be attractive or repulsive. If you have ever experienced a shock upon touching a metal doorknob after walking across a carpet in dry weather, you have experienced the macroscopic effect of electrical forces.

What about the contact forces that are all around us? What *fundamental* interaction is responsible for the force of the wind on a kite, or the force of a muscle on a tendon? What fundamental interaction binds electrons to nuclei to form atoms, and binds atoms together in molecules? What interaction is responsible for the properties of solids, liquids, and gases, and forms the basis of the sciences of chemistry and biology? Gravity is much too weak to be responsible for these forces; the strong and weak nuclear forces are too short-ranged. The electromagnetic interaction between atoms is the cause of *all* contact forces.

The electromagnetic interactions behind these macroscopic forces can be quite complicated. Let's look again at the water strider. The force holding it up is ultimately electromagnetic in nature. How does it work? All liquids have surface tension; the molecules of the liquid pull on each other, causing the surface to be stretched like an elastic film.

Making The Connection:
forces supporting
a water strider

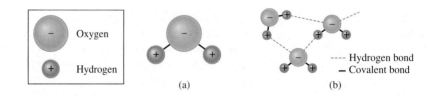

− Oxygen	
+ Hydrogen	

(a)

---- Hydrogen bond
— Covalent bond

(b)

Figure 2.4 (a) The water molecule (H_2O) is arranged in a bent chain formation with positive and negative charge separated as shown, but held together by covalent bonds (bonds in which two electrons are shared by two atoms). (b) Hydrogen bonds, which are weaker than covalent bonds, result from an electrical attraction between a positively charged hydrogen atom in one water molecule and a negatively charged oxygen atom in another water molecule. Each oxygen atom in water is bonded to four hydrogen atoms, two by covalent bonds and two by the weaker hydrogen bonds.

Figure 2.5 Hydrogen bonds make the surface of the water like an elastic membrane and enable the water strider to stand on top of the water.

The surface tension of water is particularly strong because the water molecule is strongly *polar*: its net charge is zero, but the oxygen atom has a slight negative charge and the hydrogen atoms are slightly positive (Fig. 2.4). Since unlike charges attract each other, the hydrogen atom of one water molecule attracts the oxygen atom of another molecule, forming a hydrogen bond between the molecules.

At the beginning of this chapter we asked why a water strider can stand on the surface of a pond. For the water strider's leg to break through the surface, hydrogen bonds would have to be broken (Fig. 2.5). The gravitational force acting on the tiny mass of the water strider is too small to break the hydrogen bonds. (For more detail, see Problem 55.)

A better explanation of the hydrogen bond than this simplified model would entail quantum mechanics, but it is still the electromagnetic interaction that is responsible. The formation of hydrogen bonds gives water many of its unique properties. The hydrogen bond makes the boiling temperature of water higher than it would be otherwise. Without the hydrogen bond, there would be no oceans on Earth—just water vapor.

2.3   NEWTON'S LAWS OF MOTION

In 1687, Isaac Newton published his *Philosophiae Naturalis Principia Mathematica* (or *Principia* for short), a work whose Latin title translates as *The Mathematical Principles of Natural Philosophy*. The *Principia* contains three laws of motion that form the basis of classical mechanics.

Newton's Laws of Motion

1. If no forces act on an object, then its speed and direction of motion do not change. (If the object is at rest, it remains at rest with a speed of zero; if it is moving, it moves in a straight line with constant speed.)

2. A nonzero net force acting on an object causes its state of motion to change. Quantitatively,

$$\vec{\mathbf{F}}_{net} = m\vec{\mathbf{a}} \tag{2-2}$$

where $\vec{\mathbf{F}}_{net}$ is the net force, m is the mass, and $\vec{\mathbf{a}}$ is the acceleration (which measures the rate of change of the velocity).

3. In an interaction between two objects, the forces that each exerts on the other are equal in magnitude and opposite in direction.

Newton's First Law: The Law of Inertia

Newton's first law says that a body acted on by no forces either remains at rest or moves in a straight line with constant speed. Certainly it makes sense that a body at rest remains at rest unless some force acts upon it to make it move. On the other hand, it is not at all obvious that a body can continue to move with constant speed in a straight line without forces acting to keep it moving. In our experience, moving objects come to rest because of forces that oppose motion, like friction and air drag. A hockey puck can slide the entire length of a rink with very little change in speed or direction because the ice is slippery—frictional forces are small. If we could remove *all* the resistive forces, including friction and air drag, the puck would slide without changing its speed or direction at all. No force would be needed to keep it going. The brief but large contact force exerted by a hockey stick on the puck changes the puck's state of motion, but once the puck loses contact with the stick, it slides in a straight line at nearly constant velocity.

No force is required to keep a body in motion if there are no forces opposing its motion. A force is necessary to set a body into motion, if the body was previously at rest. If you throw a ball, the contact force due to your hand sets the ball into motion, but that force stops as soon as the ball leaves your hand.

Newton's first law is also called the **law of inertia**. In physics, *inertia* means resistance to *changes* in motion, not resistance to motion or "tendency to come to rest." No force needs to act on a body to keep it moving. On the contrary, in the absence of forces, a body continues moving at the same speed and in the same direction. A body at rest has a speed of zero.

Credit for the law of inertia really goes to Galileo Galilei (1564–1642), as Newton acknowledged in the *Principia*. In a series of clever experiments in which he rolled a ball up inclines of different angles, Galileo postulated that, if he could eliminate all resistive forces, a ball rolling on a horizontal surface would never stop (Fig. 2.6). Galileo made a brilliant conceptual leap from the real world with friction to an imagined, ideal world, free of friction. The law of inertia contradicted the view of the Greek philosopher Aristotle (384–322 B.C.). Almost 2000 years before Galileo, Aristotle had formulated his view that the natural state of a body is to be at rest; and, for a body to remain in motion, a force would have to act upon it continuously. Galileo conjectured that, in the absence of friction and other resistive forces, no continued force is needed to keep an object moving.

Galileo thought that the sustained motion of a body would be in a circle around the Earth—that the horizontal motion of the ball formed part of a great circle. Shortly after Galileo's death, René Descartes (1596–1650) argued that the motion of a body free of any forces should be in a straight line rather than a circle. Newton acknowledged his debt to Galileo, Descartes, and others when he wrote: "If I have seen farther, it is because I was standing on the shoulders of giants."

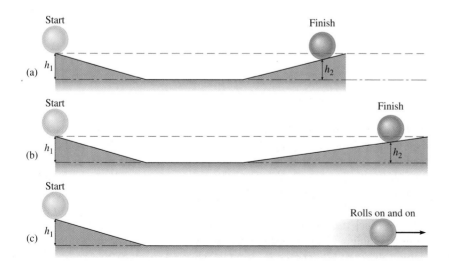

Figure 2.6 (a) Galileo found that a ball rolled down an incline stops when it reaches *almost* the same height on the second incline. He decided that it would reach the *same* height if resistive forces could be eliminated. (b) As the second incline is made less and less steep, the ball rolls farther and farther before stopping. (c) If the second incline is horizontal, and there are no resistive forces, the ball would never stop.

Conceptual Example 2.4

Snow Shoveling

The task of shoveling newly fallen snow from the driveway can be thought of as a struggle against the inertia of the snow. Without the application of a net external force, the snow remains at rest on the ground. However, there is an important way that the inertia of the snow makes it *easier* to shovel. Explain.

Strategy   Think about the physical motions used when shoveling snow. (If you live where there is no snow, think about shoveling gravel from a wheelbarrow to line a garden path.) To be "assisted by the law of inertia," there must be a time when the snow is moving on its own, without the shovel pushing it.

Solution and Discussion   Imagine scooping up a shovelful of snow and swinging the shovel forward toward the side of the driveway. The snow and the shovel are both in motion.

Then suddenly the forward motion of the shovel stops, but the snow continues to move forward because of its inertia; it slides forward off the shovel, to be pulled down to the ground by gravity. The snow does not stop moving forward when the forward force due to the shovel is removed.

This procedure works best with fairly dry snow. Wet sticky snow tends to cling to the shovel. The frictional force on the snow due to the shovel keeps it from moving forward and makes the job far more difficult.

Conceptual Practice Problem 2.4   Subway car inertia

Negar stands on a subway car, holding on to an overhead strap. As the train starts to pull out of the station, she feels thrust toward the rear of the car; as the train comes to a stop at the next station, she feels thrust forward. Explain the role played by inertia in this situation.

Physics at Home

The parlor trick in which a person pulls a tablecloth out from under dishes set for a formal dinner is a demonstration of inertia. Try a similar demonstration yourself. If you do not want to risk your dishes, place a quarter on top of an index card, or a credit card, balanced on top of a drinking glass (Fig. 2.7a). With your thumb and forefinger, flick the card so that it flies out horizontally from under the quarter. What happens to the quarter? The horizontal force on the coin due to friction is small. With a negligibly small horizontal force, the coin's inertia tends to keep it motionless while the card slides out from under it (Fig. 2.7b). Once the card is gone, gravity pulls the coin down into the glass (Fig. 2.7c).

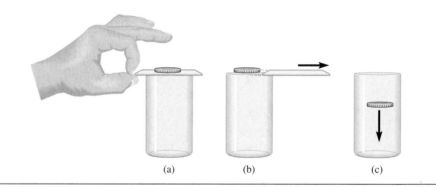

Figure 2.7   A demonstration of inertia

(a)   (b)   (c)

The law of inertia is a postulate of classical mechanics—an assumption that is used as a starting point. It is not something we can prove experimentally. No object can ever be completely free of forces acting on it. Far from any galaxies, the gravitational force on an object would exist, even if it is small. The brilliant insights of Galileo and Descartes were to imagine what would happen in an *ideal* situation they could only approximate. Newton's second and third laws of motion are valid as long as an *inertial reference frame*—a reference frame in which the law of inertia holds—is used to observe the motion of objects.

At the beginning of this chapter we asked why the Moon orbits the Earth month after month, without any engine to drive it, while an airplane that runs out of fuel falls to the ground. A continuous force—such as that provided by an engine—is not necessary to keep an object moving if there are no resistive forces acting on it. The Moon is not slowed by friction or air resistance, but an airplane does encounter significant air resistance and must compensate using the forward force due to its engines.

The Moon does not move in a straight line because the force of Earth's gravity is pulling on it, changing its direction of motion. Imagine suddenly turning off the gravitational force on the Moon. It would then move in a straight line, getting farther and farther from Earth. The gravitational force on the Moon due to the Earth *does* cause the Moon to continually fall toward Earth—just the right amount so that it stays the same distance from Earth in its circular orbit.

Newton's Second Law

In Chapters 3 and 4, the state of motion of an object is characterized by its velocity—a vector quantity that conveys both the speed and direction of motion. The law of inertia says that if no forces act on an object, it moves with constant velocity; neither its speed nor its direction change.

Newton's second law describes what happens when forces do act. When more than one force acts on an object, an extremely important quantity is the **net force** (or **resultant force**) acting on the object. The net force is the vector sum of all the forces acting on an object. Newton's second law relates changes in an object's state of motion to the net force acting on that object. When the net force on a body is not zero, the body's velocity changes; a quantity called acceleration (which is also a vector quantity) measures the way velocity changes. Newton's second law states that the acceleration is proportional to the net force acting on the body and is in the same direction as the net force. Quantitative use of Newton's second law when the net force is nonzero depends upon careful definitions of velocity and acceleration, which we defer to Chapters 3 and 4.

> The net force (or resultant force) is the vector sum of all the forces acting on an object.

For now we consider cases where the net force is zero. If the net force is zero, then the acceleration is zero and the velocity stays the same—the body moves as if there were no forces at all. Its direction and speed of motion do not change because the forces "cancel" each other. Such a body is said to be in **translational equilibrium**. *Equilibrium* conveys the idea that the forces are in balance; there is as much force upward as there is downward, as much to the right as to the left, and so forth.

> An object in translational equilibrium has a net force of zero acting on it.

An object in translational equilibrium is not necessarily at rest; it may be moving at constant speed in a straight line. If it *is* at rest, it is said to be in **static equilibrium**; otherwise, it moves at constant velocity in **dynamic equilibrium**.

Newton's Third Law

In Section 2.1, we said that an interaction always consists of *two* forces; no object can exert a force without having a force exerted on it. Nature is quite democratic when it comes to interactions: not only are there always two forces, one on each object, but the two always have the *same magnitude*. The directions of the two forces are always opposite. This is Newton's third law: an interaction between two bodies consists of two forces, one acting on each body, that are equal in magnitude and opposite in direction.

When the Earth's gravity pulls down on a satellite, the satellite pulls upward just as hard on the Earth. Though the forces are the same strength, the effect they have on the motion of the two bodies is very different. The force on the satellite keeps it moving in a circular orbit around the Earth, while the force exerted by the satellite on the Earth has an immeasurably small effect on the Earth's motion. According to Newton's second law, the effect of a force acting upon an object depends on the object's mass, which is a measure of its inertia. The Earth, with vastly greater inertia—remember that means resistance to change in motion—reacts very little to the force. The satellite's mass is much less, so the same magnitude force has a much larger effect.

Figure 2.8 Two arctic terns fight over a fish. The forces exerted on the fish by the two birds, $\vec{\mathbf{F}}_{f1}$ and $\vec{\mathbf{F}}_{f2}$, *cannot* be third-law partners since they act on the same object.

Making The Connection:
detection of
extrasolar planets

Multiplying a vector by –1 reverses its direction. The vector $-\vec{\mathbf{A}}$ has the same magnitude as $\vec{\mathbf{A}}$ but its direction is opposite.

On the other hand, if a massive planet orbits a star in a relatively small orbit, the gravitational force that the planet exerts on the star can make the star wobble enough to be observed. The wobble enables astronomers to discover planets orbiting stars other than the Sun. The planets do not reflect enough light toward Earth to be seen, but their presence can be inferred from the effect they have on the star's motion.

Do not assume that Newton's third law is involved *every* time two forces *happen* to be equal and opposite—"it ain't necessarily so." Many situations involve two equal and opposite forces acting *on a single body*. Such forces cannot be *third-law partners* because they act on the same body. The third law deals with the two forces that comprise a *single interaction*. These two forces act on *different bodies*, one on each of the two bodies that are interacting.

Two children fighting over a toy *may* be pulling on it with equal and opposite forces, but they are not *required* to do so by Newton's laws. Suppose two arctic terns fight over a fish by pulling in opposite directions with equal magnitude forces. The two forces exerted on the fish, shown in Fig. 2.8, are *not* third-law partners. The third-law partner of the force exerted on the fish by one bird ($\vec{\mathbf{F}}_{f1}$) is the force exerted on that bird by the fish ($\vec{\mathbf{F}}_{1f}$).

To correctly identify two forces that are third-law partners, one of them must be a force exerted on object 1 by object 2 and the other must be a force exerted on object 2 by object 1. To keep track of what exerts a force on what, we can use subscripts. Thus, $\vec{\mathbf{F}}_{12}$ stands for the force exerted *on* 1 *by* 2, while $\vec{\mathbf{F}}_{21}$ stands for the force exerted on 2 by 1. These two forces must be equal in magnitude and opposite in direction: $\vec{\mathbf{F}}_{21} = -\vec{\mathbf{F}}_{12}$.

Conceptual Example 2.5

A Dangling Fish

A fish is suspended by a line from a fishing pole. Choose two forces acting on the fish and describe the third-law partner of each.

Strategy For a force exerted on the fish by object A, its third-law partner is the force exerted on A by the fish. The two forces must be equal in magnitude and opposite in direction, according to Newton's third law.

Solution and Discussion The weight of the fish is the downward gravitational force exerted on the fish by the Earth, $\vec{\mathbf{F}}_{fE}$. Its third-law partner must be the upward gravitational force on

the Earth by the fish, $\vec{\mathbf{F}}_{Ef}$. The fish is suspended from a line, so there is a downward force on the line due to the fish, $\vec{\mathbf{F}}_{lf}$; its third-law partner is the upward force on the fish due to the line, $\vec{\mathbf{F}}_{fl}$. Figure 2.9a shows these two pairs of forces.

 Since third-law partners act on different objects, the two forces of the pair can never appear in a diagram showing the forces on a single object. Figure 2.9b shows the fish in isolation with the forces acting on the fish; such forces acting on a single body *cannot* be third-law partners.

Conceptual Practice Problem 2.5   The fishing rod

Identify the Newton's third-law partner of each force acting on the fishing rod.

Figure 2.9 (a) Sketch to illustrate pairs of forces and (b) forces acting on the fish.

2.4   NET FORCE: VECTOR ADDITION

To find the net force acting on an object, we sum all of the forces acting on it. As previously mentioned, vector quantities have their own rules of addition. For now we add only *collinear*—parallel or antiparallel—vectors. Chapter 4 generalizes this method to include noncollinear vectors.

Adding Collinear Vectors

Suppose we want to add two force vectors, $\vec{\mathbf{F}}_1$ and $\vec{\mathbf{F}}_2$, to find the net force acting on an object. We can add them graphically. To draw a force vector, use an arrow pointing in the direction of the force. Make the length of the arrow proportional to the magnitude of the vector. To add two vectors, first draw one of them. Then draw the second one starting where the first left off. In other words, place the tail of the second vector at the tip of the first. Next, draw an arrow starting from the tail of the first and ending at the tip of the second. This arrow represents the **resultant vector**—the vector sum of the two. This graphical method shows what happens when two collinear vectors are added.

- If the two forces are in the same direction, the resultant force is in the same direction (Fig. 2.10a). Its magnitude is the sum of the magnitudes of the two. If you and your friend push a heavy trunk with forces of 200 N and 300 N in the same direction, the net effect is that of a 500-N force in that direction.

- If two forces are in opposite directions, or **antiparallel**, the magnitude of their sum is the *difference* between the magnitudes of the two vectors (Fig. 2.10b)—the larger minus the smaller, since vector magnitudes are never negative. The direction of the vector sum is the direction of the larger of the two. Pushing that trunk with a

Figure 2.10 Addition of collinear vectors that are (a) in the same direction and (b) in opposite directions. (c) The sum of two vectors with equal magnitudes and opposite directions is zero.

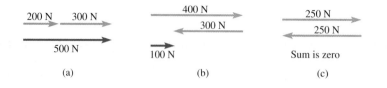

(a)   (b)   (c)

400-N force to the right and a 300-N force to the left has the net effect of a 100-N force to the right.

- The only way that two vectors can add to zero is if they are opposites: they must have the same magnitude but opposite directions (Fig. 2.10c).

Whenever quantities are added or subtracted, check whether the quantities are vectors. If so, be sure to add them correctly—do not just add their magnitudes.

Conceptual Example 2.6

Bird in Dynamic Equilibrium

A bird weighing 1 N is flying north in dynamic equilibrium. What is the force acting on the bird due to the air?

Strategy and Solution   First, we identify all the forces acting on the bird. Since the bird is in dynamic equilibrium, the vector sum of the forces acting on it must be zero.

There are only two forces acting on the bird. One of them is long-range: the gravitational pull of the Earth, which is the bird's weight $\vec{W}$. The other is a contact force due to the air ($\vec{F}_{ba}$). Nothing else is in contact with the bird, so there are no other contact forces. Since the bird is in dynamic equilibrium, the net force acting on the bird must be zero:

$$\vec{W} + \vec{F}_{ba} = 0 \quad \text{or} \quad \vec{F}_{ba} = -\vec{W}$$

So the force on the bird due to the air is 1 N, directed upward.

Discussion   The interaction between the bird and the air is complex at the microscopic level, as was the interaction between the *Alvin* and the seawater. The net effect of all the interactions between air molecules and the bird produces an upward force of 1 N.

With the framework of Newton's laws, we can now explain why a spring scale can measure weight (Section 2.1). If a melon is at rest in the pan of the scale, the net force on it must be zero. There are only two forces acting: gravity pulls down and the scale pulls up. These two forces must be equal in magnitude, so the reading of the scale is equal to the weight of the melon.

Conceptual Practice Problem 2.6   A crate of apples

An 80-N crate of apples sits at rest on a ramp that runs from the ground to the bed of a truck. The ramp is inclined at 20° to the ground. What is the force exerted on the crate by the ramp?

Making The Connection:
    forces on a
    bird in air

In Conceptual Example 2.6, we concluded that the air exerts an upward force on the bird to hold it up. How does the bird get the air to do that? By flapping its wings and pushing air downward, the bird takes advantage of Newton's third law—the air must push up on the bird. The shape of the wing is changed so that it pushes harder on the air while moving down than while moving up, so the average force on the air is downward. You might wonder whether there is a *drag* force—a force due to the air that resists the forward motion of the bird. There is, so the bird must flap its wings in such a way as to also push some air backward. The air then pushes forward on the bird. This push is called the *thrust*. If the bird is flying at constant velocity, the drag and the thrust are equal in magnitude. The total force on the bird due to the air—the sum of lift, drag, and thrust—is still 1 N upward.

Physics at Home

The next time you go swimming, you can feel the same effect that the flying bird does, except you will be in the water. Push down and backward on the water with your arms and legs. The water pushes up and forward on you. The various swimming strokes are devised so that you exert as large a force as possible backward on the water during the power part of the stroke, and then as small a force as possible forward on the water during the return part of the stroke.

Free-Body Diagrams

An essential tool used to find the net force acting on an object is a **free-body diagram** (FBD). In a free-body diagram, the body isn't free from interactions; the word *free* is taken to mean *isolated*. An FBD is a sketch of a single body with force vectors drawn to represent *every* force acting on that body. It does *not* include forces that act on other bodies. To draw an FBD:

- First draw the object. Feel free to draw it in a simplified way—you do not have to be Renoir to solve a mechanics problem. Almost any object can be represented as a box or a circle, or even a dot.

- Next, draw vector arrows representing every force acting on that object. Draw the arrows so that they correctly illustrate the directions of the forces. We usually draw the vectors as arrows that start on the object and point away from it, but remember that the vectors represent forces exerted *on* the object by something else. If you have enough information to do so, draw the lengths of the arrows so that they are proportional to the magnitudes of the forces. Figure 2.9b is an example of an FBD for the fish in Conceptual Example 2.5.

The most important thing is not to omit any forces and not to draw any extra ones. To make sure that you do not miss any, consider everything in contact with the object to see if there are contact forces; also include any long-range forces. Every force you draw must be a force exerted *on that object by something else*. Ask yourself: what is the other object that is exerting this force?

Internal and External Forces

When we say that a baseball has interactions with the Earth (gravity), with a baseball bat, and with the air, we are treating the baseball as a single entity. But the ball really consists of an enormous number of protons, neutrons, and electrons, all interacting with each other. The protons and neutrons interact with each other to form atomic nuclei; the nuclei interact with electrons to form atoms; interactions between atoms form molecules; and the molecules interact to form the structure of the thing we call a baseball. It would be difficult to have to deal with all of these interactions to predict the motion of a baseball.

Call the set of particles comprising the baseball a **system**. We can classify all the interactions as either **internal** or **external**. Internal interactions are those for which *both* interacting particles are part of the system. Suppose we want to add up all the forces acting on all of the particles comprising the baseball in order to find the net force. Every internal interaction consists of two forces, equal in magnitude and opposite in direction, so the sum of the two is zero. But for the external interactions, *only one of the two forces is exerted on the ball.* So to find the net force exerted on the ball, we only need to add external forces; the internal forces all add to zero.

The statement that internal forces always add to zero is particularly powerful because the choice of what comprises a system is completely arbitrary. We can choose *any* set of particles and think of it as a system. In one problem, it may be convenient to think of the baseball as a system; in another, we may choose a system comprising both the baseball and the bat together. The second choice might be useful if we do not have detailed information about the interaction between the bat and the ball.

Example 2.7

A Gold Crown

A gold crown of weight 17.5 N is immersed in a beaker of water while suspended from a spring scale (Fig. 2.11). When the crown is in static equilibrium, the reading on the spring scale is 9.0 N. (a) What is the force exerted on the crown by the water? (b) If the spring in the scale is 0.50 cm longer than when it is relaxed, what is its spring constant? (c) What is the

continued on next page

Example 2.7 *continued*

reading of the pan scale on which the beaker sits? The weight of the water and the beaker together is 23.5 N.

Strategy   (a) The question asks for one of the forces acting on the crown. First, we identify the forces acting on the crown. There are three: gravity, a force due to the spring scale, and a force due to the water. Since the crown is in static equilibrium, these three forces must sum to a net force of zero. A free-body diagram will help make sure we add them correctly.

(b) Assume the spring to be ideal and apply Hooke's law.

(c) The question asks for the reading of the pan scale, which is the magnitude of the force exerted on it by the beaker. Should we draw an FBD for the beaker?

Unfortunately, the weight of the beaker itself is not given, nor is the weight of the water. We do not have enough information to use an FBD for the beaker to solve for one of the forces; we would need to know the magnitude of all the other forces acting on the beaker. We do know the combined weight of the beaker including the water in it. Therefore we will consider the beaker and the water to be a single object or system throughout the problem. To solve part (c), we draw an FBD for the "water + beaker."

To keep the forces straight, use subscripts: c = crown; s = spring scale; w = water + beaker; p = pan scale; and e = Earth. The forces given in the problem are: (1) crown weight = $\vec{W}_{ce}$ = 17.5 N, downward; (2) force on the spring scale due to the crown = $\vec{F}_{sc}$ = 9.0 N, downward; and (3) weight of water + beaker = $\vec{W}_{we}$ = 23.5 N, downward.

We need to find (a) the force on the crown by the water ($\vec{F}_{cw}$), (b) the spring constant k, and (c) the pan scale reading = the magnitude of the force on the pan scale due to water + beaker = $|\vec{F}_{pw}|$.

Solution   (a) Figure 2.12a shows an FBD for the crown. The forces due to the spring scale ($\vec{F}_{cs}$) and the water ($\vec{F}_{cw}$) are directed upward, while the force of gravity ($\vec{W}_{ce}$) is downward.

The crown is in static equilibrium, so the sum of the forces acting on the crown must be zero.

$$\vec{F}_{cs} + \vec{W}_{ce} + \vec{F}_{cw} = 0$$

This is *not* ordinary addition; it is vector addition. Vectors must be added according to the rules presented earlier.

The spring scale reads 9.0 N, which means that the force on the spring scale due to the crown ($\vec{F}_{sc}$) has magnitude 9.0 N.

The force that the spring scale exerts on the crown must be equal in magnitude and opposite in direction. Therefore $\vec{F}_{cs}$ has magnitude 9.0 N and is directed upward.

The two known forces are in opposite directions, so their sum is in the direction of the larger one and has a magnitude equal to the *difference* of the two magnitudes:

$$\vec{F}_{cs} + \vec{W}_{ce} = 9.0 \text{ N up} + 17.5 \text{ N down} = 8.5 \text{ N down}$$

To this sum, the force $\vec{F}_{cw}$ must be added to give a net force of zero:

$$(\vec{F}_{cs} + \vec{W}_{ce}) + \vec{F}_{cw} = 0$$

Two vectors that add to zero must be equal and opposite, so

$$\vec{F}_{cw} = -(\vec{F}_{cs} + \vec{W}_{ce}) = 8.5 \text{ N up}$$

(b) For an ideal spring, Hooke's law says that

$$F = kx$$

where k is the spring constant, x is the extension of the spring from the relaxed length, and F is the force acting on the spring to cause the extension. The magnitude of the force acting on the spring due to the crown is 9.0 N. Solving for the spring constant,

$$k = \frac{F_{sc}}{x} = \frac{9.0 \text{ N}}{0.50 \text{ cm}} = 18 \text{ N/cm}$$

(c) Finally, what is the reading of the pan balance (the magnitude of $\vec{F}_{pw}$)? It is not necessarily equal to the weight of the object sitting on it (the water + beaker) since other forces act on that system. Is it equal to the weight of water + beaker + crown? No, because there is an upward force on the crown due to the spring scale, which is partially supporting the crown. The reading of the pan scale should fall between the weight of the water + beaker (23.5 N) and the weight of water + beaker + crown (41.0 N).

The force on the crown due to the water is $\vec{F}_{cw}$ = 8.5 N upward. The other force in the interaction between the crown and the water is the force exerted on the water by the crown, $\vec{F}_{wc}$. It must act in the opposite direction and be of the same magnitude; therefore the crown is pushing down on the water + beaker system with a force of magnitude 8.5 N. The system

Figure 2.11
A gold crown is suspended from a spring scale into a beaker of water.

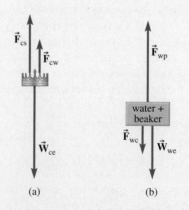

Figure 2.12
(a) A free-body diagram showing the forces acting on the crown and (b) a free-body diagram for the water + beaker system.

(a)   (b)

continued on next page

Example 2.7 continued

of water plus beaker pushes down on the pan balance with a force greater in magnitude than the weight of the water plus beaker system, $\vec{W}_{we}$, as shown in the free-body diagram (Fig. 2.12b).

The system water + beaker is in static equilibrium:

$$\vec{F}_{net} = \vec{W}_{we} + \vec{F}_{wc} + \vec{F}_{wp} = 0$$

$$\vec{F}_{wp} = -(\vec{W}_{we} + \vec{F}_{wc}) = -(23.5 \text{ N down} + 8.5 \text{ N down})$$

$$= -(32.0 \text{ N down}) = 32.0 \text{ N up}$$

The force that the water + beaker exerts on the pan scale is therefore 32.0 N down.

$$\vec{F}_{pw} = -\vec{F}_{wp} = 32.0 \text{ N down}$$

The pan scale reading, which is the magnitude of the force on the pan by the system of water + beaker, is 32.0 N, which, as expected, is between 23.5 N and 41.0 N.

Discussion Since the individual weights of the beaker and of the water are not known, it is not possible to draw an FBD for either one individually. Nor can we find out the forces that the water and beaker exert on each other. By considering the water and beaker as a single object or system, the forces that the water and beaker exert on each other become internal to the system. Therefore they are not included in an FBD for the composite object water + beaker.

Practice Problem 2.7   Force exerted by table

The weight of the pan scale in Example 2.7 is 13.0 N. What is the upward force exerted on it by the table on which it rests when it is arranged as shown in Fig. 2.11?

Example 2.8

Net Force on an Airplane

The external forces on an airplane in flight heading eastward are as follows: weight = 16.0 kN, downward; lift = 16.0 kN, upward; thrust = 1.8 kN, east; and drag = 0.8 kN, west. What is the net force on the plane?

Strategy After drawing an FBD, we must add the forces. They are not all along the same line, but we can first add those that *are* collinear and see what happens.

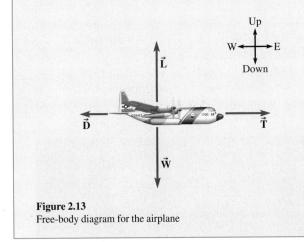

Figure 2.13
Free-body diagram for the airplane

Solution Figure 2.13 is a free-body diagram for the plane, using $\vec{W}$, $\vec{L}$, $\vec{T}$, and $\vec{D}$ for the weight, lift, thrust, and drag. There are two vertical vectors ($\vec{W}$ and $\vec{L}$). They are equal in magnitude and opposite in direction, so they add to zero:

$$\vec{W} + \vec{L} = 0$$

The two remaining vectors are horizontal. The thrust is larger, so their sum is in the direction of the thrust (east). The magnitude of their sum is the difference of their magnitudes:

$$T - D = 1.8 \text{ kN} - 0.8 \text{ kN} = 1.0 \text{ kN}$$

The net force is therefore

$$\vec{F}_{net} = \vec{W} + \vec{L} + \vec{T} + \vec{D} = (\vec{W} + \vec{L}) + (\vec{T} + \vec{D})$$

$$= 0 + (1.0 \text{ kN east}) = 1.0 \text{ kN east}$$

Discussion When faced with vectors that are not *all* along the same line, usually the easiest thing is to first add groups of vectors that are collinear.

Practice Problem 2.8   New forces on the airplane

Find the net force on the airplane if the forces are weight = 16.0 kN, downward; lift = 15.5 kN, upward; thrust = 1.2 kN, north; drag = 1.2 kN, south.

2.5   GRAVITATIONAL FORCES

It is useful to summarize our knowledge about various forces, whether fundamental or not, in **force laws**. A force law states the magnitude and direction of a particular force.

Let us start with the gravitational force. According to Newton's law of universal gravitation, two bodies exert gravitational forces on each other that are proportional to

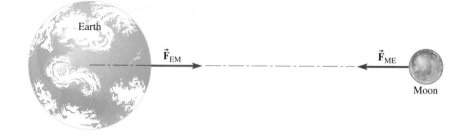

Figure 2.14 Gravity is always an attractive force. The force that each body exerts on the other is equal in magnitude, even though the masses may be very different. The force exerted on the Moon by the Earth ($\vec{F}_{ME}$) is of the same magnitude as the force exerted on the Earth by the Moon ($\vec{F}_{EM}$).

the masses (m_1 and m_2) of the two bodies and inversely proportional to the square of the distance (r) between their centers. Strictly speaking, the law of gravitation as presented here only applies to point particles and symmetric spheres. (The *point particle* is a common model in physics used when the size of an object is negligibly small and the internal structure is irrelevant.) But the law of gravitation is *approximately* true for any two bodies if the distance between their centers is large compared to their sizes.

In mathematical language, the law of universal gravitation is written:

$$F = \frac{Gm_1m_2}{r^2} \tag{2-3}$$

where the constant of proportionality ($G = 6.673 \times 10^{-11}$ N·m²/kg²) is called the universal gravitational constant. Gravity is an attractive force: each body is pulled toward the other's center (Fig. 2.14). The forces on the two bodies are equal in magnitude but the directions are opposite.

Example 2.9

Gravitational Force between Two Cars

Estimate the gravitational force exerted by one car on another when they are driving side-by-side in parallel lanes.

Strategy   The cars are not spheres and are not far apart, but as long as we want only an order-of-magnitude estimate, we can apply the law of gravity. We need to make reasonable estimates for the masses of the cars and the distance between their centers. Then the universal law of gravitation will give an estimate of the force each exerts on the other. We expect the gravitational force to be quite small—otherwise, there would be many more auto accidents.

Solution   A car might have a mass of around 1000 kg, so let $m_1 = m_2 = 1000$ kg. A reasonable approximation is that it is 4 m from the center of one car to the center of the other. Then

$$F = \frac{Gm_1m_2}{r^2} = \frac{6.673 \times 10^{-11} \text{ N·m}^2 \cdot \text{kg}^{-2} \times (1000 \text{ kg})^2}{(4 \text{ m})^2}$$

$$= 4 \times 10^{-6} \text{ N} = 4 \text{ } \mu\text{N}$$

which is about the same magnitude as the weight of a mosquito.

Discussion   The estimate shows that gravitational forces exerted by ordinary objects are so small as to be negligible in most cases. The gravitational force on one car due to the other is tiny compared to other forces acting on the car, such as the gravitational force due to the Earth (about 10 kN) and the contact forces exerted by the road and by the air. The Earth exerts appreciable gravitational forces because its mass is so large.

Writing out the units of all quantities, including the constant G, prevents a mistake such as forgetting to square r. In the calculation in the solution section here, all the units other than newtons cancel, leaving a force in units of newtons:

$$\frac{\text{N·m}^2 \cdot \text{kg}^{-2} \times \text{kg}^2}{\text{m}^2} = \text{N}$$

Practice Problem 2.9   Gravitational force between two humans

Alex is on stage playing his bass guitar. Estimate the *gravitational* attraction between Alex and Frances, a fan who is standing 8 m from Alex. Alex has a mass of 55 kg and Frances has a mass of 40 kg.

For objects near Earth's surface, we often use a simpler expression for the *weight* of an object—which is the force of gravity on the object due to the mass of the Earth. At small distances above the Earth's surface, the distance between the object and the Earth's center is approximately equal to the Earth's mean radius, $R_E = 6.37 \times 10^6$ m. The mass of the Earth is $M_E = 5.98 \times 10^{24}$ kg, so the weight of an object of mass m near Earth's surface is

$$W = \frac{GM_E m}{R_E^2} = m \times \left(\frac{GM_E}{R_E^2} \right) = mg \qquad (2\text{-}4)$$

where g is called the **gravitational field strength** at the surface of the Earth. From Eq. (2-4),

$$g = \frac{GM_E}{R_E^2} = \frac{6.673 \times 10^{-11}\ \text{N·m}^2\text{·kg}^{-2} \times (5.98 \times 10^{24}\ \text{kg})}{(6.37 \times 10^6\ \text{m})^2} = 9.8\ \text{N/kg}$$

We have rounded to two significant figures because g varies from place to place. Since the newton is defined as $1\ \text{N} = 1\ \text{kg·m/s}^2$, g can be written in SI base units as:

$$g = 9.8\ \frac{\text{N}}{\text{kg}} \times \frac{1\ \text{kg·m/s}^2}{1\ \text{N}} = 9.8\ \text{m/s}^2$$

At most locations on the surface of the Earth, the gravitational field strength g is approximately 9.8 N/kg. The Earth is not a perfect sphere; it is slightly flattened at the poles. Since the distance from the surface to the center of the Earth is smaller there, the field strength is greatest at the poles: about 9.832 N/kg, in contrast to 9.814 N/kg at the equator (for sea level). Altitude also matters; as you climb above sea level, your distance from Earth's center increases and the field strength decreases. Tiny local variations in the field strength are also caused by geologic formations. On top of dense bedrock, g is a little greater than above less dense rock. Geologists and geophysicists measure these variations to study Earth's structure and also to locate deposits of various minerals, water, and oil. The device they use, a *gravimeter*, is essentially a mass hanging on a spring. As the gravimeter is carried from place to place, the extension of the spring increases where g is larger and decreases where g is smaller. The mass hanging from the spring does not change, but its weight does ($W = mg$).

Not only is the weight per unit mass at a given point the same for any object at that point, but so is the *direction* of the gravitational force. The **gravitational field** vector $\vec{g}$ is defined as the force per unit mass; its direction is that of the gravitational force exerted on an object at that point. Then $\vec{W} = m\vec{g}$, where $\vec{g}$ is the gravitational field at some point and $\vec{W}$ is the weight of an object at that point. Note that in this equation a vector ($\vec{g}$) is multiplied by a scalar (m). Since m is positive, the magnitude of the vector is multiplied by m but its direction is not changed. Multiplying a vector by a *negative* scalar reverses its direction.

⚠️ Weight and mass are *not* the same thing. If you climb Mt. McKinley, the *weight* of your gear decreases as you get farther from the center of the Earth, but its *mass* does not change.

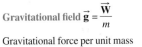

Gravitational field $\vec{g} = \dfrac{\vec{W}}{m}$

Gravitational force per unit mass

Example 2.10

Weight at High Altitude

When you are in a commercial airliner cruising at 6.40×10^3 m (21,000 ft), by what percentage has your weight (as well as the weight of the airplane) changed?

Strategy   This is a good problem to work by proportions. The masses and the universal gravitational constant are the same in both cases. Only the distance r changes.

Solution   Your weight is inversely proportional to the square of the distance r between you and Earth's center:

$$W = \frac{GM_E m}{r^2} \propto r^{-2}$$

At an altitude h above the Earth's surface, your weight is smaller by a factor

$$\frac{W}{W_{\text{surface}}} = \left(\frac{R_E + h}{R_E} \right)^{-2} = \left(1 + \frac{h}{R_E} \right)^{-2}$$

$$= \left(1 + \frac{6.40 \times 10^3\ \text{m}}{6.37 \times 10^6\ \text{m}} \right)^{-2} = 0.997994 \approx 1 - 0.0020$$

Your weight decreases by about 0.20%.

Discussion   Although 6400 m may seem like a significant altitude to us, it's a small fraction of the Earth's radius (0.100%), so the weight change is a small percentage. When

continued on next page

Example 2.10 *continued*

judging whether a quantity is small or large, always ask: small compared to what?

To find the weight of other objects on the plane, it would not be necessary to redo the whole calculation. *Every* object in the airplane—as well as the plane itself—changes weight by the same percentage. The weight of each object is still proportional to its mass, so we can still write $W = mg$, where g is the strength of the *local* gravitational field. Once g is known at a point, it can be used to find the weight of any object at that same point.

Practice Problem 2.10   A satellite put into orbit

The space shuttle carries a satellite in its cargo bay and places it into orbit around the Earth. Find the ratio of the Earth's gravitational force on the satellite when it is on a launch pad at the Kennedy Space Center to the gravitational force exerted when the satellite is orbiting 6.00×10^3 km above the launch pad.

Electrical Forces

There is another force law that is strikingly similar to Newton's law of universal gravitation. Just as all massive particles exert gravitational forces on each other, all *charged* particles exert *electrical* forces on each other. Coulomb's law says that two charged particles exert electrical forces on each other that are proportional to the magnitudes of the charges ($|q_1|$ and $|q_2|$) of the two particles and inversely proportional to the square of the distance (r) between them. In mathematical language, Coulomb's law is written

$$F = \frac{k|q_1||q_2|}{r^2} \qquad (2\text{-}5)$$

where k is a constant of proportionality analogous to G.

Unlike the gravitational force, which is always attractive, the electric force can be either attractive *or* repulsive (Fig. 2.15). The particles attract each other if their charges have opposite signs (like an electron and a proton); particles repel each other if their charges have the same sign (like two electrons). Just as for the law of universal gravitation, Coulomb's law explicitly obeys Newton's third law; the two forces are equal in magnitude and opposite in direction.

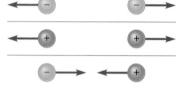

Figure 2.15 Like charges repel; opposite charges attract.

Physics at Home

Rub a balloon against your sweater and then place it against a wall. The rubbing transfers some charge between the balloon and the sweater, so they acquire net charges of opposite signs. When you place the balloon against the wall, it attracts opposite charges in the wall and repels like charges. The opposite charge in the wall is on average a little closer to the balloon than the like charge. As a result, the repulsive force due to the like charges is weaker than the attractive force due to the unlike charges. Thus there is a net attractive force on the balloon due to the wall. If there is a large enough charge on the balloon, this electrical attraction is capable of holding the balloon in place.

An **electric field** vector $\vec{E}$ can be defined in a way analogous to the gravitational field. The gravitational field is the force per unit mass, but the electric force is proportional to charge, not mass. So the electric field is defined as the force per unit *charge*:

$$\vec{E} = \frac{\vec{F}_e}{q} \quad \left(\text{compare to } \vec{g} = \frac{\vec{F}_g}{m} \right)$$

We study electric fields in detail in Part 3 of this book.

2.6   CONTACT FORCES

Contact forces are the net effect of enormous numbers of electromagnetic interactions between atoms when two objects come into contact. Electromagnetic forces are fundamentally long-range, but due to the balance of positive and negative charge in macroscopic objects, the net effect is negligible unless atoms on the two surfaces come very close to each other—what we think of as *in contact*. On a microscopic level there is no fundamental difference between the contact forces and other electromagnetic forces.

Normal Forces

Consider a book resting on a horizontal desk surface. The state of motion of the book would be very different if the desk were removed. So there must be an upward force exerted by the desk on the book. The force acting on the book due to the desk is called the **normal force** because it is perpendicular to the contact surface between the book and the desk. (In geometry, the word *normal* means *perpendicular*.) If the surface of the desk is not horizontal, the normal force would not be vertical; by definition the normal force is perpendicular to the contact surface.

According to Newton's third law, there is also a normal force exerted by the book on the desk; this normal force acts downward and is of equal magnitude. In everyday language, we might say that the desk "feels the book's weight pushing down on it." That is not an accurate statement in the language of physics. The desk cannot "feel" the gravitational force on the book; it can only feel forces exerted *on it*. What the desk does "feel" is the normal force exerted on it by the book. The normal force always pushes one object *away* from the other.

We can figure out what the magnitude of the normal force must be in various situations. In the case of the book, the normal force due to the desk must have just the right magnitude to keep the book from falling without making it shoot up into the air. If the book is in equilibrium and no other vertical forces act, the normal force on the book is equal in magnitude to the book's weight. However, if there are other vertical forces acting on the book, or if the book is not in equilibrium, then the normal force is *not* equal in magnitude to the book's weight.

How does the desk "know" how hard to push up on the book? First imagine putting the book on a bathroom scale instead of the desk. A spring inside the scale provides the upward force. The spring "knows" how hard to push because, as it is compressed, the force it exerts increases. When the book reaches equilibrium, the spring is exerting just the right amount of force, so there is no tendency to compress it further. The spring is compressed until it pushes up with a force equal to the book's weight. If the spring were stiffer, it would exert the same upward force but with less compression.

The forces that bind atoms together in a rigid solid, like the desk, act like extremely stiff springs that can provide large forces with little compression—so little that it's usually not noticed. The book makes a tiny indentation in the surface of the desk; a heavier book would make a slightly larger indentation. If the book were to be placed on a soft foam surface, the indentation would be much more noticeable.

Normal force: a contact force between two solid objects that is perpendicular to the contact surfaces. Each object pushes the other one away.

Friction

A contact force *parallel* to the surfaces of contact is called **friction**. We distinguish two types: **static friction** and **kinetic friction**. When the two surfaces are sliding across one another, as when a loose shingle slides down a roof, the friction is kinetic. When no sliding occurs, such as between the tires of a car parked on a hill and the road surface, the friction is called static. Static friction acts to prevent objects from sliding; kinetic friction acts to make sliding objects slow down and stop moving relative to each other.

Frictional forces are enormously complicated on the microscopic level and the effects of friction are not very reproducible. For dry, solid surfaces, the amount of friction depends on how smooth the surfaces are and how many contaminants are present on the surface. Does polishing two steel surfaces decrease the frictional forces when they slide across each other? Yes, up to a point. If the surfaces are extremely smooth and all surface contaminants are removed, the steel surfaces form a "cold weld"—essentially, they become one piece of steel. If there are no oxides, oils, or other contaminants, the atoms at the surfaces have no way to "know" to which piece of steel they belong. They bond as strongly with their new neighbors as they do with the old. A bit of lubricant drastically decreases the frictional forces because the two surfaces will float past each other without much contact.

Despite the complexities, we can make some approximate statements about the frictional forces between dry, solid surfaces. In the simplest model, the *maximum* force of static friction is proportional to the normal force acting between the two surfaces.

$$(f_s)_{max} \propto N$$

where $(f_s)_{max}$ is the maximum magnitude of the force of static friction and N is the magnitude of the normal force at the same contact surfaces. If you want better traction between the tires of a rear-wheel-drive car and the road, it helps to put something heavy in the trunk to increase the normal force between the tires and the road.

The constant of proportionality is called the **coefficient of static friction** (symbol μ_s):

$$(f_s)_{max} = \mu_s N \tag{2-6a}$$

The magnitude of the static frictional force must be less than or equal to the maximum possible magnitude:

$$f_s \leq \mu_s N \tag{2-6b}$$

Since it is the ratio of the magnitudes of two forces, μ_s is dimensionless. The coefficient depends on the condition and nature of the surfaces. Only the *maximum* force of static friction can be found from Eqs. (2-6a) and (2-6b). The actual force of friction in a given situation is not necessarily the maximum possible.

For sliding or kinetic friction, the force of friction is only weakly dependent on the speed and is roughly proportional to the normal force. In the simplified model we will use, the force of kinetic friction is assumed to be proportional to the normal force and independent of speed:

$$f_k = \mu_k N \tag{2-6c}$$

where f_k is the magnitude of the force of kinetic friction and μ_k is called the **coefficient of kinetic friction**. The coefficient of static friction is always larger than the coefficient of kinetic friction for an object on a given surface. On a horizontal surface, a larger force is required to start the object moving than is required to keep it moving at a constant velocity.

Equations (2-6) are only *part* of the force law for friction; we must also specify the direction of the force:

- The static frictional force acts in the direction that tends to prevent the surfaces from beginning to slide.

- The direction of the kinetic frictional force is in the direction that would tend to make the sliding stop—the direction opposite to the object's motion.

According to Newton's third law, there must be a pair of frictional forces, one acting on each object. If a book slides to the left along a desk, the desk exerts a kinetic frictional force on the book to the right; the book exerts a kinetic frictional force on the desk of equal magnitude to the left.

> Equations (2-6) relate only the *magnitudes* of the frictional and normal forces. Their directions are perpendicular to each other.

Example 2.11

Pushing on a Small Chest

In order to slide a 750-N chest across the floor at constant speed, you must push horizontally with a force of 450 N. The chest is rather short and is in no danger of toppling over. (a) What is the coefficient of kinetic friction? (b) Suppose you push with a force of 110 N and the cabinet does not budge. Now what is the force of friction on the chest? Assume that $\mu_s = 1.2 \mu_k$.

Strategy In both cases, the chest is in equilibrium—dynamic in (a) and static in (b). So in both cases, the net force acting on it is zero. In each case we draw a free-body diagram.

continued on next page

Example 2.11 *continued*

Solution (a) Since you push at constant speed, the net force on the chest is zero (dynamic equilibrium). There are four forces acting on it (Fig. 2.16a): the weight $\vec{W}$, the normal force $\vec{N}$ exerted by the floor, kinetic friction $\vec{f}_k$, and the force $\vec{F}$ you apply—which is a normal force on another surface of the chest. Now we draw an FBD (Fig. 2.16b).

The sum of these four forces must give a net force of zero. There are two pairs of collinear vectors: one pair is vertical and the other is horizontal. Since we know how to add collinear vectors, let us group the vectors in collinear pairs:

$$\vec{F}_{net} = (\vec{N} + \vec{W}) + (\vec{f}_k + \vec{F}) = \vec{0}$$

The sum $\vec{N} + \vec{W}$ is a vector in the vertical direction and the sum $\vec{f}_k + \vec{F}$ is a vector in the horizontal direction. Imagine adding the two vectors $(\vec{N} + \vec{W})$ and $(\vec{f}_k + \vec{F})$ and setting the sum equal to zero. Two nonzero vectors can add to zero only if they are equal in magnitude and opposite in direction. These two vectors cannot be opposite in direction since one is vertical and the other is horizontal. Therefore, they must *each* be zero:

$$(\vec{N} + \vec{W}) = \vec{0} \quad \text{and} \quad (\vec{f}_k + \vec{F}) = \vec{0}$$

In each pair, the vectors are in opposite directions. Their magnitudes must be equal in order to add to zero:

$$N = W \quad \text{and} \quad f_k = F$$

Now we can calculate the coefficient of kinetic friction.

$$\mu_k = \frac{f_k}{N}$$

Substituting $N = W$ and $f_k = F$,

$$\mu_k = \frac{F}{W} = \frac{450}{750} = 0.60$$

(b) It might be tempting to write $f_s = \mu_s N$, but that's the *maximum possible* static frictional force for this contact surface. The actual f_s is whatever it takes to keep the cabinet in equilibrium, so long as it does not *exceed* $\mu_s N$. The free-body diagram is the same as before, except $\vec{f}$ now represents static friction instead of kinetic friction. As before, the two horizontal forces must be equal and opposite, so $f_s = F = 110$ N.

Discussion The static frictional force is like the normal force in one way: its magnitude and direction are whatever it takes to keep the surfaces from sliding across each other—up to a *maximum* of $\mu_s N$. Do not assume that its magnitude is equal to $\mu_s N$.

Practice Problem 2.11 Net force on a moving chest

What is the net force on the chest while you slide it across the floor by applying a horizontal force of 480 N?

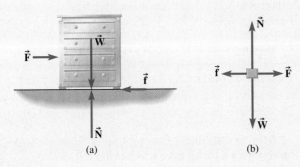

Figure 2.16
(a) Forces acting on the chest and (b) a free-body diagram for the chest.

Conceptual Example 2.12

Horse and Sleigh

To cause a sleigh to move through the snow (Fig. 2.17a), a horse pulls the sleigh with a forward horizontal force of magnitude T. (a) Draw free-body diagrams for the horse and for the sleigh when there is a net force on each in the forward direction. (According to Newton's second law, a net force in the forward direction is needed to start the sleigh moving forward.) (b) According to Newton's third law, the sleigh pulls back on the horse with a force of *the same* magnitude as the force with which the horse pulls the sleigh. Does this mean the horse can never budge the sleigh? Explain. (c) For each force acting on the sleigh, identify the other half of the interaction—the equal and opposite force exerted by the sleigh on another object.

Strategy We can draw free-body diagrams for (1) the sleigh alone, (2) the horse alone, and (3) the system of horse plus sleigh. In each, we include only the external forces acting *on* that system. Changes in the state of motion of a system are determined by the net force acting *on that system.*

Solution and Discussion (a) There are four forces acting on the sleigh: the force exerted by the horse on the sleigh $\vec{T}_{sh}$, the weight of the sleigh $\vec{W}_{se}$, the normal force of the ground on the sleigh $\vec{N}_{sg}$, and friction acting on the sleigh due to the ground $\vec{f}_{sg}$.

Figure 2.17b shows an FBD for the sleigh. In order for the net force to be in the forward direction, the force with which the horse pulls ($\vec{T}_{sh}$) must exceed the force of friction on the sleigh due to the ground ($\vec{f}_{sg}$) and the normal force ($\vec{N}_{sg}$) must be equal in magnitude to the weight of the sleigh ($\vec{W}_{se}$).

$$T_{sh} > f_{sg}$$
$$N_{sg} = W_{se}$$

continued on next page

Conceptual Example 2.12 *continued*

There are four forces acting on the horse: the force exerted by the sleigh ($\vec{\mathbf{T}}_{hs}$), the weight of the horse $\vec{\mathbf{W}}_{he}$, the normal force of the ground on the horse $\vec{\mathbf{N}}_{hg}$, and friction acting on the horse due to the ground $\vec{\mathbf{f}}_{hg}$. Newton's third law says that $\vec{\mathbf{T}}_{hs} = -\vec{\mathbf{T}}_{sh}$; the sleigh pulls back on the horse with a force equal in magnitude to the forward pull of the horse on the sleigh. Therefore $\vec{\mathbf{T}}_{hs}$ is to the left. In order for the net force on the horse to be directed to the right, the only other horizontal force, $\vec{\mathbf{f}}_{hg}$, must be larger and directed *to the right*. How does the horse get friction to push it *forward*? By pushing *backward* on the ground with its feet. We all do the same thing when taking a step; by pushing backward on the ground, we get the ground to push forward on us.

Figure 2.17c shows an FBD for the horse. The normal force ($\vec{\mathbf{N}}_{hg}$) must be equal in magnitude to the weight of the horse ($\vec{\mathbf{W}}_{he}$).

$$f_{hg} > T_{hs}$$
$$N_{hg} = W_{he}$$

(b) The forces that the sleigh and horse exert on each other must be equal in magnitude and opposite in direction no matter what happens. What determines the motion of the sleigh is the net force acting on *it*. A force acting on the horse ($\vec{\mathbf{T}}_{hs}$) is not included when finding the net force on the sleigh—the sum of all the forces *acting on the sleigh*.

The condition for the sleigh to start moving forward is that the net force on it be in the forward direction, which means that

$$T_{sh} > f_{sg}$$

For the horse to also move forward,

$$f_{hg} > T_{hs}$$

Since $T_{sh} = T_{hs}$, the horse can pull the sleigh forward if

$$f_{hg} > f_{sg}$$

The forward frictional force on the horse must exceed the backward frictional force on the sleigh. Imagine putting the horse but not the sleigh on slippery ice; it would be impossible for the horse to pull the sleigh forward.

(c)

Force Exerted on Sleigh	Third-Law Partner
Force exerted on the sleigh by the horse $\vec{\mathbf{T}}_{sh}$	Force exerted on the horse by the sleigh $\vec{\mathbf{T}}_{hs}$
The weight of the sleigh $\vec{\mathbf{W}}_{se}$ (or $\vec{\mathbf{F}}_{se}$)	Gravitational force exerted on Earth by the sleigh $\vec{\mathbf{F}}_{es}$
The normal force on the sleigh by the ground $\vec{\mathbf{N}}_{sg}$	The normal force on the ground by the sleigh $\vec{\mathbf{N}}_{gs}$
Friction on sleigh due to the ground $\vec{\mathbf{f}}_{sg}$	Friction on ground due to the sleigh $\vec{\mathbf{f}}_{gs}$

Practice Problem 2.12   Find the missing force

A car is trying to get under way on a level north-south road. The contact forces exerted *by the tires on the road* are 11.0 kN downward and 3.3 kN southward.
(a) Draw an FBD for the car. (b) What is the weight of the car? (c) What is the net force acting on the car?

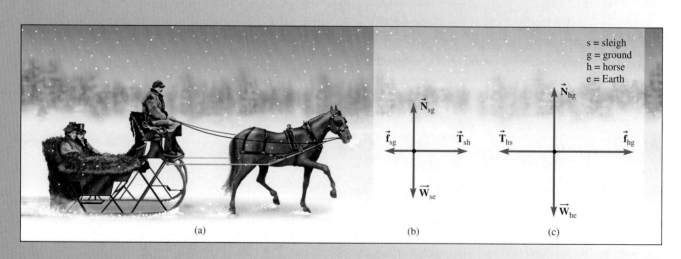

Figure 2.17 (a) Horse pulling sleigh, (b) free-body diagram for the sleigh, and (c) free-body diagram for the horse.

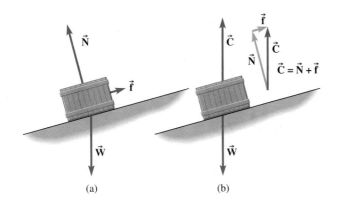

Figure 2.18 (a) Normal and frictional forces acting on a crate of weight $\vec{W}$ on an incline and (b) the total contact force $\vec{C}$ on the crate, which is directed straight up, is the vector sum of the normal and frictional forces. (In Chapter 4, we learn how to add vectors that are not collinear.)

Suppose that a crate of apples sits at rest on a ramp that is inclined at 20° to the horizontal. The normal force on the crate is perpendicular to the contact surfaces, or 20° from the vertical (Fig. 2.18a). The static frictional force points up the incline, 20° above the horizontal. In Practice Problem 2.6, we said that the contact force due to the ramp was straight up. Which is it?

Both are correct. Contact forces are the net macroscopic result of huge numbers of microscopic forces. Whether we think of a contact force as a single force or as a pair of perpendicular forces is entirely a matter of convenience. In Chapter 4, when the method of adding vectors is generalized to noncollinear vectors, we can show that the contact force $\vec{C}$ is the sum of the normal force and the frictional force (Fig. 2.18b).

2.7 TENSION

A flexible cord, rope, string, tendon, cable, or beam that is under tension exerts forces on the objects at each end, much as a spring does. There are two main differences. One is that the change in length is often negligibly small. The other is that a flexible rope or string can only pull; it cannot push. A rope may be under tension, but it cannot be compressed. A stiff cable or beam, on the other hand, can be compressed and therefore can push.

If the mass of the cord is small enough, then the forces exerted at the two ends are approximately equal in magnitude, assuming that nothing exerts forces on the cord except at the ends (see Conceptual Question 15). An *ideal cord* has zero mass and does not stretch at all for any amount of tension. The tension in an ideal cord is the same at the two ends (again assuming that nothing exerts forces on the cord except at the ends).

Pulleys

The pulley or a system of pulleys is sometimes called a *simple machine*. The name *machine* does not mean that a pulley can do work like an engine can, but a simple machine can change the direction or the magnitude of force required for performing some task. A single pulley (Fig. 2.19a) can change the direction of the force, exerted by a cord under tension, required to raise an object without changing the magnitude of the force required. If we are lifting a heavy weight it is easier to stand on the ground and pull down on the rope than to get above the weight on a platform and pull up on the rope.

An ideal pulley has no mass and no friction. The tension of a cord that runs through an ideal pulley is the same on both sides of the pulley. The proof of this statement, using the concept of torque, is found in Chapter 8. An ideal pulley exerts no forces on the cord that are *tangent* to the cord, so it is not pulling in either direction along the cord. This property enables a pulley to change the direction of a force without changing its magnitude. As long as a real pulley has a small mass and negligible amount of friction, we can approximate it as an ideal pulley.

Making The Connection:
advantages of a pulley

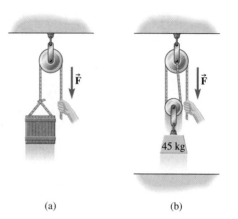

Figure 2.19 Two pulley systems: (a) a single pulley and (b) a two-pulley system

(a)   (b)

In Example 2.13, we will find that having more than one pulley enables us to pull with less force to raise an object. The pulley system of Fig. 2.19b requires only half as much force to lift the 45-kg mass as would be required to lift it directly since there are two ropes pulling up while only one is pulled down.

The pulley exerts forces on the cord and the cord exerts forces on the pulley. One way to think about the forces exerted on the pulley is to think of the part of the cord wrapped around the pulley as part of the pulley system; then there are two cords pulling on the pulley system, each with the same tension. A similar technique can be useful in other situations. Any cord can be thought of as two cords connected together at any arbitrary point; the two imagined cords pull on each other with forces whose magnitude is the tension in the cord.

Example 2.13

A Two-Pulley System

A 1804-N engine is hauled upward at constant speed (Fig. 2.20a). What are the tensions in the three ropes labeled A, B, and C? Assume ideal pulleys, L to the left and R to the right.

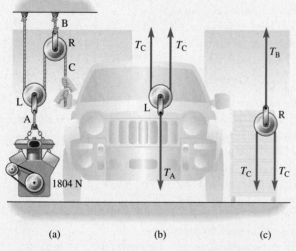

(a)   (b)   (c)

Figure 2.20
(a) A system of pulleys used to raise a heavy weight, (b) free-body diagram for the pulley on the left, and (c) free-body diagram for the pulley on the right.

Strategy   Assuming that the pulleys are ideal means that the tension of a rope is the same on both sides of the pulley. Therefore the rope that is attached to the ceiling, passes through both pulleys, and is pulled downward at the other end has the same tension throughout. Call the tensions in the three ropes T_A, T_B, and T_C. The engine and the pulley from which it hangs move up at constant speed, so the net force on each of them is zero. The other pulley is at rest, so the net force on it is also zero.

Solution   There are two forces acting on the engine: its weight (1804 N, downward) and the upward pull of rope A. These must be equal and opposite, since the net force is zero. Therefore $T_A = 1804$ N.

Figure 2.20b is an FBD for the pulley on the left. Rope A pulls down on it with a force of magnitude T_A. On *each side* of the pulley, rope C pulls upward. The rope has the same tension throughout so that all forces labeled T_C in Figs. 2.20b and 2.20c have the same magnitude. For the net force to equal zero,

$$2T_C = T_A$$

$$T_C = \tfrac{1}{2}T_A = 902.0 \text{ N}$$

Figure 2.20c is an FBD for the pulley on the right. Rope B pulls upward on it with a force of magnitude T_B. On *each side*

continued on next page

Example 2.13 *continued*

of the pulley, rope C pulls downward. For the net force to equal zero,

$$T_B = 2T_C = 1804 \text{ N}$$

Discussion The engine is raised by pulling *down* on a rope—the pulleys change the direction of the applied force needed to lift the engine. In this case, they also change the *magnitude* of the required force. They do that by making the rope pull up on the engine twice, so the person pulling the rope only needs to

exert a force equal to half the engine's weight. Where does the "extra" force come from? It comes from the ceiling, which pulls up on the two ropes (B and C) with a total force of 2706 N.

Practice Problem 2.13 System of ropes, pulleys, engine

Consider the entire collection of ropes, pulleys, and the engine to be a single system. Draw an FBD for this system and show that the net force is zero.

Changing the direction of a tensile force is of interest in the study of animal motion, or biomechanics. You do not have pulleys inside your body; instead, tendons are often used to change the direction of the force exerted by a muscle. Tendons attach to bones and are lubricated by fluids to minimize friction (Fig. 2.21). For example, the quadriceps tendon connects to the kneecap (patella) and allows you to swing your lower leg back and forth or to kick a soccer ball. Muscles are usually connected by tendons, located on either end of the muscle, to two different bones, which in turn are linked at a joint. Usually one of the bones is more easily moved than the other. As an analogy, think of the muscle as a spring that can contract or relax. The tendons correspond to ropes tied to either end of the spring and then attached at their opposite ends to the movable bones.

Making The Connection:
functions of tendons
in the body

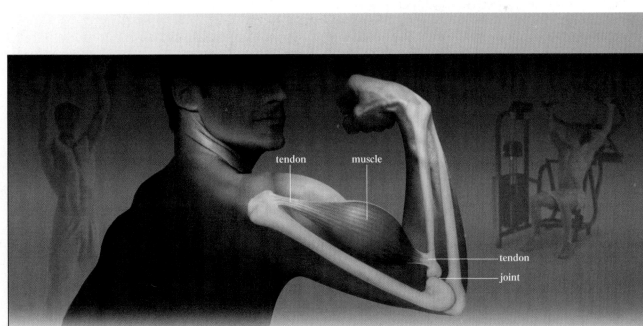

Figure 2.21 A muscle and two tendons attach two bones across a joint.

Physics at Home

Sit with your arm bent at the elbow with a heavy object on the palm of your hand. You can feel the contraction of the biceps muscle. With your other hand, feel the tendon that connects the biceps muscle to your forearm.

Now place your hand palm down on the desktop and push down. Now it is the triceps muscle that contracts, pulling up on the bone on the other side of the elbow joint. Muscles and tendons cannot push; they can only pull. The biceps muscle cannot push the forearm downward, but the triceps muscle can pull on the other side of the joint. In both cases, the arm acts as a lever.

MASTER THE CONCEPTS

Summary

Newton's laws of motion:

1. If no forces act on an object, then its state of motion does not change. If at rest, it remains at rest; if moving, it continues to move in a straight line with constant speed.

2. A nonzero net force acting on an object causes its state of motion to change.

3. In an interaction between two objects, the forces that each exerts on the other are equal in magnitude and opposite in direction.

- If the net external force acting on a body is zero, the body is in *equilibrium*. A body in equilibrium moves at a constant speed along a straight line. (The speed can be zero, in which case the body is at rest.)

- At the fundamental level, there are four interactions: gravitation, the strong and weak nuclear interactions, and the electromagnetic interaction. Contact forces are large-scale manifestations of many microscopic electromagnetic interactions.

- Gravity and electromagnetic forces are the only long-range forces. All other forces (in a macroscopic view) are contact forces.

- The SI unit of force is the newton. $1.00 \text{ N} = 0.2248 \text{ lb}$. $1 \text{ N} = 1 \text{ kg·m/s}^2$

- Force is a vector quantity—it has a direction as well as a magnitude. Vectors have their own rules for addition.

- The *net force* is the vector sum of all the forces acting on a body. Since all the internal forces come in third-law pairs, we need only sum the external forces.

- A *free-body diagram* (FBD) includes vector arrows representing every force acting on the chosen object, but no forces acting on other objects.

- The magnitude of the *gravitational force* between two objects is

$$F = \frac{Gm_1m_2}{r^2}$$

where r is the distance between their centers.

- The *weight* of an object near the Earth's surface is the gravitational force acting on the object due to the Earth. An object's weight is proportional to its mass: $\vec{W} = m\vec{g}$. Near Earth's surface, $\vec{g} = 9.8 \text{ N/kg}$ downward.

- The magnitude of the *electric force* between two charged particles is

$$F = \frac{k|q_1||q_2|}{r^2}$$

where r is the distance between the particles.

- The *normal force* is a contact force perpendicular to the contact surfaces that pushes each object away from the other.

- *Friction* is a contact force parallel to the contact surfaces. In a simplified model, the kinetic frictional force and the maximum static frictional force are proportional to the normal force acting between the same contact surfaces.

$$f_s \le \mu_s N$$

$$f_k = \mu_k N$$

The static frictional force acts in the direction that tends to keep the surfaces from beginning to slide. The direction of the kinetic frictional force is in the direction that would tend to make the sliding stop.

- Many springs obey *Hooke's law* as long as they are not stretched or compressed too far:

$$F = kx$$

- Cords and pulleys can be used to change the direction of a force without changing its magnitude.

Highlighted Figures and Tables

F2.2 Simple spring scale showing force measured by the deformation it produces in an object (p. 28)

F2.3 Schematic representation of the unification of forces (p. 31)

F2.6 Galileo postulated that if all resistive forces could be eliminated, a ball rolling on a horizontal surface would never stop. (p. 33)

F2.9 Illustration of pairs of forces (p. 37)

F2.10 Addition of collinear vectors that are in the same direction and in opposite directions (p. 38)

F2.12 Examples of free-body diagrams (p. 40)

F2.14 Gravity is always an attractive force. (p. 42)

F2.18 Contact force $\vec{C}$ is the sum of the normal force and the frictional force. (p. 49)

F2.19 Single pulley and two-pulley systems (p. 50)

F2.20 Free-body diagrams of a pulley system (p. 50)

CONCEPTUAL QUESTIONS

1. Explain the need for automobile seat belts in terms of Newton's first law.

2. An American visitor to Finland is surprised to see heavy metal frames outside of all the apartment buildings. On Saturday morning the purpose of the frames becomes evident when several apartment dwellers appear, carrying rugs and carpet beaters to each frame. How does the law of inertia aid in the rug beating process? Do you see a similarity to the role inertia plays when you throw a baseball?

3. Explain why gravitation is the only appreciable interaction between the Moon and the Earth. Why are all of the other interactions negligible?

4. You are standing on a bathroom scale in an elevator. In which of these situations must the scale read the same as when the elevator is at rest? Explain. (a) Moving up at constant speed. (b) Moving up with increasing speed. (c) In free fall (after the elevator cable has snapped).

Figure 2.22 Question 5

5. A carpenter is working on the roof of a new house. The steel head of her hammer has become loose on the tapered upper end of its wooden handle. She holds the hammer vertically, raises it up, and then brings it down rapidly, hitting the bottom end of the wood handle on the sturdy ridge pole of the roof (Fig. 2.22). Explain how this tightens the head back onto the handle.

6. When a car begins to move forward, what force makes it do so? Remember that it has to be an *external* interaction; the internal forces all add to zero. How does the engine facilitate the propelling force?

7. Two cars are headed toward each other in opposite directions along a narrow country road. The cars collide head-on, crumpling up the hoods of both. Describe what happens to the car bodies in terms of inertia. Does the rear end of the car stop at the same time as the front end?

8. In some tables of unit conversions, you are told that 1 kg = 2.2 lb. What is wrong with that statement? On the surface of the Moon, does 1 kg "equal" 2.2 lb? Why or why not?

9. (a) What assumptions do you make when you call the reading of a bathroom scale your "weight"? What does the scale really tell you? (b) Under what circumstances might the reading of the scale *not* be equal to your weight?

10. A freight train consists of an engine and several identical cars on level ground. Determine whether each of these statements is correct or incorrect and explain why. (a) If the train is moving at constant speed, the engine must be pulling with a force greater than the train's weight. (b) If the train is moving at constant speed, the engine's pull on the first car must exceed that car's backward pull on the engine. (c) If the train is coasting, its inertia makes it slow down and eventually stop.

11. (a) Does a man weigh more at the North Pole or at the equator? (b) Does he weigh more at the top of Mt. Everest or at the base of the mountain?

12. What is the distinction between a vector and a scalar quantity? Give two examples of each.

13. Which of the fundamental forces governs the motion of planets in the solar system? Is this the strongest or the weakest of the fundamental forces? Explain.

14. An old steam locomotive uses coal as a fuel. A railroad worker, called a fireman, shovels the coal onto the fire in the firebox that powers the locomotive engine. Explain how he lets Newton's first law help him move the coal onto the fire.

15. Using Newton's laws, explain why a spring *must* exert forces of equal magnitude on the objects attached to each end, assuming that the spring's mass is negligibly small. The same line of reasoning applies to a rope or cord whose mass is negligible.

16. A boy has just won his first sailing race. After mooring the sailboat, he rows his dinghy up to the dock to greet his admiring mates. He ties the line attached to the bow of his dinghy to a cleat on the dock; initially he is only 2.0 m from the dock but the line is long enough so that the boat can float 5.0 m away from the dock. Then, feeling proud of himself, he decides to make a running jump from the dinghy to the dock. What happens next? Use Newton's third law to describe the action.

17. You are standing on one end of a light wooden raft that has floated 3 m away from the pier (Fig. 2.23). If the raft is 6 m long by 2.5 m wide and you are standing on the raft end nearest to the pier, can you propel the raft back toward the pier where a friend is standing with a pole and hook trying to reach you? You have no oars. Make suggestions of what to do without getting yourself wet.

18. Tell whether each of these objects is in equilibrium. If it is, is the equilibrium static or dynamic? (a) A car driving around a curve at constant speed on a flat road. (b) A car driving straight up a 6° incline at constant speed. (c) The Moon.

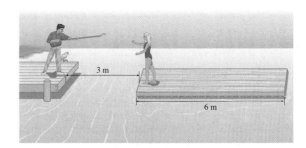

Figure 2.23 Question 17

19. A traffic light is hanging at an intersection. (a) If the air is still, is the light in equilibrium? (b) What about when a strong, steady wind blows so that the light makes an angle of 30° with its usual vertical position? Is the light in equilibrium?

20. Two boys are trying to break a cord. Gerardo says they should each pull in opposite directions on the two ends; Stefan says they should tie the cord to a pole and both pull together on the opposite end. Which plan is more likely to work?

MULTIPLE CHOICE QUESTIONS

1. A space probe leaves the solar system to explore interstellar space. Once it is far from any stars, when must it fire its rocket engines?
   (a) All the time, in order to keep moving.
   (b) Only when it wants to speed up.
   (c) When it wants to speed up or slow down.
   (d) Only when it wants to turn.
   (e) When it wants to speed up, slow down, or turn.

2. Your car won't start, so you are pushing it. You apply a horizontal force of 300 N to the car, but it doesn't budge. Which of these other forces *must* have a magnitude of 300 N?
   (a) the frictional force exerted by the road on the car
   (b) the force exerted by the car on you
   (c) the frictional force exerted on you by the road
   (d) (a), (b), and (c)
   (e) (a) and (c) but not (b)

3. The net force on a moving object suddenly becomes zero. The object then:
   (a) stops abruptly.
   (b) stops during a short time interval.
   (c) changes direction.
   (d) continues at constant velocity.
   (e) slows down gradually.

4. Which of these is *not* a long-range force?
   (a) the force that makes lightning flow between an electrically charged cloud and an oppositely charged area on the ground
   (b) the force that makes a compass point north
   (c) the force that a person exerts on a chair while sitting
   (d) the force that keeps the Moon in its orbital path around the Earth

5. The gravitational field strength on the surface of a distant planet is 1/5 that on the surface of Earth. An astronaut's mass and weight as measured on Earth are *m* and *W*, respectively. What are the astronaut's mass and weight as measured on the distant planet?
   (a) m, W (b) $\frac{1}{5}m, W$ (c) $m, \frac{1}{5}W$ (d) $\frac{1}{5}m, \frac{1}{5}W$

6. When an object is in translational equilibrium, which of these statements is *not* true?
   (a) The vector sum of the forces acting on the object is zero.
   (b) The object must be stationary.
   (c) The object has a constant velocity.
   (d) The speed of the object is constant.

7. A person is standing on a bathroom scale. Which of these is *not* a force exerted on the scale?
   (a) a contact force due to the floor
   (b) a contact force due to the person's feet
   (c) the weight of the person
   (d) the weight of the scale

8. Which of these forces bind electrons to nuclei to form atoms?
   (a) the strong nuclear force
   (b) the weak nuclear force
   (c) the electromagnetic force
   (d) the gravitational force

9. The Sun exerts a gravitational force on the Earth. Which statement is true concerning the gravitational force exerted by the Earth on the Sun?
   (a) It has the same magnitude and direction as the gravitational force exerted by the Sun on the Earth.
   (b) It has a much smaller magnitude than the gravitational force exerted by the Sun on the Earth and is opposite in direction.
   (c) It has a much smaller magnitude than the gravitational force exerted by the Sun on the Earth and is in the same direction.
   (d) It has the same magnitude as the gravitational force exerted by the Sun on the Earth and is opposite in direction.
   (e) The Earth exerts no gravitational force on the Sun; the Earth orbits the Sun but the Sun does not orbit the Earth.

10. To make an object start moving on a surface with friction requires
    (a) less force than to keep it moving on the surface.
    (b) the same force as to keep it moving on the surface.
    (c) more force than to keep it moving on the surface.
    (d) a force equal to the weight of the object.

PROBLEMS

Note: **C** indicates a combination conceptual/quantitative problem. Gold diamonds ◆, ◆◆ are used to indicate the increasing level of difficulty of each problem. Problem numbers appearing in blue, 9., denote problems that have a detailed solution available in the Student Solutions Manual. Some problems are *paired* by concept; their numbers are connected by a ruled box.

2.1 Forces

1. If the length of the Achilles tendon increases 0.50 cm when the force exerted on it by the muscle increases from 3200 N to 4800 N, what is the "spring constant" of the tendon?

2. If a force of 5.0 N causes a spring to stretch 3.5 cm from its relaxed length, how far does a force of 7.0 N cause the

same spring to stretch? What is the spring constant of this spring?

3. A block of wood is compressed 2.0 nm when inward forces of magnitude 120 N are applied to it on two opposite sides. (a) What is the effective spring constant of the block? (b) Assuming Hooke's law still holds, how much would the block be compressed by inward forces of magnitude 480 N?

4. A sack of flour has a weight of 19.8 N. What is its weight in pounds?

5. An astronaut weighs 175 lb. What is his weight in newtons?

2.3 Newton's Laws of Motion;
2.4 Net Force: Vector Addition

6. A person is standing on a bathroom scale. Identify the third-law partner of each of the forces exerted on the scale. In other words, for every interaction involving the scale, identify the force that the scale exerts on another object.

7. A skydiver is falling with her parachute open. Consider the apparatus that connects the parachute to the skydiver to be part of the parachute. (a) Identify the third-law partner of each of the forces exerted on the skydiver. (b) Identify the third-law partner of each of the forces exerted on the parachute. (c) If the skydiver is falling straight down at a constant speed, is she in equilibrium? If so, what kind of equilibrium?

8. A hummingbird is hovering motionless beside a flower. The blur of its wings shows that they are rapidly beating up and down. If the air pushes upward on the bird with a force of 0.30 N, what is the weight of the hummingbird?

9. A hanging potted plant is suspended by a cord from a hook in the ceiling. Draw a free-body diagram for each of these: (a) the system consisting of plant, soil, and pot; (b) the cord; (c) the hook; (d) the system consisting of plant, soil, pot, cord, and hook. Label each force arrow using subscripts (for example, $\vec{F}_{ch}$ would represent the force exerted on the cord by the hook).

10. A parked automobile slips out of gear, rolls unattended down a slight incline and then along a level road until it hits a stone wall. Draw a free-body diagram to show the forces acting on the car while it is in contact with the wall.

11. A sailboat, tied to a mooring with a line, weighs 820 N. The mooring line pulls horizontally toward the west on the sailboat with a force of 110 N. The sails are stowed away and the wind blows from the west. The boat is moored on a still lake—no water currents push on it. Draw a free-body diagram for the sailboat and indicate the magnitude of each force.

12. An airplane is cruising along in a horizontal level flight at a constant velocity, heading due west. (a) If the weight of the plane is 2.6×10^4 N, what is the net force on the plane? (b) With what force does the air push upward on the plane? (c) Is the plane in equilibrium? If so, what kind of equilibrium?

13. Find the magnitude and direction of the net force on the object in each of the free-body diagrams in Fig. 2.24.

14. A truck driving on a level highway is acted upon by the following forces: a downward gravitational force of 52 kN (kilonewtons); an upward contact force due to the road of 52 kN; another contact force due to the road of 7 kN, directed east; and a drag force due to air resistance of 5 kN, directed west. What is the net force acting on the truck?

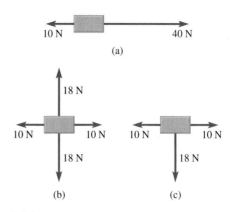

Figure 2.24 Problem 13

15. A woman who weighs 600.0 N sits on a chair with her feet on the floor and her arms resting on the chair's armrests. The chair weighs 100.0 N. Each armrest exerts an upward force of 25.0 N on her arms. The seat of the chair exerts an upward force of 500.0 N. (a) What force does the floor exert on her feet? (b) What force does the floor exert on the chair? (c) Consider the woman and the chair to be a single system. Draw a free-body diagram for this system that includes all of the *external* forces acting on it.

16. A man is lazily floating on an air mattress in a swimming pool. If the weight of the man and air mattress together is 806 N, what is the upward force of the water acting on the system of man and mattress?

17. Refer to Problem 7. Consider the skydiver and parachute to be a single system. What are the external forces acting on this system?

2.5 Gravitational Forces

18. (a) Calculate your weight in newtons. (b) What is the weight in newtons of 250 g of cheese? (c) Name a common object whose weight is about 1 N.

19. A young South African girl has a mass of 40.0 kg. Assume $g = 9.81$ N/kg. (a) What is her weight in newtons? (b) If she came to the United States, what would her weight be in pounds as measured on an American scale?

20. A man weighs 0.80 kN. What is his mass in kilograms?

21. The length of a spring increases by 7.2 cm from its relaxed length when a mass of 1.4 kg is suspended from the spring. (a) What is the spring constant? (b) A different mass is suspended and the spring length increases by 12.2 cm from its relaxed length. What is the second mass?

22. A spring scale in a French market is calibrated to show the *mass* of vegetables in grams and kilograms. (a) If the marks on the scale are 1.0 mm apart for every 25 grams, what maximum extension of the spring is required to measure up to 5.0 kg? (b) What is the spring constant of the spring? [Remember that the scale really measures *force*.]

23. Find the altitudes above the Earth's surface where Earth's gravitational field strength would be (a) two-thirds and (b) one-third of its value at the surface. [*Hint:* First find the radius for each situation; then recall that the altitude is the distance from the *surface* to a point above the surface. Use proportional reasoning.]

◆24. A spring hangs from a 2.44-m-high ceiling. The relaxed length of the spring is 0.305 m. A board, 1.98 m in length, hangs from the spring so that its lower end just reaches, but does not touch, the floor. The board has a mass of 10.4 kg and is perpendicular to the floor. Find the spring constant of the spring. Assume $g = 9.807$ N/kg.

◆25. A brick of mass 2.00 kg is "weighed under water" by hanging it from a spring scale (see Fig. 2.11). Let $g = 9.79$ N/kg. (a) If the spring scale reads 14.0 N, what is the upward force of water on the brick? (b) What is the reading of the pan scale on which the beaker sits? The weight of the water and the beaker together is 12.0 N. [*Hint:* refer to Example 2.7.]

26. The mass of the Moon is 0.0123 times that of the Earth. A spaceship is traveling along a line connecting the centers of the Earth and the Moon. At what distance from the Earth does the spaceship find the gravitational pull of the Earth equal in magnitude to that of the Moon? Express your answer as a percentage of the distance between the centers of the two bodies.

27. A 320-kg satellite is in orbit around the Earth 16,000 km above the Earth's surface. (a) What is the weight of the satellite when in orbit? (b) What was its weight when it was on the Earth's surface, before being launched? (c) While it orbits the Earth, what force does the satellite exert on the Earth?

28. At what altitude above the Earth's surface would your weight be half of what it is at the Earth's surface? [Hint: See Problem 23.]

29. (a) What is the magnitude of the gravitational force that the Earth exerts on the Moon? (b) What is the magnitude of the gravitational force that the Moon exerts on the Earth?

◆◆30. (a) If a spacecraft moves in a straight line between the Earth and the Sun, at what point would the force of gravity on the spacecraft due to the Sun be as large as that due to the Earth? (b) If the spacecraft is close to, but not at, this equilibrium point, does the net force on the spacecraft tend to push it toward or away from the equilibrium point? [*Hint:* imagine the spacecraft a small distance d closer to the Earth and find out which gravitational force is stronger.]

2.6 Contact Forces

31. Your textbook is sitting on your desk. (a) Draw a free-body diagram showing all the forces acting on the book. (b) Give the book a shove to the right. Draw a free-body diagram for the book while your hand was in contact with it. (c) Draw a free-body diagram for the book after your hand breaks contact with it, but while it is still sliding.

©◆32. A toy freight train consists of an engine and three identical cars. The train is moving to the right at constant speed along a straight, level track. Three spring scales are used to connect the cars as follows: spring scale A is located between the engine and the first car; scale B is between the first and second cars; scale C is between the second and third cars. (a) If air resistance and friction are negligible, what are the relative readings on the three spring scales A, B, and C? (b) Repeat part (a), taking air resistance and friction into consideration this time. [*Hint:* draw a free-body diagram for the car in the middle.] (c) If air resistance and friction together cause a force of magnitude 5.5 N on each car, directed toward the left, find the readings of scales A, B, and C.

33. The coefficient of static friction between a block and a horizontal floor is 0.40, while the coefficient of kinetic friction is 0.15. The mass of the block is 5.0 kg. If a horizontal force is slowly increased until it is barely enough to make the block start moving, what is the net force on the block the instant that it starts to slide?

34. A crate full of artichokes rests on a ramp that is inclined 10.0° above the horizontal. Give the direction of the normal force and the friction force acting on the crate in each of these situations. (a) The crate is at rest. (b) The crate is being pushed and is sliding up the ramp. (c) The crate is being pushed and is sliding down the ramp.

◆35. A 3.0-kg block is at rest on a horizontal floor. If you push horizontally on the 3.0-kg block with a force of 12.0 N, it just starts to move. (a) What is the coefficient of static friction? (b) A 7.0-kg block is stacked on top of the 3.0-kg block. What is the magnitude F of the force, acting horizontally on the 3.0-kg block as before, that is required to make the two blocks start to move?

36. A horse is trotting along pulling a sleigh through the snow. To move the sleigh, of mass m, straight ahead at a constant speed, the horse must pull with a force of magnitude T. (a) What is the net force acting on the sleigh? (b) What is the coefficient of kinetic friction between the sleigh and the snow?

◆37. A box full of books rests on a wooden floor. The normal force the floor exerts on the box is 250 N. (a) You push horizontally on the box with a force of 120 N, but it refuses to budge. What can you say about the coefficient of static friction between the box and the floor? (b) If you must push horizontally on the box with a force of at least 150 N to start it sliding, what is the coefficient of static friction? (c) Once the box is sliding, you only have to push with a force of 120 N to keep it sliding. What is the coefficient of kinetic friction?

38. Before hanging new William Morris wallpaper in her bedroom, Brenda sanded the walls lightly to smooth out some irregularities on the surface. Draw a free-body diagram for the sanding block as it moves straight up the wall.

39. Four separate blocks are placed side by side in a left-to-right row on a table. A horizontal force, acting toward the right, is applied to the block on the far left end of the row. Draw free-body diagrams for (a) the second block on the left and for (b) the system of four blocks.

2.7 Tension

40. A fly fisherman has hooked a Chinook salmon. The salmon is running out the line, moving at a constant velocity of 2 m/s through the water. Identify all the forces acting on the salmon.

41. A sailboat is tied to a mooring with a line. The wind is from the southwest. Identify all the forces acting on the sailboat.

42. A pulley is attached to the ceiling (Fig. 2.25a). Spring scale A is attached to the wall and a rope runs horizontally from it and over the pulley. The same rope is then attached to spring scale B. On the other side of scale B hangs a 120-N weight. What are the readings of the two scales A and B? The weights of the scales are negligible.

43. Spring scale A is attached to the floor and a rope runs vertically upward, loops over the pulley, and runs down on the other side to a 120-N weight (Fig. 2.25b). Scale B is attached

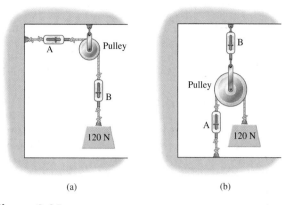

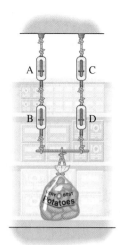

(a)   (b)

Figure 2.25 Problems 42 and 43

Figure 2.26 Problem 46

to the ceiling and the pulley is hung below it. What are the readings of the two spring scales, A and B? Neglect the weights of the pulley and scales.

✦ 44. The coefficient of static friction between block A and a horizontal floor is 0.45 and the coefficient of static friction between block B and the floor is 0.30. The mass of each block is 2.0 kg and they are connected together by a cord. (a) If a horizontal force $\vec{F}$ pulling on block B is slowly increased until it is barely enough to make the two blocks start moving, what is the magnitude of $\vec{F}$ at the instant that they start to slide? (b) What is the tension in the cord connecting blocks A and B at that same instant?

45. Two springs are connected in series so that spring scale A hangs from a hook on the ceiling and a second spring scale, B, hangs from the hook at the bottom of scale A (see springs A and B of Fig. 2.26). A 120-N weight hangs from the hook at the bottom of scale B. What are the readings on the upper scale A and the lower scale B? Neglect the weights of the scales.

✦ 46. Four *identical* spring scales, A, B, C, and D are used to hang a 220.0-N sack of potatoes (Fig. 2.26). (a) Assume the scales

have negligible weights and all four scales show the same reading. What is the reading of each scale? (b) Suppose that each scale has a weight of 5.0 N. If scales B and D show the same reading, what is the reading of each scale?

✦✦ 47. Two springs with equal spring constants k are connected first in series (one after the other) and then in parallel (side by side) with a weight hanging from the bottom of the combination. What is the effective spring constant of the two different arrangements? In other words, what would be the spring constant of a single spring that would behave exactly as (a) the series combination and (b) the parallel combination? Ignore the weight of the springs. [*Hint* for (a): *each spring stretches an amount $x = F/k$, but only one spring exerts a force on the hanging object. Hint* for (b): *each spring exerts a force $F = kx$.*]

COMPREHENSIVE PROBLEMS

Ⓒ 48. When you hold up a 100-N weight in your hand, with your forearm horizontal and your palm up, the force exerted by your biceps is much larger than 100 N—perhaps as much as 1000 N. How can that be? What other forces are acting on your arm? (See Fig. 2.27.) Draw a free-body diagram for the

forearm, showing all of the forces. Assume that all the forces exerted on the forearm are purely vertical—either up or down.

49. In the sport of curling, popular in Canada and Ireland, a player slides a 20.0-kg granite stone down a 38-m-long ice rink. Draw free-body diagrams for the stone (a) while it sits at rest on the

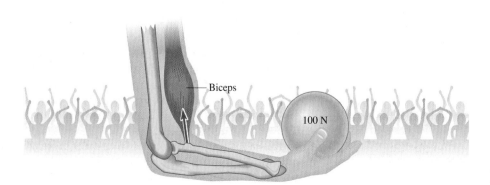

Figure 2.27 Problem 48

ice; (b) while it slides down the rink; (c) during a head-on collision with an opponent's stone that was at rest on the ice.

50. A car is driving on a straight, level road at constant speed. Draw a free-body diagram for the car, showing the significant forces that act upon it.

51. A refrigerator magnet weighing 0.14 N is used to hold up a photograph weighing 0.030 N. The magnet attracts the refrigerator door with a magnetic force of 2.10 N. (a) Identify the interactions between the magnet and other objects. (b) Draw a free-body diagram for the magnet, showing all the forces that act on it. (c) Which of these forces are long-range and which are contact forces? (d) Find the magnitudes of all the forces acting on the magnet.

52. A computer weighing 87 N sits on the surface of your desk. The coefficient of friction between the computer and the desk is 0.60. (a) Draw a free-body diagram for the computer. (b) What is the magnitude of the frictional force acting on the computer? (c) How hard would you have to push on it to get it to start to slide across the desk?

53. The readings of the two spring scales shown in Fig. 2.28 are the same. (a) Explain why they are the same. [*Hint:* draw free-body diagrams.] (b) What is the reading?

54. A 50.0-kg crate is suspended between the floor and the ceiling using two spring scales, one attached to the ceiling and one to the floor. If the lower scale reads 120 N, what is the reading of the upper scale? Ignore the weight of the scales.

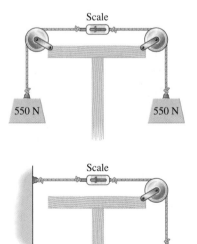

Figure 2.28
Problem 53

55. The weight of a water strider is approximately 6×10^{-6} N. (a) What is the force due to surface tension on the water strider when it is standing at rest on the water? (b) The *surface tension* γ of water is 0.07 N/m. Use dimensional analysis to determine whether the force of surface tension is proportional to the radius, the area, or the volume of the foot. In other words, to what power of r is the force proportional? (c) Explain why much larger animals cannot walk on water. Why can't they have feet big enough to make the surface tension of water support them? [*Hint:* Imagine a water strider that is twice as big in length, breadth, and height. Its volume—and therefore its weight—would be *eight* times as great.]

56. Spring scale A is attached to the ceiling. A 10.0-kg mass is suspended from the scale. A second spring scale, B, is hanging from a hook at the bottom of the 10.0-kg mass and a 4.0-kg mass hangs from the second spring scale. (a) What are the readings of the two scales if the masses of the scales are negligible? (b) What are the readings if each scale has a mass of 1.0 kg? Assume $g = 9.80$ N/kg.

57. A boy has stacked two blocks on the floor so that a 5.0-kg block is on top of a 2.0-kg block. (a) If the coefficient of static friction between the two blocks is 0.40 and the coefficient of static friction between the bottom block and the floor is 0.22, with what force should the boy push on the upper block to make both blocks in the system start to slide along the floor? (b) If he pushes too hard, the top block starts to slide off the lower block. What is the limit of the force with which he can push without that happening?

58. The forces required to extend a spring to various lengths are measured. The results are shown in the table below. Using the data in the table, plot a graph that helps you to answer the following two questions: (a) What is the spring constant? (b) What is the relaxed length of the spring?

Force (N)	1.00	2.00	3.00	4.00	5.00
Spring length (cm)	14.5	18.0	21.5	25.0	28.5

59. Plot a graph of this data for a spring resting horizontally on a table. Use your graph to find (a) the spring constant and (b) the relaxed length of the spring.

Force (N)	0.200	0.450	0.800	1.500
Spring length (cm)	13.3	15.0	17.3	22.0

60. A binary star consists of two stars of masses M_1 and $4.0M_1$ a distance d apart. Is there any point where the gravitational field due to the two stars is zero? If so, where is that point?

61. Two springs with spring constants k_1 and k_2 are connected in series. What is the effective spring constant of the combination? [See Problem 47(a).]

62. Two springs with spring constants k_1 and k_2 are connected in parallel. What is the effective spring constant of the combination? [See Problem 47(b).]

63. The tallest spot on Earth is Mt. Everest, which is 8850 m above sea level. If the radius of the Earth to sea level is 6370 km, how much does the magnitude of $\vec{g}$ change between sea level and the top of Mt. Everest?

64. In the fairy tale Rapunzel, the beautiful maiden let her long golden hair hang down from the tower in which she was held prisoner so that her prince could use her hair as a climbing rope to climb the tower and rescue her. Estimate how much force is required to pull a strand of hair out of your head. Next estimate how many hairs might be growing out of Rapunzel's head. If the prince has a mass of 60 kg, what is the force pulling on each strand of hair? Will Rapunzel be bald by the time the prince reaches the top of the 30-m tower?

65. A mass of 10 kg is hanging from a spring scale that is in turn hanging from a hook attached to the ceiling. The scale indicates a reading of 98 N for the force. Then two people hold the spring scale horizontally and pull on opposite ends until the scale again reads 98 N. With what force must each person pull to attain this result?

66. While trying to decide where to hang a framed picture, you press it against the wall to keep it from falling (Fig. 2.29). (a) What is the direction of the normal force exerted on the picture by your hand? (b) What is the direction of the normal force exerted on the picture by the wall? (c) The frictional force exerted on the picture by the wall can have two possible directions. Explain why.

Figure 2.29 Problem 66

ANSWERS TO PRACTICE PROBLEMS

2.1 The water pushes on the hull of the boat; the air pushes against the surface of the boat, including the sails; Earth's gravity pulls downward on the boat.

2.2 3.2 cm

2.3 No, the checkbook balance may be positive or negative, but there is no spatial direction associated with it. It is a scalar.

2.4 In the first case, inertia tends to keep Negar's body at rest with respect to the ground as the subway car begins to move forward. In the second case, inertia tends to keep Negar's body moving forward with respect to the ground, with constant speed as the subway car slows down.

2.5 downward force on the rod by the line and upward force on the line by the rod, downward gravitational force on the rod by the Earth and upward gravitational force on the Earth by the rod, upward force on the rod by the fisherman's hands and downward force on the hands by the rod

2.6 80 N upward

2.7 45.0 N

2.8 0.5 kN down

2.9 2 nN

2.10 3.77

2.11 30 N in the direction of the applied force

c = car
r = road
e = Earth

2.12 (a)   (b) weight of the car = 11.0 kN; (c) net force acting on the car = 3.3 kN northward

2.13

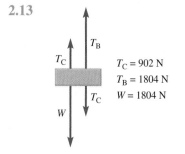

$T_C = 902$ N
$T_B = 1804$ N
$W = 1804$ N

Forces and Motion Along a Line

A sailplane (or "glider") is a small, unpowered, high-performance aircraft. A sailplane must be initially towed a few thousand feet into the air by a small airplane, after which it relies on regions of upward-moving air such as thermals and ridge currents to ascend farther. Suppose a small plane requires about 120 m of runway to take off by itself. When it is towing a sailplane, how much more runway does it need?

Making The Connection:
motion of a train

3.1   POSITION AND DISPLACEMENT

A nonzero net force acting on an object changes the object's velocity. According to Newton's second law, the net force determines the *acceleration* of the object. In order to see how interactions affect motion, we first need to carefully define the quantities used to describe motion (position, displacement, velocity, and acceleration) and we must understand the relationships among them. The description of motion using these quantities is called **kinematics**. When we combine kinematics with Newton's laws, we are in the realm of **dynamics**. Chapter 3 considers motion only along a straight line.

To describe motion clearly, we first need a way to say *where* an object is located. Suppose that at 3:00 P.M. a train stops on an east-west track as a result of an engine problem. The engineer wants to call the railroad office to report the problem. How can he tell them where to find the train? He might say something like "three kilometers east of the old trestle bridge." Notice that he uses a point of reference: the old trestle bridge. Then he states how far the train is from that point and in what direction. If he omits any of the three pieces (the reference point, the distance, and the direction) then his description of the train's whereabouts is ambiguous.

The same thing is done in physics. First, we choose a reference point, called the **origin**. Then, to describe the location of something, we give its distance from the origin and the direction. These two quantities, direction and distance, together describe the **position** of the object. Position is a vector quantity; the direction is just as important as the distance. The position vector is written symbolically as $\vec{r}$. Graphically, the position vector can be drawn as an arrow starting at the origin and ending with the arrowhead on the object. When more than one position vector is to be drawn, a scale is chosen so that the length of the vector is proportional to the distance between the object and the origin.

Once the train's engine is repaired and it gets on its way, we might want to describe its motion. At 3:14 P.M. it leaves its initial position, 3 km east of the origin (Fig. 3.1). At 3:56 P.M. the train is 26 km west of the origin, which is 29 km to the west of its initial position. **Displacement** is defined as the change of position; it is written $\Delta\vec{r}$, where the

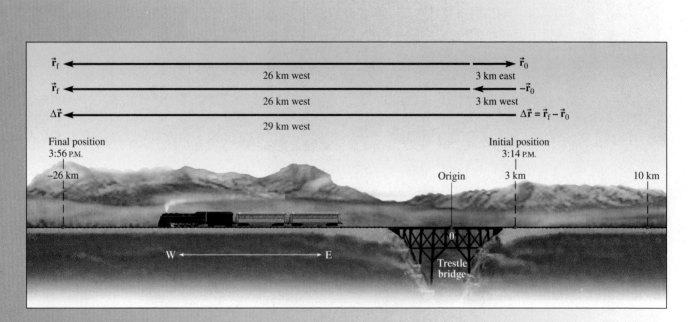

Figure 3.1   Initial ($\vec{r}_0$) and final ($\vec{r}_f$) position vectors for a train; the displacement vector $\Delta\vec{r}$ is found by subtracting the vector for the initial position from the vector for the final position.

symbol Δ means *the change in,* or the final value minus the initial value. If the initial and final positions are $\vec{\mathbf{r}}_0$ and $\vec{\mathbf{r}}_f$, then

---

Displacement

$$\Delta\vec{\mathbf{r}} = \vec{\mathbf{r}}_f - \vec{\mathbf{r}}_0 \qquad (3\text{-}1)$$

The symbol Δ stands for *the change in.* If the initial value of a quantity Q is Q_0 and the final value is Q_f, then $\Delta Q = Q_f - Q_0$. ΔQ is read "delta Q."

Since the positions are vector quantities, the operation indicated in Eq. (3-1) is a vector subtraction. To subtract a vector is to add its opposite, so $\vec{\mathbf{r}}_f - \vec{\mathbf{r}}_0 = \vec{\mathbf{r}}_f + (-\vec{\mathbf{r}}_0)$. Vectors representing the initial and final positions are shown as arrows in Fig. 3.1 and the vector subtraction is indicated. The displacement or change of position $\Delta\vec{\mathbf{r}}$ is shown; it is a vector of magnitude 29 km pointing west.

As a shortcut to subtract $\vec{\mathbf{r}}_0$ from $\vec{\mathbf{r}}_f$, instead of adding $-\vec{\mathbf{r}}_0$ to $\vec{\mathbf{r}}_f$, we can simply draw an arrow from the tip of $\vec{\mathbf{r}}_0$ to the tip of $\vec{\mathbf{r}}_f$ when the two vectors are drawn starting from the same point, as in the uppermost vector diagram of Fig. 3.1. This shortcut works because the displacement vector is the change of position; it takes us from the initial position to the final position. Since all vectors add and subtract according to the same rules, this same method can be used to subtract *any* kind of vector—as long as the vectors are drawn starting with their tails at the same point.

Notice that the magnitude of the displacement vector is not necessarily equal to the *distance traveled* by the train. Perhaps the train first travels 7 km to the east, putting it 10 km east of the origin, and then reverses direction and travels 36 km to the west. The total distance traveled in that case is 7 km + 36 km = 43 km, but the magnitude of the displacement—which is the distance between the initial and final positions—is 29 km. The displacement, since it is a vector quantity, must include the direction; it is 29 km to the west.

Displacement depends only on the starting and ending positions, not on the path taken.

Conceptual Example 3.1

A Relay Race

A relay race is run along a straight line track of length L running south to north. The first runner starts at the south end of the track and passes the baton to a teammate at the north end of the track. The teammate races back to the start line and passes the baton to a third teammate who races 1/3 of the way northward to a finish line. What is the baton's displacement during the race?

Strategy The displacement of the baton is the vector from the starting point to the finish point; it is the *vector sum* of the *displacements* of each runner.

Solution The displacement is the vector sum of the three separate displacements that occurred during the race (Fig. 3.2). Since the first two displacements are equal in magnitude but in opposite directions, their vector sum is zero. The total displacement is then equal to the 3rd displacement, from the origin to the finish line, located 1/3 of the way along the track north of the starting point.

$$\Delta\vec{\mathbf{r}} = (\Delta\vec{\mathbf{r}}_1 + \Delta\vec{\mathbf{r}}_2) + \Delta\vec{\mathbf{r}}_3$$

$$= 0 + \Delta\vec{\mathbf{r}}_3 = \tfrac{1}{3}L, \text{ directed north}$$

Discussion Note that the displacement magnitude is less than the total distance traveled by the baton $(2\tfrac{1}{3}L)$. The distance

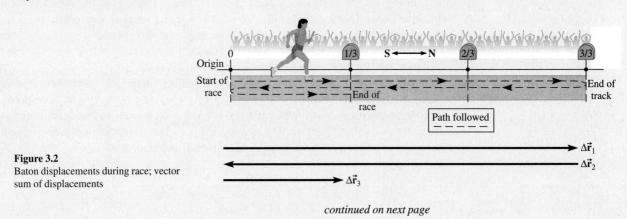

Figure 3.2
Baton displacements during race; vector sum of displacements

continued on next page

Conceptual Example 3.1 *continued*

traveled depends on the path followed, while the magnitude of the displacement is always the shortest distance between the two points of interest. The displacement vector shows the straight line path from starting point to finish point.

Conceptual Practice Problem 3.1   Around the bases

Casey hits a long fly ball over the heads of the outfielders. He runs from home plate to first base, to second base, to third base, and slides back to home plate safely (an inside-the-park home run). What is Casey's total displacement?

The sign of a vector component indicates the direction.

Vector Components

Since our universe has three spatial dimensions, positions in space are specified using a three-dimensional coordinate system, with x-, y-, and z-axes that are mutually perpendicular. Any vector can be specified either by its magnitude and direction or by its **components** along the x-, y-, and z-axes. The x-, y-, and z-components of vector $\vec{\mathbf{A}}$ are written with subscripts as follows: A_x, A_y, and A_z. One exception to this otherwise consistent notation is that the components of the position vector $\vec{\mathbf{r}}$ are usually written simply as x, y, and z rather than r_x, r_y, and r_z. The components themselves are scalars, which have a magnitude, units, and an algebraic sign. The sign indicates the direction; a positive x-component indicates the direction of the positive x-axis, while a negative x-component indicates the opposite direction (the negative x-axis). Specifying the components of a vector is as complete a description of the vector as is giving the magnitude and direction.

This chapter concentrates on motion along a straight line. If we choose one of the axes along that line, then the position, displacement, velocity, and acceleration vectors each have only one nonzero component. For horizontal linear motion, we usually choose the x-axis to lie along the direction of motion. For objects that move vertically, it is conventional to use the y-axis instead. One direction along the axis is chosen to be positive. Customarily the direction to the right of the origin is called the positive direction for a horizontal axis and upward is called the positive direction for a vertical axis. If vector $\vec{\mathbf{A}}$ points along the x-axis in the direction of $+x$ (to the right), then $A_x = +|\vec{\mathbf{A}}|$; if $\vec{\mathbf{A}}$ points in the opposite direction—in the direction of $-x$ (to the left)—then $A_x = -|\vec{\mathbf{A}}|$.

In Fig. 3.1 the compass arrows indicate that east is chosen as the direction for the positive x-axis. Using components, the train of Fig. 3.1 starts at $x = +3$ km and is later found at $x = -26$ km. The displacement (in component form) is $\Delta x = x_f - x_0 = -29$ km. The sign of the component indicates the direction. Since the positive x-axis is east, a negative x-component means that the displacement is to the west.

The rules for adding vectors translate into adding components with their algebraic signs. If $\vec{\mathbf{C}} = \vec{\mathbf{A}} + \vec{\mathbf{B}}$, then $C_x = A_x + B_x$, $C_y = A_y + B_y$, and $C_z = A_z + B_z$. Components become increasingly useful when adding vectors having more than one nonzero component. We can also subtract vectors by subtracting their components, since reversing the direction of a vector changes the sign of each of its components.

Example 3.2

A Stubborn Mule

In an attempt to get a mule moving in a horizontal direction, a farmer stands in front of the mule and pulls on the reins with a force of 320 N to the east while his son pushes on the mule's hindquarters with a force of 110 N to the east. In the meantime the mule pushes on the ground with a force of 430 N to the east (Fig. 3.3a). Use vector components to find the sum of the horizontal forces acting on the mule. Draw a vector diagram to verify the result.

Strategy We are interested in the forces acting *on the mule*. One force given in the problem statement is a force exerted *on the ground* (by the mule). From Newton's third law, the ground pushes back on the mule with a force of equal magnitude and opposite direction; therefore, we needed to know this force to determine one of the horizontal forces acting on the mule. Other forces acting on the mule that have not been explicitly stated are the mule's weight and the normal force on the mule due to the ground. These two are vertical forces: the weight is downward and the normal force is upward; these forces do not concern us since they have no effect on the horizontal motion of the mule. Therefore, we are left with three horizontal forces to sum. When finding vector components, we must first choose a direction for the positive *x*-axis. Here we choose east as the +*x*-direction.

Given: Force on <u>m</u>ule by <u>f</u>armer: $\vec{\mathbf{F}}_{mf}$ = 320 N to the east; force on <u>m</u>ule by <u>b</u>oy: $\vec{\mathbf{F}}_{mb}$ = 110 N to the east; horizontal force on <u>g</u>round by <u>m</u>ule: $\vec{\mathbf{F}}_{gm}$ = 430 N to the east.

To find: Sum of horizontal forces on mule

Solution From Newton's third law, the ground pushes on the mule with a force equal and opposite to that of the mule on the ground; therefore, the ground pushes on the mule with a force $\mathbf{F}_{mg}$ of 430 N to the west. The other two horizontal forces acting on the mule are directed to the east. Now we can drop the "m" subscripts since we deal only with the three forces acting on the mule ($\vec{\mathbf{F}}_{mf}$ becomes $\vec{\mathbf{F}}_f$, etc.). The *x*-components of forces directed to the east (F_{fx} and F_{bx}) are positive, since we chose east as the +*x*-direction. The *x*-component of the one force directed to the west (F_{gx}) is negative. Notice that the *F*'s are not in bold-faced type because we are now talking about *components* of vectors; components are scalars. Then F_{fx} = +320 N, F_{bx} = +110 N, and F_{gx} = –430 N.

$$\Sigma F_x = F_{fx} + F_{bx} + F_{gx}$$
$$\Sigma F_x = 320 \text{ N} + 110 \text{ N} + (-430 \text{ N}) = 0$$

The sum of the horizontal forces is zero. Drawing the vectors to scale (Fig. 3.3b) verifies that the vector sum is zero: $\vec{\mathbf{F}}_f + \vec{\mathbf{F}}_b + \vec{\mathbf{F}}_g = \vec{\mathbf{0}}$.

Discussion Each force component has a sign that indicates the direction along the *x*-axis in which the force acts. Adding the three *x*-components with their algebraic signs gives the *x*-component of the sum. If the sum of the *x*-components had been positive, the sum of the vectors would be to the right (east); if it had been negative, the sum of the forces would be to the left (west).

Practice Problem 3.2   At last, the mule cooperates

The mule finally starts moving and hauls the farmer's wagon along a straight road for 4.3 km directly east to the neighboring farm, where a few bushels of corn are loaded onto the wagon. Then the farmer drives the mule back along the same straight road, heading west for 7.2 km to the market. Use vector components to find the displacement of the mule from the starting point.

Figure 3.3
(a) A farmer and his son encouraging their mule to move and (b) the vector sum of the horizontal forces on the mule

East

(a)

$\vec{\mathbf{F}}_{mb}$ $\vec{\mathbf{F}}_{mf}$

$\vec{\mathbf{F}}_{mg}$

$\vec{\mathbf{F}}_{mb} + \vec{\mathbf{F}}_{mf} + \vec{\mathbf{F}}_{mg} = \vec{\mathbf{0}}$

(b)

3.2   VELOCITY

A displacement vector indicates by how much and in what direction the position has changed, but implies nothing about how long it took to move from one point to the other. If we want to describe how fast something moves, we would give its speed. But we find that a more useful quantity than speed is **velocity**. Speed is a measure of "how fast": distance traveled divided by elapsed time (distance traveled *per unit time*). Velocity includes the direction of motion as well as the speed: it is defined as *displacement*

Velocity: a vector that measures how fast and in what direction something moves

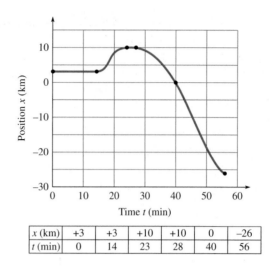

Figure 3.4 Graph of position x versus time t for the train.

x (km)	+3	+3	+10	+10	0	−26
t (min)	0	14	23	28	40	56

divided by elapsed time (displacement per unit time). Speed, the magnitude of the velocity vector, is a scalar.

Average Velocity

Making The Connection:
motion of a train

Figure 3.4 is a graph showing the position of the train considered in Section 3.1 as a function of time. The positions of the train at various times are marked with a dot. The position of the train would have to be measured at more frequent time intervals in order to accurately trace out the shape of the graph.

The graph of position versus time shows a curving line, but that does not mean the train travels along a curved path. The motion of the train is constrained along a straight line since the track runs in an east-west direction. The graph shows the x-component of the train's position as a function of time.

A horizontal portion of the graph (as from $t = 0$ to $t = 14$ min) indicates that the position is not changing during that time interval. Sloping portions of the graph indicate that the train is moving. The steeper the graph, the faster its position changes. The slope is a measure of the rate of change of position. A positive slope ($t = 14$ min to $t = 23$ min) indicates motion in the $+x$-direction while a negative slope ($t = 28$ min to $t = 56$ min) indicates motion in the $-x$-direction.

When a displacement $\Delta \vec{\mathbf{r}}$ occurs during a time interval Δt, the **average velocity** $\vec{\mathbf{v}}_{av}$ during that time interval is:

Average velocity is written $\vec{\mathbf{v}}_{av}$.

Average velocity
$$\vec{\mathbf{v}}_{av} = \frac{\Delta \vec{\mathbf{r}}}{\Delta t} \tag{3-2}$$

Average velocity is a vector because it is the product of a vector, the displacement ($\Delta \vec{\mathbf{r}}$), and a scalar, the inverse of the time ($1/\Delta t$). Since Δt is always positive, the direction of the average velocity vector must be the same as the direction of the displacement vector. In terms of x-components,

$$v_{av,x} = \frac{\Delta x}{\Delta t} \tag{3-3}$$

The symbol Δ does not stand alone and cannot be canceled in equations because it modifies the quantity that follows it; $\Delta x/\Delta t$ is *not* the same as x/t.

The *average* velocity does not convey detailed information about the motion during Δt, just the net effect. During the time interval in question, the object's motion could change direction and speed in many ways and still have the same *average* velocity. For any given displacement, the average velocity does indicate a constant speed and a straight line direction that would result in the same displacement during the same amount of time.

Suppose a butterfly flutters from a point *O* on one flower to a point *D* on another flower along the dashed line path shown in Fig. 3.5 (path *OABCD*). The displacement of the butterfly is the vector arrow $\Delta\vec{r}$ from *O* to *D*. The average velocity of the butterfly is the displacement per unit time, $\Delta\vec{r}/\Delta t$; the direction of the average velocity is the same as the direction of the displacement vector $\Delta\vec{r}$. The average velocity does not distinguish the actual path of the butterfly from any other path that begins at *O* and ends at *D* over the same time interval Δt.

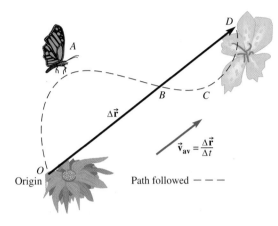

Figure 3.5 Path followed by a butterfly, fluttering from one flower to another.

Example 3.3

Average Velocity of Train

Find the average velocity of the train shown in Fig. 3.1 during the time interval between 3:14 P.M., when the train is 3 km east of the origin, and 3:56 P.M., when it is 26 km west of the origin.

Strategy We already know the displacement $\Delta\vec{r}$ from Fig. 3.1. The direction of the average velocity is the direction of the displacement.

Known: Displacement $\Delta\vec{r}$ = 29 km west; start time = 3:14 P.M.;
     finish time = 3:56 P.M.

To find: $\vec{v}_{av}$

Solution From Section 3.1, the displacement is 29 km to the west:

$$\Delta\vec{r} = 29 \text{ km west}$$

The time interval is

$$\Delta t = 56 \text{ min} - 14 \text{ min} = 42 \text{ min}$$

We convert the time interval to hours, so that we can use units of km/h.

$$\Delta t = 42 \text{ min} \times \frac{1 \text{ h}}{60 \text{ min}} = 0.70 \text{ h}$$

The average velocity is

$$\vec{v}_{av} = \frac{\text{displacement}}{\text{time interval}} = \frac{\Delta\vec{r}}{\Delta t}$$

$$\vec{v}_{av} = \frac{29 \text{ km west}}{0.70 \text{ h}} = 41 \text{ km/h to the west}$$

We can also use components to express the answer to this question. Assuming a positive *x*-axis pointing east, the *x*-component of the displacement vector is $\Delta x = -29$ km, where the negative sign indicates that the vector points to the west. Then

$$v_{av,x} = \frac{\Delta x}{\Delta t} = \frac{-29 \text{ km}}{0.70 \text{ h}} = -41 \text{ km/h}$$

The negative sign shows that the average velocity is directed along the negative *x*-axis, or to the west.

Discussion If the train had started at the same instant of time, 3:14 P.M., and had traveled directly west at a constant 41 km/h, it would have ended up in exactly the same place—26 km west of the trestle bridge—at 3:56 P.M.

Had we started measuring time from when we first spotted the motionless train at 3:00 P.M., instead of 3:14 P.M., we would have found the average velocity over a different time interval, changing the average velocity.

⚠ The average velocity depends on the time interval considered.

Practice Problem 3.3 Average velocity of the baton from the relay race

Consider the baton from the relay race of Conceptual Example 3.1. For the winning team, the magnitudes of the average velocities of the first, second, and third runners are 7.30 m/s, 7.20 m/s, and 7.80 m/s, respectively. If the length of the track is 3.00×10^2 m, what is the average velocity of the baton for the entire race? [*Hint:* Find the time spent by each runner in completing her portion of the race.]

In Example 3.3, the magnitude of the train's average velocity is *not* equal to the total distance traveled divided by the time interval for the complete trip. That quantity is called the average speed:

$$\text{average speed} = \frac{\text{distance traveled}}{\text{total time}} = \frac{43 \text{ km}}{0.70 \text{ h}} = 61 \text{ km/h}$$

The distinction arises because the average velocity is the average of a vector quantity whereas the average speed is the average of a scalar.

Instantaneous Velocity

The speedometer of a car does not indicate the average speed, but shows how fast the car is going at any instant in time—the *instantaneous speed*. When a speedometer reads 55 mi/h, it does *not* necessarily mean that the car will travel 55 miles in the next hour; the car could change its speed or stop during that hour. The speedometer reading indicates how far the car will travel during a very short time interval—short enough that the speed does not change appreciably. For instance, at 55 mi/h (= 25 m/s), we can calculate that in 0.010 s the car moves 25 m/s × 0.010 s = 0.25 m—assuming the speed does not change significantly during that 0.010 s interval.

The **instantaneous velocity** is a *vector* quantity whose magnitude is the instantaneous speed and whose direction is the direction of motion. The direction of an object's instantaneous velocity at any point is tangent to the path of the object at that point. Repeating the word *instantaneous* can get cumbersome. When we refer simply to the *velocity,* we mean the *instantaneous* velocity.

The mathematical definition of instantaneous velocity starts with the average velocity during a short time interval. Then we consider shorter and shorter time intervals Δt; the corresponding displacements $\Delta \vec{\mathbf{r}}$ that take place during Δt get smaller and smaller. We let the time interval approach—but never reach—zero (Fig. 3.6). (This mathematical process is called finding the *limit* and is written $\lim_{\Delta t \to 0}$.) As Δt approaches zero, the average velocity during the increasingly short time interval approaches the instantaneous velocity.

> The *rate of change* of a quantity Q (whether scalar or vector) is
> $$\text{rate of change of } Q = \lim_{\Delta t \to 0} \frac{\Delta Q}{\Delta t}$$
> Velocity is the rate of change of position.

Definition of instantaneous velocity

$$\vec{\mathbf{v}} = \lim_{\Delta t \to 0} \frac{\Delta \vec{\mathbf{r}}}{\Delta t} \qquad (3\text{-}4)$$

Both numerator and denominator in Eq. (3-4) approach zero, but $\Delta \vec{\mathbf{r}} / \Delta t$ approaches a limiting value that is neither undefined nor infinite; it can be but is not necessarily zero. Suppose Δt is small enough that the velocity is approximately constant during the interval. If we now cut the time interval Δt in half, the displacement during the time interval is also cut in half (or nearly so), so $\Delta \vec{\mathbf{r}} / \Delta t$ (the average velocity during the interval) changes very little.

For motion along the *x*-axis, we can rewrite the definition of velocity in terms of *x*-components:

$$v_x = \lim_{\Delta t \to 0} \frac{\Delta x}{\Delta t} \qquad (3\text{-}5)$$

Figure 3.6 (a) A small displacement $\Delta \vec{\mathbf{r}}$ of a butterfly from point *a* to point *b*. If the time to travel from point *a* to point *b* is Δt, the average velocity during that time interval is $\vec{\mathbf{v}}_{\text{av}} = \dfrac{\Delta \vec{\mathbf{r}}}{\Delta t}$. (b) As points *a* and *b* get closer and closer to point *A*, the displacement $\Delta \vec{\mathbf{r}}$ becomes tangent to the path of travel at point *A*. The average velocity during the displacement from *a* to *b* approaches the instantaneous velocity at point *A*.

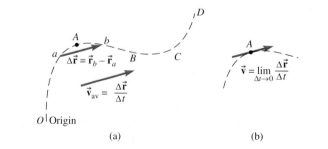

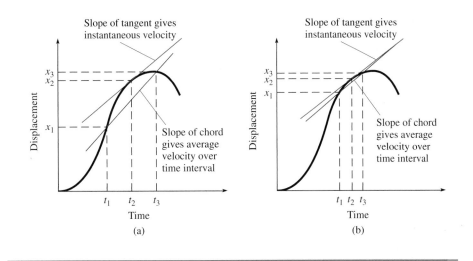

Figure 3.7 Instantaneous velocity at time t_2 is the slope of the tangent to the curve of displacement versus time at that time: (a) average velocity measured over a longer time interval and (b) average velocity measured over a shorter time interval

Table 3.1									
Velocities Determined from the Graph of Fig. 3.4									
t (min)	14	16	19	21	23	28	32	38	47	56
v_x (km/min)	0	0.5	1.8	0.7	0	0	−0.5	−1.3	−2.0	−1.3
v_x (m/s)	0	8.3	30	12	0	0	−8.3	−22	−33	−22

Graphical Relationship between Position and Velocity

How can we determine the instantaneous velocities from a graph of position versus time? Again, we look at the average velocity during a short time interval. For motion along the x-axis, the displacement (in component form) is Δx. Then the average velocity can be represented on the graph of x versus t as the slope of a line connecting two points (called a "chord"). In Fig. 3.7a, the displacement $\Delta x = x_3 - x_1$ is the "rise" of the graph and the time interval $\Delta t = t_3 - t_1$ is the "run" of the graph. The slope of the graph is the "rise over run," which is equal to the average velocity $\Delta x/\Delta t = v_{av,x}$ for that time interval.

To find the *instantaneous* velocity at some time $t = t_2$, we draw lines showing the average velocity for smaller and smaller time intervals. As the time interval is reduced (Fig. 3.7b), the average velocity changes. As $\Delta t \to 0$, the chord approaches a tangent to the graph. Therefore, v_x is the *slope of the line tangent to the graph of x versus t* at the chosen time.

Average and instantaneous velocities for the train can be found from the graph of x versus t (Fig. 3.4). Instantaneous velocities are found from the slopes of the tangents to the curve at various points. Table 3.1 shows some instantaneous velocities of the train in km/min along with equivalent velocities in m/s.

What about the other way around? Given a graph of the *velocity* as a function of time (v_x versus t), how can we determine displacements? For motion along a straight line it is most convenient to work in terms of the x-components of velocity and displacement (v_x and Δx). If v_x is constant during a time interval, then the average velocity is equal to the instantaneous velocity:

v_x is the slope of the graph of x versus t

$$v_x = v_{av,x} = \frac{\Delta x}{\Delta t}$$

and therefore $$\Delta x = v_x \Delta t \text{ (for constant } v_x)$$

The graph of Fig. 3.8 shows v_x versus t for an object moving along the x-axis with constant speed v_1 from time t_1 to t_2. The displacement Δx during the time interval $\Delta t = t_2 - t_1$ is $v_1\Delta t$. The shaded rectangle has "height" v_1 and "width" Δt. Since the area of a rectangle is the product of the height and the width, the displacement Δx is represented by the area of the rectangle between the graph of $v_x(t)$ and the time axis for the time interval considered.

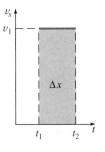

Figure 3.8 Displacement Δx between t_1 and t_2 is represented by the shaded area under the $v_x(t)$ graph.

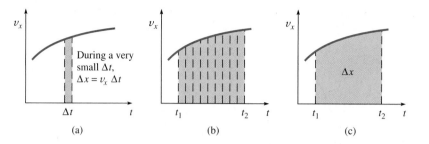

Figure 3.9 Displacement Δx is the area under the v_x versus time graph for the time interval considered.

Δx is the area under the graph of v_x versus t.
The area is negative when the graph is beneath the time axis ($v_x < 0$).

When we speak of the area under a graph, we are not talking about the literal number of square centimeters of paper or computer screen. The figurative area under a graph usually does not have dimensions of an ordinary area ($[L]^2$). In a graph of v_x versus t, v_x has dimensions $[L]/[T]$ and time has dimensions $[T]$; areas on such a graph have dimensions $\frac{[L]}{[T]} \times [T] = [L]$, which is correct for a displacement. The units of Δx are determined by the units used on the axes of the graph. If v_x is in meters per second and t is in seconds, then the displacement is in meters.

What if the velocity is not constant? The displacement Δx during a *very small* time interval Δt can be found in the same way as for constant velocity since, during a short enough time interval, the velocity does not change appreciably. Then v_x and Δt are the height and width of a narrow rectangle (Fig. 3.9a) and the displacement during that short time interval is the area of the rectangle. To find the total displacement during any time interval, the areas of all the narrow rectangles are added together.

In Fig. 3.9b the time from t_1 to t_2 is subdivided into many short time intervals so that many narrow rectangles of varying heights are formed. The width Δt is allowed to approach zero and the areas of the rectangles are added together. The total displacement Δx between t_1 and t_2 is the sum of the areas of the rectangles. Thus, *the displacement Δx during any time interval equals the area under the graph of $v_x(t)$* (Fig. 3.9c). If v_x is negative, x is decreasing and the displacement is negative, so we must count the area as negative when it is below the time axis.

The magnitude of the train's displacement from time $t = 14$ min to time $t = 23$ min is the area under the v_x versus t graph during that time interval. In Fig. 3.10 the area

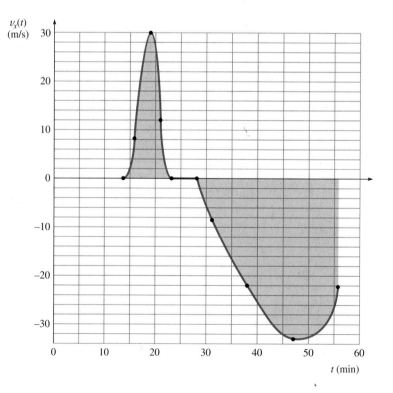

Figure 3.10 Train velocity versus time from $t = 14$ to 56 min using values from Table 3.1.

under the graph for that time interval is shaded. One way to estimate the area is to count the number of grid boxes under the curve. Each box is 2 m/s in height and 5 min (300 s) in width, so each box represents an "area" (displacement) of 2 m/s × 300 s = 600 m. When counting the number of boxes under the curve, we make our best estimate for the fraction of the boxes that are only partly below the curve.

The total number of shaded boxes for this time interval is about 12, so the displacement magnitude is 12×0.60 km ≈ 7 km, as expected—during this time interval the train went from +3 km to +10 km. To get more exact measurements of areas under a curve, we could divide the area into a finer grid.

The shaded area for the time interval $t = 28$ min to $t = 56$ min is below the x-axis, indicating a negative "area" or negative displacement. During this interval the train is headed west (in the $-x$-direction). The number of shaded grid boxes in this interval is approximately 60. The magnitude of the displacement is the product of the number of boxes times "area" of a single box and is negative since v_x is negative; $\Delta x \approx -(60) \times (0.60$ km$) \approx -36$ km. By adding this value to the displacement during the first 14 min, we have the total displacement from $t = 0$ to $t = 56$ min:

$$\Delta x = +7 \text{ km} + (-36 \text{ km}) = -29 \text{ km}$$

which agrees with Fig. 3.1.

3.3   ACCELERATION

One goal of this chapter is to quantify changes in velocity, so we can predict the effect on an object's motion of a nonzero net force. Just as we defined velocity in terms of displacements and time intervals, now we define **acceleration** in terms of changes in velocity and time intervals. The **average acceleration** during a time interval Δt is defined to be

$$\vec{\mathbf{a}}_{\text{av}} = \frac{\vec{\mathbf{v}}_{\text{f}} - \vec{\mathbf{v}}_0}{t_{\text{f}} - t_0} = \frac{\Delta \vec{\mathbf{v}}}{\Delta t} \qquad (3\text{-}6)$$

The velocity vector can change magnitude, direction, or both. Acceleration is a vector quantity because it is the product of a scalar $1/\Delta t$ and a vector $\Delta \vec{\mathbf{v}}$; *the direction of the average acceleration is the direction of the vector* $\Delta \vec{\mathbf{v}}$. The direction of the *change* in velocity $\Delta \vec{\mathbf{v}}$ is not necessarily the same as either the initial or the final velocity direction.

Just as instantaneous velocity is the limit of the average velocity as the time interval approaches zero, the **instantaneous acceleration** is defined as the limit of the average acceleration as the time interval approaches zero.

$$\vec{\mathbf{a}} = \lim_{\Delta t \to 0} \frac{\Delta \vec{\mathbf{v}}}{\Delta t} \qquad (3\text{-}7)$$

Acceleration is the rate of change of velocity.

The instantaneous acceleration is the rate of change of the velocity.

The SI units of acceleration are m/s^2, read as "meters per second squared." Just as with instantaneous velocity, the word *instantaneous* is not always repeated; acceleration without the adjective means *instantaneous* acceleration.

In physics, the word acceleration does not necessarily mean speeding up. A velocity can also change by decreasing speed or by changing direction. A car going around a curve at constant speed has a nonzero acceleration because its velocity is changing; the change is in the *direction* of the velocity rather than in the magnitude. Of course, both the magnitude and direction of the velocity can be changing simultaneously, as when a skateboarder goes up a curved ramp.

Conceptual Example 3.4

Direction of Acceleration While Slowing Down

When Damon approaches a stop sign on his motor scooter, he "decelerates" before coming to a full stop. If he moves in the negative x-direction while he slows down, is the scooter's acceleration component, a_x, positive or negative?

Strategy and Solution The term "decelerate" is not a scientific term. In common usage it means the scooter is slowing: the scooter's velocity is decreasing in magnitude. The acceleration vector $\vec{a}$ points in the direction of the *change* in the velocity vector $\Delta\vec{v}$ during a short time interval. Since the velocity is decreasing in magnitude, the direction of $\Delta\vec{v}$ is opposite to $\vec{v}$ (Fig. 3.11). The acceleration vector is in the same direction as $\Delta\vec{v}$—the positive x-direction. Thus a_x is positive (see Fig. 3.12).

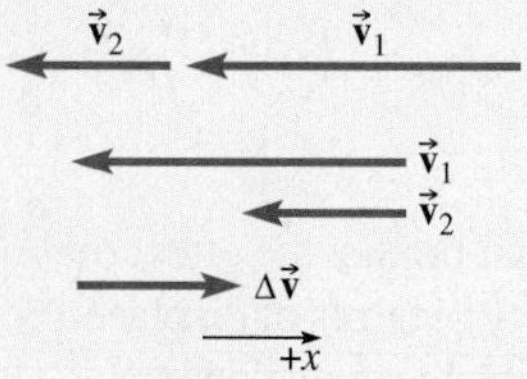

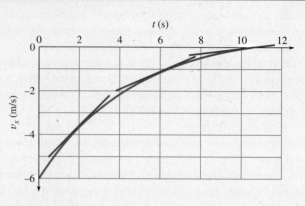

Figure 3.12
In this graph of v_x versus t, as Damon is stopping, v_x is negative, while a_x (the slope) is positive. The value of v_x is increasing, but—since it is less than zero to begin with and is getting closer to zero as time goes on—the speed is *decreasing*.

Figure 3.11
The velocity vectors at two different times as the scooter moves to the left with decreasing speed. The change in velocity $\Delta\vec{v} = \vec{v}_2 - \vec{v}_1$ is to the right.

Conceptual Practice Problem 3.4 Continuing on his way

As Damon pulls away from the stop sign, continuing in the $-x$-direction, his speed gradually increases. What is the sign of a_x?

---

a_x is the slope of a $v_x(t)$ graph.
Δv_x is the area under an $a_x(t)$ graph.

Both velocity and acceleration measure rates of change: velocity is the rate of change of position and acceleration is the rate of change of velocity. Therefore the graphical relationship of acceleration to velocity is the same as the graphical relationship of velocity to position: a_x is the slope of a $v_x(t)$ graph and Δv_x is the area under an $a_x(t)$ graph. On a graph of *any* quantity Q as a function of time, the slope of the graph represents the instantaneous rate of change of Q. On a graph of the *rate of change of Q* as a function of time, the area under the graph represents ΔQ.

Example 3.5

Acceleration of a Sports Car

A sports car can accelerate from 0 to 30.0 m/s in 4.7 s according to the advertisements. Figure 3.13 shows data for the speed of the car as a function of time as the sports car starts from rest and accelerates to 60.0 m/s while traveling in a straight line headed east.

(a) What is the average acceleration of the sports car from 0 to 30.0 m/s?

(b) What is the maximum acceleration of the car?

(c) How far has the car traveled when it reaches 60.0 m/s?

(d) What is the car's average velocity during the entire 19.2-s interval?

Strategy The velocity of the car is always in the same direction. If we choose that direction as the $+x$-axis, the x-component of the car's velocity is always positive. Therefore, v_x is equal to

the car's speed (which is the *magnitude* of the velocity). We can reinterpret the graph as a graph of v_x versus t.

(a) To find the average acceleration, the change in velocity for the time interval is divided by the time interval. (b) The instantaneous acceleration is the slope of the velocity graph, so it is maximum where the graph is steepest. At that point, the velocity is changing at a high rate. We expect the maximum acceleration to take place early on; the magnitude of acceleration must decrease as the velocity gets higher and higher—there is a maximum velocity for the car, after all. (c) The displacement is the area under the $v_x(t)$ graph. The graph is not a simple shape such as a triangle or rectangle, so an estimate of the area is made. (d) Once we have a value for the displacement, we can apply the definition of average velocity.

continued on next page

Example 3.5 *continued*

Given: Graph of $v(t)$ in Fig. 3.13

To find: (a) $a_{av,x}$ for $v_x = 0$ to 30.0 m/s; (b) a_{max}; (c) Δx from $v_x = 0$ to 60.0 m/s; (d) $v_{av,x}$ from $t = 0$ to 19.2 s

Solution (a) During the first 4.7 s, the car is moving in a straight line to the east, so the average component of acceleration in the east direction is, by definition,

$$a_{av,x} = \frac{\Delta v_x}{\Delta t} = \frac{30.0 \text{ m/s}}{4.7 \text{ s}} = 6.4 \text{ m/s}^2$$

The positive sign indicates the acceleration is directed to the east. On average, then, the car's velocity increases 6.4 m/s each second during the first 4.7 s.

(b) The acceleration, at any instant of time, is the slope of the tangent line to the $v_x(t)$ graph, at that time. To find the maximum acceleration, notice where the graph is steepest. In this case, the largest slope occurs near $t = 0$, just as the car is starting out. In Fig. 3.13, a tangent line to the $v_x(t)$ graph at $t = 0$ passes through $t = 0$. Values for the change in velocity and change in time are read from the graph; the tangent line passes through the two points ($t = 0$, $v_x = 0$ and $t = 6.0$ s, $v_x = 55$ m/s) on the graph so that the change in velocity is 55 m/s for a change in time of 6.0 s. The slope of this line is

$$a_{max} = \frac{\text{rise}}{\text{run}} = \frac{55 \text{ m/s}}{6.0 \text{ s}} = +9.2 \text{ m/s}^2$$

Since the slope is positive, the direction of the acceleration is east.

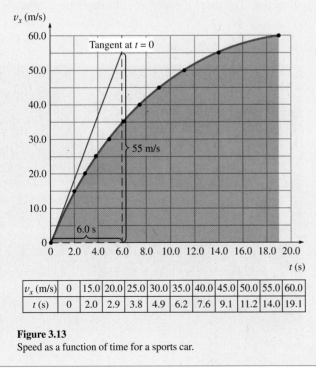

Figure 3.13
Speed as a function of time for a sports car.

v_x (m/s)	0	15.0	20.0	25.0	30.0	35.0	40.0	45.0	50.0	55.0	60.0
t (s)	0	2.0	2.9	3.8	4.9	6.2	7.6	9.1	11.2	14.0	19.1

(c) Displacement is the area under the $v_x(t)$ graph shown shaded in Fig. 3.13. The area can be estimated by counting the number of grid boxes under the curve. Each box is 5.0 m/s in height and 2.0 s in width, so each represents an "area" (displacement) of 10 m. When counting the number of boxes under the curve, a best estimate is made for the fraction of the boxes that are only partly below the curve. Approximately 75 boxes lie below the curve, so the displacement magnitude is $(75 \times 10 \text{ m}) = 750$ m.

(d) The average velocity for the 19.2-s interval is

$$v_{av,x} = \frac{\Delta x}{\Delta t}$$

$$= \frac{750 \text{ m}}{19.2 \text{ s}} = 39 \text{ m/s}$$

This result is reasonable; if the acceleration were constant, the average velocity would be $\frac{1}{2}(0 + 60.0 \text{ m/s}) = 30.0$ m/s. The actual average velocity is somewhat higher because the acceleration is greater at the start so less of the time interval is spent going (relatively) slowly and more is spent going fast. The speed is less than 30.0 m/s for only 4.7 s, but is greater than 30.0 m/s for 14.5 s.

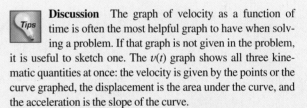 **Discussion** The graph of velocity as a function of time is often the most helpful graph to have when solving a problem. If that graph is not given in the problem, it is useful to sketch one. The $v(t)$ graph shows all three kinematic quantities at once: the velocity is given by the points or the curve graphed, the displacement is the area under the curve, and the acceleration is the slope of the curve.

Practice Problem 3.5 Braking a car

An automobile is traveling along a straight road heading to the southeast at 24 m/s when the driver sees a deer begin to cross the road ahead of her. She steps on the brake and brings the car to a complete stop in an elapsed time of 8.0 s. A data recording device, triggered by the sudden braking action, records the following velocities and times as the car slows (Table 3.2). Let the positive x-axis be directed to the southeast. Plot a graph of v_x versus t and find (a) the average acceleration as the car comes to a stop and (b) the instantaneous acceleration at a time of 2.0 s after braking begins.

Table 3.2

Velocities and Times as Car Slows

v_x (m/s)	24	17.3	12.0	8.7	6.0	3.5	2.0	0.75	0
t (s)	0	1.0	2.0	3.0	4.0	5.0	6.0	7.0	8.0

When the acceleration and the velocity of an object are both pointing in the same direction, the object is speeding up. In terms of components, a_x and v_x must either both be positive or both be negative when speed is increasing. When they are both positive, the object is moving in the $+x$-direction and is speeding up. If they are both negative, the object is moving in the $-x$-direction and is speeding up.

When the velocity and acceleration point in opposite directions, so that their components have opposite signs, the object is slowing down. When v_x is positive and a_x is negative, the body is moving in the positive x-direction, but it is slowing down. When v_x is negative and a_x is positive, the body is moving in the negative x-direction and is slowing down (its speed is decreasing).

The acceleration does not contain any information about the initial velocity. Acceleration is the *rate of change* of velocity. Can an object have a velocity of zero and a nonzero acceleration at the same time? Yes, but the velocity does not *remain* zero. A ball thrown straight up into the air has a velocity of zero at its highest point. Its acceleration is *not* zero at that point; if it were, the ball would not fall back down. On the way up, the ball's velocity is upward and decreasing in magnitude. At the highest point, the velocity is zero but then the ball starts moving downward. On the way down, the velocity is downward and increasing in magnitude. For the entire flight of the ball, its acceleration is downward.

We can define the upward direction as positive or negative as we like; the choice is arbitrary, but we must remain consistent throughout a particular problem. For vertical motion, upward is usually defined as the positive y-direction. For a ball thrown into the air, v_y is positive on the way up and negative on the way down; a_y is negative and constant throughout the motion—on the way up, at the highest point, and on the way down. Since v_y and a_y have opposite signs on the way up, the ball is slowing down (the velocity is decreasing in magnitude). On the downward trip, v_y and a_y have the same sign so the ball's speed is increasing.

3.4 NEWTON'S SECOND LAW: FORCE AND ACCELERATION

According to Newton's second law, the acceleration of an object is proportional to the net force on it and is in the same direction. The larger the net force, the larger the acceleration.

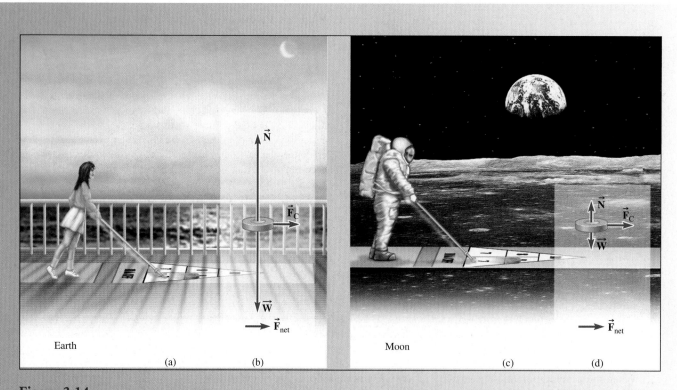

Figure 3.14 An astronaut playing shuffleboard (a) on Earth and (c) on the Moon. Free-body diagrams for a puck being given the same acceleration on a *frictionless* court on (b) Earth and (d) on the Moon. The contact force on the puck due to the *pushing stick* ($\vec{\mathbf{F}}_C$) must be the same since the mass of the puck is the same.

If the net force is zero, the acceleration is zero and the object moves with constant velocity—possibly zero velocity, but not necessarily. Newton's second law also says that the acceleration is inversely proportional to the object's mass. The same net force acting on two different objects causes a smaller acceleration on the object with greater mass. Mass is a measure of an object's inertia—the amount of resistance to *changes in velocity*.

Mass and weight measure different physical properties. The mass of a body is a measure of its inertia, while weight is a measure of the gravitational force acting on it. Imagine taking a shuffleboard puck to the Moon. Since the Moon's gravitational field is weaker than the Earth's, the puck's weight $\vec{\mathbf{W}}$ would be smaller. A smaller normal force $\vec{\mathbf{N}}$ would be required to hold it up. On the other hand, the puck's mass, an intrinsic property, is the same. Neglecting the effects of friction, an astronaut playing shuffleboard on the Moon would have to exert the same horizontal force on the puck as on Earth to give it the same acceleration (Fig. 3.14).

Newton's law relating net force and acceleration is

Newton's Second Law

$$\vec{\mathbf{F}}_{net} = m\vec{\mathbf{a}} \qquad\qquad (3\text{-}8)$$

---

or

$$\Sigma\vec{\mathbf{F}} = m\vec{\mathbf{a}}$$

where Σ, the Greek capital letter sigma, stands for *the sum of*. $\Sigma\vec{\mathbf{F}}$ means the sum of all the forces acting on a system. The order of the symbols in Eq. (3-8) does not reflect a cause-and-effect relationship; the net force causes the acceleration, not the other way around. The SI unit of force, the newton, is defined in terms of SI base units so that a 1-N net force acting on a 1-kg mass produces an acceleration of 1 m/s^2; therefore,

$$1\text{ N} = 1\text{ kg·m/s}^2$$

When calculating the net force on a system, only *external* forces need be considered. According to Newton's third law, internal forces always add to zero.

Tips

Example 3.6

Coupling Force on First and Last Freight Cars

A train engine pulls out of a station along a straight track with five identical freight cars behind it, each of which weigh 90.0 kN. The train reaches a speed of 15.0 m/s within 5.00 min of starting out. Assuming the acceleration is constant, with what magnitude of force must

the coupling between cars pull forward on the first and last of the freight cars? Ignore friction and air resistance. Assume $g = 9.80$ N/kg.

Strategy   A sketch of the situation is shown in Fig. 3.15a. We can calculate the acceleration of the train from the initial and final velocities and the elapsed time. Then we can relate the acceleration to the net force using Newton's second law.

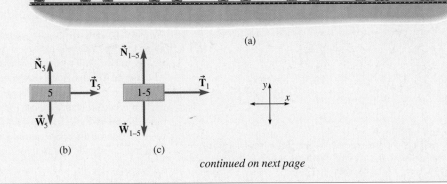

Figure 3.15
(a) An engine pulling five identical freight cars. The entire train has a constant acceleration $\vec{\mathbf{a}}$ to the right.
(b) FBD for car 5. (c) FBD for cars 1–5.

continued on next page

Example 3.6 *continued*

To find the force of the first coupling, we can consider all five cars to be one system so that we do not have to worry about the force exerted on the first car by the second car. Once we identify a system, we draw a free-body diagram before applying Newton's second law.

Given: W = weight of each freight car = 90.0 kN = 9.00×10^4 N;
$v_x = 15.0$ m/s at $t = 5.00$ min = 300 s; $v_{0x} = 0$ since the train starts from rest; a_x = constant

To find: Tensions T_1 and T_5

Solution   The acceleration of the train is

$$a_x = \frac{\Delta v_x}{\Delta t} = \frac{15.0 \text{ m/s}}{300 \text{ s}} = 0.0500 \text{ m/s}^2$$

First consider the last freight car (car 5). If we ignore friction and air resistance, the only forces acting are the force $\vec{T}_5$ due to the tension in the coupling, the normal force $\vec{N}_5$, and the car's weight $\vec{W}_5$; an FBD is shown in Fig. 3.15b. The normal force and the weight are vertical and act in opposite directions. They must be equal in magnitude; the vertical component of the net force is zero since the vertical component of the acceleration is zero. Then the net force is equal to the tension in the coupling. The mass of the car is $m = W/g$, where g = 9.80 N/kg = 9.80 m/s^2. Then, from Newton's second law,

$$T_5 = \Sigma F_x = ma_x = \frac{W}{g} a_x$$

$$T_5 = 9.00 \times 10^4 \text{ N} \times \frac{0.0500 \text{ m/s}^2}{9.80 \text{ m/s}^2} = 459 \text{ N}$$

For the tension in the first coupling, consider the five cars as *one system*. Fig. 3.15c shows an FBD in which cars 1–5 are treated as a single object. Again, the vertical forces on the system add to zero. The only external horizontal force is the force $\vec{T}_1$ due to the tension in the first coupling. The mass of the system is five times the mass of one car. Therefore,

$$T_1 = \Sigma F_x = ma_x = (5 \times 9.00 \times 10^4 \text{ N}) \times \frac{0.0500 \text{ m/s}^2}{9.80 \text{ m/s}^2} = 2.30 \text{ kN}$$

Discussion   The solution to this problem is much simpler when Newton's second law is applied to a system comprised of all five cars, rather than to each car individually. Although the problem can be solved by looking at individual cars, to find the tension in the first coupler you would have to draw five free-body diagrams (one for each car) and apply Newton's second law five times. That's because each car, except the fifth, is acted on by the unequal tensions in the couplers on either side. You'd have to first find the tension in the fifth coupler, then in the fourth, then the third, and so on.

Practice Problem 3.6   Coupling force between first and second freight cars

With what force does the coupling between the first and second cars pull forward on the second car? [*Hint:* Try two methods. One of them is to draw an FBD for the first car and apply Newton's *third* law as well as the second.]

Example 3.7

Two Blocks Hanging on a Pulley

In Fig. 3.16a, two blocks are connected by a massless, flexible cord that does not stretch; the cord passes over a massless, frictionless pulley. If the masses are m_1 = 26.0 kg and m_2 = 42.0 kg,

what are the accelerations of each block and the tension in the cord? The gravitational field strength is 9.80 N/kg.

Strategy   Since m_2 is greater than m_1, the downward pull of gravity is stronger on the right side than on the left. We expect m_2 to accelerate downward and m_1 to accelerate upward.

Since the cord does not stretch, the accelerations of the two blocks are equal in magnitude. If the accelerations had different magnitudes, then soon the two blocks would be moving with different speeds. That could only happen if the cord either stretches or contracts. The fixed length of the cord constrains the blocks to move with equal speeds (in opposite directions) at all times, so the magnitudes of their accelerations must be equal.

The tension in the cord must be the same everywhere along the cord since the masses of the cord and pulley are negligible and the pulley turns without friction.

We treat each block as a separate system, draw free-body diagrams for each, and then apply Newton's second law to each. It is convenient to choose the positive y-direction differently for the two blocks. For each, we choose the $+y$-axis in the direction of the acceleration. Doing so means that a_y has the same magnitude *and sign* for the two.

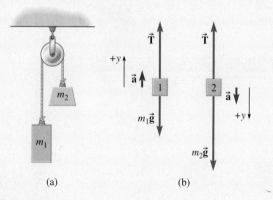

(a)   (b)

Figure 3.16
(a) Two hanging blocks connected on either side of a frictionless pulley by a massless, flexible cord that does not stretch. (b) Free-body diagrams for the hanging blocks.

continued on next page

Example 3.7 *continued*

Given: $m_1 = 26.0$ kg and $m_2 = 42.0$ kg
To find: a_y; T

Solution Figure 3.16b shows free-body diagrams for the two blocks. Two forces act on each: gravity and the pull of the cord. The acceleration vectors are drawn next to the free-body diagrams. Thus we know the direction of the net force: it is always the same as the direction of the acceleration. Then we know that the tension must be greater than m_1g to give block 1 an upward acceleration and less than m_2g to give block 2 a downward acceleration. The +y-axes are drawn for each block to be in the direction of the acceleration.

From the free-body diagram of block 1, the pull of the cord is in the +y-direction and the gravitational force is in the −y-direction. Then

$$\Sigma F_y = T - m_1g = m_1 a_y$$

For block 2, the pull of the cord is in the −y-direction and the gravitational force is in the +y-direction. Therefore,

$$\Sigma F_y = m_2g - T = m_2 a_y$$

Both T and a_y are identical in these two equations. We then have a system of two equations with two unknowns. Adding the equations, we obtain

$$m_2g - m_1g = m_2 a_y + m_1 a_y$$

Solving for a_y we find

$$a_y = \frac{(m_2 - m_1)g}{m_2 + m_1}$$

Substituting numerical values,

$$a_y = \frac{(42.0 \text{ kg} - 26.0 \text{ kg}) \times 9.80 \text{ N/kg}}{42.0 \text{ kg} + 26.0 \text{ kg}} = \frac{16.0 \text{ kg}}{68.0 \text{ kg}} \times 9.80 \text{ N/kg}$$

$$a_y = 2.31 \text{ m/s}^2$$

since

$$1 \frac{\text{N}}{\text{kg}} = 1 \frac{\text{kg·m/s}^2}{\text{kg}} = 1 \text{ m/s}^2$$

The blocks have the same magnitude acceleration. For block 1 the acceleration points upward and for block 2 it points downward.

To find T we can substitute the expression for a_y into either of the two original equations. Using the first equation,

$$T - m_1g = m_1 \frac{(m_2 - m_1)g}{m_2 + m_1}$$

Solving for T yields

$$T = \frac{2m_1 m_2}{m_1 + m_2}g$$

Substituting,

$$T = \frac{2 \times 26.0 \text{ kg} \times 42.0 \text{ kg}}{68.0 \text{ kg}} \times 9.80 \text{ N/kg} = 315 \text{ N}$$

Discussion A few quick checks:

- a_y is positive, which means that the accelerations are in the directions we expect.
- The tension (315 N) is between m_1g (255 N) and m_2g (412 N) as expected.
- The units and dimensions are correct for all equations.

It is also instructive to examine what happens to the expressions for a_y and T for special cases of hanging blocks with: equal masses, masses just slightly unequal, or one mass much greater than the other. We often have intuition about what should happen in such special cases. See Practice Problem 3.7 for some examples.

⚠ Note that we did *not* find out which way the blocks move. We found the directions of their *accelerations*. If the blocks start out at rest, then the block of mass m_2 moves downward and the block of mass m_1 moves upward. However, if initially m_2 is moving up and m_1 down, they continue to move in those directions, slowing down since their accelerations are opposite to their velocities. Eventually, they come to rest and then reverse directions.

Conceptual Practice Problem 3.7   Equal and slightly unequal masses

Suppose that the blocks attached to the pulley are of equal mass ($m_1 = m_2$). What do the expressions for a_y and T yield in that case? Can you explain why that must be correct? What if the blocks are only slightly unequal so that their masses $m_2 - m_1 \ll m_2 \approx m_1$? What is the tension in that case? What about the magnitude of the acceleration?

These steps are helpful in most problems that involve Newton's second law.

Problem-Solving Strategies for Newton's Second Law

- Decide what objects will have Newton's second law applied to them.
- Identify all the interactions affecting that object.
- Draw a free-body diagram to show all the forces acting on the object.
- Find the net force by adding the forces as vectors.
- Use Newton's second law to relate the net force to the acceleration.

Forces acting on an object determine the motion of the object. The sum of the forces is the net force, which causes an acceleration. The acceleration is a measure of the *rate of change* of the object's velocity. To describe the motion, one other piece of information is required: the object's initial velocity. From the initial velocity and the changes in velocity caused by the forces acting, the velocity at any later time can be calculated. The velocity is the *rate of change* of the position. This strategy for finding the velocity is a triumph of Newton's method of analyzing and then predicting the motion based on the initial conditions and the forces acting on the object.

Example 3.8

Forward Acceleration of a Grocery Cart

Alfredo is shopping at the supermarket (Fig. 3.17a). Alfredo's mass is 72 kg and the mass of his shopping cart plus groceries is 46 kg. At some particular instant, Alfredo's foot pushes backward on the floor with a force of 147 N. What is the acceleration of the cart at that instant? The forces that oppose the forward motion of the cart—friction in the rotation of the cart wheels, air resistance, and so on—add to 5 N.

Strategy   To simplify this problem we may choose a system composed of Alfredo and the cart; they move together with the same acceleration. Two horizontal external forces act on the system: the forces opposing the motion of the cart, $\vec{f}$, and a force of the floor pushing forward on Alfredo's foot. There is no motion in the vertical direction so the net vertical force is zero; the weight of the system is equal to the normal force with which the floor pushes up on the system. We apply both Newton's third law and Newton's second law to solve this problem.

Given: Alfredo's mass: $m_1 = 72$ kg
cart + groceries mass: $m_2 = 46$ kg
force on floor by Alfredo: $\vec{F}_{fA} = 147$ N to the left
force opposing motion of cart: $\vec{f} = 5$ N to the left

To find: $\vec{a}$

Solution   From Newton's third law, we know that the floor pushes forward on the system with the same force that Alfredo pushes backward on the floor.

$$\vec{F}_{Af} = -\vec{F}_{fA}$$

Since $\vec{F}_{fA}$ is to the left, $\vec{F}_{Af}$ is to the right and of magnitude 147 N. We draw a free-body diagram (Fig. 3.17b), showing the external forces acting on the system. Since the vertical acceleration component is zero, the free-body diagram indicates that the weight of the system is balanced by an equal normal force. There may be a nonzero horizontal acceleration component. The +x-axis direction is chosen to the right, in the forward direction of the cart.

From Newton's second law for the vertical direction,

$$\Sigma F_y = N - (m_1 + m_2)g = ma_y = 0$$

and for the horizontal direction,

$$\Sigma F_x = F_{Af} - f = ma_x$$

Solving for the acceleration,

$$a_x = \frac{F_{Af} - f}{m_1 + m_2}$$

Substituting the given values,

$$a_x = \frac{147\ \text{N} - 5\ \text{N}}{72\ \text{kg} + 46\ \text{kg}} = \frac{142\ \text{N}}{118\ \text{kg}} = 1.20\ \text{m/s}^2$$

The direction of the acceleration is to the right.

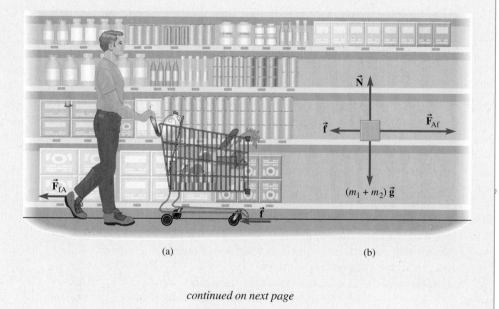

Figure 3.17
(a) Alfredo pushes backward on the floor with force $\vec{F}_{fA}$ and a force $\vec{f}$ opposes the forward motion of the cart; (b) a free-body diagram for the system of Alfredo and cart + groceries.

(a)

(b)

continued on next page

Example 3.8 continued

Discussion By choosing the system to be composed of Alfredo and the cart and groceries, we did not have to worry about the forces internal to the system. Alfredo pushes on the cart handle while the cart handle pushes back on Alfredo's hands—these are two internal forces within the system that cancel each other. The only forces that are of concern for the motion of the system as a whole are the external forces, shown in Fig. 3.17b, that act on the system.

Practice Problem 3.8   Alfredo's pushing force

What is the force with which Alfredo's hands push on the grocery cart? [*Hint:* Choose a new system so that Alfredo is external to the system and consider forces acting external to the chosen system.]

3.5   MOTION WITH CONSTANT ACCELERATION

If the net force acting on an object is constant, then the acceleration of the object is also constant, both in magnitude and direction. *Uniform acceleration* is a synonym for constant acceleration.

Three essential relationships between position, velocity, and acceleration can be used to solve *any* constant acceleration problem. We write these relationships using these conventions:

- For motion along the x-axis, write position, velocity, and acceleration in terms of their x-components (x, v_x, a_x).
- At a time $t = 0$, we designate the initial position and velocity as x_0 and v_{0x}.
- At a later time $t > 0$, the position and velocity are x and v_x.

Just as the origin of a coordinate system can be chosen at any convenient point in space, the time at which $t = 0$ can be freely chosen to be any instant in time—whatever makes the problem easiest to solve.

The three essential relationships are: First, since the acceleration is constant, the change in velocity over a given time interval Δt is just the acceleration—the rate of change of velocity—times the elapsed time:

$$\Delta v_x = v_x - v_{0x} = a_x \, \Delta t \quad \text{for } a_x \text{ constant} \tag{3-9}$$

Second, the displacement is the average velocity times the time interval:

$$\Delta x = x - x_0 = v_{\text{av},x} \, \Delta t \tag{3-10}$$

Equation (3-10) is true whether the acceleration is constant or not; it comes directly from the definition of average velocity.

Third, since the velocity changes linearly with time, the average velocity is given by

$$v_{\text{av},x} = \frac{v_{0x} + v_x}{2} \quad \text{for } a_x \text{ constant} \tag{3-11}$$

Equation (3-11) is *not* true in *general,* but it is true for constant acceleration. To see why, refer to the velocity versus time graph in Fig. 3.18a. The graph is linear because the acceleration is constant. The displacement during any time interval is represented by the area under the $v(t)$ graph. The average velocity is found by forming a rectangle with an area equal to the area under the curve in Fig. 3.18a, because the average velocity should give the same displacement in the same time interval. Fig. 3.18b shows that, to make the area of one colored triangle equal to the area of the other, the average velocity must be exactly halfway between the initial and final velocities.

If the acceleration is *not* constant, there is no reason why the average velocity would have to be exactly halfway between the initial and the final velocity. As an illustration, imagine a trip where you drive along a straight highway at 80 km/h for 50 min and then at 60 km/h for 30 min. Your acceleration is zero for the entire trip *except* during the few seconds while you slowed from 80 km/h to 60 km/h. The magnitude of your average velocity would *not* be 70 km/h. You spent more time going 80 km/h than you did going

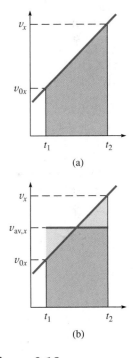

Figure 3.18 Finding average velocity with the aid of a graph

60 km/h, so the magnitude of your average velocity would be greater than 70 km/h (see Problem 16).

To summarize:

When the time interval is from $t = 0$ until a later time t, $\Delta t = t - 0 = t$

> If a_x is constant during the entire time interval from $t = 0$ until a later time t,
>
> $$\Delta v_x = v_x - v_{0x} = a_x \Delta t$$
>
> $$\Delta x = x - x_0 = v_{av,x} \Delta t$$
>
> $$v_{av,x} = \frac{v_{0x} + v_x}{2}$$

Other relationships can be formed between the various quantities (displacement, velocity, acceleration, and time interval), but it is usually better to start with the three basic relationships and work from there. Any constant acceleration problem can be solved using just these three, and they are not difficult to remember—two of them are really definitions.

Two such relationships that are often useful shortcuts can be derived from Eqs. (3-9) through (3-11) (see Problems 37 and 38). They are

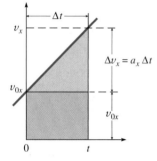

Figure 3.19 Graphical interpretation of Eq. (3-12)

> $$\Delta x = x - x_0 = v_{0x}t + \frac{1}{2}a_x t^2 \qquad (3\text{-}12)$$
>
> $$v_x^2 - v_{0x}^2 = 2a_x \Delta x \qquad (3\text{-}13)$$

Note that Δt is replaced by t because the time interval begins at $t = 0$ and $\Delta t = t$. Equation (3-12) is useful when the final velocity is not known, while Eq. (3-13) is useful when the elapsed time is not known.

We can interpret Eq. (3-12) graphically. Figure 3.19 shows a $v_x(t)$ graph for constant-acceleration motion. The displacement that occurs between $t = 0$ and a later time t is the area under the graph for that time interval. Partition this area into a rectangle plus a triangle. The area of the rectangle is

$$\text{height} \times \text{base} = v_{0x}t$$

The height of the triangle is the change in velocity, which is equal to $a_x t$. The area of the triangle is

$$\tfrac{1}{2} \times \text{height} \times \text{base} = \tfrac{1}{2} \times a_x t \times t = \tfrac{1}{2}a_x t^2$$

Adding these "areas" gives Eq. (3-12).

Equations (3-9) through (3-13) are called *kinematic* equations for motion with constant acceleration, since they deal with the relationships between position, velocity, acceleration, and time, but not with forces.

Example 3.9

Displacement of Motorboat

A motorboat accelerates from rest at a dock with a constant acceleration of magnitude 2.8 m/s². After traveling directly to the east for 140 m the motor is throttled down so that the boat slows down at 1.2 m/s² while still moving east until its speed is 16 m/s. Just as the boat attains the velocity of 16 m/s, it passes a buoy due east of the dock. What is the total displacement of the motorboat from the dock at that time?

Strategy This problem involves two different values of acceleration, so it must be divided into two subproblems. The kinematic equations for constant acceleration cannot be applied to a time interval during which the acceleration changes. But for each of two time intervals, the acceleration of the boat is constant. The two subproblems are connected by the position and velocity of the boat at the instant that the acceleration changes.

For the first subproblem, the boat speeds up with a constant acceleration of 2.8 m/s² to the east. We know the acceler-

continued on next page

Example 3.9 *continued*

ation, the displacement (140 m east), and the initial velocity: the boat starts from rest, so the initial velocity is zero. We need to calculate the final velocity, which then becomes the initial velocity for the second subproblem. The boat is always headed to the east, so we let east be the positive x-direction.

> First subproblem: $v_{0x} = 0$; $a_x = +2.8$ m/s^2
>
> $\Delta x = 140$ m; find v_x

For the second subproblem, we know acceleration, final velocity, and we have just found initial velocity from the first subproblem. Since the boat is slowing down, its acceleration is in the direction opposite its velocity; therefore a_x is negative. From these three quantities we can find the displacement of the boat during the second time interval.

> Second subproblem: v_{0x} comes from first subproblem
>
> $a_x = -1.2$ m/s^2; $v_x = +16$ m/s; find Δx

Adding the displacements for the two time intervals gives the total displacement.

Solution (1) To find v_x without knowing the time, the most convenient kinematic equation is

$$v_x^2 - v_{0x}^2 = 2a_x\Delta x$$

Solving for v_x,

$$v_x = \sqrt{v_{0x}^2 + 2a_x\Delta x}$$

$$v_x = \sqrt{0 + 2 \times 2.8 \text{ m/s}^2 \times 140 \text{ m}} = +28 \text{ m/s}$$

(2) For the second interval the *initial* velocity is the final velocity for the first interval: $v_{0x} = +28$ m/s. Then

$$v_x^2 - v_{0x}^2 = 2a_x\Delta x$$

This time we solve for the displacement.

$$\Delta x = \frac{v_x^2 - v_{0x}^2}{2a_x}$$

$$\Delta x = \frac{(16 \text{ m/s})^2 - (28 \text{ m/s})^2}{2 \times (-1.2 \text{ m/s}^2)} = +220 \text{ m}$$

And the total displacement is

$$\Delta x = \Delta x_1 + \Delta x_2 = 140 \text{ m} + 220 \text{ m} = +360 \text{ m}$$

The boat is 360 m east of the dock.

Discussion This problem is solved by applying the same kinematic equation twice, once to find a velocity and once to find a displacement. The natural division of the problem into two parts occurs because the boat has two different constant accelerations during two different time periods.

In problems that can be subdivided in this way, the final velocity found in the first part becomes the initial velocity for the second part. The same is true for position.

Practice Problem 3.9 Time to reach buoy and average velocities

(a) What is the time required by the boat in the previous example to reach the buoy? (b) Find the average velocity for the entire trip from the dock to the buoy.

Visualizing Motion with Constant Acceleration

In Fig. 3.20 three carts move in the same direction with three different values of constant acceleration. The position of each cart is depicted as it would appear in a stroboscopic photograph with one picture taken every second.

The yellow cart has zero acceleration and therefore constant velocity. During each 1.0-s time interval its displacement is the same: 1.0 m/s × 1.0 s = 1.0 m to the right.

The red cart has a constant acceleration of 0.2 m/s^2 to the right. Although m/s^2 is normally read "meters per second squared," it can be useful to think of it as "m/s per second": the cart's velocity changes by 0.2 m/s during each 1.0-s time interval. In this case,

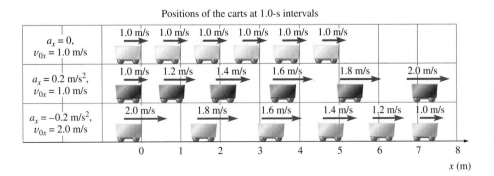

Positions of the carts at 1.0-s intervals

Figure 3.20 Each cart is shown as if stroboscopic photographs were taken with time intervals of 1.0 s between flashes. The arrows above each cart indicate velocity vectors as the strobe flashes occur.

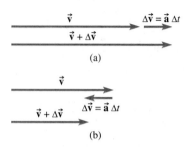

Figure 3.21 (a) If the acceleration is parallel to the velocity, then the change in velocity ($\Delta\vec{v} = \vec{a}\,\Delta t$) is also parallel to the velocity. The result is an increase in the magnitude of the velocity: the object speeds up. (b) If the acceleration is antiparallel (parallel but pointed in the opposite direction) to the velocity, then the change in velocity ($\Delta\vec{v} = \vec{a}\,\Delta t$) is also antiparallel to the velocity. The result is a decrease in the magnitude of the velocity: the object slows down.

acceleration is in the same direction as the velocity, so the velocity increases (Fig. 3.21a). The displacement of the cart during successive 1.0-s time intervals gets larger and larger.

The blue cart experiences a constant acceleration of 0.2 m/s² in the –x-direction—the direction *opposite* to the velocity. The magnitude of the velocity then decreases (Fig. 3.21b); during each one-second interval the speed decreases by 0.2 m/s. Now the displacements during one-second intervals get smaller and smaller.

Figure 3.22 shows graphs of $x(t)$, $v_x(t)$, and $a_x(t)$ for each of the carts. The acceleration graphs are horizontal since each of the carts has a constant acceleration. All three v_x graphs are straight lines. Since a_x is the rate of change of v_x, the slope of the v_x graph at any value of t is a_x at that value of t. With constant acceleration, the slope is the same everywhere and the graph is linear. Note that a positive a_x does mean that v_x is increasing, but not necessarily that the *speed* is increasing; if v_x is negative then a positive a_x indicates a *decreasing* speed. (See Conceptual Example 3.4 and Fig. 3.12.) Speed is increasing when the acceleration and velocity are in the same direction (a_x and v_x both positive *or* both negative). Speed is decreasing when acceleration and velocity are in opposite directions—when a_x and v_x have opposite signs.

The position graph is linear for the yellow cart because it has constant velocity. For the red cart the slope of the $x(t)$ graph increases, showing that v_x is increasing; for the blue cart the slope of the $x(t)$ graph decreases, showing that v_x is decreasing.

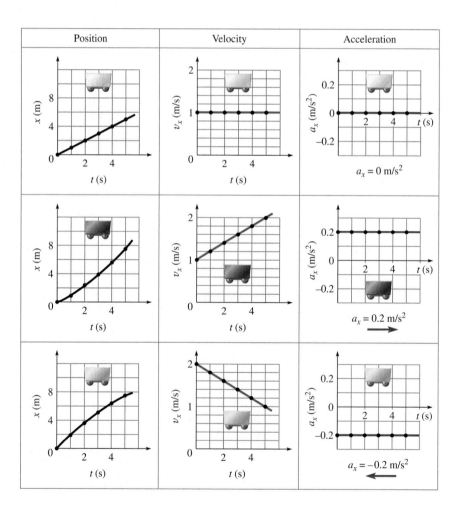

Figure 3.22 Plots of position, velocity, and acceleration along the x-axis for the carts of Fig. 3.20.

Example 3.10

Drag-Racing Spaceships

Two spaceships are moving from the same starting point in the $+x$-direction with constant accelerations. In component form, the silver spaceship starts with an initial velocity of $+2.00$ km/s and has an acceleration of $+0.400$ km/s^2. The black spaceship starts with a velocity of $+6.00$ km/s and has an acceleration of -0.400 km/s^2. Find the time at which the silver spaceship just overtakes the black spaceship.

Strategy We can find the positions of the spaceships at later times from the initial velocities and the accelerations. At first, the black spaceship is moving faster, so it pulls out ahead. Later, the silver ship overtakes the black ship when their *positions are equal*.

Solution The position of either spaceship at a later time is given by

$$x = x_0 + v_{0x}t + \tfrac{1}{2}at^2$$

We set the positions of the spaceships equal to each other ($x_{silver} = x_{black}$) and solve algebraically for the time at which this occurs. The initial positions are the same: $x_{0s} = x_{0b} = x_0$.

$$x_0 + v_{0sx}t + \tfrac{1}{2}a_{sx}t^2 = x_0 + v_{0bx}t + \tfrac{1}{2}a_{bx}t^2$$

Subtracting x_0 from each side, moving all terms to one side, and factoring out one power of t yields

$$t(v_{0sx} + \tfrac{1}{2}a_{sx}t - v_{0bx} - \tfrac{1}{2}a_{bx}t) = 0$$

One solution of the equation is $t = 0$ (the two spaceships at the same *initial* position). That is not the solution we seek, so the expression in the parentheses must be equal to zero. Solving for t,

$$t = \frac{2(v_{0sx} - v_{0bx})}{a_{bx} - a_{sx}} = \frac{2(2.00 \text{ km/s} - 6.00 \text{ km/s})}{-0.400 \text{ km/s}^2 - 0.400 \text{ km/s}^2} = 10.0 \text{ s}$$

The silver spaceship overtakes the black spaceship at $t = 10.0$ s.

Discussion Quick check: the two ships must have the same displacement at $t = 10.0$ s.

$$\Delta x_s = v_{0sx}t + \tfrac{1}{2}a_{sx}t^2$$
$$= 2.00 \text{ km/s} \times 10.0 \text{ s} + \tfrac{1}{2} \times 0.400 \text{ km/s}^2 \times (10.0 \text{ s})^2 = 40.0 \text{ km}$$
$$\Delta x_b = v_{0bx}t + \tfrac{1}{2}a_{bx}t^2$$
$$= 6.00 \text{ km/s} \times 10.0 \text{ s} + \tfrac{1}{2} \times (-0.400 \text{ km/s}^2) \times (10.0 \text{ s})^2$$
$$= 40.0 \text{ km}$$

Practice Problem 3.10 Time to reach same velocity

Find the time at which the two spaceships have the same *velocity*.

Example 3.11

Two Blocks, One Sliding and One Hanging

A block of mass $m_1 = 3.0$ kg rests on a frictionless horizontal surface. A second block of mass $m_2 = 2.0$ kg hangs from a flexible cord of negligible mass that runs over an ideal pulley and then is connected to the first block (Fig. 3.23a). The blocks are released from rest. (a) Find the accelerations of the two blocks after they are released. (b) Find the tension in the cord connecting the blocks. (c) What is the velocity of the first block 1.2 s after the release of the blocks, assuming the first block does not run out of room on the table and the second block does not land on the floor? (d) What is the minimum distance from the pulley for block 1 to be located so that it can attain the velocity found in part (c)?

Strategy We consider each block as a separate system and draw a free-body diagram for each. The tension in the cord is the same at both ends of the cord since the cord and pulley are ideal. We choose the $+x$-axis to the right and the $+y$-axis up. To find the accelerations of the blocks (which are equal in magnitude if the cord length is fixed) and the tension in the cord, we apply Newton's second law. Then we use kinematic equations for constant acceleration to answer (c) and (d).

Given: $m_1 = 3.0$ kg; $m_2 = 2.0$ kg; $\vec{v}_0 = 0$ for both; $\Delta t = 1.2$ s

Solution (a) Figure 3.23b shows free-body diagrams for the two blocks. Block 1 slides along the table surface, so the vertical component of acceleration is zero; the normal force must

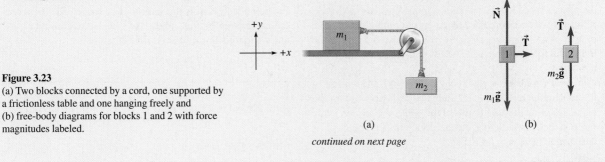

Figure 3.23
(a) Two blocks connected by a cord, one supported by a frictionless table and one hanging freely and
(b) free-body diagrams for blocks 1 and 2 with force magnitudes labeled.

continued on next page

Example 3.11 *continued*

be equal in magnitude to the weight. With the two vertical forces canceling, the only remaining force is horizontal: the pull of the cord.

$$\Sigma F_{1x} = T = m_1 a_{1x}$$

Only vertical forces act on block 2. From the free-body diagram for block 2, we write Newton's second law:

$$\Sigma F_{2y} = T - m_2 g = m_2 a_{2y}$$

The two accelerations have the same magnitude. The acceleration of block 1 is in the $+x$-direction while that of block 2 is in the $-y$-direction. Therefore we substitute

$$a_{1x} = a \quad \text{and} \quad a_{2y} = -a$$

where a is the magnitude of the acceleration.

Substituting $T = m_1 a$ and $a_{2y} = -a$ into the second equation,

$$m_1 a - m_2 g = -m_2 a$$

Now we solve for a:

$$a = \frac{m_2}{m_1 + m_2} g$$

Substituting the known quantities

$$a = \frac{2.0 \text{ kg}}{3.0 \text{ kg} + 2.0 \text{ kg}} \times 9.8 \text{ m/s}^2 = 3.9 \text{ m/s}^2$$

The acceleration of block 1 is 3.9 m/s² to the right and that of block 2 is 3.9 m/s² downward.

(b) The tension is now found from the acceleration:

$$T = m_1 a = 3.0 \text{ kg} \times 3.9 \text{ m/s}^2 = 12 \text{ N}$$

(c) Next we find the velocity of block 1 after 1.2 s. The problem gives the initial velocity, $v_{0x} = 0$ at $t = 0$, and the elapsed time, $\Delta t = t = 1.2$ s.

$$v_x = v_{0x} + a_x t$$

$$v_x = 0 + 3.9 \text{ m/s}^2 \times 1.2 \text{ s} = 4.7 \text{ m/s}$$

The positive sign indicates that block 1 moves to the right.

(d) To find the minimum distance from the pulley, we find the distance traveled during 1.2 s after starting from rest

$$\Delta x = v_{0x} t + \tfrac{1}{2} a_x t^2$$

$$\Delta x = 0 + \tfrac{1}{2} \left[3.9 \text{ m/s}^2 \times (1.2 \text{ s})^2 \right] = 2.8 \text{ m}$$

Discussion   An algebraic expression for the tension is

$$T = m_1 a = \frac{m_1 m_2}{m_1 + m_2} g$$

An algebraic expression can lead to insights that are lost when a numerical answer is calculated (see Practice Problem 3.11).

Conceptual Practice Problem 3.11   A quick check

Check the expressions for acceleration and tension in the special case $m_1 \gg m_2$. [*Hint:* What is the sum $m_1 + m_2$ approximately equal to if $m_1 \gg m_2$?]

Example 3.12

Towing a Glider

 A small plane of mass 760 kg requires 120 m of runway to take off by itself (120 m is the horizontal displacement of the plane just before it lifts off the runway, not the entire length of the runway). (a) When the plane is towing a 330-kg glider, how much runway does it need? (b) If the final speed of the plane just before it lifts off the runway is 28 m/s, what is the tension in the tow cable while the plane and glider are moving along the runway?

Strategy   The plane's engines produce thrust—the forward force on the plane due to the air. There is also a backward force on the plane due to the air: drag. The drag force increases as the plane's speed increases, but it is much less than the thrust. A small backward force is exerted on the rolling tires by the runway. As a simplified model we assume that the sum of the horizontal forces on the plane (thrust plus drag plus runway) is constant.

As the glider is towed along the runway, the tension in the cable pulls forward on the sailplane and backward on the plane. Ignoring the small mass of the cable, the tension is the same at both ends. Drag on the glider is negligible—it is designed to have very little drag.

Since the plane and glider are moving along the runway, the vertical component of acceleration is zero. We need not be concerned with gravity and with lift (the upward force on the aircraft's wings due to the air) since they add to zero to produce zero vertical acceleration.

Solution   (a) When the plane takes off by itself, we assume a constant horizontal net force. From Newton's second law,

$$\Sigma F_x = m_1 a_{1x}$$

where m_1 is the plane's mass and a_{1x} is its horizontal acceleration component.

When the glider is towed, we can consider the plane, glider, and cable to be a single system. The net horizontal force on the system is the same as ΣF_x above. The tension in the cable is an internal force; the glider produces no thrust; and we ignore drag on the glider. Therefore

$$m_1 a_{1x} = (m_1 + m_2) a_{2x}$$

where m_2 is the glider's mass and a_{2x} is the horizontal acceleration component of the plane and glider system. The same net force applied to a larger mass produces a smaller acceleration:

continued on next page

Example 3.12 *continued*

$a_{2x} < a_{1x}$. Rearranging the last equation shows that the acceleration is inversely proportional to the total mass:

$$\frac{a_{2x}}{a_{1x}} = \frac{m_1}{m_1 + m_2}$$

How is the acceleration related to the runway distance? The plane must get to the same final speed in order to lift off the runway. From

$$v_x^2 - v_{0x}^2 = 2a_x \Delta x$$

with the same values of v_x and v_{0x} in both cases, we see that Δx is inversely proportional to a_x; a smaller acceleration means a longer runway distance is required. Since the acceleration is inversely proportional to the total mass, and the runway distance is inversely proportional to the acceleration, the runway distance is directly proportional to the total mass:

$$\frac{\Delta x_2}{\Delta x_1} = \frac{a_{1x}}{a_{2x}} = \frac{m_1 + m_2}{m_1} = \frac{1090 \text{ kg}}{760 \text{ kg}} = 1.43$$

$$\Delta x_2 = 1.43 \times 120 \text{ m} = 172 \text{ m} \approx 170 \text{ m}$$

The plane uses 170 m of runway when towing the glider.

(b) The final speed given enables us to find the acceleration:

$$v_x^2 - v_{0x}^2 = 2a_x \Delta x$$

With $v_x = 28$ m/s, $v_{0x} = 0$, and $\Delta x = 172$ m,

$$a_x = \frac{v_x^2}{2 \Delta x} = \frac{(28 \text{ m/s})^2}{2 \times 172 \text{ m}} = 2.28 \text{ m/s}^2$$

The tension in the cable is the only horizontal force acting on the glider. Therefore

$$\Sigma F_x = T = m_2 a_x = 330 \text{ kg} \times 2.28 \text{ m/s}^2 = 752 \text{ N}$$

The tension in the cable is approximately 750 N.

Discussion This solution is based on a simplified model, so we can only regard the answers as approximate. Nevertheless, it illustrates Newton's second law. The same net force produces an acceleration inversely proportional to the mass of the object on which it acts. Here we have the same net force acting on two different objects: first the plane alone, then the plane and glider together.

Alternatively, we can look at forces acting only on the plane. When towing the glider, the cable pulls backward on the plane. The net force *on the plane* is smaller, so its acceleration is smaller. The smaller acceleration means that it takes more time to reach takeoff speed and travels a longer distance before lifting off the runway.

Practice Problem 3.12   Engine thrust

Neglecting air resistance, what is the thrust provided by the airplane's engines in the preceding example?

Example 3.13

Hauling a Crate Up to a Third-Floor Window

 A student is moving into a dorm room on the third floor and he decides to use a block and tackle arrangement (Fig. 3.24) to move a crate of mass 91 kg from the ground up to his window. If the breaking strength of the available cable is 550 N, what is the minimum time required to haul the crate to the level of the window, 30.0 m above the ground?

Strategy The tension in the cable is T and is the same at both ends or anywhere along the cable, assuming the cable and pulleys are ideal. Two cable strands support the crate, each pulling upward with a force of magnitude T. The weight of the crate acts downward. We draw a free-body diagram and set the tension equal to the breaking force of the rope to find the maximum possible acceleration of the crate. Then from the maximum acceleration, we use kinematic relationships to find the minimum time to move the required distance to the third-floor window with that acceleration.

Given: $m = 91$ kg; $\Delta y = 30.0$ m; $g = 9.8$ m/s^2; $T_{max} = 550$ N; $v_{0y} = 0$

To find: Δt, the minimum time to raise the crate 30.0 m

continued on next page

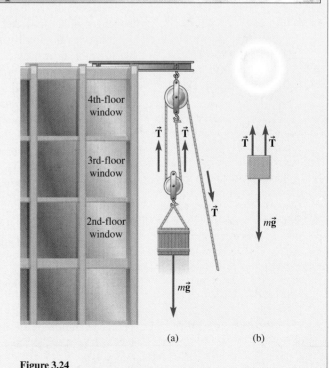

Figure 3.24
(a) Block and tackle setup and (b) free-body diagram for the crate

Example 3.13 *continued*

Solution   From the free-body diagram (Fig. 3.24b), if the forces acting up are greater than the force acting down, the net force is upward and the crate's acceleration is upward. In terms of components, with the +y-direction chosen to be upward,

$$\Sigma F_y = T + T - mg = ma_y$$

Solving for the acceleration,

$$a_y = \frac{T + T - mg}{m}$$

Setting $T = 550$ N, the maximum possible value before the cable breaks, and substituting the other known values,

$$a_y = \frac{550 \text{ N} + 550 \text{ N} - 91 \text{ kg} \times 9.8 \text{ m/s}^2}{91 \text{ kg}}$$

$$= 2.3 \text{ m/s}^2$$

The minimum time to move the crate up a distance Δy starting from rest can be found from

$$\Delta y = v_{0y}t + \tfrac{1}{2}a_y t^2$$

Setting $v_{0y} = 0$ and solving for t, we find

$$t = \pm\sqrt{\frac{2\,\Delta y}{a_y}}$$

Our equation applies only for $t \geq 0$ (the crate reaches the window *after* it leaves the ground). Taking the positive root and substituting numerical values,

$$t = \sqrt{\frac{2 \times 30.0 \text{ m}}{2.3 \text{ m/s}^2}} = 5.1 \text{ s}$$

This is the minimum time if a_y is the maximum acceleration.

Discussion   If the crate is accelerated upward, the tension of the cable is greater than if the crate is moved at a constant velocity. For the crate to accelerate, there must be an upward net force. For motion with a constant velocity, the tension would be equal to half the weight of the crate, 450 N.

Practice Problem 3.13   Hauling the crate with a single pulley

If only a single pulley, attached to the pole above the fourth-floor, were available and if the student had a few friends to help him pull on the cable, could they haul the crate up to the third-floor window? If so, what is the minimum time required to do so?

3.6   FALLING OBJECTS

Suppose you are standing on a bridge over a deep gorge. If you drop a stone into the gorge, how fast does it fall? You know from experience that it does not fall at a constant velocity; the longer it falls, the faster it goes. A better question is: what is the stone's acceleration?

First, let us simplify the problem. If the stone were moving very fast, an appreciable force of air resistance would oppose its motion. When it is not falling so fast, air resistance is negligibly small. If air resistance is negligible, the only appreciable force is that of gravity. **Free fall** is a situation in which no forces act on an object other than the gravitational force that makes the object fall. On Earth, free fall is an idealization since there is always *some* air resistance.

What is the acceleration of an object in free fall? More massive objects are harder to accelerate: the acceleration of an object subjected to a given force is inversely proportional to its mass. However, the stronger gravitational force on a more massive object compensates for its greater inertia, giving it the same acceleration as a less massive object. The gravitational force on an object is

$$\vec{\mathbf{W}} = m\vec{\mathbf{g}}$$

From Newton's second law,

$$\vec{\mathbf{F}}_{\text{net}} = m\vec{\mathbf{g}} = m\vec{\mathbf{a}}$$

Dividing by the mass yields

$$\vec{\mathbf{a}} = \vec{\mathbf{g}} \tag{3-14a}$$

The acceleration of an object in free fall is $\vec{\mathbf{g}}$, regardless of the object's mass. Since $1 \text{ N} = 1 \text{ kg·m/s}^2$, an object in free fall near the Earth's surface has an acceleration of magnitude

$$a = g = 9.8 \, \frac{\text{N}}{\text{kg}} = 9.8 \, \frac{\text{N}}{\text{kg}} \times 1 \, \frac{\text{kg·m/s}^2}{\text{N}} = 9.8 \text{ m/s}^2$$

Thus any object in free fall near the Earth's surface has a constant downward acceleration of magnitude 9.8 m/s². For this reason, $\vec{\mathbf{g}}$ is sometimes called *the freefall acceleration*—the acceleration of an object near the surface of the Earth when the *only* force acting is gravity.

When dealing with vertical motion, the y-axis is usually chosen to be positive pointing upward. In two-dimensional motion, the x-axis is often used for the horizontal direction and the y-axis for the vertical direction. The direction of the acceleration is down, so in free fall

$$a_y = -g \qquad \text{(3-14b)}$$

The same techniques and equations used for other constant acceleration situations are used with free fall. The only change is that the constant acceleration in free fall is always directed toward the center of the Earth and has a known constant magnitude of approximately 9.8 m/s² for objects near the surface of the Earth.

Earth's gravity always pulls downward, so the acceleration of an object in free fall is always downward, regardless of whether the object is moving up, down, or is at rest. If the object is moving downward, the downward acceleration makes it speed up; if it is moving upward, the downward acceleration makes it slow down; and if it is at rest, the downward acceleration makes it start moving downward.

If an object is thrown straight up, its velocity is zero at the highest point of its flight. Why? On the way up, the y-component of its velocity v_y is positive if the positive y-axis is pointing up. On the way down, v_y is negative. Since v_y changes continuously, at a rate of 9.8 m/s², it must pass through zero to change sign. At the one instant of time at its highest point, the object is neither moving up nor down. The object's acceleration is *not* zero at the top of flight. If the acceleration were to suddenly become zero at the top of flight, the velocity would no longer change; the object would get *stuck* at the top rather than fall back down! The velocity is zero at the top but it does not stay zero; it is still changing at the same rate.

> In free fall near the Earth's surface, $a_y = -g$ (if the y-axis points up).

Example 3.14

Throwing Stones

Standing on a bridge, you throw a stone straight upward. The stone hits a stream, 44.1 m below the point at which you release it, 4.00 s later. Assume $g = 9.81$ m/s². (a) What is the speed of the stone just after it leaves your hand? (b) What is the speed of the stone just before it hits the water?

Strategy Ignoring air resistance, the acceleration is constant. Choose the positive y-axis pointing up. Let the stone be thrown at $t = 0$ and hit the stream at a later time t.

Known: $a_y = -9.81$ m/s²; $\Delta y = -44.1$ m at $t = 4.00$ s.
To find: $|v_{0y}|$ (speed at $t = 0$) and $|v_y|$ (speed at $t = 4.00$ s)

Solution (a) Equation (3-12) can be used to solve for v_{0y}, since all the other quantities in it (Δy, t, and a_y) are known.

$$\Delta y = v_{0y}t + \tfrac{1}{2}a_yt^2$$

Solving for v_{0y},

$$v_{0y} = \frac{\Delta y}{t} - \frac{1}{2}a_yt = \frac{-44.1 \text{ m}}{4.00 \text{ s}} - \frac{1}{2}(-9.81 \text{ m/s}^2 \times 4.00 \text{ s}) \qquad \text{(1)}$$

$$= -11.0 \text{ m/s} + 19.6 \text{ m/s} = 8.6 \text{ m/s}$$

The initial speed is 8.6 m/s.

(b) The change in v_y is a_yt from Eq. (3-9):

$$v_y = v_{0y} + a_yt$$

Substituting the expression for v_{0y} found in Eq. (1),

$$v_y = \left(\frac{\Delta y}{t} - \frac{1}{2}a_yt\right) + a_yt \qquad \text{(2)}$$

$$= \frac{\Delta y}{t} + \frac{1}{2}a_yt$$

$$= \frac{-44.1 \text{ m}}{4.00 \text{ s}} + \frac{1}{2}(-9.81 \text{ m/s}^2 \times 4.00 \text{ s})$$

$$v_y = -11.0 \text{ m/s} - 19.6 \text{ m/s} = -30.6 \text{ m/s}$$

The final speed is 30.6 m/s.

Discussion The final speed is greater than the initial speed, as expected.

Equations (1) and (2) have a direct interpretation, which is a good check on their validity. The first term, $\Delta y/t$, is the average velocity of the stone during the 4.00 s of free fall. The second term, $\frac{1}{2}a_yt$, is *half* the change in v_y since $(v_y - v_{0y}) = a_yt$. Because the acceleration is constant, the average velocity is halfway between the initial and final velocities. Therefore, the initial velocity is the average velocity minus half of the change, while the final velocity is the average velocity plus half of the change.

Practice Problem 3.14 Height attained by stone

(a) How high above the bridge does the stone go? [*Hint:* What is v_y at the highest point?] (b) If you dropped the stone instead of throwing it, how long would it take to hit the water?

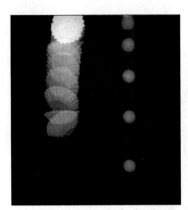

Figure 3.25 A stroboscopic photograph shows two objects falling through the air with very different terminal speeds. The exposures are taken at fixed time intervals of 1/15 s.

Table 3.3

Some Typical Terminal Speeds

Object	Terminal Speed (m/s)
Feather	0.5
Snowflake	1
Raindrop	7
Skydiver (open parachute)	5–9
Basketball	20
Baseball	40
Skydiver (spread-eagle)	50–60
Skydiver (dive)	100
Bullet	100

Air Resistance—Falling with Varying Acceleration

A skydiver relies on a parachute to provide a large drag force of air resistance. Even with the parachute closed, drag is not negligible when the skydiver is falling rapidly. The drag force exerted on a body falling through air increases dramatically with speed; it is proportional to the *square* of the speed:

$$F_d = bv^2$$

where b is a constant that depends on the size and shape of the object. The direction of the drag force is opposite to the direction of motion.

Since the drag force increases as the speed increases, a falling object may eventually reach equilibrium when the drag force is equal in magnitude to the weight. The speed at which the drag force is equal in magnitude to the weight is called the object's **terminal speed**. As the speed gets near the terminal speed, the acceleration gets smaller and smaller. The acceleration is zero when the object falls at its terminal speed.

In Fig. 3.25, a baseball and a coffee filter are released from rest and fall through the air. The strobe photograph shows the positions of the two at equal time intervals. The baseball has a terminal velocity of about 40 m/s, so air resistance is negligible for the speeds shown in the photo. The displacement of the baseball in equal time intervals increases linearly, showing that its acceleration is constant. The coffee filter has a very large surface area for its small mass. As a result, its terminal speed is much smaller—about 1 m/s. The displacement of the filter barely changes from one strobe flash to the next, showing that it is falling at a small, nearly constant velocity.

At terminal speed v_t, the drag force is equal in magnitude to the weight. Therefore, $F_d = mg = bv_t^2$ and

$$b = \frac{mg}{v_t^2}$$

Therefore, at any speed v,

$$F_d = mg \frac{v^2}{v_t^2} \tag{3-15}$$

The terminal speed of an object depends on its size, shape, and mass (see Table 3.3). A skydiver with the parachute closed will reach a terminal speed of about 50 m/s (≈ 110 mi/h) in the spread-eagle position or as much as 100 m/s (≈ 220 mi/h) in a dive. When the parachute is opened, the drag force increases dramatically—the larger surface area of the parachute means that more air has to be pushed out of the way. The terminal speed with the parachute open is typically about 9 m/s (20 mi/h). When the parachute is opened, the skydiver is initially moving *faster* than the new terminal speed. For $v > v_t$, the drag force is larger in magnitude than the weight and the acceleration is *upward*. The skydiver slows down, approaching the new terminal speed. Note that the terminal speed is not the maximum possible speed; it is the speed that the falling object approaches, regardless of initial conditions, when the only forces acting are drag and gravity.

Example 3.15

Skydivers Falling Freely

Two skydivers have identical parachutes. Their masses (including parachutes) are 62.0 kg and 82.0 kg. Which of the skydivers has the larger terminal speed? What is the ratio of their terminal speeds?

Strategy With identical parachutes, we expect the same amount of drag at a given speed. The more massive skydiver must fall faster in order for the drag force to equal his weight, so the 82.0-kg skydiver should have a larger terminal speed. For the ratio of the terminal speeds, we first find how the terminal speed depends on mass, all other things being equal. Then we work by proportions.

Solution At terminal speed v_t, the drag force must be equal in magnitude to the weight.

continued on next page

Example 3.15 *continued*

$$mg = F_d = bv_t^2$$

Since the parachutes are identical, we expect the constant b to be the same for the two divers. Therefore,

$$v_t \propto \sqrt{m}$$

The more massive skydiver has a larger terminal speed—he must move faster in order for the drag force to equal his larger weight. The ratio of the terminal speeds is

$$\frac{v_{t2}}{v_{t1}} = \sqrt{\frac{m_2}{m_1}} = \sqrt{\frac{82.0}{62.0}} = 1.15$$

The terminal speed of the 82.0-kg diver is 1.15 times that of the less massive skydiver, or 15% faster.

Discussion The 82.0-kg skydiver is 32% more massive:

$$\frac{82.0 \text{ kg}}{62.0 \text{ kg}} = 1.32$$

but his terminal speed is only 15% greater. That is because the drag force is proportional to the *square* of the speed. It only takes a 15% greater speed to make the drag force 32% greater:

$$(1.15)^2 = 1.32$$

Practice Problem 3.15 Air resistance at terminal speed

A pilot has bailed out of her airplane at a height of 2000 m above the surface of the Earth. The mass of pilot plus parachute is 112 kg. What is the force of air resistance when the pilot reaches terminal speed?

Example 3.16

Dropping the Ball

A basketball is dropped off a tall building. (a) What is the initial acceleration of the ball, just after it is released? (b) What is the acceleration of the ball when it is falling at its terminal speed? (c) What is the acceleration of the ball when falling at half its terminal speed?

Strategy We choose the positive y-axis to point upward as usual. The ball is dropped from rest so *initially* the only force acting is gravity—the drag force is zero when the velocity is zero. Once the ball is moving, air drag contributes to the net force on the basketball.

Solution (a) The initial acceleration is the free-fall acceleration ($\vec{a} = \vec{g}$) since the drag force is zero.

(b) Once the ball reaches terminal speed, the drag force is equal in magnitude to the weight of the ball but acts in the opposite direction. The net force on the ball is zero, so the acceleration is zero. At terminal speed, $\vec{a} = \vec{0}$.

(c) When the ball is falling at half its terminal speed, the drag force is significant but it is smaller than the weight. The net force is down and therefore the acceleration is still downward, though with a smaller magnitude. The drag force at any speed is given by

$$F_d = mg \frac{v^2}{v_t^2}$$

and this drag force acts in the opposite direction to the weight; it acts upward.

The net vertical force is

$$\Sigma F_y = F_d - mg = mg \frac{v^2}{v_t^2} - mg = mg \left(\frac{v^2}{v_t^2} - 1 \right)$$

Now we apply Newton's second law.

$$\Sigma F_y = ma_y$$

Solving for the acceleration yields

$$a_y = g \left(\frac{v^2}{v_t^2} - 1 \right)$$

continued on next page

Example 3.16 *continued*

At a time when the velocity is at half the terminal speed,

$$v = \frac{1}{2}v_t \quad \text{and} \quad \frac{v^2}{v_t^2} = \frac{1}{4}$$

$$a_y = g\left(\frac{1}{4} - 1\right) = -\frac{3}{4}g$$

so that the acceleration of the ball is

$$\vec{a} = \frac{3}{4}\vec{g}$$

where $\vec{a}$ and $\vec{g}$ both point downward.

Discussion   How do we know when air resistance is negligible? If we know the approximate terminal speed of an object, then air resistance is negligible as long as its speed moving through the air is small compared to the terminal speed.

Practice Problem 3.16   Acceleration graph sketch

Sketch a qualitative graph of $v_y(t)$ for the basketball using a y-axis that is positive pointing upward. [*Hint:* At first air resistance is negligible. After a long time the basketball is in equilibrium. Figure out what the beginning and end of the graph look like and then connect them smoothly.]

Physics at Home

Go to a balcony or climb up a ladder and drop a basket-style paper coffee filter (or a cupcake paper) and a penny simultaneously. Air resistance on the penny is negligible unless it is dropped from a very high balcony. At the other extreme, the effect of air resistance on the coffee filter is very noticeable; it reaches its terminal speed almost immediately. Stack several (two to four) coffee filters together and drop them simultaneously with a single coffee filter. Why is the terminal speed higher for the stack? Crumple a coffee filter into a ball and drop it simultaneously with the penny. Air resistance on the coffee filter is now reduced, but still noticeable.

3.7   APPARENT WEIGHT

Imagine being in an elevator when the cable snaps. Assume that some safety mechanism brings you to rest after you have been in free fall for a while. While you are in free fall, you *seem* to be "weightless," but your weight has not changed; the Earth still pulls downward with the same gravitational force. In free fall, gravity makes the elevator and everything in it accelerate downward at 9.8 m/s². The floor of the elevator stops pushing up on you, as it does when the elevator is at rest. If you jump up from the elevator floor, you seem to "float" up to the ceiling of the elevator. Your *weight* hasn't changed, but your **apparent weight** is zero while you are in free fall.

You don't need a disastrous elevator mishap to notice an apparent weight that differs from your true weight. A normally operating elevator will do quite nicely. Step in the elevator and push a button for a higher floor. When the elevator accelerates upward, you can feel your apparent weight increase. When the elevator slows down to stop, the elevator's acceleration is downward and your apparent weight is less than your true weight.

What is happening in the body while the elevator accelerates? Blood tends to collect in the lower extremities during acceleration upward and in the upper body during acceleration downward. The internal organs shift position within the body cavity resulting in a funny feeling in the gut as the elevator starts and stops. To avoid this problem, high-speed express elevators in skyscrapers keep the acceleration relatively small, but maintain that acceleration long enough to reach high speeds. That way, the elevator can travel quickly to the upper floors without making the passengers feel too uncomfortable.

Imagine an object resting on a bathroom scale. The scale measures the object's *apparent* weight W', which is equal to the true weight only if the object has zero acceleration. Newton's second law requires that

$$\vec{F}_{net} = \vec{N} + m\vec{g} = m\vec{a}$$

where $\vec{N}$ is the normal force of the scale pushing up. The apparent weight is the reading of the scale—the magnitude of $\vec{N}$:

$$W' = |\vec{N}| = N$$

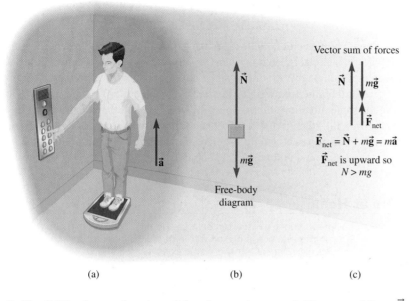

(a)   (b)   (c)

Figure 3.26 (a) Apparent weight
in an elevator with acceleration upward
(b) Free-body diagram for the passenger
(c) Normal force must be greater than the
weight to have an upward net force.

In Fig. 3.26a, the acceleration of the elevator is upward. The normal force $\vec{N}$ must
be larger than the weight $m\vec{g}$ in order for the net force to be upward (Fig. 3.26c). Writing
the forces in component form where the $+y$-direction is upward,

$$\Sigma F_y = N - mg = ma_y$$

or

$$N = mg + ma_y$$

Therefore,

$$W' = N = m(g + a_y) \tag{3-16}$$

Since the elevator accelerates upward, $a_y > 0$; the apparent weight is greater than the
true weight (Fig. 3.26c).

In Fig. 3.27a, the acceleration is downward. Then the net force must also point
downward. The normal force is still upward, but it must be smaller than the weight in
order to produce a downward net force (Fig. 3.27c). It is still true that

$$W' = m(g + a_y)$$

but now the acceleration is downward ($a_y < 0$). The apparent weight is less than the true
weight.

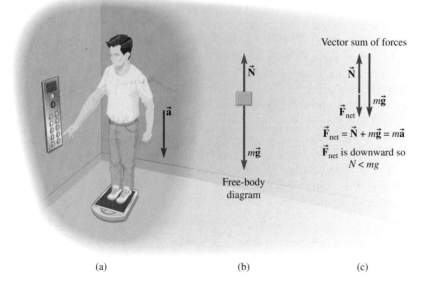

(a)   (b)   (c)

Figure 3.27 (a) Apparent weight
in an elevator with acceleration down
(b) Free-body diagram for the passenger
(c) Normal force must be less than the
weight to have a downward net force.

In both these cases—and in general—the apparent weight is given by

$$W' = m|\vec{g} - \vec{a}| \qquad (3\text{-}17)$$

where $\vec{g}$ and $\vec{a}$ are both vector quantities. When $\vec{g}$ and $\vec{a}$ point in opposite directions, the magnitude $|\vec{g} - \vec{a}|$ is greater than it is when they point in the same direction. If the acceleration of an elevator is upward, then $\vec{g}$ and $\vec{a}$ have opposite directions and the apparent weight is greater than the true weight. In an elevator with a downward acceleration, $\vec{g}$ and $\vec{a}$ have the same direction and the apparent weight is less than the true weight. If an elevator is in free fall, then $\vec{g}$ and $\vec{a}$ are in the same direction and are equal in magnitude; in free fall the apparent weight is zero. Draw the vectors and perform the vector subtraction to satisfy yourself that Eq. (3-17) is correct.

Physics at Home

Take an empty half-gallon paper milk carton or a plastic milk jug and poke one hole in the bottom and another in the side of the carton with a pencil. Fill the carton with water while sealing the holes with your fingers so the water does not pour out. Throw the carton straight up into the air and watch what happens at the holes. Does water start to pour out of the unsealed holes as the carton ascends? What about when the carton is falling downward? Can you explain your observations using the ideas of free fall and apparent weight?

Example 3.17

Apparent Weight in an Elevator

A passenger weighing 598 N rides in an elevator. The gravitational field strength is 9.80 N/kg. What is the apparent weight of the passenger in each of the following situations? In each case, the magnitude of the elevator's acceleration is 0.500 m/s². (a) The passenger is on the 1st floor and has pushed the button for the 15th floor; the elevator is beginning to move upward. (b) The elevator is slowing down as it nears the 15th floor.

Strategy   Let the +y-axis be upward. The apparent weight is equal to the magnitude of the normal force exerted by the floor on the passenger. Newton's second law lets us find the normal force from the weight and the acceleration.

Given: $W = 598$ N; magnitude of the acceleration is
$\qquad a = 0.500$ m/s²

Find: W'

Solution   (a) When the elevator starts up from the 1st floor it accelerates in the upward direction as its speed increases. Since the elevator accelerates upward, $a_y > 0$ (as in Fig. 3.26). We expect the apparent weight $W' = N$ to be greater than the true weight—the floor must push up with a force greater than W to cause an upward acceleration.

$$\Sigma F_y = N - W = ma_y$$

Since $m = W/g$,

$$W' = N = W + ma_y = W + \frac{W}{g}a_y = W\left(1 + \frac{a_y}{g}\right)$$

$$= 598\text{ N} \times \left(1 + \frac{0.500\text{ m/s}^2}{9.80\text{ m/s}^2}\right) = 629\text{ N}$$

(b) When the elevator approaches the 15th floor, it slows down while still moving upward; its acceleration is downward ($a_y < 0$) as in Fig. 3.27. The apparent weight is less than the true weight. Again, $\Sigma F_y = N - W = ma_y$, but this time $a_y = -0.500$ m/s².

$$N = W\left(1 + \frac{a_y}{g}\right)$$

$$= 598\text{ N} \times \left(1 + \frac{-0.500\text{ m/s}^2}{9.80\text{ m/s}^2}\right) = 567\text{ N}$$

Discussion   The apparent weight is greater when the direction of the elevator's acceleration is upward. That can happen in two cases: either the elevator is moving up with increasing speed, or it is moving down with decreasing speed.

Practice Problem 3.17   Elevator descending

What is the apparent weight of a passenger of mass 42.0 kg traveling in an elevator in each of the following situations? The gravitational field strength is 9.80 N/kg. In each case, the magnitude of the elevator's acceleration is 0.460 m/s². (a) The passenger is on the 15th floor and has pushed the button for the 1st floor; the elevator is beginning to move downward. (b) The elevator is slowing down as it nears the 1st floor.

MASTER THE CONCEPTS

Summary

- Position (symbol $\vec{r}$) is a vector from the origin to an object's location. Its magnitude is the distance from the origin and its direction points from the origin to the object.

- Displacement is the change in position: $\Delta\vec{r} = \vec{r}_f - \vec{r}_0$. The displacement depends only on the starting and ending positions, not on the path taken. The magnitude of the displacement vector is not necessarily equal to the total distance traveled; it is the straight line distance from the initial position to the final position.

- Vector components: If $\vec{A}$ points along the $+x$-axis, then $A_x = +A$; if $\vec{A}$ points in the opposite direction—along the negative x-axis—then $A_x = -A$. When adding or subtracting vectors, we can add or subtract their components.

- The average velocity states at what constant speed and in what direction to travel to cause that same displacement in the same amount of time.
$$\vec{v}_{av} = \frac{\Delta\vec{r}}{\Delta t}$$

- Velocity is a vector that states how fast and in what direction something moves. Its direction is the direction of the object's motion and its magnitude is the instantaneous speed.
$$\vec{v} = \lim_{\Delta t \to 0} \frac{\Delta\vec{r}}{\Delta t} \qquad (3\text{-}4)$$

- Average acceleration is the constant acceleration that would give the same velocity change in the same amount of time. In terms of changes in velocity and time,
$$\vec{a}_{av} = \frac{\Delta\vec{v}}{\Delta t} \qquad (3\text{-}6)$$

- Acceleration is the instantaneous rate of change of velocity:
$$\vec{a} = \lim_{\Delta t \to 0} \frac{\Delta\vec{v}}{\Delta t} \qquad (3\text{-}7)$$
Acceleration does not necessarily mean speeding up. A velocity can also change by decreasing speed or by changing directions.

- Interpreting graphs: On a graph of $x(t)$, the slope at any point is v_x. On a graph of $v_x(t)$, the slope at any point is a_x and the area under the graph during any time interval is the displacement during that time interval. If v_x is negative, the displacement is also negative, so we must count the area as negative when it is below the time axis. On a graph of $a_x(t)$, the area under the curve is the change in v_x during that time interval.

- Newton's second law:
$$\vec{F}_{net} = m\vec{a} \qquad (3\text{-}8)$$

- The SI unit of force is the newton; $1 \text{ N} = 1 \text{ kg·m/s}^2$. One newton is the magnitude of the net force that gives a 1-kg object an acceleration of magnitude 1 m/s^2.

- Essential relationships for solving any constant acceleration problem: if a_x is constant during the entire time interval from $t = 0$ until a later time t,
$$\Delta v_x = v_x - v_{0x} = a_x t \qquad (3\text{-}9)$$
$$\Delta x = x - x_0 = v_{av,x} t \qquad (3\text{-}10)$$
$$v_{av,x} = \frac{v_{0x} + v_x}{2} \qquad (3\text{-}11)$$
$$\Delta x = x - x_0 = v_{0x}t + \frac{1}{2}a_x t^2 \qquad (3\text{-}12)$$
$$v_x^2 - v_{0x}^2 = 2a_x \Delta x \qquad (3\text{-}13)$$

Highlighted Figures and Tables

F3.1 Displacement is the change in position: $\Delta\vec{r} = \vec{r}_f - \vec{r}_0$ (p. 64)

F3.5 Displacement and average velocity (p. 67)

F3.6 As Δt approaches zero, the average velocity during the increasingly short time interval approaches the instantaneous velocity (p. 68)

F3.7 Graphical relationship between position and velocity (p. 69)

F3.9 Displacement Δx is the area under the v_x versus time graph for the time interval considered (p. 70)

F3.14 Mass versus weight (p. 76)

F3.18 Finding average velocity with the aid of a graph (p. 79)

F3.19 Graphical interpretation of Eq. (3-12) (p. 80)

F3.20 Visualizing motion with constant acceleration (p. 81)

T3.3 Some typical terminal speeds (p. 88)

CONCEPTUAL QUESTIONS

1. Explain the difference between distance traveled, displacement, and displacement magnitude.

2. Explain the difference between speed and velocity.

3. On a graph of v_x versus time, what quantity does the area under the graph represent?

4. On a graph of v_x versus time, what quantity does the slope of the graph represent?

5. On a graph of a_x versus time, what quantity does the area under the graph represent?

6. On a graph of x versus time, what quantity does the slope of the graph represent?

7. What is the relationship between average velocity and instantaneous velocity? An object can have different instantaneous velocities at different times. Can the same object have different average velocities? Explain.

8. If an object is traveling at a constant velocity, is it necessarily traveling in a straight line? Explain.

9. Can the average speed and the magnitude of the average velocity ever be equal? If so, under what circumstances?

10. If a feather and a lead brick are dropped simultaneously from the top of a ladder, the lead brick hits the ground first. What would happen if the experiment is repeated on the surface of the Moon?

11. Name a situation where the speed of an object is constant while the velocity is not.

12. Can the velocity of an object be zero and the acceleration be nonzero at the same time? Explain.

13. Why does a 1-kg sandbag fall with the same acceleration as a 5-kg sandbag? Explain in terms of Newton's second law and his law of gravitation.

14. If an object is acted on by a single constant force, is it possible for the object to remain at rest? Is it possible for the object to move with constant velocity? Is it possible for the object's speed to be decreasing? Is it possible for it to change direction?

15. If an object is acted on by two constant forces is it possible for the object to move at constant velocity? If so, what must be true about the two forces?

16. An object is placed on a scale. Under what conditions does the scale read something other than the object's weight, even though the scale is functioning properly and is calibrated correctly? Explain.

17. What is meant by the terminal speed of a falling object? Can an object ever move through air faster than the object's terminal speed? If so, give an example.

18. What force(s) act on a parachutist descending to Earth with a constant velocity? What is the acceleration of the parachutist?

19. A baseball is tossed straight up. Taking into consideration the force of air resistance, is the magnitude of the baseball's acceleration zero, less than g, equal to g, or greater than g on the way up? At the top of the flight? On the way down? Explain.

20. What is the acceleration of an object thrown straight up into the air at the highest point of its motion? Does the answer depend on whether air resistance is negligible or not? Explain.

21. You are bicycling along a straight north-south road. Let the x-axis point north. Describe your motion in each of the following cases. Example: $a_x > 0$ and $v_x > 0$ means you are moving north and speeding up. (a) $a_x > 0$ and $v_x < 0$. (b) $a_x = 0$ and $v_x < 0$. (c) $a_x < 0$ and $v_x = 0$. (d) $a_x < 0$ and $v_x < 0$. (e) Based on your answers, explain why it is not a good idea to use the expression "negative acceleration" to mean slowing down.

MULTIPLE CHOICE QUESTIONS

1. A go-kart travels around a circular track at a constant speed. Which of these is a true statement?
   (a) The go-kart has a constant velocity.
   (b) The go-kart has zero acceleration.
   (c) Both (a) and (b) are true.
   (d) Neither (a) nor (b) is true.

2. A ball is thrown straight up into the air. Neglect air resistance. While the ball is in the air its acceleration
   (a) increases   (b) is zero   (c) remains constant
   (d) decreases on the way up and increases on the way down
   (e) changes direction

3. A stone is thrown upward and reaches a height Δy. After an elapsed time Δt, measured from the time the stone was first thrown, the stone has fallen back down to the ground. The magnitude of the average velocity of the stone during this time is
   (a) zero   (b) $2\dfrac{\Delta y}{\Delta t}$ (c) $\dfrac{\Delta y}{\Delta t}$ (d) $\dfrac{1}{2}\dfrac{\Delta y}{\Delta t}$

4. A stone is thrown upward and reaches a height Δy. After an elapsed time Δt, measured from the time the stone was first thrown, the stone has fallen back down to the ground. The average speed of the stone during this time is
   (a) zero   (b) $2\dfrac{\Delta y}{\Delta t}$ (c) $\dfrac{\Delta y}{\Delta t}$ (d) $\dfrac{1}{2}\dfrac{\Delta y}{\Delta t}$

5. A ball is thrown straight up. At the top of its trajectory the ball is
   (a) instantaneously at rest.
   (b) instantaneously in equilibrium.
   (c) Both (a) and (b) are true.
   (d) Neither (a) nor (b) is true.

Multiple Choice Questions 6–15 refer to Fig. 3.28.

6. What distance does the jogger travel during the first 10.0 min ($t = 0$ to 10.0 min)?
   (a) 8.5 m   (b) 510 m   (c) 900 m   (d) 1020 m

7. What is the displacement of the jogger from $t = 18.0$ min to $t = 24.0$ min?
   (a) 720 m, south   (b) 720 m, north
   (c) 2160 m, south   (d) 3600 m, north

8. What is the displacement of the jogger for the entire 30.0 min?
   (a) 3120 m, south   (b) 2400 m, north
   (c) 2400 m, south   (d) 3840 m, north

9. What is the total distance traveled by the jogger in 30.0 min?
   (a) 3840 m   (b) 2340 m   (c) 2400 m   (d) 3600 m

10. What is the average velocity of the jogger during the 30.0 min?
    (a) 1.3 m/s, north   (b) 1.7 m/s, north
    (c) 2.1 m/s, north   (d) 2.9 m/s, north

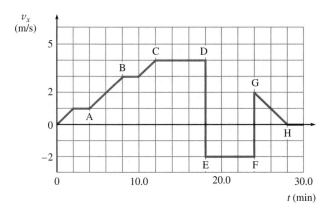

Figure 3.28 Multiple Choice Questions 6–15. A jogger is exercising along a long, straight road that runs north-south. She starts out heading north.

11. What is the average speed of the jogger for the 30 min?
(a) 1.4 m/s   (b) 1.7 m/s   (c) 2.1 m/s   (d) 2.9 m/s

12. In what direction is she running at time $t = 20$ min?
(a) south   (b) north   (c) not enough information

13. In which region of the graph is a_x positive?
(a) A to B   (b) C to D   (c) E to F   (d) G to H

14. In which region is a_x negative?
(a) A to B   (b) C to D   (c) E to F   (d) G to H

15. In which region is the velocity directed to the south?
(a) A to B   (b) C to D   (c) E to F   (d) G to H

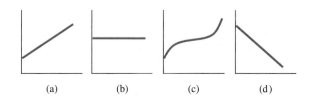

Figure 3.29 Multiple Choice Questions 16–20

Multiple Choice Questions 16–20 refer to Fig. 3.29.

16. If Fig. 3.29 shows four graphs of x versus time, which graph shows a constant, positive, nonzero velocity?

17. If Fig. 3.29 shows four graphs of v_x versus time, which graph shows a constant velocity?

18. If Fig. 3.29 shows four graphs of v_x versus time, which graph shows a_x constant and positive?

19. If Fig. 3.29 shows four graphs of v_x versus time, which graph shows a_x constant and negative?

20. If Fig. 3.29 shows four graphs of v_x versus time, which graph shows a changing a_x that is always positive?

21. Two blocks are connected by a light string passing over a pulley (see Fig. 3.23a). The block with mass m_1 slides on the frictionless horizontal surface, while the block with mass m_2 hangs vertically. ($m_1 > m_2$.) The tension in the string is:
(a) zero   (b) less than $m_2 g$
(c) equal to $m_2 g$ (d) greater than $m_2 g$, but less than $m_1 g$
(e) equal to $m_1 g$ (f) greater than $m_1 g$

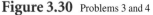

PROBLEMS

Note: **C** indicates a combination conceptual/quantitative problem. Gold diamonds ◆, ◆◆ are used to indicate the increasing level of difficulty of each problem. Problem numbers appearing in blue, 9., denote problems that have a detailed solution available in the Student Solutions Manual. Some problems are *paired* by concept; their numbers are connected by a ruled box.

3.1 Position and Displacement; 3.2 Velocity

1. Two cars, a Porsche and a Honda, are traveling in the same direction, although the Porsche is 186 m behind the Honda. The speed of the Porsche is 24.4 m/s and the speed of the Honda is 18.6 m/s. How much time does it take for the Porsche to catch the Honda? [*Hint:* What must be true about the displacements of the two cars when they meet?]

2. To get to a concert in time, a harpsichordist has to drive 121 mi in 2.00 h. If he drove at an average speed of 55 mi/h for the first 1.20 h, what must be his average speed for the remaining 0.80 h?

3. Figure 3.30 shows the vertical velocity component of an elevator versus time. How high is the elevator above the starting point ($t = 0$ s) after 20 s have elapsed?

4. Figure 3.30 shows the vertical velocity component of an elevator versus time. If the elevator starts to move at $t = 0$ s, when is the elevator at its highest location above the starting point?

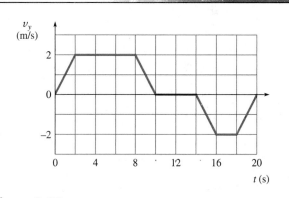

Figure 3.30 Problems 3 and 4

5. Figure 3.31 shows a graph of speedometer readings obtained as a car comes to a stop along a straight-line path. How far does the car move between $t = 0$ and $t = 16$ s?

6. Figure 3.32 shows a graph of speedometer readings, in meters per second (on the vertical axis), obtained as a car travels along a straight-line path. How far does the car move between $t = 3.00$ s and $t = 8.00$ s?

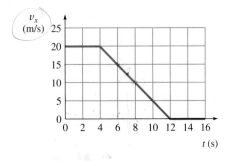

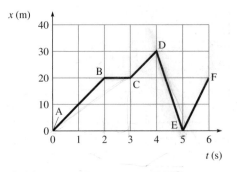

Figure 3.31 Problems 5, 17, and 79

Figure 3.33 Problem 10

3.3 Acceleration

7. Figure 3.32 shows a graph of $x(t)$ in meters, on the vertical axis, for an object traveling in a straight line. (a) What is $v_{\mathrm{av},x}$ for the interval from $t = 0$ to $t = 4.0$ s? (b) from $t = 0$ to $t = 5.0$ s?

8. Figure 3.32 shows a graph of $x(t)$ in meters for an object traveling in a straight line. What is v_x at $t = 2.0$ s?

12. If a car traveling at 28 m/s is brought to a full stop in 4.0 s after the brakes are applied, find the average acceleration during braking.

13. If a pronghorn antelope accelerates from rest in a straight line with a constant acceleration of 1.7 m/s², how long does it take for the antelope to reach a speed of 22 m/s?

14. An airtrack glider, 8.0 cm long, blocks light as it goes through a photocell gate (Fig. 3.34). The glider is released from rest on a frictionless inclined track and the gate is positioned so that the glider has traveled 96 cm when it is in the middle of the gate. The timer gives a reading of 333 ms for the glider to pass through this gate. Friction is negligible. What is the acceleration (assumed constant) of the glider along the track?

15. Figure 3.35 shows a graph of v_x versus t for a body moving along a straight line. (a) What is a_x at $t = 11$ s? (b) What is a_x at $t = 3$ s? (c) How far does the body travel from $t = 12$ to $t = 14$ s?

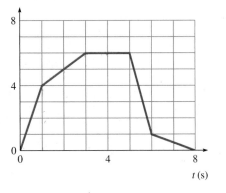

16. Figure 3.36 shows a plot of $v_x(t)$ for a car. (a) What is $a_{\mathrm{av},x}$ between $t = 6$ s and $t = 11$ s? (b) What is $v_{\mathrm{av},x}$ for the same time interval? (c) What is $v_{\mathrm{av},x}$ for the interval $t = 0$ to $t = 20$ s? (d) What is the increase in the car's speed between 10 and 15 s? (e) How far does the car travel from time $t = 10$ s to time $t = 15$ s?

17. Figure 3.31 shows a graph of speedometer readings as a motorcycle comes to a stop. What is the magnitude of the acceleration at $t = 7.0$ s?

18. At 3:00 P.M., a bank robber is spotted driving north on I-15 at milepost 126. His speed is 112.0 mi/h. At 3:37 P.M. he is spotted at milepost 185 doing 105.0 mi/h. During this time interval, what are the bank robber's displacement, average velocity, and average acceleration? (Assume a straight highway.)

Figure 3.32 Problems 6, 7, and 8.

9. A rabbit, nervously trying to cross a road, first moves 80 cm to the right, then 30 cm to the left, then 90 cm to the right, and then 310 cm to the left. (a) What is the rabbit's total displacement? (b) If the elapsed time was 18 s, what was the rabbit's average speed? (c) What was its average velocity?

10. An object is moving along a straight line. The graph in Fig. 3.33 shows its position from the starting point as a function of time. (a) In which section(s) of the graph does the object have the highest speed? (b) At which time(s) does the object reverse its direction of motion? (c) How far does the object move from $t = 0$ to $t = 3$ s?

11. A motor scooter travels east at a speed of 12 m/s. The driver then reverses direction and heads west at 15 m/s. What was the change in velocity of the scooter? Give magnitude and direction.

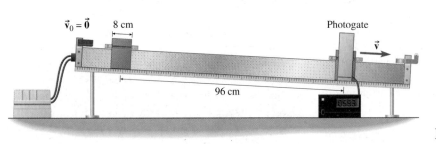

Figure 3.34 Problem 14

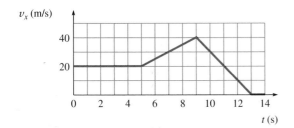

Figure 3.35 Problems 15, 33, and 34

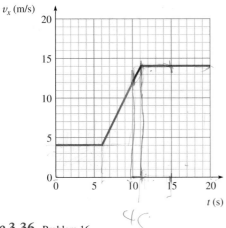

Figure 3.36 Problem 16

3.4 Newton's Second Law: Force and Acceleration

19. An engine accelerates a train of 20 freight cars, each having a mass of 5.0×10^4 kg, from rest to a speed of 4.0 m/s in 20.0 s on a straight track. Neglecting friction and assuming constant acceleration, what is the force with which the 10th car pulls the 11th one (at the middle of the train)?

20. In Fig. 3.16a, two blocks are connected by a lightweight, flexible cord that passes over a frictionless pulley. (a) If $m_1 = 3.0$ kg and $m_2 = 5.0$ kg, what are the accelerations of each block? (b) What is the tension in the cord?

21. A 2.0-kg toy locomotive is pulling a 1.0-kg caboose. The frictional force of the track on the caboose is 0.50 N backward along the track. If the train is accelerating forward at 3.0 m/s², what is the magnitude of the force exerted by the locomotive on the caboose?

22. A 2010-kg elevator accelerates upward at 1.5 m/s². What is the tension in the cable that supports the elevator?

23. A 2010-kg elevator accelerates downward at 1.5 m/s². What is the tension in the cable that supports the elevator?

24. A model sailboat is slowly sailing west across a pond. A gust of wind gives the sailboat a constant acceleration of magnitude 0.30 m/s² during a time interval of 2.0 s. If the net force on the sailboat during the 2.0-s interval has magnitude 0.375 N, what is the sailboat's mass?

25. The vertical component of the acceleration of a sailplane is zero when the air pushes up against its wings with a force of 3.0 kN. (a) Assuming that the only forces on the glider are that due to gravity and that due to the air pushing against its wings, what is the gravitational force on the Earth due to the glider? (b) If the wing stalls and the upward force decreases to 2.0 kN, what is the acceleration of the glider?

26. A man lifts a 2.0-kg stone vertically with his hand at a constant velocity of 1.5 m/s. What is the force exerted by his hand on the stone?

27. A man lifts a 2.0-kg stone vertically with his hand at a constant upward *acceleration* of 1.5 m/s². What is the magnitude of the total force of the stone on the man's hand?

✦ 28. A crate of oranges weighing 180 N rests on a flatbed truck 2.0 m from the back of the truck. The coefficients of friction between the crate and the bed are $\mu_s = 0.30$ and $\mu_k = 0.20$. The truck drives on a straight, level highway at a constant 8.0 m/s. (a) What is the force of friction acting on the crate? (b) If the truck speeds up with an acceleration of 1.0 m/s², what is the force of friction on the crate? (c) What is the maximum acceleration the truck can have without the crate starting to slide?

3.5 Motion with Constant Acceleration

29. A trolley car in New Orleans starts from rest at the St. Charles Street stop and accelerates uniformly at 1.20 m/s² for 12.0 s. (a) How far has the train traveled at the end of the 12.0 s? (b) What is the speed of the train at the end of the 12.0 s?

30. A train, traveling at a constant speed of 22 m/s, comes to an incline with a constant slope. While going up the incline, the train slows down with a constant acceleration of magnitude 1.4 m/s². (a) What is the speed of the train after 8.0 s on the incline? (b) How far has the train traveled up the incline after 8.0 s?

31. A car is speeding up and has an instantaneous speed of 10.0 m/s when a stopwatch reads 10.0 s. It has a constant acceleration of 2.0 m/s². (a) What change in speed occurs between $t = 10.0$ s and $t = 12.0$ s? (b) What is the speed when the stopwatch reads 12.0 s?

32. A train is traveling south at 24.0 m/s when the brakes are applied. It slows down at a constant rate to a speed of 6.00 m/s in a time of 9.00 s. (a) What is the acceleration of the train during the 9.00 s interval? (b) How far does the train travel during the 9.00 s?

33. Figure 3.35 shows the graph of v_x versus time for a body moving along the x-axis. How far does the body go between $t = 9.0$ s and $t = 13.0$ s? Solve using two methods: a graphical analysis and an algebraic solution using kinematic equations.

34. Figure 3.35 shows the graph of v_x versus time for a body moving along the x-axis. What is the average acceleration between $t = 5.0$ s and $t = 9.0$ s? Solve using two methods: a graphical analysis and an algebraic solution using kinematic equations.

35. An airplane lands and starts down the runway at a southwest velocity of 55 m/s. What constant acceleration allows it to come to a stop in 1.0 km?

✦ 36. The minimum stopping distance of a car moving at 30.0 mi/h is 12 m. Under the same conditions (so that the maximum braking force is the same), what is the minimum stopping distance for 60.0 mi/h? Work by proportions to avoid converting units.

37. Derive Eq. (3-12) using these steps. (a) Start with $\Delta x = v_{av,x}t$. Rewrite this equation with the average velocity expressed in terms of the initial and final velocities. (b) Replace the final velocity with the initial velocity plus the change in velocity $(v_{0x} + \Delta v_x)$. (c) Find the change in velocity in terms of a_x and t and substitute. Rearrange to form Eq. (3-12).

38. Derive Eq. (3-13) using these steps. (a) Start with $\Delta x = v_{av,x}t$. Rewrite this equation with the average velocity expressed in terms of the initial and final velocities. (b) Use $v_x - v_{0x} = a_x t$ to eliminate t. (c) Rearrange algebraically to form Eq. (3-13). [*Hint*: $A^2 - B^2 = (A + B)(A - B)$.]

39. A train of mass 55,200 kg is traveling along a straight, level track at 26.8 m/s (60.0 mi/h). Suddenly the engineer sees a truck stalled on the tracks 184 m ahead. If the maximum possible braking force has magnitude 84.0 kN, can the train be stopped in time?

40. In a television tube, electrons are accelerated from rest by a constant electric force of magnitude 6.4×10^{-17} N during the first 2.0 cm of the tube's length; then they move at essentially constant velocity another 45 cm before hitting the screen. (a) Find the speed of the electrons when they hit the screen. (b) How long does it take them to travel the length of the tube?

3.6 Falling Objects

In the problems, please assume $g = 9.8$ m/s^2 unless a more precise value is given in the problem.

41. A penny is dropped from the observation deck of the Empire State building (369 m above ground). With what velocity does it strike the ground? Ignore air resistance.

42. (a) How long does it take for a golf ball to fall from rest for a distance of 12.0 m? (b) How far would the ball fall in twice that time?

43. Grant Hill jumps 1.3 m straight up into the air to slam-dunk a basketball into the net. With what speed did he leave the floor?

44. A student, looking toward his fourth-floor dormitory window, sees a flowerpot with nasturtiums (originally on a window sill above) pass his 2.0-m-high window in 0.093 s. The distance between floors in the dormitory is 4.0 m. From a window on which floor did the flowerpot fall?

45. A balloonist, riding in the basket of a hot air balloon that is rising vertically with a constant velocity of 10.0 m/s, releases a sandbag when the balloon is 40.8 m above the ground. Neglecting air resistance, what is the bag's speed when it hits the ground? Assume $g = 9.80$ m/s^2.

46. A 55-kg lead ball is dropped from the leaning tower of Pisa. The tower is 55 m high. (a) How far does the ball fall in the first 3.0 s of flight? (b) What is the speed of the ball after it has traveled 2.5 m downward? (c) What is the speed of the ball 3.0 s after it is released?

47. During a walk on the Moon, an astronaut accidentally drops his camera over a 20.0-m cliff. It leaves his hands with zero speed, and after 2.0 s it has attained a velocity of 3.3 m/s downward. How far has the camera fallen after 4.0 s?

48. A stone is launched straight up by a slingshot. Its initial speed is 19.6 m/s and the stone is 1.50 m above the ground when launched. Assume $g = 9.80$ m/s^2. (a) How high above the ground does the stone rise? (b) How much time elapses before the stone hits the ground?

49. How far must something fall for its speed to get close to its terminal speed? Use dimensional analysis to come up with an estimate based on the object's terminal speed v_t and the gravitational field g. Estimate this distance for (a) a raindrop ($v_t \approx 7$ m/s) and (b) a skydiver in a dive ($v_t \approx 100$ m/s).

50. A paratrooper with a fully loaded pack has a mass of 120 kg. The force due to air resistance on him when falling with an unopened parachute has magnitude $F_d = bv^2$, where $b = 0.14 \; \dfrac{\text{N} \cdot \text{s}^2}{\text{m}^2}$. (a) If he is falling with an unopened parachute at 64 m/s, what is the force of air resistance acting on him? (b) What is his acceleration? (c) What is his terminal speed?

51. A bobcat weighing 72 N jumps out of a tree. (a) What is the drag force on the bobcat when it falls at its terminal velocity? (b) What is the drag force on the bobcat when it falls at 75% of its terminal velocity? (c) What is the acceleration of the bobcat when it falls at its terminal velocity? (d) What is its acceleration when it falls at 75% of its terminal velocity?

52. In free fall, we assume the acceleration to be constant. Not only is air resistance neglected, but the gravitational field strength is assumed to be constant. (a) From what height can an object fall to the Earth's surface such that the gravitational field strength changes less than 1.000% during the fall? (b) In most cases, which do we have to worry about first: air resistance becoming significant or g changing?

3.7 Apparent Weight

In the problems, please assume $g = 9.8$ m/s^2 unless a more precise value is given in the problem.

53. Refer to Example 3.17. What is the apparent weight of the same passenger (weighing 598 N) in the following situations? In each case, the magnitude of the elevator's acceleration is 0.50 m/s^2. (a) After having stopped at the 15th floor, the passenger pushes the 8th floor button; the elevator is beginning to move downward. (b) Elevator is slowing down as it nears the 8th floor.

54. You are standing on a bathroom scale inside an elevator. Your weight is 140 lb, but the reading of the scale is 120 lb. (a) What is the magnitude and direction of the acceleration of the elevator? (b) Can you tell whether the elevator is speeding up or slowing down?

55. Felipe is going for a physical before joining the swim team. He is concerned about his weight, so he carries his scale into the elevator to check his weight while heading to the doctor's office on the 21st floor of the building. If his scale reads 750 N while the elevator is accelerating upward at 2.0 m/s^2, what does the nurse measure his weight to be?

56. Luke stands on a scale in an elevator that has a constant acceleration upward. The scale reads 0.960 kN. When Luke picks up a box of mass 20.0 kg, the scale reads 1.200 kN. (The acceleration remains the same.) Let $g = 9.80$ m/s^2. (a) Find the acceleration of the elevator. (b) Find Luke's weight.

COMPREHENSIVE PROBLEMS

In the problems, please assume $g = 9.8 \text{ m/s}^2$ unless a more precise value is given in the problem statement.

57. The conduction electrons in a copper wire have instantaneous speeds of around 10^6 m/s. The electrons keep colliding with copper atoms and reversing direction, so that even when a current is flowing in the wire, they only drift along with average velocities on the order of centimeters per hour. If the drift velocity in a particular wire has magnitude 10 cm/h, calculate the distance an electron travels in order to move 1 m down the wire.

58. Imagine a trip where you drive along a straight east-west highway at 80.0 km/h for 50.0 min and then at 60.0 km/h for 30.0 min. What is your average velocity for the trip (a) if you drive east for both parts of the trip? (b) if you drive east for the first 50.0 min and then west for the last 30.0 min?

59. A rocket is launched from rest. After 8.0 min, it is 160 km above the Earth's surface and is moving at a speed of 7.6 km/s. Assuming the rocket moves up in a straight line, what are its (a) average velocity and (b) average acceleration?

60. Based on the information given in Problem 59, is it possible that the rocket moves with constant acceleration? Explain.

61. An unmarked police car starts from rest just as a speeding car passes at a speed of v_0. If the police car accelerates at a constant value of a, what is the speed of the police car when it catches up to the speeder, who does not realize she is being pursued and does not vary her speed?

62. In Fig. 3.23a, the block of mass m_1 slides to the right with coefficient of kinetic friction μ_k on a horizontal surface. The block is connected to a hanging block of mass m_2 by a light cord that passes over a light, frictionless pulley. (a) Find the acceleration of each of the blocks and the tension in the cord. (b) Check your answers in the special cases $m_1 \ll m_2$, $m_1 \gg m_2$, and $m_1 = m_2$. (c) For what value of m_2 (if any) do the two blocks slide at constant velocity? What is the tension in the cord in that case?

63. While passing a slower car on the highway, you accelerate uniformly from 39 mi/h to 61 mi/h in a time of 10.0 s. (a) How far do you travel during this time? (b) What is your acceleration in (mi/h)/s and m/s²?

64. A cheetah can accelerate from rest to 24 m/s in 2.0 s. Assuming the acceleration is uniform, (a) what is the acceleration of the cheetah? (b) What is the distance traveled by the cheetah in these 2.0 s? (c) A runner can accelerate from rest to 6.0 m/s in the same time, 2.0 s. What is the acceleration of the runner? By what factor is the cheetah's average acceleration magnitude greater than that of the runner?

65. Neglecting air resistance, (a) from what height must a hockey puck drop if it is to attain a speed of 30.0 m/s (approximately 67 mi/h) before striking the ground? (b) If the puck comes to a full stop in a time of 1.00 s from initial impact, what acceleration (assumed constant) is experienced by the puck during the impact with the ground?

66. Locusts can jump to heights of 0.30 m. (a) Assuming the locust jumps straight up, and ignoring air resistance, what is the takeoff speed of the locust? (b) The locust actually jumps at an angle of about 55° to the horizontal, and air resistance is not negligible. The result is that the takeoff speed is about 40% higher than the value you calculated in part (a). If the

mass of the locust is 2.0 g and its body moves 4.0 cm in a straight line while accelerating from rest to the takeoff speed, calculate the acceleration of the locust (assumed constant). (c) Ignore the locust's weight and estimate the force exerted on the hind legs by the ground. Compare this force to the locust's weight. Was it OK to ignore the locust's weight?

67. In Fig. 3.16a, two blocks are connected by a lightweight, flexible cord that passes over a frictionless pulley. If m_1 is 3.6 kg and m_2 is 9.2 kg, and block 2 is initially at rest 140 cm above the floor, how long does it take block 2 to reach the floor?

68. A locomotive pulls a train of 10 identical cars with a force of 2.0×10^6 N directed east. What is the force with which the *last* car pulls on the rest of the train?

69. A woman of mass 51 kg is standing in an elevator. If the elevator floor pushes up on her feet with a force of 408 N, what is the acceleration of the elevator?

70. A drag racer crosses the finish line of a 400.0-m track with a final speed of 104 m/s. (a) Assuming constant acceleration during the race, find the racer's time and the minimum coefficient of static friction between the tires and the road. (b) If, because of bad tires or wet pavement, the acceleration were 30.0% smaller, how long would it take to finish the race?

71. The graph in Fig. 3.37 shows the position x of a switch engine in a rail yard as a function of time t. At which of the labeled times t_0 to t_7 is (a) $a_x < 0$, (b) $a_x = 0$, (c) $a_x > 0$, (d) $v_x = 0$, (e) the speed decreasing?

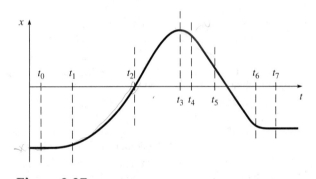

Figure 3.37   Problem 71

72. An elevator starts at rest on the ninth floor. At $t = 0$, a passenger pushes a button to go to another floor. The graph in Fig. 3.38 shows the acceleration a_y of the elevator as a function of time. Let the y-axis point upward. (a) Has the passenger gone to a higher or lower floor? (b) Sketch a graph of the velocity

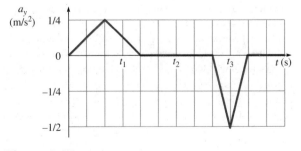

Figure 3.38   Problem 72

v_y of the elevator versus time. (c) Sketch a graph of the position y of the elevator versus time. (d) If a 1.40×10^2 lb person stands on a scale in the elevator, what does the scale read at times t_1, t_2, and t_3?

73. A streetcar named Desire travels between two stations 0.60 km apart. Leaving the first station, it accelerates for 10.0 s at 1.0 m/s^2 and then travels at a constant speed until it is near the second station, when it brakes at 2.0 m/s^2 in order to stop at the station. How long did this trip take? [*Hint:* What's the average velocity?]

© ✦74. The graph in Fig. 3.39 is the vertical velocity component v_y of a bouncing ball as a function of time. The y-axis points up. Answer these questions based on the data in the graph. (a) At what time does the ball reach its maximum height? (b) For how long is the ball in contact with the floor? (c) What is the maximum height of the ball? (d) What is the acceleration of the ball while in the air? (e) What is the average acceleration of the ball while in contact with the floor?

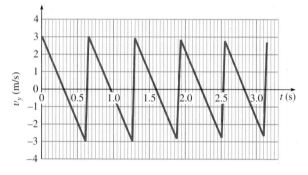

Figure 3.39 Problem 74

✦75. A rocket engine can accelerate a rocket launched from rest vertically up with an acceleration of 20.0 m/s^2. However, after 50.0 s of flight the engine fails. Neglect air resistance. (a) What is the rocket's altitude when the engine fails? (b) When does it reach its maximum height? (c) What is the maximum height reached? [*Hint:* a graphical solution may be easiest.] (d) What is the velocity of the rocket just before it hits the ground?

76. Two blocks lie side by side on a frictionless table. The block on the left is of mass m; the one on the right is of mass $2m$. The block on the right is pushed to the left with a force of magnitude F, pushing the other block in turn. What force does the block on the left exert on the block to its right?

✦77. A helicopter of mass M is lowering a truck of mass m onto the deck of a ship. (a) At first, the helicopter and the truck move downward together (the length of the cable doesn't change). If their downward speed is decreasing at a rate of $0.10g$, what is the tension in the cable? (b) As the truck gets close to the deck, the helicopter stops moving downward. While it hovers, it lets out the cable so that the truck is still moving downward. If the truck's downward speed is decreasing at a rate of $0.10g$, while the helicopter is at rest, what is the tension in the cable?

© ✦78. Fish don't move as fast as you might think. A small trout has a top swimming speed of only about 2 m/s, which is about the speed of a brisk walk (for a human, not a fish!). It may seem to move faster because it is capable of large *accelerations*—it

can dart about, changing its speed or direction very quickly. (a) If a trout starts from rest and accelerates to 2 m/s in 0.05 s, what is the trout's average acceleration? (b) During this acceleration, what is the average net force on the trout? Express your answer as a multiple of the trout's weight. (c) Explain how the trout gets the water to push it forward.

79. Figure 3.31 shows a graph of speedometer readings obtained as an automobile came to a stop. What is the displacement along a straight-line path in the first 16 s?

✦80. In Fig. 3.16a, two blocks are connected by a lightweight, flexible cord that passes over a frictionless pulley. If $m_1 \gg m_2$, find (a) the acceleration of each block and (b) the tension in the cord.

✦✦81. Find the point of no return for an airport runway of 1.50 mi in length if a jet plane can accelerate at 10.0 ft/s^2 and decelerate at 7.00 ft/s^2. The point of no return occurs when the pilot can no longer abort the takeoff without running out of runway. What length of time is available from the start of the motion in which to decide on a course of action?

82. Two blocks, masses m_1 and m_2, are connected by a massless cord (Fig. 3.40). If the two blocks are accelerated uniformly on a frictionless surface by applying a force of magnitude T_2 to a second cord connected to m_2, what is the ratio of the tensions in the two cords in terms of the masses? $T_1/T_2 = ?$

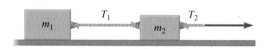

Figure 3.40 Problem 82

83. In the human nervous system, signals are transmitted along neurons as *action potentials* that travel at speeds of up to 100 m/s. (An action potential is a traveling influx of sodium ions through the membrane of a neuron.) The signal is passed from one neuron to another by the release of neurotransmitters in the synapse. Suppose someone steps on your toe. The pain signal travels along a 1.0-m-long sensory neuron to the spinal column, across a synapse to a second 1.0-m-long neuron, and across a second synapse to the brain. Suppose that the synapses are each 100 nm wide, that it takes 0.10 ms for the signal to cross each synapse, and that the action potentials travel at 100 m/s. (a) At what average speed does the signal cross a synapse? (b) How long does it take the signal to reach the brain? (c) What is the average speed of propagation of the signal?

84. A car traveling at 65 mi/h runs into a bridge abutment after the driver falls asleep at the wheel. (a) If the driver is wearing a seat belt and comes to rest within a 1.0-m distance, what is his acceleration (assumed constant)? (b) A passenger who isn't wearing a seat belt is thrown into the windshield and comes to a stop in a distance of 10.0 cm. What is the acceleration of the passenger? (c) Express these accelerations in terms of the free fall acceleration, $g = 9.8$ m/s^2, and find "how many g's" are felt in each case. (Test pilots can black out at accelerations of $4g$ or greater; pressure suits are designed to help force blood to the brain and prevent such a loss of consciousness.)

ANSWERS TO PRACTICE PROBLEMS

3.1 His displacement is zero since he ends up at the same place from which he started (home plate).

3.2 $\Delta x = -2.9$ km, so the displacement is 2.9 km to the west.

3.3 1.05 m/s to the north

3.4 The velocity is increasing in magnitude, so $\Delta\vec{v}$ and $\vec{a}$ are in the same direction as the velocity (the $-x$-direction). Thus a_x is negative.

3.5 (a) $a_{av,x} = -3.0$ m/s^2, where the negative sign means the acceleration is directed to the northwest; (b) $a_x = -4.2$ m/s^2 (northwest)

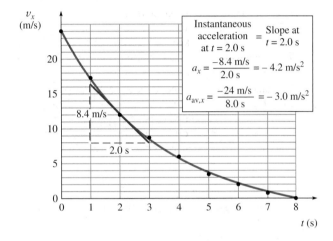

3.6 1.84 kN

3.7 For equal masses, $a_y = 0$ and $T = mg$. The pull of gravity is equal on the two sides. For slightly unequal masses, $m_1 \approx m_2$, so again $T \approx mg$; the acceleration is very small since there is only a slight excess pull of gravity on the heavier side but plenty of inertia.

3.8 60 N

3.9 (a) 20 s; (b) 18 m/s east

3.10 5.00 s

3.11 If $m_1 \gg m_2$, a is very small and $T \approx m_2 g$. Block 1 is so massive that it barely accelerates; block 2 is essentially just hanging there, being supported by the cord.

3.12 2500 N

3.13 It is impossible to pull the crate up with a single pulley. The entire weight of the crate would be supported by a single strand of cable and that weight exceeds the breaking strength of the cable.

3.14 (a) 3.8 m; (b) 3.00 s

3.15 1.1 kN

3.16 At first, constant acceleration: straight line with slope $-g$. After a long time, constant velocity.

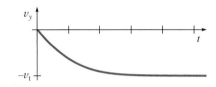

3.17 (a) 392 N; (b) 431 N

Forces and Motion in Two Dimensions

A seagull scoops up a clam and takes it high above the ground. While flying parallel to the ground, the seagull lets go so that the clam lands on a rock below and cracks open. Then the seagull alights and enjoys lunch. A beachcomber on the beach sees the clam fall along a parabolic path, just as a projectile would. Why does the clam not drop straight down? What does the path of the falling clam look like to the bird?

4.1   ADDITION AND SUBTRACTION OF VECTORS IN TWO DIMENSIONS

This chapter is about analyzing motion in two dimensions. Before we delve into the topic, we must acquire the necessary mathematical tools. We start by extending the principles introduced in Chapter 3 to two-dimensional motion. Suppose we plan a trip in Ireland from Killarney to Cork in order to visit Blarney Castle. There are many possible routes. One route is to travel due east from Killarney for 39 km to Mallow and then south from Mallow for 22 km to Cork. Figure 4.1 shows the displacement vectors for the two legs of the trip. The total distance traveled is 39 km + 22 km = 61 km. (The direction doesn't matter; distance is a scalar quantity.) To find the *displacement*, a vector quantity, the displacement *vectors* are added. We have already learned the rules for adding vectors graphically. To add two or more vectors, place the tail of each successive vector at the tip of the preceding vector. The value of a vector is not changed by moving it as long as its direction and magnitude are not changed, so a vector can be drawn starting at any point. Moreover, it does not matter in what order the vectors are added since vector addition is commutative (that is, $\vec{\mathbf{A}} + \vec{\mathbf{B}} = \vec{\mathbf{B}} + \vec{\mathbf{A}}$; see Problem 1). The resultant vector—the sum of the two vectors—is then drawn from the tail of the first vector to the tip of the last vector. If the lengths and directions of the vectors are drawn accurately to scale, using a ruler and a protractor, then the length and direction of the resultant can be determined with the ruler and protractor.

The displacement vectors (Fig. 4.1) are labeled $\Delta\vec{\mathbf{r}}_e$ for the trip from Killarney to Mallow and $\Delta\vec{\mathbf{r}}_s$ for the trip from Mallow to Cork; the total displacement is labeled $\Delta\vec{\mathbf{r}}$.

$$\Delta\vec{\mathbf{r}} = \Delta\vec{\mathbf{r}}_e + \Delta\vec{\mathbf{r}}_s$$

The two separate displacements to the east and south are at right angles and the displacement $\Delta\vec{\mathbf{r}}$ forms the hypotenuse of a right triangle so the magnitude can be found using the Pythagorean theorem. The magnitude of the displacement is

$$|\Delta\vec{\mathbf{r}}| = \sqrt{(39 \text{ km})^2 + (22 \text{ km})^2} = 45 \text{ km}$$

The vector arrow drawn from Killarney directly to Cork represents the total displacement for the trip; it indicates the route that an airplane would take in flying *directly* from one city to the other along a straight line. The magnitude of the displacement is the straight-line distance from Killarney to Cork.

Now suppose we take a different route from Killarney to Cork. We might drive at a compass heading of 27° west of south for 18 km to Kenmare, then directly south for 17 km to Glengariff, then at a compass heading of 13° north of east for 48 km to Cork, a total travel distance of 83 km. The total displacement (Fig. 4.3) is the same as for the previous route, since the trip begins and ends at the same two places.

Concepts & Skills to Review

- net force: vector addition (Section 2.4)
- gravitational forces (Section 2.5)
- Newton's second law: force and acceleration (Section 3.4)
- motion with constant acceleration (Section 3.5)
- falling objects (Section 3.6)

The order does not matter in vector addition: $\vec{\mathbf{A}} + \vec{\mathbf{B}} = \vec{\mathbf{B}} + \vec{\mathbf{A}}$.

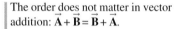

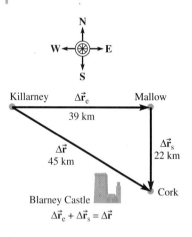

$\Delta\vec{\mathbf{r}}_e + \Delta\vec{\mathbf{r}}_s = \Delta\vec{\mathbf{r}}$

Figure 4.1 Displacement vectors for the trip from Killarney to Cork via Mallow

Compass headings are often used to specify vector directions. Any direction in the horizontal plane can be specified by giving an angle with respect to the directions of the compass. For example, the direction of the vector in Fig. 4.2 is "20° north of east," which means that the vector makes a 20° angle with the east direction and is on the north (rather than the south) side of east. The same direction could be described as "70° east of north," although it is customary to use the smaller angle. Northeast means "45° north of east" or, equivalently, "45° east of north."

Blarney Castle

In Fig. 4.4a, we select an origin and then draw position vectors from the origin to the starting and ending points of the trip. To subtract $\vec{r}_0$ from $\vec{r}_f$, the vectors are drawn with their tails at the same point. Then the vector drawn from the tip of $\vec{r}_0$ to the tip of $\vec{r}_f$ is the vector $\vec{r}_f - \vec{r}_0$ since it takes us from $\vec{r}_0$ to $\vec{r}_f$. To double-check, note that the initial position plus the change in position gives the final position:

$$\vec{r}_0 + \Delta\vec{r} = \vec{r}_f$$

We draw $\Delta\vec{r}$ so that, when added to $\vec{r}_0$, it gives $\vec{r}_f$. This procedure is equivalent to adding $-\vec{r}_0$ to $\vec{r}_f$ (Fig. 4.4b):

$$\Delta\vec{r} = \vec{r}_f - \vec{r}_0 = \vec{r}_f + (-\vec{r}_0)$$

The same methods of vector addition and subtraction are used for *any* vectors measuring like quantities, such as two force vectors or two displacement vectors. Example 4.1 considers the addition of two force vectors.

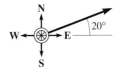

Figure 4.2 Measuring angles with respect to compass headings. The direction of this vector is 20° north of east (20° N of E).

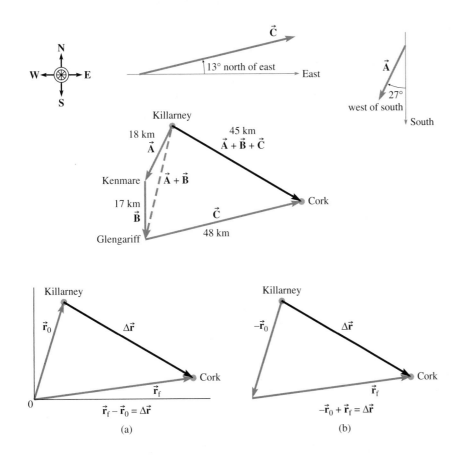

Figure 4.3 Graphical addition of the displacement vectors for the trip from Killarney to Cork via Kenmare and Glengariff.

Figure 4.4 (a) Two position vectors, $\vec{r}_0$ and $\vec{r}_f$, drawn from an *arbitrary origin* to the starting point (Killarney) and to the ending point (Cork) of a trip. The final position vector minus the initial position vector is the displacement $\Delta\vec{r}$, drawn from the tip of $\vec{r}_0$ to the tip of $\vec{r}_f$. (b) Another way to find $\vec{r}_f - \vec{r}_0$ is to add $-\vec{r}_0$ to $\vec{r}_f$.

Example 4.1

Bringing Home the Maple Syrup

Two draft horses, Sam and Bob, are dragging a sled loaded with jugs of maple syrup. They pull with forces of equal magnitude 1.50 kN on two ropes attached to the front of the sled. Each rope makes an angle of 15° with the forward direction (Fig. 4.5). Use the graphical method of vector addition to find the magnitude and direction of the sum of the forces exerted by the two horses.

Strategy Forces are vector quantities, so they add just like displacement vectors. To add vectors graphically, we need to use a ruler and a protractor. The protractor is used to draw the vector arrows in the correct directions and the ruler is used to draw them with the correct lengths. We must choose a convenient scale for the lengths of the vector arrows. Here we choose to represent 200 N as a length of

continued on next page

Example 4.1 *continued*

1 cm, so that the 1.50 kN forces are drawn as vector arrows of length

$$1.50 \text{ kN} \times \frac{1 \text{ cm}}{200 \text{ N}} = 7.50 \text{ cm}$$

Once the resultant is drawn, its direction and magnitude are determined with the ruler and protractor.

Solution   We can redraw any vector starting at any point, as long as we do not change its direction or magnitude. Thus, we can either add vector $\vec{F}_B$ to vector $\vec{F}_S$ or add vector $\vec{F}_S$ to vector $\vec{F}_B$. Both possibilities are shown in Fig. 4.6.

The direction of the resultant vector is straight ahead since it lies on a horizontal ruling of the grid. Measurement of the length of the resultant vector arrow yields 14.5 cm. Therefore, the magnitude of the resultant is

$$14.5 \text{ cm} \times \frac{200 \text{ N}}{1 \text{ cm}} = 2.90 \text{ kN}$$

Discussion   That the resultant is straight ahead is not surprising given the symmetry of the situation: the two forces are equal in magnitude and pull at equal angles on either side of the forward direction.

A quick check on the magnitude of the resultant: if the two vectors being added were parallel, their sum would have magnitude 3.00 kN. Since they are not parallel, their sum must have a magnitude *smaller* than 3.00 kN. The magnitude is only slightly smaller because the angle between the two vectors is relatively small.

The accuracy of the graphical method depends on the accuracy with which the directions and magnitudes of the arrows are drawn. In Section 4.2 we introduce an algebraic method of vector addition that is both easier and more accurate.

Practice Problem 4.1   Pulling at other angles

(a) If Sam and Bob were to pull with forces of the same magnitude but angled at 10° to the forward direction, would the sum of the two be larger, smaller, or the same magnitude? Illustrate with a sketch. (b) If Sam pulls at 10° while Bob pulls at 15° to the forward direction, is it still possible for the sum of the two to be in the forward direction? If so, which force must have the larger magnitude? Illustrate with a sketch.

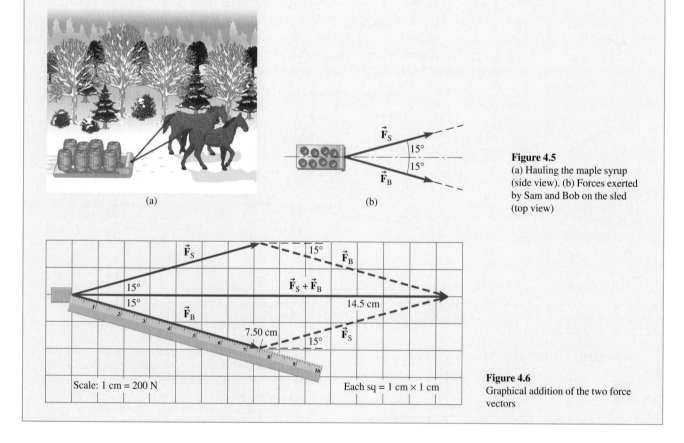

(a)

(b)

Figure 4.5
(a) Hauling the maple syrup (side view). (b) Forces exerted by Sam and Bob on the sled (top view)

Scale: 1 cm = 200 N

Each sq = 1 cm × 1 cm

Figure 4.6
Graphical addition of the two force vectors

4.2 COMPONENTS OF VECTORS IN TWO DIMENSIONS

It is generally easier and more accurate to add vectors algebraically rather than graphically. The algebraic method of vector addition relies on adding the components of the vectors. Vector components were introduced in Chapter 3, but only in cases where the vectors were parallel to one of the axes. If a vector lies in the xy-plane but *not* along one of the axes, then it has both x- and y-components.

A problem can be made easier to solve with a good choice of axes. We can choose any direction we want for the x- and y-axes, as long as they are perpendicular to each other. The two most common choices when near the surface of the Earth are

- x-axis horizontal and y-axis vertical, when the vectors all lie in a vertical plane; and

- x-axis east and y-axis north, when the vectors all lie in a horizontal plane.

Some problems are considerably easier to solve for a choice of axes other than the two common choices above.

Consider vector $\vec{A}$ in Fig. 4.7. Note that $\vec{A}$ is the sum of two vectors $\vec{A}_x$ and $\vec{A}_y$, the former along the x-axis and the latter along the y-axis:

$$\vec{A}_x + \vec{A}_y = \vec{A}$$

$\vec{A}_x$ and $\vec{A}_y$ are called the **vector components** of $\vec{A}$. Using these two vectors we can find the **scalar components** (A_x and A_y) of $\vec{A}$ exactly as in Chapter 3:

$$A_x = \pm \left| \vec{A}_x \right|$$

where the choice of sign reflects whether $\vec{A}_x$ points in the positive or negative x-direction.

In Fig. 4.7, A_x is positive and A_y is negative. Using the grid, we can measure the components of $\vec{A}$: $A_x = +5.0$ cm and $A_y = -8.0$ cm.

The process of finding the scalar components of a vector is called **resolving** or decomposing the vector into its components. From now on, when we refer simply to "components," we mean the *scalar* components.

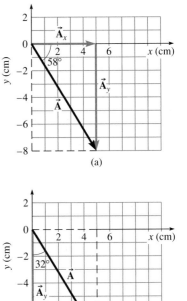

Figure 4.7 Resolving vector $\vec{A}$ into components along two axes in two equivalent ways

---

Finding the *x*- and *y*-Components of a Vector from Its Magnitude and Direction

1. Draw a right triangle with the vector as the hypotenuse and the other two sides parallel to the x- and y-axes.
2. Determine one of the angles in the triangle.
3. Use trigonometric functions to find the magnitudes of the components.
4. Supply the correct algebraic sign for each component.

Suppose that the magnitude of vector $\vec{A}$ is 9.4 cm and the direction is 58° below the horizontal. Choosing the x-axis to be horizontal and positive to the right and the y-axis to be vertical and positive upward, the vector is drawn as in Fig. 4.7 with the 58° angle labeled. Now, using trigonometry,

$$\cos 58° = \frac{\text{adjacent}}{\text{hypotenuse}} = \frac{\left| A_x \right|}{A}$$

where $A = \left| \vec{A} \right|$ is the magnitude of the vector $\vec{A}$. (See Appendix A.7 for a review of the trigonometric functions.) The x-component is positive so

$$A_x = +A \cos 58° = +(9.4 \text{ cm}) \times 0.530 = +5.0 \text{ cm}$$

For the y-component,

$$\sin 58° = \frac{\text{opposite}}{\text{hypotenuse}} = \frac{\left| A_y \right|}{A}$$

The *y*-component is *negative,* so

$$A_y = -A \sin 58° = -(9.4 \text{ cm}) \times 0.848 = -8.0 \text{ cm}$$

We must also know how to reverse the process:

Finding the Magnitude and Direction of a Vector from Its *x*- and *y*-Components

1. Sketch the vector on a set of *xy*-axes in the correct quadrant according to the signs of the components.
2. Draw a right triangle with the vector as the hypotenuse and the other two sides parallel to the *x*- and *y*-axes.
3. In the right triangle, choose which of the unknown angles you want to determine.
4. Use the inverse tangent function to find the angle. The lengths of the sides of the triangle represent $|A_x|$ and $|A_y|$. If θ is opposite the side parallel to the *x*-axis, then since $\tan \theta$ = opposite/adjacent,

$$\theta = \tan^{-1} \left| \frac{A_x}{A_y} \right|$$

If θ is opposite the side parallel to the *y*-axis, then

$$\theta = \tan^{-1} \left| \frac{A_y}{A_x} \right|$$

5. Interpret the angle: specify whether it is the angle below the horizontal, or the angle west of south, or the angle clockwise from the negative *y*-axis, and so on.
6. Use the Pythagorean theorem to find the magnitude:

$$A = \sqrt{A_x^2 + A_y^2}$$

For the vector $\vec{A}$ in Fig. 4.7, let us choose to find the angle between $\vec{A}$ and the +*x*-axis. Since $\tan \theta$ is the ratio of the opposite side to the adjacent side,

$$\theta = \tan^{-1} \frac{\text{opposite}}{\text{adjacent}} = \tan^{-1} \left| \frac{A_y}{A_x} \right| = \tan^{-1} \frac{8.0}{5.0} = 58°$$

The magnitude of $\vec{A}$ is

$$A = \sqrt{A_x^2 + A_y^2} = \sqrt{(+5.0 \text{ cm})^2 + (-8.0 \text{ cm})^2} = 9.4 \text{ cm}$$

To add vectors algebraically, we work with the components of the vectors:

Adding Vectors Algebraically

1. Find the *x*- and *y*-components of each vector to be added.
2. Add the *x*-components (with their algebraic signs) of the vectors to find the *x*-component of the resultant.
3. Add the *y*-components (with their algebraic signs) of the vectors to find the *y*-component of the resultant.
4. Use the *x*- and *y*-components of the resultant to find the magnitude and direction of the resultant.

Example 4.2

Adding Displacements Algebraically

Use the algebraic method to add the displacement vectors for the trip from Killarney to Cork via Kenmare and Glengariff (Fig. 4.8a).

Strategy   First we find the *x*- and *y*-components for the three displacements $\vec{A}$, $\vec{B}$, and $\vec{C}$ using the *x*-axis to the east and the *y*-axis to the north. Then the *x*- and *y*-components Δx and Δy of the total displacement $\Delta \vec{r}$ are

$$\Delta x = A_x + B_x + C_x$$

continued on next page

Example 4.2 *continued*

and
$$\Delta y = A_y + B_y + C_y$$

Finally, we use Δx and Δy to find the magnitude and direction of $\Delta \vec{r}$.

Solution   The first displacement ($\vec{A}$) is directed 27° west of south. Both of its components are negative since west is the $-x$-direction and south is the $-y$-direction. Using the triangle in Fig. 4.8b, the side of the triangle opposite the 27° angle is parallel to the x-axis. The sine function relates the opposite side to the hypotenuse

$$A_x = -A \sin 27° = -18 \text{ km} \times 0.454 = -8.2 \text{ km}$$

while the cosine relates the adjacent side to the hypotenuse

$$A_y = -A \cos 27° = -18 \text{ km} \times 0.891 = -16 \text{ km}$$

Vector $\vec{B}$ has no x-component, since its direction is south. Therefore,

$$B_x = 0$$
$$B_y = -B = -17 \text{ km}$$

The direction of $\vec{C}$ is 13° north of east. Both its components are positive. From Fig. 4.8b, the side of the triangle opposite the 13° angle is parallel to the y-axis, so

$$C_x = +C \cos 13° = +48 \text{ km} \times 0.974 = +47 \text{ km}$$
$$C_y = +C \sin 13° = +48 \text{ km} \times 0.225 = +11 \text{ km}$$

Now we sum the x- and y-components separately to find the x- and y-components of the total displacement:

$$\Delta x = A_x + B_x + C_x = (-8.2 \text{ km}) + 0 + 47 \text{ km} = +39 \text{ km}$$
$$\Delta y = A_y + B_y + C_y = (-16 \text{ km}) + (-17 \text{ km}) + 11 \text{ km} = -22 \text{ km}$$

The magnitude and direction of $\Delta \vec{r}$ can be found from the triangle in Fig. 4.8c. The magnitude is represented by the hypotenuse:

$$\Delta r = \sqrt{(\Delta x)^2 + (\Delta y)^2} = \sqrt{(39 \text{ km})^2 + (-22 \text{ km})^2} = \sqrt{2005 \text{ km}^2} = 45 \text{ km}$$

The angle θ is

$$\theta = \tan^{-1} \frac{\text{opposite}}{\text{adjacent}} = \tan^{-1} \frac{22 \text{ km}}{39 \text{ km}} = 29°$$

Since $+x$ is east and $-y$ is south, the direction of the displacement is 29° south of east. The magnitude and direction of the resultant found algebraically agree with the resultant found graphically in Fig. 4.8a.

Discussion   Note that A_x was found using the sine function, while C_x was found using the cosine. The x-component (or the y-component) of the vector can be related to *either* the sine or the cosine, depending on which angle in the triangle is used.

When finding the components of a vector, it is crucial to write them with the correct algebraic signs. Otherwise, when the components are added to find the components of the resultant, the resultant vector found will be wrong.

Practice Problem 4.2   Rotating the coordinate axes

Find the x- and y-components of $\vec{A}$, $\vec{B}$, and $\vec{C}$ if the x-axis points south and the y-axis points east.

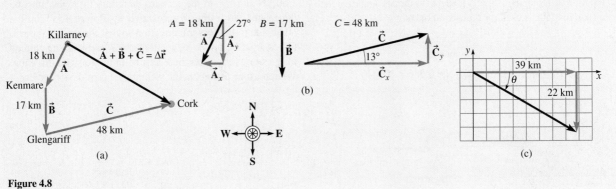

(a)   (b)   (c)

Figure 4.8
Addition of displacement vectors using components

Tips Even though the algebraic method is the preferred way to add vectors, the graphical method is an important first step. Making a rough sketch of vector addition has two important benefits.

1. Drawing the vectors makes it much easier to get the angles and the signs of the components right.

2. The graphical addition serves as a check on the answer. Even a rough sketch of the vectors $\vec{A}$, $\vec{B}$, and $\vec{C}$ being added graphically, without carefully measuring their lengths or the angles, would give an approximate idea of the magnitude and direction of the resultant, which can be used to check the algebraic answer. Graphical addition gives you a mental picture of what is going on and an intuitive feel for the algebraic calculations.

Example 4.3

Exercise Is Good for You

Suppose you are standing on the floor doing your daily exercises. For one exercise, you lift your arms up and out until they are horizontal. In this position, assume that the deltoid muscle exerts a force of 270 N at an angle of 15° above the horizontal on the humerus (Fig. 4.9a). What are the vertical and horizontal components of this force?

Strategy   This problem gives the magnitude and direction of a force and asks for the components of the force. Let the *x*-axis be horizontal and the *y*-axis be vertical. To find the components, we draw a right triangle with the force vector as the hypotenuse and the sides parallel to the *x*- and *y*-axes.

Solution   Figure 4.9b shows a triangle that can be used to find the components. With the choice of *x*- and *y*-axes shown, the *x*-component is negative; the +*x*-axis points to the right while the *x*-component of the force is to the left. The side of the triangle along the *x*-axis is adjacent to the 15° angle, so the cosine function is used for the *x*-component.

$$\cos 15° = \frac{\text{adjacent}}{\text{hypotenuse}} = \frac{|F_x|}{F}$$

$$F_x = -F \cos 15° = -270 \text{ N} \times 0.966 = -260 \text{ N}$$

The *y*-component is up, along the +*y*-axis. The sine function is used for the *y*-component, since the side representing F_y is opposite the 15° angle.

$$\sin 15° = \frac{\text{opposite}}{\text{hypotenuse}} = \frac{|F_y|}{F}$$

$$F_y = +F \sin 15° = 270 \text{ N} \times 0.259 = 70 \text{ N}$$

Discussion   The vertical component of the muscle force supports the arm against the downward force of gravity. The horizontal component of the force stabilizes the joint by pulling the humerus in toward the scapula. The horizontal component is much larger than the vertical component because the force makes a small angle with the horizontal.

Practice Problem 4.3   Adding forces algebraically

In Practice Problem 4.1(b), suppose Sam pulls with a force of 1.50 kN at 10.0° to the forward direction. (a) What must be the magnitude of the force exerted by Bob at 15.0° on the other side of the forward direction so that the sum of the two forces is in the forward direction? (b) What is the magnitude of the sum? [*Hint:* Let the *x*-axis be in the forward direction. What must be true about the *y*-components of the two forces?]

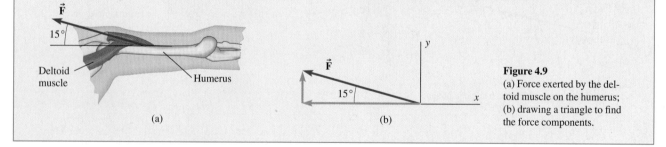

Figure 4.9
(a) Force exerted by the deltoid muscle on the humerus; (b) drawing a triangle to find the force components.

4.3   EQUILIBRIUM

According to Newton's second law, the acceleration of an object is determined by the net force acting on it:

$$\Sigma \vec{F} = m\vec{a}$$

This is a vector equation that can be rewritten in terms of perpendicular components:

$$\Sigma F_x = ma_x \quad \text{and} \quad \Sigma F_y = ma_y$$

When an object is in equilibrium (either at rest or in motion at constant velocity), the net force acting on it is zero. Since the object's acceleration is zero, both of its components must be zero.

For an object in equilibrium,

$$\Sigma F_x = ma_x = 0 \quad \text{and} \quad \Sigma F_y = ma_y = 0 \qquad (4\text{-}1)$$

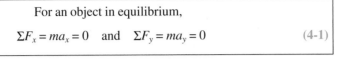

Example 4.4

Archery Practice

Figure 4.10a shows the bowstring of a bow and arrow just before it is released. The archer is pulling back on the midpoint of the bowstring with a horizontal force of 162 N. What is the tension in the bowstring?

Strategy   The bowstring, arrow, bow, and the archer's hand are all in equilibrium. We must choose some object or system, draw a free-body diagram, and apply Newton's second law. The only object about which we have enough information is the bowstring. We know the force exerted on the string by the archer's fingers. The bow exerts two forces on the string—one at each end; the magnitude of each is equal to the tension in the string. (At either end, the magnitude of the force with which the string pulls on the bow is the tension in the string. From Newton's third law, the bow pulls on the string with a force equal in magnitude and opposite in direction.) We neglect the weight of the string, since it is small compared to the other forces on the bowstring.

Solution   Figure 4.10b is a free-body diagram for the bowstring. The weight of the string is negligibly small compared to the other forces, so it is omitted. To find the components of the forces exerted by the bow, we draw a triangle as in Fig. 4.10c. From the measurements given, we can find the angle θ.

$$\sin \theta = \frac{\text{opposite}}{\text{hypotenuse}} = \frac{35 \text{ cm}}{72 \text{ cm}} = 0.486$$

$$\theta = \sin^{-1} 0.486 \approx 29°$$

The x-component of the force exerted on the upper end of the string is

$$F_x = -T \sin \theta$$

The x-component of the force exerted on the lower end of the string is the same. Therefore,

$$\Sigma F_x = -2T \sin \theta + F_a = 0$$

where F_a is the magnitude of the force applied by the archer. Solving for T,

$$T = \frac{F_a}{2 \sin \theta}$$

Substituting the value of $\sin \theta$ from the ratio of lengths (35 cm/72 cm = 0.486),

$$T = \frac{162 \text{ N}}{2 \times 0.486} = 170 \text{ N}$$

Discussion   In this problem, only the x-components of the forces had to be used. The y-components also add to zero. At the upper end of the string, the y-component of the force exerted by the bow is $+T \cos \theta$ while at the lower end it is $-T \cos \theta$. Therefore, $\Sigma F_y = 0$.

The expression $T = F_a/(2 \sin \theta)$ can be evaluated for limiting values of θ to make sure that the expression is correct. As θ approaches 90°, the tension approaches $F_a/(2 \sin 90°) = F_a/2$. That is correct because the archer would be pulling to the right with a force F_a, while each side of the bowstring would pull to the left with a force of magnitude T. For equilibrium, $F_a = 2T$ or $T = F_a/2$.

As θ gets smaller, $\sin \theta$ decreases and the tension increases (for a fixed value of F_a). That agrees with our intuition. The larger the tension, the smaller the angle the string needs to make in order to supply the necessary horizontal force.

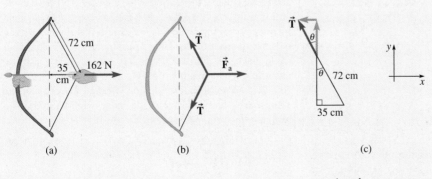

Figure 4.10
(a) The force applied to the bowstring by an archer. (b) Free-body diagram for a point on the bowstring with the forces labeled. The magnitudes of the two forces exerted by the bow are equal to the tension in the string T. (c) The angle θ is used to find the x- and y-components of the forces exerted at each end of the bowstring.

(a)   (b)   (c)

continued on next page

Example 4.4 *continued*

Practice Problem 4.4   Tightrope practice

Jorge decides to rig up a tightrope in the backyard so that his children can develop a good sense of balance (Fig. 4.11). For safety reasons he positions a horizontal cable only 0.60 m above the ground. If the 6.00 m long cable sags by 0.12 m from its taut horizontal position when Denisha (weight 250 N) is standing on the middle of it, what is the force on the eye-bolt due to the cable?

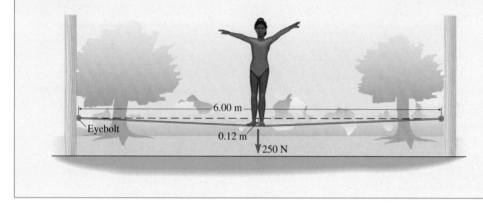

Figure 4.11
Tightrope for balancing practice

Example 4.5

Traction on a Foot

Figure 4.12 shows a traction apparatus designed to apply a force to stretch the leg in a controlled fashion. What is the force exerted on the foot by this traction apparatus?

Strategy   Assuming ideal pulleys, the tension of the cord is the same everywhere along the cord ($T = 22.0$ N). The force exerted on the foot by the attached pulley is equal in magnitude and opposite in direction to the force that the foot exerts on the pulley according to Newton's third law. Since this pulley is in equilibrium, the net force acting on it is zero.

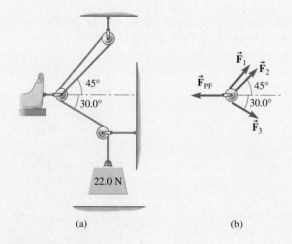

(a)   (b)

Figure 4.12
(a) A foot in traction; (b) forces acting on the pulley.

Solution   Four forces act on the pulley: three forces due to the cord and one due to the patient's foot. (The pulley's weight is negligible.) Then

$$\vec{\mathbf{F}}_{PF} + \vec{\mathbf{F}}_1 + \vec{\mathbf{F}}_2 + \vec{\mathbf{F}}_3 = 0$$

where $\vec{\mathbf{F}}_{PF}$ is the force exerted on the pulley by the foot and $\vec{\mathbf{F}}_1$, $\vec{\mathbf{F}}_2$, and $\vec{\mathbf{F}}_3$ are the forces on the pulley due to the three cords. Solving for $\vec{\mathbf{F}}_{PF}$,

$$\vec{\mathbf{F}}_{PF} = -(\vec{\mathbf{F}}_1 + \vec{\mathbf{F}}_2 + \vec{\mathbf{F}}_3)$$

The problem asks for the force exerted on the foot by the pulley ($\vec{\mathbf{F}}_{FP}$). From Newton's third law,

$$\vec{\mathbf{F}}_{FP} = -\vec{\mathbf{F}}_{PF} = \vec{\mathbf{F}}_1 + \vec{\mathbf{F}}_2 + \vec{\mathbf{F}}_3$$

This equation confirms what intuition tells us: the resultant force exerted on the foot by the apparatus is the sum of the three forces due to the cord.

Figure 4.13a shows the graphical addition of the forces exerted on the pulley by the cord. The magnitudes of $\vec{\mathbf{F}}_1$, $\vec{\mathbf{F}}_2$, and $\vec{\mathbf{F}}_3$ are equal to the tension in the cord (22.0 N). Two cords pull upward at a 45° angle; they are represented by a single force vector of magnitude 44.0 N. From this sketch, we can tell that the sum of the three forces is at a relatively small angle above the horizontal (roughly half of 45°) and has a magnitude a bit larger than 44.0 N.

To find an algebraic solution, we find the components along the x- and y-axes and add them (Fig. 4.13b-d). The resultant force component along the x-axis is

$$\Sigma F_x = (44.0\text{ N})\cos 45° + (22.0\text{ N})\cos 30.0° = 31.11\text{ N} + 19.05\text{ N}$$
$$= 50.16\text{ N}$$

continued on next page

Example 4.5 *continued*

and along the *y*-axis,

$$\Sigma F_y = 44.0\ \text{N} \sin 45.0° + (-22.0\ \text{N} \sin 30.0°) = 31.11\ \text{N} - 11.0\ \text{N}$$

$$= 20.11\ \text{N}$$

The resultant force is found from the Pythagorean theorem (Fig. 4.13e):

$$F = \sqrt{(20.11\ \text{N})^2 + (50.16\ \text{N})^2} = 54.0\ \text{N}$$

and the direction of the force is found from

$$\tan \theta = \frac{\text{opposite}}{\text{adjacent}} = \left| \frac{F_y}{F_x} \right| = \frac{20.11\ \text{N}}{50.16\ \text{N}} = 0.4009$$

$$\theta = \tan^{-1} 0.4009 = 21.8°$$

The force exerted on the foot is 54.0 N at an angle 21.8° above the horizontal.

Discussion The result of the algebraic addition is what we expect from doing the graphical addition: the total force

(54.0 N) is somewhat larger than 44.0 N and is at an angle just about half of 45° above the horizontal.

A shortcut to the beginning of the solution would have been to consider the foot and the attached pulley as a single system. Then we immediately see that the total force acting on the foot–pulley system due to the three cords is the quantity to be found.

Practice Problem 4.5 Effect of changing the pulley angles

The pulleys are moved so that the two cords above the foot make an angle of 30.0° with the horizontal and the cord below the foot makes an angle of 60.0° with the horizontal. (a) What are the components of the traction force on the foot? (b) What is the magnitude of the traction force? (c) At what angle with the horizontal does the traction force act?

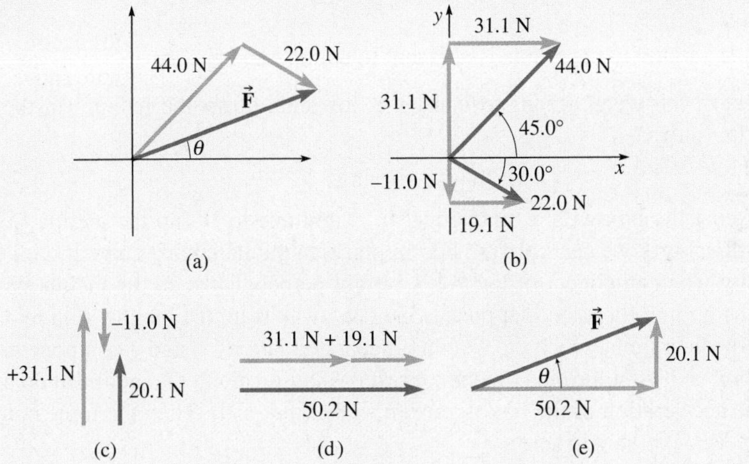

Figure 4.13
(a) Graphical sum of the forces on the pulley due to the cord, (b) finding the *x*- and *y*-components of the forces due to the cord, (c) adding the *y*-components, (d) adding the *x*-components, and (e) finding the resultant from its components.

Equilibrium on an Inclined Plane

Making The Connection:
an inclined plane

Suppose we wish to move a large box up to a loading dock platform. If we pick the box up, we have to support its entire weight. If we slide it along an incline, we can move it up to the same loading dock by exerting a force smaller than the box's weight. The inclined plane enables us to move an object up an incline by using less force than would be needed if we just picked up the object and lifted it. Part of the weight of the object is supported by the inclined plane itself. The inclined plane is one example of a *simple machine* that reduces the magnitude of the applied force needed for some task.

Figure 4.14a shows the three forces acting on a box being pushed up a frictionless incline. $\vec{F}_a$ represents the applied force. If we choose the *x*- and *y*-axes to be horizontal and vertical respectively, then two of the three forces have both *x*- and *y*-components. On the other hand, if we choose the *x*-axis parallel to the incline and the *y*-axis perpendicular to it, then only one of the three forces has both *x*- and *y*-components (the weight). In an equilibrium problem, choose *x*- and *y*-axes so that the fewest number of force vectors have both *x*- and *y*-components.

With axes chosen, the weight of the box is then resolved into two perpendicular components. In order to find the *x*- and *y*-components of $\vec{W}$, we must determine the

Tips

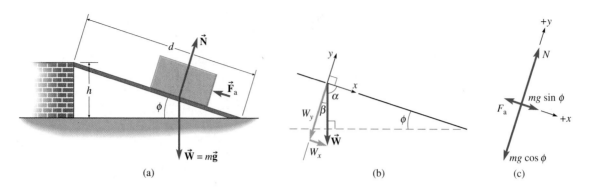

(a)   (b)   (c)

Figure 4.14 (a) A box of mass m pushed up an incline, (b) resolving the weight into components parallel to and perpendicular to the incline, and (c) free-body diagram for the box on the inclined plane

angle that $\vec{W}$ makes with one of the axes. In Fig. 4.14b, $\alpha = 90° - \phi$, since the three angles inside the large right triangle add up to $180°$. The x- and y-axes are perpendicular, so $\alpha + \beta = 90°$ and, therefore, the angle between $\vec{W}$ and the $-y$-axis is $\beta = \phi$.

The y-component of the weight is perpendicular to the surface of the incline. From the triangle in Fig. 4.14b,

$$W_y = -mg \cos \phi$$

since the side parallel to the y-axis is adjacent to angle ϕ and $\cos \phi = \dfrac{\text{adjacent}}{\text{hypotenuse}}$. The x-component of the weight tends to make the box slide down the incline (in the $+x$-direction). Its value is

$$W_x = +mg \sin \phi$$

By pushing the box with a force equal in magnitude to W_x up the incline (in the negative x-direction), we can hold the box in place on the incline or move it with constant velocity. The component of the box's weight perpendicular to the incline is supported by the normal force, $\vec{N}$, that pushes the box away from the incline. Figure 4.14c is a free-body diagram in which the weight is separated into its x- and y-components.

If the box is in equilibrium, whether at rest or moving along the incline at constant velocity, the acceleration of the box is zero ($a_x = 0$ and $a_y = 0$). Then the force components along each axis sum to zero:

$$\Sigma F_x = (-F_a) + mg \sin \phi = 0$$

and

$$\Sigma F_y = N + (-mg \cos \phi) = 0$$

The normal force is *not* equal in magnitude to the weight and it does not point straight up. If the applied force is of magnitude $mg \sin \phi$, we can push the box up the incline at constant velocity. If friction acts on the box, we must push with a force greater than $mg \sin \phi$ to move the box up the incline at constant velocity.

Example 4.6

Pushing a Safe Up an Incline

A new safe is being delivered to the Corner Book Store. It is to be placed in the wall at a height of 1.5 m above the floor. The delivery people have a portable ramp that they plan to use to help them push the safe up and into position.

The mass of the safe is 510 kg, the coefficient of static friction along the incline is $\mu_s = 0.42$, and the coefficient of kinetic friction along the incline is $\mu_k = 0.33$. The ramp forms an angle $\theta = 15°$ above the horizontal. (a) Find the minimum applied force parallel to the incline needed to start the safe moving up

continued on next page

Example 4.6 *continued*

the incline. (b) If the movers continue to push with this same force, what is the safe's acceleration? (c) To move the safe up at a constant speed, with what force must the movers push?

Strategy   First we draw a free-body diagram for the safe. Before resolving the forces into components, we must choose *x*- and *y*-axes. The coefficient of friction relates the frictional force to the normal force. To use the coefficient of friction, we have to resolve the contact force on the safe due to the incline into components *parallel and perpendicular to the incline*— friction and the normal force, respectively—rather than into horizontal and vertical components. Therefore, we choose *x*- and *y*-axes parallel and perpendicular to the incline so that friction is along the *x*-axis and the normal force is along the *y*-axis. The applied force and the safe's acceleration are parallel to the incline as well. Using this choice of axes, the *only* vector that is not parallel to one of the axes is the weight. The problem would be much harder to solve using a horizontal *x*-axis and a vertical *y*-axis since only one force would be parallel to the axes; all the other forces and the acceleration would have to be resolved into horizontal and vertical components.

 When the direction of the acceleration is known in a nonequilibrium problem, it is usually easiest to choose axes parallel and perpendicular to the acceleration.

Solution   First we draw a diagram to show the forces acting (Fig. 4.15a) and sketch in the *x*- and *y*-axes. The weight $\vec{\mathbf{W}}$ must be resolved into two components: the component perpendicular to the incline, $W_y = -mg \cos \theta$, and the component along the incline, $W_x = -mg \sin \theta$ (Fig. 4.15b). Then

we draw the FBD with the weight replaced by its components (Fig. 4.15c).

(a) Suppose that the safe is initially at rest. For the safe to *begin* moving up the ramp, a nonzero net force in the +*x*-direction must act on it. As the movers start to push, $F_{applied}$ gets larger and the force of static friction gets larger to "try" to keep the safe from sliding. Eventually, at some value of $F_{applied}$, static friction reaches its maximum possible value $\mu_s N$. If the movers push any harder than that, the safe starts to slide.

To *start* the box moving up the incline, it must have an acceleration directed up the incline:

$$\Sigma F_x = F_{applied} + (-mg \sin \theta) + (-f_s) = ma_x > 0$$

The maximum force of static friction is

$$(f_s)_{max} = \mu_s N$$

The normal force is *not* equal in magnitude to the weight of the safe. To find the normal force, sum the *y*-components of the forces:

$$\Sigma F_y = N + (-mg \cos \theta) = ma_y = 0$$

The safe slides along the surface of the incline; it never moves in the ±*y*-direction. Since v_y never changes (it is always zero), $a_y = 0$. Then

$$N = mg \cos \theta$$

The normal force is less than the weight since $\cos \theta < 1$.

In order for a_x to be positive,

$$F_{applied} > mg \sin \theta + f_s = mg \sin 15° + \mu_s mg \cos 15°$$

$$F_{applied} > (510 \text{ kg} \times 9.8 \text{ m/s}^2 \times 0.2588) + (0.42 \times 510 \text{ kg} \times 9.8 \text{ m/s}^2 \times 0.9659)$$

$$F_{applied} > 1293 \text{ N} + 2028 \text{ N} = 3321 \text{ N}$$

Rounding to two significant figures, we conclude that an applied force that exceeds 3300 N starts the box moving up the incline.

(b) Once the safe starts moving, the static friction force stops acting against the motion and the kinetic friction force replaces it. Since the kinetic friction force is less than the static friction force, the movers are pushing with more force than is necessary to maintain motion at constant speed. The net force along the incline, ΣF_x, determines the acceleration of the safe up the incline. If the movers continue to push with a force of 3321 N,

$$\Sigma F_x = F_{applied} + (-mg \sin 15°) + (-\mu_k mg \cos 15°) = ma_x$$

$$a_x = \frac{F_{applied} - mg \sin 15° - \mu_k mg \cos 15°}{m}$$

$$= \frac{3321 \text{ N} - (510 \text{ kg} \times 9.8 \text{ m/s}^2 \times 0.2588) - (0.33 \times 510 \text{ kg} \times 9.8 \text{ m/s}^2 \times 0.9659)}{510 \text{ kg}}$$

$$= \frac{3321 \text{ N} - 1293 \text{ N} - 1593 \text{ N}}{510 \text{ kg}}$$

$$= \frac{435 \text{ N}}{510 \text{ kg}} = 0.9 \text{ m/s}^2$$

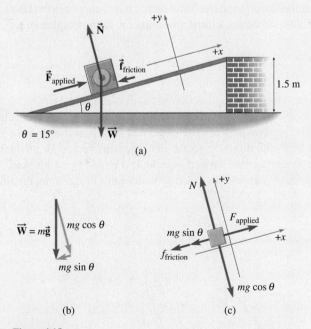

$\theta = 15°$

(a)

(b)   (c)

Figure 4.15
(a) Forces acting on the safe as it is moved up the incline, (b) resolving the weight into *x*- and *y*-components, and (c) a free-body diagram in which the weight is replaced with its *x*- and *y*-components.

continued on next page

Example 4.6 *continued*

(c) Once the safe starts moving, the movers need only push hard enough to make the net force on the safe equal to zero if they want the safe to slide at constant velocity.

$$\Sigma F_x = F_{\text{applied}} + (-mg \sin 15°) + (-\mu_k mg \cos 15°) = 0$$

$$F_{\text{applied}} = mg \sin 15° + \mu_k mg \cos 15°$$

$$= 1293 \text{ N} + 1593 \text{ N} = 2900 \text{ N}$$

Discussion   The force required to lift the safe straight up at constant speed would be equal to the weight.

$$F_{\text{lift}} = mg = (510 \text{ kg} \times 9.8 \text{ m/s}^2) = 5000 \text{ N}$$

The incline enables us to use a smaller force, but we must push over a longer distance.

Practice Problem 4.6   Smoothing the infield dirt

During the seventh-inning stretch of a baseball game, groundskeepers drag mats across the infield dirt to smooth it for the remainder of the game. A groundskeeper is pulling a mat at a constant velocity across the dirt surface of the infield at Fenway Park. If the force pulling the mat is 120 N at an angle of 22° above the ground, find the frictional force between the dirt and the mat and the weight of the mat assuming a coefficient of kinetic friction of 0.60.

Physics at Home

To estimate the coefficient of static friction between a penny and the cover of your physics book, place the penny on the book and slowly lift the cover. Note the angle of the cover when the penny starts to slide. Explain how you can use this angle to find the coefficient of static friction. Can you devise an experiment to find the coefficient of kinetic friction?

4.4   VELOCITY AND ACCELERATION

The definitions of average velocity, instantaneous velocity, average acceleration, and instantaneous acceleration in Chapter 3 still apply when the motion is not in a straight line. Suppose we want to know the instantaneous velocity of a race car at point P as it goes around a curved section of the Indy 500 (Fig. 4.16a). At a slightly later time the race car is at point Q. Let $\vec{\mathbf{r}}_0$ be the position of the car at P and $\vec{\mathbf{r}}_f$ be the position at point Q.

The displacement $\Delta\vec{\mathbf{r}}$ is $\vec{\mathbf{r}}_f - \vec{\mathbf{r}}_0$ (Fig. 4.16b). To subtract $\vec{\mathbf{r}}_0$ from $\vec{\mathbf{r}}_f$, the two vectors are drawn with their tails at the same point; the difference is the vector from the tip of $\vec{\mathbf{r}}_0$ to the tip of $\vec{\mathbf{r}}_f$. The average velocity for this time interval is the displacement $\Delta\vec{\mathbf{r}}$ divided by the time interval:

$$\vec{\mathbf{v}}_{\text{av}} = \frac{\vec{\mathbf{r}}_f - \vec{\mathbf{r}}_0}{t_f - t_0} = \frac{\Delta\vec{\mathbf{r}}}{\Delta t} \tag{3-2}$$

The direction of the average velocity is the direction of the displacement $\Delta\vec{\mathbf{r}}$.

The instantaneous velocity at P is the limit of the average velocity as Δt approaches zero. As we shorten the time interval between the initial and final positions by moving point Q closer and closer to P, the direction of the displacement vector $\Delta\vec{\mathbf{r}}$ gradually changes, approaching the tangent to the curved path at P (Fig. 4.16c). Expressed in mathematical terminology, the instantaneous velocity is the limit of $\Delta\vec{\mathbf{r}}/\Delta t$ as the time interval approaches zero:

$$\vec{\mathbf{v}} = \lim_{\Delta t \to 0} \frac{\Delta\vec{\mathbf{r}}}{\Delta t} \tag{3-4}$$

Figure 4.16 (a) Position vectors for two points on the curve. (b) The displacement $\Delta\vec{\mathbf{r}}$ from point P to point Q. (c) As time interval is decreased, the final point moves closer and closer to P; the direction of the displacement $\Delta\vec{\mathbf{r}}$ approaches the tangent to the curve at P. (d) Instantaneous velocity can be resolved into components along perpendicular axes.

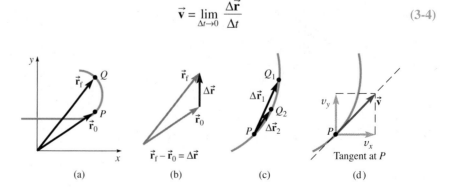

With this definition, the instantaneous velocity at P becomes tangent to the curve at P.

Since velocity is a vector quantity, it can be resolved into components (Fig. 4.16d). The vector equations defining average and instantaneous velocity can be rewritten in terms of x- and y-components:

$$v_{\mathrm{av},x} = \frac{\Delta x}{\Delta t} \quad \text{and} \quad v_{\mathrm{av},y} = \frac{\Delta y}{\Delta t}$$

$$v_x = \lim_{\Delta t \to 0} \frac{\Delta x}{\Delta t} \quad \text{and} \quad v_y = \lim_{\Delta t \to 0} \frac{\Delta y}{\Delta t}$$

Similarly, the average acceleration $\vec{\mathbf{a}}_{\mathrm{av}}$ is the change in velocity divided by the elapsed time:

$$\vec{\mathbf{a}}_{\mathrm{av}} = \frac{\vec{\mathbf{v}}_{\mathrm{f}} - \vec{\mathbf{v}}_0}{t_{\mathrm{f}} - t_0} = \frac{\Delta \vec{\mathbf{v}}}{\Delta t} \tag{3-6}$$

This vector equation is equivalent to two component equations:

$$a_{\mathrm{av},x} = \frac{\Delta v_x}{\Delta t} \quad \text{and} \quad a_{\mathrm{av},y} = \frac{\Delta v_y}{\Delta t}$$

Instantaneous acceleration is the limit of the average acceleration as the time interval approaches zero and point Q approaches point P:

$$\vec{\mathbf{a}} = \lim_{\Delta t \to 0} \frac{\Delta \vec{\mathbf{v}}}{\Delta t} \tag{3-7}$$

or

$$a_x = \lim_{\Delta t \to 0} \frac{\Delta v_x}{\Delta t} \quad \text{and} \quad a_y = \lim_{\Delta t \to 0} \frac{\Delta v_y}{\Delta t}$$

In straight-line motion the acceleration is always either parallel or antiparallel to the velocity. For more general motion in two dimensions, the acceleration can be in *any* direction, as illustrated in Example 4.7.

Example 4.7

Skating Uphill

An inline skater is traveling on a level road with a speed of 8.94 m/s. Then she comes to a long hill at a 15.0° angle of incline. 120.0 s later she is still climbing the hill while her speed has decreased to 7.15 m/s. (a) What is the change in her velocity? (b) What is her average acceleration during the 120.0-s time interval?

Strategy The change in velocity is *not* 8.94 m/s − 7.15 m/s = 1.79 m/s. That is the change in *speed*. The change in velocity is found by subtracting the initial velocity *vector* from the final velocity *vector*. After first making a graphical sketch, we use the component method. The average acceleration is the change in velocity divided by the elapsed time.

Solution (a) Figure 4.17a shows the initial and final velocity vectors and the slope of the hill. The initial velocity is horizontal as she skates on level ground. The final velocity is 15.0° above the horizontal. To subtract the two velocity vectors graphically, we place the tails of the vectors together. The change in velocity, $\Delta \vec{\mathbf{v}}$, is found by drawing a vector arrow from the tip of $\vec{\mathbf{v}}_0$ to the tip of $\vec{\mathbf{v}}_{\mathrm{f}}$. Judging by the graphical subtraction in Fig. 4.17b, the change in velocity is roughly at a 45° angle above the −x-axis. Its magnitude is smaller than the magnitudes of the initial and final velocity vectors—something like 2 or 3 m/s.

The components v_{fx} and v_{fy} can be found from a right triangle (Fig. 4.17c):

$$v_{\mathrm{fx}} = v_{\mathrm{f}} \cos \theta = 7.15 \text{ m/s} \times 0.9659 = 6.91 \text{ m/s}$$

$$v_{\mathrm{fy}} = v_{\mathrm{f}} \sin \theta = 7.15 \text{ m/s} \times 0.2588 = 1.85 \text{ m/s}$$

continued on next page

Example 4.7 *continued*

Since v_0 has only an x-component,

$$v_{0y} = 0 \quad \text{and} \quad v_{0x} = v_0 = 8.94 \text{ m/s}$$

Now we subtract the components to find the components of $\Delta\vec{v}$:

$$\Delta v_x = v_{fx} - v_{0x} = (6.91 - 8.94) \text{ m/s} = -2.03 \text{ m/s}$$

and

$$\Delta v_y = v_{fy} - v_{0y} = (1.85 - 0) \text{ m/s} = +1.85 \text{ m/s}$$

To find the magnitude of $\Delta\vec{v}$, we apply the Pythagorean theorem (Fig. 4.17d):

$$(|\Delta\vec{v}|)^2 = (\Delta v_x)^2 + (\Delta v_y)^2 = (-2.03 \text{ m/s})^2 + (1.85 \text{ m/s})^2 = 7.54 \text{ (m/s)}^2$$

$$|\Delta\vec{v}| = 2.75 \text{ m/s}$$

The angle is found from

$$\tan\phi = \frac{\text{opposite}}{\text{adjacent}} = \left|\frac{\Delta v_y}{\Delta v_x}\right| = \frac{1.85 \text{ m/s}}{2.03 \text{ m/s}} = 0.911$$

$$\phi = \tan^{-1} 0.911 = 42.3°$$

The direction of the change in velocity $\Delta\vec{v}$ is 42.3° above the negative x-axis.

(b) The magnitude of the average acceleration is

$$|\vec{a}_{av}| = \frac{|\Delta\vec{v}|}{\Delta t} = \frac{2.75 \text{ m/s}}{120.0 \text{ s}} = 0.0229 \text{ m/s}^2$$

The direction of the average acceleration is the same as the direction of $\Delta\vec{v}$: 42.3° above the negative x-axis.

Discussion   Checking back with the graphical subtraction in Fig. 4.17b, the magnitude of $\Delta\vec{v}$ appears to be roughly $\frac{1}{4}$ to $\frac{1}{3}$ the magnitude of $\vec{v}_0$. Since $\frac{1}{4} \times 8.94$ m/s = 2.24 m/s and $\frac{1}{3} \times 8.94$ m/s = 2.98 m/s, the answer of 2.75 m/s is reasonable.

Figure 4.17b also shows the direction of $\Delta\vec{v}$ to be roughly midway between the +y- and −x-axes. We found the direction of $\Delta\vec{v}$ to be 42.3° above the −x-axis and therefore, 47.7° from the +y-axis. So the direction we calculated is also reasonable based on the graphical subtraction.

Practice Problem 4.7   Change in sailboat velocity

A C&C 30 sailboat is sailing at 12.0 knots heading directly east across the harbor. When a gust of wind comes up, the boat changes its heading to 11.0° north of east and its speed increases to 14.0 knots. What is the magnitude and direction of the change in velocity of the sailboat? [A boat's speed is customarily expressed in knots, which means nautical miles per hour. A nautical mile (6076 ft) is a little longer than a statute mile (5280 ft).]

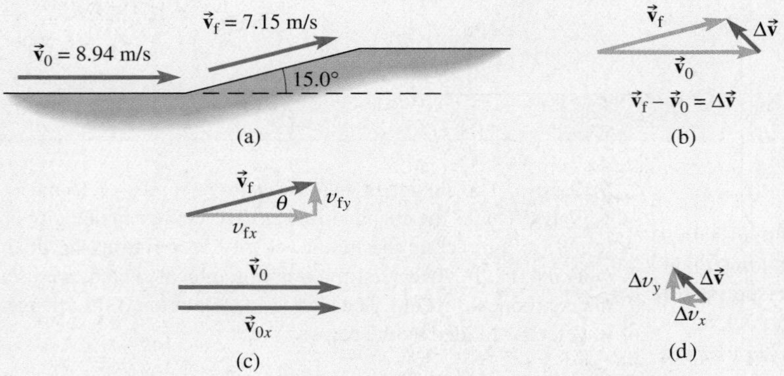

Figure 4.17
(a) Change in velocity as the skater slows going uphill, (b) graphical subtraction of velocity vectors, (c) initial and final velocity vectors resolved into components, and (d) reconstruction of $\Delta\vec{v}$ from its components.

4.5   MOTION OF PROJECTILES

If an object moves in the xy-plane with constant acceleration, then both a_x and a_y are constant. By looking separately at the motion along two perpendicular axes, the y-direction and the x-direction, each component becomes a one-dimensional problem, which we already know how to solve. We can apply any of the constant-acceleration kinematic relations from Chapter 3 separately to the x-components and to the y-components. *It is generally easiest to choose the axes so that the acceleration has only one nonzero component.* Suppose we choose the axes so that the acceleration is in the positive or negative y-direction. Then $a_x = 0$ and v_x is constant.

As in Chapter 3, it is often convenient to write the kinematic relations for constant acceleration assuming that $t_0 = 0$.

Kinematic Equations	
for constant acceleration along the y-axis and $t_0 = 0$	

$a_x = 0$:

$$v_x = v_{0x} \qquad\qquad\qquad (4\text{-}2)$$

$$x - x_0 = v_{0x}t \qquad\qquad\qquad (4\text{-}3)$$

$a_y = $ constant:

$$v_y = v_{0y} + a_y t \qquad\qquad\qquad (4\text{-}4)$$

$$\Delta y = v_{\text{av},y}t \qquad\qquad\qquad (4\text{-}5)$$

$$v_{\text{av},y} = \tfrac{1}{2}(v_{0y} + v_y) \qquad\qquad\qquad (4\text{-}6)$$

$$y - y_0 = v_{0y}t + \tfrac{1}{2}a_y t^2 \qquad\qquad\qquad (4\text{-}7)$$

$$v_y^2 - v_{0y}^2 = 2a_y \Delta y \qquad\qquad\qquad (4\text{-}8)$$

Note that there is no mixing of components in Eqs. (4-2) through (4-8). Each equation pertains either to the x-components or to the y-components; none contains the x-component of one vector quantity and the y-component of another. The only quantity that appears in both x- and y-component equations is the time—a scalar.

An object in free fall near the Earth's surface has a constant acceleration. As long as air resistance is negligible, the constant downward pull of gravity gives the object a constant downward acceleration of 9.8 m/s². In Chapter 3 we considered objects in free fall, but only when they moved straight up or straight down. An object in free fall that has a nonzero horizontal velocity component is called a **projectile**; its motion takes place in a vertical plane. For such an object, Eqs. (4-2) through (4-8) describe projectile motion with

$$a_y = -g$$

assuming the y-axis points upward.

Suppose some medieval marauders are attacking a castle. They have a catapult that can propel large stones into the air to bombard the walls of the castle (Fig. 4.18). Picture a stone leaving the catapult with an initial velocity $\vec{v}_0$. The **angle of elevation** is the angle of the initial velocity above the horizontal. Once the stone is in the air, the only force acting on it is the downward gravitational force, provided that air resistance has a negligible effect on the motion. The **trajectory** (path) of the stone is shown in Fig. 4.19. The positive x-axis is chosen in the horizontal direction (to the right) and the positive y-axis is upward.

If the initial velocity is at an angle θ above the horizontal, then resolving it into components gives:

$$v_{0x} = v_0 \cos\theta$$

$$v_{0y} = v_0 \sin\theta$$

Since the gravitational force pulls downward, only the y-component of the velocity changes. Notice in Fig. 4.19 that the horizontal velocity component v_x never changes. With zero net force in the x-direction, $a_x = 0$; the stone moves along the horizontal axis at constant speed. The vertical velocity component v_y changes at a constant rate, exactly as if the stone were propelled straight up with an initial speed of v_{0y}. The initially positive v_y decreases until, at the top of flight, $v_y = 0$. Then the pull of gravity makes the projectile fall back downward. During the downward trip, velocity v_y is still changing at the same constant rate with which it changed on the way up and at the top of the path. The acceleration has the same constant value—magnitude *and* direction—for the entire path.

The displacement of the projectile at any instant is the vector sum of the displacements in the two mutually perpendicular directions. The motion of a projectile when air resistance is negligible is the superposition of horizontal motion with constant velocity and vertical

Tips

Figure 4.18   A medieval catapult

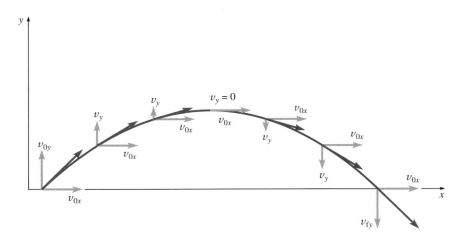

Figure 4.19 Trajectory of a projectile. The velocity vectors are drawn at equal time intervals along with the x- and y-components of $\vec{\mathbf{v}}$.

> The horizontal and vertical motions of a projectile can be treated separately; they are independent of each other.

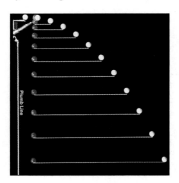
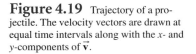

Figure 4.20 Independence of horizontal and vertical motion of a projectile in the absence of air resistance.

motion with constant acceleration. The vertical and horizontal motions each proceed independently, as if the other motion were not present. In the experiment of Fig. 4.20, one ball was dropped and, at the same instant, another was projected horizontally. The strobe photo shows snapshots of the two balls at closely spaced time intervals. The *vertical* motion of the two is identical; at every instant the two are at the same height. The fact that they have different horizontal motion does not affect their vertical motion. (This statement would *not* be true if air resistance were significant.)

Physics at Home

Take two golf balls and a friend to an open window or balcony on the second or third floor of a building. After making sure that no one is below, you drop one ball while your friend, *at the same time*, throws the other ball horizontally. Which ball hits the ground first?

Conceptual Example 4.8

Trajectory of a Projectile

The graph of an equation of the form

$$y = kx^2, \quad k = \text{a nonzero constant}$$

is a parabola. Show that the trajectory of a projectile is a parabola. [*Hint:* Choose the origin at the highest point of the trajectory and let $t = 0$ at that instant.]

Strategy and Solution   We start at the high point of the path and look at displacements from there. The horizontal displacement is proportional to the elapsed time, since the horizontal velocity is constant. The vertical displacement is the average vertical velocity component times the elapsed time. The average vertical velocity component is itself proportional to t since v_y changes at a constant rate. Therefore, the vertical displacement is proportional to t^2. Thus, the vertical displacement y is proportional to the square of the horizontal displacement x and $y = kx^2$, where k is a constant of proportionality. The path followed by a projectile in free fall is parabolic.

Discussion   The same conclusion can be drawn algebraically. If we choose the origin and $t = 0$ at the top of flight, then x_0, y_0, and v_{0y} are all zero. We can write $y = v_{\text{av},y}t$, but y is *not* proportional to t. From Eq. (4-6)

$$v_{\text{av},y} = \tfrac{1}{2}(v_{0y} + v_y)$$

Substituting $v_{0y} = 0$ and $v_y = v_{0y} + a_y t = at$ yields

$$v_{\text{av},y} = \tfrac{1}{2}a_y t$$

Then $y = v_{\text{av},y}t = \tfrac{1}{2}a_y t^2$. From $x = v_{0x}t$, we can substitute $t = x/v_{0x}$ into this expression for y:

$$y = \tfrac{1}{2}a_y\left(\frac{x}{v_{0x}}\right)^2$$

Thus, $y = kx^2$ where $k = a_y/(2v_{0x}^2)$ and $a_y = -g$.

Conceptual Practice Problem 4.8   Throwing stones

You stand at the edge of a cliff and throw stones horizontally into the river below. To double the horizontal displacement of a stone from the cliff to where it lands, by what factor must you increase the stone's initial speed? Neglect air resistance.

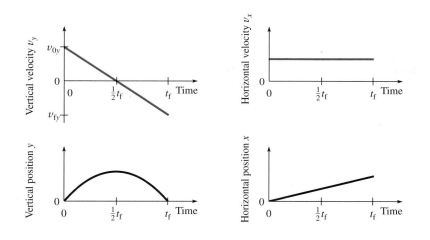

Figure 4.21 Projectile motion: separate vertical and horizontal quantities versus time.

Figure 4.21 shows graphs of the x- and y-components of the velocity and position of a projectile as functions of time. In this case, the projectile is launched above flat ground at $t = 0$ and returns to the same elevation at a later time t_f. The kinematic equations are valid only while the object is in free fall; once the projectile hits the ground, forces other than gravity act on it and its acceleration changes.

Note that the y-component graphs are symmetrical about the vertical line through the highest point in the trajectory. The y-component of velocity decreases linearly from its initial value; the slope of the line is $a_y = -g = -9.8$ m/s^2. When $v_y = 0$ the projectile is at the apex of its trajectory. Then v_y continues to decrease at the same rate and is now negative with its magnitude getting larger and larger. At t_f, when the projectile has returned to its original altitude, the y-component of the velocity has the same magnitude as at $t = 0$ but with the opposite sign ($v_y = -v_{0y}$).

The graph of y versus t indicates that the projectile moves upward, quickly at first and then gradually slowing, until it reaches the maximum height. The slope of the tangent to the y versus t graph at any particular moment of time is v_y at that instant. At the highest point of the y versus t graph, the tangent is horizontal and $v_y = 0$. After that gravity makes the projectile start to fall downward.

The horizontal velocity is constant because the projectile is not acted upon by any horizontal forces ($a_x = \Sigma F_x/m = 0$). Thus, the graph of v_x versus t is a horizontal line. The horizontal position x increases uniformly in time because the object is moving with a constant v_x.

Example 4.9

Attacking the Castle Walls

The catapult used by the marauders hurls a stone of mass 32.0 kg with a velocity of 50.0 m/s at an angle of elevation of 30.0° (Fig. 4.22). Assume that $g = 9.80$ m/s^2. (a) What is the maximum height reached by the stone? (b) What is its range (defined as the horizontal distance traveled when the stone returns to its original height)? (c) How long has the stone been in the air when it returns to its original height?

Strategy   The problem gives both the magnitude and direction of the initial velocity of the stone. Ignoring air resistance,

the stone has a constant downward acceleration once it has been launched—until it hits the ground or some obstacle. We choose the positive y-axis upward and the positive x-axis in the direction of horizontal motion of the stone (toward the castle). When the stone reaches its maximum height, the velocity component in the y-direction is zero, since the stone goes no higher. When the stone returns to its original height, $\Delta y = 0$ and $v_y = -v_{0y}$. The range can be found once the time of flight t_f is known—time is the quantity that connects the x-component equations to the y-component equations. Therefore, we solve part (c) before part (b). One way to find t_f is to find the time to

continued on next page

Example 4.9 *continued*

reach maximum height and then double it (see Fig. 4.21). (Other methods include finding the time when $\Delta y = 0$ (the stone returns to its original height) or the time when $v_y = -v_{0y}$.)

Solution (a) First we find the x- and y-components of the initial velocity for an angle of elevation $\theta = 30.0°$.

$$v_{0y} = v_0 \sin \theta$$

$$v_{0x} = v_0 \cos \theta$$

The maximum height is the vertical displacement Δy when $v_y = 0$. From Eq. (4-8),

$$v_y^2 - v_{0y}^2 = 2a_y \, \Delta y$$

Solving for Δy,

$$\Delta y = \frac{v_y^2 - v_{0y}^2}{2a_y}$$

Substituting $v_y = 0$ and $v_{0y} = v_0 \sin \theta$ yields

$$\Delta y = \frac{-(v_0 \sin \theta)^2}{2a_y}$$

$$= \frac{-[(50.0 \text{ m/s}) \sin 30.0°]^2}{2 \times (-9.80 \text{ m/s}^2)} = 31.9 \text{ m}$$

The maximum height of the projectile is 31.9 m above its launch height.

(c) The time of flight (t_f) is *twice* the time it takes the projectile to reach its maximum height. The time to reach the maximum height can be found from

$$v_y = v_{0y} + a_y t$$

Solving for t,

$$t = \frac{v_y - v_{0y}}{a_y} = \frac{-v_{0y}}{a_y}$$

since $v_y = 0$. The time of flight is

$$t_f = 2 \times \frac{-v_{0y}}{a_y} = 2 \times \frac{-(50.0 \text{ m/s}) \sin 30.0°}{-9.80 \text{ m/s}^2} = 5.10 \text{ s}$$

(b) The range is

$$\Delta x = v_{0x} t_f = (50.0 \text{ m/s}) \cos 30.0° \times 5.10 \text{ s} = 221 \text{ m}$$

Discussion Other kinematic relations can be used to check the answers. For example,

$$y - y_0 = v_{0y} t + \tfrac{1}{2} a_y t^2$$

can be used to check that $\Delta y = 31.9$ m at $t = \tfrac{1}{2} \times 5.10$ s and that $\Delta y = 0$ at $t = 5.10$ s. Here we check the first of these:

$$\Delta y = (50.0 \text{ m/s}) \sin 30.0° \times 2.55 \text{ s} + \tfrac{1}{2} \times (-9.80 \text{ m/s}^2) \times (2.55 \text{ s})^2$$

$$= 63.8 \text{ m} + (-31.9 \text{ m}) = 31.9 \text{ m}$$

which is correct. This is not an *independent* check, since this equation can be derived from the others, but it can reveal algebra or calculation errors.

Since we analyze the horizontal motion independently from the vertical motion, we start by resolving the initial velocity given into x- and y-components. Time is what connects the two.

Notice that we did not use the mass of the stone in the solution. The mass does not affect the maximum height, the time of flight, or the range, if we are given the initial velocity. Where mass does matter is in the ability of the catapult to accelerate the stone from rest to its launch velocity. The catapult (or, in other problems, a cannon, a human arm, or some other propelling machine) has an easier time imparting a particular velocity to a less massive projectile. A projectile that is too massive for the machine can only be given a small launch speed.

Practice Problem 4.9   Maximum height for arrows

Archers have joined in the attack on the castle and are shooting arrows over the walls. If the angle of elevation for an arrow is 45°, find an expression for the maximum height of the arrow in terms of v_0 and g. [*Hint:* Simplify the expression using $\sin 45° = \cos 45° = 1/\sqrt{2}$.]

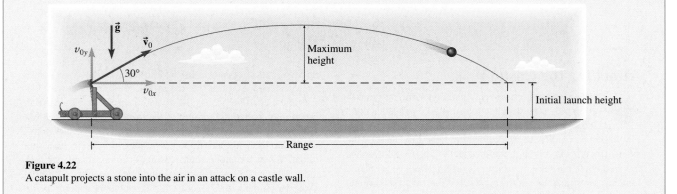

Figure 4.22
A catapult projects a stone into the air in an attack on a castle wall.

Conceptual Example 4.10

Monkey and Hunter

A thoughtless child aims and shoots an arrow, which fortunately has a rubber, suction-cup tip, straight at a monkey who is sitting in a tree and holding a coconut (Fig. 4.23). At the same instant the arrow leaves the bow, the monkey drops the coconut. Does the arrow hit the monkey?

Strategy and Solution If there were no gravitational field, the arrow would fly straight to the monkey and coconut (along the dashed line from the bow to the monkey on the branch in Fig. 4.23). Since gravity gives the dropped coconut and the released arrow the same constant acceleration downward, they each fall the same vertical distance below the positions they would have had with no gravity. The coconut falls as shown by the dashed red line; the distance fallen at 0.25-s intervals is marked along a vertical axis. At the same time, the arrow drops below the gray dashed line by the amounts marked along its indicated trajectory at 0.25-s intervals.

The arrow ends up hitting the coconut no matter what the initial velocity of the arrow. The higher the velocity of the arrow, the sooner they meet and the shorter the vertical distance that the coconut falls before being hit.

Discussion An experienced hunter would have aimed above the monkey on the branch to compensate for the amount his arrow would drop during the time of flight; he would have missed the dropping coconut but might have hit the monkey unless the monkey jumped down to retrieve the coconut.

Conceptual Practice Problem 4.10   Changes in position and velocity for consecutive arrows

An arrow is shot into the air. One second later, a second arrow is shot with the same initial velocity. While the two are both in the air, does the difference in their positions $(\vec{r}_2 - \vec{r}_1)$ stay constant or does it change with time? Does the difference in their velocities $(\vec{v}_2 - \vec{v}_1)$ stay constant or does it change with time?

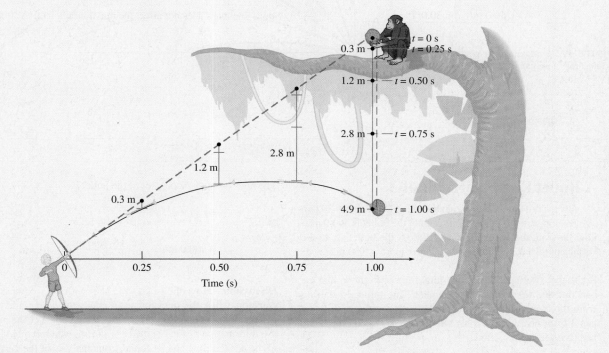

Figure 4.23
A monkey drops a coconut at the very instant an arrow is shot toward the monkey. In each quarter second, the coconut and arrow have fallen the same distance below where their positions would be if there were no gravity.

Physics at Home

On a warm day, take a garden hose and aim the nozzle so that the water streams upward at an angle above the horizontal. Set the nozzle for a fast, narrow stream for best effect. Once the water leaves the nozzle, it becomes a projectile subject only to the force of gravity (neglecting the small effect of air resistance). The continuous stream of water lets us see the parabolic path easily. Stand in one place and try aiming the nozzle at different angles of elevation to find an angle that gives the maximum range. Aim for a particular spot on the ground (at a distance less than the maximum range) and see if you can find two different angles of elevated nozzle position that allow the stream to hit the target spot (see Fig. 4.24).

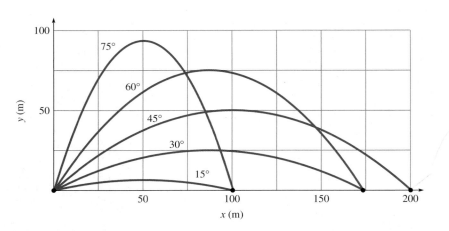

Figure 4.24 Maximum height versus range for various angles of elevation ($v_0 = 44.3$ m/s).

Example 4.11

A Bullet Fired Horizontally

A bullet is fired horizontally from the top of a cliff that is 20.0 m above a long lake. If the muzzle speed of the bullet is 500.0 m/s, how far from the bottom of the cliff does the bullet strike the surface of the lake? Ignore air resistance. Let $g = 9.80$ m/s^2.

Strategy We need to find the total time of flight so that we can find the horizontal displacement. The bullet is starting from the high point of the parabolic path because $v_{0y} = 0$. As usual in projectile problems, we choose the y-axis to be the positive vertical direction.

Given: $\Delta y = -20.0$ m; $v_{0y} = 0$; $v_{0x} = 500.0$ m/s
To find: Δx

Solution The vertical displacement through which the bullet falls is 20.0 m. The time to fall can be found from

$$\Delta y = v_{0y}t + \tfrac{1}{2}a_y t^2$$

Solving for t with $v_{0y} = 0$ yields

$$t = \sqrt{\frac{2\Delta y}{a_y}}$$

The horizontal displacement of the bullet is

$$\Delta x = v_{0x}t = v_{0x}\sqrt{\frac{2\Delta y}{a_y}}$$

$$= 500.0 \text{ m/s} \times \sqrt{\frac{2 \times (-20.0 \text{ m})}{-9.80 \text{ m/s}^2}} = 1.01 \text{ km}$$

Discussion How did we know to start with the y-component equation when the question asks about the *horizontal* displacement? The question gives v_{0x} and asks for Δx. The missing information needed is the time during which the bullet is in the air; the time can be found from analysis of the *vertical* motion.

We neglected air resistance in this problem, which is not very realistic. The actual distance would be less than 1.01 km.

Practice Problem 4.11   Bullet velocity

Find the horizontal and vertical components of the bullet's velocity just before it hits the surface of the lake. At what angle does it strike the surface?

At the beginning of the chapter, we asked why the clam does not fall straight down when the seagull lets go. The seagull is flying horizontally with the clam, so the clam has the same horizontal velocity as the seagull. When the seagull lets go, the net force on the clam is downward due to gravity. The clam falls toward Earth, but since $a_x = 0$, the clam retains the same horizontal component of velocity as the seagull. Therefore, the clam is a projectile starting at the top of its parabolic trajectory.

Making The Connection:
seagull dropping
a clam

4.6   OTHER EXAMPLES OF CONSTANT ACCELERATION

Projectiles moving with negligible air resistance in a uniform gravitational field are not the only objects with constant acceleration. Any object moves with constant acceleration if the net force acting on it is constant. Once we have added all the forces and used Newton's second law to find the acceleration of the object, we can use the same kinematics relations as for projectiles. The only difference is that the acceleration is not necessarily 9.8 m/s² downward; the acceleration is determined by Newton's second law.

If the direction of the acceleration is known, it is easiest to choose x- and y-axes so that the acceleration has one component that is zero. Then we can view the motion as a superposition of constant velocity along one axis and constant acceleration along the other, as we do for projectiles.

Tips

Example 4.12

The Broken Suitcase

The wheels fell off of Beatrice's suitcase so she tied a rope to it and dragged it along the floor of the airport terminal (Fig. 4.25a). The rope makes an angle of 40.0° with the horizontal. The suitcase has a mass of 36.0 kg and Beatrice pulls on the rope with a force of 65.0 N. (a) What is the magnitude of the normal force acting on the suitcase due to the floor? (b) If the coefficient of kinetic friction between the suitcase and the marble floor is $\mu_k = 0.13$, find the frictional force acting on the suitcase. (c) What is the acceleration of the suitcase? (d) For how long a time must Beatrice pull with this force until the suitcase reaches a comfortable walking speed of 0.40 m/s?

Strategy   Since the suitcase is dragged along the floor, it can have no vertical acceleration component. (If it did have a vertical acceleration component, the suitcase would begin to move either down through the floor or up into the air.) Therefore, if we choose the $+y$-axis up and the $+x$-axis to be horizontal, $a_y = 0$. We resolve the forces acting on the suitcase into their components, draw a free-body diagram for the suitcase, and apply Newton's second law.

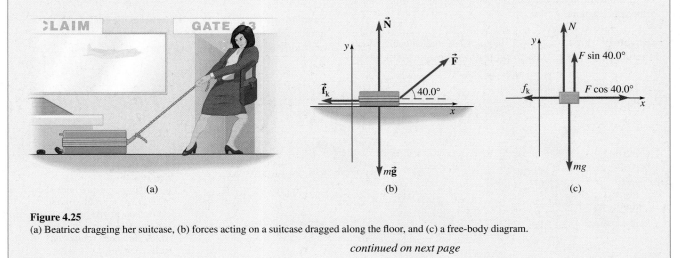

Figure 4.25
(a) Beatrice dragging her suitcase, (b) forces acting on a suitcase dragged along the floor, and (c) a free-body diagram.

continued on next page

Example 4.12 *continued*

Solution   (a) Figure 4.25b shows the forces acting on the suitcase, where $\vec{F}$ is the force exerted by Beatrice. All the other forces are either parallel or perpendicular to the floor, so only $\vec{F}$ needs to be resolved into x- and y-components.

$$F_x = F \cos 40.0° = 65.0 \text{ N} \times 0.766 = 49.8 \text{ N}$$
$$F_y = F \sin 40.0° = 65.0 \text{ N} \times 0.643 = 41.8 \text{ N}$$

We don't round to two significant figures yet because these intermediate results are used in subsequent calculations.

Figure 4.25c is a free-body diagram where $\vec{F}$ is replaced by its two components. The vertical force components add to zero since $a_y = 0$.

$$\Sigma F_y = ma_y = 0$$
$$N + F \sin 40.0° - mg = 0$$
$$N = mg - F \sin 40.0° = (36.0 \text{ kg} \times 9.8 \text{ m/s}^2) - (65.0 \text{ N} \times \sin 40.0°)$$
$$= 353 \text{ N} - 41.8 \text{ N} = 311.2 \text{ N}$$

Now we round the final answer: the normal force is 310 N.

(b) The kinetic frictional force is

$$f_k = \mu_k N = 0.13 \times 311.2 \text{ N} = 40.46 \text{ N}$$

The frictional force is 40 N in the –x-direction (opposite the motion of the suitcase).

(c) The y-component of the acceleration is zero. To find the x-component, we apply Newton's second law to the x-components of the forces:

$$\Sigma F_x = ma_x$$
$$\Sigma F_x = +F \cos 40.0° + (-f_k)$$
$$= 49.8 \text{ N} - 40.46 \text{ N} = 9.34 \text{ N}$$
$$a_x = 9.34 \text{ N}/36.0 \text{ kg} = 0.259 \text{ m/s}^2$$

The acceleration is 0.26 m/s^2 in the +x-direction.

(d) With constant a_x,

$$v_x = v_{0x} + a_x t$$

The suitcase starts from rest so $v_{0x} = 0$. Then

$$t = \frac{v_x}{a_x} = \frac{0.40 \text{ m/s}}{0.259 \text{ m/s}^2} = 1.5 \text{ s}$$

Discussion   What Beatrice probably wants to do is to drag the suitcase along at constant velocity. To do that, she must first accelerate the suitcase from rest. Once the suitcase is moving, she pulls a little less hard so the net force is zero and the suitcase slides at constant speed. She would do so without thinking much about it, of course!

Practice Problem 4.12   The continuing suitcase story

How hard must Beatrice pull at a 40.0° angle so that the suitcase slides at constant velocity? [*Warning:* Do *not* assume that the normal force is the same as in the previous discussion.]

Example 4.13

A Sliding Brick

A brick of mass 1.0 kg slides down an icy roof inclined at 30.0° with respect to the horizontal. If the brick starts from rest 2.0 m from the edge of the roof, how fast is it moving at the edge? Ignore friction.

Strategy   First we draw a diagram and indicate the forces acting on the brick. Then we write down Newton's second law

and solve it by breaking it into components along two mutually perpendicular directions. The acceleration is directed along the roof, so we choose the x-axis parallel to the roof (toward the edge) and the y-axis perpendicular to the roof.

Solution   In Fig. 4.26a, $\vec{W}$ is the weight of the brick and $\vec{N}$ is the normal force exerted on the brick by the roof. The weight of the brick must be resolved into components in the x- and y-directions

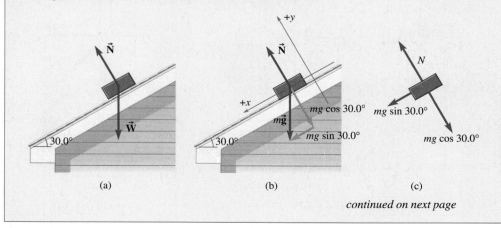

Figure 4.26
(a) Forces acting on a brick sliding down an icy roof, (b) resolving the weight into components, and (c) a free-body diagram in which the weight is replaced by its components.

(a)   (b)   (c)

continued on next page

Example 4.13 *continued*

(Fig. 4.26b). Figure 4.26c is a free-body diagram in which the weight is separated into two perpendicular components.

Ignoring friction, the roof exerts no force in the *x*-direction on the brick. Then only the weight has an *x*-component. From Newton's second law,

$$\Sigma F_x = mg \sin 30.0° = ma_x$$

Solving for a_x,

$$a_x = g \sin 30.0°$$

Now that we have the acceleration, we use kinematic relationships to find the final velocity. We know that $\Delta x = 2.00$ m and $v_{0x} = 0$. Then,

$$v_x^2 - v_{0x}^2 = v_x^2 = 2a_x \Delta x$$

Solving for v_x,

$$v_x = \sqrt{2a_x \Delta x} = \sqrt{2g \sin 30.0° \times \Delta x}$$
$$= \sqrt{2 \times 9.8 \text{ m/s}^2 \times \tfrac{1}{2} \times 2.0 \text{ m}} = 4.4 \text{ m/s}$$

Discussion   In this problem it was not necessary to apply Newton's second law to the *y*-components of the forces. Doing so would enable us to find the normal force. If the roof were not frictionless, we would need to find the normal force in order to calculate the force of kinetic friction.

Practice Problem 4.13   Effect of friction on sliding brick

Repeat Example 4.13 assuming that the coefficient of kinetic friction is 0.20.

Example 4.14

A Pulley, an Incline, and Two Blocks

A block of mass $m_1 = 2.6$ kg rests upon an incline that is angled at 30.0° above the horizontal (Fig. 4.27a). A lightweight, flexible cord is connected from block 1 over an ideal, frictionless pulley to another block of mass $m_2 = 2.2$ kg that is hanging freely 2.00 m above the ground. The coefficient of kinetic friction between the incline and block 1 is 0.18. If the blocks are initially at rest, how long does it take for block 2 to reach the ground? Assume the cord is of negligible mass and does not stretch.

Strategy   It is not obvious which way the blocks start to move once they are released. Fortunately the problem tells us that the blocks start from rest and that block 2 hits the floor, so block 2 accelerates downward and block 1 accelerates up the incline. For block 1, we choose axes parallel and perpendicular to the incline so that its acceleration has only one nonzero component. The magnitudes of the accelerations of the two blocks are equal, since they are connected by a cord that does not stretch. The fixed length of the cord means that during any time interval, the blocks move the same distance. Since that is

true even for tiny time intervals, the speeds of the blocks are equal at any instant. Since the speeds change the same amount during any time interval, their accelerations have equal magnitudes. Since the cord's mass is neglected, the tension is the same at the two ends.

Solution   We start by drawing separate free-body diagrams for each block (Figs. 4.27b and 4.27c). Since block 1 slides up the incline, the frictional force f_k acts down the incline to oppose the sliding. The weight of block 1 is resolved into two components, one along the incline and one perpendicular to the incline.

Using the free-body diagrams we write Newton's second law in component form for each block. Block 1 has no acceleration in the *y*-direction. It does not sink into the incline or rise above it; it can only slide along the incline. Thus, the net force in the *y*-direction is zero.

$$\Sigma F_y = N - m_1 g \cos \theta = 0$$

or

$$N = m_1 g \cos \theta$$

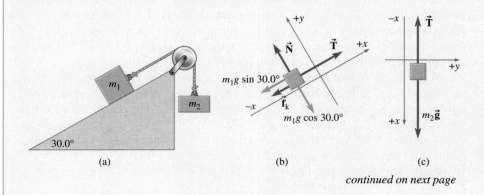

Figure 4.27
(a) Block on an incline connected to a freely hanging block by a cord passing over a pulley, (b) free-body diagram for block 1, and (c) free-body diagram for block 2.

continued on next page

Example 4.14 *continued*

where $\theta = 30.0°$. Along the incline, in the x-direction, the acceleration is nonzero:

$$\Sigma F_x = T - m_1 g \sin\theta - f_k = m_1 a_x$$

The kinetic frictional force is related to the normal force:

$$f_k = \mu_k N = \mu_k m_1 g \cos\theta$$

By substitution,

$$T - m_1 g \sin\theta - \mu_k m_1 g \cos\theta = m_1 a_x \qquad (1)$$

For block 2, we choose an x-axis pointing downward. Doing so simplifies the solution since then the two blocks have the same a_x. Applying Newton's second law,

$$\Sigma F_x = m_2 g - T = m_2 a_x \qquad (2)$$

The tension in the cord, T, and the x-component of acceleration, a_x, are both unknown in Eqs. (1) and (2). We solve for T in Eq. (2) and substitute into Eq. (1),

$$T = m_2 g - m_2 a_x = m_2 (g - a_x)$$

$$m_2 (g - a_x) - m_1 g \sin\theta - \mu_k m_1 g \cos\theta = m_1 a_x$$

Rearranging and solving for a_x yields

$$a_x = \frac{m_2 g - m_1 g (\sin\theta + \mu_k \cos\theta)}{m_1 + m_2} \qquad (3)$$

Substituting the known and given values,

$$a_x = \frac{2.2 \text{ kg} \times 9.8 \text{ m/s}^2 - 2.6 \text{ kg} \times 9.8 \text{ m/s}^2 \times (0.500 + 0.18 \times 0.866)}{2.6 \text{ kg} + 2.2 \text{ kg}}$$

$$= 1 \text{ m/s}^2$$

Block 2 has a distance of 2.00 m to travel starting from rest with a constant downward acceleration of 1.01 m/s².

$$\Delta x = \tfrac{1}{2} a_x t^2$$

The time to travel that distance is

$$t = \sqrt{\frac{2\Delta x}{a_x}} = \sqrt{\frac{2 \times 2.00 \text{ m}}{1 \text{ m/s}^2}} = 2 \text{ s}$$

Discussion   There are several advantages to solving for a_x algebraically in Eq. (3) before substituting numerical values.

Dimensional analysis can easily be used to check for errors. In Eq. (3), the quantity in parentheses is dimensionless—the values of trigonometric functions are pure numbers as are coefficients of friction. Therefore the numerator is the sum of two quantities with dimensions of force, the denominator is the sum of two masses, and force divided by mass gives an acceleration.

Equation (3) also enables us to ask many "what if" questions. These both deepen our understanding of the physics behind the solution and serve as further checks on the validity of the solution. For one example, what if m_2 is much larger than m_1? In the numerator, the term $m_2 g$ would be much larger than $m_1 g$ ($\sin\theta + \mu_k \cos\theta$). In the denominator, m_2 would be much larger than m_1. Then Eq. (3) says that a_x would be approximately equal to g. If m_1 is negligibly small, then the tension in the cord would be small and block 2 would be almost in free fall.

Other such "what if" questions can be posed regarding the magnitudes of the masses, the angle of the incline, and the coefficient of friction.

What if the problem did not tell us the direction of the blocks' acceleration? We could figure it out by comparing the force with which gravity pulls down on block 2 ($m_2 g$) with the component of the gravitational force pulling block 1 down the incline ($m_1 g \sin\theta$). Whichever is greater "wins the tug-of-war." Once we know the direction of the acceleration, then we can determine the direction of the kinetic frictional force. If the blocks are not initially at rest, the kinetic frictional force opposes the direction of sliding, even though that may be opposite the direction of the acceleration.

Practice Problem 4.14   More fun with a pulley and an incline

Suppose that $m_1 = 3.8$ kg and $m_2 = 1.2$ kg and the coefficient of kinetic friction is 0.18. The blocks are released from rest and block 1 starts to slide. (a) Does block 1 slide up or down the incline? (b) In which direction does the kinetic frictional force act? (c) Find the acceleration of block 1.

4.7   RELATIVE VELOCITY

Suppose Wanda is walking down the aisle of a moving train (Fig. 4.28). Imagine asking, "How fast is she walking?" This question is not well defined. Do we mean her speed as measured by Tim, a passenger on the train, or her speed as measured by Greg, who is standing on the ground and looking into the train as it passes by? The answer to the question "How fast?" depends on the observer.

Figure 4.28b shows Wanda walking from one end of the car to the other during a time interval Δt. The displacement of Wanda as measured by Tim—her displacement relative to the train—is $\Delta\vec{\mathbf{r}}_{WT} = \vec{\mathbf{v}}_{WT}\Delta t$. During the same time interval, the *train's* displacement relative to the ground is $\Delta\vec{\mathbf{r}}_{TG} = \vec{\mathbf{v}}_{TG}\Delta t$. As measured by Greg, Wanda's displacement is partly due to her motion relative to the train and partly due to the motion of the train relative to the ground. Figure 4.28b shows that $\Delta\vec{\mathbf{r}}_{WT} + \Delta\vec{\mathbf{r}}_{TG} = \Delta\vec{\mathbf{r}}_{WG}$. Dividing by the time interval Δt gives the relationship between the three velocities:

$$\vec{\mathbf{v}}_{WT} + \vec{\mathbf{v}}_{TG} = \vec{\mathbf{v}}_{WG} \qquad (4\text{-}9)$$

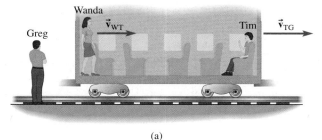

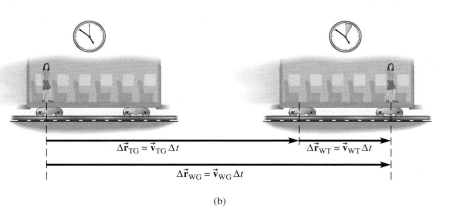

Figure 4.28 (a) Tim and Greg watch Wanda walk down the aisle of a train. Wanda's velocity with respect to Tim (or with respect to the train) is $\vec{\mathbf{v}}_{WT}$; Tim's velocity with respect to Greg (or with respect to the ground) is $\vec{\mathbf{v}}_{TG}$. (b) Wanda's displacement relative to the ground is the sum of her displacement relative to the train and the displacement of the train relative to the ground.

To be sure that you are adding the velocity vectors correctly, think of the subscripts as if they were fractions that get multiplied when the velocity vectors are added. In Eq. (4-9), *Tips*

$\dfrac{W}{T} \times \dfrac{T}{G} = \dfrac{W}{G}$, so the equation is correct.

Example 4.15

Flight from Denver to Chicago

An airplane flies from Denver to Chicago (1770 km) in 4.4 h when no wind blows. On a day with a tailwind, the plane makes the trip in 4.0 h. (a) What is the wind speed? (b) If a headwind blows with the same speed, how long does the trip take?

Strategy We assume the plane has the same *airspeed*—the same speed relative to the air—in both cases. Once the plane is up in the air, the behavior of the wings and the control surfaces depends on how fast the air is rushing by; the ground speed is irrelevant. But it is not irrelevant for the passengers, who are interested in a displacement relative to the ground.

Solution Let $\vec{\mathbf{v}}_{PG}$ and $\vec{\mathbf{v}}_{PA}$ represent the velocity of the plane relative to the ground and air, respectively. The wind velocity—the velocity of the air relative to the ground—can be written

$\vec{\mathbf{v}}_{AG}$. Then $\vec{\mathbf{v}}_{PA} + \vec{\mathbf{v}}_{AG} = \vec{\mathbf{v}}_{PG}$. The equation is correct since P/A × A/G = P/G. With no wind,

$$v_{PA} = v_{PG} = \frac{1770 \text{ km}}{4.4 \text{ h}} = 400 \text{ km/h}$$

(a) On the day with the tailwind,

$$v_{PG} = \frac{1770 \text{ km}}{4.0 \text{ h}} = 440 \text{ km/h}$$

We expect v_{PA} to be the same regardless of whether there is a wind or not. Since we are dealing with a tailwind, $\vec{\mathbf{v}}_{PA}$ and $\vec{\mathbf{v}}_{AG}$ are in the same direction, so their magnitudes add (Fig. 4.29a):

$$|\vec{\mathbf{v}}_{PA}| + |\vec{\mathbf{v}}_{AG}| = |\vec{\mathbf{v}}_{PG}|$$

$$|\vec{\mathbf{v}}_{AG}| = |\vec{\mathbf{v}}_{PG}| - |\vec{\mathbf{v}}_{PA}| = 440 \text{ km/h} - 400 \text{ km/h} = 40 \text{ km/h}$$

(b) With a 40 km/h headwind, $\vec{\mathbf{v}}_{PA}$ and $\vec{\mathbf{v}}_{AG}$ are in opposite directions (Fig. 4.29b); the ground speed of the plane is

$$|\vec{\mathbf{v}}_{PG}| = |\vec{\mathbf{v}}_{PA}| - |\vec{\mathbf{v}}_{AG}| = 400 \text{ km/h} - 40 \text{ km/h} = 360 \text{ km/h}$$

continued on next page

Example 4.15 *continued*

Then the trip takes

$$\frac{1770 \text{ km}}{360 \text{ km/h}} = 4.9 \text{ h}$$

Discussion   If we choose the *x*-axis to the right, then the vector addition can be done by components: $v_{PAx} + v_{AGx} = v_{PGx}$. The vectors v_{PAx} and v_{PGx} are both positive. Vector v_{AGx} is positive in (a) and negative in (b).

Practice Problem 4.15   Rowing across the bay

Jamil, practicing to get on the crew team at school, rows a one-person racing shell to the north shore of the bay for a distance of 3.6 km to his friend's dock. On a day when the water is still (no current flowing), it takes him 20 min (1200 s) to reach his friend. On another day when a current flows southward, it takes him 30 min (1800 s) to row the same course. Ignore air resistance. (a) What is the speed of the current in m/s? (b) How long does it take Jamil to return home with that same current flowing?

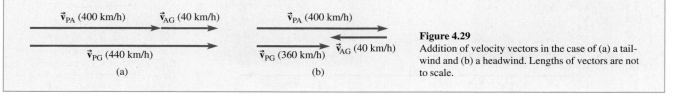

Figure 4.29
Addition of velocity vectors in the case of (a) a tailwind and (b) a headwind. Lengths of vectors are not to scale.

Equation (4-9) applies to situations where the velocities are not all along the same line, as illustrated in Example 4.16.

Example 4.16

Rowing to the Dock

Jack is enjoying a pleasant day at the ocean and is rowing his friend Kristina in a small boat in the harbor. A sudden squall comes up and the wind blows steadily along the beach, causing a current to travel from north to south at a uniform velocity of 0.61 m/s. Jack drops Kristina off at her sailboat and then proceeds 250 m directly west to the dock

(Fig. 4.30a). If Jack gets to the dock in a time of 4.2 min after leaving Kristina, in what direction did he head his rowboat? At what speed with respect to still water is Jack able to row?

Strategy   To keep the various velocities straight, we choose subscripts as follows: R = rowboat, W = water, and D = dock. The velocity of the current given is the velocity of the water relative to the dock ($\vec{v}_{WD}$). The velocity of the rowboat relative to the dock ($\vec{v}_{RD}$) is due west. The question asks for the magnitude and direction of the velocity of the rowboat relative to the water ($\vec{v}_{RW}$). The three velocities are related by

$$\vec{v}_{RW} + \vec{v}_{WD} = \vec{v}_{RD}$$

To compensate for the current's tendency to push the rowboat south of the dock, the rowboat is headed "against the current" at some angle to the north of west.

Given: $\vec{v}_{WD} = 0.61$ m/s; displacement from sailboat to dock = 250 m due west, time to travel from sailboat to dock = 4.2 min

To find: $\vec{v}_{RW}$

Solution   We start with a sketch of the vector addition (Fig. 4.30b). The velocity of the rowboat with respect to the water is at an angle θ north of west. With respect to the dock, Jack travels 250 m in 4.2 min, so his speed with respect to the dock is

$$v_{RD} = \frac{250 \text{ m}}{4.2 \text{ min} \times 60 \text{ s/min}} = 0.992 \text{ m/s}$$

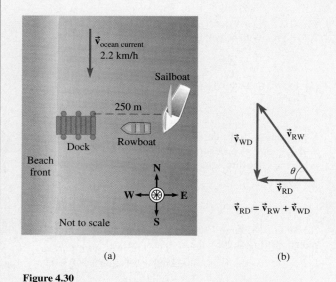

Figure 4.30
(a) Seaside scene and (b) graphical addition of velocities.

continued on next page

Example 4.16 continued

We can find the angle at which the rowboat should be headed into the current by finding the tangent of the angle between $\vec{v}_{RW}$ and $\vec{v}_{RD}$.

$$\tan \theta = \frac{v_{WD}}{v_{RD}} = \frac{0.61 \text{ m/s}}{0.992 \text{ m/s}}$$

$$\theta = 32° \text{ N of W}$$

The speed at which Jack is able to row with respect to still water is the magnitude of $\vec{v}_{RW}$. Since $\vec{v}_{RD}$ and $\vec{v}_{WD}$ are perpendicular, the Pythagorean theorem yields

$$|\vec{v}_{RW}| = \sqrt{v_{WD}^2 + v_{RD}^2} = \sqrt{(0.61 \text{ m/s})^2 + (0.992 \text{ m/s})^2}$$

$$= 1.16 \text{ m/s}$$

Jack rows at a speed of 1.16 m/s with respect to the water.

Discussion If Jack rowed directly west toward the dock, the ocean current would push him to the south, so he would end up traveling in a direction south of west. He has to compensate by heading against the ocean current by just the amount needed to have the velocity relative to the dock be directly west. By looking at the shore, Jack can see that he is headed straight west, but he knows that the prow of his boat is pointed to the north of west and that he is constantly fighting against the current.

Practice Problem 4.16 Paddling the canoe

A canoeist is attempting to paddle directly from east to west across a river that is flowing south at 5.0 km/h. If she can paddle the canoe so that it moves at 7.0 km/h with respect to the river, at what angle should she head the canoe to travel directly west as she intends?

At the beginning of this chapter, we asked what the path followed by the falling clam looks like as seen by the seagull flying through the air. With respect to a beachcomber on the ground, the clam has a constant horizontal velocity component given to it by the seagull and a changing vertical component of velocity due to the gravitational force (Fig. 4.31a); the clam moves in a parabolic path. If the gull continues to fly at the same horizontal velocity after dropping the clam, it is directly overhead when the clam hits the rock because they both have the same constant horizontal component of velocity with respect to Earth.

In its own reference frame—that is, using its own position as the origin of the coordinate axes—the seagull sees the clam drop straight down toward the ground while rocks and other objects on the beach are moving horizontally (Fig. 4.31b). The bird sees a collision between the horizontally moving rocks and the vertically falling clam. At any instant, if the velocity of the clam with respect to the seagull is $\vec{v}_{CS}$, the velocity of the clam with respect to the beach is $\vec{v}_{CB}$, and the velocity of the seagull with respect to the beach is $\vec{v}_{SB}$, then $\vec{v}_{CS} + \vec{v}_{SB} = \vec{v}_{CB}$.

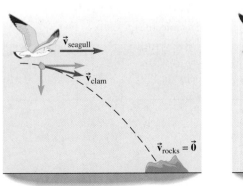

(a)

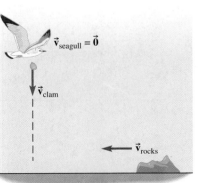

(b)

Figure 4.31 (a) Beachcomber view: The bird flies along a horizontal line while the clam follows a parabolic path. (b) Bird's-eye view: The gull sees the ground and objects upon the ground moving while the clam drops straight down, landing upon the rocks just as the rocks move under the clam.

MASTER THE CONCEPTS

Summary

- Addition of vectors is commutative: $\vec{A} + \vec{B} = \vec{B} + \vec{A}$.

- Vectors are added graphically by moving each vector so that its tail is placed at the tip of the previous vector. The sum is drawn as a vector arrow from the tail of the first vector to the tip of the last.

- Vectors are subtracted graphically by drawing the vectors with their tails at the same point. Then the difference $\vec{A} - \vec{B}$ is a vector drawn from the tip of $\vec{B}$ to the tip of $\vec{A}$.

- Addition and subtraction of vectors algebraically using components is generally easier and more accurate than the graphical method. The graphical method is still a useful first step to get an approximate answer.

- To add vectors algebraically, add their components to find the components of the resultant: if $\vec{A} + \vec{B} = \vec{C}$, then $A_x + B_x = C_x$ and $A_y + B_y = C_y$.

- Components are found by drawing a right triangle with the vector as the hypotenuse and the other two sides parallel to the x- and y-axes. Trigonometric functions are used to find the magnitudes of the components. The correct algebraic sign must be supplied for each component. The same triangle can be used to find the magnitude and direction of a vector if its components are known.

- The x- and y-axes are chosen to make the problem easiest to solve. Any choice is valid as long as the two are perpendicular. In an equilibrium problem, choose x- and y-axes so that the fewest number of force vectors have to be resolved into both x- and y-components. In a nonequilibrium problem, if the direction of the acceleration is known, choose x- and y-axes so that the acceleration vector is parallel or antiparallel to one of the axes.

- The instantaneous velocity vector is tangent to the path of motion.

- The instantaneous acceleration vector does *not* have to be tangent to the path of motion, since velocities can change both in direction and in magnitude.

- The kinematic equations for an object moving in two dimensions with constant acceleration along the y-axis and $t_0 = 0$ are:

$$v_x = v_{0x} \tag{4-2}$$

$$x - x_0 = v_{0x}t \tag{4-3}$$

$$v_y = v_{0y} + a_y t \tag{4-4}$$

$$\Delta y = v_{av,y}t \tag{4-5}$$

$$v_{av,y} = \tfrac{1}{2}(v_{0y} + v_y) \tag{4-6}$$

$$y - y_0 = v_{0y}t + \tfrac{1}{2}a_y t^2 \tag{4-7}$$

$$v_y^2 - v_{0y}^2 = 2a_y \Delta y \tag{4-8}$$

- Problems involving Newton's second law—whether equilibrium or nonequilibrium—can be solved by treating the x- and y-components of the forces and the acceleration separately. The vector equation $\vec{F}_{net} = m\vec{a}$ is equivalent to the two scalar equations

$$\Sigma F_x = ma_x \quad \text{and} \quad \Sigma F_y = ma_y \tag{4-1}$$

- For a projectile or any object moving with constant acceleration in the $\pm y$-direction, the motion in the x- and y-directions can be treated separately. Since $a_x = 0$, v_x is constant. Thus, the motion is a superposition of constant velocity motion in the x-direction on constant acceleration motion in the y-direction.

- To relate the velocities of objects measured in different reference frames, use the equation

$$\vec{v}_{AC} = \vec{v}_{AB} + \vec{v}_{BC}$$

where $\vec{v}_{AC}$ represents the velocity of A relative to C, and so forth.

Highlighted Figures and Tables

F4.3 Graphical addition of displacement vectors (p. 105)

F4.4 The final position vector minus the initial position vector is the displacement $\Delta\vec{r}$ ($\vec{r}_f - \vec{r}_0 = \Delta\vec{r}$) (p. 105)

F4.7 Resolving vector $\vec{A}$ into components along two axes in two equivalent ways (p. 107)

F4.8 Addition of displacement vectors using components (p. 109)

F4.14 Forces acting on an object of mass m pushed up an inclined plane; resolving the weight into x- and y-components; and the free-body diagram (p. 114)

F4.16 Instantaneous velocity resolved into components along perpendicular axes (p. 116)

F4.19 Trajectory of a projectile (p. 120)

F4.21 Projectile motion: vertical and horizontal quantities versus time (p. 121)

F4.28 Relative velocity (p. 129)

CONCEPTUAL QUESTIONS

1. Why is the muzzle of a rifle not aimed directly at the center of the target?

2. Give an example of an object whose acceleration is (1) in the same direction as its velocity, (2) opposite its velocity, and (3) perpendicular to its velocity.

3. Does the monkey, coconut, and hunter demonstration still work if the gun is pointed *downward* at the monkey and coconut? Explain.

4. Can a body in free fall be in equilibrium? Explain.

5. Can the *x*-component of a vector ever be greater than the magnitude of the vector? Explain.

6. Explain how to combine two displacement vectors of magnitudes 3*L* and 4*L* so that the resultant vector has magnitude (a) *L*; (b) 7*L*; (c) 5*L*.

7. Compare the advantages and disadvantages of the two methods of vector addition (graphical and algebraic).

8. An object is subjected to two constant forces that are perpendicular to each other. Can a set of *x*- and *y*-axes be chosen so that the acceleration of the object has only one nonzero component? If so, how? Explain.

9. Is it possible for two identical projectiles with identical initial speeds, but with two different angles of elevation, to land in the same spot? Explain.

10. Pulleys and inclined planes are often called simple machines. Explain why in terms of the forces exerted with and without the use of the "machine."

11. Demonstrate with a vector diagram that a displacement is the same when measured in two different reference frames that are at rest with respect to each other.

12. If the trajectory is parabolic in one reference frame, is it always, never, or sometimes parabolic in another reference frame that moves at constant velocity with respect to the first reference frame? If the trajectory can be other than parabolic, what else can it be?

13. In free fall, neglecting air resistance, the *x*- and *y*-components of the motion are independent. Are the components independent when air resistance is significant? Why or why not? [*Hint:* The drag force has a magnitude determined by the *speed* of the projectile and a direction opposite to the velocity.]

MULTIPLE CHOICE QUESTIONS

1. Two balls, identical except for color, are projected horizontally from the roof of a tall building at the same instant. The initial speed of the red ball is twice the initial speed of the blue ball. Ignoring air resistance,

   (a) the red ball reaches the ground first.
   (b) the blue ball reaches the ground first.
   (c) both balls land at the same instant with different speeds.
   (d) both balls land at the same instant with the same speed.

2. A person stands on the roof garden of a tall building with one ball in each hand. If the red ball is thrown horizontally off the roof and the blue ball is simultaneously dropped over the edge, which statement is true?

   (a) Both balls hit the ground at the same time, but the red ball has a higher speed just before it strikes the ground.
   (b) The blue ball strikes the ground first, but with a lower speed than the red ball.
   (c) The red ball strikes the ground first with a higher speed than the blue ball.
   (d) Both balls hit the ground at the same time with the same speed.

3. A ball is thrown into the air and follows a parabolic trajectory. At the highest point in the trajectory,

   (a) the velocity is zero, but the acceleration is not zero.
   (b) both the velocity and the acceleration are zero.
   (c) the acceleration is zero, but the velocity is not zero.
   (d) neither the acceleration nor the velocity are zero.

4. A ball is thrown into the air and follows a parabolic trajectory. Point *A* is the highest point in the trajectory and point *B* is a point as the ball is falling back to the ground. Choose the correct relationship between the speeds and the magnitudes of the accelerations at the two points.

   (a) $v_A > v_B$ and $a_A = a_B$ (b) $v_A < v_B$ and $a_A > a_B$
   (c) $v_A = v_B$ and $a_A \neq a_B$ (d) $v_A < v_B$ and $a_A = a_B$

5. A small plane is flying directly west with an airspeed of 30 m/s. The plane flies into a region where the wind is blowing at 10 m/s directly to the southwest. If the pilot does not change the heading of the plane, the ground speed of the airplane

   (a) is greater than the airspeed.
   (b) is less than the airspeed.
   (c) is the same as the airspeed.
   (d) depends on the height of the plane.

6. A small plane is flying directly west with an airspeed of 30 m/s. The plane flies into a region where the wind is blowing at 10 m/s directly to the southwest. In that region, the pilot changes the directional heading to maintain the due west heading. The ground speed of the airplane

   (a) is greater than the airspeed.
   (b) is less than the airspeed.
   (c) is the same as the airspeed.
   (d) depends on the height of the plane.

7. A boy plans to cross a river in a rubber raft. The current flows from north to south at 1 m/s. In what direction should he head to get across the river to the east bank in the least amount of time if he is able to paddle the raft at 1.5 m/s in still water?

   (a) directly to the east
   (b) south of east
   (c) north of east
   (d) The three directions require the same time to cross the river.

8. A boy plans to paddle a rubber raft across a river to the east bank while the current flows downriver from north to south at 1 m/s. He is able to paddle the raft at 1.5 m/s in still water. In what direction should he head the raft to go straight east across the river to the opposite bank?

   (a) directly to the east   (b) south of east
   (c) north of east   (d) north
   (e) south

9. Vector $\vec{A}$ in Fig. 4.32 is equal to

   (a) $\vec{C} + \vec{D}$ (b) $\vec{C} + \vec{D} + \vec{E}$ (c) $\frac{1}{2}(\vec{D} - \vec{C})$
   (d) $\vec{B} + \vec{C}$ (e) $\vec{B} - \vec{D}$

10. Vector $\vec{E}$ in Fig. 4.32 is equal to

    (a) $-2(\vec{B} + \vec{C})$ (b) $-(\vec{C} + \vec{D})$ (c) $-2\vec{A}$
    (d) $-(\vec{A} + \vec{B} + \vec{C})$ (e) all of the above

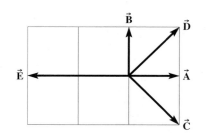

Figure 4.32   Multiple Choice Questions 9 and 10

PROBLEMS

Note: ☉ indicates a combination conceptual/quantitative problem. Gold diamonds ◆, ◆◆ are used to indicate the increasing level of difficulty of each problem. Problem numbers appearing in blue, 9., denote problems that have a detailed solution available in the Student Solutions Manual. Some problems are *paired* by concept; their numbers are connected by a ruled box.

4.1 Addition and Subtraction of Vectors in Two Dimensions

1. Vectors $\vec{A}$ and $\vec{B}$ are drawn in Fig. 4.33. (a) Draw vectors $\vec{C}$ and $\vec{D}$, where $\vec{C} = \vec{A} + \vec{B}$ and $\vec{D} = \vec{A} - \vec{B}$. (b) Show that $\vec{A} + \vec{B} = \vec{B} + \vec{A}$ by graphical means.

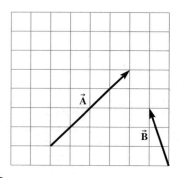

Figure 4.33   Problem 1

2. A scout troop is practicing its orienteering skills with map and compass. First they walk due east for 1.2 km. Next, they walk 45° west of north for 2.7 km. In what direction must they walk to go directly back to their starting point? How far will they have to walk? Use graph paper, ruler, and protractor to find a geometrical solution.

☉ 3. We have assumed that the displacement for a trip is equal to the sum of the displacements for each leg of the trip. Prove that this is true. [*Hint:* Imagine a trip that consists of n segments. The trip starts at position $\vec{r}_0$, proceeds to $\vec{r}_1$, then to $\vec{r}_2$, . . . , then to $\vec{r}_n - 1$, then finally to $\vec{r}_n$. Write an expression for each displacement as the difference of two position vectors and then add them.]

4.2 Components of Vectors in Two Dimensions

4. The pilot of a small plane finds that the airport where he intended to land is fogged in. He flies 55 miles west to another airport to find that conditions there are too icy for him to land. He flies 25 miles south and is finally able to land at the third airport. (a) How far and in what direction must he fly the next day to go directly to his original destination? (b) How many extra miles beyond his original flight plan has he flown?

5. A vector is 20.0-m long and makes an angle of 60.0° counterclockwise from the y-axis (on the side of the −x-axis). What are the x- and y-components of this vector?

6. Vector $\vec{A}$ has magnitude 4.0 units; vector $\vec{B}$ has magnitude 6.0 units. The angle between $\vec{A}$ and $\vec{B}$ is 60.0°. What is the magnitude of $\vec{A} + \vec{B}$?

7. Vector $\vec{A}$ is directed along the positive y-axis and has magnitude $\sqrt{3.0}$ units. Vector $\vec{B}$ is directed along the negative x-axis and has magnitude 1.0 unit.

   (a) What are the magnitude and direction of $\vec{A} + \vec{B}$?
   (b) What are the magnitude and direction of $\vec{A} - \vec{B}$?
   (c) What are the x- and y-components of $\vec{B} - \vec{A}$?

8. Vector $\vec{a}$ has components $a_x = -3.0$ m/s^2 and $a_y = +4.0$ m/s^2. (a) What is the magnitude of $\vec{a}$? (b) What is the direction of $\vec{a}$? Give an angle with respect to one of the coordinate axes.

9. A roller coaster is towed up an incline at a steady speed of 0.50 m/s by a chain parallel to the surface of the incline. The slope is 3.0%, which means that the elevation increases by 3.0 m for every 100.0 m of distance along the surface. The mass of the roller coaster is 400.0 kg. Neglecting friction, what is the magnitude of the force exerted on the roller coaster by the chain?

10. Find the x- and y-components of the four vectors shown in Fig. 4.34.

11. In each of these, the x- and y-components of a vector are given. Find the magnitude and direction of the vector. (a) $x = -5.0$ cm,

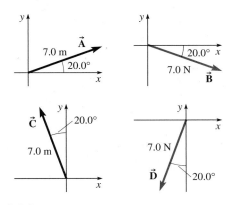

Figure 4.34 Problem 10

$y = +8.0$ cm. (b) $F_x = +120$ N, $F_y = -60.0$ N. (c) $v_x = -13.7$ m/s, $v_y = -8.8$ m/s. (d) $a_x = 2.3$ m/s^2, $a_y = 6.5$ cm/s^2.

12. Vector $\vec{\mathbf{b}}$ has magnitude 7.1 and direction 14° below the +x-axis. Vector $\vec{\mathbf{c}}$ has x-component $c_x = -1.8$ and y-component $c_y = -6.7$. Compute (a) the x- and y-components of $\vec{\mathbf{b}}$; (b) the magnitude and direction of $\vec{\mathbf{c}}$; (c) the magnitude and direction of $\vec{\mathbf{c}} + \vec{\mathbf{b}}$; (d) the x- and y-components of $\vec{\mathbf{c}} - \vec{\mathbf{b}}$.

13. Repeat Problem 2 using the component (algebraic) method.

14. A runner runs three-quarters of the way around a circular track, of radius 60.0 m, when she collides with another runner and trips. (a) How far had the runner traveled on the track before the collision? (b) What was the magnitude of the displacement of the runner from her starting position when the accident occurred?

15. When a particle undergoes diffusion, it follows a random path, bumping into obstacles here and there as it travels along. Suppose an electron is diffusing through a crystal. The black dots in Fig. 4.35 represent the regularly spaced atoms in the crystal. When the electron bumps into an atom, the direction of its motion is changed. If Fig. 4.35 shows a path taken by an electron, find the length of the path taken and the displacement of the electron.

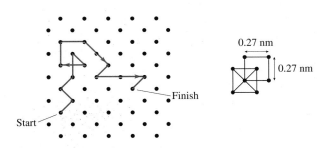

Figure 4.35 Problem 15: random path of diffusing electron

4.3 Equilibrium

16. (a) In Example 4.6, if the movers stop pushing on the safe, can the frictional force hold the safe in place without having it slide back down? (b) If not, what force needs to be applied to hold the safe in place?

17. Mechanical advantage is the ratio of the force required without the use of a simple machine to that needed when using the simple machine. Compare the force to lift an object to that needed to slide the same object up a frictionless incline and show that the mechanical advantage of the inclined plane is the length of the incline divided by the height of the incline (d/h in Fig. 4.14a).

18. A crow sits on a clothesline midway between two poles (Fig. 4.36). Each end of the rope makes an angle of θ below the horizontal where it connects to the pole. If the combined weight of the crow and the rope is W, what is the tension in the rope?

Figure 4.36 Problem 18

19. Tamar wants to cut down a large, dead poplar tree with her chain saw, but she does not want it to fall onto the nearby gazebo. Yoojin comes to help with a long rope. Yoojin, a physicist, suggests they tie the rope taut from the poplar to the oak tree and then pull *sideways* on the rope (Fig. 4.37). If the rope is 40.0 m long and Yoojin pulls sideways at the midpoint of the rope with a force of 360.0 N, causing a 2.00-m sideways displacement of the rope at its midpoint, what force will the rope exert on the poplar tree? Compare this with pulling the rope directly away from the poplar with a force of 360.0 N. [*Hint:* Until the poplar is cut through enough to start falling, the rope is in equilibrium.] Explain where the "extra" force comes from.

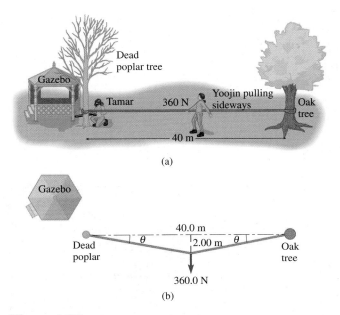

Figure 4.37 Problem 19: (a) side view of trees and rope and (b) overhead view of rope tied to trees.

20. A pulley is hung from the ceiling by a rope. A block of mass M is suspended by another rope that passes over the pulley and is attached to the wall (Fig. 4.38). The rope fastened to the wall makes a right angle with the wall. Neglect the masses of the rope and the pulley. Find (a) the tension in the rope from which the pulley hangs and (b) the angle θ that the rope makes with the ceiling.

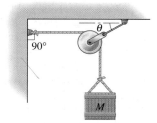

Figure 4.38 Problem 20

21. A 2.0-kg ball tied to a string fixed to the ceiling is pulled to one side by a force $\vec{F}$ (Fig. 4.39). Just before the ball is released and allowed to swing back and forth, (a) how large is the force $\vec{F}$ that is holding the ball in position and (b) what is the tension in the string?

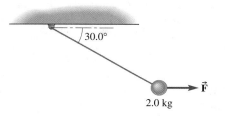

Figure 4.39 Problem 21

22. A young boy with a broken leg is undergoing traction (Fig. 4.40). (a) Find the magnitude of the total force of the traction apparatus applied to the leg, assuming the weight of the leg is 22 N and the weight hanging from the traction apparatus is also 22 N. (b) What is the horizontal component of the traction

force acting on the leg? (c) What is the magnitude of the force exerted on the femur by the lower leg?

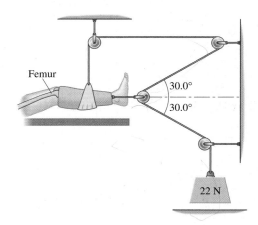

Figure 4.40 Problem 22

23. A large wrecking ball of mass m is resting against a wall. It hangs from the end of a cable that is attached at its upper end to a crane that is just touching the wall. The cable makes an angle of θ with the wall (Fig. 4.41). Neglecting friction between the ball and the wall, find the tension in the cable.

24. A crumb of bread is pulled upon by several ants from rival anthills. They exert the following forces: 0.06 N to the north, 0.08 N to the east, 0.02 N to the west, and $0.06\sqrt{2}$ N to the southeast (45° S of E). What additional force should be applied by a fifth ant to keep the crumb in equilibrium?

25. An 80.0-N crate of apples sits at rest on a ramp that runs from the ground to the bed of a truck. The ramp is inclined at 20.0° to the ground. (a) What is the normal force exerted on the crate by the ramp? (b) What is the static frictional force exerted on the crate by the ramp? (c) What is the minimum possible value of the coefficient of static friction? (d) The normal and frictional forces are perpendicular components of the contact force exerted on the crate by the ramp. Find the magnitude and direction of the contact force. Compare your answer to that of Practice Problem 2.6.

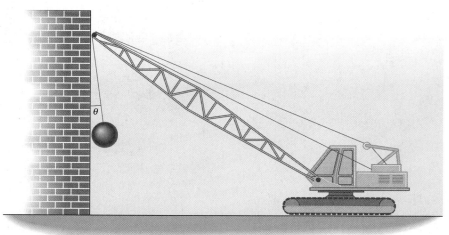

Figure 4.41 Problem 23

26. A 45-N lithograph is supported by two wires. One wire makes a 25° angle with the vertical and the other makes a 15° angle with the vertical. Find the tension in each wire.

27. A tire swing hangs at a 12° angle to the vertical when a stiff breeze is blowing. In terms of the tire's weight W, (a) what is the magnitude of the horizontal force exerted on the tire by the wind? (b) What is the tension in the rope supporting the tire? Ignore the weight of the rope.

4.4 Velocity and Acceleration

28. A car travels east at 96 km/h for 1 h. It then travels north at 128 km/h for 1 h. (a) What is the average speed for the trip? (b) What is the average velocity for the trip?

29. A car travels 3.2 km due east in 0.10 h, then 4.8 km due north in 0.15 h, and finally another 3.2 km due east in 0.10 h to reach its destination. The time lost in turning is negligible. What is the average velocity for the entire trip?

30. A car travels west at 108 km/h for 20.0 min. It then travels south at 90.0 km/h for 10.0 min. (a) What is the average speed for the trip? (b) What is the average velocity for the trip?

31. A car travels three-quarters of the way around a circle of radius 20.0 m in a time of 3.0 s at a constant speed. The initial velocity is west and the final velocity is south. (a) Find its average velocity for this trip. (b) What is the car's average acceleration during these 3.0 s? (c) Explain how a car moving at constant speed has a nonzero average acceleration.

32. At $t = 0$, an automobile traveling north begins to make a turn. It follows one-quarter of the arc of a circle of radius 10.0 m until, at $t = 1.60$ s, it is traveling east. The car does not alter its speed during the turn. Find (a) the car's speed, (b) the change in its velocity during the turn, and (c) its average acceleration during the turn.

33. Peggy drives from Cornwall to Atkins Glen in 45 min. Cornwall is 73.6 km from Illium in a direction 25° west of south. Atkins Glen is 27.2 km from Illium in a direction 15° south of west. Using Illium as your origin, (a) draw the initial and final position vectors, (b) find the displacement during the trip, and (c) find Peggy's average velocity for the trip.

34. A jetliner flies east for 600.0 km, then turns 30.0° toward the south and flies another 300.0 km. (a) How far is the plane from its starting point? (b) In what direction could the jetliner have flown directly to the same destination (in a straight-line path)? (c) If the jetliner flew at a constant speed of 400.0 km/h, how long did the trip take? (d) Moving at the same speed, how long would the direct flight have taken?

35. At the beginning of a 3.0-h plane trip, you are traveling due north at 192 km/h. At the end, you are traveling 240 km/h in the northwest direction (45° west of north). (a) Draw your initial and final velocity vectors. (b) Find the change in your velocity. (c) What is your average acceleration during the trip?

4.5 Motion of Projectiles

36. A ball is thrown from a point 1.0 m above the ground. The initial velocity is 19.6 m/s at an angle of 30.0° above the horizontal. (a) Find the maximum height of the ball above the ground. (b) Calculate the speed of the ball at the highest point in the trajectory.

37. The invention of the cannon in the fourteenth century made the catapult unnecessary and ended the safety of castle walls. Stone walls were no match for balls shot from cannons. Suppose a cannonball of mass 5.00 kg is launched from a height of 1.10 m, at an angle of elevation of 30.0° with an initial velocity of 50.0 m/s, toward a castle wall of height 30 m and located 215 m away from the cannon. (a) The range of a projectile is defined as the horizontal distance traveled when the projectile returns to its original height. Derive an equation for the range in terms of v_0, g, and angle of elevation θ. (b) What will be the range reached by the projectile, if it is not intercepted by the wall? (Let $g = 9.80$ m/s^2.) (c) If the cannonball travels far enough to hit the wall, find the height at which it strikes.

38. An arrow is shot into the air at an angle of 60.0° above the horizontal with a speed of 20.0 m/s. (a) What are the x- and y-components of the velocity of the arrow 3.0 s after it leaves the bowstring? (b) What are the x- and y-components of the displacement of the arrow during the 3.0-s interval?

39. Show that for a projectile launched at an angle of 45° the maximum height of the projectile is one quarter of the range (the distance traveled on flat ground).

40. Two ballplayers are practicing how to catch fly balls. One stands by home plate and hits balls into the air for the other, some distance away, to catch. A ball is hit with an initial velocity of 22.0 m/s at an angle of 60.0° above the horizontal. (a) How high will the ball rise? (b) How much time will elapse from the time the ball leaves the bat until it reaches the fielder? (c) At what distance from home plate will the fielder be when he catches the ball?

41. A ball is thrown horizontally off the edge of a cliff with an initial speed of 20.0 m/s. (a) How long does it take for the ball to fall to the ground 20.0 m below? (b) How long would it take for the ball to reach the ground if it were dropped from rest off the cliff edge? (c) How long would it take the ball to fall to the ground if it is thrown at an initial velocity of 20.0 m/s but 18° below the horizontal?

42. A marble is rolled so that it is projected horizontally off the top landing of a staircase. The initial speed of the marble is 3.0 m/s. Each step is 0.18 m high and 0.30 m wide. Which step does the marble strike first?

43. A suspension bridge is 60.0 m above the level base of a gorge. A stone is thrown or dropped from the bridge. Neglect air resistance. Let $g = 9.81$ m/s^2. (a) If you drop the stone, how long does it take for it to fall to the base of the gorge? (b) If you *throw* the stone straight down with a speed of 20.0 m/s, how long before it hits the ground? (c) If you throw the stone with a velocity of 20.0 m/s at 30.0° above the horizontal, how far from the point directly below the bridge will it hit the level ground?

44. A beanbag is thrown horizontally from a dorm room window a height h above the ground. It hits the ground a horizontal distance h (the *same* distance h) from the dorm directly below the window from which it was thrown. Neglecting air resistance, find the direction of the beanbag's velocity just before impact.

45. From the edge of the rooftop of a building, a boy throws a stone at an angle 25.0° above the horizontal. The stone hits the ground 4.20 s later, 105 m away from the base of the building. (Ignore air resistance.) (a) For the stone's path through the air, sketch graphs of x, y, v_x, and v_y as functions of time. These need to be only *qualitatively* correct—you

need not put numbers on the axes. (b) Find the initial velocity of the stone. (c) Find the initial height h from which the stone was thrown. (d) Find the maximum height H reached by the stone.

46. The citizens of Paris were terrified during World War I when they were suddenly bombarded with shells fired from a long-range gun known as Big Bertha. The barrel of the gun was 36.6 m long and it had a muzzle speed of 1.46 km/s. When the gun's angle of elevation was set to 55°, what would be the range? For the purposes of solving this problem, neglect air resistance. (The actual range at this elevation was 121 km; air resistance cannot be neglected for the high muzzle speed of the shells.)

✦ 47. A projectile is launched from the origin with initial speed v_0 at an angle θ above the horizontal. Show that the equation of the trajectory followed by the projectile is

$$ y = \left(\frac{v_{0y}}{v_{0x}}\right)x + \left(\frac{-g}{2v_{0x}^2}\right)x^2 $$

4.6 Other Examples of Constant Acceleration

48. A particle experiences a constant acceleration that is south at 2.50 m/s². At $t = 0$, its velocity is 40.0 m/s east. What is its velocity at $t = 8.00$ s?

49. A particle experiences a constant acceleration that is north at 100 m/s². At $t = 0$, its velocity vector is 60 m/s east. At what time will the magnitude of the velocity be 100 m/s?

50. In the physics laboratory, a glider is released from rest on a frictionless air track inclined at an angle similar to the one shown in Fig. 3.34. If the glider has gained a speed of 25.0 cm/s in traveling 50.0 cm from the starting point, what was the angle of inclination of the track?

51. A 10.0-kg block is released from rest on a frictionless track inclined at an angle of 45°. (a) What is the net force on the block after it is released? (b) What is the acceleration of the block? (c) If the block is released from rest, how long will it take for the block to attain a speed of 10.0 m/s?

52. A 6.0-kg block, starting from rest, slides down a frictionless incline of length 2.0 m. When it arrives at the bottom of the incline, its speed is v_f. At what distance from the top of the incline is the speed of the block $\frac{1}{2}v_f$?

53. A crate of books is to be put on a truck by rolling it up an incline of angle θ using a dolly. The total mass of the crate and the dolly is m. Assume that rolling the dolly up the incline is the same as sliding it up a frictionless surface. (a) What is the magnitude of the *horizontal* force that must be applied just to hold the crate in place on the incline? (b) What horizontal force must be applied to roll the crate up at constant speed? (c) In order to start the dolly moving, it must be accelerated from rest. What horizontal force must be applied to give the crate an acceleration up the incline of magnitude a?

✦ 54. A toy cart of mass m_1 moves on frictionless wheels as it is pulled by a string under tension T. A block of mass m_2 rests on top of the cart. The coefficient of friction between the cart and the block is μ. Find the maximum tension T that will not cause the block to slide on the cart if the cart rolls on (a) a horizontal surface; (b) up a ramp of angle θ above the horizontal. In both cases, the string is parallel to the surface on which the cart rolls.

55. A solar sailplane is going from Earth to Mars (Fig. 4.42). Its sail is oriented to give a solar radiation force of 8.00×10^2 N. The gravitational force due to the Sun is 173 N and the gravitational force due to the Earth and Moon is 1.00×10^2 N. All forces are in the plane formed by Earth, Sun, and sailplane. The mass of the sailplane is 14,500 kg. (a) What is the net force (magnitude and direction) acting on the sailplane? (b) What is the acceleration of the sailplane?

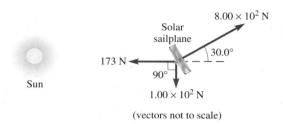

(vectors not to scale)

Figure 4.42 Problem 55

✦ 56. In a playground, two slides have different angles of incline θ_1 and θ_2 ($\theta_2 > \theta_1$). A child slides down the first at constant speed; on the second, his acceleration down the slide is a. Assume the coefficient of kinetic friction is the same for both slides. (a) Find a in terms of θ_1, θ_2, and g. (b) Find the numerical value of a for $\theta_1 = 45°$ and $\theta_2 = 61°$.

✦ 57. A 15-kg crate starts at rest at the top of a 60.0° incline. The coefficients of friction are $\mu_s = 0.40$ and $\mu_k = 0.30$. The crate is connected to a hanging 8.0-kg box by an ideal rope and pulley. (a) As the crate slides down the incline, what is the tension in the rope? (b) How long does it take the crate to slide 2.00 m down the incline? (c) To push the crate back up the incline at constant speed, with what force P should Pauline push on the crate (parallel to the incline)? (d) What is the smallest mass that you could substitute for the 8.0-kg box to keep the crate from sliding down the incline?

58. An accelerometer—a device to measure acceleration—can be as simple as a small pendulum hanging in the cockpit. Suppose you are flying a small plane in a straight horizontal line and your accelerometer hangs 12° behind the vertical (Fig. 4.43). What is your acceleration?

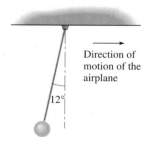

Figure 4.43 Problem 58

4.7 Relative Velocity

59. Two cars are driving toward each other on a straight, flat Kansas road. The Jeep Wrangler is traveling at 82 km/h north and the Ford Taurus is traveling at 48 km/h south, both measured relative to the road. What is the velocity of the Jeep relative to an observer in the Ford?

60. Two cars are driving toward each other on a straight and level road in Alaska. The BMW is traveling at 100.0 km/h north and the VW is traveling at 42 km/h south, both velocities measured relative to the road. At a certain instant, the distance between the cars is 10.0 km. Approximately how long will it take from that instant for the two cars to meet? [*Hint:* Consider a reference frame in which one of the cars is at rest.]

61. A Nile cruise ship takes 20.8 h to go upstream from Luxor to Aswan, a distance of 208 km, and 19.2 h to make the return trip downstream. Assuming the ship's speed relative to the water is the same in both cases, calculate the speed of the current in the Nile.

62. An airplane has a velocity relative to the ground of 210 m/s toward the east. The pilot measures his airspeed (the speed of the plane relative to the air) to be 160 m/s. What is the minimum wind velocity possible?

63. A boat that can travel at 4.0 km/h in still water crosses a river with a current of 1.8 km/h. At what angle must the boat be pointed upstream to travel straight across the river? In other words, in what direction is the velocity of the boat relative to the water?

64. A boy is attempting to swim directly across a river; he is able to swim at a speed of 0.500 m/s relative to the water. The river is 25.0 m wide and the boy ends up at 50.0 m downstream from his starting point. (a) How fast is the current flowing in the river? (b) What is the speed of the boy relative to a friend standing on the riverbank?

65. An aircraft has to fly between two cities, one of which is 600.0 km north of the other. The pilot starts from the southern city and encounters a steady 100.0 km/h wind that blows from the northeast. The plane has a cruising speed of 300.0 km/h in still air. (a) In what direction (relative to east) must the pilot head her plane? (b) How long does the flight take?

66. A Stanley Steamer automobile travels north at 40 km/h and a Pierce Arrow automobile travels east at 50 km/h. Relative to an observer riding in the Stanley Steamer, what are the *x*- and *y*-components of the velocity of the Pierce Arrow car?

67. A person climbs from a Paris metro station to the street level by walking up a stalled escalator in 94 s. It takes 66 s to ride the same distance when standing on the escalator when it is operating normally. How long would it take for him to climb from the station to the street by walking up the moving escalator?

68. Sheena can row a boat at 3.00 mi/h in still water. She needs to cross a river that is 1.20 mi wide with a current flowing at 1.60 mi/h. Not having her calculator ready, she guesses that to go straight across, she should head 60.0° upstream. (a) What is her speed with respect to the bank? (b) How long does it take her to cross the river? (c) How far upstream or downstream from her starting point will she reach the opposite bank? (d) In order to go straight across, what angle upstream should she have headed?

69. A dolphin wants to swim directly back to its home bay, which is 0.80 km due west. It can swim at a speed of 4.00 m/s relative to the water, but a uniform water current flows with speed 2.83 m/s in the southeast direction. (a) What direction should the dolphin head? (b) How long does it take the dolphin to swim the 0.80-km distance home?

70. A pilot starting from Athens, New York, wishes to fly to Sparta, New York, which is 320 km from Athens in the direction 20.0° N of E. The pilot heads directly for Sparta and flies at an airspeed of 160 km/h. After flying for 2.0 h, the pilot expects to be at Sparta, but instead he finds himself 20 km due west of Sparta. He has forgotten to correct for the wind. (a) What is the velocity of the plane relative to the air? (b) Find the velocity (magnitude and direction) of the plane relative to the ground. (c) Find the wind speed and direction.

COMPREHENSIVE PROBLEMS

71. A sailboat sails from Marblehead Harbor directly east for 45 nautical miles, then south for 20.0 nautical miles, returns to an easterly heading for 30.0 nautical miles, and sails north for 15.0 nautical miles, then west for 63 nautical miles. At that time the boat becomes becalmed and the auxiliary engine fails to start. The crew decides to notify the Coast Guard of their position. Using graph paper, ruler, and protractor, find the sailboat's displacement from the harbor. Then add the displacement vectors using the component method to obtain a more accurate answer.

72. A vector has magnitude 2.00 units and its *y*-component is +1.00 unit. What is the *x*-component of this vector? Is there more than one possibility? Explain.

73. A bike rider follows a trail 100.0 m south, then 200.0 m east, and finally 75.0 m northeast (45.0° north of east). (a) What total distance has he traveled from his starting point? (b) He decides to ride off across a field directly back to his starting point. How far must he travel to get back and in what direction?

74. A barge is hauled along a straight-line section of canal by two horses harnessed to tow ropes and walking along the tow paths on either side of the canal. Each horse pulls with a force of 560 N at an angle of 15° with the centerline of the canal. Find the resultant force on the barge.

75. A 200.0-N sign is suspended from a horizontal strut of negligible weight (Fig. 4.44). The force exerted on the strut by the

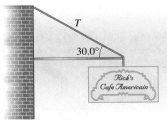

Figure 4.44   Problem 75

wall is horizontal. Draw a free-body diagram to show the forces acting on the strut. Find the tension T in the diagonal cable supporting the strut.

✦ 76. Peter is collecting paving stones from a quarry. He harnesses two dogs, Sandy and Rufus, in tandem to the loaded cart. Sandy pulls with force $\vec{F}$ at a 15° angle to the north of east; Rufus pulls with 1.5 times the force of Sandy and at an angle of 30.0° south of east. Use a ruler and protractor to draw the force vectors to scale (choose a simple scale, such as 2.0 cm ↔ F). Find the resultant net force vector graphically. Measure its length and find the magnitude of the net force from the scale used and the direction with the protractor. Will the cart stay on the road that runs directly west to east?

77. The seagull discussed in this chapter is flying horizontally 8.00 m above the ground at 6.00 m/s. Assume $g = 9.80$ m/s². Ignoring air resistance, (a) what is the horizontal distance to the rocks at the moment that the seagull should let go of the clam? (b) With what speed relative to the rocks does the clam smash into the rocks? (c) With what speed relative to the seagull does the clam smash into the rocks?

78. An airplane is traveling from New York to Paris, a distance of 5.80×10^3 km. Neglect the curvature of the Earth. (a) If the cruising speed of the airplane is 350.0 km/h, how much time will it take for the airplane to make the round-trip on a calm day? (b) If a steady wind blows from New York to Paris at 60.0 km/h, how much time will the round-trip take? (c) How much time will it take if there is a crosswind of 60.0 km/h?

79. A helicopter is flying horizontally at 8.0 m/s and an altitude of 18 m when a package of emergency medical supplies is ejected horizontally backward with a speed of 12 m/s *relative to the helicopter*. Ignoring air resistance, what is the horizontal distance between the package and the helicopter when the package hits the ground?

80. A particle has a constant acceleration of 5.0 m/s² to the east. At time $t = 0$, it is 2.0 m east of the origin and its velocity is 20 m/s north. What are the components of its position vector at $t = 2.0$ s?

ⓒ 81. A toboggan is sliding down a snowy slope. Table 4.1 shows the speed of the toboggan at various times during its trip. (a) Make a graph of the speed as a function of time. (b) Judging by the graph, is it plausible that the toboggan's acceleration is constant? If so, what is the acceleration? (c) Ignoring friction, what is the angle of incline of the slope? (d) If friction is significant, is the angle of incline larger or smaller than that found in (c)? Explain.

82. The range R of a projectile is defined as the magnitude of the horizontal displacement of the projectile *when it returns to its original altitude*. In other words, the range is the distance the projectile will travel on flat ground. A projectile is launched at $t = 0$ with initial speed v_0 at an angle θ above the horizontal. (a) Find the time t at which the projectile returns to its original altitude. (b) Show that the range is

$$R = \frac{v_0^2 \sin 2\theta}{g}$$

[*Hint:* Use the trigonometric identity $\sin 2\theta = 2 \sin \theta \cos \theta$.] (c) What value of θ gives the maximum range? What is this maximum range?

83. A projectile is launched at $t = 0$ with initial speed v_0 at an angle θ above the horizontal. (a) What are v_x and v_y at the projectile's highest point? (b) Find the time t at which the projectile reaches its maximum height. (c) Show that the maximum height H of the projectile is

$$H = \frac{(v_0 \sin \theta)^2}{2g}$$

✦✦ 84. An airplane of mass 2800 kg has just lifted off the runway. It is gaining altitude at a constant 2.3 m/s while the horizontal component of its velocity is increasing at a rate of 0.86 m/s². Assume $g = 9.81$ m/s². (a) Find the direction of the force exerted on the airplane by the air. (b) Find the horizontal and vertical components of the plane's acceleration if the force due to the air has the same magnitude but has a direction 2.0° closer to the vertical than its direction in part (a).

85. A locust jumps at an angle of 55.0° and lands 0.800 m from where it jumped. (a) What is the maximum height of the locust during its jump? Ignore air resistance and assume $g = 9.80$ m/s². (b) If it jumps with the same initial speed at an angle of 45.0°, would the maximum height be larger or smaller? (c) What about the range? (d) Calculate the maximum height and range for this angle.

✦ 86. A spotter plane sees a school of tuna swimming at a steady 5.00 km/h northwest. The pilot informs a fishing trawler; which is just then 100.0 km due south of the fish. The trawler sails along a straight-line course and intercepts the tuna after 4.0 h. How fast did the trawler move? [*Hint:* First find the velocity of the trawler relative to the tuna.]

Table 4.1
Problem 81

Time Elapsed, t (s)	Speed of Toboggan, v (m/s)
0	0
1.14	2.8
1.62	3.9
2.29	5.6
2.80	6.8

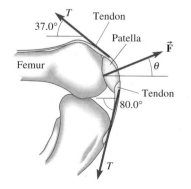

Figure 4.45 Problem 88

87. A baseball batter hits a long fly ball that rises to a height of 44 m. An outfielder on the opposing team can run at 7.6 m/s. What is the farthest the fielder can be from where the ball will land so that it is possible for him to catch the ball?

88. Figure 4.45 shows the quadriceps and the patellar tendons attached to the patella (the kneecap). If the tension T in each tendon is 1.30 kN, what is (a) the magnitude and (b) the direction of the contact force $\vec{F}$ exerted on the patella by the femur?

✦ 89. One of the tricky things about learning to sail is distinguishing the true wind from the apparent wind. When you are on a sailboat and you feel the wind on your face, you are experiencing the *apparent wind*—the motion of the air relative to you. The true wind is the speed and direction of the air relative to the water while the apparent wind is the speed and direction of the air relative to the *sailboat*. Fig. 4.46 shows three different directions for the true wind along with one possible sail orientation as indicated by the position of the boom attached to the mast. (a) In each case, draw a vector diagram to establish the magnitude and direction of the apparent wind. (b) In which of the three cases is the apparent wind speed greater than the true wind speed? (Assume that the speed of the boat relative to the water is less than the true wind speed.) (c) In which of the three cases is the direction of the apparent wind direction forward of the true wind? ("Forward" means coming from a direction more nearly straight ahead. For example, (1) is forward of (2) which is forward of (3).)

90. Figure 4.47 shows an elastic cord attached to two back teeth and stretched across a front tooth. The purpose of this arrangement is to apply a force $\vec{F}$ to the front tooth. (The figure has been simplified by running the cord straight from the front tooth to the back teeth.) If the tension in the cord is 1.2 N, what are the magnitude and direction of the force $\vec{F}$ applied to the front tooth?

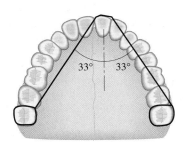

Figure 4.47   Problem 90

91. Figure 4.48 shows a student's head bent over her physics book. The head weighs 50.0 N and is supported by the muscle force $\vec{F}_m$ exerted by the neck extensor muscles and by the contact force $\vec{F}_c$ exerted at the atlanto-occipital joint. Given that the magnitude of $\vec{F}_m$ is 60.0 N and is directed 35° below the horizontal, find (a) the magnitude and (b) the direction of $\vec{F}_c$.

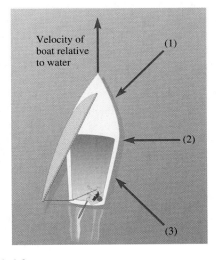

Figure 4.46   Problem 89

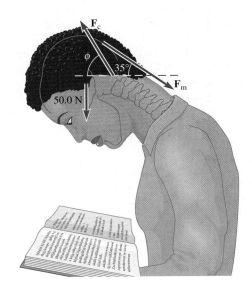

Figure 4.48   Problem 91

92. When salmon head upstream to spawn, they often must make their way up a waterfall. If the water is not moving too fast, the salmon can swim right up through the falling water. If the water is falling with too great a speed, the salmon jump out of the water to get to a place in the waterfall where the water is not falling so fast. When humans build dams that interrupt the usual route followed by the salmon, artificial fish ladders must also be built to allow the salmon to get back uphill to the spawning area. These fish ladders consist of a series of small waterfalls with still pools of water in between them (Fig. 4.49). Assume that the water is at rest in the pools at the top and bottom of one "rung" of the fish ladder, that water falls straight down from one pool to the next, and that salmon can swim at 5.0 m/s with respect to the water. (a) What is the maximum height of a waterfall up which the salmon can swim without having to jump? (b) If a waterfall is 1.5 m high, how high must the salmon jump to get to water through which it can swim? Assume that they jump straight up. (c) What initial speed must a salmon have to jump the height found in part (b)? (d) For a 1.0-m-high waterfall, how fast will the salmon be swimming with respect to the ground when it starts swimming up the waterfall?

93. Jesse James and his gang are planning to rob the train carrying the payroll for the Lost Gulch mine. A Wells Fargo guard on the train spots Jesse astride his horse at a perpendicular distance of 0.300 km from the train tracks. The train is moving at 25.0 m/s on a straight track. The guard wants to frighten Jesse and the desperadoes away by shooting a hole in Jesse's ten-gallon hat, which happens to be at the same height as the guard's gun (Fig. 4.50). The muzzle velocity of the guard's gun is 0.350 km/s. Ignore air resistance in parts (a)–(c). (a) Should he aim the gun directly at Jesse's hat, in front of the hat, or behind the hat? Explain. (b) Unsure of the correct angle to aim the gun, the guard decides to aim his gun at a right angle to the track and fire the bullet a little before the time when he will be directly opposite to Jesse. At what distance before the point where the guard is directly opposite Jesse James should the guard fire? (c) Does the guard need to worry about the force of gravity on the bullet? If so, how far above the hat should he aim? (d) Given that the terminal speed of the bullet is about 120 m/s, should the guard fire earlier or later than the answer found in part (b)?

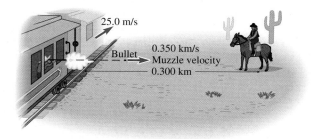

Figure 4.50   Problem 93 (figure not to scale)

Figure 4.49   Problem 92. A fish ladder

ANSWERS TO PRACTICE PROBLEMS

4.1 (a) larger; (b) yes, $F_S > F_B$

4.2 $A_x = +16$ km; $A_y = -8.2$ km; $B_x = +17$ km; $B_y = 0$; $C_x = -11$ km; $C_y = +47$ km

4.3 (a) 1.01 kN; (b) 2.45 kN

4.4 3.1 kN

4.5 (a) $F_x = 49.1$ N; $F_y = 2.9$ N; (b) $F = 49.2$ N; (c) 3.4° above the horizontal

4.6 $f_k = 110$ N; $W = 230$ N

4.7 3.19 knots directed 56.9° north of east

4.8 2

4.9 $y_{max} = \dfrac{v_0^2}{4g}$

4.10 $\vec{v}_2 - \vec{v}_1$ stays the same, since both velocities change at the same rate—both have the same acceleration. The difference in position ($\vec{r}_2 - \vec{r}_1$) changes, since their velocities have different y-components and, therefore, y_2 changes at a different rate from y_1. $x_2 - x_1$ stays constant, though, since both have the same constant horizontal velocity component.

4.11 $v_{fx} = 500.0$ m/s; $v_{fy} = -19.8$ m/s; the bullet enters the water at an angle of 2.27° below the horizontal.

4.12 54 N

4.13 $a_x = 3.2$ m/s^2 and $v_x = 3.6$ m/s

4.14 (a) down the incline; (b) up the incline; (c) 0.2 m/s^2 down the incline

4.15 (a) 1.0 m/s; (b) 15 min

4.16 46° north of west

Circular Motion

Chapter 5

Alekos, a Greek goatherd, notices a wolf sneaking up toward his goats. After a moment of panic, he picks up a stone and puts it in his sling. He begins to whirl the stone faster and faster in a circle above his head, as the wolf inches closer and closer. To save the goats, Alekos must release the stone at precisely the right point in its circular path so it hits the wolf and drives it away. At what point should he release the stone?

5.1　DESCRIPTION OF UNIFORM CIRCULAR MOTION

It might seem overly dramatic to introduce circular motion in life and death terms, but your life depends every day on *someone's* understanding of circular motion. Ask someone to name the most important machine ever invented by humans and you are likely to get *the wheel* as a response. Rotating objects are so essential to modern—and even not-so-modern—technology that we barely notice them. Examples include wheels on cars, bicycles, trains, and lawnmowers; propellers on airplanes and helicopters; CDs and DVDs; hard and floppy computer disks; the gears and hands of an analog watch; amusement park rides and centrifuges—the list is endless.

As Alekos whirls the stone, it travels in a circle. A nonzero net force must act on any object moving in a circle since its velocity is changing in direction (and, perhaps, also in magnitude). The stone rests in a leather pouch attached to two cords, which Alekos holds in his hand. The stone is held in its circular path by an inward force due to the pouch. If Alekos releases one of the cords, the pouch no longer holds the stone in its circular path; what is the subsequent path of the stone?

Initially, let us neglect the weight of the stone. Once the force due to the sling stops acting on the stone, the stone's velocity stops changing direction. The stone then continues to move in the direction of its velocity at the instant the force due to the sling stops acting. Newton's first law applies now: a body with no external forces acting on it moves at a constant speed along a straight line. Since the velocity is tangent to the stone's circular path, the stone initially heads off along a straight line tangent, to the circle (Fig. 5.1).

A satellite orbiting Earth may also travel in a circular path. Instead of being held in its path by the tension in a cord, it is held by the gravitational force on the satellite due to Earth's gravitational field. One of the main goals of this chapter is to find out what net force pulling inward is required to keep an object moving in a circle. The net force required turns out to depend both on the radius of the circular path and the speed at which the object moves.

To describe the circular motion of the stone or the satellite, we could use the familiar definitions of displacement, velocity, and acceleration. But much of the circular motion around us occurs in the rotation of a rigid object. A **rigid body** is one for which the distance between any two points of the body remains the same when the body is translated or rotated. When such an object rotates, every point on the object moves in a circular path. The radius of the path for any point is the distance between that point and the axis of rotation. When a compact disk spins inside a CD player, each point on the CD moves through a different distance and has a different velocity and acceleration. The velocity and acceleration of a given point keep changing direction as the CD spins. It would be clumsy to describe the rotation of the CD by talking about the motion of arbitrary points on it. However, some quantities are *the same* for every point on the CD. It is much simpler, for instance, to say "the CD spins at 210 rpm" instead of saying "a point 6.0 cm from the rotation axis of the CD is moving at 1.3 m/s."

To simplify the description of circular motion, we concentrate on angles instead of distances. If a CD spins through 1/4 of a turn, every point moves through the same angle (90°), but different points move different linear distances. On the CD shown in Fig. 5.2a, point 1 near the axis of rotation moves through a smaller distance than point 4 on the circumference. For this reason we define a set of variables that are analogous to displacement, velocity, and acceleration, but use angular measure instead of linear distance. Instead of displacement, we speak of **angular displacement** $\Delta\theta$, the angle through which the CD turns. A point on the CD moves along the circumference of a circle. As the point moves from the angular position θ_1 to angular position θ_2, a radial line drawn between the center of the circle and that point sweeps out an angle $\Delta\theta = \theta_2 - \theta_1$, which is the angular displacement of the CD during that time interval (Fig. 5.2b).

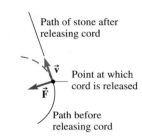

Making The Connection:
whirling a stone in a sling

Path of stone after releasing cord

$\vec{v}$

Point at which cord is released

$\vec{F}$

Path before releasing cord

Figure 5.1　Path of a stone before and after a cord is released.

The abbreviation rpm means "revolutions per minute."

Definition of angular displacement

$$\Delta\theta = \theta_2 - \theta_1$$

(5-1)

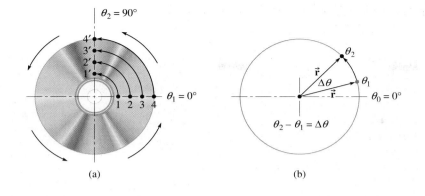

(a)   (b)

Figure 5.2 (a) A CD rotates through $\frac{1}{4}$ turn; points 1, 2, 3, and 4 travel through the same angle but different distances to reach their new positions, marked 1′, 2′, 3′, and 4′, respectively. (b) Angular positions such as θ_1 and θ_2 are measured from a reference axis (here marked $\theta_0 = 0°$).

Just as with displacement, angular displacement tells us in what direction the rotation occurred. The usual convention is that a positive angular displacement represents counterclockwise rotation and a negative angular displacement represents clockwise rotation. Counterclockwise and clockwise are only well defined for a particular viewing direction; counterclockwise rotation viewed from above is clockwise when viewed from below.

The **average angular velocity** ω_{av} is the average rate of change of the angular displacement.

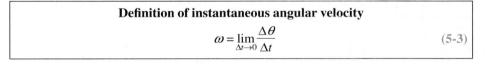

Definition of average angular velocity

$$\omega_{av} = \frac{\Delta\theta}{\Delta t} \tag{5-2}$$

If we let the time interval Δt become shorter and shorter, we are averaging over smaller and smaller time intervals. In the limit $\Delta t \rightarrow 0$, ω_{av} becomes the **instantaneous angular velocity** ω.

Definition of instantaneous angular velocity

$$\omega = \lim_{\Delta t \rightarrow 0} \frac{\Delta\theta}{\Delta t} \tag{5-3}$$

The angular velocity also indicates—through its algebraic sign—in what direction the CD is spinning. Since angular displacements can be measured in degrees or radians, angular velocities have units such as degrees/second, radians/second, degrees/day, and the like. The SI unit is rad/s, which stands for radians per second.

You may be most familiar with measuring angles in degrees, but in many situations the most convenient measure is the *radian*. One such situation is when we relate the angular displacement or angular velocity of a rotating object with the distance traveled by, or the speed of, some point on the object.

In Fig. 5.3, an angle θ between two radii of a circle define an arc of length s. We say that θ is the angle *subtended* by the arc. The angle θ in radians is defined as

$$\theta \text{ (in radians)} = \frac{s}{r} \tag{5-4}$$

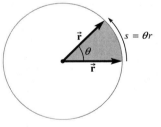

Figure 5.3 Definition of the radian: angle θ in radians is defined by the arc length s divided by the radius r.

where r is the radius of the circle. Since an angle in radians is defined by the ratio of two lengths, it is dimensionless (a pure number). We use the term radians, abbreviated "rad" to keep track of the angular measure used. Since "rad" is not a physical unit like meters or kilograms, it does not have to balance in Eq. (5-4). For the same reason, we can drop "rad" whenever there is no chance of being misunderstood. We may write $\omega = 23 \text{ s}^{-1}$ as long as context makes it clear that we mean 23 radians per second.

Since the arc length for an angle of 360° is the circumference of the circle, the radian measure of an angle of 360° is

$$\theta = \frac{s}{r} = \frac{2\pi r}{r} = 2\pi \text{ rad}$$

Therefore, the conversion factor between degrees and radians is

$$360° = 2\pi \text{ rad}$$

Example 5.1

Angular Speed of Earth

Earth is rotating about its axis. What is its angular speed in rad/s? (The question asks for angular *speed*, so we do not have to worry about the direction of rotation.)

Strategy The Earth's angular velocity is constant, or nearly so. Therefore, we can calculate the average angular velocity for any convenient time interval and, in turn, the Earth's instantaneous angular speed $|\omega|$.

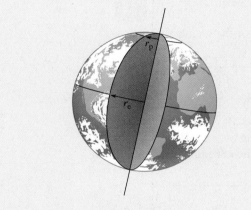

Solution It takes the Earth 1 day to complete one rotation, during which the angular displacement is 2π rad. More formally, during a time interval $\Delta t = 1$ day, the angular displacement of the Earth is $\Delta\theta = 2\pi$ rad. So the angular speed of the Earth is 2π rad/day and then convert days to seconds.

$$1 \text{ day} = 24 \text{ h} = 24 \text{ h} \times 3600 \text{ s/h} = 86{,}400 \text{ s}$$

$$|\omega| = \frac{2\pi \text{ rad}}{86{,}400 \text{ s}} = 7.27 \times 10^{-5} \text{ rad/s}$$

Discussion Notice that this problem is analogous to a problem in linear motion such as: "A car travels in a straight line at constant speed. In 3 h, it has traveled 192 mi. What is its velocity in m/s?" Just about everything in circular motion and rotation has this kind of analog—which means we can draw heavily on what we have already learned.

The *angular* velocity of a person near one of the poles due to Earth's rotation is the same as that of someone at the equator, although the linear speeds are very different. Both move through the same *angle* in the same time interval, but the person at the equator moves through a much larger distance.

Practice Problem 5.1 Angular speed of Venus

Given that a day on Venus is 243 Earth days, what is the angular speed of the rotation of Venus in rad/s?

Relation between Linear and Angular Speed

For a point moving in a circular path of radius r, the linear distance traveled along the circular path during an angular displacement of $\Delta\theta$ (in radians) is the arc length s where

$$s = r\Delta\theta = r(\theta_2 - \theta_1) \quad \text{(angles in radians)} \tag{5-5}$$

The point in question could be a point particle moving in a circular path, or it could be any point on a rotating rigid object. Since Eq. (5-5) comes directly from the definition of the radian, any equation derived from Eq. (5-5) is valid only when the angles are measured in radians.

What is the linear speed at which the point moves? The average linear speed is the distance traveled divided by the time interval,

$$v_{av} = \frac{s}{\Delta t} = \frac{r\Delta\theta}{\Delta t} \quad (\Delta\theta \text{ in radians})$$

We recognize $\Delta\theta/\Delta t$ as the average angular velocity ω_{av}. If we take the limit as Δt goes to zero, both average quantities (v_{av} and ω_{av}) become instantaneous quantities. Therefore, the relationship between linear speed and angular speed is

$$v = r\omega \quad (\omega \text{ in radians per unit time}) \tag{5-6}$$

For a rotating object, points farther from the axis move at higher linear speeds; they have a circle of bigger radius to travel and therefore cover more distance in the same time interval.

When the speed of a point moving in a circle is constant, its motion is called **uniform circular motion**. Even though the speed of the point is constant, the velocity is not: the direction of the velocity is changing. This distinction is important when we find the acceleration of an object in uniform circular motion (Section 5.2). The time for the point to travel completely around the circle is called the **period** of the motion, T. The **frequency** of the motion, which is the number of revolutions per unit time, is defined as

$$f = \frac{1}{T} \tag{5-7}$$

since

$$\frac{\text{revolutions}}{\text{second}} = \frac{1}{\text{seconds/revolution}}$$

The speed is the total distance traveled divided by the time taken,

$$v = \frac{2\pi r}{T} = 2\pi r f \tag{5-8}$$

Then, for uniform circular motion

$$\omega = \frac{v}{r} = 2\pi f \quad (\omega \text{ in radians per unit time}) \tag{5-9}$$

where, in SI units, angular velocity ω is measured in rad/s and frequency f is measured in hertz (Hz). The hertz is a derived unit equal to 1 rev/s. The dimensions of Eq. (5-9) are correct since both revolutions and radians are pure numbers. The physical dimensions on both sides are a number per second (s^{-1}).

In uniform circular motion, speed is constant but velocity is *not* constant because the *direction* of the velocity is changing.

SI unit of frequency: 1 Hz = 1 rev/s

Example 5.2

Speed in a Centrifuge

A centrifuge is spinning at 5400 rpm. (a) Find the period (in s) and frequency (in Hz) of the motion. (b) If the radius of the centrifuge is 14 cm, how fast (in m/s) is an object at the outer edge moving?

Strategy Remember that rpm means *revolutions per minute*. 5400 rpm *is* the frequency, but in a unit other than Hz. After a unit conversion, the other quantities can be found using the relations already discussed.

Solution (a) First convert rpm to Hz:

$$f = 5400 \, \frac{\text{rev}}{\text{min}} \times \frac{1 \, \text{min}}{60 \, \text{s}} = 90 \, \text{rev/s}$$

so the frequency is $f = 90$ Hz $= 90 \, \text{s}^{-1}$. The period is

$$T = 1/f = 0.011 \, \text{s}$$

(b) To find the linear speed, we first find the angular speed in rad/s:

$$\omega = 90 \, \frac{\text{rev}}{\text{s}} \times 2\pi \, \frac{\text{rad}}{\text{rev}} = 180\pi \, \text{rad/s}$$

So $\omega = 2\pi f = 180\pi$ rad/s. The linear speed is

$$v = \omega r = 180\pi \, \text{s}^{-1} \times 0.14 \, \text{m} = 79 \, \text{m/s}$$

Discussion Notice that much of this problem was done with unit conversions. Instead of memorizing a formula such as $\omega = 2\pi f$, an understanding of where the formula came from (in this case, that 2π radians correspond to one revolution) is more useful and less prone to error.

Practice Problem 5.2   Clothing in the drier

An automatic clothing drier spins at 51.6 rpm. If the radius of the drier drum is 30.5 cm, how fast is the outer edge of the drum moving?

Rolling: Rotation and Translation Combined

When an object is rolling, it is both rotating and translating. The wheel rotates about an axle, but the axle is not at rest; it moves forward or backward. What is the relationship between the angular speed of the wheel and the linear speed of the axle? You might guess that $v = \omega r$ is the answer. You would be right, as long as the object rolls without slipping or skidding.

There is no fixed relationship between the linear and angular speeds of a wheel if it is allowed to skid or slip. When an impatient driver guns the engine the instant a traffic light turns green, the automobile wheels are likely to slip. The rubber sliding against the road surface makes the squealing sound and leaves tracks on the road. The driver could actually make the acceleration of the car greater by giving the engine *less* gas. When the wheels are skidding or slipping, *kinetic* friction propels the car forward instead of the potentially larger force of *static* friction.

For a wheel that rolls *without* slipping, as the wheel turns through one complete rotation, the axle moves a distance equal to the circumference of the wheel. Think of a paint roller leaving a line of paint as it rolls along a wall. After one complete rotation, the same point on the roller wheel is touching the wall as was initially touching it. The length of the line of paint is $2\pi r$. The elapsed time is T, so the axle's speed is

$$v_{\text{axle}} = \frac{2\pi r}{T}$$

while the angular velocity of the roller is

$$\omega = \frac{2\pi}{T}$$

Thus,

$$v_{\text{axle}} = \omega r \quad (\omega \text{ in radians per unit time})$$

Example 5.3

Angular Speed of a Rolling Wheel

Kevin is riding his motorcycle at a speed of 13.0 m/s. If the diameter of the rear tire is 65.0 cm, what is the angular speed of the rear wheel?

Strategy   The given diameter of the tire enables us to find the circumference and thus the distance traveled in one revolution of the wheel. From the speed of the motorcycle we can find how many revolutions the tire must make per second.

Given: $v = 13.0$ m/s; $d_{\text{wheel}} = 65.0$ cm.
To find: angular velocity ω.

Solution   During one revolution of the wheel, the motorcycle travels a distance equal to the tire's circumference (Fig. 5.4). Then the time T to make one revolution is the time to travel a linear distance equal to $2\pi r$,

$$T = \frac{\text{distance}}{\text{speed}} = \frac{2\pi r}{v}$$

Since T is the time per revolution, the number of revolutions per second is $1/T$.

$$f = \text{number of revolutions/s} = \frac{v}{2\pi r}$$

For each revolution there is an angular displacement of 2π radians,

$$\omega = 2\pi \text{ radians} \times \text{number of revolutions/s} = 2\pi f = 2\pi \frac{v}{2\pi r}$$

so, as expected,

$$\omega = \frac{v}{r} = \frac{13.0 \text{ m/s}}{0.325 \text{ m}} = 40.0 \text{ rad/s}$$

Discussion   Check: time for one revolution is $\frac{2\pi \text{ rad}}{40.0 \text{ rad/s}} = $ 0.157 s. Time to travel a distance $2\pi r = 2.04$ m is $\frac{2.04 \text{ m}}{13.0 \text{ m/s}} = $ 0.157 s. Looks good.

You could have obtained this answer immediately by looking back through the text for the equation $\omega = v/r$ and plugging in numbers, but the solution here shows that you can re-create that equation. Here, and in many cases, there is no need to memorize a formula if you understand the concepts behind the formula. You are also less apt to make a mistake by forgetting a factor or constant in the equation, or by using an inappropriate formula. For another example, if an object moves along a straight line at a constant velocity, you know instantly that the displacement is the velocity times the time interval—

continued on next page

Figure 5.4   Position of a motorcycle wheel after one revolution.

Example 5.3 *continued*

not because you have memorized an equation ($\Delta \vec{r} = \vec{v} \Delta t$) but because you understand the concepts of displacement and velocity. This is the sort of internalization of scientific thinking that you will develop with more and more practice in problem solving.

Practice Problem 5.3   Rolling drum

A cylindrical steel drum is tipped over and rolled along the floor of a warehouse. If the drum has a radius of 0.40 m and makes one complete turn every 8.0 s, how long does it take to roll the drum 36 m?

5.2   CENTRIPETAL ACCELERATION

For a particle undergoing uniform circular motion, the magnitude of the velocity is constant, but the direction is continuously changing. At any instant of time, the direction of the instantaneous velocity is tangent to the path, as discussed in Section 4.4. Since the *direction* of the velocity continually changes, the particle has a nonzero acceleration.

In Fig. 5.5a, two velocity vectors of equal magnitude are drawn tangent to a circular path of radius r, representing the velocity at two different times of an object moving around a circular path with constant speed. At any instant, the velocity vector is perpendicular to a radius drawn from the center of the circle to the position of the object. As the time between velocity measurements approaches zero, the radii become closer together (Fig. 5.5b). To find the acceleration, $\vec{a} = \lim\limits_{\Delta t \to 0} \dfrac{\Delta \vec{v}}{\Delta t}$, we must first find the change in velocity $\Delta \vec{v}$ for a very short time interval. Figure 5.5c shows that as the time interval Δt approaches zero, the angle between the two velocities also approaches zero and $\Delta \vec{v}$ becomes perpendicular to the velocity.

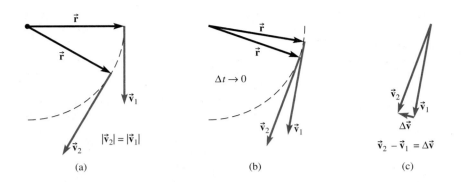

Figure 5.5 Uniform circular motion at constant speed. (a) The velocity is always tangent to the circular path and perpendicular to the radius at that point. (b) As the time interval between two velocity measurements decreases, the angle between the velocity vectors decreases. (c) The change in velocity ($\Delta\vec{\mathbf{v}}$) is found by placing the tails of the two velocity vectors together. Then $\Delta\vec{\mathbf{v}}$ is drawn from the tip of the initial velocity ($\vec{\mathbf{v}}_1$) to the tip of the final velocity ($\vec{\mathbf{v}}_2$).

Since $\Delta\vec{\mathbf{v}}$ is perpendicular to the velocity, it is directed along a radius of the circle. Inspection of Figs. 5.5b and 5.5c shows that $\Delta\vec{\mathbf{v}}$ is radially *inward* (toward the center of the circle). Since the acceleration $\vec{\mathbf{a}}$ has the same direction as $\Delta\vec{\mathbf{v}}$ (in the limit $\Delta t \rightarrow 0$), the acceleration is also directed radially inward.

The word **centripetal** is a synonym for *radially inward*. Literally it means *toward the center*. The acceleration of an object undergoing *uniform* circular motion is often called the **centripetal acceleration, $\vec{\mathbf{a}}_c$.** The word "centripetal" here just reminds us of the direction of the acceleration.

Since the acceleration of an object in uniform circular motion is radially inward, the net force acting on that object must also be directed radially inward, perpendicular to the path of motion. For the stone Alekos whirls about in a sling, the tension in the cords pulls inward on the stone to hold it in its circular path. For a satellite orbiting Earth in a circular orbit, the gravitational force always pulls inward on the satellite toward the center of the Earth, which is the center of its circular orbit.

> In uniform circular motion, the direction of the acceleration is radially inward.

Magnitude of the Centripetal Acceleration

To find the magnitude of the centripetal acceleration for uniform circular motion, we must find the change in velocity $\Delta\vec{\mathbf{v}}$ for a time interval Δt in the limit $\Delta t \rightarrow 0$. The velocity keeps the same magnitude but changes direction at a steady rate, equal to the angular velocity ω. In a time interval Δt, the velocity $\vec{\mathbf{v}}$ rotates through an angle equal to the angular displacement $\Delta\theta = \omega\,\Delta t$. During this time interval, the velocity vector sweeps out an arc of a circle of "radius" v (Fig. 5.6). In the limit $\Delta t \rightarrow 0$, the magnitude of $\Delta\vec{\mathbf{v}}$ becomes equal to the arc length, since a very short arc approaches a straight line. Then

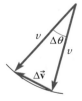

Figure 5.6 The velocity vector sweeps out an arc of a circle whose magnitude is equal to the magnitude of $\Delta\vec{\mathbf{v}}$ in the limit $\Delta t \rightarrow 0$.

$$|\Delta\vec{\mathbf{v}}| = \text{arc length} = \text{radius of circle} \times \text{angle subtended}$$
$$= v\Delta\theta = v\omega\Delta t$$

Acceleration is the rate of change of velocity, so the magnitude of the centripetal acceleration is

$$a_c = |\vec{\mathbf{a}}| = \frac{|\Delta\vec{\mathbf{v}}|}{\Delta t} = v\omega \quad (\omega \text{ in radians per unit time}) \tag{5-10}$$

where absolute value symbols are used with the vector quantities to indicate their magnitudes. Velocity and angular velocity are not independent; $v = \omega r$. It is usually most convenient to write the magnitude of the centripetal acceleration in terms of one or the other of these two quantities. So we write the centripetal acceleration in two other equivalent ways using $v = \omega r$:

$$a_c = \frac{v^2}{r} \quad \text{or} \quad a_c = \omega^2 r \quad (\omega \text{ in radians per unit time}) \tag{5-11}$$

Note that Eqs. (5-10) and (5-11) are valid only when ω is measured in *radians* per unit time (normally rad/s, but rad/min or rad/h would be correct).

Example 5.4

A Spinning CD

If a CD spins at 210 rpm, what is the centripetal acceleration of a point on the outer rim of the CD? The CD is 12 cm in diameter.

Strategy From the number of revolutions per minute, we can find the frequency and the angular velocity. The angular velocity and the radius of the CD enable us to calculate the centripetal acceleration.

Solution We convert 210 rpm into a frequency in revolutions per second (Hz).

$$f = 210 \, \frac{\text{rev}}{\text{min}} \times \frac{1}{60} \, \frac{\text{min}}{\text{s}} = 3.5 \, \frac{\text{rev}}{\text{s}} = 3.5 \, \text{Hz}$$

For each revolution, the CD rotates through an angle of 2π radians. The angular velocity is

$$\omega = 2\pi f = 2\pi \, \frac{\text{radians}}{\text{rev}} \times 3.5 \, \frac{\text{rev}}{\text{s}} = 7.0\pi \, \text{rad/s}$$

Then using Eq. (5-11), the centripetal acceleration is

$$a_c = \omega^2 r = (7.0\pi \, \text{rad/s})^2 \times 0.060 \, \text{m} = 29 \, \text{m/s}^2$$

Discussion When finding the centripetal acceleration, use whichever form of Eq. (5-11) is more convenient. For rotating objects such as the spinning CD, it's usually easiest to think in terms of the angular velocity. For an object moving around a circle, such as a satellite in orbit whose speed is known, it might be easier to use v^2/r. Since the two equations are equivalent, either can be used in any situation.

Practice Problem 5.4 Centripetal acceleration of a point on an old record

What is the centripetal acceleration of a point 25.4 cm from the center of a record that is rotating at 78 rpm on a turntable?

Now that we know the magnitude and direction of the acceleration of any object in uniform circular motion, we can use Newton's second law to relate the net force acting on the object to the speed and radius of its motion. The net force is found in the usual way: each of the individual forces acting on the object is identified and then the forces are added as vectors. Every force acting must be exerted *by some other object*. Resist the temptation to add in a new, separate force just because something moves in a circle. For an object to move in a circle at constant speed, real, physical forces such as gravity, tension, normal forces, and friction must act on it; these forces combine to produce a net force that has the correct magnitude and is always perpendicular to the velocity of the object.

Problem-Solving Strategy for an Object in Uniform Circular Motion

1. Begin as for any Newton's second law problem: identify all the forces acting on the object and draw a free-body diagram.
2. Choose perpendicular axes at the point of interest so that one is centripetal and the other is tangent to the circular path.
3. Find the centripetal component of each force.
4. Apply Newton's second law as follows:

$$\Sigma F_c = ma_c$$

where ΣF_c is the centripetal component of the net force and the centripetal component of the acceleration is

$$a_c = \frac{v^2}{r} = \omega^2 r$$

(For uniform circular motion, neither the net force nor the acceleration has a tangential component.)

Now let's consider Alekos and the whirling stone again. The path of the stone, once it is released, is tangent to the circular path it had been following. We had neglected the weight of the stone in comparison with the other forces acting. However, once the cord is released, the weight of the stone can no longer be neglected; the gravitational force pulls the stone down toward the ground. The stone then moves in a parabolic path with an initial velocity tangent to the original circular path and parallel to the ground (assuming the plane of the circular path is parallel to the plane of the ground). The stone is also at the high point of the parabolic path, on its journey to the ground, because it has no initial component of velocity in the vertical direction.

Example 5.5

Alekos and the Wolf

Alekos, practicing his wolf deterrence strategies, is whirling a 0.50-kg stone about in a sling. The sling consists of two 80.0-cm-long cords tied to either side of a pouch that holds the stone. To release the stone, Alekos lets go of one cord. The sling makes 4.8 revolutions per second in a horizontal plane 2.0 m above the ground. (a) What is the speed of the stone when it leaves the sling? (b) What is the tension in each of the two cords (assumed equal) before one is released? (c) How far from the release point does the stone land? Ignore air resistance. (d) Sketch an overhead view of Alekos and the wolf and indicate where along the circular path he should release the cord.

Strategy We can find the linear speed of the stone at the time Alekos releases it from the radius and angular velocity. The tension in the cords can be found from the centripetal acceleration and the mass of the stone. Once the stone flies off, it follows the path of a projectile launched horizontally. We use the equations of motion for constant acceleration to find the landing distance of the projectile.

Solution (a) We find the angular velocity from the number of revolutions per second.

$$\omega \,(\text{rad/s}) = 4.8 \text{ rev/s} \times 2\pi \text{ rad/rev} = 30 \text{ rad/s}$$

The tangential velocity component is

$$v_t = \omega r = 30 \text{ rad/s} \times 0.800 \text{ m} = 24 \text{ m/s}$$

so that is the speed with which the stone leaves the sling.

(b) The net force on the stone and the leather pouch can be found from Newton's second law. We neglect the weight of the stone and the pouch, assuming that it is small compared to the tension in the cords. Then the tension in the cords is the only significant force acting on the pouch. Assuming that the tensions are equal and the cords are parallel,

$$F_{net} = 2T$$

For uniform circular motion, the net force must act radially inward. We write the centripetal acceleration component in terms of either ω or v_t. Choosing the former,

$$2T = F_{net} = ma_c = mr\omega^2$$

Substituting numerical values,

$$T = \tfrac{1}{2} \times 0.50 \text{ kg} \times 0.800 \text{ m} \times (30 \text{ rad/s})^2 = 180 \text{ N}$$

The tension is much larger than the weight of the stone (≈ 5 N), so the assumption that we could ignore the weight is justified.

(c) Once the cord is released, we treat the stone as a projectile with an initial horizontal velocity of 24 m/s. The time before the stone hits the ground is found by looking at the vertical motion. Initially the stone has no velocity component in the vertical direction, so it is acted on only by the downward force of gravity.

We can use a kinematic equation for constant acceleration to find the time it takes the stone to reach the ground:

$$\Delta y = v_{0y}t + \tfrac{1}{2} a_y t^2 \qquad (4\text{-}7)$$

Everything in this equation is given except for the time. The vertical displacement of the stone, starting 2.0 m above the ground and ending on the ground, is

$$\Delta y = -2.0 \text{ m}$$

Since $v_{0y} = 0$ and $a_y = -g$,

$$\Delta y = \tfrac{1}{2} a_y t^2 = -\tfrac{1}{2} g t^2$$

so that the time before the stone lands is

$$t = \sqrt{\frac{2\,\Delta y}{-g}} = \sqrt{\frac{2 \times (-2.0 \text{ m})}{-9.8 \text{ m/s}^2}} = 0.64 \text{ s}$$

The horizontal displacement of the stone during this time is

$$\Delta x = v_{0x}t = 24 \text{ m/s} \times 0.64 \text{ s} = 15 \text{ m}$$

15 m is the horizontal distance between the release point of the stone and the point where it hits the ground—if it misses the wolf!

(d) The stone flies away at a tangent to the circle, so it must be let go at the position shown in Fig. 5.7 to be aimed directly at the wolf.

Discussion This example demonstrates the cumulative nature of physics concepts. The basic concepts keep reappearing, to be used over and over and to be extended for use in new contexts. Part of the problem involves new concepts (angular velocity and centripetal acceleration); the rest of the problem involves old material (Newton's second law, linear velocity and acceleration, projectile motion, and tension in a cord).

Practice Problem 5.5 Rotating carousel

A carousel is rotating at such a rate that a horse located 8.0 m from the central axis moves at a speed of 6.0 m/s. (a) What is the centripetal acceleration of the horse? (b) At what speed does a horse located 4.0 m from the center move?

Figure 5.7   The stone hits the wolf only if it is released at the correct point in its orbit.

Example 5.6

Conical Pendulum

Suppose you whirl a stone in a horizontal circle at a slower speed so that the weight of the stone is *not* negligible compared to the tension in the cord. Then the cord cannot be horizontal—the tension must have a vertical component to cancel the weight and leave a horizontal net force (Fig. 5.8). If the cord has length L, the stone has mass m, and the cord makes an angle ϕ with the vertical direction, what is the constant angular speed of the stone?

Strategy   The net force must point toward the center of the circle, since the stone is in uniform circular motion. With the

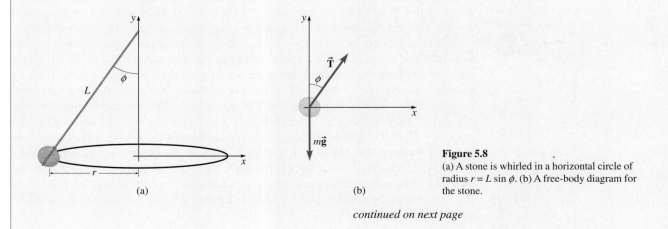

Figure 5.8
(a) A stone is whirled in a horizontal circle of radius $r = L \sin \phi$. (b) A free-body diagram for the stone.

continued on next page

Example 5.6 *continued*

stone in the position depicted in Fig. 5.8a, the direction of the net force is along the +x-axis. This time the tension in the cord does not pull toward the center, but the *net* force does.

Solution Start by drawing a free-body diagram (Fig. 5.8b). Now apply Newton's second law in component form. The acceleration has components $a_x = \omega^2 r$ and $a_y = 0$. For the x-components,

$$\Sigma F_x = T \sin \phi = ma_x = m\omega^2 r$$

Since the problem does not specify r, we must express r in terms of L and ϕ. In Fig. 5.8a, the radius forms a right triangle with the cord and the y-axis. Then

$$r = L \sin \phi$$

and

$$T \cancel{\sin \phi} = m\omega^2 L \cancel{\sin \phi}$$

$$\Sigma F_y = T \cos \phi - mg = ma_y = 0 \Rightarrow T \cos \phi = mg$$

Now we eliminate the tension:

$$(m\omega^2 L) \cos \phi = mg$$

Solving for ω,

$$\omega = \sqrt{\frac{g}{L \cos \phi}}$$

Discussion We should check the dimensions of the final expression. Since $\cos \phi$ is dimensionless,

$$\sqrt{\frac{[L]/[T]^2}{[L]}} = \frac{1}{[T]}$$

which is correct for ω (SI unit rad/s).

Another check is to ask how ω and ϕ are related for a given length cord. As ϕ increases toward 90°, the cord gets closer to horizontal and the radius increases. In our expression, as ϕ increases, $\cos \phi$ decreases and therefore ω increases, in accordance with experience: the stone would have to be whirled faster and faster to make the cord more nearly horizontal.

Conceptual Practice Problem 5.6 Conical pendulum on the Moon

Examine the result of Example 5.6 to see how ω depends on g, all other things being equal. Where the gravitational field is weaker, do you have to whirl the stone faster or more slowly to keep the cord at the same angle ϕ? Is that in accord with your intuition?

5.3 BANKED CURVES

When you drive an automobile in a circular path along an unbanked roadway, friction acting on the tires due to the pavement acts to keep the automobile moving in a curved path. This frictional force acts *sideways*, toward the center of the car's circular path (Fig. 5.9) The frictional force might also have a tangential component; for example, if the car is braking, a component of the frictional force makes the car slow down by acting backward (opposite to the car's velocity). For now we assume that the car's speed is constant and that the tangential component of friction is negligible. As long as the tires roll without slipping, there is no relative motion between the bottom of the tire and the road, so it is the force of *static* friction that acts. If the car is in a skid, then it is the smaller force of kinetic friction that acts as the bottom portion of the tire slides along the pavement. As the speed of the car increases, or for slippery surfaces with low coefficients of friction, the static frictional force may not be enough to hold the car in its curved path.

Making The Connection:
banked roadways

Tips

To help prevent cars from going into a skid or losing control, the roadway is often banked (tilted at a slight angle) around curves so that the outer portion of the road—the part farthest from the center of curvature—is higher than the inner portion. Banking changes the angle and magnitude of the normal force, $\vec{N}$, so that it has a horizontal component N_x directed toward the center of curvature (in the centripetal direction—see Fig. 5.10). Then we need no longer rely solely on friction to keep the car moving in a circular path as it negotiates the curve; this component of the normal force acts to help the car remain on the curved path. Figure 5.10 shows an exaggerated illustration of a banked road with the normal force, the car's weight, and, in parts (c) and (d), the centripetal component of the normal force N_x. We choose the axes so that the x-axis is in the direction of the acceleration, which is to the left; the axes are *not* parallel and perpendicular to the incline.

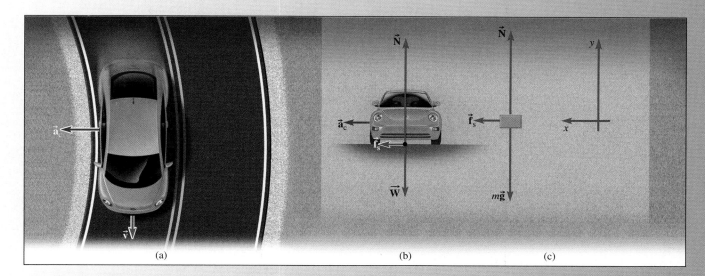

Figure 5.9 A car negotiating a curve on an unbanked roadway: (a) top view and (b) head-on view showing the frictional force acting to the left. (c) A free-body diagram for the car, in which we choose the x-axis in the direction of the acceleration.

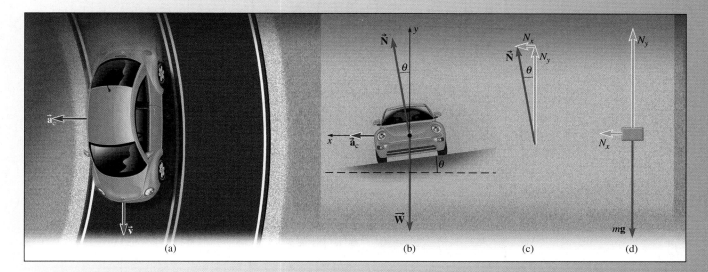

Figure 5.10 A car negotiating a curve on a banked roadway: (a) top view and (b) head-on view showing banked roadway. (c) Components of the normal force. (d) Free-body diagram for the car.

From Section 5.2, any object traveling at a constant speed along a circular path has centripetal acceleration component

$$a_c = \frac{v^2}{r} \tag{5-11}$$

To keep the car safely on the road and to prevent a skid, even under slippery conditions, we want the road to be banked so that the horizontal component of the normal force *alone* gives rise to the centripetal acceleration. With no friction,

$$N_x = \Sigma F_x = ma_x = m\frac{v^2}{r}$$

From Fig. 5.10c, the components of the normal force are

$$N_y = N \cos \theta$$

and

$$N_x = N \sin \theta$$

Since the y-component of the acceleration is zero,

$$\Sigma F_y = N_y - mg = ma_y = 0$$

Therefore, the y-component of the normal force is equal to mg.

Now we can determine the correct banking angle of the roadway. Note that if we divided N_x by N_y, the unknown magnitude of the normal force cancels out and leaves an expression for the banking angle:

$$\frac{N_x}{N_y} = \frac{N \sin \theta}{N \cos \theta} = \tan \theta$$

$$\frac{N_x}{N_y} = \frac{mv^2/r}{mg} = \frac{v^2}{rg}$$

Therefore,

$$\theta = \tan^{-1} \frac{v^2}{rg} \tag{5-12}$$

where v is the speed of the car, r is the radius of curvature of the circular path, and g is the gravitational field strength. Notice that the mass of the car does not appear in this equation; the same banking angle holds for a scooter, motorcycle, car, or tractor-trailer. Notice also that the banking angle depends on the square of the speed. Automobile racetracks and bicycle racetracks have highly banked road surfaces at hairpin curves to minimize skidding of the high-speed vehicles.

If there is no friction between the road and the tires, then there is only one speed—the speed that satisfies Eq. (5-12)—at which it is safe to drive around a given curve:

$$v = \sqrt{rg \tan \theta}$$

With friction, there is a *range* of safe speeds. The static frictional force can have any magnitude from 0 to μN and it can be directed either up or down the bank of the road.

Example 5.7

A Possible Skid

A car is going around an unbanked curve at the recommended speed of 11 m/s (25 mi/h). (a) If the radius of curvature of the path is 25 m and the coefficient of static friction between the rubber and the road is $\mu_s = 0.70$, does the car skid as it goes around the curve? (b) What happens if the driver ignores the highway speed limit sign and travels at 18 m/s (40 mi/h)? (c) What speed is safe for traveling around the curve if the road surface is wet from a recent rainstorm and the coefficient of static friction between the wet road and the rubber tires is $\mu_s = 0.50$? (d) For a car to safely negotiate the curve in icy conditions at a speed of 13 m/s (29 mi/h), what banking angle would be required?

Strategy The force of static friction is the only horizontal force acting on the car when the curve is not banked. The maximum force of static friction, which depends on road conditions, determines the maximum possible centripetal acceleration of the car. Therefore, we can compare the centripetal acceleration necessary to go around the curve at the specified speeds with the maximum possible centripetal acceleration determined by the coefficient of static friction. For part (d), in icy conditions we cannot rely much on friction, but the normal force has a horizontal component when the road is banked.

Solution (a) We find the centripetal acceleration required for a speed of 11 m/s:

$$a_c = \frac{v^2}{r} = \frac{(11 \text{ m/s})^2}{25 \text{ m}} = 4.8 \text{ m/s}^2$$

In order to have that acceleration, the component of the net force acting toward the center of curvature must be

$$\Sigma F_c = ma_c = m\frac{v^2}{r}$$

continued on next page

Example 5.7 *continued*

The only force with a horizontal component is the static frictional force acting on the tires due to the road (see the FBD in Fig. 5.10d). Therefore,

$$\Sigma F_c = f_s = m\frac{v^2}{r}$$

We must check to make sure that the maximum frictional force is not exceeded:

$$f_s \leq \mu_s N$$

Since $N = mg$, the car can go around the curve without skidding as long as

$$m\frac{v^2}{r} \leq \mu_s mg$$

Thus, the centripetal acceleration cannot exceed $\mu_s g$. That limits the car to speeds satisfying

$$v \leq \sqrt{\mu_s gr}$$

Substituting numerical values,

$$v \leq \sqrt{0.70 \times 9.8 \text{ m/s}^2 \times 25 \text{ m}} = 13 \text{ m/s}$$

Since 11 m/s is less than the maximum safe speed of 13 m/s, the car safely negotiates the curve.

(b) At 18 m/s, the car moves at a speed higher than the maximum safe speed of 13 m/s. The frictional force cannot supply the centripetal acceleration needed for the car to go around the curve—the car goes into a skid.

(c) The maximum safe speed occurs when the static frictional force has its maximum possible magnitude. For the wet road conditions, the static frictional force acting between the tires and the road is given by

$$f_s \leq \mu_s N = \mu_s mg$$

where $\mu_s = 0.50$. The maximum possible acceleration is therefore

$$a_c = \mu_s g = 0.50 \times 9.8 \text{ m/s}^2 = 4.9 \text{ m/s}^2$$

Now we use the centripetal acceleration to find the maximum speed:

$$a_c = v^2/r$$

or

$$v = \sqrt{ra_c} = \sqrt{25 \text{ m} \times 4.9 \text{ m/s}^2} = 11 \text{ m/s}$$

which is the same speed recommended by the road sign. The highway engineer knew what he was doing when he had the sign placed along the road.

(d) Finally, we find the banking angle that would enable cars to travel around the curve at 13 m/s in icy conditions. Assuming that friction is negligible, the horizontal component of the normal force is the only horizontal force. With the *x*-axis pointing toward the center of curvature and the *y*-axis vertical,

$$\Sigma F_x = N \sin \theta = m v^2/r \qquad (1)$$

and

$$\Sigma F_y = N \cos \theta - mg = 0 \qquad (2)$$

Dividing Eq. (1) by Eq. (2) gives

$$\frac{N \sin \theta}{N \cos \theta} = \tan \theta = \frac{mv^2/r}{mg} = \frac{v^2}{rg}$$

$$\theta = \tan^{-1} \frac{v^2}{rg} = \tan^{-1} \frac{(13 \text{ m/s})^2}{25 \text{ m} \times 9.8 \text{ m/s}^2} = 35°$$

Discussion A banking angle of 35° is far greater than those used in practice along roadways. Careful drivers would not try to drive around this curve in icy conditions at 13 m/s. What do you think might happen in icy conditions to a car that is traveling *very slowly* along a road banked at such a steep angle?

Highway curves are banked at slight angles to help drivers who are driving at reasonable speeds for the road conditions. They are not banked to save speed demons from their folly.

Practice Problem 5.7 A bobsled race

A bobsled races down an icy hill and then comes upon a horizontal curve, located 60.0 m from the bottom of the hill. The sled is traveling at 22.4 m/s (50 mph) as it approaches the curve that has a radius of curvature of 50.0 m. The curve is banked at an angle of 45° and the frictional force on the sled runners is negligible. Does the sled make it safely around the curve?

When an airplane pilot makes a turn in the air, the pilot makes use of a banking angle. The airplane itself is tilted as if it were traveling over an inclined surface. Because of the shape of the wings, an aerodynamic force called *lift* acts upward when the plane is in level flight. To go around a turn, the wings are tilted; the lift force stays perpendicular to the wings and, therefore, now has a horizontal component, just as the normal force has a horizontal component for a car on a banked curve. This component supplies the necessary centripetal acceleration, while the vertical component of the lift holds the plane up. Therefore,

$$L_x = ma_c = \frac{mv^2}{r} \quad \text{and} \quad L_y = mg$$

where the *x*-axis is horizontal and the *y*-axis is vertical. The lift force is different in its physical origin from the normal force, but its components split up the same way, so a plane in a turn banks its wings at the same angle that a road would be banked for the same speed and radius of curvature.

Making The Connection:
banking angle
of an airplane

5.4   CIRCULAR ORBITS

A satellite orbiting Earth travels in a circular path because of the long-range gravitational force between the satellite and the Earth. The same gravitational interaction that makes Earth-bound objects fall keeps satellites and planets from flying off into space. The law of universal gravitation is

$$F = \frac{Gm_1m_2}{r^2} \tag{2-3}$$

where the universal gravitational constant is

$$G = 6.67 \times 10^{-11} \text{ N·m}^2/\text{kg}^2$$

We can use Newton's second law to find the speed of a satellite in circular orbit at constant speed. Let m be the mass of the satellite and M_E be the mass of the Earth. The direction of the gravitational force on the satellite is always toward the center of the Earth, which is the center of the orbit. Since gravity is the only force acting on the satellite,

$$\Sigma F_c = G\frac{mM_E}{r^2}$$

where r is the distance from the center of the Earth to the satellite. Then, from Newton's second law,

$$\Sigma F_c = ma_c = \frac{mv^2}{r}$$

Setting these equal,

$$G\frac{\cancel{m}M_E}{r^2} = \frac{\cancel{m}v^2}{\cancel{r}}$$

Solving for the speed yields

$$v = \sqrt{\frac{GM_E}{r}} \tag{5-13}$$

Notice that the mass of the satellite does not appear in the equation for speed; it has been algebraically canceled. The greater inertia of a more massive satellite is overcome by a proportionally greater gravitational force acting on it. Thus, the speed of a satellite in a circular orbit does not depend on the mass of the satellite. Equation (5-13) also shows that satellites in lower orbits (smaller radii) have greater speeds.

We have been discussing satellites orbiting Earth, but the same principles apply to the circular orbits of satellites around other planets and to the orbits of the planets around the Sun. For planetary orbits, the mass of the Sun would appear in Eq. (5-13) instead of the Earth's mass. Gravity is the force that keeps planets orbiting about the Sun. The paths of planetary orbits are actually ellipses instead of circles, although for most of the planets in the solar system the ellipses are nearly circular. Pluto and Mercury are the exceptions; their orbits are markedly different from circles.

Example 5.8

Speed of Satellite

The Hubble telescope is in a circular orbit 613 km above Earth's surface. The average radius of the Earth is 6.37×10^3 km and the mass of Earth is 5.98×10^{24} kg. What is the speed of the telescope in its orbit?

Strategy We first need to find the orbital radius of the telescope. It is not 613 km; that is the distance from the *surface* of Earth to the telescope. We must add the radius of the Earth to

613 km to find the orbital radius, which is measured from the center of the Earth to the telescope. Then we use Newton's second law, along with what we know about centripetal acceleration.

Solution The radius of the telescope's orbit is

$$r = 6.13 \times 10^2 \text{ km} + 6.37 \times 10^3 \text{ km} = (0.613 + 6.37) \times 10^3 \text{ km}$$
$$= 6.98 \times 10^3 \text{ km}$$

continued on next page

Example 5.8 continued

The net force on the telescope is equal to the gravitational force, given by Newton's law of gravity. Newton's second law relates the net force to the acceleration. Both are directed radially inward.

$$\Sigma F_c = \frac{GmM_E}{r^2} = \frac{mv^2}{r}$$

where m is the mass of the telescope. Solving for the speed, we find

$$v = \sqrt{\frac{GM_E}{r}}$$

$$v = \sqrt{\frac{6.67 \times 10^{-11} \text{ N·m}^2/\text{kg}^2 \times 5.98 \times 10^{24} \text{ kg}}{6.98 \times 10^6 \text{ m}}}$$

$$v = 7560 \text{ m/s} = 27{,}200 \text{ km/h}$$

Discussion *Any* satellite orbiting Earth at an altitude of 613 km has this same speed, regardless of its mass.

Practice Problem 5.8 Speed of Earth in its orbit

What is the speed of Earth in its approximately circular orbit about the Sun? The average Earth–Sun distance is 1.496×10^{11} m and the mass of the Sun is 1.99×10^{30} kg. Once you find the speed, use it along with the distance traveled by the Earth during one revolution about the Sun to calculate the time in seconds for one orbit. Compare your answer with the number of seconds in one year.

Kepler's Laws of Planetary Motion

At the beginning of the seventeenth century, Johannes Kepler (1571–1630) proposed three laws to describe the motion of the planets. These laws predated Newton's laws of motion and his law of gravity. They offered a far simpler description of planetary motion than anything that had been proposed previously. We turn history on its head and look at one of Kepler's laws as a consequence of Newton's laws. The fact that Newton could derive Kepler's laws from his own work on gravity was seen as a confirmation of Newtonian mechanics.

Kepler's laws of planetary motion are

- The planets travel in elliptical orbits with the Sun at one focus of the ellipse.
- A line drawn from a planet to the Sun sweeps out equal areas in equal time intervals. (Fig. 5.11).
- The square of the orbital period is proportional to the cube of the average distance from the planet to the Sun.

Kepler's first law can be derived from the inverse square law of gravitational attraction. The derivation is a bit complicated, but for any two objects that have such an attraction, the orbit of one about the other is an ellipse, with the stationary object located at one focus. The circle is a special case of an ellipse where the two foci coincide.

We can derive Kepler's third law from Newton's law of universal gravitation for a circular orbit. The gravitational force gives rise to the centripetal acceleration:

$$\Sigma F_c = \frac{GmM_{sun}}{r^2} = \frac{mv^2}{r}$$

Solving for v yields

$$v = \sqrt{\frac{GM_{sun}}{r}}$$

Making The Connection:
planetary motion

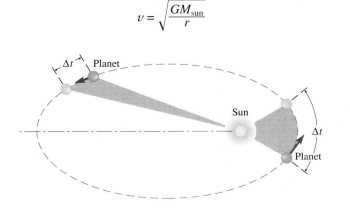

Figure 5.11 A planet orbiting the Sun. When the planet is closer to the Sun, it moves faster; when it is farther away, it moves more slowly. No matter where the planet is in its orbit, the radial line sweeps out the same area during the same time interval Δt. (The planetary orbits are nearly circular, unlike the elliptical orbit shown here, which is exaggerated for clarity.)

The distance traveled during one revolution is the circumference of the circle, which is equal to $2\pi r$. The speed is the distance traveled during one orbit divided by the period:

$$v = \sqrt{\frac{GM_{sun}}{r}} = \frac{2\pi r}{T}$$

Now we solve for T:

$$T = 2\pi \sqrt{\frac{r^3}{GM_{sun}}}$$

Squaring both sides yields

$$T^2 = \frac{4\pi^2}{GM_{sun}} r^3 = \text{constant} \times r^3 \qquad (5\text{-}14)$$

Equation (5-14) is Kepler's third law: the square of the period of a planet is directly proportional to the cube of the average orbital radius.

Planetary orbits are affected by gravitational interactions with other planets; Kepler's laws ignore these small effects. Although Kepler's laws were derived for the motion of planets, they apply to satellites orbiting the Earth as well. Many satellites, such as those used for communications, are placed in a *geostationary* (or *geosynchronous*) orbit—a circular orbit in Earth's equatorial plane whose period is equal to Earth's rotational period (Fig. 5.12). A satellite in geostationary orbit remains directly above a particular point on the equator; to observers on the ground, it seems to hover above that point without moving. Due to their fixed positions with respect to Earth's surface, geostationary satellites are used as relay stations for communication signals. In Example 5.9, we find the speed of a geostationary satellite.

Example 5.9

Geostationary Satellite

A 300.0-kg communications satellite is placed in a geostationary orbit 35,800 km above a relay station located in Kenya. What is the speed of the satellite in orbit?

Strategy   The period of the satellite is one day or approximately 24 h. To find the speed of the satellite in orbit we use Newton's law of gravity and his second law of motion along with what we know about centripetal acceleration.

Solution   Let m be the mass of the satellite and let M_E be the mass of the Earth. Gravity is the only force acting on the satellite in its orbit. From Newton's law of universal gravitation, Newton's second law, and the expression for centripetal acceleration,

$$\Sigma F_c = \frac{GmM_E}{r^2} = \frac{mv^2}{r}$$

Solving for the speed yields

$$v = \sqrt{\frac{GM_E}{r}}$$

We must add the mean radius of the Earth, $R_E = 6.37 \times 10^6$ m, to the height of the satellite above the Earth's surface to find the orbital radius.

$$r = h + R_E = 3.58 \times 10^7 \text{ m} + 0.637 \times 10^7 \text{ m}$$

$$r = 4.217 \times 10^7 \text{ m}$$

Substituting numerical values into the speed equation,

$$v = \sqrt{\frac{6.67 \times 10^{-11} \text{ N·m}^2/\text{kg}^2 \times 5.98 \times 10^{24} \text{ kg}}{4.217 \times 10^7 \text{ m}}}$$

$$= \sqrt{9.459 \times 10^6 \text{ m}^2/\text{s}^2}$$

$$v = 3.08 \times 10^3 \text{ m/s}$$

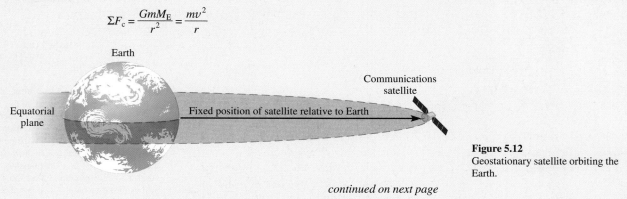

Figure 5.12
Geostationary satellite orbiting the Earth.

continued on next page

Example 5.9 continued

Discussion This result, an orbital speed of 3.08 km/s and a distance above Earth's surface of 35,800 km, applies to *all* geosynchronous satellites. The mass of the satellite does not matter; it cancels out of the equations for orbital radius and for speed.

If we were actually putting a satellite into orbit, we would use a more accurate value for the period. We should use a time of 23 h and 56 min, which is the length of a sidereal day—the time for Earth to complete one rotation about its axis relative to the fixed stars. The solar day, 24 h, is the period of time between the daily appearances of the Sun at its highest point in the sky. The fact that Earth moves around the Sun is what causes the difference between these two ways of measuring the length of a day. The error introduced by using the longer time is negligible in this problem.

We can use Kepler's third law to check the result. Examples 5.8 and 5.9 both concern circular orbits around the Earth. Is the square of the period proportional to the cube of the orbital radius? From Example 5.8, $r_1 = 6.98 \times 10^3$ km and

$$T_1 = \frac{2\pi r_1}{v} = \frac{2\pi \times 6.98 \times 10^3 \text{ km}}{7.56 \text{ km/s}} = 5800 \text{ s}$$

and, from the present example, $r_2 = 4.22 \times 10^7$ m and

$$T_2 = 24 \text{ h} \times \frac{3600 \text{ s}}{1 \text{ h}} = 86,400 \text{ s}$$

The ratio of the squares of the periods is

$$\left(\frac{T_2}{T_1}\right)^2 = \left(\frac{86,400 \text{ s}}{5800 \text{ s}}\right)^2 = 221$$

The ratio of the cubes of the radii is

$$\left(\frac{r_2}{r_1}\right)^3 = \left(\frac{4.22 \times 10^7 \text{ m}}{6.98 \times 10^6 \text{ m}}\right)^3 = 221$$

Practice Problem 5.9 Orbital radius of Venus

The period of the orbit of Venus around the Sun is 0.615 Earth years. Using this information, find the radius of its orbit in terms of R, the radius of Earth's orbit around the Sun.

Example 5.10

Orbiting Satellites

A satellite revolves about Earth with an orbital radius of r_1 and speed v_1. If an identical satellite were set into circular orbit with the same speed about a planet of mass three times that of Earth, what would its orbital radius be?

Strategy We can apply Newton's law of universal gravitation and set up a ratio to solve for the new orbital radius.

Solution From Newton's second law, the magnitude of the gravitational force on the satellite is equal to the satellite's mass times the magnitude of its centripetal acceleration:

$$\frac{GmM_E}{r_1^2} = \frac{mv_1^2}{r_1}$$

where M_E and m are the masses of Earth and of the satellite, respectively. Solving for r_1 yields

$$r_1 = \frac{GM_E}{v_1^2}$$

Now we apply Newton's second law to the orbit of the second satellite about the planet of mass $3M_E$:

$$\frac{Gm \times 3M_E}{r_2^2} = \frac{mv_1^2}{r_2}$$

$$r_2 = \frac{G \times 3M_E}{v_1^2}$$

The ratio of r_2 to r_1 is

$$\frac{r_2}{r_1} = \frac{G \times 3M_E/v_1^2}{GM_E/v_1^2} = 3$$

Thus $r_2 = 3r_1$.

Discussion Notice that we did not rush to substitute numerical values for the constants G and M_E into the equations. We took the ratio r_2/r_1 so that these constants would cancel.

Practice Problem 5.10 Period of lunar lander

A lunar lander is orbiting about the Moon. If the radius of its orbit is $\frac{1}{3}$ the radius of Earth, what is the period of its orbit?

5.5 NONUNIFORM CIRCULAR MOTION

So far we have focused on *uniform* circular motion. Now we can extend the discussion to nonuniform circular motion, where the angular velocity changes with time.

Figure 5.13a shows the velocity vectors $\vec{v}_1$ and $\vec{v}_2$ at two different times for an object moving in a circle with changing speed. In this case the speed is increasing ($v_2 > v_1$). In Fig. 5.13b we subtract $\vec{v}_1$ from $\vec{v}_2$ to find the change in velocity. In the limit $\Delta t \to 0$, $\Delta\vec{v}$

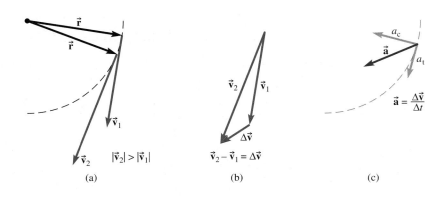

Figure 5.13 Motion along a curved path with a changing speed: (a) the magnitude of velocity $\vec{v}_2$ is greater than the magnitude of velocity $\vec{v}_1$, (b) the direction of $\Delta\vec{v}$ is not radial when the speed is changing, and (c) components of $\vec{a}$ can be taken along a tangent to the curved path (a_t) and along a radius (a_c).

does *not* become perpendicular to the velocity, as it did for uniform circular motion. Thus the direction of the acceleration is *not* centripetal if the speed is changing. However, we can resolve the acceleration into tangential and centripetal components (Fig. 5.13c). The centripetal component a_c changes the *direction* of the velocity, while the tangential component a_t changes the *magnitude* of the velocity. Since these are perpendicular components of the acceleration, the magnitude of the acceleration is

$$a = \sqrt{a_c^2 + a_t^2}$$

Using the same method as in Section 5.2 to find the centripetal acceleration, but working here with only the centripetal *component* of the acceleration, we find that

$$a_c = \frac{v^2}{r} = \omega^2 r \quad (\omega \text{ in radians per unit time}) \tag{5-11}$$

For circular motion, whether uniform or nonuniform, the centripetal component of the acceleration is given by Eq. (5-11). However, in *uniform* circular motion the centripetal component of the acceleration a_c is constant in magnitude, while for nonuniform circular motion a_c changes as the speed changes.

Also still true for nonuniform circular motion is the relationship between speed and angular speed:

$$v = r\omega \tag{5-6}$$

Many problems involving nonuniform circular motion are solved in the same way as for uniform circular motion. We find the *centripetal component of the net force* and then apply Newton's second law along the centripetal direction:

$$\Sigma F_c = ma_c$$

Problem-Solving Strategy for an Object in Nonuniform Circular Motion

1. Begin as for any Newton's second law problem: identify all the forces acting on the object and draw a free-body diagram.

2. Choose perpendicular axes at the point of interest so that one axis is centripetal and the other is tangent to the circular path.

3. Resolve each force into centripetal and tangential components.

4. Apply Newton's second law

$$\Sigma F_c = ma_c$$

where

$$a_c = \frac{v^2}{r} = \omega^2 r$$

and

$$\Sigma F_t = ma_t$$

The tangential acceleration component a_t determines how the speed of the body changes.

Example 5.11

Vertical Loop-the-Loop

A roller coaster includes a vertical circular loop of radius 20.0 m (Fig. 5.14a). What is the minimum speed at which the car must move at the top of the loop so that it doesn't lose contact with the track?

Strategy   A roller coaster car moving around a vertical loop is in nonuniform circular motion; its speed decreases on the way up and increases on the way back down. Nevertheless, it is moving in a circle and has a centripetal acceleration component as given in Eq. (5-11) as long as it moves in a circle. The only forces acting on the car are gravity and the normal force of the track pushing the car. Even if frictional or drag forces are present, at the top of the loop they act in the tangential direction and thus do not contribute to the centripetal component of the net force. At the top of the loop, the track exerts a normal force on the car as long as the car moves with a speed great enough to stay on the track. If the car moves too slowly, it loses contact with the track and the normal force is then zero.

Solution   The normal force exerted by the track on the car at the top pushes the car *away* from the track (downward); the normal force cannot pull up on the car. Then, at the top of the loop, the gravitational force and the normal force both point straight down toward the center of the loop. Figure 5.14b is a free-body diagram for the car. From Newton's second law,

$$\Sigma F_c = N + mg = ma_c = \frac{mv_t^2}{r}$$

or

$$N = \frac{mv_t^2}{r} - mg$$

where v_t stands for the speed at the top. In this expression, N stands for the magnitude of the normal force. Since $N \geq 0$,

$$m\left(\frac{v_t^2}{r} - g\right) \geq 0$$

or

$$v_t \geq \sqrt{gr}$$

Imagine sending a roller coaster car around the loop many times with a slightly smaller speed at the top each time. As v_t approaches $\sqrt{gr}$, the normal force at the top gets smaller and smaller. When $v_t = \sqrt{gr}$, the normal force just becomes zero at the top of the loop. Any slower and the car loses contact with the track *before* getting to the highest point and would fall off the track unless prevented from falling by a backup safety mechanism. Therefore, the minimum speed at the top is

$$v_t = \sqrt{gr} = \sqrt{9.8 \text{ m/s}^2 \times 20.0 \text{ m}} = 14 \text{ m/s}$$

Discussion   If the car is going faster than 14 m/s at the top, its centripetal acceleration is larger. The track pushing on the car provides the additional net force component that results in a larger centripetal acceleration. The minimum speed occurs when gravity alone provides the centripetal acceleration at the top of the loop. In other words, $a_c = g$ at the top of the loop for minimum speed.

Practice Problem 5.11   Normal force at bottom of track

If the speed of the roller coaster at the *bottom* of the loop is 25 m/s, what is the normal force exerted on the car by the track in terms of the car's weight mg? (See Fig. 5.14c.)

Physics at Home

Go outside on a warm day and fill a bucket with water. Swing the bucket around in a vertical circle over your head. What, if anything, keeps the water in the bucket when the bucket is upside down over your head? Why doesn't the water spill out? Do any upward forces act on the water at that point? [*Hint:* An FBD for the water when it is directly overhead is identical to the FBD for a roller coaster car at the top of a loop.] If it is warm enough and you don't mind getting wet, try swinging the bucket at lower and lower speeds to show that there is a minimum speed at the top if the water is to stay in the bucket.

Figure 5.14   (a) A roller coaster on a circular vertical loop, (b) free-body diagram for the car at the top of the loop, and (c) free-body diagram for the car at the bottom of the loop.

Conceptual Example 5.12

Acceleration of Pendulum Bob

A pendulum is released from rest at point A (Fig. 5.15). Sketch qualitatively the acceleration vector of the pendulum bob at points B and C.

Strategy   The pendulum bob moves along the arc of a circle, but not at constant speed. To find the centripetal component, we use

$$a_c = \frac{v^2}{r}$$

Unless $v = 0$, the centripetal component is nonzero. As the pendulum bob swings toward the bottom, its speed is increasing; as it rises on the other side, its speed is decreasing. When the speed

is increasing, the tangential component of the acceleration a_t is in the same direction as the velocity; when the speed is decreasing, a_t is in the direction *opposite* to the velocity. At the lowest point, the speed is neither increasing nor decreasing and $a_t = 0$.

Solution and Discussion   The bob's speed increases on the way down until it reaches point B. Therefore, between A and B, the tangential acceleration is parallel to the velocity. On the way back up, between points B and D, the speed is decreasing and the tangential acceleration is opposite to the velocity. At point B, where the bob has its maximum speed, the speed is neither increasing nor decreasing. At that point the tangential

continued on next page

Conceptual Example 5.12 *continued*

acceleration is zero. Therefore, at point *B*, the acceleration points in the centripetal direction: straight up (Fig. 5.15b).

The acceleration at point *C* has both tangential and centripetal components. The tangential acceleration is opposite to the velocity. Figure 5.15c shows the tangential and centripetal components added to form the acceleration vector $\vec{\mathbf{a}}$. We do not know the exact direction since we don't know the relative magnitudes of a_c and a_t.

Conceptual Practice Problem 5.12 Acceleration of bob at later time

Sketch the acceleration vector of the pendulum bob at point *D*, the highest point in its swing to the right.

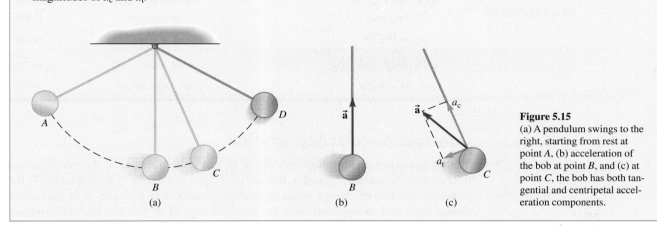

Figure 5.15
(a) A pendulum swings to the right, starting from rest at point *A*, (b) acceleration of the bob at point *B*, and (c) at point *C*, the bob has both tangential and centripetal acceleration components.

5.6   ANGULAR ACCELERATION

An object in nonuniform circular motion has a changing speed and a changing angular velocity. To describe how the angular velocity changes, we define an angular acceleration. If the angular velocity is ω_1 at time t_1 and is ω_2 at time t_2, the change in angular velocity is

$$\Delta\omega = \omega_2 - \omega_1$$

The time interval during which the angular velocity changes is $\Delta t = t_2 - t_1$. The average rate at which the angular velocity changes is called the **average angular acceleration, α_{av}**.

$$\alpha_{av} = \frac{\omega_2 - \omega_1}{t_2 - t_1} = \frac{\Delta\omega}{\Delta t} \tag{5-15}$$

As we let the time interval become shorter and shorter, α_{av} approaches the **instantaneous angular acceleration, α**.

$$\alpha = \lim_{\Delta t \to 0} \frac{\Delta\omega}{\Delta t} \tag{5-16}$$

If ω is in units of rad/s, α is in units of rad/s^2.

The angular acceleration is closely related to the tangential component of the acceleration. The tangential component of velocity is

$$v_t = r\omega \tag{5-6}$$

Equation (5-6) gives us a way to relate tangential acceleration to the angular acceleration. The tangential acceleration is the rate of change of the tangential velocity, so

$$a_t = \frac{\Delta v_t}{\Delta t} = r\frac{\Delta\omega}{\Delta t} \quad \text{(in the limit } \Delta t \to 0)$$

Therefore,

$$a_t = r\alpha \tag{5-17}$$

$$\frac{W_{orbit}}{W_{surface}} = \frac{(R_E + h)}{\frac{GM_E m}{R_E^2}} = \frac{R_E}{(R_E + h)^2} = \frac{(6400 \text{ km})}{(7000 \text{ km})^2} = 0.84$$

The weight in orbit is 0.84 times the weight on the surface. The astronaut weighs less but certainly isn't *weightless!* Then why does the astronaut *seem* to be weightless?

Recall Section 3.7 on the apparent weightlessness of someone unfortunate enough to be in an elevator when the cable snaps. In that situation, the elevator and the passenger both have the same acceleration ($\vec{\mathbf{a}} = \vec{\mathbf{g}}$). Similarly, the astronaut has the same acceleration as the space shuttle, which is equal to the *local* gravitational field $\vec{\mathbf{g}}$. The

Table 5.1

Relationships between θ, ω, and α for Constant Angular Acceleration Are Analogous to the Relationships between x, v_x, and a_x for Constant Acceleration Along the x-Axis.

Linear Variables: x, v_x, a_x;	Angular Variables: θ, ω, α;
~~Constant Linear Acceleration~~	~~Constant Angular Acceleration~~

magnitude of the apparent weight (the feeling of "weight" as experienced by the astronaut) is the magnitude of the normal force acting on the body:

$$W' = m|\vec{\mathbf{g}} - \vec{\mathbf{a}}| \tag{3-17}$$

Apparent weightlessness occurs when $\vec{\mathbf{a}} = \vec{\mathbf{g}}$, where $\vec{\mathbf{g}}$ is the *local* gravitational field.

Stunt pilots have to be careful about the accelerations to which they subject their bodies. An acceleration of about $3g$ can cause temporary blindness due to an inadequate supply of oxygen to the retina; the heart has difficulty pumping blood up to the head due to the blood's increased apparent weight. Larger accelerations can cause unconsciousness. Pressurized flight suits enable pilots to sustain accelerations up to about $5g$.

Example 5.14

Stunt Pilot

Dave wants to practice vertical circles for a flying show exhibition (Fig. 5.16). (a) What must the minimum radius of the circle be to ensure that his acceleration at the bottom does not exceed $3.0g$? The speed of the plane is 78 m/s at the bottom of the circle. (b) What is Dave's apparent weight at the bottom of the circular path? Express your answer in terms of his true weight W.

Strategy   For the *minimum* radius, we use the maximum possible centripetal acceleration since $a_c = v^2/r$. For the maximum centripetal acceleration, the *tangential* acceleration must be zero—the magnitude of the acceleration is $a = \sqrt{a_c^2 + a_t^2}$. Therefore, the centripetal acceleration component has magnitude $3.0g$ at the bottom. To find Dave's apparent weight, we do not need to use the numerical value of the radius found in part (a); we already know that his acceleration is upward and has magnitude $3.0g$.

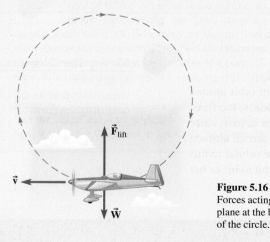

Figure 5.16
Forces acting on the plane at the bottom of the circle.

Solution   (a) The magnitude of the centripetal acceleration is

$$a_c = v^2/r$$

Solving for the radius,

$$r = \frac{v^2}{a_c} = \frac{v^2}{3.0g}$$

$$= \frac{(78 \text{ m/s})^2}{3.0 \times 9.8 \text{ m/s}^2} = 210 \text{ m}$$

(b) Dave's apparent weight is the magnitude of the normal force of the plane pushing up on him. Let the y-axis point upward. The normal force is up and the gravitational force is down. Then

$$\Sigma F_y = N - mg = ma_y$$

where $a_y = +3.0g$. Therefore,

$$W' = N = m(g + a_y) = 4.0mg$$

His apparent weight is 4.0 times his true weight.

Discussion   It might have been tempting to jump to the conclusion that an acceleration of $3.0g$ means that his apparent weight is $3.0mg$. But is his apparent weight zero when his acceleration is zero? No.

According to Eq. (3-17), the apparent weight is $W' = m|\vec{\mathbf{g}} - \vec{\mathbf{a}}|$. In the solution we *added* $g + a_y$. Is there a contradiction? No; because $\vec{\mathbf{g}}$ is *down* while $\vec{\mathbf{a}}$ is *up*. Therefore,

$$\vec{\mathbf{g}} - \vec{\mathbf{a}} = \vec{\mathbf{g}} + (-\vec{\mathbf{a}}) = (g \text{ downward}) + (3.0g \text{ downward})$$

$$= (4.0g \text{ downward}).$$

Practice Problem 5.14   Astronaut's apparent weight

What is the apparent weight of a 730-N astronaut when her spaceship has an acceleration of magnitude $2.0g$ in the following two situations: (a) just above the surface of Earth, acceleration straight up; (b) far from any stars or planets?

In order for astronauts to spend long periods of time living in a space station without the deleterious effects of apparent weightlessness, *artificial gravity* must be created on the station. Many science fiction novels and movies feature ring-shaped space stations that rotate in order to create artificial gravity for the occupants.

The centrifuge is a device that creates artificial gravity on a smaller scale. Centrifuges are common not only in scientific and medical laboratories but also in everyday life. The first successful centrifuge was used to separate cream from milk in the 1880s. Water drips out of sopping wet clothes due to the pull of gravity when the clothes are hung on a clothesline, but the water is removed much faster by the artificial gravity created in the spin cycle of a washing machine.

If we attribute the apparent weight of an object to an apparent gravitational field $\vec{\mathbf{g}}'$, then

$$\vec{\mathbf{W}}' = m(\vec{\mathbf{g}} - \vec{\mathbf{a}}) = m\vec{\mathbf{g}}'$$

and

$$\vec{\mathbf{g}}' = \vec{\mathbf{g}} - \vec{\mathbf{a}} \qquad (5\text{-}23)$$

In many cases, $g \ll a$ so that $\vec{\mathbf{g}}' \approx -\vec{\mathbf{a}}$. The apparent gravitational field is then equal in magnitude to the acceleration, but oppositely directed. In a rotating space station, the acceleration of an astronaut is inward (toward the rotation axis), but the apparent gravitational field is outward. Therefore, the ceiling of rooms on the station are closest to the rotation axis and the floor is farthest away (Fig. 5.17).

Figure 5.17 A rotating space station from the movie *2001: A Space Odyssey.*

MASTER THE CONCEPTS

Summary

- The angular displacement $\Delta\theta$ is the angle through which an object has turned. Positive and negative angular displacements indicate rotation in different directions. Conventionally, positive represents counterclockwise motion.

- Average angular velocity:

$$\omega_{av} = \frac{\theta_2 - \theta_1}{t_2 - t_1} = \frac{\Delta\theta}{\Delta t} \qquad (5\text{-}2)$$

- Average angular acceleration:

$$\alpha_{av} = \frac{\omega_2 - \omega_1}{t_2 - t_1} = \frac{\Delta\omega}{\Delta t} \qquad (5\text{-}15)$$

- The instantaneous angular velocity and acceleration are the limits of the average quantities as $\Delta t \to 0$.

- A useful measure of angle is the radian:

$$2\pi \text{ radians} = 360°$$

Using radian measure for θ, the arc length s of a circle of radius r subtended by an angle θ is

$$s = \theta r \quad (\theta \text{ in radian measure}) \qquad (5\text{-}5)$$

- Using radian measure for ω, the speed of an object in circular motion (including a point on a rotating object) is

$$v = r\omega \quad (\omega \text{ in radians per unit time}) \qquad (5\text{-}6)$$

- Using radian measure for α, the tangential acceleration component is related to the angular acceleration by

$$a_t = r\alpha \quad (\alpha \text{ in radians per time}^2) \qquad (5\text{-}17)$$

- An object moving in a circle has a centripetal acceleration component given by

$$a_c = \frac{v^2}{r} = \omega^2 r \quad (\omega \text{ in radians per unit time}) \qquad (5\text{-}11)$$

The tangential and centripetal accelerations are two perpendicular components of the acceleration. The centripetal acceleration component changes the direction of the velocity and the tangential acceleration component changes the speed.

- Uniform circular motion means that v and ω are constant. In uniform circular motion, the time to complete one revolution is constant and is called the period T. The frequency f is the number of revolutions completed per second.

$$f = 1/T \qquad (5\text{-}7)$$
$$v = 2\pi r/T = 2\pi rf \qquad (5\text{-}8)$$
$$\omega = v/r = 2\pi f \qquad (5\text{-}9)$$

where the SI unit of angular velocity is rad/s and that of frequency is rev/s = Hz.

- A rolling object is both rotating and translating. If the object rolls without skidding or slipping, then

$$v_{axle} = r\omega$$

- Kepler's third law says that the square of the period of a planetary orbit is proportional to the cube of the orbital radius:

$$T^2 = \text{constant} \times r^3 \qquad (5\text{-}14)$$

MASTER THE CONCEPTS *continued*

- In the case of constant angular acceleration, we can solve rotational kinematics problems using relationships analogous to those we developed for linear motion:

$$\Delta\theta = \theta - \theta_0 = \omega_{av}\,\Delta t \qquad (5\text{-}18)$$

$$\omega - \omega_0 = \alpha t \qquad (5\text{-}19)$$

$$\omega_{av} = \frac{\omega_0 + \omega}{2} \qquad (5\text{-}20)$$

$$\theta - \theta_0 = \omega_0 t + \tfrac{1}{2}\alpha t^2 \qquad (5\text{-}21)$$

$$\omega^2 - \omega_0^2 = 2\alpha\,\Delta\theta \qquad (5\text{-}22)$$

- When an object is accelerating, its apparent weight differs from its true weight:

$$W' = m|\vec{\mathbf{g}} - \vec{\mathbf{a}}| \qquad (3\text{-}17)$$

where $\vec{\mathbf{g}}$ is the local gravitational field. The apparent gravitational field is

$$\vec{\mathbf{g}}' = \vec{\mathbf{g}} - \vec{\mathbf{a}} \qquad (5\text{-}23)$$

Highlighted Figures and Tables

F5.2 Angular displacement of a spinning CD (p. 147)

F5.3 Definition of the radian (p. 147)

F5.5 Uniform circular motion at a constant speed (p. 152)

F5.6 To find the magnitude of the centripetal acceleration for uniform circular motion, you must find the change in velocity $\Delta\vec{\mathbf{v}}$ for a time interval Δt in the limit $\Delta t \to 0$. (p. 152)

F5.9 Car negotiating a curve on an unbanked roadway (p. 157)

F5.10 Car negotiating a curve on a banked roadway (p. 157)

F5.13 Nonuniform circular motion (p. 164)

T5.1 Relationships between θ, ω, and α for constant angular acceleration are analogous to the relationships between x, v_x, and a_x for constant acceleration along the x-axis. (p. 168)

CONCEPTUAL QUESTIONS

1. Is depressing the "accelerator" (gas pedal) of a car the only way that the driver can make the car accelerate (in the physics sense of the word)? If not, what else can the driver do to give the car an acceleration?

2. Two children ride on a merry-go-round. One is 2 m from the axis of rotation and the other is 4 m from it. Which child has the larger (a) linear speed, (b) acceleration, (c) angular speed, and (d) angular displacement?

3. Explain why the orbital radius and the speed of a satellite in circular orbit are not independent.

4. In uniform circular motion, is the velocity constant? Is the acceleration constant? Explain.

5. In uniform circular motion, the net force is perpendicular to the velocity and changes the direction of the velocity but not the speed. If a projectile is launched horizontally, the net force (ignoring air resistance) is perpendicular to the initial velocity, and yet the projectile gains speed as it falls. What is the difference between the two situations?

6. The speed of a satellite in circular orbit around a planet does not depend on the mass of the satellite. Does it depend on the mass of the planet? Explain.

7. A flywheel (a massive disk) rotates with constant angular acceleration. For a point on the rim of the flywheel, is the tangential acceleration constant? Is the centripetal acceleration constant?

8. Explain why the force of gravity due to the Earth does not pull the Moon in closer and closer on an inward spiral until it hits Earth's surface.

9. When a roller coaster takes a sharp turn to the right, it feels as if you are pushed toward the left. Does a force push you to the left? If so, what is it? If not, why does there *seem* to be such a force?

10. Is there anywhere on Earth where a bathroom scale reads your true weight? If so, where? Where does your apparent weight due to Earth's rotation differ most from your true weight?

MULTIPLE CHOICE QUESTIONS

1. A spider sits on a turntable that is rotating at a constant 33 rpm. The acceleration $\vec{\mathbf{a}}$ of the spider is

   (a) greater the closer the spider is to the central axis.
   (b) greater the farther the spider is from the central axis.
   (c) nonzero and independent of the location of the spider on the turntable.
   (d) zero.

Questions 2–5. A satellite in orbit travels around the Earth in uniform circular motion. In Fig. 5.18, the satellite moves counterclockwise (*ABCDA*). Answer choices:

   (a) $+x$ (b) $+y$ (c) $-x$ (d) $-y$
   (e) 45° above $+x$ (toward $+y$)   (f) 45° below $+x$ (toward $-y$)
   (g) 45° above $-x$ (toward $+y$)   (h) 45° below $-x$ (toward $-y$)

2. What is the direction of the satellite's instantaneous velocity at point *D*?

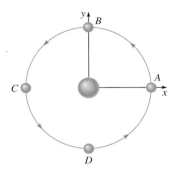

Figure 5.18 Multiple Choice Questions 2–5 and Problem 30

3. What is the direction of the satellite's average velocity for one quarter of an orbit, starting at C and ending at D?

4. What is the direction of the satellite's average acceleration for one half of an orbit, starting at C and ending at A?

5. What is the direction of the satellite's instantaneous acceleration at point C?

6. Two satellites are in orbit around Mars with the same orbital radius. Satellite 2 has twice the mass of satellite 1. The centripetal acceleration of satellite 2 has

   (a) twice the magnitude of the centripetal acceleration of satellite 1.

   (b) the same magnitude as the centripetal acceleration of satellite 1.

(c) half the magnitude of the centripetal acceleration of satellite 1.

(d) four times the magnitude of the centripetal acceleration of satellite 1.

Questions 7–8. A boy swings in a tire swing. Answer choices:

   (a) At the highest point of the motion
   (b) At the lowest point of the motion
   (c) At a point neither highest nor lowest
   (d) It is constant.

7. When is the tension in the rope the greatest?

8. When is the tangential acceleration the greatest?

Questions 9–10 concern these three statements:

   (1) Its acceleration is constant.
   (2) Its centripetal acceleration component is constant in magnitude.
   (3) Its tangential acceleration component is constant in magnitude.

9. An object is in uniform circular motion. Identify the correct statement(s).

   (a) 1 only   (b) 2 only   (c) 3 only
   (d) 1, 2, and 3   (e) 2 and 3   (f) 1 and 2
   (g) 1 and 3   (h) None of them

10. An object is in nonuniform circular motion with constant angular acceleration. Identify the correct statement(s). (Use the same answer choices as Question 9.)

PROBLEMS

Note: **C** indicates a combination conceptual/quantitative problem. Gold diamonds ✦, ✦✦ are used to indicate the increasing level of difficulty of each problem. Problem numbers appearing in blue, 9., denote problems that have a detailed solution available in the Student Solutions Manual. Some problems are *paired* by concept; their numbers are connected by a ruled box.

5.1 Description of Uniform Circular Motion

1. A carnival swing is fixed on the end of an 8.0-m-long beam. If the swing and beam sweep through an angle of 120°, what is the distance through which the riders move?

2. A soccer ball of diameter 31 cm rolls without slipping at a linear speed of 2.8 m/s. Through how many revolutions has the soccer ball turned as it moves a linear distance of 18 m?

3. Find the average angular speed of the second hand of a clock.

4. Convert these to radian measure: (a) 30.0°, (b) 135°, (c) 1/4 revolution, (d) 33.3 revolutions.

5. A bicycle is moving at 9.0 m/s. What is the angular speed of its tires if their radius is 35 cm?

6. An elevator cable winds on a drum of radius 90.0 cm that is connected to a motor. (a) If the elevator is moving down at 0.50 m/s, what is the angular speed of the drum? (b) If the elevator moves down 6.0 m, how many revolutions has the drum made?

7. Grace is playing with her dolls and decides to give them a ride on a merry-go-round. She places one of them on a record player and sets the angular speed at 33.3 rpm. (a) What is their angular speed in rad/s? (b) If the doll is 13 cm from the center of the spinning turntable platform, how fast (in m/s) is the doll moving?

8. A wheel is rotating at a rate of 2.0 revolutions every 3.0 s. Through what angle, in radians, does the wheel rotate in 1.0 s?

✦ 9. In the construction of railroads, it is important that curves be gentle, so as not to damage passengers or freight. Curvature is not measured by the radius of curvature, but in the following way. First a 100.0-ft-long chord is measured. Then the curvature is reported as the angle subtended by two radii at the endpoints of the chord (Fig. 5.19). (The angle is measured by determining the angle between two tangents 100 ft apart; since each tangent is perpendicular to a radius, the angles are the same.) In modern railroad construction, track curvature is kept below 1.5°. What is the radius of curvature of a "1.5° curve"? [*Hint:* Since the angle is small, the length of the chord is approximately equal to the arc length along the curve.]

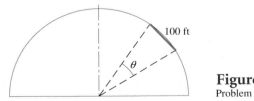

Figure 5.19 Problem 9

5.2 Centripetal Acceleration

10. Verify that all three expressions for centripetal acceleration ($v\omega$, v^2/r, and $\omega^2 r$) have the correct dimensions for an acceleration.

11. The apparatus of Fig. 5.20 is designed to study insects at an acceleration of magnitude 980 m/s² ($= 100g$). The apparatus consists of a 2.0-m rod with insect containers at either end.

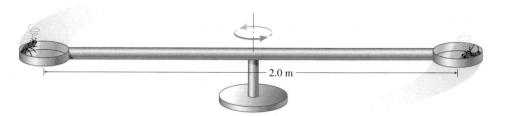

Figure 5.20 Problem 11

The rod rotates about an axis perpendicular to the rod and at its center. (a) How fast does an insect move when it experiences a centripetal acceleration of 980 m/s²? (b) What is the angular speed of the insect?

12. The rotor (Fig. 5.21) is an amusement park ride where people stand against the inside of a cylinder. Once the cylinder is spinning fast enough, the floor drops out. (a) What force keeps the people from falling out the bottom of the cylinder? (b) If the coefficient of friction is 0.40 and the cylinder has a radius of 2.5 m, what is the minimum angular speed of the cylinder so that the people don't fall out? (Normally the operator runs it considerably faster as a safety measure.)

Figure 5.21 Problem 12

13. Objects that are at rest relative to Earth's surface are in circular motion due to Earth's rotation. What is the centripetal acceleration of an African baobab tree located at the equator?

C 14. A curve in a stretch of highway has radius R. The road is unbanked. The coefficient of static friction between the tires and road is μ_s. (a) What is the fastest speed that a car can safely travel around the curve? (b) Explain what happens when a car enters the curve at a speed greater than the maximum safe speed. Illustrate with a free-body diagram.

C ✦ 15. Earth's orbit around the Sun is nearly circular. The period is 1 yr = 365.25 days. (a) In an elapsed time of 1 day, what is Earth's angular displacement? (b) What is the change in Earth's velocity, $\Delta\vec{v}$? (c) What is Earth's average acceleration during 1 day? (d) Compare your answer for (c) to Earth's instantaneous centripetal acceleration. Explain.

✦ 16. A child swings a rock of mass m in a horizontal circle using a rope of length L. The rock moves at constant speed v. (a) Ignoring gravity, find the tension in the rope. (b) Now include gravity (the weight of the rock is no longer negligible, although the weight of the rope still is negligible). What is the tension in the rope? What angle does the rope make with the horizontal?

✦ 17. A *conical pendulum* consists of a bob (mass m) attached to a string (length L) swinging in a horizontal circle (Fig. 5.8). As the string moves, it sweeps out the area of a cone. The angle

that the string makes with the vertical is ϕ. (a) What is the tension in the string? (b) What is the period of the pendulum?

5.3 Banked Curves

18. A highway curve has a radius of 122 m. At what angle should the road be banked so that a car traveling at 26.8 m/s (60 mph) has no tendency to skid sideways on the road? [*Hint:* No tendency to skid means the frictional force is zero.]

19. A curve in a highway has radius of curvature 120 m and is banked at 3.0°. On a day when the road is icy, what is the safest speed to go around the curve?

20. A roller coaster car of mass 320 kg (including passengers) travels around a horizontal curve of radius 35 m. Its speed is 16 m/s. What is the magnitude and direction of the total force exerted on the car by the track?

21. A velodrome is built for use in the Olympics (Fig. 5.22). The radius of curvature of the surface is 20.0 m. At what angle should the surface be banked for cyclists moving at 18 m/s? (Choose an angle so that no frictional force is needed to keep the cyclists in their circular path. Large banking angles *are* used in velodromes.)

Figure 5.22 Problem 21

C 22. A car drives around a curve with radius 410 m at a speed of 32 m/s. The road is not banked. The mass of the car is 1400 kg. (a) What is the frictional force on the car? (b) Does the frictional force necessarily have magnitude $\mu_s N$? Explain.

✦ 23. A car drives around a curve with radius 410 m at a speed of 32 m/s. The road is banked at 5.0°. The mass of the car is 1400 kg. (a) What is the frictional force on the car? (b) At what speed could you drive around this curve so that the force of friction is zero?

✦✦ 24. A curve in a stretch of highway has radius R. The road is banked at angle θ to the horizontal. The coefficient of static friction between the tires and road is μ_s. What is the fastest speed that a car can travel through the curve?

25. An airplane is flying at constant speed v in a horizontal circle of radius r. The lift (aerodynamic force) on the wings is perpendicular to the wings. At what angle to the vertical must the wings be banked to fly in this circle?

5.4 Circular Orbits

26. A spy satellite is in circular orbit around Earth. It makes one revolution in 6.00 hours. (a) How high above Earth's surface is the satellite? (b) What is the satellite's acceleration?

✦ 27. Two satellites are in circular orbits around Jupiter. One, with orbital radius r, makes one revolution every 16 h. The other satellite has orbital radius $4.0r$. How long does the second satellite take to make one revolution around Jupiter?

28. The Hubble Space Telescope orbits Earth 613 km above Earth's surface. What is the period of the telescope's orbit?

29. The orbital speed of Earth about the Sun is 3.0×10^4 m/s and its distance from the Sun is 1.5×10^{11} m. The mass of Earth is approximately 6.0×10^{24} kg and that of the Sun is 2.0×10^{30} kg. What is the magnitude of the force exerted by the Sun on Earth? [*Hint:* Two different methods are possible. Try both.]

30. A satellite travels around Earth in uniform circular motion at altitude 35,800 km above Earth's surface. The satellite is in geosynchronous orbit (that is, the time for it to complete one orbit is exactly one day). In Fig. 5.18, the satellite moves counterclockwise (*ABCDA*). State directions in terms of the *x*- and *y*-axes. (a) What is the satellite's instantaneous velocity at point *C*? (b) What is the satellite's average velocity for one quarter of an orbit, starting at *A* and ending at *B*? (c) What is the satellite's average acceleration for one quarter of an orbit, starting at *A* and ending at *B*? (d) What is the satellite's instantaneous acceleration at point *D*?

✦✦ 31. A spacecraft is in orbit around Jupiter. The radius of the orbit is 3.0 times the radius of Jupiter (which is $R_J = 71,500$ km). The gravitational field at the surface of Jupiter is 23 N/kg. What is the period of the spacecraft's orbit? [*Hint:* You don't need to look up any more data about Jupiter to solve the problem.]

5.5 Nonuniform Circular Motion

32. A roller coaster has a vertical loop with radius 20.0 m. With what minimum speed should the roller coaster car be moving at the top of the loop so that the passengers do not lose contact with the seats?

ⓒ 33. A pendulum is 0.80 m long and the bob has a mass of 1.0 kg. At the bottom of its swing, the bob's speed is 1.6 m/s. (a) What is the tension in the string at the bottom of the swing? (b) Explain why the tension is greater than the weight of the bob.

✦ 34. A pendulum (Fig. 5.23) is 0.800 m long and the bob has a mass of 1.00 kg. When the string makes an angle of $\theta = 15.0°$ with the vertical, the bob is moving at 1.40 m/s. Let $g = 9.80$ m/s^2. Find the tangential and centripetal acceleration components and the tension in the string. [*Hint:* Draw an FBD for the bob. Choose the *x*-axis to be tangential to the motion of the bob and the *y*-axis to be centripetal. Apply Newton's second law.]

✦ 35. Find the tangential acceleration of a freely swinging pendulum when it makes an angle θ with the vertical (Fig. 5.23).

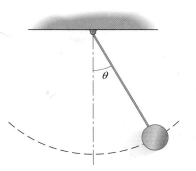

Figure 5.23
Problems 34, 35, and 48

5.6 Angular Acceleration

36. A wheel is subjected to uniform angular acceleration. Initially its angular velocity is zero. During the first 1.0-s time interval, it rotates through an angle of 90.0°. (a) Through what angle does it rotate during the next 1.0-s time interval? (b) Through what angle during the third 1.0-s time interval?

37. A cyclist starts from rest and pedals so that the wheels make 8.0 revolutions in the first 5.0 s. What is the angular acceleration of the wheels (assumed constant)?

38. During normal operation, a computer's hard disk spins at 7200 rpm. If it takes the hard disk 4.0 s to reach this angular velocity starting from rest, what is the average angular acceleration of the hard disk in rad/s^2?

39. The turntable of a record player reaches its rated frequency of rotation, 33.3 rpm, in 2.0 s, starting from rest. (a) Assuming the angular acceleration is constant, what is its magnitude? (b) How many revolutions does the turntable make during this time interval?

40. In a Beams ultracentrifuge, the rotor is suspended magnetically in a vacuum. Since there is no mechanical connection to the rotor, the only friction is the air resistance due to the few air molecules in the vacuum. If the rotor is spinning with an angular speed of 5.0×10^5 rad/s and the driving force is turned off, its spinning slows down at an angular rate of 0.40 rad/s^2. (a) How long does the rotor spin before coming to rest? (b) During this time, through how many revolutions does the rotor spin?

41. The rotor of the Beams ultracentrifuge (see Problem 40) is 20.0 cm long. For a point at the end of the rotor, find the (a) initial speed, (b) tangential acceleration component, and (c) maximum centripetal acceleration component.

5.7 Artificial Gravity

ⓒ ✦ 42. Objects that are at rest relative to the Earth's surface are in circular motion due to Earth's rotation. (a) What is the centripetal acceleration of an object at the equator? (b) Is the object's apparent weight greater or less than its weight? Explain. (c) By what percentage does the apparent weight differ from the weight at the equator? (d) Does a bathroom scale measure your true weight or your apparent weight?

43. A space station is shaped like a ring and rotates to simulate gravity. If the radius of the space station is 120 m, at what frequency must it rotate so that it simulates Earth's gravity? [*Hint:* The apparent weight of the astronauts must be the same as their weight on Earth.]

44. A person of mass M stands on a bathroom scale inside a Ferris wheel compartment. The Ferris wheel has radius R and angular velocity ω. What is the apparent weight of the person (a) at the top and (b) at the bottom?

45. A person rides a Ferris wheel that turns with constant angular velocity. Her weight is 520.0 N. At the top of the ride her apparent weight is 1.5 N different from her true weight. (a) Is her apparent weight at the top 521.5 N or 518.5 N? Why? (b) What is her apparent weight at the bottom of the ride? (c) If the angular speed of the Ferris wheel is 0.025 rad/s, what is its radius?

The Millenium Wheel in London.

46. If a washing machine's drum has a radius of 25 cm and spins at 4.0 rev/s, what is the strength of the artificial gravity to which the clothes are subjected? Express your answer as a multiple of g.

♦♦ 47. Objects that are at rest relative to Earth's surface are in circular motion due to Earth's rotation. What is the centripetal acceleration of a painting hanging in the Prado Museum in Madrid, Spain, at a latitude of 40.2° North (Fig. 5.24)? (Note that the object's centripetal acceleration is not directed toward the center of the Earth.)

♦♦ 48. A little girl is having fun on a swing of length 8.0 m. If her weight is 180 N, find her apparent weight (a) at the lowest point, where her speed is 9.7 m/s; and (b) at the highest point, where the swing is at an angle $\theta = 75°$ to the vertical (Fig. 5.23).

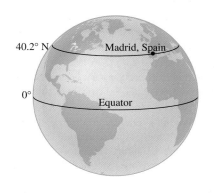

(a) Three-dimensional view

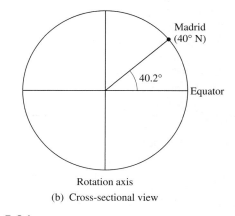

(b) Cross-sectional view

Figure 5.24 Problem 47

COMPREHENSIVE PROBLEMS

49. The Earth rotates on its own axis once per day (24.0 h). What is the tangential speed of the summit of Mt. Kilimanjaro (elevation 5895 m above sea level), which is located approximately on the equator, due to the rotation of the Earth? The equatorial radius of Earth is 6378 km.

50. A trimmer for cutting weeds and grass near trees and borders has a nylon cord of 0.23-m length that whirls about an axle at 660 rad/s. What is the linear speed of the tip of the nylon cord?

51. A high-speed dental drill is rotating at 3.14×10^4 rad/s. Through how many degrees does the drill rotate in 1.00 s?

52. A jogger runs counterclockwise around a path of radius 90.0 m at constant speed. He makes 1.00 revolution in 188.4 s. At $t = 0$, he is heading due east. (a) What is the jogger's instantaneous velocity at $t = 376.8$ s? (b) What is his instantaneous velocity at $t = 94.2$ s?

♦ 53. The time to sunset can be estimated by holding out your arm, holding your fingers horizontally in front of your eyes, and counting the number of fingers that fit between the horizon and the setting Sun. (a) What is the angular speed, in rad/s, of

the Sun's apparent circular motion around the Earth? (b) Estimate the angle subtended by one finger held at arm's length. (c) How long in minutes does it take the Sun to "move" through this same angle?

© 54. In the professional videotape recording system known as quadriplex, four tape heads are mounted on the circumference of a drum of radius 2.5 cm that spins at 1500 rad/s. (a) At what speed are the tape heads moving? (b) Why are moving tape heads used instead of stationary ones, as in audio tape recorders? [*Hint:* How fast would the tape have to move if the heads were stationary?]

55. The Milky Way galaxy rotates about its center with a period of about 200 million years. The Sun is 2×10^{20} m from the center of the galaxy. How fast is the Sun moving with respect to the center of the galaxy?

56. A small body of mass 0.50 kg is attached by a 0.50-m-long cord to a pin set into the surface of a frictionless table top. The body moves in a circle on the horizontal surface with a speed of 2.0π m/s. (a) What is the magnitude of the centripetal acceleration of the body? (b) What is the tension in the cord?

✦ 57. What's the fastest way to make a U-turn at constant speed? Suppose that you need to make a 180° turn on a circular path. The minimum radius (due to the car's steering system) is 5.0 m, while the maximum (due to the width of the road) is 20.0 m. Your acceleration must never exceed 3.0 m/s² or else you will skid. Should you use the smallest possible radius, so the distance is small, or the largest, so you can go faster without skidding, or something in between? What is the minimum possible time for this U-turn?

58. The Milky Way galaxy rotates about its center with a period of about 200 million years. The Sun is 2×10^{20} m from the center of the galaxy. (a) What is the Sun's centripetal acceleration? (b) What is the net gravitational force on the Sun due to the other stars in the Milky Way?

59. A rotating flywheel slows down at a constant rate due to friction in its bearings. After 1 min its angular velocity has diminished to 0.80 of its initial value ω. At the end of the third minute what is the angular velocity in terms of the initial value?

60. A coin is placed on a record that is rotating at 33.3 rpm. If the coefficient of static friction between the coin and the record is 0.1, how far from the center of the record can the coin be placed without having it slip off?

61. Grace gives her dolls a merry-go-round ride on an old phonograph set at 33.3 rpm. The dolls are 5.0 in. from the central axis. She changes the setting to 45 rpm. (a) For this new setting, what is the linear speed of a point on the turntable at the location of the dolls? (b) If the coefficient of static friction between the dolls and the turntable is 0.13, do the dolls stay on the record?

62. Your car's wheels are 65 cm in diameter and the wheels are spinning at an angular velocity of 101 rad/s. How fast is your car moving in kilometers per hour (assume no slippage)?

63. In an amusement park rocket ride, cars are suspended from 4.25-m cables attached to rotating arms at a distance of 6.00 m from the axis of rotation (Fig. 5.25). The cables swing out at an angle of 45.0° when the ride is operating. What is the angular speed of rotation?

64. Centrifuges are commonly used in biological laboratories for the isolation and maintenance of cell preparations. For cell separation, the centrifugation conditions are typically 1.0×10^3 rpm using an 8.0-cm-radius rotor. (a) What is the acceleration of material in the centrifuge under these conditions? Express your answer as a multiple of g. (b) At 1.0×10^3 rpm (and with a 8.0-cm rotor) what is the net force on a red blood cell whose mass is 9.0×10^{-14} kg? (c) What is the net force on a virus particle of mass 5.0×10^{-21} kg under the same conditions? (d) To pellet out virus particles and even to separate large molecules such as proteins, super-high-speed centrifuges called ultracentrifuges are used in which the rotor spins in a vacuum to reduce heating due to friction. What is the centripetal acceleration inside an ultracentrifuge at 75,000 rpm with an 8.0-cm rotor? Express your answer as a multiple of g.

65. Bacteria swim using a corkscrew-like helical flagellum that rotates. For a bacterium with a flagellum that has a pitch of 1.0 μm that rotates at 110 rev/s, how fast could it swim if there were no "slippage" in the medium in which it is swimming? The pitch of a helix is the distance between "threads."

66. Massimo, a machinist, is cutting threads for a bolt on a lathe. He wants the bolt to have 18 threads per inch. If the cutting tool moves parallel to the axis of the would-be bolt at a linear velocity of 0.080 in./s, what must the rotational speed of the lathe chuck be to ensure the correct thread density? [*Hint:* One thread is formed for each complete revolution of the chuck.]

✦ 67. In Chapter 19 we will see that a charged particle can undergo uniform circular motion when acted on by a magnetic force and no other forces. (a) For that to be true, what must be the angle between the magnetic force and the particle's velocity? (b) The magnitude of the magnetic force on a charged particle is proportional to the particle's speed, $F = kv$. Show that two identical charged particles moving in circles at different speeds in the same magnetic field must have the same period. (c) Show that the radius of the particle's circular path is proportional to the speed.

✦✦ 68. Find the orbital radius of a geosynchronous satellite. Do not assume the speed found in Example 5.9. Start by writing an equation that relates the period, radius, and speed of the orbiting satellite. Then apply Newton's second law to the satellite. You will have two equations with two unknowns (the speed and radius). Eliminate the speed algebraically and solve for the radius.

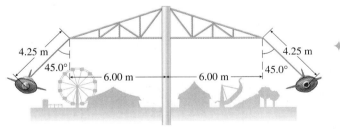

Figure 5.25   Problem 63

ANSWERS TO PRACTICE PROBLEMS

5.1 2.99×10^{-7} rad/s **5.2** 1.65 m/s **5.3** 1.9 min **5.4** 17 m/s² **5.5** (a) 4.5 m/s²; (b) 3.0 m/s **5.6** More slowly **5.7** No **5.8** 29.8 km/s; 3.16×10^7 s **5.9** 0.723R **5.10** 2.44 h **5.11** 4.2mg **5.12** Acceleration is purely tangential:
5.13 (a) 1.7 rad/s²; (b) 3.5 rad **5.14** (a) 2200 N; (b) 1500 N

Energy

How high can a kangaroo jump? A kangaroo can jump about 0.70 m high when stretching its hind leg tendons approximately half of the maximum amount. How high can the same kangaroo jump when the tendons are stretched by an additional 10%?

Conservation law: a physical law phrased in terms of a quantity that does not change with time.

6.1   A CONSERVATION LAW

The hopping kangaroo is an example of the transformation of energy in a living thing. A kangaroo has long elastic tendons in its hind legs. As the kangaroo lands, these tendons store *elastic energy* to assist in making the next jump. Energy is then alternately expended and stored by the kangaroo as it propels itself forward in one jump after another. The storage of energy in the tendons reduces the amount of energy that the muscles must supply for subsequent jumps.

Energy comes in many different forms, including chemical energy, nuclear energy, mechanical energy, electrical energy, kinetic energy, and elastic energy. Energy can be changed from one form to another, or transferred from one place to another, but the total quantity of energy in any closed system cannot change. (A *closed system* is a set of interacting bodies upon which no net external force acts and that does not transfer energy to or from anything outside the system.) With energy we encounter our first **conservation law**: a physical law phrased in terms of a quantity that does not change with time. Conservation laws are often considered to be the most fundamental laws of physics.

The law of conservation of energy says that energy can neither be created nor destroyed. To apply this law, we need to learn how to calculate the amount of each kind of energy. There isn't one formula that applies to all.

Conservation of energy is a powerful tool in the search to understand nature. It is a great simplification to look at energy transformations in the complex interactions that occur, for instance, in the human body. The body converts chemical energy stored in food into various other forms of energy. Some of the energy is available to fuel physical activity (Fig. 6.1). Conservation of energy requires that when the amount of one form of energy decreases, the amount of other forms of energy increases such that the total energy remains the same. Over the years, scientists have faced situations where the conservation of energy seemed to be violated, only to find that energy had been converted into a previously unrecognized form.

6.2   WORK DONE BY CONSTANT FORCES

Since conservation of energy is an important law of nature, we need to define various forms of energy and learn how energy can be transformed from one form to another. We begin with an example. Suppose the trunk in Fig. 6.2a weighs 640 N and must be lifted a height $h = 1.5$ m. To lift it at constant speed, Rosie must exert a force of 640 N on the rope (assuming an ideal pulley and rope). As discussed in Example 2.13, the force required would be half as much (320 N) if she were to use the two-pulley system of Fig. 6.2b

Figure 6.1 These rock climbers must convert chemical energy into other forms of energy.

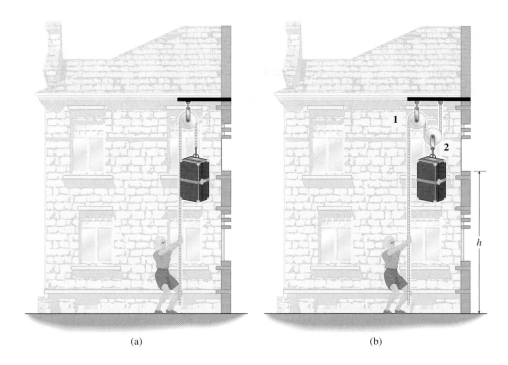

(a)   (b)

Figure 6.2 (a) Rosie is trying to move a trunk into her dorm room through the window. As long as she holds the trunk motionless in one position, she is doing no work on the trunk (in the physics sense of the word *work*). (b) The two-pulley system makes it easier for Rosie to lift the trunk: the force she must exert is halved. Is she getting something for nothing, or does she still have to do the same amount of work to lift the trunk?

instead. She doesn't get something for nothing, though. To lift the trunk 1.5 m, the sections of rope on *both* sides of pulley 2 must be shortened by 1.5 m, so Rosie must pull a 3.0-m length of rope. The two-pulley system enables her to pull with half the force, but now she must pull the rope twice as far.

Notice that the *product* of the magnitude of the force and the distance is the same for both cases:

$$640 \text{ N} \times 1.5 \text{ m} = 320 \text{ N} \times 3.0 \text{ m} = 960 \text{ N·m} = W$$

This product is called the **work** (*W*) done. Work is a scalar quantity; it does not have a direction. The same symbol *W* is also commonly used for the magnitude of the weight of an object. We can distinguish these two uses of the same symbol by context. Weight has dimensions of force, while work has dimensions of force times distance. On the other hand, the symbol $\vec{\mathbf{W}}$ can never represent work, since work is not a vector quantity.

As with velocity, force, and acceleration, the word *work* has a precise definition in physics; don't be misled by the many different meanings it has in ordinary conversation. We talk about doing homework, or going to work, or having too much work to do to have time to go to the movies. Not everything we call "work" in conversation is *work* as defined in physics.

The SI unit of work and energy is the newton-meter (N·m), which is given the name joule (symbol: J). Using either method, Rosie must do 960 J of work to lift the trunk. When we say that Rosie does 960 J of work, we mean that Rosie supplies 960 J of energy—the amount of energy required to lift the trunk 1.5 m. *Work is an energy transfer due to the application of a force.* Simple machines such as levers, pulleys, and inclined planes can reduce the *force* needed to accomplish a task, but they cannot reduce the *work* that must be done.

According to the definition, Rosie does no work while she holds the trunk in one place. She can just as well fasten the rope to a tie-down and walk away (Fig. 6.3). Work is done only when she *moves* the trunk by moving the rope. *If there is no displacement, no work is done and no energy is transferred.* A more exact statement is that Rosie does no work *on the rope* or *on the trunk*. Work *is* being done inside her body by the muscle fibers, which continually contract and expand to hold a load in one place. Thus, Rosie does expend chemical energy to hold the trunk in place and she gets tired if she holds it there very long, but the energy expended is not transferred to the trunk.

SI unit of work and energy is the joule: 1 J = 1 N·m

Definition of work: an energy transfer due to the application of a force.

Figure. 6.3 While the trunk is held in place by tying the rope, no work is done and no energy transfers occur.

The force that Rosie exerts on the rope is in the same direction as the displacement of that end of the rope. How much work is done by a force that is at some other angle to the displacement? Only the component of the force in the direction of the displacement does work. So, in general, the work done by a constant force is defined as the product of the magnitude of the displacement and the *component* of the force *parallel to the displacement*. If the displacement is in the *x*-direction, then $W = F_x \Delta x$. If θ is the angle between the force and displacement vectors, then the force component in the direction of the displacement is $F \cos \theta$ (Fig. 6.4a). Therefore, work done by a constant force can be written $W = F \Delta r \cos \theta$, where F is the magnitude of the force and Δr is the magnitude of the displacement of the point of application of the force.

Work done by a constant force:
$$W = F \Delta r \cos \theta \qquad (6\text{-}1)$$
For a displacement along the *x*-axis,
$$W = F_x \Delta x \qquad (6\text{-}1a)$$

There is another way to calculate work. We identified $F \cos \theta$ as the component of the force along the direction of the displacement. We can just as well identify $\Delta r \cos \theta$ as the component of the *displacement* in the direction of the *force* (Fig. 6.4b). Therefore, the work done by a force is equal to the magnitude of the force times the parallel component of the displacement. For example, gravity pulls downward, so the work done by gravity is the weight times the *downward component* of the displacement. If the *y*-axis is taken to be upward, then Δy is the upward component of the displacement. The *downward* component of the displacement is $-\Delta y$.

$$W_g = F_g \times (-\Delta y) = -mg\, \Delta y \qquad (6\text{-}2)$$

(in a uniform gravitational field with the *y*-axis pointing up)

Equation (6-2) holds even if the object does not move in a straight-line path.

The application of a force can decrease as well as increase the energy of the object to which the force is applied. If a force has a component parallel to the displacement of the object, then the work done by the force increases the object's energy. If a force has a component *antiparallel* to the displacement of the object, then the work done by the force decreases the object's energy. When the angle between $\vec{F}$ and $\Delta\vec{r}$ is less than 90°, $\cos \theta$ in Eq. (6-1) is positive. The force does positive work and the object's energy increases. If the angle between $\vec{F}$ and $\Delta\vec{r}$ is greater than 90°, $\cos \theta$ is negative. The force does negative work and the object's energy decreases. Pay careful attention to the algebraic sign when calculating work: negative work done on a system means that the energy of the system decreases, while positive work represents an energy increase.

If the force is perpendicular to the displacement, then there is no force component along the direction of the displacement and therefore no work is done. Equation (6-1) includes this case: if the angle between the force and the displacement is $\theta = 90°$, then the work done is zero since $\cos 90° = 0$. The normal force does no work on a sliding

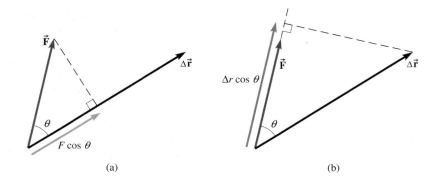

Figure 6.4 (a) The component of $\vec{F}$ in the direction of $\Delta\vec{r}$ is $F \cos \theta$. (b) The component of $\Delta\vec{r}$ in the direction of $\vec{F}$ is $\Delta r \cos \theta$.

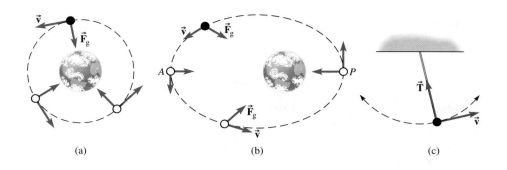

Figure 6.5 (a) No matter where the satellite is in its circular orbit, it experiences a gravitational force directed toward the center of the Earth. This force is always perpendicular to the satellite's velocity; thus, gravity does no work on the satellite. (b) In an elliptical orbit, the gravitational force is *not* always perpendicular to the velocity. As the satellite moves counterclockwise in its orbit from point *P* to point *A*, gravity does negative work; from *A* to *P*, gravity does positive work. (c) The tension in the string of a pendulum is always perpendicular to the velocity of the pendulum bob.

object since it is perpendicular to the displacement of the object, even if the surface is not horizontal. The direction of the normal force is always perpendicular to the velocity of an object sliding along the surface. During a short time interval, the normal force is perpendicular to the displacement $\vec{v}\,\Delta t$ of the object. Since the normal force and displacement are perpendicular for *every* small time interval, the normal force does no work.

The normal force is not the only instance of a force whose direction is always perpendicular to the velocity. No work is done by the Earth's gravitational force on a satellite in circular orbit (Fig. 6.5a). In a circular orbit, the gravitational force is always directed along a radius from the satellite to the center of the Earth. At every point in the orbit, the gravitational force is perpendicular to the velocity of the satellite (which is tangent to the circular orbit).

By contrast, gravity does work on a satellite in a noncircular orbit (Fig. 6.5b). Everywhere except at points *A* and *P*, there *is* a component of the gravitational force along the direction of the satellite's velocity. Wherever the gravitational force has a component parallel to the velocity, gravity does positive work; where the gravitational force has a component antiparallel to the velocity, gravity does negative work.

As a third example, no work is done by the tension in the string on a swinging pendulum bob because the tension is always perpendicular to the velocity (Fig. 6.5c).

Making The Connection:
**work done on an
orbiting satellite**

No work is done on an object by a force that is perpendicular to the object's velocity.

Example 6.1

Antique Chest Delivery

A rare and valuable antique chest, made in 1907 by Gustav Stickley, is being moved into a truck using a 4.0-m-long ramp. The weight of the chest plus packing material is 1400 N. To get the chest from the ground onto the truck bed, which is 1.0 m higher, the movers must decide what to do. Should they lift it straight up, or should they push it up the ramp? Assume they push the chest on a dolly, which in a simplified model is equivalent to sliding it up a *frictionless* ramp.

(a) Find the work done by the movers on the chest if they lift it straight up 1.0 m at constant speed.

(b) Find the work done if they slide the chest up the 4.0-m-long *frictionless* ramp at constant speed by pushing parallel to the ramp.

(c) Find the work done by gravity on the chest in each case.

(d) Find the work done by the normal force of the ramp on the chest. Assume that all forces are constant.

Strategy   In each case we must calculate the force required to move the chest, calculate the displacement, and then find

the work done by multiplying the magnitude of one by the parallel component of the other. Drawing a free-body diagram helps us calculate the forces. The ramp is a simple machine—just as for Rosie's pulleys, the ramp cannot reduce the amount of work that must be done, so we expect the work done by the movers to be the same in both cases (neglecting friction). We expect the work done by gravity to be negative in both cases, since the chest is moving up rather than down. Gravity is *not* helping the movers lift the chest up into the truck. The normal force is perpendicular to the ramp so it does zero work on the chest.

Given: Weight of chest = 1400 N; length of ramp = 4.0 m; height of ramp = 1.0 m

To find: W_{movers} lifting straight up and W_{movers} using the ramp; W_{g}; W_{N}

Tips   When several forces act on a body, use subscripts to clarify which work is being calculated.

continued on next page

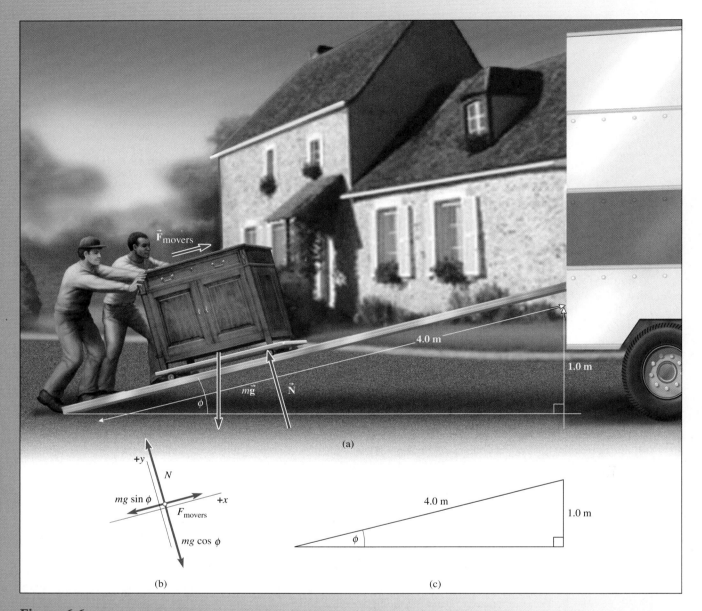

Figure 6.6 (a) An antique chest is pushed up a ramp into a truck, (b) free-body diagram for the chest, and (c) finding the angle of the incline.

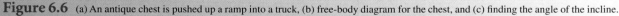

Example 6.1 *continued*

Solution (a) The displacement is 1.0 m straight up. The movers must exert an upward force equal in magnitude to the weight of the chest to move it at constant speed. The work done to lift it 1.0 m is

$$W_{movers} = F\,\Delta r\cos\theta = 1400\text{ N} \times 1.0\text{ m} \times \cos 0 = +1400\text{ J}$$

where $\theta = 0$ because the force and the displacement are in the same direction (upward).

(b) Figure 6.6a shows a sketch of the situation. We draw an FBD for the chest as it is pushed up the ramp (Fig. 6.6b). We take the x-axis along the inclined ramp and the y-axis perpendicular to the ramp and replace the gravitational force with its x- and y-components. Sliding along at constant speed, the chest's acceleration is zero, so the x-components of the forces add to zero.

continued on next page

Example 6.1 *continued*

The *x*-component of the gravitational force acts in the −*x*-direction and the force from the movers acts in the +*x*-direction.

$$\Sigma F_x = F_{\text{movers}} - mg \sin \phi = 0$$

From the right triangle formed by the ramp, the ground, and the truck bed in Fig. 6.6c:

$$\sin \phi = \frac{\text{height of truck bed}}{\text{distance along ramp}}$$

$$= \frac{h}{d} = \frac{1.0 \text{ m}}{4.0 \text{ m}} = 0.25$$

We can now solve for F_{movers}:

$$F_{\text{movers}} = mg \sin \phi = mg \times \frac{h}{d}$$

$$= 0.25 mg = 0.25 \times 1400 \text{ N} = 350 \text{ N}$$

The force and displacement are parallel, so the work done by the movers is F_{movers} times the distance *d*.

Solving for the work done by the movers yields

$$W_{\text{movers}} = F_{\text{movers}} d$$

$$= 350 \text{ N} \times 4.0 \text{ m}$$

$$= +1400 \text{ J}$$

The work done is the same as in (a). The movers apply 1/4 the force over a displacement that is four times larger.

(c) In both cases, the force of gravity is the weight of magnitude *mg* acting downward. Also, in both cases, the upward component of the displacement in the direction of the gravitational force is $\Delta y = h = 1.0$ m. The work done by gravity is the same for the two cases.

$$W_g = -mg \, \Delta y$$

$$= -1400 \text{ N} \times 1.0 \text{ m} = -1400 \text{ J}$$

(d) The normal force of the ramp on the chest does no work because it acts in a direction perpendicular to the displacement of the chest.

$$W_N = N \Delta r \cos 90° = 0$$

Discussion   In (b), we could have worked with algebraic symbols as

$$W_{\text{movers}} = F_{\text{movers}} d = \frac{mgh}{d} \times d = mgh$$

Since *d*, the length of the ramp, cancels when multiplying the force times the distance, the work done by the movers is the same for *any* length ramp (as long as the height is the same). Since we found the work by multiplying 350 N by 4.0 m, we might not have noticed that *d* does not affect the final answer.

Using a frictionless ramp, the movers exert a smaller force than if they lift the chest straight up, but they exert it over a longer distance; the total work done is the same. With a *real* ramp, friction acts to oppose the motion of the chest, so the movers would have to do *more* than 1400 J of work to slide the chest up the ramp. There's no getting around it; if the movers want to get that chest into the truck, they're going to have to do *at least* 1400 J of work.

Practice Problem 6.1   Bicycling uphill

A bicyclist climbs a 2.0-km-long hill that makes an angle of 7.0° with the horizontal. The total weight of the bike and the rider is 750 N. How much work is done on the bike and rider by gravity?

Example 6.2

Fun on a Sled

Diane pulls a sled along a snowy path on level ground with her little brother Jasper riding on the sled (Fig. 6.7a). The total mass of the sled and Jeremy is 26 kg. The cord makes a 20.0° angle with the ground. As a rough approximation, assume that the force of friction on the sled is determined by $\mu_k = 0.16$, even though the surfaces are not dry (some snow melts as the runners slide along it). Find the work done by Diane and the work done by the ground on the sled while the sled moves 120 m along the path at a constant 3 km/h.

Strategy   The sled's acceleration is zero, so the vector sum of all the external forces (gravity, friction, rope tension, and the normal force) is zero. We draw an FBD and use Newton's second law to find the tension in the rope.

Solution   The FBD is shown in Fig. 6.7b. The *x*- and *y*-axes are parallel and perpendicular to the ground, respectively. From Newton's second law with zero acceleration,

$$\Sigma F_x = +T \cos \theta - f_k = 0 \qquad (1)$$

$$\Sigma F_y = +T \sin \theta - mg + N = 0 \qquad (2)$$

where *T* is the tension and $\theta = 20.0°$. The force of kinetic friction is the coefficient of friction times the normal force,

$$f_k = \mu_k N$$

Substituting this into (1)

$$T \cos \theta - \mu_k N = 0 \qquad (3)$$

To find the tension, *T*, we need to eliminate the unknown normal force *N*. Equation (2) also involves the normal force *N*. We multiply Eq. (2) by μ_k,

$$\mu_k T \sin \theta - \mu_k mg + \mu_k N = 0 \qquad (4)$$

continued on next page

Example 6.2 *continued*

Adding Eqs. (3) and (4) to eliminate the normal force, N, and solving for T yields

$$T \cos \theta + \mu_k T \sin \theta - \mu_k mg = 0$$

$$T = \frac{\mu_k mg}{\mu_k \sin \theta + \cos \theta}$$

$$= \frac{0.16 \times 26 \text{ kg} \times 9.8 \text{ m/s}^2}{(0.16 \times \sin 20.0°) + \cos 20.0°} = 41.0 \text{ N}$$

Now the tension is used to find the work done by Diane. The component of the tension T acting parallel to the displacement is the force that does work. That component is $T \cos \theta$ and the displacement is $\Delta x = 120$ m. So the work done by Diane is

$$W_T = (T \cos \theta) \Delta x$$

$$= 41.0 \text{ N} \times \cos 20.0° \times 120 \text{ m} = +4600 \text{ J}$$

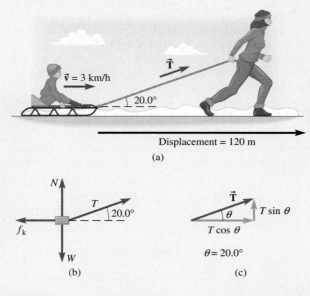

$\vec{v} = 3$ km/h

$\vec{T}$

20.0°

Displacement = 120 m

(a)

N

T

20.0°

f_k

W

(b)

$\vec{T}$

θ

$T \sin \theta$

$T \cos \theta$

$\theta = 20.0°$

(c)

Figure 6.7
(a) Jasper being pulled on a sled; (b) free-body diagram; (c) resolving the tension into x- and y-components.

The force on the sled due to the ground has two components: N and f_k. The normal force does no work since it acts perpendicular to the displacement of the sled. Friction acts in a direction opposite to the displacement, so the work done by friction is

$$W_f = f_k \Delta x \cos 180° = -f_k \Delta x$$

From Eq. (1),

$$f_k = T \cos \theta$$

Therefore, the work done by the ground—the work done by the frictional force—is

$$W_f = -f_k \Delta x = - (T \cos \theta) \Delta x$$

Except for the minus sign, W_f is the same as W_T. Therefore, $W_f = -4600$ J.

Discussion   A force can be resolved into components parallel and perpendicular to the displacement. The perpendicular component does no work. Here, since the sled's displacement is in the x-direction, the work done is the x-component of the force times the displacement. The x-component of the tension is

$$T \cos \theta = 41.0 \text{ N} \cos 20.0° = 38.5 \text{ N}$$

Then the work done is

$$W = T_x \Delta x = 38.5 \text{ N} \times 120 \text{ m} = 4600 \text{ J}$$

Note that the speed (3 km/h) was not used in the solution. Assuming that the frictional force on the sled is independent of speed, Diane would exert the same force to pull the sled at *any* constant speed. Then the work she does is the same for a 120-m displacement. At a higher speed, though, she would have to do that amount of work in a shorter time interval.

Practice Problem 6.2   A different angle

Find the tension if Diane pulls at an angle $\theta = 30.0°$ instead of 20.0°, assuming the same coefficient of friction. What is the work done by Diane on the sled in this case for a 120-m displacement? Explain how the tension can be greater but the work done by Diane smaller.

Work Done by Dissipative Forces

The work done by kinetic friction was calculated in Example 6.2 according to a simplified model of friction. If friction truly does −4600 J of work on the sled, then 4600 J of energy would have to be transferred from the sled to the ground. What really happens is that 4600 J of energy is converted into thermal energy *shared* between the ground and the sled—both the ground and the sled warm up a little. So the 4600 J is not all transferred to the ground; some stays in the sled. A more accurate statement is that friction *dissipates* 4600 J of energy. **Dissipation** is the conversion of energy from an organized form to a disorganized form.

Kinetic friction is a complicated phenomenon. The friction force acts all along the bottom of the sled. The transfer of energy by friction is due to huge numbers of microscopic forces acting at many different points of contact. Each of these forces acts through some small displacement as the surfaces continually deform against one another. The work done is not the same as if a single force of kinetic friction were to act at a single point that moves through the same displacement as the sled.

As a practical matter, though, we usually are not concerned with *where* the thermal energy appears. In our simplified model of kinetic friction, we assume that there *is* a single force of kinetic friction moving through the same displacement as the object. Then we can calculate the work done by friction using Eq. (6-1) and get the correct amount of energy dissipated; we just won't know how much of it is transferred to the stationary surface and how much remains in the sliding object. This is how we apply the term *work* to kinetic friction or to other dissipative forces such as air resistance.

6.3   KINETIC ENERGY

When a single, constant force $\vec{F}$ directed along the x-axis acts on a point particle of mass m during a displacement $\Delta\vec{r}$, the work done is

$$W = F_x \Delta x$$

where Δx is the x-component of the displacement. If $\vec{F}$ is the only force acting, then Newton's second law tells us that $\vec{F} = \vec{F}_{net} = m\vec{a}$, so

$$W = F_x \Delta x = m a_x \Delta x \tag{6-3}$$

Since the acceleration is constant, we can use any of the kinematic equations for constant acceleration from Chapter 3. The product of acceleration and displacement components occurs in Eq. (3-13):

$$a_x \Delta x = \tfrac{1}{2}(v_x^2 - v_{0x}^2)$$

Substituting Eq. (3-13) into Eq. (6-3) yields

$$W = m(a_x \Delta x) = \tfrac{1}{2}m(v_x^2 - v_{0x}^2)$$

Since the net force is along the x-axis, a_y and a_z are both zero. Then only the x-component of the velocity changes; v_y and v_z are constant. As a result,

$$v^2 - v_0^2 = (v_x^2 + \cancel{v_y^2} + \cancel{v_z^2}) - (v_{0x}^2 + \cancel{v_{0y}^2} + \cancel{v_{0z}^2}) = v_x^2 - v_{0x}^2$$

Therefore, the work done is

$$W = \tfrac{1}{2}m(v^2 - v_0^2)$$

The work done is equal to the change in the quantity $\tfrac{1}{2}mv^2$. This quantity is called the **kinetic energy** of the object (symbol K). Kinetic energy is a scalar quantity and is always nonnegative. The kinetic energy of an object moving with speed v is equal to the work that must be done on the object to accelerate it to that speed starting from rest. For a point particle, work done on the particle always appears as a change in the particle's kinetic energy. When a force does positive work on a point particle, it acts in a direction that tends to increase the particle's speed and thereby increase the object's kinetic energy. Negative work done on a point particle means that the force doing the work tends to slow the particle down and decrease its kinetic energy.

Kinetic energy:

$$K = \tfrac{1}{2}mv^2 \tag{6-4}$$

$$W = \Delta K \quad \text{(point particle)} \tag{6-5}$$

Equation (6-5) is a special case: it applies only to a **point particle**—an object with no internal structure. A point particle cannot be deformed, cannot change temperature, and cannot rotate. By assuming no internal structure, we can be sure that every part of the object goes through the same displacement while the force is applied and that the energy transferred all appears as kinetic energy. We must be careful when applying Eq. (6-5) to ordinary objects since some or all of the work done on the object may go into changing some other form of the object's energy, such as rotational kinetic energy or internal energy. Internal forces can do internal work, changing energy from one form to another within the object (as happens in our muscles when we hold something up without moving it). The internal energy of an object changes if it is deformed (its size or shape is altered), if its temperature changes, or if it undergoes a phase change such as melting. If we are sure that changes in these other forms of energy can be neglected, then we can apply Eq. (6-5) to ordinary objects.

Equation (6-4) defines the *translational* kinetic energy of an object—the kinetic energy due to the object's motion as a whole. Translational kinetic energy is the only kind of kinetic energy that a point particle can have. Extended objects have internal structure; they can have *rotational* and *vibrational* kinetic energy as well. For now, we consider only translational kinetic energy.

If more than one force acts on a point object, the change in kinetic energy is equal to the *total* work done by all the forces [Eq. (6-6)]. There are two ways to calculate the total work. The total work is the sum of the work done by each force individually, so we can calculate the work done by each force and add them [Eq. (6-7)]. An alternative is to calculate the work done by the *net* force [Eq. (6-8)]. Both methods give the same result as long as we consider only point particles, so that the point of application of every force moves through the same displacement.

$$W_{total} = \Delta K \quad \text{(point particle)} \tag{6-6}$$

$$W_{total} = \sum_i W_i \quad \text{(general)} \tag{6-7}$$

where W_i = work done by each individual force and Σ means *the sum of*.

$$W_{total} = F_{net}\, \Delta r \cos \theta \quad \text{(point particle)} \tag{6-8}$$

Conceptual Example 6.3

Collision Damage

Why is the damage caused by an automobile collision so much worse when the vehicles involved are moving at high speeds?

Strategy   When a collision occurs, the kinetic energy of the automobiles gets converted into other forms of energy. Thus, we can use the kinetic energy as a rough measure of how much damage can be done in a collision.

Solution and Discussion   Suppose we compare the kinetic energy of a car at two different speeds: 60.0 mph and 72.0 mph (which is 20.0% greater than 60.0 mph). If kinetic energy were proportional to speed, then a 20.0% increase in speed would mean a 20.0% increase in kinetic energy. However, since kinetic energy is proportional to the *square* of the speed, a 20.0% speed increase causes an increase in kinetic energy much greater than 20.0%. Working by proportions, we can find the percent increase in kinetic energy:

$$K \propto v^2$$

$$\frac{K'}{K} = \left(\frac{v'}{v}\right)^2 = (1.200)^2 = 1.440$$

Therefore, a 20.0% increase in speed causes a 44.0% increase in kinetic energy. What seems like a relatively modest difference in speed makes a lot of difference when a collision occurs.

Practice Problem 6.3   Two different cars collide with a stone wall

Suppose a sports utility vehicle and a small electric car both collide with a stone wall and come to a dead stop. If the SUV mass is 2.5 times that of the small car and the speed of the SUV is 60.0 mph while that of the other car is 40.0 mph, what is the ratio of the kinetic energy changes for the two cars (SUV to small car)? Does this ratio necessarily indicate the relative damage done to the two cars? Explain.

Example 6.4

Bungee Jumping

A bungee jumper makes a jump in the Gorge du Verdon in southern France. The jumping platform is 182 m above the bottom of the gorge. The jumper weighs 780 N. If the jumper falls to within 68 m of the bottom of the gorge, how much work is done by the bungee cord on the jumper during his descent? Ignore air resistance.

Strategy   Ignoring air resistance, only two forces act on the jumper during the descent: gravity and the tension in the cord. Since the jumper has zero kinetic energy at both the highest and lowest points of the jump, the change in kinetic energy for the descent is zero. Therefore, the total work done by the two forces on the jumper must equal zero.

Solution   Let W_g and W_c represent the work done on the jumper by gravity and by the cord. Then

$$W_{\text{total}} = W_g + W_c = \Delta K = 0$$

The work done by gravity is

$$W_g = -mg\,\Delta y$$

where the weight of the jumper is

$$mg = 780 \text{ N}$$

With $y = 0$ at the bottom of the gorge, the vertical component of the displacement is

$$\Delta y = y_f - y_i = 68 \text{ m} - 182 \text{ m} = -114 \text{ m}$$

Then the work done by gravity is

$$W_g = -(780 \text{ N}) \times (-114 \text{ m}) = +89 \text{ kJ}$$

The work done by the cord is therefore

$$W_c = -W_g = -89 \text{ kJ}$$

Discussion   The work done by gravity is positive, since the force and the displacement are in the same direction (downward). If not for the negative work done by the cord, the jumper would have a kinetic energy of 89 kJ after falling 114 m.

The length of the bungee cord is not given, but it does not affect the answer. At first the jumper is in free fall as the cord plays out to its full length; only then does the cord begin to stretch and exert a force on the jumper, ultimately bringing him to rest again. Regardless of the length of the cord, the total work done by gravity and by the cord must be zero since the change in the jumper's kinetic energy is zero.

Practice Problem 6.4   The bungee jumper's speed

Suppose that during the jumper's descent, at a height of 110 m above the bottom of the gorge, the cord has done −21.7 kJ of work on the jumper. What is the jumper's speed at that point?

6.4   WORK DONE BY VARIABLE FORCES

So far we have considered only constant forces when calculating work. How can we calculate the work done by a variable force? Consider an archer drawing back a compound bow (Fig. 6.8a). The compound bow is designed to make it easier to draw the string back and hold it back because, at a certain point, the force required to draw the string farther stops increasing. A convenient way to describe how the force varies with string position is to plot a graph. Figure 6.8b shows the force required to hold the string back as a function of distance. How can we calculate the work done by the archer as he draws the string back 40 cm?

We've asked analogous questions in previous chapters. Recall how we find the change in velocity Δv_x when the acceleration a_x is not constant (Section 3.3). During a short time interval Δt the acceleration change is negligible and $\Delta v_x = a_x \Delta t$. To find the velocity change for a longer time interval, we divide the time interval into a series of short time intervals and sum up all the velocity changes that occur during each one. On a graph of $a_x(t)$, each of these velocity changes is the area of a rectangle of height a_x and width Δt, so the area under a graph of a_x versus t is Δv_x.

To find the work done by a variable force F_x, we divide the displacement into a series of small displacements Δx. The displacements are small enough that the change

Making The Connection:
work done in drawing a bow

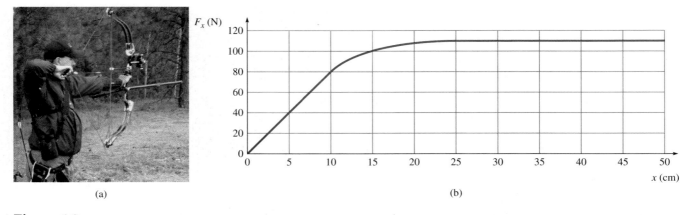

Figure 6.8 (a) A compound bow. (b) The force to draw back the compound bow depends on how far it is drawn. In this graph, the "area" represented by each rectangle is (0.050 m × 20.0 N) = 1.0 J.

in F_x during each one of them can be ignored. Then, during each small displacement, the work done is

$$\Delta W = F_x \, \Delta x \qquad (6\text{-}9)$$

Each ΔW is the area of a rectangle of height F_x and width Δx on a graph of $F_x(x)$. The total work done is the sum of these areas. In the limit that the small displacements are infinitesimally small, *the total work done is the area under the graph of $F_x(x)$.*

In Fig. 6.8b, the "area" of each rectangle represents (0.050 m × 20.0 N) = 1.0 J of work. There are approximately 36 rectangles under the force curve between $x = 0$ and $x = 40$ cm, so the work done by the archer is +36 J.

Example 6.5

Archery Practice

To draw back a simple bow, the force the archer exerts on the string continues to increase as the displacement of the string increases and the bow bends slightly. The force-versus-position graph of Fig. 6.9 describes such a bow, which, like an ideal spring, requires a force proportional to the displacement. Calculate the work done to draw the string back 40.0 cm.

Strategy The work done is the area under the force-versus-position graph. This time, instead of counting rectangles, we can calculate the triangular area formed by the force-versus-position graph.

Solution We want to find the work done to draw the string back 40.0 cm, so the base of the triangle is 40.0 cm. The altitude of the triangle is the force at 40.0 cm: 160 N. Therefore, the work done is

$$W = \tfrac{1}{2}(0.400 \text{ m} \times 160 \text{ N}) = +32 \text{ J}$$

Discussion When the arrow is released from this displacement of 40.0 cm, the string and bow do +32 J of work on the arrow. The work done on the arrow is equal to the increase in the arrow's kinetic energy. Therefore, the arrow has a kinetic energy equal to +32 J as it leaves the string.

This simple bow is *initially* easier to pull back than the compound bow considered previously. The spring constant of

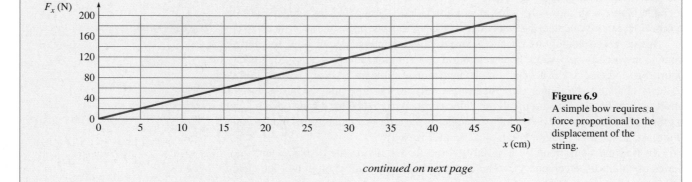

Figure 6.9
A simple bow requires a force proportional to the displacement of the string.

continued on next page

Example 6.5 *continued*

this bow is (200 N)/(50 cm) = 4 N/cm, while the initial slope of the $F(x)$ graph for the compound bow is (80 N)/(10 cm) = 8 N/cm. The compound bow is stiffer initially, but the force then levels out at 110 N, while the force required to pull the simple bow continues to increase to 160 N (at 40 cm). The work done to pull the compound bow back 40 cm is greater than the work done to pull the simple bow back 40 cm, but the work done to pull the compound bow back 50 cm is less than the work done to pull the simple bow back 50 cm.

Practice Problem 6.5   A gentle pull
How much work would you do to draw the string of each of the two bows back 10.0 cm instead of 40.0 cm?

We can generalize Example 6.5. For small amounts of stretch, springs tend to obey Hooke's law: $F = kx$, where F is the magnitude of the external force applied to the spring, x is the distance the spring is stretched or compressed from its equilibrium length, and k is the spring constant. The work done to stretch the spring from equilibrium a distance x is the area of a right triangle whose base is x and height is kx, or

$$W = \tfrac{1}{2}kx^2$$

Think of $\tfrac{1}{2}kx^2$ as the average force ($\tfrac{1}{2}kx$) times the displacement (x). The work done *by* the spring during the same displacement is $-\tfrac{1}{2}kx^2$, because the spring exerts the same magnitude force in the opposite direction.

More generally, if a spring starts at some displacement x_i, not necessarily at the equilibrium point, the work done to stretch or compress it more, to a new displacement x_f, is $\tfrac{1}{2}kx_f^2 - \tfrac{1}{2}kx_i^2$. Imagine the spring starting at equilibrium and ultimately ending up at a displacement x_f after passing through x_i. The total work done is $\tfrac{1}{2}kx_f^2$; then we subtract the work that was done to get the spring to position x_i.

Work done to compress or stretch an ideal spring:

$$W = \tfrac{1}{2}kx_f^2 - \tfrac{1}{2}kx_i^2$$

Example 6.6

The Dart Gun

In a dart gun (Fig. 6.10), a spring with $k = 400.0$ N/m is compressed 8.0 cm when the dart (mass 20.0 g) is loaded (Fig. 6.10a). What is the muzzle speed of the dart when the spring is released (Fig. 6.10b)?

Strategy   As the spring expands, it does positive work on the dart because it exerts a force on the dart parallel to the dart's motion. Assuming that no other forces do work, the work done on the dart is equal to the change in the dart's kinetic energy. The dart loses contact with the spring when the spring is at its relaxed length, so $x_f = 0$ and $x_i = 8.0$ cm.

Solution   Originally the dart was forced into the gun barrel, cocking the gun by compressing the spring 8.0 cm. Once the gun is fired, the compressed spring is released and the force exerted on the dart by the spring is equal in magnitude and opposite in direction to the force originally exerted on the spring by the dart during the compression. Both forces move through the same displacement. Therefore, the work done by the spring on the dart is the opposite of the work first done by the dart on the spring. As the spring does work it expands from its compressed position $x_i = 8.0$ cm to $x_f = 0$:

$$W_{spring} = -\tfrac{1}{2}k(x_f^2 - x_i^2) = \tfrac{1}{2}kx_i^2$$

The work done by the spring is positive because the spring pushes the dart in the direction of its motion and increases the

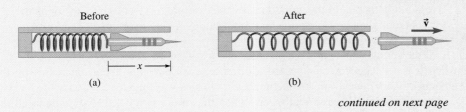

Before   After

(a)   (b)

Figure 6.10
Dart gun (a) before and (b) after firing; note that the spring is compressed by a distance x when the gun is cocked before firing.

continued on next page

Example 6.6 *continued*

dart's kinetic energy. The work done equals the dart's gain in kinetic energy:

$$W_{\text{spring}} = \tfrac{1}{2}kx_i^2 = \tfrac{1}{2}mv^2$$

Solving for v,

$$v = \sqrt{\frac{kx_i^2}{m}} = \sqrt{\frac{400.0 \text{ N/m} \times (0.080 \text{ m})^2}{0.0200 \text{ kg}}} = 11 \text{ m/s}$$

Discussion   Notice that the muzzle speed is proportional to the distance the spring is compressed when the gun is cocked. If the spring is compressed halfway, it does only *one-quarter* as much work on the dart. The dart then acquires one-quarter

the kinetic energy, which means its speed is only half as much. A more massive dart fired from the same gun would have a smaller muzzle speed, but the *same* kinetic energy.

Practice Problem 6.6   A misfire

The same dart gun is cocked by compressing the spring the same distance (8.0 cm). This time the spring gets caught inside the gun, stopping at the point where it is still compressed by 4.0 cm. The dart is not caught inside the gun but is released. Find the muzzle speed of the dart. [*Hint:* Identify x_i and x_f and make use of both in finding the work.]

6.5   POTENTIAL ENERGY

An example of energy transfer occurs when the archer of Example 6.5 releases the string; the bent bow transfers energy to the arrow by doing work on it. The arrow acquires kinetic energy and flies off toward the target. Where did the bow get this energy?

The energy is transferred to the bow when the archer draws the string back. As the string is drawn back, the archer does positive work on it and therefore transfers energy to the string. What happens to this energy? It must be stored by the bow in a form other than kinetic energy, since nothing is moving. The farther back the string is drawn, the more the bow bends, and the more energy is stored in the bow. This kind of energy is called **potential energy** (symbol U). Kinetic energy is due to an object's motion, while potential energy is *stored* energy due to an object's position or configuration. The bow's potential energy depends on how far the string is drawn and thus how much the bow is bent.

The symbol for potential energy is U.

There are different kinds of potential energy. The kind stored in the bow is **elastic potential energy** since it is associated with the ability of the bent bow to spring back and straighten out. Other forms of potential energy include gravitational and electrical potential energy. Not every kind of interaction has an associated potential energy; for instance, there is no such thing as "frictional potential energy." Friction converts energy into a form that is not easily recoverable.

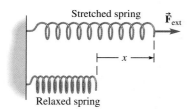

Figure 6.11   An ideal spring is stretched a distance x beyond its relaxed length by the application of an external force $\vec{\mathbf{F}}_{\text{ext}}$.

Let's calculate the potential energy for a spring that obeys Hooke's law (Fig. 6.11). In order to stretch or compress the spring a distance x starting from its relaxed position, we must do an amount of work equal to $+\tfrac{1}{2}kx^2$. In other words, the amount of energy we transfer to the spring is $+\tfrac{1}{2}kx^2$. Therefore, the potential energy of the spring increases by this same amount. The change in potential energy of the spring equals the work done by the external force to compress or stretch it. The change in *any* kind of potential energy is calculated in the same way: the potential energy change equals the work done by the external force in changing the object's position or configuration, assuming that all of the energy goes into potential energy and none into other forms.

Potential energy:

$$\Delta U = W_{\text{ext}} \qquad \text{(6-10)}$$

(assuming that all of the work done by the external force goes into changing the potential energy and none goes into changing kinetic or internal energy)

If the spring starts with zero potential energy when it is relaxed ($U = 0$ at $x = 0$), then when it is stretched or compressed a distance x, its potential energy increases by the amount of work done to stretch or compress it—which was found in Section 6.4 to be $\tfrac{1}{2}kx^2$.

Is the potential energy really zero when the spring is relaxed? What about all the forces binding the atoms together? They are storing some potential energy even when the

---

Potential energy of an ideal spring:

$$U = +\tfrac{1}{2}kx^2 \qquad (6\text{-}11)$$

spring is relaxed. We really don't know what the absolute total amount of potential energy in the spring is. Luckily we don't have to; only *changes* in potential energy ever enter our calculations. The potential energy in Eq. (6-10) is the *additional* potential energy the spring acquires when it is stretched. As long as a problem concerns only stretching and compressing the spring—not burning it or putting it into the core of a nuclear reactor—we can use Eq. (6-10) to calculate the potential energy changes that occur.

Since our calculations only involve *changes* in potential energy, we can always choose the potential energy for any *one* configuration to be an arbitrary amount. Most often, we choose one configuration to have zero potential energy. Once that choice is made, the potential energy of every other configuration is determined by Eq. (6-11).

For example, consider the bow (Fig. 6.12). We can decide that the potential energy is zero before the string is pulled. Then the bow acquires *additional* potential energy when the string is pulled back and the bow is bent more. The potential energy is then greater than zero. When an arrow is shot, this additional potential energy is converted into other forms of energy, mostly the kinetic energy of the arrow. The potential energy of the bow is now zero again. It is possible for potential energy to be negative. If the bowstring is cut, the bow straightens out and its potential energy decreases from zero to negative values. We would have to do positive work to restring the bow and restore its original potential energy.

Gravitational Potential Energy

For another example, we calculate the potential energy of an object of mass m as it moves in the Earth's gravitational field near Earth's surface (Fig. 6.13). How much work is done to lift an object a height h? The force that must be applied to move the object up at constant speed—so that the kinetic energy stays constant—is mg upward; therefore, the work done by the external force is $+mgh$. Thus, the change in gravitational potential energy is $+mgh$. If we choose the potential energy to be zero initially, then $U = mgh$.

Potential energy for object of mass m in uniform gravitational field:

$$U = mgh \qquad (6\text{-}12)$$

where h is the height above a reference level at which $U = 0$

The potential energy depends only on changes in height; no work is done by gravity when an object moves horizontally since it then moves perpendicular to the gravitational force. Since only changes in potential energy are significant, we can choose $U = 0$ at any convenient place: on the floor, on a table, or at the top of a ladder. Potential energy is then positive above the chosen zero point and negative below it. If we put the origin at the point where $U = 0$, then h is the same as y.

The gravitational force is only *approximately* uniform near the surface of the Earth; the gravitational force between two bodies sufficiently far apart varies inversely as the square of the distance r between the centers of the two bodies:

$$F = \frac{Gm_1m_2}{r^2} \qquad (2\text{-}3)$$

The *general* expression for gravitational potential energy as a function of the distance between two bodies is

---

Gravitational potential energy:

$$U(r) = -\frac{Gm_1m_2}{r} \qquad (6\text{-}13)$$

where $U = 0$ when $r = \infty$

Only *changes* in potential energy are significant.

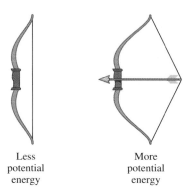

Figure 6.12 The elastic potential energy stored in a bow increases as the string is pulled back. We can define the potential energy of the bow on the left to be zero.

Less potential energy — More potential energy

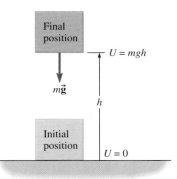

Figure 6.13 An object of mass m moving in the Earth's gravitational field near the surface of the Earth.

Convention: The gravitational potential energy of two bodies infinitely far apart is taken to be zero in Eq. (6-13). For finite separation, gravitational potential energy is always negative.

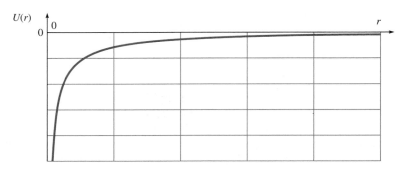

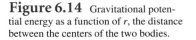

Figure 6.14 Gravitational potential energy as a function of r, the distance between the centers of the two bodies.

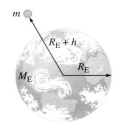

Figure 6.15 An object at a distance h above the Earth's surface is a distance $R_E + h$ from the Earth's center.

Equation (6-13) applies whenever the force law [Eq. (2-3)] does. A graph showing the variation of gravitational potential energy as a function of r is shown in Fig. 6.14.

For any finite separation r, the gravitational potential energy given by Eq. (6-13) is negative because $U = 0$ at infinite separation and the potential energy *decreases* as the bodies get closer together. But if we choose the Earth's *surface* to be the point where $U = 0$, as in Eq. (6-12), then the potential energy a little above the Earth's surface is *positive*. Which is it: positive or negative? That depends on the choice of zero; what is important is that potential energy *changes* are the same using either choice.

Consider an object of mass m at a height h above the surface of the Earth (Fig. 6.15). Let us write the potential energy at height h minus the potential energy at height 0:

$$\Delta U = U(R_E + h) - U(R_E) = \left(-\frac{GM_E m}{R_E + h}\right) - \left(-\frac{GM_E m}{R_E}\right)$$

Rearranging and factoring out the common factors $GM_E m$,

$$\Delta U = GM_E m \left(\frac{1}{R_E} - \frac{1}{R_E + h}\right)$$

Rewriting with a common denominator,

$$\Delta U = GM_E m \frac{R_E + h - R_E}{R_E(R_E + h)} \tag{1}$$

For values of h that are small compared to the radius of the Earth, $R_E + h \approx R_E$. Making that approximation in the denominator of Eq. (1),

$$\Delta U = \frac{GM_E mh}{R_E^2} \tag{2}$$

Recall that the gravitational field strength at the Earth's surface is

$$g = \frac{F_g}{m} = \frac{GM_E}{R_E^2}$$

Replacing GM_E/R_E^2 with g in Eq. (2) reduces that equation to

$$\Delta U = mgh$$

Therefore, Eqs. (6-12) and (6-13) *agree*: the difference between the potential energy for a mass m at a height h and the same mass on the Earth's surface is $\Delta U = mgh$ as long as $h \ll R_E$.

Example 6.7

Communications Satellite in Orbit

Calculate the change in potential energy of a 300.0-kg communications satellite when it is put into geosynchronous orbit from a launch pad on the surface of the Earth (see Example 5.9).

Strategy In Example 5.9 we found that a satellite in geosynchronous orbit at altitude $h = 35,800$ km above the Earth's surface moves at speed $v = 3.07$ km/s. We can calculate the initial

continued on next page

Example 6.7 continued

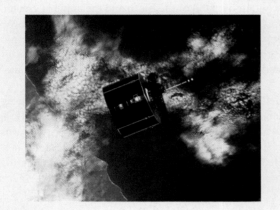

and final potential energies using the more general expression [Eq. (6-13)], where r stands for the distance between the satellite and the Earth's *center*. We cannot use Eq. (6-12) because the satellite's change in altitude is *not* small compared to the Earth's radius.

Solution   The initial potential energy is

$$U_i = -\frac{GM_E m}{R_E} = -\frac{6.67 \times 10^{-11}\ \text{N·m}^2\text{·kg}^{-2} \times 5.98 \times 10^{24}\ \text{kg} \times 300.0\ \text{kg}}{6.37 \times 10^6\ \text{m}}$$

$$= -1.88 \times 10^{10}\ \text{J}$$

The final potential energy is

$$U_f = -\frac{GM_E m}{R_E + h} = -\frac{6.67 \times 10^{-11}\ \text{N·m}^2\text{·kg}^{-2} \times 5.98 \times 10^{24}\ \text{kg} \times 300.0\ \text{kg}}{6.37 \times 10^6\ \text{m} + 3.58 \times 10^7\ \text{m}}$$

$$= -2.84 \times 10^9\ \text{J}$$

The change in potential energy is the final value minus the initial value:

$$\Delta U = U_f - U_i = (-0.284 \times 10^{10}\ \text{J}) - (-1.88 \times 10^{10}\ \text{J})$$

$$= +1.60 \times 10^{10}\ \text{J}$$

Discussion   What does this answer mean? It means that we would have to do a minimum of $+1.60 \times 10^{10}$ J of work to put the satellite 35,800 km above the Earth's surface *without changing its kinetic energy*. In order for the satellite to stay in a circular orbit, it must have a kinetic energy of

$$K = \tfrac{1}{2}mv^2 = \tfrac{1}{2}(300.0\ \text{kg}) \times (3070\ \text{m/s})^2 = +1.41 \times 10^9\ \text{J}$$

The increase in kinetic energy must also come from the work done on the satellite. To launch a satellite it is necessary to lift it up to the correct altitude *and* to get it *moving* at the correct speed and direction in its orbit.

Note the relationship between the satellite's potential and kinetic energies when it is in orbit:

$$U = -2.84 \times 10^9\ \text{J} \approx -2 \times (1.41 \times 10^9\ \text{J}) = -2K$$

It is *not* a coincidence that $U \approx -2K$. Problem 26 proves that $U = -2K$ for any gravitational circular orbit.

Practice Problem 6.7   Potential energy of the bungee jumper

What is the gravitational potential energy of the bungee jumper of Example 6.4 while he stands on the jumping platform? Take $U = 0$ at the bottom of the gorge.

Conservative Forces

Since potential energy is energy of configuration, potential energy can depend on only the positions of various objects, not on the *path* they took to get to those positions. The increase in the gravitational potential energy of an airplane that starts in New Orleans and ends in Denver is the same regardless of how many stops the plane makes on the way, or what its maximum cruising altitude is, or any other detail of the path it takes. Thus, the work done by the force associated with the potential energy *cannot depend on the path taken* by an object, but only on the initial and final positions of the object. Forces such as those we have considered in this section (ideal springs and gravity) that can have potential energies associated with them are called **conservative forces**.

The change in potential energy is equal in magnitude but opposite in sign to the work done by the conservative force associated with the potential energy:

$$\Delta U = -W_c \qquad (6\text{-}14)$$

Why the opposite sign? If a conservative force does positive work W_c on an object, then the object's energy is increased by an amount W_c. Where does that energy come from? It is taken from stored potential energy. The conservative force dips into its "potential energy bank account" and gives the energy to the object. So the potential energy must decrease when the force does positive work. Contrast Eq. (6-14) with Eq. (6-10), which concerns the work done by an *external* force. For example, if we lift a bowling ball from the floor, we (the *external* force) do positive work while gravity does *negative* work; the change in potential energy is positive.

Work done on an object by a conservative force depends only on the initial and final positions of the object, not on the path taken.

What *is* potential energy? Is it just a bookkeeping trick? No, *potential energy is field energy*. When an apple of weight *mg* hangs in a tree at a height *h* above the ground, the energy of the combined gravitational field of the Earth and the apple is greater by an amount *mgh* than when the apple is on the ground. Luckily we concern ourselves only with *changes* in potential energy, so we need not try to calculate the *total* energy of the Earth's gravitational field just to find the speed at which the apple hits the ground. Elastic potential energy is stored in the *electromagnetic* fields that bind atoms and molecules together.

There are two types of field energy. *Static* field energy—the potential energy that we discuss in this chapter—depends only on the configuration of the system. If the configuration is changed and then returns to its original state, the field energy returns to its original value. Field energy can also propagate as *waves*. For example, light is a wave that consists of oscillating electromagnetic fields. For now we are considering only static or nearly static fields, so we can ignore the possibility of energy being carried away by waves.

The work done by friction, air resistance, and other contact forces does depend on path, so these forces cannot have potential energies associated with them. We cannot use friction to store energy in a form that is completely recoverable. These forces are **nonconservative**. In a sense, though, it depends on your point of view. From a macroscopic point of view, these forces are certainly nonconservative. On the microscopic level, these contact forces are the result of large numbers of electromagnetic forces, which are *conservative*. On the microscopic level, every *one* of the electromagnetic interactions between two atoms is conservative. A nonconservative force like kinetic friction converts ordered energy (kinetic energy of a sliding object) into disordered energy (kinetic and potential energies of randomly vibrating atoms).

6.6   CONSERVATION OF MECHANICAL ENERGY

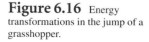

Making The Connection:
energy
transformations
in a jumping
grasshopper

What is "conservative" about a conservative force? Imagine the jump of a grasshopper. Just before the jump (Fig. 6.16a), the grasshopper stores potential energy for the jump by stretching an elastic protein that acts like a spring. When the "spring" is released, the work done by the spring transforms the stored elastic energy into kinetic energy (Fig. 6.16b). Then, as the grasshopper moves upward, the gravitational force transforms kinetic energy into gravitational potential energy (Fig. 6.16c,d). As long as friction and air resistance are negligible, the sum of the kinetic and potential energies does not change during the jump.

We call the sum of the kinetic and potential energies the **mechanical energy**. Whenever nonconservative forces do no net work, mechanical energy is conserved (that is, the mechanical energy does not change). When a conservative force does positive

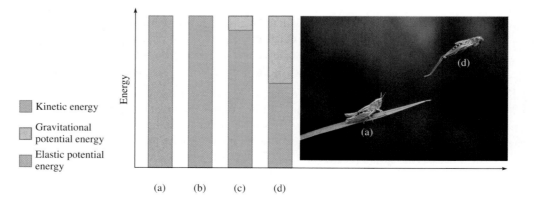

Figure 6.16 Energy transformations in the jump of a grasshopper.

Kinetic energy

Gravitational potential energy

Elastic potential energy

work, it transforms some of its potential energy into kinetic energy or other forms of potential energy; when it does negative work, the transformation goes the other way. A conservative force cannot change the mechanical energy because it never transforms energy into nonmechanical forms such as thermal energy. Now we see why forces associated with potential energies are called forces—they conserve mechanical energy.

Mechanical energy:

$$E = K + U \qquad (6\text{-}15\text{a})$$

Conservation of mechanical energy:

If no nonconservative forces act, then the mechanical energy is constant.

$$E_i = E_f$$

$$\Delta E = \Delta K + \Delta U = 0 \qquad (6\text{-}15\text{b})$$

Mechanical energy: the sum of the kinetic and potential energies

Example 6.8

A Quick Descent

 A ski trail makes a vertical descent of 78 m (Fig. 6.17). A novice skier, unable to control his speed, skis down this trail and is lucky enough not to hit any trees. What is his speed at the bottom of the trail?

Strategy A skilled skier can control his speed by controlling the frictional force. For a worst-case scenario, we'll assume there's no friction at all; the only forces acting on the skier are the normal force and gravity. The normal force does no work, since it is always perpendicular to the skier's velocity, so mechanical energy is conserved.

Solution As the skier descends, his change in potential energy is

$$\Delta U = -mgh$$

where $h = 78$ m. If he starts with zero kinetic energy, then

$$\Delta K = +\tfrac{1}{2}mv_f^2$$

Mechanical energy is conserved. Potential energy is converted to kinetic energy, but the total energy doesn't change:

$$\Delta K + \Delta U = \tfrac{1}{2}mv_f^2 - mgh = 0$$

We can solve for the final speed v_f:

$$v_f = \sqrt{2gh} = \sqrt{2 \times 9.8 \text{ m/s}^2 \times 78 \text{ m}} = 39 \text{ m/s}$$

Discussion As an alternative solution, instead of looking at the energy changes directly, we could equate the initial and final mechanical energies:

$$E_i = E_f$$

which means

$$U_i + K_i = U_f + K_f$$

Initially, the kinetic energy is zero and the potential energy is mgh. At the bottom, the potential energy is zero and the kinetic energy is $\tfrac{1}{2}mv_f^2$. Substituting these energy values,

$$mgh + 0 = 0 + \tfrac{1}{2}mv_f^2$$

The two methods are equivalent, so we can use whichever is most convenient.

A speed of 39 m/s (87 mi/h) is dangerously fast. In reality, there would be some friction and air resistance, so the final speed would be smaller. How much smaller would depend on the length and contour of the trail, among other things. Without friction, the final speed doesn't depend on the detailed shape of the trail: it could have been a steep descent, or a long gradual one, or have a complicated profile with varying slope. It could even be a vertical descent—the final speed is the same for free-fall off a 78-m high building.

Practice Problem 6.8 Speeding roller coaster

A roller coaster is hauled to the top of the first hill of the ride by a motorized chain drive. After that, the train of cars is released and no more energy is supplied by an external motor. If the cars start from rest 35 m above the ground at the top of the first hill, how fast are they moving at the top of the second hill, which is 22 m above the ground? Ignore friction and air resistance.

78 m

Figure 6.17   The final speed of the skier depends only on the initial and final altitudes if no friction acts.

Example 6.9

Escape Speed

What is the minimum speed that a projectile must have at the Earth's surface so that the projectile can escape the Earth's gravitational attraction? Ignore air resistance.

Strategy   To just escape the Earth's gravity the projectile must have enough initial kinetic energy so that it can reach an infinite distance from the Earth with no leftover kinetic energy to spare. It is going to end up at the place where we have defined the potential energy to be zero. The projectile must be given enough kinetic energy at the beginning to overcome the initial negative value of potential energy it possesses by virtue of its position on the surface of the Earth—it has enough kinetic energy to climb the potential energy hill of Fig. 6.18.

Solution   By equating the initial mechanical energy at the Earth's surface to the final mechanical energy at the maximum value of r ($r = \infty$), we obtain an equation with initial speed v_i as the unknown variable.

Initial conditions: $K_i = \dfrac{1}{2}mv_i^2$

Since $r_i = R_E,\; U_i = -\dfrac{GM_E m}{R_E}$

Therefore, $E_i = \dfrac{1}{2}mv_i^2 + \left(-\dfrac{GM_E m}{R_E}\right)$

Final conditions:

$K_f = 0$ (the projectile is momentarily at rest at the maximum r)

At an infinite distance, $\dfrac{1}{r_f} = 0$, so $U_f = 0$

continued on next page

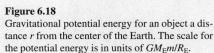

Example 6.9 *continued*

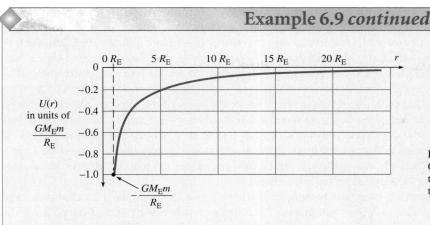

Figure 6.18
Gravitational potential energy for an object a distance r from the center of the Earth. The scale for the potential energy is in units of GM_Em/R_E.

Mechanical energy is conserved, so we equate the initial and final mechanical energies:

$$E_f = E_i$$

Substituting the kinetic and potential energies, we find

$$0 + 0 = \frac{1}{2}mv_i^2 + \left(-\frac{GM_Em}{R_E}\right)$$

Solving for v_i,

$$v_i = \sqrt{\frac{2GM_E}{R_E}}$$

This speed is called the **escape speed** of the Earth because it is the minimum initial speed for a projectile launched from the surface that can escape the influence of Earth's gravity.

$$v_{esc} = \sqrt{\frac{2GM_E}{R_E}} = 11.2 \text{ km/s (a little over 25,000 mi/hr)}$$

Discussion Note that the answer is independent of the mass of the projectile because the mass cancels out in the solution for v_i. No matter what size projectile is launched from the

Earth, the same initial speed must be supplied to have the energy needed to escape the gravitational pull of the Earth.

We ignored air drag in the calculation. The friction of the Earth's atmosphere would actually cause so much heating that the projectile would burn up long before it got to a height of two Earth radii above the surface. This frictional heating caused by the dense lower Earth atmosphere makes it necessary for space scientists to gradually increase the speed of rockets launched from Earth by using more than one stage of rocket boosters. The rocket is eventually given enough kinetic energy to escape the Earth's gravity.

The concept of escape speed helps explain why there is little hydrogen or helium in Earth's atmosphere, even though these are by far the most common elements in the universe. We will see in Chapter 13 that the molecules in a gas have an average kinetic energy determined by the temperature of the gas. In a mixture of gases, the molecules with the smallest mass have the highest average speeds. The average speeds of hydrogen and helium in our atmosphere are large enough that a significant fraction of the molecules have speeds greater than the escape speed, and can therefore escape the atmosphere. A negligible fraction of the nitrogen, oxygen, or water molecules have speeds greater than the escape speed.

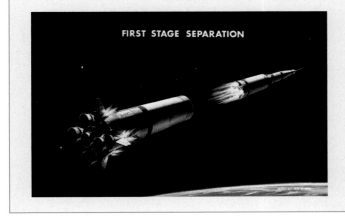

FIRST STAGE SEPARATION

Practice Problem 6.9 Protons streaming away from the Sun

Particles such as protons and electrons are continually streaming away from the Sun in all directions. They carry off some of the energy released in the thermonuclear reactions occurring in the Sun. How fast must a proton be moving at a distance of 7.00×10^9 m from the center of the Sun in order for it to escape the Sun's gravitational pull and leave the solar system? The Sun's mass is 1.99×10^{30} kg.

When a human jumps (Fig. 6.19), the muscles provide the energy to propel the body upward. Try jumping as high as you can from a standing start. You no doubt start by crouching down. Then you accelerate upward, straightening your legs and your body; your muscles convert chemical energy into the mechanical energy of your jump. You might be able to jump about 1 m above the floor.

The kangaroo uses a slightly different mechanism. It has long, elastic tendons and small muscles, in contrast to the relatively large muscles and short, stiffer tendons found in humans. The kangaroo slowly folds its legs before a jump, stretching the tendons to store elastic potential energy just as energy is stored in a stretched spring. The kangaroo then quickly extends its legs, relaxing the tendon like a released spring. The tendon supplies much of the energy needed for the jump; the rest is supplied by the kangaroo's leg muscles, which convert some chemical energy into mechanical energy. When the kangaroo lands on the ground, the tendons are stretched again as its legs bend, and it decelerates, storing up energy for the next jump. As the kangaroo takes off again and accelerates back up into the air, the stored energy from the previous jump's landing is used. The storage of elastic energy reduces the amount of chemical energy that must be supplied for each successive jump.

Insects use a catapult technique. The knee joint of a flea contains an elastic material called resilin (a rubber-like protein). The flea slowly bends its knee, stretching out the resilin and storing energy, and then locks its knee in place. When the flea is ready to jump, the knee is unlocked and the resilin quickly contracts with a sudden release of the stored elastic energy.

Figure 6.19 Shane Battier leaps to make a basket for Duke University.

Example 6.10

The Hopping Kangaroo

Let's consider the hopping kangaroo (Fig. 6.20). If we assume that its tendons are like springs, a fair assumption in this case, then the potential energy stored is

$$U = \tfrac{1}{2}kx^2$$

If the kangaroo can jump 0.70 m straight up into the air when the tendons have been stretched by about half their maximum possible stretched length, how high can the kangaroo jump if the tendons are stretched 10% more?

Strategy By equating the initial mechanical energy before the jump to the final mechanical energy at the highest point of the jump, we can set up a ratio and find the relative change in height jumped. At first we imagine a kangaroo jumping straight up; then we try to generalize to more realistic hopping where there is forward motion as well.

Solution When the kangaroo is crouched, it is at rest momentarily and has zero kinetic energy. For convenience, we choose the initial gravitational potential energy to be zero. We must still account for the elastic potential energy stored in the tendon. The initial energies are

$$K_i = 0 \quad \text{and} \quad U_i = \tfrac{1}{2}kx_i^2$$

where x_i represents the initial stretch of the tendon. With the kangaroo at the high point of the jump, the kinetic energy is again zero if it jumped straight up. The tendon is no longer stretched, so the elastic potential energy is zero. But now there is gravitational potential energy. The final energies are

$$K_f = 0 \quad \text{and} \quad U_f = mgh$$

We equate initial and final mechanical energies to express the conservation of energy

$$E_i = E_f$$

$$\tfrac{1}{2}kx_i^2 = mgh$$

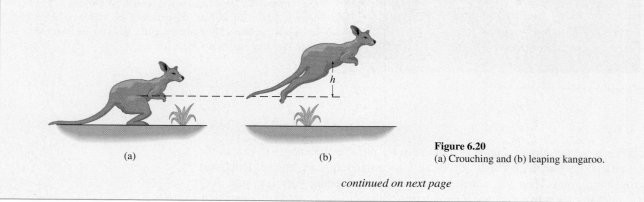

(a)

(b)

Figure 6.20
(a) Crouching and (b) leaping kangaroo.

continued on next page

Example 6.10 *continued*

We don't know all of the constants (mass, spring constant), so we set up a ratio,

$$\frac{E_{i1}}{E_{i2}} = \frac{E_{f1}}{E_{f2}} \quad \text{or} \quad \frac{x_1^2}{x_2^2} = \frac{h_1}{h_2}$$

For a 10% increase in stretch, $x_2 = 1.1x_1$ and

$$h_2 = (1.1)^2\, h_1 = 1.21 \times 0.70\text{ m} = 0.85\text{ m}$$

Using a 10% increase in the stretch of the tendon, the kangaroo jumps about 20% higher.

When the kangaroo is hopping along, it does not jump straight up. Will the kangaroo's jump still be 20% higher when jumping at another angle? Imagine the kangaroo hopping along so that it leaves the ground at a 45° angle, which gives the maximum horizontal range per hop in the absence of air resistance. The elastic energy in the tendon is first converted to kinetic energy. This time, not all of the kinetic energy is converted to gravitational potential energy. The kinetic energy at the highest point of the jump is *not* zero because the kangaroo is still moving forward. The initial velocity can be resolved into components:

$$v^2 = v_x^2 + v_y^2 = 2v_x^2 \quad \text{(since } v_x = v_y \text{ for a 45° angle)}$$

At the highest point of the jump, the kinetic energy is $\frac{1}{2}mv_x^2$, which is half of the initial kinetic energy. Overall, *half* of the elastic energy of the tendon is converted to gravitational potential energy:

$$\tfrac{1}{2} \times \left(\tfrac{1}{2}kx_i^2\right) = mgh$$

Since h is still proportional to x_i^2, the height of the jump still increases by 20% if the stretch of the tendon is increased by 10%.

Discussion   The storage of elastic energy in the tendon is a clever way for the kangaroo to get more "miles per gallon." Without such an energy storage system, most of the kangaroo's kinetic energy would be "lost" at the end of each hop (converted to unrecoverable thermal energy). The tendons store some of the energy that would otherwise be lost and then release it to help the next jump. Since less energy is lost on each hop, the average power supplied by the kangaroo's body is less than it would be without the long, elastic tendons. Humans use a similar energy-saving mechanism when running (see Problem 81).

The analogy that the tendon behaves like an ideal spring fails for large changes in the amount of stretch because the tendon does not obey Hooke's law. For small changes in the amount of stretch, Hooke's law is a reasonable approximation.

Practice Problem 6.10   Jumping with joey

Suppose the kangaroo has a baby kangaroo (a *joey*) riding in her pouch. If the joey has grown to be one-sixth the mass of its mother, how high can the kangaroo jump with the additional load? Assume that, without the joey, she can jump 2.8 m.

Work Done by Nonconservative Forces

In general, a body can be acted on by both conservative and nonconservative forces, and both can do work. Then the total work done on the body can be written

$$W_{\text{total}} = W_c + W_{\text{nc}}$$

where W_c and W_{nc} represent the work done by conservative and nonconservative forces, respectively. For instance, a falling ball has work done on it by both gravity (conservative) and air resistance (nonconservative).

The work done by the conservative forces is accounted for by the change in potential energy [Eq. (6-14)]. This work doesn't change the mechanical energy. Work done by nonconservative forces does change the mechanical energy. If the work is done on a point particle, so that its internal energy cannot change, then the total work done is equal to the change in kinetic energy [see Eq. (6-5)]:

$$W_{\text{total}} = W_c + W_{\text{nc}} = \Delta K$$

Substituting $W_c = -\Delta U$,

$$W_{\text{nc}} = \Delta K - W_c = \Delta K + \Delta U = \Delta E \qquad \text{(6-16)}$$

The work done by the nonconservative forces is equal to the change in the mechanical energy. When nonconservative forces do no work, $W_{\text{nc}} = \Delta E = 0$ and mechanical energy is conserved.

When finding the change in mechanical energy, do not include the work done by conservative forces. Work done by conservative forces is already accounted for by the change in potential energy.

Example 6.11

Rock Climbing in Yosemite

A team of climbers is rappelling down steep terrain in the Yosemite valley (Fig. 6.21). Mei-Ling (mass 60.0 kg) slides down a line starting from rest 12.0 m above a horizontal shelf. If she lands on the shelf below with a speed of 2.0 m/s, calculate the energy dissipated by the kinetic frictional forces acting between her and the line. The local value of g is 9.78 N/kg. Ignore air resistance.

Strategy   Since we know Mei-Ling's initial and final speeds as well as her mass, we can calculate the change in her kinetic energy. From the change in height, we can calculate the change in potential energy. Mechanical energy is *not* conserved here; mechanical energy decreases by an amount equal to the thermal energy generated by friction.

Given: mass of climber, $m = 60.0$ kg; $\Delta h = -12.0$ m;
$\quad\quad\quad v_i = 0$ m/s and $v_f = 2.0$ m/s, just before stopping;
$\quad\quad\quad$ local field strength $g = 9.78$ N/kg

To find: change in mechanical energy, ΔE

Solution   Mei-Ling's kinetic energy is initially zero since she starts at rest. The change in her kinetic energy is

$$\Delta K = \tfrac{1}{2}mv_f^2 - \tfrac{1}{2}mv_i^2 = \tfrac{1}{2}mv_f^2 - 0 = \tfrac{1}{2}(60.0 \text{ kg}) \times (2.0 \text{ m/s})^2 = +120 \text{ J}$$

The change in her potential energy is

$$\Delta U = mgh_f - mgh_i$$

$$= (60.0 \text{ kg} \times 9.78 \text{ m/s}^2) \times (0 - 12.0 \text{ m}) = -7040 \text{ J}$$

Therefore, the change in her mechanical energy is

$$\Delta E = \Delta K + \Delta U = 120 \text{ J} + (-7040 \text{ J}) = -6920 \text{ J}$$

The energy dissipated by friction is 6920 J.

Discussion   One reason for wearing gloves while rappelling is to keep the hands from getting burned due to the thermal energy generated by kinetic friction.

If the line had broken when Mei-Ling was at the top, her final kinetic energy would have been +7040 J—disastrously large since it corresponds to a final speed of

$$v = \sqrt{\frac{K}{\tfrac{1}{2}m}} = \sqrt{\frac{7040 \text{ J}}{30.0 \text{ kg}}} = 15.3 \text{ m/s}$$

Instead, kinetic friction reduces her final kinetic energy to a manageable +120 J (which corresponds to a final speed of 2.0 m/s). Mei-Ling can absorb this much kinetic energy safely by landing on the shelf while bending her knees.

Practice Problem 6.11   Energy dissipated by air resistance

A ball thrown straight up at an initial speed of 14.0 m/s reaches a maximum height of 7.6 m. What fraction of the ball's initial kinetic energy is dissipated by air resistance as the ball moves upward?

6.7   GENERAL LAW OF ENERGY CONSERVATION

So far we have treated objects, no matter how large, as *point particles* that have all their mass located at a single point. We have ignored internal motion and rotation. A point particle can only have mechanical energy (translational kinetic energy plus the potential energy due to external interactions). A point particle has no rotational kinetic energy; nor does it have any way to change its internal energy. The principle of energy conservation, which we have applied only to point particles so far, is a fundamental pillar of physics that has much more general application.

In Chapter 8, we learn about the rotation of rigid objects and how to calculate the kinetic energy of rotation. Then we will be able to extend the conservation of energy to rotating bodies. After that, we will see how conservation of energy applies to fluid flow, heat flow, electromagnetic fields, and radioactivity. Many devices function by converting energy from one form to another. For instance, electric generators convert mechanical energy into electrical energy, while electric motors convert electrical energy into mechanical energy. All waves, whether sound, light, water waves, or earthquakes, transport energy from one place to another.

We can do little in the natural sciences—or for that matter, in life—without encountering the principle of energy conservation. Think about the energy conversions that make life possible. Green plants use photosynthesis to convert the energy they receive from the Sun into stored chemical energy. When animals eat the plants, that stored energy enables motion, growth, and maintenance of body temperature. Energy conservation governs every one of these processes.

Figure 6.21   Mei-Ling rappelling downward from a position 12.0 m above a shelf.

Since there are so many different forms of energy, we might wonder just what energy really *is*. Energy is sometimes defined as the ability to do work, but this definition is misleading. The availability of energy to do work depends on whether that energy exists in an ordered or a disordered form. Energy is a fundamental conserved quantity; work is just a way to *transfer* energy.

Another kind of energy transfer is heat flow. When two objects of different temperatures are put in contact, heat flows from the hotter object to the cooler one until the temperatures are equal. Before the nineteenth century, heat was thought to be the flow of an invisible fluid, called *caloric,* from one body to another. In the mid-nineteenth century, the English physicist James Prescott Joule (1818–1889) provided definitive evidence that heat flow is *energy* in transit; in other words, heat is energy. Before that time, heat and energy were measured in different units (the calorie was a unit of heat but *not* of energy). Joule determined the conversion factor, sometimes called "the mechanical equivalent of heat." More precisely, we might call this factor the mechanical energy equivalent of a given quantity of thermal energy. In recognition of Joule's experiments, the SI unit of energy was named for him.

Mechanical equivalent of heat

1 cal = 4.186 J

6.8   POWER

Sometimes the *rate* of energy transfer is important. When shopping for a sports car, you wouldn't ask the salesman how much work the engine can do. A tiny economy car like the Geo Metro does more work than a Ferrari if the Metro is used for daily commuting while the Ferrari sits in the garage most of the time. But the Ferrari can do work *at a much faster rate* than the Metro can. In other words, it can change chemical energy in the gasoline into mechanical energy of the car at a much faster rate—it has a larger maximum power output. We give the name **power** (symbol *P*) to the rate of energy transfer. The average power is the amount of energy transferred divided by the time the transfer takes.

Power is the rate of energy transfer.

Average power

$$P_{av} = \frac{\Delta E}{\Delta t} \qquad (6\text{-}17)$$

SI unit of power: watt
1 W = 1 J/s

The SI unit of power, the joule per second, is given the name watt (1 W = 1 J/s). Remember that the unit symbol W stands for *watt*, not *work*. In the United States, the maximum power output of an automobile engine is usually specified in horsepower, which is a non-SI unit of power (1 hp = 746 W).

The *kilowatt-hour* (kW·h) is sometimes mistaken for a unit of power, but it is a unit of *energy*—the amount of energy transferred at a constant rate of 1 kW during a time interval of 1 h. The kilowatt-hour is commonly used by utility companies to measure the amount of electrical energy used by consumers.

Since the work done by a constant force during a small time interval Δt is

$$\Delta W = F \, \Delta r \cos \theta$$

where the magnitude of the displacement is

$$\Delta r = v \, \Delta t$$

then power—the rate at which the force does work—can be found from the force and the velocity.

$$P = \frac{\Delta W}{\Delta t} = \frac{F \, \Delta r \cos \theta}{\Delta t} = F \frac{\Delta r}{\Delta t} \cos \theta = Fv \cos \theta$$

Instantaneous power

$$P = Fv \cos \theta \qquad (6\text{-}18)$$

Again, only the component of force in the direction of motion does work.

Example 6.12

Air Resistance on a Hill-Climbing Car

A 1000.0-kg car climbs a hill with a 4.0° incline at a constant 12.0 m/s (Fig. 6.22a). (a) At what rate is the gravitational potential energy increasing? (b) If the mechanical power output of the engine is 20.0 kW, find the force of air resistance on the car.

Strategy   (a) We can find the rate of gravitational potential energy increase in two ways. One is to find the potential energy change during a time interval Δt and divide it by the time interval, which is equivalent to using the definition of average power [Eq. (6-17)]. The other possibility is to use Eq. (6-18) to find the rate at which the gravitational force does work.

(b) The car moves at constant speed, so its kinetic energy is not changing. Therefore, during any time interval, the work done by the engine plus the (negative) work done by air resistance is equal to the increase in the gravitational potential energy.

Given: car mass = 1000.0 kg; v = 12.0 m/s; 4.0° incline

To find: (a) rate of potential energy change, $\Delta U/\Delta t$; (b) force of air resistance, F_a

Solution   (a) For a small change in elevation Δy, the change in potential energy is

$$\Delta U = mg \, \Delta y$$

continued on next page

Example 6.12 *continued*

Then the *rate* of potential energy change is

$$\frac{\Delta U}{\Delta t} = \frac{mg\,\Delta y}{\Delta t} = mg\,\frac{\Delta y}{\Delta t}$$

We recognize $\Delta y/\Delta t$ as the y-component of the velocity (v_y). Therefore,

$$\frac{\Delta U}{\Delta t} = mgv_y$$

From Fig. 6.22b, $v_y = v \sin \phi$, where $\phi = 4.0°$. Then

$$\frac{\Delta U}{\Delta t} = mgv \sin \phi$$

$$= 1000.0 \text{ kg} \times 9.8 \text{ m/s}^2 \times 12.0 \text{ m/s} \times \sin 4.0°$$

$$= 8200 \text{ W}$$

(b) During any time interval Δt, the work done by the engine must equal the increase in the gravitational potential energy plus the amount of energy dissipated by air resistance:

$$W_e + W_a = \Delta U$$

Dividing each term by Δt, we find

$$\frac{W_e}{\Delta t} + \frac{W_a}{\Delta t} = \frac{\Delta U}{\Delta t}$$

or

$$P_e + P_a = \frac{\Delta U}{\Delta t}$$

where P_e and P_a represent the power output of the engine and the rate at which air resistance dissipates energy, respectively. Then

$$P_a = \frac{\Delta U}{\Delta t} - P_e = 8.2 \text{ kW} - 20.0 \text{ kW} = -11.8 \text{ kW}$$

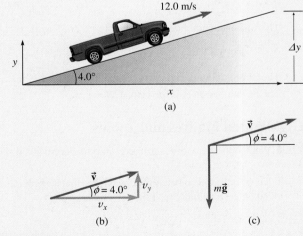

Figure 6.22
(a) Car climbing a hill at constant speed; (b) resolving the velocity into x- and y-components; and (c) the angle between the force and the velocity is 94°.

So, of the 20.0 kJ of mechanical work that the engine does each second, 8.2 kJ goes into gravitational potential energy and 11.8 kJ goes into stirring up the air.

The direction of the force of air resistance $\vec{F}_a$ on the car is opposite to the car's velocity, so

$$P_a = F_a v \cos 180° = -F_a v$$

Solving for F_a,

$$F_a = -\frac{P_a}{v} = -\frac{-11,800 \text{ W}}{12.0 \text{ m/s}} = 983 \text{ N}$$

Discussion   We can check (a) by using Eq. (6-18) to find the rate at which the gravitational force does work:

$$P = Fv \cos \theta$$

where $F = mg$. The angle θ is *not* the same as ϕ. In Eq. (6-18), θ is the angle between the force and velocity vectors, which is 94.0° (Fig. 6.22c). Then

$$P = mgv \cos 94.0°$$

$$= 1000.0 \text{ kg} \times 9.8 \text{ m/s}^2 \times 12.0 \text{ m/s} \times \cos 94.0°$$

$$= -8200 \text{ W}$$

Gravity does work on the car at a rate of −8200 W, which means the potential energy is *increasing* at a rate of +8200 W.

We can also figure out what mechanical power the engine must supply to go 12.0 m/s on level ground. With no change in potential energy, all of the mechanical power output of the engine goes into stirring up the air, so

$$P_e + P_a = 0$$

The magnitude of the force of air resistance is the same (983 N) since the speed is the same. Then air resistance dissipates energy at the same rate as before:

$$P_a = -Fv = -(983 \text{ N} \times 12.0 \text{ m/s}) = -11.8 \text{ kW}$$

Therefore, $P_e = 11.8$ kW. On level ground, the engine isn't increasing gravitational potential energy, so it only needs to do enough work to counteract the tendency of air resistance to slow down the car.

In this example we have assumed that all of the mechanical power output of the engine is delivered to the wheels to propel the car forward. In reality, some of the engine's power output is used to run auxiliary devices such as headlights, radios, and windshield wipers. Friction (in the moving parts of the engine, transmission, and drivetrain) also reduces the amount of power that is actually delivered to the wheels.

Practice Problem 6.12   Mechanical power output of an automobile engine

Assuming the force of air resistance on the car is proportional to the square of the car's speed, what mechanical power must the engine supply in order to drive on level ground at 15 m/s? What must the mechanical power output of the engine be to go up a 4.0° incline at 15 m/s?

When animals move, they depend on their muscles to supply the necessary power. We can think of muscles as engines for animal locomotion. The body supplies chemical energy to the muscle; the muscle requires oxygen to convert this chemical energy into mechanical energy. When an animal runs, jumps, or hops, it uses more oxygen than when at rest. A runner whose cardiovascular system is in excellent condition can run faster and farther than the rest of us partly because his body is much more efficient at delivering oxygen to the muscle tissues. The mechanical power output of an animal as it moves is limited by how fast it can deliver oxygen to the muscles.

The concept of energy can help us understand why the rat kangaroo (which is about the size of a rabbit) can jump just about as high as the larger wallaby and the much larger kangaroo. The work that can be done by the muscles of an animal is proportional to the animal's mass. If the work is used to propel the animal into the air in a jump, then the work done by the muscle (or some fraction of it) is equal to the increase in potential energy, mgh. Since both the work done by the muscles and the potential energy change are proportional to mass, we expect h to be *independent* of the animal mass. The larger animal can jump slightly higher mainly because it *starts out* higher above the ground. Similarly, the tiny flea and the much larger locust also have about the same jumping height (typically about 0.2 m).

MASTER THE CONCEPTS

Summary

- The principle of conservation of energy states that energy can be converted from one form to another, but the total energy in an isolated system never changes.
- The work done by a constant force is
$$W = F \, \Delta r \cos \theta \qquad (6\text{-}1)$$
where Δr is the displacement of the point of application of the force and θ is the angle between the force and displacement vectors.
- The kinetic energy of an object of mass m moving with speed v is
$$K = \tfrac{1}{2}mv^2 \qquad (6\text{-}4)$$
- Work done during a small displacement (variable forces) is
$$\Delta W = F_x \, \Delta x \qquad (6\text{-}9)$$
- The change in potential energy is equal to the negative of the work done by the conservative force that stores the potential energy
$$\Delta U = -W_c \qquad (6\text{-}14)$$
Why the minus sign? If the force does positive work, it takes energy *from* its stored potential energy and gives it away in some other form; therefore for a positive W_c, ΔU must be negative.
- The elastic potential energy of an ideal spring of spring constant k and stretched or compressed a distance x from its relaxed length is
$$U = +\tfrac{1}{2}kx^2 \qquad (6\text{-}11)$$
- The gravitational potential energy for an object of mass m raised a distance h above some reference level near Earth's surface, where h is small relative to the radius of the Earth, is
$$U = mgh \qquad (6\text{-}12)$$
if we choose $U = 0$, where $h = 0$.
- The gravitational potential energy as a function of the distance between the two bodies of masses m_1 and m_2 whose centers are separated by a distance r is
$$U(r) = -\frac{Gm_1m_2}{r} \qquad (6\text{-}13)$$
where $U = 0$ at $r =$ infinity.

- The work done by a conservative force on an object does not depend on the path taken, but only on the initial and final positions of the object.
- The principle of conservation of mechanical energy states that for a system on which only conservative forces act, the sum of the kinetic and potential energies is constant.
$$E = K + U \qquad (6\text{-}15a)$$
$$\Delta E = \Delta K + \Delta U = 0 \qquad (6\text{-}15b)$$
- The work done on a point particle by nonconservative forces is equal to the change in the mechanical energy:
$$W_{nc} = \Delta E = \Delta K + \Delta U \qquad (6\text{-}16)$$
- Average power is the average rate of energy transfer.
$$P_{av} = \frac{\Delta E}{\Delta t} \qquad (6\text{-}17)$$
The instantaneous rate at which a force does work is the instantaneous power P:
$$P = Fv \cos \theta \qquad (6\text{-}18)$$
- The SI unit of work and energy is the joule. 1 J = 1 N·m. The SI unit of power is the watt. 1 W = 1 J/s.

Highlighted Figures and Tables

F6.4 Work done by a constant force and displacement in the direction of a force (p. 182)

F6.5 No work is done on an object by a force that is perpendicular to the object's velocity (p. 183)

F6.6 Work done on an antique chest (p. 185)

F6.13 An object of mass m moving in the Earth's gravitational field near the surface of the Earth (p. 193)

F6.14 The variation of gravitational potential energy as a function of r (p. 194)

F6.16 Conservation of mechanical energy (p. 196)

F6.17 Energy conservation applied to a skier (p. 198)

F6.21 Work done by friction on a rappeller (p. 203)

CONCEPTUAL QUESTIONS

1. An object moves in a circle. Is the total work done on the object by external forces necessarily zero? Explain.

2. You are walking to class with a backpack full of books. As you walk at constant speed on flat ground, does the force exerted on the backpack by your back and shoulders do any work? If so, is it positive or negative? Answer the same questions in two other situations: (1) you are walking down some steps at constant speed; (2) you start to run faster and faster on a level sidewalk to catch a bus.

3. Why do roads leading to the top of a mountain wind back and forth? [*Hint:* Think of the road as an inclined plane.]

4. A mango falls to the ground. During the fall, does the Earth's gravitational field do positive or negative work W_m on the mango? Does the mango's gravitational field do positive or negative work W_E on the Earth? Compare the signs and the magnitudes of W_m and W_E.

5. Can static friction do work? If so, give an example. [*Hint:* Static friction acts to prevent *relative* motion along the contact surface.]

6. In the design of a roller coaster, is it possible for any hill of the ride to be higher than the first one? If so, how?

7. When a ball is dropped to the floor from a height h, it strikes the ground and briefly undergoes a change of shape before rebounding to a maximum height less than h. Explain why it does not return to the same height h.

8. A gymnast is swinging in a vertical circle about a crossbar. In terms of energy conservation, explain why the speed of the gymnast's body is slowest at the top of the circle and fastest at the bottom.

9. A bicycle rider notices that he is approaching a steep hill. Explain, in terms of energy, why the bicyclist pedals hard to gain as much speed as possible on level road before reaching the hill.

10. You need to move a heavy crate by sliding it across a smooth floor. The coefficient of sliding friction is 0.2. You can either push the crate horizontally or pull the crate using an attached rope. When you pull on the rope, it makes a 30° angle with the floor. Which way should you choose to move the crate so that you will do the least amount of work? How can you answer this question without knowing the weight of the crate or the displacement of the crate?

11. The main effort of running is the work done by the muscles to accelerate and decelerate the legs. When a foot strikes the ground, it is momentarily brought to rest while the remainder of the animal's body continues to move forward. When the foot is picked up, it is accelerated forward by one set of muscles in order to move ahead of the rest of the body. Then the foot is slowed down by a second set of muscles until it is brought to rest on the ground again. The muscles expend energy both when accelerating and when decelerating the leg. How are thoroughbred horses, deer, and greyhounds adapted so that they can run at great speed?

MULTIPLE CHOICE QUESTIONS

1. After getting on the Santa Monica Freeway, a sports car accelerates from 30 mi/h to 90 mi/h. The ratio of its final to its initial kinetic energy is
   (a) $\sqrt{3}$ (b) $1/\sqrt{3}$ (c) 3
   (d) 1/3   (e) 9   (f) 1/9
   (g) Depends on the car's mass

2. If a kangaroo on Earth can jump from a standing start so that its feet reach a height h above the surface, approximately how high can the same kangaroo jump from a standing start on the Moon's surface? $g_{Moon} \approx \frac{1}{6}g_{Earth}$. (Assume the kangaroo has an oxygen tank and pressure suit with negligible mass.)
   (a) h (b) $6h$ (c) $\frac{1}{6}h$
   (d) $36h$ (e) $\frac{1}{36}h$ (f) $\sqrt{6}\,h$

Questions 3–6. The orbits of Mercury and Pluto are much more eccentric than the orbits of the other planets. That is, instead of being nearly circular, the orbits are noticeably elliptical. The point in the orbit nearest the Sun is called the *perihelion* and the point farthest from the Sun is called the *aphelion* (Fig. 6.23).

Answer choices for Questions 3–5:
   (a) its maximum value   (b) its minimum value
   (c) the same value as at every other point in the orbit

3. At perihelion, the gravitational potential energy of Pluto's orbit has

4. At perihelion, the kinetic energy of Pluto has

5. At perihelion, the mechanical energy of Pluto's orbit has

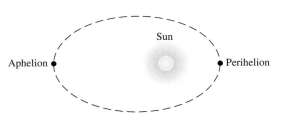

Figure 6.23 Multiple Choice Questions 3–6.

6. As Pluto moves from the perihelion to the aphelion, the work done by gravity on Pluto is
   (a) zero   (b) positive   (c) negative

7. Two balls are thrown from the roof of a building with the same initial speed. One is thrown horizontally while the other is thrown at an angle of 20° above the horizontal. Which hits the ground with the greater speed? Ignore air resistance.
   (a) the one thrown horizontally
   (b) the one thrown at 20°
   (c) they hit the ground with the same speed
   (d) the answer cannot be determined with the given information

8. A hiker descends from the South Rim of the Grand Canyon to the Colorado River. During this hike, the work done by gravity on the hiker is

   (a) positive and depends on the path taken
   (b) negative and depends on the path taken
   (c) positive and independent of the path taken
   (d) negative and independent of the path taken
   (e) zero

9. A stone is thrown straight down from the edge of a cliff. It hits the river below with a speed v_1. A second, identical stone is thrown straight up with the same initial speed from the same point at the edge of the cliff. When it hits the river, how does its speed v_2 compare to v_1? Do not ignore air resistance. [*Hint:* compare the work done by air resistance on the two stones.]

   (a) $v_2 > v_1$ (b) $v_2 < v_1$ (c) $v_2 = v_1$
   (d) The answer cannot be determined with the given information.

10. A simple catapult, consisting of a leather pouch attached to rubber bands tied to two forks of a wooden Y, has a spring constant k and is used to shoot a pebble horizontally. When the catapult is stretched by a distance d, it gives the pebble a speed v. What speed does it give the same pebble when it is stretched to a distance $3d$?

    (a) $\sqrt{3}v$ (b) $3v$ (c) $3\sqrt{3}v$ (d) $9v$ (e) $27v$

11. A projectile is launched at an angle θ above the horizontal. Ignoring air resistance, what fraction of its initial kinetic energy does the projectile have at the top of its trajectory?

    (a) $\cos\theta$ (b) $\sin\theta$ (c) $\tan\theta$ (d) $\dfrac{1}{\tan\theta}$ (e) $\frac{1}{2}$
    (f) $\cos^2\theta$ (g) $\sin^2\theta$ (h) 0   (i) 1

PROBLEMS

Note: **C** indicates a combination conceptual/quantitative problem. Gold diamonds ◆, ◆◆ are used to indicate the increasing level of difficulty of each problem. Problem numbers appearing in blue, 9., denote problems that have a detailed solution available in the Student Solutions Manual. Some problems are *paired* by concept; their numbers are connected by a ruled box.

Section 6.2 Work Done by Constant Forces

1. How much work must Denise do to drag her basket of laundry of mass 5.0 kg a distance of 5.0 m along a floor, if the force she exerts is 30.0 N at an angle of 60.0° with the horizontal?

2. A sled is dragged along a horizontal path at a constant speed of 1.5 m/s by a rope that is inclined at an angle of 30.0° with respect to the horizontal (Fig. 6.24). The total weight of the sled is 470 N. The tension in the rope is 240 N. How much work is done by the rope on the sled in a time interval of 10.0 s?

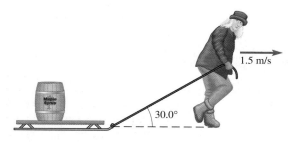

1.5 m/s

30.0°

Figure 6.24 Problem 2

3. Hilda holds a gardening book of weight 10 N at a height of 1.0 m above her patio for 50 s. How much work does she do *on the book* during that 50 s?

4. The tension in the horizontal towrope pulling a water-skier is 240 N while the skier moves due west a distance of 54 m. How much work does the towrope do on the water-skier?

5. A barge of mass 5.0×10^4 kg is pulled along the Erie Canal by two mules, walking along towpaths parallel to the canal on either side of it. The ropes harnessed to the mules make angles of 45° to the canal. Each mule is pulling on its rope with a force of 1.0 kN. How much work is done on the barge by both of these mules together as they pull the barge 150 m along the canal?

6. A 402-kg pile driver (Fig. 6.25) is raised 12 m above ground. (a) How much work must be done to raise the pile driver? (b) How much work does gravity do on the driver as it is raised? (c) The driver is now dropped. How much work does gravity do on the driver as it falls?

Figure 6.25
Problem 6

Section 6.3 Kinetic Energy

7. An automobile with a mass of 1600 kg has a speed of 30.0 m/s. What is its kinetic energy?

8. A record company executive is on his way to a TV interview and is carrying a promotional CD in his briefcase. The mass of the briefcase and its contents is 5.00 kg. The executive realizes that he is going to be late. Starting from rest, he starts to run, reaching a speed of 2.50 m/s. What is the work done by the executive on the briefcase during this time?

9. Sam pushes a 10.0-kg sack of bread flour on a frictionless horizontal surface with a constant horizontal force of 2.0 N starting from rest. (a) What is the kinetic energy of the sack after Sam has pushed it a distance of 35 cm? (b) What is the speed of the sack after Sam has pushed it a distance of 35 cm?

10. Josie and Charlotte push a 12-kg bag of playground sand for a sandbox on a frictionless, horizontal, wet polyvinyl surface with a constant, horizontal force for a distance of 8.0 m, starting from rest. If the final speed of the sand bag is 0.40 m/s, what is the magnitude of the force with which they pushed?

11. A ball of mass 0.10 kg moving with speed of 2.0 m/s hits a wall and bounces back with the same speed in the opposite direction. What is the change in the ball's kinetic energy?

12. A plane weighing 220 kN (25 tons) lands on an aircraft carrier. The plane is moving horizontally at 67 m/s (150 mi/h) when its tailhook grabs hold of the arresting cables. The cables bring the plane to a stop in a distance of 84 m. (a) How much work is done on the plane by the arresting cables? (b) What is the force (assumed constant) exerted on the plane by the cables? (Both answers will be *underestimates*, since the plane lands with the engines full throttle forward; in case the tailhook fails to grab hold of the cables, the pilot must be ready for immediate takeoff.)

13. A shooting star is a meteor that burns up when it reaches Earth's atmosphere. Many of these meteors are quite small. Calculate the kinetic energy of a meteor of mass 5.0 g moving at a speed of 48 km/s and compare it to the kinetic energy of a 1100-kg car moving at 29 m/s (65 mi/h).

Section 6.4 Work Done by Variable Forces

14. How much work is done on the bowstring of Example 6.5 to draw it back by 20.0 cm? [*Hint:* Rather than recalculate from scratch, use proportional reasoning.]

15. An ideal spring has a spring constant $k = 20.0$ N/m. What is the amount of work that must be done to stretch the spring 0.40 m from its relaxed length?

16. A spring is compressed from its relaxed position by a distance of 0.20 m. Figure 6.26 is a graph of the applied external force, F_x, versus the compression, x, of the spring. (a) Find the work done by the external force in compressing the spring 0.20 m starting from its relaxed position. (b) Find the work done by the external force to compress the spring from 0.10 m to 0.20 m.

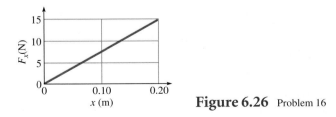

Figure 6.26 Problem 16

17. Rhonda keeps a 2.0-kg model airplane moving at constant speed in a horizontal circle at the end of a string of length 1.0 m. The tension in the string is 18 N. How much work does the string do on the plane during each revolution?

18. Figure 6.27 shows the force exerted on an object versus the position of that object along the *x*-axis. The force has no components other than along the *x*-axis. What is the work

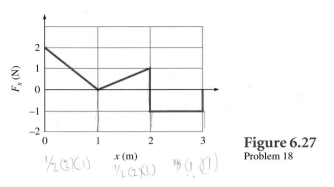

Figure 6.27
Problem 18

done by the force on the object as the object is displaced from 0 to 3.0 m?

19. The force that must be exerted to drive a nail into a wall is roughly as shown in the graph of Fig. 6.28. The first 1.2 cm are through soft drywall; then the nail enters the solid wooden stud. How much work must be done to hammer the nail a horizontal distance of 5.0 cm into the wall?

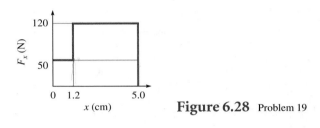

Figure 6.28 Problem 19

20. The bungee jumper of Example 6.4 made a jump into the Gorge du Verdon in southern France from a platform 182 m above the bottom of the gorge. The jumper weighed 780 N and came within 68 m of the bottom of the gorge. The cord's unstretched length is 30.0 m. (a) Assuming that the bungee cord follows Hooke's law when it stretches, find its spring constant. [*Hint:* The cord does not begin to stretch until the jumper has fallen 30.0 m.] (b) At what speed is the jumper falling when he reaches a height of 92 m above the bottom of the gorge?

Section 6.5 Potential Energy

21. Justin moves a desk 5.0 m across a level floor by pushing on it with a constant horizontal force of 340 N. (It slides for a negligibly small distance before coming to a stop when the force is removed.) Then, changing his mind, he moves it back to its starting point, again by pushing with a constant force of 340 N. (a) What is the change in the desk's gravitational potential energy during the round-trip? (b) How much work has Justin done on the desk? (c) If the work done by Justin is not equal to the change in gravitational potential energy of the desk, then where has the energy gone?

22. A spring used in an introductory physics laboratory stores 10.0 J of elastic potential energy when it is compressed 0.20 m. Suppose the spring is cut in half. When one of the halves is compressed by 0.20 m, how much potential energy is stored in it? [*Hint:* Does the half spring have the same *k* as the original uncut spring?]

23. Emil is learning to juggle with three oranges, each of mass 0.30 kg. (a) Emil throws one orange straight up and then catches it, throwing and catching it at the same point. How much work is done by gravity during the orange's free fall? (b) Emil throws another orange straight up, starting 1.0 m above the ground. He fails to catch it. How much work is done by gravity during this orange's free fall?

24. An airline executive decides to economize by reducing the energy, and thus the amount of fuel, required for long distance flights. He orders the ground crew to remove the paint from the outer surface of each plane. The paint removed from a single plane has a mass of approximately 100 kg. (a) If the airplane cruises at an altitude of 12,000 m, how much energy is saved in not having to lift the paint to that altitude? (b) How much energy is saved by not having to move that amount of paint from rest to a cruising speed of 250 m/s?

25. A satellite is placed in a noncircular orbit about the Earth. The farthest point of its orbit (*apogee*) is 4 Earth radii from the center of the Earth, while its nearest point (*perigee*) is 2 Earth radii from the Earth's center. If we define the gravitational potential energy U to be zero for an infinite separation of Earth and satellite, find the ratio $U_{perigee}/U_{apogee}$.

26. Prove that $U = -2K$ for any gravitational circular orbit. [*Hint:* Use Newton's second law to relate the gravitational force to the acceleration required to maintain uniform circular motion.]

Section 6.6 Conservation of Mechanical Energy; Section 6.7 General Law of Energy Conservation

27. How much energy is expended by an 80.0-kg person in climbing a vertical distance of 15 m? Assume that muscles have an efficiency of 22%; that is, the work done by the muscles to climb is 22% of the energy expended.

28. When a 0.20-kg mass is suspended from a vertically hanging spring, it stretches the spring from its original length of 5.0 cm to a total length of 6.0 cm. The spring with the same mass attached is then placed on a horizontal frictionless surface. The mass is pulled so that the spring stretches to a *total* length of 10.0 cm; then the mass is released and it oscillates back and forth (Fig. 6.29). What is the maximum speed of the mass as it oscillates?

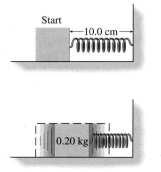

Figure 6.29   Problem 28

29. A cart starts from position 4 in Fig. 6.30 with a velocity of 15 m/s to the left. Find the speed with which the cart reaches positions 1, 2, and 3. Neglect friction.

30. A cart moving to the *right* passes point 1 in Fig. 6.30 at a speed of 20.0 m/s. Let $g = 9.81$ m/s^2. (a) What is the speed of the cart as it passes point 3? (b) Will the cart reach position 4? Neglect friction.

31. In an emergency, the passengers on an airplane slide down a curved escape ramp to safety (Fig. 6.31). The ramp has a 10.0-m radius of curvature and the passengers exit the plane 2.5 above the ground. (a) What is the change in the gravitational potential energy of a passenger of mass 72 kg as she slides down the ramp? (b) Neglecting friction, at what speed does the passenger reach the bottom of the ramp? (c) What is the apparent weight of the passenger near the bottom of the ramp? (d) If 33% of the gravitational potential energy change

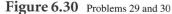

Figure 6.30   Problems 29 and 30

Figure 6.31   Problem 31

is changed into thermal energy by friction, what is the passenger's speed at the bottom?

32. Bruce stands on a bank beside a pond, grasps the end of a 20.0-m long rope attached to a nearby tree and swings out to drop into the water. If the rope starts at an angle of 35.0° with the vertical, what is Bruce's speed at the bottom of the swing?

33. The maximum speed of a child on a swing is 4.9 m/s. The child's height above the ground is 0.70 m at the lowest point in his motion. How high above the ground is he at his highest point?

34. An object slides down an inclined plane of angle 30.0° and of incline length 2.0 m. If the initial speed of the object is 4.0 m/s directed down the incline, what is the speed at the bottom? Neglect friction.

© 35. Rachel is on the roof of a building, h meters above ground. She throws a heavy ball into the air with a speed v, at an angle θ with respect to the horizontal. Ignore air resistance. (a) Find the speed of the ball when it hits the ground in terms of h, v, θ, and g. (b) For what value(s) of θ is the speed of the ball greatest when it hits the ground?

36. The escape speed from the surface of Planet Zoroaster is 12.0 km/s. The planet has no atmosphere. A meteor far away from the planet moves at speed 5.0 km/s on a collision course with Zoroaster. How fast is the meteor going when it hits the surface of the planet?

37. The escape speed from the surface of the Earth is 11.2 km/s. What would be the escape speed from another planet of the same density (mass per unit volume) as Earth but with a radius twice that of Earth?

♦ 38. The orbit of Halley's Comet around the Sun is a long thin ellipse. At its aphelion (point farthest from the Sun), the comet is 5.3×10^{12} m from the Sun and moves with a speed of 10.0 km/s. What is the comet's speed at its perihelion (closest approach to the Sun) where its distance from the Sun is 8.9×10^{10} m?

39. A spring with $k = 40.0$ N/m is at the base of a frictionless 30.0° inclined plane. A 0.50-kg object is pressed against the spring, compressing it 0.20 m from its equilibrium position. The object is then released. If the object is not attached to the spring, how far up the incline does it travel before coming to rest and then sliding back down? (See Fig. 6.32.)

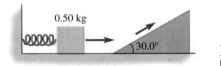

0.50 kg

30.0°

Figure 6.32
Problems 39 and 78

40. A pendulum consists of a bob of mass m attached to the end of a cord of length L. The pendulum is released from a point at a height of $L/2$ above the lowest point of the swing. What is the tension in the cord as the bob passes the lowest point?

♦ 41. A pendulum, consisting of a bob of mass M on a cord of length L, is interrupted in its swing by a peg a distance d directly below its point of suspension (Fig. 6.33). (a) If the bob is to travel in a full circle of radius $(L - d)$ around the peg, what is the minimum possible speed it can have at the lowest point in its motion, just before it starts to go around? [*Hint:* Ignore any decrease in the length of the string due to the peg's circumference.] (b) From what minimum angle θ must the

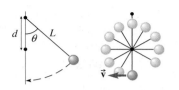

Figure 6.33   Problem 41

pendulum be released so that the bob attains the speed calculated in (a)?

42. A roller coaster car (mass = 988 kg including passengers) is about to roll down a track (Fig. 6.34). The diameter of the circular loop is 20.0 m and the car starts out from rest 40.0 m above the lowest point of the track. Ignore friction and air resistance and assume $g = 9.81$ m/s². (a) At what speed does the car reach the top of the loop? (b) What is the force exerted on the car by the track at the top of the loop? (c) From what minimum height above the bottom of the loop can the car be released so that it does not lose contact with the track at the top of the loop?

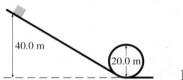

40.0 m

20.0 m

Figure 6.34   Problem 42

43. You shoot a 51-g pebble straight up with a catapult whose spring constant is 320 N/m. The catapult is initially stretched by 0.20 m. How high above the starting point does the pebble fly? Ignore air resistance.

44. A spring gun ($k = 28$ N/m) is used to shoot a 56-g ball horizontally. Initially the spring is compressed by 18 cm. The ball loses contact with the spring and leaves the gun when the spring is still compressed by 12 cm. What is the speed of the ball when it hits the ground, 1.4 m below the spring gun?

♦ 45. A block (mass m) hangs from a spring (spring constant k). The block is released from rest a distance d above its *equilibrium* position. (a) What is the speed of the block as it passes through the equilibrium point? (b) What is the maximum distance below the equilibrium point that the block will reach?

46. A hang glider moving at speed 9.5 m/s dives to an altitude 8.2 m lower. Neglecting drag, how fast is it then moving?

47. In Example 6.1, find the work done by the movers as they slide the chest up the ramp if the coefficient of friction between the chest and the ramp is 0.20.

48. A gymnast of mass 52 kg is jumping on a trampoline. She jumps so that her feet reach a maximum height of 2.5 m above the trampoline and, when she lands, her feet stretch the trampoline down 75 cm. How far does the trampoline stretch when she stands on it at rest? [*Hint:* Assume the trampoline obeys Hooke's law when it is stretched.]

Section 6.8 Power

49. Show that 1 kilowatt-hour (kW·h) is equal to 3.6 MJ.

50. If a man has an average useful power output of 40.0 W, what minimum time would it take him to lift fifty 10.0-kg boxes to a height of 2.00 m? The value of g is 9.80 m/s².

51. A bicycle and its rider together have a mass of 75 kg. What power output of the rider is required to maintain a constant

speed of 4.0 m/s (about 9 mph) up a 5.0% grade (a road that rises 5.0 m for every 100 m along the pavement)? Assume that frictional losses of energy are negligible.

52. A car with mass of 1000.0 kg accelerates from 0 m/s to 40.0 m/s in 10.0 s. Ignore air resistance. The engine has a 22% efficiency, which means that 22% of the thermal energy released by the burning gasoline is converted into mechanical energy. (a) What is the average mechanical power output of the engine? (b) What volume of gasoline is consumed? Assume that the burning of 1.0 L of gasoline releases 46 MJ of thermal energy.

53. A motorist driving a 1200-kg car on level ground accelerates from 20.0 m/s to 30.0 m/s in a time of 5.0 s. Neglecting friction and air resistance, determine the *average* mechanical power in watts the engine must supply during this time interval.

54. A 62-kg woman takes 6.0 s to run up a flight of stairs. The landing at the top of the stairs is 5.0 m above her starting place. (a) What is the woman's average power output while she is running? (b) Would that be equal to her average power *input*—the rate at which chemical energy in food or stored fat—is used? Why or why not?

55. How many grams of carbohydrate does a person of mass 74 kg need to metabolize to climb five flights of stairs (15 m height increase)? Each gram of carbohydrate provides 17.6 kJ of energy. Assume 10.0% efficiency—that is, 10.0% of the available chemical energy in the carbohydrate is converted to mechanical energy. What happens to the other 90% of the energy?

56. A race car with a mass of 500.0 kg completes a quarter-mile (402 m) drag race in a time of 4.2 s. The car's final speed is 125 m/s. What is the engine's average power output? Neglect friction and air resistance.

57. The power output of a cyclist moving at a constant speed of 6.0 m/s on a level road is 120 W. (a) What is the force exerted on the cyclist and the bicycle by the air? (b) By bending low over the handlebars, the cyclist reduces the air resistance to 18 N. If she maintains a power output of 120 W, what will her speed be?

58. Use this method to find how the speed with which animals of similar shape can run up a hill depends on the size of the animal. Let L represent some characteristic length such as the height or diameter of the animal. Assume that the maximum rate at which the animal can do work is proportional to the animal's surface area:

$$P_{max} \propto L^2$$

Set the maximum power output equal to the rate of increase of gravitational potential energy and determine how the speed v depends on L.

59. A 1500-kg car coasts in neutral down a 2.0° hill. The car attains a terminal speed of 20.0 m/s. (a) How much power must the engine deliver to drive the car on a *level* road at 20.0 m/s? (b) If the maximum useful power that can be delivered by the engine is 40.0 kW, what is the steepest hill the car can climb at 20.0 m/s?

COMPREHENSIVE PROBLEMS

60. Rosie lifts a trunk weighing 600 N up 1.5 m. If it takes her 15 s to lift the trunk, at what average rate does she do work?

61. If the skier of Example 6.8 is moving at 12 m/s at the bottom of the trail, calculate the total work done by friction and air resistance during the run. The skier's mass is 75 kg.

62. A child's playground swing is supported by chains that are 4.0 m long. If the swing is 0.50 m above the ground and moving at 6.0 m/s when the chains are vertical, what is the maximum height of the swing?

63. (a) Calculate the change in potential energy of 1 kg of water as it passes over Niagara Falls (a vertical descent of 50 m). (b) At what rate is gravitational potential energy lost by the water of the Niagara River? The rate of flow is 5.5×10^6 kg/s. (c) If 10% of this energy can be converted into electrical energy, how many households would the electricity supply? (An average household uses an average electrical power of about 1 kW.)

64. A 750-kg automobile is moving at 20.0 m/s at a height of 5.0 m above the bottom of a hill when it runs out of gasoline (Fig. 6.35). The car coasts down the hill and then continues coasting up the other side until it comes to rest. Ignoring frictional forces and air resistance, what is the value of h, the highest position the car reaches above the bottom of the hill?

65. A planet with a radius of 6.00×10^7 m has a gravitational field of magnitude 30.0 m/s^2 at the surface. What is the escape speed from the planet?

66. Suppose a satellite is in a circular orbit 3.0 Earth radii above the surface of the Earth (4.0 Earth radii from the center of the Earth). By how much does it have to increase its speed in order to be able to escape Earth? [*Hint:* You need to calculate the orbital speed and the escape speed.]

67. What is the minimum speed with which a meteor strikes the top of the Earth's stratosphere (about 40 km above Earth's surface), assuming that the meteor begins as a bit of interplanetary debris far from Earth?

68. A projectile with mass of 500 kg is launched straight up from the Earth's surface with an initial speed v_0. What magnitude of v_0 enables the projectile to just reach a maximum height of $5R_E$, measured from the *center* of the Earth? Ignore air friction as the projectile goes through the Earth's atmosphere.

69. The escape speed from Earth is 11.2 km/s, but that is only the minimum speed needed to escape *Earth's* gravitational pull; it does not give the object enough energy to leave the solar system. What is the minimum speed for an object near the Earth's surface so that the object escapes both the Earth's and the Sun's gravitational pulls? Ignore drag due to the atmosphere and the

750 kg

20.0 m/s

5.0 m

$h = ?$

Figure 6.35 Problem 64

gravitational forces due to the Moon and the other planets. Also ignore the rotation and the orbital motion of the Earth.

70. A car moving at 30 mi/h is stopped by jamming on the brakes and locking the wheels. The car skids 50 ft before coming to rest. How far would the car skid if it were initially moving at 60 mi/h? [*Hint:* You will not have to do any unit conversions if you set up the problem as a proportion.]

C ✦ 71. An elevator can carry a maximum load of 1202 kg (including the mass of the elevator car). The elevator has an 801-kg counterweight that always moves with the same speed but in the *opposite direction* to the car (Fig. 6.36). Let $g = 9.81$ m/s^2. (a) What is the average power that must be delivered by the motor to carry the maximum load up 40.0 m in 60.0 s? (b) How would your answer be different if there were no counterweight?

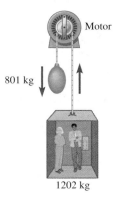

Motor

801 kg

1202 kg

Figure 6.36 Problem 71

72. If a high jumper needs to make his center of gravity rise 1.2 m, how fast must he be able to sprint? (For an extended object, the gravitational potential energy is $U = mgh$, where h is the height of the center of gravity.)

73. A pole-vaulter converts the kinetic energy of running to elastic potential energy in the pole, which is then converted to gravitational potential energy. If a pole-vaulter's center of gravity is 1.0 m above the ground while he sprints at 10.0 m/s, what is the maximum height of his center of gravity during the vault? For an extended object, the gravitational potential energy is $U = mgh$, where h is the height of the center of gravity. (In 1988, Sergei Bubka was the first pole-vaulter ever to clear 6 m.)

74. (a) How much work does a major-league pitcher do on the baseball when he throws a 90.0 mi/h (40.2 m/s) fastball? The mass of a baseball is 145 g. (b) How many fastballs would a pitcher have to throw to "burn off" a 1520 Calorie meal? (1 Calorie = 1000 cal = 1 kcal.) Assume that 80.0% of the chemical energy in the food is converted to thermal energy and only 20.0% becomes the kinetic energy of the fastballs.

75. The number of kilocalories per day required by a person resting under standard conditions is called the basal metabolic rate (BMR). (a) To generate 1 kcal, Jermaine's body needs approximately 0.010 mol of oxygen. If Jermaine's net intake of oxygen through breathing is 0.015 mol/min while he is resting, what is his BMR in kcal/day? (b) If Jermaine fasts for 24 h, how many pounds of fat does he lose? Assume that only fat is consumed. Each gram of fat consumed generates 9.3 kcal.

76. A kangaroo decides to see how high it can hop on *one leg*. Assuming the elastic energy stored in the tendon is the same as for Example 6.10, how high can it jump using a single leg?

C ✦ 77. A 4.0-kg block is released from rest at the top of a frictionless plane of length 8.0 m that is inclined at an angle of 15° to the horizontal. A cord is attached to the block and trails along behind it (Fig. 6.37). When the block reaches a point 5.0 m along the incline from the top, someone grasps the cord and exerts a constant tension parallel to the incline. The tension is such that the block just comes to rest when it reaches the bottom of the incline. (The person's force is a nonconservative force.) What is this constant tension? Solve the problem twice, once using work and energy and again using Newton's laws and constant-acceleration kinematics. Which method do you prefer?

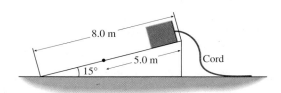

8.0 m

5.0 m

Cord

15°

Figure 6.37 Problem 77

78. A 0.50-kg block, starting at rest, slides down a 30.0° incline with kinetic friction coefficient 0.25 (Fig. 6.32). After sliding 85 cm down the incline, it slides across a frictionless horizontal surface and encounters a spring ($k = 35$ N/m). (a) What is the maximum compression of the spring? (b) After the compression of part (a), the spring rebounds and shoots the block back up the incline. How far along the incline does the block travel before coming to rest?

79. The extinction of the dinosaurs and the majority of species on Earth in the Cretaceous period is thought to have been caused by an asteroid striking the Earth on the Yucatán peninsula about 65 million years ago (see Fig. 6.38). The collision put so much debris into the atmosphere that the amount of sunlight reaching Earth's surface was reduced enough to cause severe changes in climate. The asteroid was about 16 km (10 mi) in diameter, had a mass of about 10^{16} kg, and was moving at about 30 km/s (about 20 miles per *second*) when it struck the Earth. Calculate the energy released when the asteroid struck Earth and compare it to the energy released by a thermonuclear bomb ($\sim 10^{17}$ J).

Figure 6.38 The Barringer Meteor Crater in Arizona. The asteroid discussed in Problem 79 had a kinetic energy about 100 million times as large as that of the meteor that caused the Barringer crater.

✦ 80. The solar wind is a continuous stream of particles, mostly protons and electrons, moving away from the Sun. When the particles encounter Earth's upper atmosphere, they are moving at speeds of several hundred km/s. If such a particle reaches Earth's upper atmosphere with a speed of 92 km/s, how fast was it moving when it crossed Mercury's orbit at a distance of 5.8×10^7 km from the Sun? [*Hint:* Assume that the particles are slowed by the Sun's gravitational field as they move away from the Sun. Ignore the slight *increase* in speed that occurs close to Earth due to the Earth's gravitational field.]

81. Human feet and legs store elastic energy when walking or running. They are not nearly as efficient at doing so as kangaroo legs, but the effect is significant nonetheless. If not for the storage of elastic energy, a 70-kg man running at 4 m/s would lose about 100 J of mechanical energy each time he sets down a foot. Some of this energy is stored as elastic energy in the Achilles tendon and in the arch of the foot; the elastic energy is then converted back into the kinetic and gravitational potential energy of the leg, reducing the expenditure of metabolic energy. If the maximum tension in the Achilles tendon when the foot is set down is 4.7 kN and the tendon's spring constant is 350 kN/m, calculate how far the tendon stretches and how much elastic energy is stored in it.

82. The graph in Fig. 6.39 shows the tension in a rubber band as it is first stretched and then allowed to contract. As you stretch a rubber band, the tension force at a particular length (on the way to a maximum stretch) is larger than the force at that same length as you let the rubber band contract. That is why Fig. 6.39 shows two separate lines, one for stretching and one for contracting; the lines are not superimposed as you might have thought they would be. (a) Make a rough estimate of the total work done by the external force applied to the rubber band for the entire process. (b) If the rubber band obeyed Hooke's law, what would the answer to (a) have to be? (c) While the rubber band is stretched, is all of the work done on it accounted for by the increase in elastic potential energy? If not, what happens to the rest of it? [*Hint:* Take a rubber band and stretch it rapidly several times. Then hold it against your wrist or your lip.]

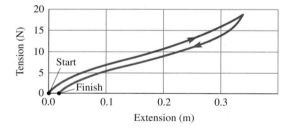

Figure 6.39   Problem 82

83. A wind turbine (Fig. 6.40) converts some of the kinetic energy of the wind into electrical energy. Suppose that the blades of a small wind turbine have length $L = 4.0$ m. (a) When a 10 m/s (22 mi/h) wind blows head-on, what volume of air (in m³) passes through the circular area swept out by the blades in 1.0 s? (b) What is the mass of this much air?

Each m³ of air has a mass of 1.2 kg. (c) What is the translational kinetic energy of this mass of air? (d) If the turbine can convert 40% of this kinetic energy into electrical energy, what is its electrical power output? (e) What happens to the power output if the wind speed decreases to $\frac{1}{2}$ of its initial value? What can you conclude about electrical power production by wind turbines?

84. Use dimensional analysis to show that the electrical power output of a wind turbine (Fig. 6.40) is proportional to the *cube* of the wind speed. The relevant quantities on which the power can depend are the length L of the rotor blades, the density ρ of air (SI units kg/m³), and the wind speed v.

Figure 6.40   Problems 83 and 84

85. Figure 6.41 is a graph of force versus extension for a wallaby tendon as it is stretched (upper curve) and then allowed to return back to its relaxed length (lower curve). The maximum extension of the tendon in the graph is similar to what occurs when a wallaby jumps. Estimate the percentage of the elastic energy stored in the tendon when the wallaby lands that is dissipated (and therefore cannot be recovered to assist the next jump).

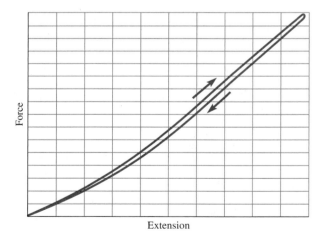

Figure 6.41   Problem 85

ANSWERS TO PRACTICE PROBLEMS

6.1 −180 kJ

6.2 43 N; 4500 J; she pulls with a greater force but its component in the direction of the displacement is smaller.

6.3 $\dfrac{(2.5m) \times (1.50v)^2}{mv^2} = 5.6$; it is difficult to judge the relative damage because the lighter car is made from lighter, less sturdy materials. It might be that both vehicles suffer the same amount of damage.

6.4 29 m/s

6.5 simple bow, 2.0 J; compound bow, 4.0 J

6.6 9.8 m/s

6.7 140 kJ

6.8 16 m/s

6.9 195 km/s

6.10 2.4 m

6.11 0.24

6.12 23 kW; 33 kW

Linear Momentum

Why were cannons on battleships mounted on wheels instead of being attached to the deck? According to a cautionary pirate tale, which may or may not be true, a ship called the *Ocean Queen* was captured in about 1820 by a French pirate, Captain la Bouche, and adapted for use as a pirate ship. The ship was relatively small, a little under 100 tons, and was equipped with 28 guns. As was the custom in the pirating trade, Captain la Bouche would sail along-side a merchant ship, run up the Jolly Roger pirate flag, and fire a broadside at the other ship if it did not surrender at once. (A "broadside" is the firing of all guns along one side of the ship at the same time.) When the *Ocean Queen* set off to hunt for unsuspecting ships, a young mate was put in charge because Captain la Bouche was sailing another ship. The ocean was rough, so the mate decided to bolt the guns firmly to the deck to prevent them from rolling about as the ship rocked from side to side. What happened when the *Ocean Queen* fired a broadside at another ship with the guns still bolted to the deck?

7.1   A VECTOR CONSERVATION LAW

In Chapter 3 we learned how to determine the acceleration of an object by finding the net force acting on it and applying Newton's second law of motion. If the forces happen to be constant, then the resulting constant acceleration enables us to calculate changes in velocity and position. Calculating velocity and position changes when the forces are not constant is much more difficult. In many cases the forces cannot even be easily determined. Conservation of energy is one tool that enables us to draw conclusions about motion without knowing all the details of the forces acting. Recall, for example, how easily we can calculate the escape speed of a projectile using conservation of energy, without even knowing the path the object takes. Now imagine how difficult the same calculation would be using Newton's second law, with a gravitational force that changes magnitude and direction depending on the path taken.

In this chapter we develop another conservation law. Conservation laws are powerful tools. If a quantity is conserved, then no matter how complicated the situation, we can set the value of the conserved quantity at one time equal to its value at a later time. The "before-and-after" aspect of a conservation law enables us to draw conclusions about the results of a complicated set of interactions without knowing all of the details.

The new conserved quantity, *momentum*, is a vector quantity, in contrast to energy, which is a scalar. When momentum is conserved, both the magnitude and the direction of the momentum must be constant. Equivalently, the *x*- and *y*-components of momentum are constant. When we find the total momentum of more than one object, we must add the momentum vectors according to the procedure by which vectors are always added.

7.2   MOMENTUM

The word *momentum* is often heard in broadcasts of sporting events. A sports broadcaster might say, "The home team has won five consecutive games; they have the momentum in their favor." The team with "momentum" is hard to stop; they are moving forward on a winning streak. A football player, running for the goal line with a football tucked under his arm, has momentum; he is hard to stop. This use of the word *momentum* is closer to the physics usage. In physics we would agree that the runner has momentum, but we have a precise definition in mind.

In everyday use, momentum has something to do with mass as well as with velocity. Would you rather have a running child bump into you, or a defensive lineman running with the same velocity? The child has much less momentum than the lineman, even though their velocities are the same.

Could a quantity combining mass and velocity be useful in physics? Imagine a collision between two spaceships (Fig. 7.1). Let the spaceships be so far from planets and stars that we can neglect gravitational interactions with celestial bodies. The spaceships exert forces on each other while they are in contact. According to Newton's third law, these forces are equal and opposite. The force on ship 2 exerted by ship 1 is equal and opposite to the force exerted on ship 1 by ship 2:

$$\vec{\mathbf{F}}_{21} = -\vec{\mathbf{F}}_{12}$$

The changes in velocities of the two spaceships are *not* equal and opposite if the masses are different. Suppose a large spaceship (mass m_1) collides with a much smaller ship (mass $m_2 \ll m_1$). Assume for now that the forces are constant during the time interval Δt that the spaceships are in contact. Although the forces have the same magnitude, the magnitudes of the accelerations of the two ships are different because their masses are different. The ship with the larger mass has the smaller acceleration.

The acceleration of either spaceship causes its velocity to change by

$$\Delta \vec{\mathbf{v}} = \vec{\mathbf{a}}\, \Delta t = \frac{\vec{\mathbf{F}}}{m}\, \Delta t$$

$\vec{\mathbf{F}}_{21}$ is the force exerted *on* object 2 *by* object 1.

Figure 7.1 Two spaceships about to collide.

The time interval Δt is the duration of the interaction between the two ships, so it must be the same for both ships.

Since the changes in velocity are inversely proportional to the masses, the changes in the *products* of mass and velocity are equal and opposite for the two bodies involved in the interaction:

$$m_1 \Delta \vec{\mathbf{v}}_1 = \vec{\mathbf{F}}_{12} \Delta t$$

$$m_2 \Delta \vec{\mathbf{v}}_2 = \vec{\mathbf{F}}_{21} \Delta t$$

Since $\vec{\mathbf{F}}_{21} = -\vec{\mathbf{F}}_{12}$,

$$m_2 \Delta \vec{\mathbf{v}}_2 = -m_1 \Delta \vec{\mathbf{v}}_1$$

This is a useful insight, so we give the product of mass and velocity a name and symbol: **linear momentum** (symbol $\vec{\mathbf{p}}$). Linear momentum (or just *momentum*) is a vector quantity having the same direction as the velocity.

Definition of linear momentum:

$$\vec{\mathbf{p}} = m\vec{\mathbf{v}} \qquad (7\text{-}1)$$

The collision of the two spaceships causes changes in their momenta that are equal in magnitude and opposite in direction:

$$\Delta \vec{\mathbf{p}}_1 = m_1 \vec{\mathbf{v}}_{1f} - m_1 \vec{\mathbf{v}}_{1i} = \vec{\mathbf{F}}_{12} \Delta t$$

$$\Delta \vec{\mathbf{p}}_2 = m_2 \vec{\mathbf{v}}_{2f} - m_2 \vec{\mathbf{v}}_{2i} = \vec{\mathbf{F}}_{21} \Delta t \qquad (7\text{-}2)$$

Since $\vec{\mathbf{F}}_{21} = -\vec{\mathbf{F}}_{12}$,

$$\Delta \vec{\mathbf{p}}_2 = -\Delta \vec{\mathbf{p}}_1$$

The momentum changes are equal and opposite.

Our new model of an interaction is this: in an interaction, an exchange of momentum takes place. Some momentum is transferred from one body to another. The momentum changes of the two bodies are equal and opposite, so the total momentum of the

During an interaction, momentum is transferred from one body to another.

isolated system composed of the two bodies is unchanged. The momentum of a system is the vector sum of the individual momenta of the bodies composing the system.

$$\text{change in momentum of system} = \Delta\vec{\mathbf{p}} = \Delta\vec{\mathbf{p}}_1 + \Delta\vec{\mathbf{p}}_2$$

Since $\Delta\vec{\mathbf{p}}_2 = -\Delta\vec{\mathbf{p}}_1$, the change in the total momentum of the system is

$$\Delta\vec{\mathbf{p}} = \Delta\vec{\mathbf{p}}_1 + \Delta\vec{\mathbf{p}}_2 = \Delta\vec{\mathbf{p}}_1 + (-\Delta\vec{\mathbf{p}}_1) = \vec{\mathbf{0}}$$

The momentum of either ship changes during the collision, but the total momentum of the two ships does *not* change.

Example 7.1

Change of Momentum of a Moving Car

A car weighing 12 kN is driving due north at 30.0 m/s. After driving around a sharp curve, the car is moving east at 13.6 m/s. What is the change in momentum of the car?

Strategy The definition of momentum is $\vec{\mathbf{p}} = m\vec{\mathbf{v}}$. We can start by finding the car's mass. There are two potential pitfalls:

(1) momentum depends not on weight but on mass, and

(2) momentum is a vector, so we must take its direction into consideration as well as its magnitude. To find the change in momentum, we need to do a *vector* subtraction.

Solution The car's mass is

$$m = \frac{W}{g} = \frac{1.2 \times 10^4\ \text{N}}{9.8\ \text{m/s}^2} = 1220\ \text{kg}$$

The car's initial velocity is

$$\vec{\mathbf{v}}_i = 30.0\ \text{m/s, north}$$

The car's initial momentum is then

$$\vec{\mathbf{p}}_i = m\vec{\mathbf{v}}_i = 1220\ \text{kg} \times 30.0\ \text{m/s north} = 3.66 \times 10^4\ \text{kg·m/s north}$$

After the curve, the final velocity is

$$\vec{\mathbf{v}}_f = 13.6\ \text{m/s, east}$$

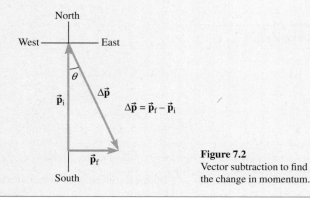

Figure 7.2
Vector subtraction to find the change in momentum.

The final momentum is

$$\vec{\mathbf{p}}_f = m\vec{\mathbf{v}}_f = 1220\ \text{kg} \times 13.6\ \text{m/s east} = 1.66 \times 10^4\ \text{kg·m/s east}$$

Momentum vectors are added and subtracted according to the same methods used for other vectors. To find the change in the momentum, we draw vector arrows representing the initial and final momenta with their tails at the same point (Fig. 7.2). Then the change in momentum $\Delta\vec{\mathbf{p}} = \vec{\mathbf{p}}_f - \vec{\mathbf{p}}_i$ is represented by the arrow drawn from the tip of $\vec{\mathbf{p}}_i$ to the tip of $\vec{\mathbf{p}}_f$. Since the three vectors in Fig. 7.2 form a right triangle, the magnitude of $\Delta\vec{\mathbf{p}}$ can be found from the Pythagorean theorem

$$|\Delta\vec{\mathbf{p}}| = \sqrt{p_i^2 + p_f^2} = \sqrt{(3.66 \times 10^4\ \text{kg·m/s})^2 + (1.66 \times 10^4\ \text{kg·m/s})^2}$$
$$= 4.02 \times 10^4\ \text{kg·m/s}$$

From the vector diagram, $\Delta\vec{\mathbf{p}}$ is directed at an angle θ east of south. Using trigonometry,

$$\tan\theta = \frac{\text{opposite}}{\text{adjacent}} = \frac{p_f}{p_i} = \frac{1.66 \times 10^4\ \text{kg·m/s}}{3.66 \times 10^4\ \text{kg·m/s}} = 0.454$$

$$\theta = \tan^{-1} 0.454 = 24.4°$$

Since the weight is given with two significant figures, we report the change in momentum of the car as 4.0×10^4 kg·m/s directed 24° east of south.

Discussion As with displacements, velocities, accelerations, and forces, it is crucial to remember that momentum is a vector. When finding changes in momentum, we must find the difference between final and initial momentum *vectors*. If the initial and final momenta had not been perpendicular, we would have had to resolve the vectors into *x*- and *y*-components in order to subtract them.

Practice Problem 7.1   Falling apple

(a) What is the momentum of an apple weighing 1.0 N just before it hits the ground, if it falls out of a tree from a height of 3.0 m? (b) The apple falls because of the gravitational interaction between the apple and the Earth. How much does this interaction change the *Earth's* momentum? How much does it change the Earth's velocity?

7.3   THE IMPULSE-MOMENTUM THEOREM

We found that the change in momentum of an object due to an interaction is equal to the product of the force acting on the object and the time interval during which the force acts [Eq. (7-2)]:

$$\Delta \vec{p} = \vec{F}\, \Delta t$$

The product $\vec{F}\Delta t$ is given the name **impulse**. Since the impulse is the product of a vector (the force) and a positive scalar (the time), impulse is a vector quantity having the same direction as that of the force. In words, $\Delta \vec{p} = \vec{F}\,\Delta t$ can be read as *"the change in momentum equals the impulse."* The SI units of impulse are newton-seconds (N·s) and those of momentum are kilogram-meters per second (kg·m/s). These are equivalent units, as can be demonstrated using the definition of the newton (Problem 16). The relation between impulse and momentum change is called the *impulse-momentum theorem* and is especially useful in solving problems that involve collisions and impacts.

Impulse $= \vec{F}\Delta t$

Impulse-Momentum Theorem:

$$\Delta \vec{p} = \vec{F}\,\Delta t \qquad (7\text{-}2)$$

If an object is involved in more than one interaction, then its change in momentum during any time interval is equal to the *total* impulse during that time interval. The total impulse is the vector sum of the impulses due to each force. The total impulse is also equal to the net force times the time interval:

$$\text{total impulse} = \vec{F}_1\,\Delta t + \vec{F}_2\,\Delta t + \cdots$$
$$= (\vec{F}_1 + \vec{F}_2 + \cdots)\,\Delta t$$
$$= \vec{F}_{\text{net}}\,\Delta t$$

Our discussion so far has assumed that the force acting during an interaction is constant. That is a rather unusual situation; the concept of momentum would be of limited use if it were applicable only when forces are constant. However, everything we have said still applies to situations where the forces are not constant, as long as we use the *average* force to calculate the impulse. The average force divided by the mass gives the average acceleration:

$$\vec{a}_{\text{av}} = \frac{\vec{F}_{\text{av}}}{m}$$

The change in velocity is always equal to the average acceleration times the time interval:

$$\Delta \vec{v} = \vec{a}_{\text{av}}\,\Delta t = \frac{\vec{F}_{\text{av}}\,\Delta t}{m}$$

Then we conclude that impulse equals the change in momentum when impulse is the product of the average force and the time interval:

$$\Delta \vec{p} = m\,\Delta \vec{v} = \vec{F}_{\text{av}}\,\Delta t \qquad (7\text{-}3)$$

When the force is not constant, the impulse is found using the average force.

Conceptual Example 7.2

Big Force–Short Time versus Small Force–Long Time

Which causes the larger change in momentum of an object, an average force of 5 N acting for 4 s or an average force of 2 N acting for 10 s? How might this principle be used when designing products to protect the human body from injury? Give an example.

Solution and Discussion   The change in momentum is found from the impulse-momentum theorem. The product of the force and the time interval gives the momentum change of the object. Over a period of 4 s, the 5-N force causes a momentum change of magnitude (5 N × 4 s) = 20 N·s, while the 2-N force acting for 10 s also causes a momentum change of magnitude (2 N × 10 s) = 20 N·s. The smaller force causes the same change in momentum because it acts for a longer time interval.

continued on next page

Conceptual Example 7.2 *continued*

Figure 7.3
A stuntman lands safely in an air bag to break his fall. The air bag reduces the risk of injury in two ways. It changes the stuntman's momentum more gradually, so that forces of smaller magnitude act on his body. It also spreads these forces over a larger area so they are less likely to cause serious injury.

When designing products to protect the human body, one goal is to lengthen the time period during which a velocity change occurs. For example, when a movie stuntman falls from a great height to simulate a victim being pushed off a balcony, he lands on a large air bag, which changes his momentum much more gradually than if he were to fall onto concrete (Fig. 7.3). The average force exerted by the air bag on the stuntman is much smaller than the average force exerted by concrete would be. Nets used under circus acrobats serve the same purpose. The net gives and dips downward when the acrobat falls into it, gradually reducing the speed of the fall over a longer time interval than if she fell directly onto the ground.

Practice Problem 7.2   Pole-vaulter landing on padded surface

A pole-vaulter vaults over the pole and falls onto thick padding. He lands with a speed of 9.8 m/s; the padding then brings him to a stop in a time of 0.40 s. What is the average force on his body *due to the padding* during that time interval? Express your answer as a fraction or multiple of his weight. [*Hint:* The force due to the padding is not the only force acting on the vaulter during the 0.40-s interval.]

Example 7.3

Collision between an Automobile and a Tree

A car moving at 20.0 m/s (44.7 mi/h) crashes into a tree. Find the magnitude of the average force acting on a passenger of mass 65 kg in each of the following cases. (a) The passenger is not wearing a seat belt. He is brought to rest by a collision with the windshield and dashboard that lasts 3 ms. (b) The passenger is wearing a seat belt. The force exerted by the seat belt for 0.3 s brings him to rest. (c) The car is equipped with a passenger-side air bag. The force due to the air bag acts for 1 s, bringing the passenger to rest.

Strategy In each of the three cases, the passenger's momentum change is the same. What differs is the time interval during which the change occurs. It takes a larger force to change the momentum in a shorter time interval.

Solution The magnitude of the passenger's initial momentum is

$$|\vec{\mathbf{p}}_i| = |m\vec{\mathbf{v}}_i| = (65 \text{ kg} \times 20.0 \text{ m/s}) = 1300 \text{ kg·m/s}$$

His final momentum is zero, so the magnitude of the momentum change is

$$|\Delta\vec{\mathbf{p}}| = 1300 \text{ kg·m/s}$$

This momentum change divided by the time interval gives the magnitude of the average force in each case.

(a) No seat belt: $|\vec{\mathbf{F}}_{av}| = \dfrac{|\Delta\vec{\mathbf{p}}|}{\Delta t} = \dfrac{1300 \text{ kg·m/s}}{0.003 \text{ s}} = 4 \times 10^5 \text{ N}$

(b) With seat belt: $|\vec{\mathbf{F}}_{av}| = \dfrac{|\Delta\vec{\mathbf{p}}|}{\Delta t} = \dfrac{1300 \text{ kg·m/s}}{0.3 \text{ s}} = 4 \times 10^3 \text{ N}$

(c) Air bag: $|\vec{\mathbf{F}}_{av}| = \dfrac{|\Delta\vec{\mathbf{p}}|}{\Delta t} = \dfrac{1300 \text{ kg·m/s}}{1 \text{ s}} = 1 \times 10^3 \text{ N}$

Discussion Without a seat belt (or air bag), the average force is roughly 700 times the passenger's weight; with an air bag (and seat belt), the average force is only about 2 times the passenger's weight. The average forces required to bring the passenger to rest are inversely proportional to the time interval over which those forces act. It is a far happier situation to have the momentum change over as long a period as possible

continued on next page

Example 7.3 *continued*

to make the forces smaller. Automotive safety engineers design cars to minimize the average forces on the passengers during sudden stops and collisions.

Practice Problem 7.3   Catching a fastball

A baseball catcher is catching a fastball that is thrown at 43 m/s (96 mi/h) by the pitcher. If the mass of the ball is 0.15 kg and if the catcher moves his mitt backward toward his body by 8.0 cm as the ball lands in the glove, what is the magnitude of the average force acting on the catcher's mitt? Estimate the time interval required for the catcher to move his hands.

One automotive design change implemented to minimize injury upon collision is the foam padding built into automobile dashboards (Fig. 7.4). Some steering wheels are collapsible in order to give way in a crash. Automobile bumpers have shock absorbers built in to lessen damage to the car body in small collisions. The structure of the car itself is often a single piece of metal with reinforced supports (*unibody* construction) so that the entire body can crumple and absorb the change in momentum more slowly than it would if it were made of separate sections of metal that would slide into or over each other or fall off the car. The safety glass in a windshield has two advantages. One is that it does not shatter and send sharp shards of glass into human tissue, but the other is that it distorts when struck by solid objects like human bones or a human head. The glass doesn't give much, but in a crash every little bit helps.

The use of seat belts plus the air bag is better than either alone. Without a seat belt, the body continues moving with the same speed the car had before the crash. The rapidly inflating air bag moves toward the body and the effective velocity is then the sum of the two velocities (air bag velocity + body velocity) when the two collide. The body flying

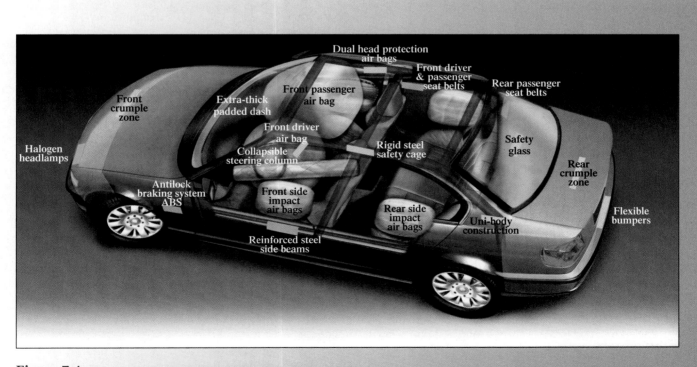

Figure 7.4 Some safety features of the modern automobile. Many of these features serve to lengthen the time interval during which a momentum change occurs in a crash, thereby lessening the forces acting on the passengers.

into the air bag can be injured more than a restrained body making more gradual contact with the air bag. An adult should sit at least 12 in. from the air bag container to avoid injury from the deploying air bag itself. Small children should always be placed in the back seat, in proper car seats for their size, to ensure their safety.

Physics at Home

Try playing catch with a friend (on the lawn) while using a raw egg or a water balloon as a ball. How do you move your hands to minimize the chance of breaking the egg or balloon when you catch it? What is likely to happen if you forget that the "ball" is an egg or balloon and catch it as you would a ball?

Impulse Due to a Changing Force

When a force is changing, how can we find the impulse? We've asked similar questions in previous chapters. For simplicity we consider components along the x-axis. Recall:

- displacement = $\Delta x = v_{av,x} \, \Delta t$ = area under $v_x(t)$ graph
- change in velocity = $\Delta v_x = a_{av,x} \, \Delta t$ = area under $a_x(t)$ graph

> A graph of $v_x(t)$ means the quantity v_x is plotted as a function of the variable t.

In both cases, the mathematical relationship is that of a rate of change. Velocity is the rate of change of position with time and acceleration is the rate of change of velocity with time. Now we have force as the rate of change of momentum with time. By analogy:

- impulse = $\vec{\mathbf{F}}_{av,x} \, \Delta t$ = area under $F_x(t)$ graph

So to find the impulse for a variable force, we find the area under the $F_x(t)$ graph. Then, if we wish to know the average force, we can divide the impulse by the time interval during which the force is applied.

The variable force of Fig. 7.5a increases linearly from 0 to 4 N in a time of 2 s; then it decreases from 4 N to 0 N in 2 s. The area under the $F(t)$ graph is found from the triangular area

$$\text{area} = \tfrac{1}{2} \, \text{base} \times \text{height} = 2 \text{ s} \times 4 \text{ N} = 8 \text{ N·s} = \text{impulse}$$

The average force during the 4-s time interval is

$$\text{average force} = \frac{\text{impulse}}{\text{time interval}} = \frac{8 \text{ N·s}}{4 \text{ s}} = 2 \text{ N}$$

Figure 7.5b shows the average force over the 4-s time interval; the area under the curve (the impulse) is the same as in Fig. 7.5a.

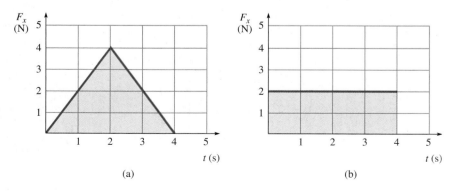

(a)   (b)

Figure 7.5   Areas under $F(t)$ graphs for (a) variable and (b) average force are a measure of the impulse.

Example 7.4

Hitting the Wall

A toy car of mass 10.2 kg moving at 1.2 m/s in the $+x$-direction crashes into a brick wall and rebounds. A force sensor on the car's bumper records the force that the wall exerts on the car as a function of time. These data are shown in graphical form in Fig. 7.6. (a) What is the maximum magnitude of the force exerted on the car? (b) What is the average force on the car during the collision? (c) At what speed does the car rebound from the wall?

Strategy The maximum force can be read directly from the graph. To solve parts (b) and (c) of this problem, we must find the impulse exerted on the car. Since impulse is the area under the $F(t)$ curve, we'll make an estimate of the area. The impulse is then equal to the average force times the time interval and also to the car's change in momentum. Once we find the change in momentum, we use it to find the car's final speed.

Given: $m = 10.2$ kg; $v_{ix} = 1.2$ m/s; graph of $F_x(t)$
To find: (a) F_{max}; (b) $F_{av,x}$; (c) v_{fx}

Solution (a) From Fig. 7.6, the maximum force is approximately 750 N in magnitude.

(b) Each division on the horizontal axis represents 0.01 s and each vertical division represents 200 N. Then the area of each grid box represents (200 N × 0.01 s) = 2 N·s. Counting the number of grid boxes between the $F(x)$ curve and the time axis, estimating fractions of boxes, yields about 10 boxes. Then the magnitude of the impulse is approximately

$$10 \text{ boxes} \times 2 \text{ N·s/box} = 20 \text{ N·s}$$

The collision is underway when the force is nonzero. So the collision begins at $t = 0.025$ s and ends at $t = 0.095$ s. The duration of the collision is

$$\Delta t = 0.07 \text{ s}$$

The magnitude of the average force is therefore approximately

$$|F_{av,x}| = \frac{\text{impulse}}{\Delta t} = \frac{20 \text{ N·s}}{0.07 \text{ s}} = 300 \text{ N}$$

(c) The impulse gives us the momentum change. The force exerted by the wall is in the $-x$-direction. Thus, the x-component of the impulse is negative. In the graph of F_x versus t, the area lies under the time axis and so is counted as negative. So, working with x-components,

$$\Delta p_x = mv_{fx} - mv_{ix} = F_{av,x} \Delta t = -20 \text{ N·s}$$

Solving for v_{fx},

$$v_{fx} = \frac{\Delta p_x + mv_{ix}}{m} = \frac{\Delta p_x}{m} + v_{ix}$$

Substituting numerical values in this expression yields

$$v_{fx} = \frac{-20 \text{ N·s}}{10.2 \text{ kg}} + 1.2 \text{ m/s} = -0.8 \text{ m/s}$$

The car rebounds at a speed of 0.8 m/s.

Discussion As a check, we compare the average force to the maximum force. The average force is a bit less than half of the maximum force. If the force were a linear function of time, the average would be exactly half the maximum. Here, the average force is less than that because more time is spent at smaller values of force than at the larger values.

Practice Problem 7.4 Car-van collision

A car weighing 13.6 kN is moving at 10.0 m/s in the $+x$-direction when it collides head-on with a van weighing 33.0 kN. The horizontal force exerted on the car before, during, and after the collision is shown in Fig. 7.7. What is the car's velocity just after the collision?

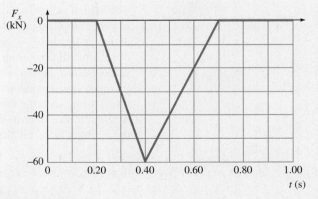

Figure 7.7
Varying force on a car during a car-van collision.

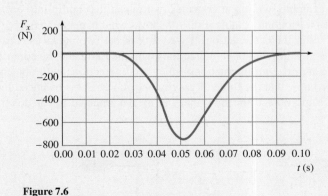

Figure 7.6
Force versus time for toy car colliding with a wall.

Newton's Second Law Revisited

We can use the relationship between impulse and momentum ($\Delta \vec{\mathbf{p}} = \vec{\mathbf{F}}_{av} \, \Delta t$) to find a new way to understand Newton's second law. Let's rewrite the impulse-momentum theorem this way:

$$\vec{\mathbf{F}}_{av} = \frac{\Delta \vec{\mathbf{p}}}{\Delta t}$$

What happens if we let the time interval Δt get smaller and smaller, approaching zero? Then the average force is taken over a smaller and smaller time interval, approaching the instantaneous force:

$$\vec{\mathbf{F}} = \lim_{\Delta t \to 0} \frac{\Delta \vec{\mathbf{p}}}{\Delta t}$$

If more than one force acts, we must replace $\vec{\mathbf{F}}$ with $\vec{\mathbf{F}}_{net}$. Then our restatement of Newton's second law becomes

Figure 7.8 The space shuttle is propelled upward as hot gases are exhausted downward at high speeds.

Newton's second law:

$$\vec{\mathbf{F}}_{net} = \lim_{\Delta t \to 0} \frac{\Delta \vec{\mathbf{p}}}{\Delta t} \tag{7-4}$$

In words, *the net force is the rate of change of momentum.*

Equation (7-4) is *more general* than $\vec{\mathbf{F}}_{net} = m\vec{\mathbf{a}}$, the form of Newton's second law used in Chapters 2 through 6, which holds only when mass is constant. One situation in which mass is not constant is the rocket engine. In a rocket engine, fuel combustion produces hot gases that are then expelled at high speeds (Fig. 7.8). The rocket's mass decreases as the exhaust gases are expelled.

When mass is constant, then it can be factored out:

$$\vec{\mathbf{F}}_{net} = \lim_{\Delta t \to 0} \frac{\Delta \vec{\mathbf{p}}}{\Delta t} = \lim_{\Delta t \to 0} \frac{\Delta (m\vec{\mathbf{v}})}{\Delta t} = m \lim_{\Delta t \to 0} \frac{\Delta \vec{\mathbf{v}}}{\Delta t} = m\vec{\mathbf{a}}$$

Thus Eq. (7-4) reduces to the familiar form of Newton's second law from Chapters 2–6 when mass is constant.

7.4   CONSERVATION OF MOMENTUM

The idea of an interaction as a transfer of momentum is most useful when the forces acting are complicated and we are interested in knowing the state of motion after the interaction is finished. As an example, we might consider two pucks that bump into each other after sliding along on a frictionless table. Fig. 7.9 shows what happens to the two pucks before, during, and after their interaction. If we think of the two pucks as comprising a single system, then the gravitational interactions with the Earth and the contact interactions with the table are *external* interactions—they are interactions with objects external to the system. We know that the force of gravity on each object is balanced by the normal force on the same object and thus there is no net impulse up or down.

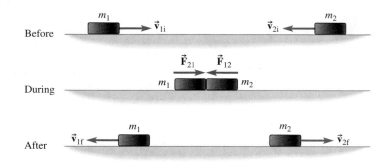

Figure 7.9 Two sliding pucks before, during, and after collision.

Together, these forces produce a net external force of zero, so they leave the system's momentum unchanged. Since these two always cancel, we can ignore these "external" interactions and just focus on the interaction between the pucks. Therefore, we omit the normal and gravitational forces in Fig. 7.9.

Until contact is made, there is no interaction between the pucks (ignoring the small gravitational interaction between the two). During the collision, the pucks exert forces on each other. Force $\vec{\mathbf{F}}_{12}$ is the contact force acting on mass m_1 and force $\vec{\mathbf{F}}_{21}$ is the contact force acting on mass m_2. If we continue to regard the two pucks as parts of a single interacting system, then those forces are *internal* forces of this system. When they collide, some momentum is transferred from one puck to the other. The changes in momentum of the two are equal and opposite:

$$\Delta\vec{\mathbf{p}}_1 = -\Delta\vec{\mathbf{p}}_2$$

Since the change in momentum is the final momentum minus the initial momentum, we write:

$$\vec{\mathbf{p}}_{1f} - \vec{\mathbf{p}}_{1i} = -(\vec{\mathbf{p}}_{2f} - \vec{\mathbf{p}}_{2i})$$

Moving the initial momenta to the right side and the final momenta to the left:

$$\vec{\mathbf{p}}_{1f} + \vec{\mathbf{p}}_{2f} = \vec{\mathbf{p}}_{1i} + \vec{\mathbf{p}}_{2i} \tag{7-5}$$

The sum of the momenta of the pucks after the interaction is equal to the sum of the momenta before the interaction; or, more simply, the total momentum of the objects is unchanged by the collision. This isn't surprising since, if some momentum is just transferred from one to the other, the total hasn't changed. We say that momentum is *conserved* for this collision. The interaction between the pucks changes the momentum of each puck, but the total momentum of the system is unchanged.

To summarize:

- The total momentum of a system is the vector sum of the momenta of each object in the system.

- External interactions can change the total momentum of a system.

- Internal interactions do not change the total momentum of a system.

In the absence of external interactions, momentum is conserved:

Law of Conservation of Linear Momentum
If the net external force acting on a system is zero,
then the momentum of the system is conserved.

$$\text{if } \vec{\mathbf{F}}_{net} = 0, \quad \vec{\mathbf{p}}_i = \vec{\mathbf{p}}_f \tag{7-6}$$

By definition, an isolated, or closed, system is subject to no external interactions; thus *linear momentum is always conserved for an isolated system.* Remember that momentum is a vector quantity, so both the magnitude and the direction of the momentum at the beginning and end of the interaction must be the same.

Example 7.5

Adrift on a Raft

Diana is standing on a raft of mass 100.0 kg that is floating on a still lake. She decides to walk the length of the raft (Fig. 7.10). If Diana's mass is 55 kg and she walks with a velocity of 0.91 m/s *relative to the water,* how fast and in what direction does the raft move while Diana is walking? Assume the raft is stationary with respect to the shore before Diana starts walking.

Strategy Diana and the raft can be considered to be a single *isolated system:* as long as frictional forces on the raft due to the water and air are small enough to ignore, the net external force on the system is zero. Then the momentum of this system (raft + Diana) is conserved. We let the subscripts D stand for Diana and r for the raft and set the change in momentum of the system equal to zero.

continued on next page

Example 7.5 *continued*

Solution   To walk, Diana must exert a force on the raft: the static frictional force between her feet and the raft. This is an internal interaction within the isolated system, so it cannot change the total momentum of the system. Only something acting from outside the system could do that. As Diana walks in one direction, she acquires some momentum. The rest of the system (the raft) must acquire an equal and opposite momentum, because the momentum of the isolated system (Diana + raft) is conserved, which means that the change in momentum of the system is zero.

First we set the change in momentum of the system equal to zero:

$$\Delta\vec{p} = \vec{0} = \Delta\vec{p}_D + \Delta\vec{p}_r$$

or

$$\Delta\vec{p}_D = -\Delta\vec{p}_r$$

This means that the momentum changes of Diana and of the raft are equal and opposite. Since momentum is the product of mass and velocity and the masses of the raft and Diana do not change,

$$m_D \,\Delta\vec{v}_D = -m_r \,\Delta\vec{v}_r$$

Figure 7.10
Diana walking along a raft.

Solving for the change in velocity of the raft gives

$$\Delta\vec{v}_r = -\frac{m_D \,\Delta\vec{v}_D}{m_r}$$

Finally we substitute numerical values from the given information in the statement of the problem. Let Diana walk in the +x-direction.

$$\Delta\vec{v}_r = -\frac{55 \text{ kg} \times 0.91 \text{ m/s (in } +x\text{-direction)}}{100.0 \text{ kg}}$$

$$= 0.50 \text{ m/s in } -x\text{-direction}$$

The negative sign indicates that the raft moves in a direction opposite to Diana's movement to keep the momentum unchanged and thus conserved. Since the raft was originally stationary, this is the new velocity of the raft.

Discussion   In any momentum conservation problem there are two equivalent ways to proceed. In this example we set the momentum change of the system equal to zero. We could just as well write an equation that sets the initial total momentum equal to the final momentum of the system. The raft and Diana are initially at rest, so the initial momentum is zero:

$$\vec{0} = m_D\vec{v}_D + m_r\vec{v}_r$$

where $\vec{v}_D$ and $\vec{v}_r$ are the final velocities of Diana and the raft.

Practice Problem 7.5   Skaters pushing apart

Two skaters on in-line skates, Alexia and Bart, are initially at rest. They push apart and start moving in opposite directions. If Alexia's speed just after they push apart is 2.0 m/s and her mass is 85% of Bart's mass, how fast is Bart moving at that time?

When a bullet is fired from a rifle, the system of rifle plus bullet must conserve momentum. Suppose the rifle is at rest before the bullet is fired. The momentum of the system is zero. When the bullet is fired, part of the system's mass breaks away and travels in one direction with a certain momentum. The rifle, which is the remaining mass of the system, moves in the exact opposite direction such that the total momentum of the system is still zero. The rifle has a much larger mass than the bullet so it has a much smaller speed. The backward motion of the rifle is the *recoil* felt by anyone who has held a rifle against her shoulder and squeezed the trigger.

Jet engines and rockets operate by conservation of momentum. Hot combustion gases are forced out of nozzles at high speed by the engines. The increased backward momentum of the hot gases as they are expelled must be accompanied by an increased forward momentum of the engines.

Conceptual Example 7.6

Escape on Slippery Ice

A pilot parachutes from his disabled aircraft and lands on the frozen surface of a lake. There is no breeze blowing and the lake surface is too slippery to walk on. What can the pilot do to reach the shore?

Strategy and Solution Since the person in jeopardy is a pilot, he begins to think about how hot gases forced backward from a jet engine cause the plane to move forward. That gives him an idea: he bundles the parachute into a package and pushes it as hard as possible in a direction away from the nearest point of the shore. If the net external force on the system of pilot plus parachute is zero, then the total momentum of the system cannot change. The momentum of the parachute plus the momentum of the pilot must still equal zero. By conservation of momentum, the pilot begins sliding in the opposite direction and glides toward the shore.

Discussion If friction brings the pilot to rest before he reaches the shore, he can search his pockets and belt loops for other items to throw away. Once he reaches shore, he can tie one end of a rope to a tree and, holding onto the other end, venture back out onto the ice to retrieve any essential items. The rope provides him with an external force so he can get back to shore.

Conceptual Practice Problem 7.6   Bank robbery in the Old West

After tying the bank manager to his office chair, which has wheels, the James brothers begin emptying the bank safe. The manager faces away from his desk. His legs are securely tied, but he manages to free one arm, with which he retrieves a heavy gold nugget from the secret compartment below the armrest. How might the manager reach the telegraph on his desk to send a message for help? What might make this scheme fail?

Physics at Home

In case you and a friend ever end up stuck in the middle of the ice, practice the technique the pilot used to escape to the lakeshore. Bring a heavy medicine ball out to the middle of the ice rink and face each other with your skates aligned parallel. Toss the ball to your friend. What happens to you? What happens to your friend when he catches the ball? Can you both be "saved" by tossing the ball back and forth? (The same technique works using in-line skates.)

Example 7.7

A Broadside and Its Aftermath

The *Ocean Queen* has 14 guns mounted along the starboard side and the same number on the port side. The mass of the *Ocean Queen* is 8.5×10^4 kg. The cannonballs are each of mass 5.0 kg and leave the cannons with a horizontal velocity of 144 m/s. If all 14 cannons on the starboard side are fired in unison, with the firing taking place within a 0.50-s time interval, what is the average force exerted on the ship by the cannons during the firing?

Strategy We can find the momentum change of the cannonballs. Then, using the impulse-momentum relation, we can find the average force on the ship.

Given: $M_{ship} = 8.5 \times 10^4$ kg; $m_{ball} = 5.0$ kg; $v_{ball} = 144$ m/s; $\Delta t = 0.50$ s

To find: F_{av}

Solution The momentum of the cannonballs just after they are fired is

$$p = mv = (14 \text{ cannonballs} \times 5.0 \text{ kg/cannonball}) \times 144 \text{ m/s}$$

$$= 10,000 \text{ kg·m/s}$$

The cannonballs must exert an impulse of equal magnitude on the ship as they are fired:

$$F_{av} \times 0.50 \text{ s} = 1.0 \times 10^4 \text{ kg·m/s}$$

$$F_{av} = 2.0 \times 10^4 \text{ N}$$

The magnitude of the average force is 2.0×10^4 N—more than 2 tons. The direction is opposite to the momentum of the cannonballs as they leave the cannons.

Discussion If the ship were free to recoil, it would do so at a speed given by momentum conservation:

$$M_{ship} v_{ship} = 1.0 \times 10^4 \text{ kg·m/s}$$

$$v_{ship} = \frac{1.0 \times 10^4 \text{ kg·m/s}}{8.5 \times 10^4 \text{ kg}} = 0.12 \text{ m/s}$$

The ship is prevented from moving sideways at 0.12 m/s by the resistance of the ocean water in which it is partially immersed, but it is relatively free to rotate and reacts to the impulse felt by rocking to one side in the water. Unfortunately for the young pirate, his ship tilted enough that water poured

continued on next page

Example 7.7 *continued*

through the open cannon portholes on the port side, scuttling the ship, and forcing the pirates to appeal to their intended victims for rescue.

One reason that cannons were mounted on wheels and held loosely in place by ropes is that the ropes allowed the cannons to roll backward as they were fired. The impulse exerted on the ship is thereby spread over a longer time interval, reducing the average force. The total impulse on the ship is *not* reduced; if the ship were free to recoil, it would do so at the same speed either way. However, the water exerts large forces on the ship, so it does not recoil as if it were free. The impor-

tant quantity is the *average force* exerted on the deck of the ship, which determines how much the ship tips in the water.

Practice Problem 7.7   Recoil of a rifle

During an afternoon of target practice, you fire a Winchester .308 rifle of mass 3.8 kg. The bullets have a mass of 9.72 g and leave the rifle at a muzzle velocity of 860 m/s. If you are sloppy and fire a round when the butt of the rifle is not firmly up against your shoulder, at what speed does the rifle butt smash into your shoulder? (Ouch!)

7.5   CENTER OF MASS

We have seen that the momentum of an isolated system is conserved even though parts of the system may interact with other parts; internal interactions transfer momentum between parts of the system but do not change the total momentum of the system. We can define a point called the **center of mass** (CM) that serves as an average location of the system. In Section 7.6 we prove that the center of mass of an isolated system must move with constant velocity, regardless of how complicated the motions of parts of the system may be. Then we can treat the mass of the system as if it were all concentrated at the center of mass, like a point particle. The center of mass of an object is not necessarily located within the object; for some objects, such as a boomerang, the center of mass is located outside of the object itself (Fig. 7.11a).

Figure 7.11 (a) The center of mass of a boomerang is a point outside of the boomerang. (b) The path followed by the center of mass when a hammer is tossed through the air. (c) Ben Challenger's center of mass actually passes *beneath* the bar as his body passes over the bar.

What if a system is not isolated, but has external interactions? Again imagine all of the mass of the system concentrated into a single point particle located at the center of mass. The motion of this fictitious point particle is determined by Newton's second law, where the net force is the sum of all of the external forces acting on *any part* of the system. In the case of a complex system composed of many parts interacting with each other, the motion of the center of mass is considerably simpler than the motion of an arbitrary particle of the system (Fig. 7.11b,c).

For a system composed of two particles, the center of mass lies somewhere on a line between the two particles. In Fig. 7.12, particles of masses m_1 and m_2 are located at positions x_1 and x_2, respectively. We define the location of the center of mass for these two particles as

$$x_{CM} = \frac{m_1 x_1 + m_2 x_2}{m_1 + m_2} \qquad (7\text{-}7)$$

The center of mass is a *weighted average* of the positions of the two particles. Here we use the word *weighted* in its statistical sense. The position of a particle with more mass counts more—carries more *statistical* weight—than does the position of a particle with a smaller mass. We can rewrite Eq. (7-7) as a weighted average:

$$x_{CM} = \frac{m_1}{M} x_1 + \frac{m_2}{M} x_2 \qquad (7\text{-}8)$$

Here $M = m_1 + m_2$ represents the total mass of the system. The statistical weight used for the location of each particle is the mass of that particle as a fraction of the total mass of the system.

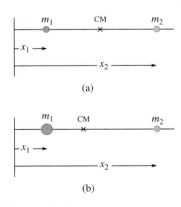

Figure 7.12 (a) Two particles of equal mass located at positions x_1 and x_2 from the origin. The center of mass is midway between the two. (b) Two particles of unequal mass. The center of mass is closer to the more massive particle.

Suppose masses m_1 and m_2 are equal. Then we expect the center of mass to be located midway between the two particles (Fig. 7.12a). If $m_1 = 2m_2$, as in Fig. 7.12b, then the center of mass is closer to the particle of mass m_1. Figure 7.12b shows that, in this case, the center of mass is twice as far from m_2 as from m_1.

For a system of N particles, at arbitrary locations in three-dimensional space, the definition of the center of mass is a generalization of Eq. (7-7).

Definition of center of mass

Vector form:
$$\vec{\mathbf{r}}_{CM} = \frac{\Sigma m_i \vec{\mathbf{r}}_i}{M}$$
(7-9)

Component form:
$$x_{CM} = \frac{\Sigma m_i x_i}{M} \qquad y_{CM} = \frac{\Sigma m_i y_i}{M} \qquad z_{CM} = \frac{\Sigma m_i z_i}{M}$$

where $i = 1, 2, 3, \ldots, N$ and $M = \Sigma m_i$

The symbol Σ stands for *sum*. The shorthand notation $\Sigma m_i x_i$ is interpreted as

$$\Sigma m_i x_i = m_1 x_1 + m_2 x_2 + \cdots + m_N x_N$$

For particles in two-dimensional space, we use only two of these equations for the x-y plane and find the x- and y-components of the center of mass.

Example 7.8

Center of Mass of a Binary Star System

The two stars in a binary star system are positioned as shown in Fig. 7.13a. Due to the gravitational interaction between the two, the stars each move in a circular orbit around their center of mass. One star has a mass of 15.0×10^{30} kg; its center is located at $x = 1.0$ AU and $y = 5.0$ AU. The other has a mass of 3.0×10^{30} kg; its center is at $x = 4.0$ AU and $y = 2.0$ AU. Find the CM of the system composed of the two stars. (AU stands for *astronomical unit*. 1 AU = the average distance between the Earth and the Sun = 1.5×10^8 km.)

Strategy   We treat the stars as point particles located at their centers. Since we are given x- and y-coordinates, the easiest way to proceed is to find the x- and y-coordinates of the CM. There is no particular advantage here in finding the position vector of the CM in terms of its length and direction.

Given: $m_1 = 15.0 \times 10^{30}$ kg $x_1 = 1.0$ AU $y_1 = 5.0$ AU
$\qquad\quad m_2 = 3.0 \times 10^{30}$ kg $x_2 = 4.0$ AU $y_2 = 2.0$ AU

To find: x_{CM}; y_{CM}

Solution   The total mass of the system is the sum of the individual masses:

$$M = m_1 + m_2 = 15.0 \times 10^{30} \text{ kg} + 3.0 \times 10^{30} \text{ kg} = 18.0 \times 10^{30} \text{ kg}$$

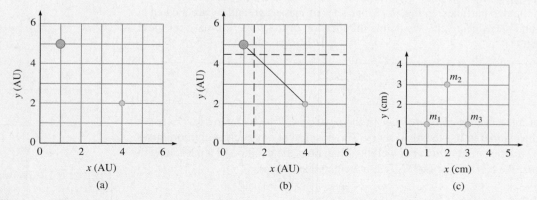

Figure 7.13
(a) Positions of two stars; (b) finding the CM for the system of two stars; and (c) three spheres located at x, y positions (1.0 cm, 1.0 cm), (2.0 cm, 3.0 cm), and (3.0 cm, 1.0 cm).

continued on next page

Example 7.8 *continued*

For the *x*-position, we find

$$x_{CM} = \frac{m_1}{M}x_1 + \frac{m_2}{M}x_2$$

$$= \frac{15.0 \times 10^{30}\,\text{kg}}{18.0 \times 10^{30}\,\text{kg}} \times 1.0\,\text{AU} + \frac{3.0 \times 10^{30}\,\text{kg}}{18.0 \times 10^{30}\,\text{kg}} \times 4.0\,\text{AU} = 1.5\,\text{AU}$$

and for the *y*-position, we find

$$y_{CM} = \frac{m_1}{M}y_1 + \frac{m_2}{M}y_2$$

$$= \frac{15.0}{18.0} \times 5.0\,\text{AU} + \frac{3.0}{18.0} \times 2.0\,\text{AU} = 4.5\,\text{AU}$$

Discussion   In Fig. 7.13b we mark the position of the center of mass. As we expect for the case of two particles, it is located closer to the larger mass and on a line connecting the two. Once the CM position is found in a problem, check to be sure its location is reasonable. Suppose we had made an error in this example and found the CM to be at $x = 1.5$ AU and $y = 1.7$ AU. This is not a reasonable location for the CM since it is not along the line connecting the two and is closer to the less massive star; we then would go back to look for the error.

Practice Problem 7.8   Three balls with unequal masses

Three spherical objects are shown in Fig. 7.13c. Their masses are $m_1 = m_3 = 1.0$ kg and $m_2 = 4.0$ kg. Find the location of the center of mass for the three objects.

Most objects we deal with in real life are not composed of a small set of point particles or spherically symmetric objects. In Example 7.8 we used the location of the center of each star to find the center of mass. Due to spherical symmetry, the center of mass of either star (by itself) is at its geometric center. The same technique can be applied to other shapes with symmetry. A standard 2 by 4, which is an 8-ft-long uniform piece of wood 1.5" deep × 3.5" high, has its center of mass at its geometric center: at half the length, half the depth, and half the height of the board. This assumes the object is uniform in construction. A "loaded" die does *not* have its CM at its geometric center, since a small metal plug has been inserted near one face. The definition of the CM [Eq. (7-9)] still holds as long as (x_i, y_i, z_i) are the coordinates of the center of mass of a part of the system with mass m_i.

7.6   MOTION OF THE CENTER OF MASS

Now that we know how to find the position of the CM of a system, we turn our attention to the motion of the CM. How is the velocity of the CM related to the velocities of the various parts of the system?

During a short time interval Δt, the displacement of the *i*th particle is

$$\Delta \vec{\mathbf{r}}_i = \vec{\mathbf{v}}_i \, \Delta t$$

and the displacement of the center of mass is

$$\Delta \vec{\mathbf{r}}_{CM} = \vec{\mathbf{v}}_{CM} \, \Delta t$$

From the definition of the center of mass [Eq. (7-9)], the displacements must be related as follows:

$$\Delta \vec{\mathbf{r}}_{CM} = \frac{\Sigma m_i \, \Delta \vec{\mathbf{r}}_i}{M}$$

or

$$M \, \Delta \vec{\mathbf{r}}_{CM} = m_1 \, \Delta \vec{\mathbf{r}}_1 + m_2 \, \Delta \vec{\mathbf{r}}_2 + \cdots + m_N \, \Delta \vec{\mathbf{r}}_N$$

We can replace each displacement by the product of a velocity and the time interval Δt.

$$M \vec{\mathbf{v}}_{CM} \, \Delta t = m_1 \vec{\mathbf{v}}_1 \, \Delta t + m_2 \vec{\mathbf{v}}_2 \, \Delta t + \cdots + m_N \vec{\mathbf{v}}_N \, \Delta t$$

The time interval Δt is the same in each term. Dividing both sides by Δt,

$$M \vec{\mathbf{v}}_{CM} = m_1 \vec{\mathbf{v}}_1 + m_2 \vec{\mathbf{v}}_2 + \cdots + m_N \vec{\mathbf{v}}_N \qquad (7\text{-}10)$$

The total momentum of a system is defined as the sum of the individual momenta of the particles comprising the system:

$$\vec{\mathbf{p}} = \Sigma \vec{\mathbf{p}}_i = \vec{\mathbf{p}}_1 + \vec{\mathbf{p}}_2 + \cdots + \vec{\mathbf{p}}_N$$

Equation (7-10) therefore says that the total momentum of a system is equal to the total mass of the system times the velocity of the center of mass:

$$\vec{p} = M\vec{v}_{CM} \tag{7-11}$$

For two-dimensional motion it is usually easiest to work with components of momenta in the x- and y-directions.

$$p_x = Mv_{CM,x} \quad \text{and} \quad p_y = Mv_{CM,y} \tag{7-12}$$

where p_x and p_y are the x- and y-components of the total momentum of the system.

In Section 7.4, we showed that, for an isolated system, the total linear momentum is conserved. In such a system, Eq. (7-11) implies that the center of mass must move with constant velocity regardless of the motions of the individual particles. On the other hand, what if the system is not isolated? If a net external force acts on a system, the center of mass does not move with constant velocity. Instead, it moves as if all the mass were concentrated there into a fictitious point particle with all the external forces acting on that point. The motion of the center of mass obeys the following statement of Newton's second law:

$$\Sigma\vec{F}_{ext} = M\vec{a}_{CM} \tag{7-13}$$

where M is the total mass of the system, $\Sigma\vec{F}_{ext}$ is the net external force, and $\vec{a}_{CM}$ is the acceleration of the center of mass. [Eq. (7-13) is proved in Problem 29.]

Example 7.9

An Exploding Rocket

A model rocket is fired from the ground in a parabolic trajectory. At the top of the trajectory, a horizontal distance of 260 m from the launch point, an explosion occurs within the rocket, breaking it into two fragments. One fragment, having one-third of the mass of the rocket, falls straight down to Earth as if it had been dropped from rest at that point. At what horizontal distance from the launch point does the other fragment land? Ignore air resistance.

Strategy   There are two different strategies that can be used to solve this problem.

Strategy 1: We apply conservation of momentum to the explosion. The momentum of the rocket *just before* the explosion is equal to the total momentum of the two fragments *just after* the explosion. Why can momentum conservation be assumed here? There is an external force—gravity—acting on the system. External forces change momentum. However, the explosion takes place in a *very short time interval*. From the impulse-momentum theorem [Eq. (7-2)], the momentum change of the system is the force of gravity multiplied by the time interval. As long as the time interval considered is sufficiently short, the momentum change of the system can be ignored.

Strategy 2: The explosion is caused by an *internal* interaction between two parts of the rocket. The motion of the center of mass of the system is unaffected by internal interactions, so it continues in the same parabolic path. The two pieces of the rocket land simultaneously; at that same instant, the center of mass also reaches the ground.

Solution 1   First we make a sketch of the situation (Fig. 7.14). At the top of the trajectory, where the explosion occurs, $v_y = 0$; the rocket is moving in the x-direction. The initial momentum just before the explosion is entirely in the x-direction. If M is the mass of the rocket, then

$$p_{ix} = Mv_{ix}$$

Just after the explosion, one-third of the mass of the rocket is at rest; it then drops straight down under the influence of the gravitational force. This piece has zero momentum just after the explosion. To conserve momentum, the other two-thirds of the rocket must have a momentum equal to the momentum just before the explosion.

$$p_{ix} = p_{1x} + p_{2x}$$

$$Mv_{ix} = 0 + (\tfrac{2}{3}M)v_{2x}$$

Solving for v_{2x}, we find

$$v_{2x} = \tfrac{3}{2}v_{ix}$$

The y-component of momentum must also be conserved:

$$p_{iy} = p_{1y} + p_{2y}$$

We know that both p_{iy} and p_{1y} are zero; therefore p_{2y} is zero as well. Just after the explosion both parts of the rocket have zero vertical components of velocity. Then both parts take the same time to fall to the ground as if the rocket had not exploded. With a horizontal velocity larger by a factor of 3/2, the second piece of the rocket travels a horizontal distance from the explosion a factor of 3/2 larger than 260 m (see Fig. 7.14). The distance from the launch point where this piece lands is

$$\Delta x = 260 \text{ m} + \tfrac{3}{2} \times 260 \text{ m} = 650 \text{ m}$$

continued on next page

Example 7.9 continued

Solution 2   The piece with mass $\frac{1}{3}M$ falls straight down and lands 260 m from the launch point. After the explosion, the CM continues to travel just as the rocket itself would have done if it had not broken apart. From the symmetry of the parabola, the CM touches the ground at a distance of 2×260 m = 520 m from the launch point. Since we know the location of the CM and that of one of the pieces, we can find where the second piece lands:

$$Mx_{CM} = \tfrac{1}{3}Mx_1 + \tfrac{2}{3}Mx_2$$

After canceling the common factor of M,

$$x_{CM} = \tfrac{1}{3}x_1 + \tfrac{2}{3}x_2$$

Solving for x_2 yields

$$x_2 = \frac{3x_{CM} - x_1}{2} = \frac{3 \times 520 \text{ m} - 260 \text{ m}}{2} = 650 \text{ m}$$

which is the same answer that we found in Solution 1.

Discussion   The insight that the motion of the CM is unaffected by internal interactions can be of enormous help. Note,

however, that Solution 2 would not be so simple if the two fragments did not land simultaneously. As soon as one fragment (fragment 1) hits the ground, the external force on the system is no longer due exclusively to gravity, so the CM doesn't continue to follow the same parabolic path. The normal and frictional forces acting on fragment 1 affect its subsequent motion and the subsequent motion of the CM even though the motion of fragment 2 is unaffected.

Practice Problem 7.9   Diana and the raft revisited

In Example 7.5, Diana (mass 55 kg) walks at 0.91 m/s (relative to the water) on a raft of mass 100.0 kg. The raft moves in the opposite direction at 0.50 m/s. Suppose it takes her 3.0 s to walk from one end of the raft to the other. (a) How far does Diana walk (relative to the water)? (b) How far does the raft move while Diana is walking? (c) How far does the center of mass of Diana and the raft move during the 3.0 s?

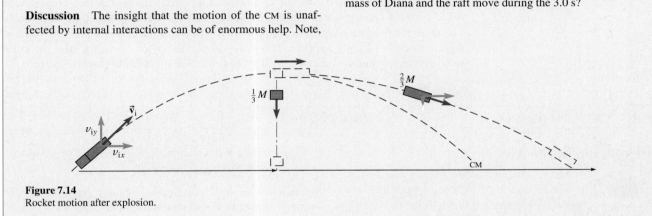

Figure 7.14
Rocket motion after explosion.

7.7   COLLISIONS IN ONE DIMENSION

What is a collision? In the macroscopic world, a moving body bumps into another body that may be at rest or in motion. The two bodies exert forces on each other while they are in contact; as a result, their velocities change. In the microscopic and submicroscopic world, our picture of a collision is different. When atoms collide, they don't "touch" each other: the atom doesn't have a definite spatial boundary, so there are no surfaces to make "contact." However, the collision model is still useful for atoms and subatomic particles whenever there is an interaction in which the forces are strong over a short time interval, so that there is a clear "before collision" and a clear "after collision."

The forces exerted during a collision between macroscopic objects generally vary with time in a way that is hard to predict. What we *are* able to do is look at the motion of the bodies before and after the interaction. Thanks to the law of conservation of linear momentum, we can make predictions of the motion after an interaction by knowing the conditions of motion before the interaction. The same techniques that are used for collisions in the macroscopic world (car crashes, billiard ball collisions, baseball bats hitting balls) are also used in collisions in the microscopic world (gas molecules colliding with each other and with surfaces, radioactive decays of nuclei). First, we study collisions limited to motion along a line; later, we consider collisions limited to motion in a plane (in two dimensions).

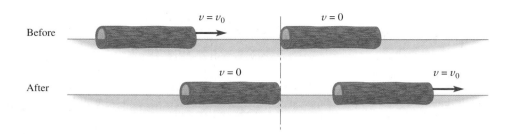

Figure 7.15 Colliding logs shown before and after a collision.

Suppose we observe a log traveling along a flume at speed v_0 toward a second log that is floating at rest in the water. The masses of the two logs are equal. When the first log hits the second, what happens?

Based on momentum considerations *alone*, there are many possible outcomes. One possibility is that the first log stops moving and the second log moves off with the same velocity that the first one had to begin with (Fig. 7.15). This possibility satisfies conservation of momentum because the total momentum is the same before and after.

Another possibility is that the two logs stick together, moving away together. With what speed do they move after the collision? If the momentum is to be the same with twice as much mass moving, the speed must be half the initial speed of the first object. There are many other possibilities. Conservation of momentum doesn't tell us which of these outcomes actually happens, but if we know one object's velocity after the collision, we can use momentum conservation to determine the other object's velocity.

Example 7.10

Collision in the Air

A krypton atom (mass 83.9 u) moving with a velocity of 0.80 km/s to the right and a water molecule (mass 18.0 u) moving with a velocity of 0.40 km/s to the left collide head-on. The water molecule has a velocity of 0.60 km/s to the right after the collision. What is the velocity of the krypton atom after the collision? (The symbol "u" stands for the atomic mass unit.)

Strategy Since we know both initial velocities and one of the final velocities, we can find the second final velocity by applying momentum conservation. Let the subscript "1" refer to the krypton atom and let the subscript "2" refer to the water

molecule. Let the x-axis point to the right. Figure 7.16 shows before and after pictures of the collision.

Solution Momentum conservation requires that the final momentum be equal to the initial momentum:

$$\vec{\mathbf{p}}_{1f} + \vec{\mathbf{p}}_{2f} = \vec{\mathbf{p}}_{1i} + \vec{\mathbf{p}}_{2i}$$

Now we substitute $\vec{\mathbf{p}} = m\vec{\mathbf{v}}$ for each momentum. It is easiest to work in terms of components. For simplicity we drop the "x" subscripts, remembering that all quantities refer to x-components:

$$m_1 v_{1f} + m_2 v_{2f} = m_1 v_{1i} + m_2 v_{2i}$$

Since $m_1/m_2 = 83.9/18.0 = 4.661$, we can substitute $m_1 = 4.661 m_2$:

$$4.661 m_2 v_{1f} + m_2 v_{2f} = 4.661 m_2 v_{1i} + m_2 v_{2i}$$

The common factor m_2 cancels out. Solving for v_{1f},

$$
\begin{aligned}
v_{1f} &= \frac{4.661 v_{1i} + v_{2i} - v_{2f}}{4.661} \\
&= \frac{4.661 \times 0.80 \text{ km/s} + (-0.40 \text{ km/s}) - 0.60 \text{ km/s}}{4.661} = 0.59 \text{ Km/s}
\end{aligned}
$$

After the collision, the krypton atom moves to the right with a speed of 0.59 Km/s.

Discussion To check this result, we calculate the total momentum (x-component) before and after the collision:

$$m_1 v_{1i} + m_2 v_{2i} = (83.9 \text{ u})(0.80 \text{ km/s}) + (18.0 \text{ u})(-0.40 \text{ km/s}) = 60 \text{ u·km/s}$$

$$m_1 v_{1f} + m_2 v_{2f} = (83.9 \text{ u})(0.59 \text{ km/s}) + (18.0 \text{ u})(0.60 \text{ km/s}) = 60 \text{ u·km/s}$$

continued on next page

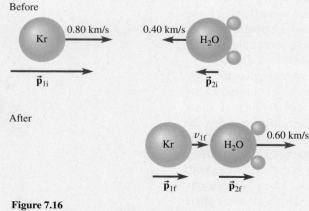

Figure 7.16
Before and after snapshots of a collision.

Before

After

Example 7.10 *continued*

Momentum is conserved. There is no need to convert u to kg since we only need to compare these two values.

⚠ If we made the mistake of thinking of momentum as a scalar, we would get the wrong answer. The sum of the *magnitudes* of the momenta before the collision is *not* equal to the sum of the *magnitudes* of the momenta after the collision. Conservation of energy is perhaps easier to understand intuitively since energy is a scalar quantity. Converting kinetic energy to potential energy is analogous to moving money from a checking account to a savings account; the total amount of money is the same before and after. This sort of analogy does *not* work with momentum!

Practice Problem 7.10   Head-on collision

A 5.0-kg ball is at rest when it is struck head-on by a 2.0-kg ball moving along a track at 10.0 m/s. If the 2.0-kg ball is at rest after the collision, what is the speed of the 5.0-kg ball after the collision?

Elastic and Inelastic Collisions

Collisions are often classified based on what happens to the kinetic energy of the colliding objects. A ball dropped from a height h does not rebound to the same height. The kinetic energy of the ball just after the collision with the floor or ground is less than it was just before the collision; the amount of the kinetic energy decrease depends on the makeup of the ball and the ground. A racquetball dropped onto a hard wooden floor may rebound nearly to its original height, but a watermelon rebounds very little or not at all. Why do some objects rebound much better than others?

Imagine a racquetball colliding with the floor (Fig. 7.17). The bottom of the ball is flattened. What makes the ball rebound from the floor? The forces holding the ball together are like springs; the kinetic energy of the ball has been transformed largely into potential energy in these springs. When the ball bounces back up, this energy is transformed back into kinetic energy. Then why does the watermelon not rebound? The watermelon, too, is deformed when it collides with the floor, but this deformation is not reversible. The kinetic energy of the watermelon is changed mostly into thermal energy rather than into potential energy.

A collision in which the total kinetic energy is the same before and after is called **elastic**. Not all collisions are elastic; when the final kinetic energy is less than the initial kinetic energy, the collision is said to be **inelastic**. A stick-together collision is a **perfectly inelastic collision**. The decrease of kinetic energy in a perfectly inelastic collision is as large as *possible* (consistent with the conservation of momentum). Now that we have defined elastic and inelastic collisions, we can put together a problem-solving strategy for collision problems.

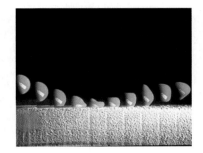

Figure 7.17 Deformation of a racquetball during its collision with the floor.

Problem-Solving Strategy for Collisions Involving Two Objects

1. Draw before and after diagrams of the collision.

2. Collect and organize information on the masses and velocities of the two objects before and after the collision. Express the velocities in component form (with correct algebraic signs).

3. Set the sum of the momenta of the two before the collision equal to the sum of the momenta after the collision. Write one equation for each nonzero component:

$$m_1 v_{1ix} + m_2 v_{2ix} = m_1 v_{1fx} + m_2 v_{2fx}$$
$$m_1 v_{1iy} + m_2 v_{2iy} = m_1 v_{1fy} + m_2 v_{2fy}$$

4. If the collision is known to be perfectly inelastic, set the final velocities equal:

$$v_{1fx} = v_{2fx}$$
$$v_{1fy} = v_{2fy}$$

5. If the collision is known to be perfectly elastic, then set the final kinetic energy equal to the initial kinetic energy:

$$\tfrac{1}{2}m_1 v_{1i}^2 + \tfrac{1}{2}m_2 v_{2i}^2 = \tfrac{1}{2}m_1 v_{1f}^2 + \tfrac{1}{2}m_2 v_{2f}^2$$

6. Solve for the unknown quantities.

There is no *conservation law* for kinetic energy by itself. Total energy is always conserved, but that does not preclude some kinetic energy being transformed into another type of energy. The elastic collision is just a special kind of collision in which no kinetic energy is changed into other forms of energy.

In Practice Problem 7.11, the relative speed of the cars was the same before and after the collision. It can be proved (see Problem 43) that for *any* elastic collision between two objects, the relative speed is the same before and after the collision. This

Example 7.11

Collision at the Highway Entry Ramp

At a Route 128 highway on-ramp, a car of mass 1.50×10^3 kg is stopped at a stop sign, waiting for a break in traffic before merging with the cars on the highway. Another car of mass 2.00×10^3 kg comes up from behind, hits the stopped car, and locks bumpers with it. How fast was the moving car going just before the collision if the two cars move together at 10.0 m/s just after the collision?

Strategy   The problem says that, after the collision, the two cars are locked together by their bumpers. The two separate cars then behave as one single, larger mass vehicle. We treat the cars as separate bodies before the collision and as a single object after the collision. The collision is perfectly inelastic since the cars "stick together." Let "1" refer to the car stopped at the stop sign and "2" refer to the other car. Fig. 7.18 shows a before and after diagram for the collision.

Given: $m_1 = 1.50 \times 10^3$ kg, $m_2 = 2.00 \times 10^3$ kg; before the collision, $v_1 = 0$; after the collision, $v_f = 10.0$ m/s (both cars have the same velocity)

To find: v_2 (speed of car 2 before the collision)

Solution   The momentum of the pair of locked cars after the collision is equal to the sum of the momenta of the two cars before the collision. All motions are in one direction so we need not include subscripts; let all p's and v's refer to x-components. From conservation of momentum,

$$p_f = p_i$$

The initial momentum is the sum of the momenta of the two cars.

$$p_f = p_1 + p_2$$

Since car 1 is initially at rest, $p_1 = 0$. Then

$$p_f = p_2$$

or

$$(m_1 + m_2)v_f = m_2 v_2$$

Solving for v_2,

$$v_2 = \frac{(m_1 + m_2)v_f}{m_2}$$

$$v_2 = \frac{(1500 \text{ kg} + 2000 \text{ kg}) \times 10.0 \text{ m/s}}{2000 \text{ kg}} = 17.5 \text{ m/s (or about 39 mph)}$$

Discussion   The entry ramp speed limit is 20 mph. By measuring the length of the skid marks from the stop sign, the investigating officer is able to determine that the stopped car was pushed onto the highway at a velocity of 10 m/s. Then the investigator calculates the speed of the other car before adding speeding to the charges against the offending driver.

Here the net external force acting on the cars wasn't zero because the friction from the road acts on the cars during the collision. We still use conservation of momentum because during the short time interval of the collision, friction doesn't have time to change the system's momentum significantly. The change of momentum of the system is equal to the external impulse, which is the product of the net external force with the time interval. For a short time interval, the impulse is small.

Practice Problem 7.11   Elastic collision between the cars

Instead of locking bumpers, suppose the two cars collided *elastically*. With the same initial conditions ($v_{1i} = 0$ and $v_{2i} = 17.5$ m/s), an elastic collision would push the stopped car onto the highway at 20.0 m/s (as you will confirm in this problem). (a) Given that $v_{1f} = 20.0$ m/s, calculate the velocity of car 2 just after the collision (v_{2f}) using momentum conservation. (b) Find the total kinetic energy of the two cars both before and after the collision and verify that the collision is elastic. (c) Find the relative speed of the two cars before and after the collision.

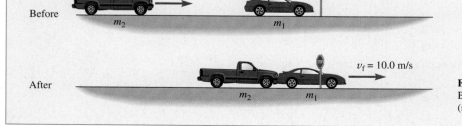

Before v_2 $v_1 = 0$
m_2 m_1

After $v_f = 10.0$ m/s
m_2 m_1

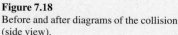
Figure 7.18
Before and after diagrams of the collision (side view).

fact is most useful in one-dimensional collisions; in two-dimensional collisions the *direction* of the relative velocity changes due to the collision. Since the relative velocity is in the opposite direction after the collision—first the objects move together, then they move apart—we can write:

$$v_{2ix} - v_{1ix} = -(v_{2fx} - v_{1fx}) \tag{7-14}$$

for a one-dimensional elastic collision along the *x*-axis. For a one-dimensional elastic collision, Eq. (7-14) is a useful alternative to setting the final kinetic energy equal to the initial kinetic energy.

7.8   COLLISIONS IN TWO DIMENSIONS

Most collisions are not limited to motion in one dimension in the absence of a track or other device to constrain motion to a single line. In many collisions the initial velocities of the objects are neither parallel nor antiparallel (Fig. 7.19a). Even if the initial velocities are parallel or antiparallel, if the centers of mass do not move along a single straight line, then the final velocities do not have to be parallel or antiparallel to the initial velocities (Figs. 7.19b and c).

In a two-dimensional collision, we use the same techniques we used for one-dimensional collisions, as long as we remember that momentum is a vector. To apply conservation of momentum, it is usually easiest to work with *x*- and *y*-components.

See the Problem-Solving Strategy on p. 237.

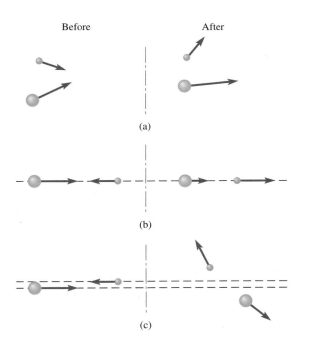

Before   After

(a)

(b)

(c)

Figure 7.19 (a) A two-dimensional collision, (b) a head-on collision, and (c) a glancing collision; even though the initial velocities are in the +x- and –x-directions, the final velocities do not lie along the x-axis.

Example 7.12

Colliding Pucks on an Air Table

A small puck (mass $m_1 = 0.10$ kg) is sliding to the right with an initial speed of 8.0 m/s on an air table (Fig. 7.20a). An air table has many tiny holes through which air is blown; the resulting air cushion allows objects to slide with very little friction. The puck collides with a larger puck (mass $m_2 = 0.40$ kg), which is initially at rest. Figure 7.20b shows the outcome of the collision: the pucks move off at angles $\phi_1 = 60.0°$ above and $\phi_2 = 30.0°$

continued on next page

Example 7.12 *continued*

below the initial direction of motion of the small puck. (a) What are the final speeds of the pucks? (b) Is this an elastic collision or an inelastic collision? (c) If inelastic, what fraction of the initial kinetic energy is converted to other forms of energy in the collision?

Strategy The system of two pucks is an isolated system because the net external force is zero. Therefore we can apply conservation of momentum. Since motions in two dimensions are involved, we treat the horizontal and vertical components of momentum separately.

Figure 7.20 shows the pucks before and after the collision. Now we collect information on the known quantities, writing velocities in component form.

Masses: $m_1 = 0.10$ kg; $m_2 = 0.40$ kg

Before collision: $v_{1ix} = 8.0$ m/s; $v_{1iy} = v_{2ix} = v_{2iy} = 0$

After collision: $v_{1fx} = v_{1f} \cos \phi_1$; $v_{1fy} = v_{1f} \sin \phi_1$;

$v_{2fx} = v_{2f} \cos \phi_2$; $v_{2fy} = -v_{2f} \sin \phi_2$

($\phi_1 = 60.0°$ and $\phi_2 = 30.0°$)

To find: v_{1f} and v_{2f}; total kinetic energy before and after the collision

Solution (a) Working with components means that we set the total x-component of momentum before the collision equal to the total x-component of momentum after the collision. We treat the y-components in the same way. The initial momentum is in the x-direction only. Thus, the total momentum y-component after the collision must be zero.

First we set the x-component of the total momentum after the collision equal to the x-component of the total momentum before the collision:

$$P_{fx} = P_{ix}$$

$$p_{1fx} + p_{2fx} = p_{1ix} + p_{2ix}$$

Each momentum component is now rewritten using $p_x = mv_x$:

$$m_1 v_{1f} \cos \phi_1 + m_2 v_{2f} \cos \phi_2 = m_1 v_{1i} + 0$$

Since $m_2 = 4m_1$,

$$m_1 v_{1f} \cos 60.0° + 4m_1 v_{2f} \cos 30.0° = m_1 v_{1ix}$$

After canceling the common factor m_1 and substituting numerical values for $\cos 60.0°$ and $\cos 30.0°$, this reduces to

$$0.500v_{1f} + 3.46v_{2f} = 8.0 \text{ m/s} \qquad (1)$$

For conservation of the y-component of the momentum:

$$P_{fy} = P_{iy}$$

$$p_{1fy} + p_{2fy} = p_{1iy} + p_{2iy} = 0$$

The y-component of $\vec{\mathbf{p}}_{2f}$ is negative because the y-component of $\vec{\mathbf{v}}_{2f}$ is negative.

$$m_1 v_{1f} \sin \phi_1 + (-4m_1 v_{2f} \sin \phi_2) = 0$$

$$v_{1f} \sin 60.0° - 4v_{2f} \sin 30.0° = 0$$

We solve for v_{2f} in terms of v_{1f}:

$$v_{2f} = \frac{\sin 60.0°}{4 \sin 30.0°} v_{1f} = 0.433v_{1f} \qquad (2)$$

Equations (1) and (2) contain two unknowns. To eliminate one unknown, we substitute $0.433v_{1f}$ for v_{2f} in Eq. (1):

$$0.500v_{1f} + 3.46 (0.433v_{1f}) = 2.00v_{1f} = 8.0 \text{ m/s}$$

Solving this equation gives the value of v_{1f}:

$$v_{1f} = 4.0 \text{ m/s}$$

Then by substitution into Eq. (2), we find the value of v_{2f}:

$$v_{2f} = 0.433v_{1f} = 1.7 \text{ m/s}$$

(b) Now that we have the final velocities, we can compare the initial and final kinetic energies.

$$K_i = \tfrac{1}{2}m_1 v_{1i}^2$$

$$K_i = \tfrac{1}{2}(0.10 \text{ kg}) \times (8.0 \text{ m/s})^2 = 3.2 \text{ J}$$

and

$$K_f = \tfrac{1}{2}m_1 v_{1f}^2 + \tfrac{1}{2}m_2 v_{2f}^2$$

$$= \tfrac{1}{2}(0.10 \text{ kg}) \times (4.0 \text{ m/s})^2 + \tfrac{1}{2}(0.40 \text{ kg}) \times (1.7 \text{ m/s})^2$$

$$K_f = 0.80 \text{ J} + 0.58 \text{ J} = 1.38 \text{ J}$$

The final kinetic energy is less than the initial kinetic energy so the collision is inelastic.

(c) The amount of kinetic energy converted to other forms of energy is

$$3.2 \text{ J} - 1.38 \text{ J} = 1.82 \text{ J}$$

We divide by the initial kinetic energy to find the fraction of the initial kinetic energy converted to other forms:

$$\frac{1.82 \text{ J}}{3.2 \text{ J}} = 0.57$$

continued on next page

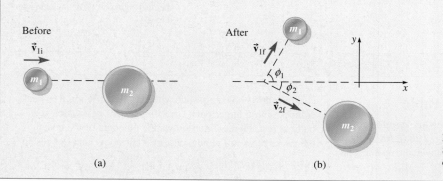

Figure 7.20
Snapshots in time, (a) before and (b) after a collision.

Example 7.12 *continued*

Less than half of the kinetic energy of the incident puck therefore survives the collision as the kinetic energies of the two pucks.

Discussion   Although a two-dimensional collision problem tends to require more complicated algebra than a one-dimensional problem, the physical principles are the same. As long as the net external force on the system is zero (or negligibly small), the total vector momentum must be conserved.

Practice Problem 7.12   Colliding balls

A ball of mass m_1 moves at speed v_0 along the $+x$-axis toward a second ball of mass m_2, which is initially at rest. The second ball has five times the mass of the first ball. After the collision between these two objects, m_1 moves along the $+y$-axis at a speed v_1, and m_2 moves at a speed $v_2 = \frac{1}{4}v_0$ at an angle of $36.9°$ from the x-axis. Find v_1 in terms of v_0.

Example 7.13

Eric at the Pool Table

Playing a game of pool, Eric is trying to decide whether to attempt a shot to sink the 4-ball in the pocket at corner B without *scratching* (sinking the cue ball "C" in corner A). The lines from the 4-ball to the two corner pockets make a right angle (Fig. 7.21a). The collision of the balls is nearly elastic. Assume Eric is an amateur player and does not know how to do fancy things, like putting sidespin on a ball. Should he attempt the shot?

Strategy   We assume a perfectly elastic collision between the balls. They have the same mass. The cue ball moves with an initial velocity $\vec{v}_0$ and strikes the 4-ball, which is initially at rest. The 4-ball falls in pocket B if its velocity after the collision, $\vec{v}_4$, points toward the pocket. Assuming it does, we use conservation of momentum and kinetic energy to find the direction of the cue ball velocity, $\vec{v}_c$, after the collision.

continued on next page

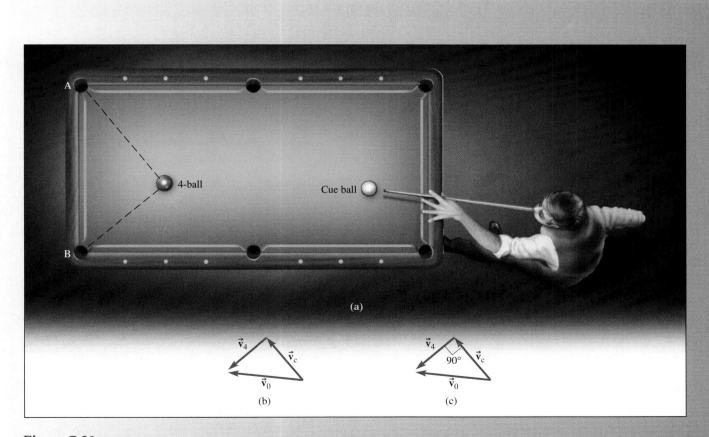

Figure 7.21   (a) Should Eric try to sink the 4-ball? (b) Graphical addition of velocity vectors as required by the conservation of momentum. (c) Since $v_0^2 = v_c^2 + v_4^2$, the three velocities form a *right* triangle.

Example 7.13 *continued*

Solution   Conservation of momentum requires that

$$m\vec{v}_0 = m\vec{v}_c + m\vec{v}_4$$

or

$$\vec{v}_0 = \vec{v}_c + \vec{v}_4$$

This vector addition is shown graphically in Fig. 7.21b. Since the collision is elastic, the total kinetic energy doesn't change:

$$\tfrac{1}{2}mv_0^2 = \tfrac{1}{2}mv_c^2 + \tfrac{1}{2}mv_4^2$$

or

$$v_0^2 = v_c^2 + v_4^2$$

Since v_0, v_c, and v_4 are the sides of a triangle, this is a statement of the Pythagorean theorem—the triangle must be a *right* triangle with v_0 as the hypotenuse (Fig. 7.21c). Therefore, the velocities of the 4-ball and the cue ball after the collision are perpendicular to each other.

If Eric sinks the 4-ball, the cue ball falls into pocket A. He shouldn't attempt this shot until he learns how to put some spin on the ball.

Discussion   Note that we did not resolve the velocities into x- and y-components. Doing so would have made the solution longer in this case.

We found that the two balls must move at right angles after the collision if the collision is elastic. This is true in general only if the two objects colliding have equal masses and if one is initially at rest. In Example 7.12 the two pucks move at right angles after the collision, but the collision is inelastic—the masses are unequal.

Practice Problem 7.13   Finding the speed ratio

Suppose that the cue ball initially moves in the $-x$-direction. After the collision, the cue ball moves at 52.0° above the $-x$-axis and the 4-ball moves at 38.0° below the $-x$-axis. Find the ratio of the balls' speeds v_c/v_4 after the collision.

MASTER THE CONCEPTS

Summary

- Definition of linear momentum:

$$\vec{p} = m\vec{v} \tag{7-1}$$

- During an interaction, momentum is transferred from one body to another, but the total momentum of the two is unchanged.

$$\Delta\vec{p}_2 = -\Delta\vec{p}_1$$

- Impulse is the average net force times the time interval. The impulse equals the change in momentum:

$$\Delta\vec{p} = \vec{F}\,\Delta t \tag{7-2}$$

- A conserved quantity is one that remains unchanged as time passes.
- Impulse is the area under a graph of force versus time.
- Force is the rate of change of momentum.

$$\vec{F} = \lim_{\Delta t \to 0} \frac{\Delta\vec{p}}{\Delta t} \tag{7-4}$$

- External interactions may change the total momentum of a system.
- Internal interactions do not change the total momentum of a system.
- Conservation of linear momentum: if the net external force acting on a system is zero, then the momentum of the system is conserved.

- The position of the center of mass of a system of N particles is

$$x_{CM} = \frac{m_1 x_1 + m_2 x_2 + \cdots + m_N x_N}{M}$$

and

$$y_{CM} = \frac{m_1 y_1 + m_2 y_2 + \cdots + m_N y_N}{M} \tag{7-9}$$

where M is the total mass of the particles:

$$M = m_1 + m_2 + \cdots + m_N$$

- The total momentum of a system is equal to the total mass times the velocity of the center of mass:

$$\vec{p} = \vec{p}_1 + \vec{p}_2 + \cdots + \vec{p}_N = M\vec{v}_{CM} \tag{7-11}$$

- No matter how complicated a system is, the center of mass moves as if all the mass of the system were concentrated to a point particle with all the external forces acting on it:

$$\Sigma\vec{F}_{ext} = M\vec{a}_{CM} \tag{7-13}$$

- The center of mass of an isolated system moves at constant velocity.
- Conservation of momentum is used to solve problems involving collisions, explosions, and the like. Even when external forces are acting, the momentum of the system just before a collision is nearly equal to the momentum just after if the collision interaction is brief. The impulse, and therefore the change in momentum of the system, is small since the time interval is small.

MASTER THE CONCEPTS *continued*

Highlighted Figures and Tables

F7.2 Vector subtraction to find the change in momentum (p. 220)

F7.4 Safety features of the modern automobile that serve to lengthen the time interval of a momentum change during a crash (p. 223)

F7.5 Impulse on a graph of F_x versus t (p. 224)

F7.9 Conservation of momentum before, during, and after collision (p. 226)

F7.11 Center of mass (p. 231)

F7.14 Motion of the center of mass (p. 235)

F7.15 Collisions in one dimension (p. 236)

F7.19 Collisions in two dimensions (p. 239)

CONCEPTUAL QUESTIONS

1. You are trapped on the second floor of a burning building. The stairway is impassable, but there is a balcony outside your window. Describe what might happen in the following situations. (a) You jump from the second-story balcony to the pavement below, landing stiff-legged on your feet. (b) You jump into a privet hedge, landing on your back and rolling to your feet. (c) You jump into a firefighters' net, landing on your back. What happens to the net as you land in it? What do the firefighters do to cushion your fall even more?

2. A force of 30 N is applied for 5 s to each of two bodies of different masses. (a) Which one has the greatest momentum change? (b) The greatest velocity change? (c) The greatest acceleration?

3. If you take a rifle and saw off part of the barrel, the muzzle speed (the speed at which bullets emerge from the barrel) will be smaller. Why?

4. A firecracker at rest explodes, sending fragments off in all directions. Initially the firecracker has zero momentum, but after the explosion the fragments flying off each have quite a lot of momentum. Hasn't momentum been created? If not, explain why not.

5. An astronaut in deep space is taking a space walk when the tether connecting him to his spaceship breaks. How can he get back to the ship? He doesn't have a rocket propulsion backpack, unfortunately, but he is carrying a big wrench.

6. An astronaut hits a golf ball on the surface of the Moon. Is the momentum of the ball conserved while it is in flight? Is there a *component* of its momentum that is conserved?

7. Which would be more effective: a hammer that collides *elastically* with a nail, or one that collides perfectly *inelastically*? Assume that the mass of the hammer is much larger than that of the nail.

8. Squid are the fastest swimmers among invertebrates. A cavity within the squid is filled with water. The *mantle*, a powerful muscle, squeezes the cavity and expels the water through a narrow opening (the *siphon*)

at high speed. Using momentum conservation, explain how this propels the squid forward. How is the squid's swimming mechanism like a rocket engine?

9. In your own words, phrase each of Newton's three laws of motion as a statement about momentum.

10. Two objects with different masses have the same kinetic energy. Which has the larger magnitude of momentum?

11. A woman is 1.60 m tall. When standing straight, is her center of mass necessarily 0.80 m above the floor? Explain.

12. The momentum of a system can only be changed by an external force. What is the external force that changes the momentum of a bicycle (with its rider) as it speeds up, slows down, or changes direction? Is it true that changes in the bicycle's kinetic energy must come from an external force? Explain.

13. In an egg toss, two people try to toss a raw egg back and forth without breaking it as they move farther and farther apart. Discuss a strategy in terms of impulse and momentum for catching the egg without breaking it.

14. In the "executive toy" (Fig. 7.22), two balls are pulled back and then released. After the collision, two balls move away on the opposite side. Why do we never see three balls move

Figure 7.22
Conceptual Question 14

away following this action, although with a lower velocity so that linear momentum is still conserved?

15. A baseball batting coach emphasizes the importance of "follow-through" when a batter is trying for a home run. The coach explains that the follow-through allows the bat to remain in contact with the ball for a longer time so that the ball will travel a greater distance. Explain the reasoning behind this statement in terms of the impulse-momentum theorem.

MULTIPLE CHOICE QUESTIONS

1. A ball of mass m with initial speed v collides with another ball of mass M, initially at rest. After the collision the two balls stick together, moving with speed V. The ratio of the final speed V to the initial speed v is $V/v =$

    (a) $\dfrac{M}{M+m}$ (b) $\dfrac{M+m}{M}$

    (c) $\dfrac{m}{M+m}$ (d) $\dfrac{M+m}{m}$

    (e) $\sqrt{\dfrac{M}{M+m}}$ (f) $\sqrt{\dfrac{m}{M+m}}$

2. Two particles A and B of equal mass are located at some distance from each other. Particle A is at rest while B moves away from A at speed v. What happens to the center of mass of the system of two particles?

    (a) It does not move.
    (b) It moves with a speed v away from A.
    (c) It moves with a speed v toward A.
    (d) It moves with a speed $\frac{1}{2}v$ away from A.
    (e) It moves with a speed $\frac{1}{2}v$ toward A.

3. Two uniform spheres with equal mass per unit volume are in contact with one another. The mass of sphere A is five times that of sphere B. The center of mass of the system is

    (a) at the point where A and B touch.
    (b) inside sphere B somewhere on the line joining the centers of A and B.
    (c) inside sphere A somewhere on the line joining the centers.
    (d) at the center of sphere A.
    (e) outside of both spheres.

4. An object at rest suddenly explodes into three parts of equal mass (Fig. 7.23). Two of the parts move away at right angles to each other and with equal speeds v. What is the initial velocity of the third part?

    (a) Direction of vector 1 and magnitude $2v$

    (b) Direction of vector 2 and magnitude $\sqrt{2}v$

    (c) Direction of vector 3 and magnitude $\dfrac{1}{\sqrt{2}}v$

    (d) Direction of vector 2 and magnitude $\dfrac{1}{\sqrt{2}}v$

    (e) Direction of vector 1 and magnitude $\frac{1}{2}v$

5. A 3.0-kg object is initially at rest. It then receives an impulse of magnitude 15 N·s. After the impulse, the object has

    (a) a speed of 45 m/s.
    (b) a momentum of magnitude 5.0 kg·m/s.
    (c) a speed of 7.5 m/s.
    (d) a momentum of magnitude 15 kg·m/s.

6. An object of mass m drops from rest a little above the Earth's surface for a time t. Ignore air resistance. After time t the magnitude of its momentum is

    (a) mgt^2
    (b) mgt
    (c) $mg\sqrt{t}$
    (d) $\sqrt{mgt}$
    (e) $\dfrac{mgt^2}{2}$

Multiple Choice Questions 7–12 refer to a situation in which a golf ball is projected straight upward in the +y-direction. Ignore air resistance. The answer choices are found in Fig. 7.24.

7. Which graph shows the acceleration a_y of the ball as a function of time?

8. Which graph shows the vertical position y of the ball as a function of time?

9. Which graph shows the momentum p_y of the ball as a function of time?

10. Which graph shows the kinetic energy of the ball as a function of time?

11. Which graph shows the potential energy of the ball as a function of time?

12. Which graph shows the total energy of the ball as a function of time?

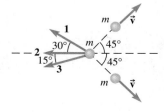

Figure 7.23 Multiple Choice Question 4

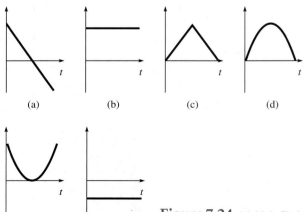

Figure 7.24 Multiple Choice Questions 7–12

PROBLEMS

Note: **C** indicates a combination conceptual/quantitative problem. Gold diamonds ✦, ✦✦ are used to indicate the increasing level of difficulty of each problem. Problem numbers appearing in blue, 9., denote problems that have a detailed solution available in the Student Solutions Manual. Some problems are *paired* by concept; their numbers are connected by a ruled box.

7.2 Momentum; 7.3 The Impulse-Momentum Theorem

1. A 2.0-kg block is moving to the right at 1.0 m/s just before it strikes and sticks to a 1.0-kg block initially at rest. What is the total momentum of the two blocks after the collision?

2. What is the momentum of an automobile (weight = 9800 N) when it is moving at 35 m/s to the south?

✦ 3. A pole-vaulter of mass 60.0 kg vaults to a height of 6.0 m before dropping to thick padding placed below to cushion her fall. (a) Find the speed with which she lands. (b) If the padding brings her to a stop in a time of 0.50 s, what is the average force on her body due to the padding during that time interval?

4. A cue stick hits a cue ball with an average force of 24 N for a duration of 0.028 s. If the mass of the ball is 0.16 kg, how fast is it moving after being struck?

5. A system consists of three particles with these masses and velocities: mass 3.0 kg, moving north at 3.0 m/s; mass 4.0 kg, moving south at 5.0 m/s; and mass 7.0 kg, moving north at 2.0 m/s. What is the total momentum of the system?

6. A sports car traveling along a straight line increases its speed from 20.0 mi/h to 60.0 mi/h. (a) What is the ratio of the final to the initial magnitude of its momentum? (b) What is the ratio of the final to the initial kinetic energy?

7. A ball of mass 5.0 kg moving with speed of 2.0 m/s in the $+x$-direction hits a wall and bounces back with the same speed in the $-x$-direction. What is the change of momentum of the ball?

8. An object of mass 3.0 kg is projected into the air at a 55° angle. It hits the ground 3.4 s later. What is its change in momentum while it is in the air? Ignore air resistance.

9. An object of mass 3.0 kg is allowed to fall from rest under the force of gravity for 3.4 s. What is the change in its momentum? Ignore air resistance.

10. What average force is necessary to bring a 50.0-kg sled from rest to a speed of 3.0 m/s in a period of 20.0 s? Assume frictionless ice.

11. For a safe re-entry into the Earth's atmosphere, the pilots of a space capsule must reduce their speed from 2.6×10^4 m/s to 1.1×10^4 m/s. The rocket engine produces a backward force on the capsule of 1.8×10^5 N. The mass of the capsule is 3800 kg. For how long must they fire their engine? [*Hint:* Ignore the change in mass of the capsule due to the expulsion of exhaust gases.]

✦ 12. Redo Problem 11, this time taking into account that 6.8 kg of exhaust gases are expelled per second. [*Hint:* The momentum change of the exhaust gases is already accounted for in the given force on the capsule; the force on the capsule is the rate of change of *its* momentum.]

13. An automobile traveling at a speed of 30.0 m/s applies its brakes and comes to a stop in 5.0 s. If the automobile has a mass of 1.0×10^3 kg, what is the average horizontal force exerted on it during braking? Assume the road is level.

14. A 3.0-kg body is initially moving northward at 15 m/s. Then a force of 15 N, toward the east, acts on it for 4.0 s. (a) At the end of the 4.0 s, what is the body's final velocity? (b) What is the change in momentum during the 4.0 s?

✦ 15. A boy of mass 60.0 kg is rescued from a hotel fire by leaping into a firefighters' net. The window from which he leapt was 8.0 m above the net. The firefighters lower their arms as he lands in the net so that he is brought to a complete stop in a time of 0.40 s. (a) What is his change in momentum during the 0.40-s interval? (b) What is the impulse on the net due to the boy during the interval? [*Hint:* Do not ignore gravity.] (c) What is the average force on the net due to the boy during the interval?

16. Verify that the SI unit of impulse is the same as the SI unit of momentum.

7.4 Conservation of Momentum

17. A rifle has a mass of 4.5 kg and it fires a bullet of mass 10.0 g at a muzzle speed of 820 m/s. What is the recoil speed of the rifle as the bullet leaves the gun barrel?

18. A submarine of mass 2.5×10^6 kg and initially at rest fires a torpedo of mass 250 kg. The torpedo has an initial speed of 100.0 m/s. What is the initial recoil speed of the submarine? Neglect the drag force of the water.

19. A uranium nucleus (mass 238 u), initially at rest, undergoes radioactive decay. After an alpha particle (mass 4.0 u) is emitted, the remaining nucleus is thorium (mass 234 u). If the alpha particle is moving at 0.050 times the speed of light, what is the recoil speed of the thorium nucleus? (Note: "u" is a unit of mass; it is *not* necessary to convert it to kg.)

✦✦ 20. A cannon on a railroad car is facing in a direction parallel to the tracks (Fig. 7.25). It fires a 98-kg shell at a speed of 105 m/s (relative to the ground) at an angle of 60.0° above the horizontal. If the cannon plus car have a mass of 5.0×10^4 kg, what is the recoil speed of the car if it was at rest before the cannon was fired? [*Hint:* A *component* of a system's momentum along an axis is conserved if the net external force acting on the system has no component along that axis.]

✦✦ 21. A marksman standing on a motionless railroad car fires a gun into the air at an angle of 30.0° from the horizontal (Fig. 7.26).

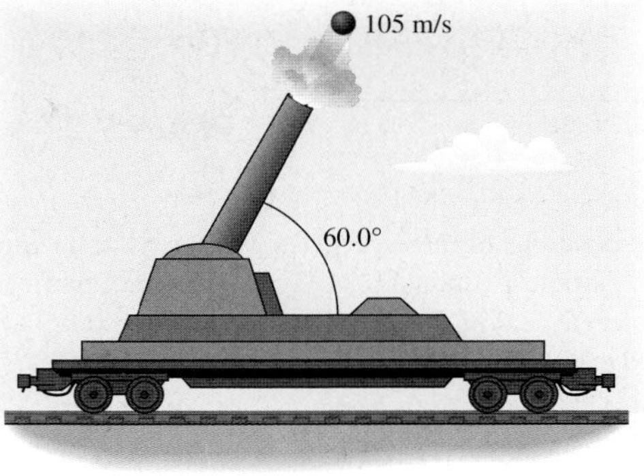

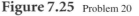

Figure 7.25   Problem 20

Figure 7.26 Problem 21

The bullet has a speed of 173 m/s (relative to the ground) and a mass of 0.010 kg. The man and car move to the left at a speed of 1.0×10^{-3} m/s after he shoots. What is the mass of the man and car? (See the hint in Problem 20.)

7.5 Center of Mass; 7.6 Motion of the Center of Mass

22. Particle A is at the origin and has a mass of 30.0 g. Particle B has a mass of 10.0 g. Where must particle B be located if the coordinates of the center of mass are $(x, y) = (2.0 \text{ cm}, 5.0 \text{ cm})$?

23. Particle A has a mass of 5.0 g, particle B has a mass of 1.0 g. Particle A is located at the origin, particle B at the point $(x, y) = (25 \text{ cm}, 0)$. What is the location of the center of mass?

24. The three bodies in Fig. 7.27 each have the same mass. If one of the bodies is moved 12 cm in the positive x-direction, by how much does the center of mass move?

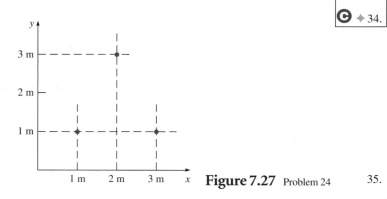

Figure 7.27 Problem 24

25. The positions of three particles, written as (x, y) coordinates, are: particle 1 (mass 4.0 kg) at (4.0 m, 0 m); particle 2 (mass 6.0 kg) at (2.0 m, 4.0 m); particle 3 (mass 3.0 kg) at (−1.0 m, −2.0 m). What is the location of the center of mass?

26. Consider two falling bodies. Their masses are 3.0 kg and 4.0 kg. At time $t = 0$, the two are released from rest. What is the velocity of their center of mass at $t = 10.0$ s? Ignore air resistance.

27. Body A of mass 3 kg is moving in the $+x$-direction with a speed of 14 m/s. Body B of mass 4 kg is moving in the $-y$-direction with a speed of 7 m/s. What are the x- and y-components of the velocity of the center of mass of the two bodies?

28. If a particle of mass 5.0 kg is moving east at 10 m/s and a particle of mass 15 kg is moving west at 10 m/s, what is the velocity of the center of mass of the pair?

29. Prove Eq. (7-13) $\Sigma \vec{F}_{ext} = M \vec{a}_{CM}$. [*Hint:* Start with $\Sigma \vec{F}_{ext} = \lim_{\Delta t \to 0} (\Delta \vec{p}/\Delta t)$, where $\Sigma \vec{F}_{ext}$ is the net external force acting on a system and $\vec{p}$ is the total momentum of the system.]

7.7 Collisions in One Dimension

30. In the railroad freight yard, an empty freight car of mass m rolls along a straight level track at 1.0 m/s and collides with an initially stationary, fully loaded boxcar of mass $4.0m$. The two cars couple together upon collision. (a) What is the speed of the two cars after the collision? (b) Suppose instead that the two cars are at rest after the collision. With what speed was the loaded boxcar moving before the collision if the empty one was moving at 1.0 m/s?

31. A 0.020-kg bullet traveling at 200.0 m/s east hits a motionless 2.0-kg block and bounces off it, retracing its original path with a velocity of 100.0 m/s west. What is the final velocity of the block? Assume the block rests on a perfectly frictionless horizontal surface.

32. A block of wood of mass 0.95 kg is initially at rest. A bullet of mass 0.050 kg traveling at 100.0 m/s strikes the block and becomes embedded in it. With what speed do the block of wood and the bullet move just after the collision?

33. A spring of negligible mass is compressed between two blocks, A and B, which are at rest on a frictionless horizontal surface at a distance of 1.0 m from a wall on the left and 3.0 m from a wall on the right. The sizes of the blocks and spring are small. When the spring is released, body A moves toward the left wall and strikes it at the same instant that body B strikes the right wall. The mass of A is 0.60 kg. What is the mass of B?

34. Two identical gliders on an air track are held together by a piece of string, compressing a spring between the gliders. While they are moving to the right at a common speed of 0.50 m/s, someone holds a match under the string and burns it, letting the spring force the gliders apart. One glider is then observed to be moving to the right at 1.30 m/s. (a) What velocity does the other glider have? (b) Is the total kinetic energy of the two gliders after the collision greater than, less than, or equal to the total kinetic energy before the collision? If greater, where did the extra energy come from? If less, where did the "lost" energy go?

35. A BMW of mass 2.0×10^3 kg is traveling at 42 m/s. It approaches a 1.0×10^3 kg Volkswagen going 25 m/s in the same direction and strikes it in the rear. Neither driver applies the brakes. Neglect the relatively small frictional forces on the cars due to the road and due to air resistance. If the collision slows the BMW down to 33 m/s, what is the speed of the VW after the collision?

36. A projectile of 1.0-kg mass approaches a stationary body of 5.0 kg at 10.0 m/s and, after colliding, rebounds in the reverse direction along the same line with a speed of 5.0 m/s. What is the speed of the 5.0-kg body after the collision?

37. A 2.0-kg object is at rest on a perfectly frictionless surface when it is hit by a 3.0-kg object moving at 8.0 m/s. If the two objects are stuck together after the collision, what is the speed of the combination?

38. Two cars, each of mass 1300 kg, are approaching each other on a head-on collision course. Each speedometer reads 19 m/s. What is the magnitude of the total momentum of the system?

39. A 75-kg man is at rest on ice skates. A 0.20-kg ball is thrown to him. The ball is moving horizontally at 25 m/s just before the man catches it. How fast is the man moving just after he catches the ball?

✦ 40. A 0.010-kg bullet traveling horizontally at 400.0 m/s strikes a 4.0-kg block of wood sitting at the edge of a table. The bullet is lodged into the wood. If the table height is 1.2 m, how far from the table does the block hit the floor?

✦ 41. Two objects with masses m_1 and m_2 approach each other with equal and opposite momenta so that the total momentum is zero. Show that, if the collision is elastic, the final *speed* of each object must be the same as its initial speed. (The final *velocity* of each object is *not* the same as its initial velocity, however.)

42. A helium atom (mass 4.0 u) moving at 600.0 m/s to the right collides with an oxygen molecule (mass 32 u) moving in the same direction at 400.0 m/s. After the collision, the oxygen molecule moves at 500.0 m/s to the right. What is the velocity of the helium atom after the collision?

✦✦ 43. Use the result of Problem 41 to show that in *any* elastic collision between two objects, the relative speed of the two is the same before and after the collision. [*Hints:* Look at the collision in its *center of mass frame*—the reference frame in which the center of mass is at rest. The *relative* speed of two objects is the same in any inertial reference frame.]

7.8 Collisions in Two Dimensions

44. Body A of mass M has an original velocity of 6.0 m/s in the $+x$-direction toward a stationary body (body B) of the same mass. After the collision, body A has velocity components of 1.0 m/s in the $+x$-direction and 2.0 m/s in the $+y$-direction. What is the magnitude of body B's velocity after the collision?

✦ 45. (a) With reference to Practice Problem 7.12, find the momentum change of the ball of mass m_1 during the collision. Give your answer in x- and y-component form; express the components in terms of m_1 and v_0. (b) Repeat for the ball of mass m_2. How are the momentum changes related?

46. A hockey puck moving at 0.45 m/s collides into another puck that was at rest. The pucks have equal mass. The first puck is deflected 37° to the right and moves off at 0.36 m/s. Find the speed and direction of the second puck after the collision.

✦ 47. Puck 1 sliding along the x-axis strikes stationary puck 2 of the same mass. After the collision, puck 1 moves off at speed v_{1f} in the direction 60.0° above the x-axis; puck 2 moves off at speed v_{2f} in the direction 30.0° below the x-axis. Find v_{2f} in terms of v_{1f}.

48. A projectile of mass 2.0 kg approaches a stationary target body at 5.0 m/s. The projectile is deflected through an angle of 60.0° and its speed after the collision is 3.0 m/s. What is the magnitude of the momentum of the target body after the collision?

✦ 49. In a nuclear reactor, a neutron moving at speed v_0 in the positive x-direction strikes a deuteron, which is at rest. The neutron is deflected by 90.0° and moves off with speed $v_0/\sqrt{3}$ in the positive y-direction. Find the x- and y-components of the deuteron's velocity after the collision. (The mass of the deuteron is twice the mass of the neutron.)

50. A firecracker is tossed straight up into the air. It explodes into three pieces of equal mass just as it reaches the highest point. Two pieces move off at 120 m/s at right angles to each other. How fast is the third piece moving?

COMPREHENSIVE PROBLEMS

51. An intergalactic spaceship is traveling through space far from any planets or stars, where no human has gone before. The ship carries a crew of 30 people (of total mass 2.0×10^3 kg). If the speed of the spaceship is 1.0×10^5 m/s and its mass (excluding the crew) is 4.8×10^4 kg, what is the magnitude of the total momentum of the ship and the crew?

52. A baseball player pitches a fastball toward home plate at a speed of 41 m/s. The batter swings, connects with the ball of mass 145 g, and hits it so that the ball leaves the bat with a speed of 37 m/s. Assume that the ball is moving horizontally just before and just after the collision with the bat. (a) What is the magnitude of the change in momentum of the ball? (b) What is the impulse delivered to the ball by the bat? (c) If the bat and ball are in contact for 3.0 ms, what is the magnitude of the average force exerted on the ball by the bat?

✦ 53. A tennis ball of mass 0.060 kg is served. It strikes the ground with a velocity of 54 m/s (120 mph) at an angle of 22° below the horizontal. Just after the bounce it is moving at 53 m/s at an angle of 18° above the horizontal. If the interaction with the ground lasts 0.065 s, what average force did the ground exert on the ball?

54. A uniform rod of length 30.0 cm is bent into the shape of an inverted U (Fig. 7.28). Each of the 3 sides is of length 10.0 cm. Find the location, in x- and y-coordinates, of the center of mass as measured from the origin.

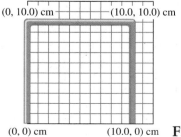

(0, 10.0) cm   (10.0, 10.0) cm

(0, 0) cm   (10.0, 0) cm   **Figure 7.28** Problem 54

55. To contain some unruly demonstrators, the riot squad approaches with fire hoses. Suppose that the rate of flow of water through a fire hose is 24 kg/s and the stream of water from the hose moves at 17 m/s. What force is exerted by such a stream on a person in the crowd? Assume that the water comes to a dead stop against the demonstrator's chest.

56. An inexperienced catcher catches a 130 km/h fastball of mass 140 g within 1 ms, whereas an experienced catcher slightly retracts his hand during the catch, extending the stopping time to 10 ms. What are the average forces imparted to the two gloved hands during the catches?

✦ 57. A stationary 0.1-g fly encounters the windshield of a 1000-kg automobile traveling at 100 km/h. (a) What is the change in momentum of the car due to the fly? (b) What is the change of momentum of the fly due to the car? (c) Approximately how many flies does it take to reduce the car's speed by 1 km/h?

58. For a system of three particles moving along a line, an observer in a laboratory measures the following masses and velocities:

Mass (kg)	v_x (m/s)
3.0	+290
5.0	−120
2.0	+52

What is the velocity of the center of mass of the system?

✦ 59. A projectile of mass 2.0 kg approaches a stationary target body at 8.0 m/s. The projectile is deflected through an angle of 90.0° and its speed after the collision is 6.0 m/s. What is the speed of the target body after the collision if the collision is perfectly elastic?

60. An automobile weighing 13.6 kN is moving at 17.0 m/s when it collides with a stopped car weighing 9.0 kN. If they lock bumpers and move off together, what is their speed just after the collision?

61. A sled of mass 5.0 kg is coasting along on a frictionless ice-covered lake at a constant speed of 1.0 m/s. A 1.0-kg book is dropped vertically onto the sled. At what speed does the sled move once the book is on it?

62. A radioactive nucleus is at rest when it spontaneously decays by emitting an electron and neutrino (Fig. 7.29). The momentum of the electron is 8.20×10^{-19} kg·m/s and it is directed at right angles to that of the neutrino. The neutrino momentum is of magnitude 5.00×10^{-19} kg·m/s. (a) In what direction does the newly formed (daughter) nucleus recoil? Let the electron direction be along the positive x-axis and find direction of nucleus recoil with respect to the electron's direction. (b) What is its momentum?

✦ 63. A 60.0-kg woman stands at one end of a 120-kg raft that is 6.0 m long. The other end of the raft is 0.50 m from a pier. (a) The woman walks toward the pier until she gets to the other end of the raft and stops there. Now what is the distance between the raft and the pier? (b) In (a), how far did the woman walk (relative to the pier)?

✦✦ 64. A jet plane is flying at 130 m/s relative to the ground. There is no wind. The engines take in 81 kg of air per second. Hot gas (burned fuel and air) is expelled from the engines at high speed. The engines provide a forward force on the plane of magnitude 6.0×10^4 N. At what speed relative to the ground is the gas being expelled? [*Hint:* Look at the momentum change of the air taken in by the engines during a time interval Δt.] This calculation is approximate since we are ignoring the 3.0 kg of fuel consumed and expelled with the air each second.

✦ 65. Within cells, small organelles containing newly synthesized proteins are transported along microtubules by tiny molecular motors called kinesins. What force does a kinesin molecule need to deliver in order to accelerate an organelle with mass 0.01 pg (10^{-17} kg) from 0 to 1 μm/s within a time of 10 μs?

✦ 66. The pendulum bobs in Fig. 7.30 are made of soft clay so that they stick together after impact. The mass of bob A is half that of bob B. Bob B is initially at rest. What is the ratio of the kinetic energy of the combined bobs, just after impact, to the kinetic energy of bob A just before impact?

✦✦ 67. The pendulum bobs in Fig. 7.30 are made of soft clay so that they stick together after impact. The mass of bob A is half that of bob B. Bob B is initially at rest. If bob A is released from a height h above its lowest point, what is the maximum height attained by bobs A and B after the collision?

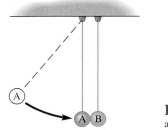

Figure 7.30 Problems 66 and 67

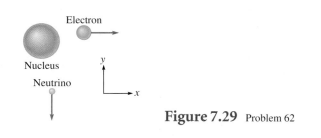

Figure 7.29 Problem 62

✦✦ 68. A flat, circular metal disk of uniform thickness has a radius of 3.0 cm. A hole is drilled in the disk that is 1.5 cm in radius. The hole is tangent to one side of the disk (Fig. 7.31). Where is the CM of the disk now that the hole has been drilled? [*Hint:* The original disk (before the hole is drilled) can be thought of as having two pieces—the disk with the hole plus the smaller disk of metal drilled out. Write an equation that expresses x_{CM} of the original disk in terms of the x_{CM}'s of the two pieces. Since the thickness is uniform, the mass of any piece is proportional to its area.]

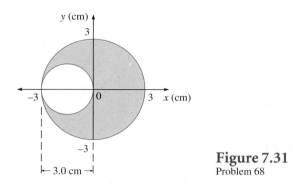

Figure 7.31
Problem 68

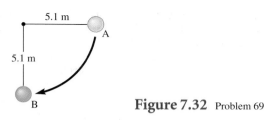

Figure 7.32 Problem 69

✦ 69. Two pendulum bobs have equal masses and lengths (5.1 m). Bob A is initially held horizontally while bob B hangs vertically at rest (Fig. 7.32). Bob A is released and collides elastically with bob B. How fast is bob B moving immediately after the collision?

70. Two identical gliders, each with elastic bumpers and mass 0.10 kg, are on a horizontal air track. Friction is negligible. Glider 2 is stationary. Glider 1 moves toward glider 2 from the left with a speed of 0.20 m/s. They collide. After the collision, what are the velocities of glider 1 and glider 2?

✦ 71. A radium nucleus (mass 226 u) at rest decays into a radon nucleus (symbol Rn, mass 222 u) and an alpha particle (symbol α, mass 4 u). (a) Find the ratio of the speeds v_α/v_{Rn} after the decay. (b) Find the ratio of the magnitudes of the momenta p_α/p_{Rn}. (c) Find the ratio of the kinetic energies K_α/K_{Rn}. (Note: "u" is a unit of mass; it is *not* necessary to convert it to kg.)

ANSWERS TO PRACTICE PROBLEMS

7.1 (a) 0.78 kg·m/s downward; (b) 0.78 kg·m/s toward the apple, 1.3×10^{-25} m/s

7.2 3.5 times his weight

7.3 1700 N; 0.0037 s

7.4 0.8 m/s in the $-x$-direction

7.5 1.7 m/s

7.6 The manager throws the nugget in a direction away from the desk. The scheme might fail if frictional forces bring the chair to rest before he reaches the telegraph.

7.7 2.2 m/s

7.8 (2.0 cm, 2.3 cm)

7.9 (a) 2.7 m; (b) 1.5 m in the other direction; (c) the center of mass does not move

7.10 4.0 m/s

7.11 (a) 2.5 m/s; (b) 306 kJ (same before and after, since the collision is elastic); (c) both are 17.5 m/s

7.12 $0.751v_0$

7.13 0.781

Torque and Angular Momentum

In gymnastics, the iron cross is a notoriously difficult feat that requires incredible strength. Why does it require such great strength? To perform the iron cross, the forces exerted by the gymnast's muscles must be much greater in magnitude than the gymnast's weight. How does the design of the human body make such large forces necessary?

The symbol Σ stands for a sum. $\overset{N}{\underset{i=1}{\Sigma}}$ means the sum for $i = 1, 2, \ldots, N$.

8.1   ROTATIONAL KINETIC ENERGY AND ROTATIONAL INERTIA

When a rigid object is rotating in place, it has kinetic energy because each particle of the object is moving in a circle around the axis of rotation. In principle, we can calculate the kinetic energy of rotation by summing the kinetic energy of each particle. To say the least, that sounds like a laborious task. We need a simpler way to express the rotational kinetic energy of such an object so that we don't have to calculate this sum over and over. Our simpler expression exploits the fact that the speed of each particle is proportional to the angular speed of rotation ω.

If a rigid object consists of N particles, the sum of the kinetic energies of the particles can be written mathematically using a subscript to label the mass and speed of each particle:

$$K_{rot} = \tfrac{1}{2}m_1 v_1^2 + \tfrac{1}{2}m_2 v_2^2 + \cdots + \tfrac{1}{2}m_N v_N^2 = \overset{N}{\underset{i=1}{\Sigma}} \tfrac{1}{2}m_i v_i^2$$

The speed of each particle is related to its distance from the axis of rotation. Particles that are farther from the axis move faster. In Section 5.1 we found that the speed of a particle moving in a circle is

$$v = r\omega \tag{5-6}$$

where ω is the angular speed and r is the distance between the rotation axis and the particle. By substitution, the rotational kinetic energy can be written

$$K_{rot} = \overset{N}{\underset{i=1}{\Sigma}} \tfrac{1}{2}m_i r_i^2 \omega^2$$

The entire object rotates at the same angular velocity ω, so the constants $\tfrac{1}{2}$ and ω^2 can be factored out of each term of the sum:

$$K_{rot} = \tfrac{1}{2}\left(\overset{N}{\underset{i=1}{\Sigma}} m_i r_i^2\right)\omega^2$$

The quantity in the parentheses *cannot change* since the distance between each particle and the rotation axis stays the same if the object is rigid and doesn't change shape. However difficult it may be to compute the sum in the parentheses, we only need to do it *once* for any given mass distribution and axis of rotation.

Let's give the quantity in the brackets the symbol I. In Chapter 5, we found it useful to draw analogies between translational variables and their rotational equivalents. By using the symbol I, we can see that translational and rotational kinetic energy have similar forms: translational kinetic energy is

$$K_{tr} = \tfrac{1}{2}mv^2$$

and **rotational kinetic energy** is

Rotational kinetic energy
$K_{rot} = \tfrac{1}{2}I\omega^2$ $\qquad$ (8-1)

The quantity I is called the **rotational inertia**:

Rotational inertia
$I = \overset{N}{\underset{i=1}{\Sigma}} m_i r_i^2$ $\qquad$ (8-2)
(SI unit: kg·m^2)

Comparing the expressions for translational and rotational kinetic energy, we see that angular speed ω takes the place of speed v and rotational inertia I takes the place of

Tips Since $v = r\omega$ was used to derive Eq. (8-1), ω must be expressed in radians per unit time (normally rad/s).

mass m. Mass is a measure of the inertia of an object, or in other words how difficult it is to change the object's velocity. Similarly, for a rigid rotating object, I is a measure of its rotational inertia—how hard it is to change its angular velocity. That is why the quantity I is called the rotational inertia; it is also sometimes called the **moment of inertia**.

When a problem requires you to find a rotational inertia, there are four principles to follow.

Finding the Rotational Inertia

1. If the object consists of a *small* number of particles, calculate the sum $I = \sum_{i=1}^{N} m_i r_i^2$ directly.

2. For symmetric objects with simple geometric shapes, advanced mathematical methods can be used to perform the sum in Eq. (8-2). For your convenience, Table 8.1 lists the results of these calculations for the shapes most commonly encountered.

3. Since the rotational inertia is a sum, you can always mentally decompose the object into several parts, find the rotational inertia of each part, and then add them. This is an example of the *divide-and-conquer* problem-solving technique.

4. Since the rotational inertia involves distances from the axis of rotation, you can move mass parallel to the rotation axis without changing I. For example, you might want to calculate the rotational inertia of a door about the axis through its hinges. Mentally "compress" the door vertically (parallel to the axis) into a horizontal rod with the same mass, as in Fig. 8.1. The rotational inertia is unchanged by the compression, since every particle maintains the same distance from the axis of rotation. Thus, the formula for the rotational inertia of a rod (listed in Table 8.1) can be used for the door.

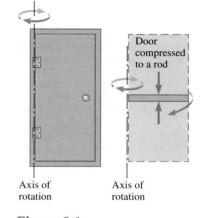

Figure 8.1 The rotational inertias of a door and a rod are similar.

Table 8.1

Rotational Inertia for Uniform Objects with Various Geometrical Shapes

Shape		Axis of Rotation	Rotational Inertia	Shape		Axis of Rotation	Rotational Inertia
Thin hollow cylindrical shell (or hoop)		Central axis of cylinder	MR^2	Solid sphere		Through center	$\frac{2}{5}MR^2$
Solid cylinder (or disk)		Central axis of cylinder	$\frac{1}{2}MR^2$	Thin hollow spherical shell		Through center	$\frac{2}{3}MR^2$
Hollow cylindrical shell or disk		Central axis of cylinder	$\frac{1}{2}M(a^2+b^2)$	Thin rod		Perpendicular to rod through end	$\frac{1}{3}ML^2$
				Rectangular plate		Perpendicular to plate through center	$\frac{1}{12}M(a^2+b^2)$

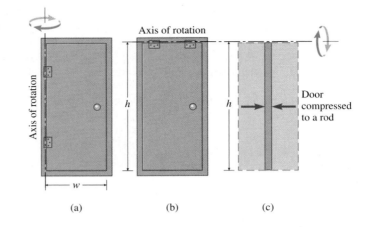

Figure 8.2 The rotational inertia of a door depends on the rotation axis. (a) The door with hinges at the side has a smaller rotational inertia, $I = \frac{1}{3}Mw^2$, than (b) the rotational inertia, $I = \frac{1}{3}Mh^2$, of the same door with hinges at the top. (c) The door can be mentally compressed horizontally into a rod.

Keep in mind that the rotational inertia of an object depends on the location of the rotation axis. For instance, imagine taking the hinges off the side of a door and putting them on the top, so that the door swings about a horizontal axis like a cat flap door (Fig. 8.2b). The door now has a considerably larger rotational inertia than before the hinges were moved. The door has the same mass as before, but its mass now lies on average much farther from the axis of rotation than that of the door in Fig. 8.2a. In applying Eq. (8-2) to find the rotational inertia of the door, the values of r_i range from 0 to the height of the door (h), whereas with the hinges in the normal position the values of r_i range from 0 only to the width of the door (w). If we think of compressing the door with hinges on top into a rod, the door would have to be compressed *horizontally* into a rod of length h (Fig. 8.2c).

Example 8.1

Rotational Inertia of a Rod About Its Midpoint

Table 8.1 gives the rotational inertia of a thin rod with the axis of rotation perpendicular to its length and passing through one end ($I = \frac{1}{3}ML^2$). Write an expression for the rotational inertia of a rod with mass M and length L that rotates about a perpendicular axis through its *midpoint* (Fig. 8.3).

Strategy In general, the same object rotating about a different axis has a different rotational inertia. With the axis at the midpoint, the rotational inertia is smaller than for the axis at the end, since the mass is closer to the axis. Imagine performing the

sum $\sum\limits_{i=1}^{N} m_i r_i^2$; for the axis at the end, the values of r_i range from 0 to L, whereas with the axis at the midpoint, r_i is never larger than $\frac{1}{2}L$.

Since the answer is not in Table 8.1, there are two options: either calculate the sum $\sum\limits_{i=1}^{N} m_i r_i^2$ directly, or decompose the object into parts. Imagine cutting the rod in half; then there are two rods, each with its axis of rotation at one of its *ends*. The rotational inertia of a rod about its end is listed in Table 8.1. Then, since rotational inertia is additive, the rotational inertia for two such rods is just twice the value for one rod.

Solution Each of the halves has mass $\frac{1}{2}M$ and length $\frac{1}{2}L$ and rotates about an axis at its endpoint. Table 8.1 gives $I = \frac{1}{3}ML^2$ for a rod with mass M and length L rotating about its end, so each of the halves has

$$I_{\text{half}} = \frac{1}{3} \times \text{mass of half} \times (\text{length of half})^2$$
$$= \frac{1}{3} \times (\frac{1}{2}M) \times (\frac{1}{2}L)^2 = \frac{1}{3} \times \frac{1}{2} \times \frac{1}{4} \times ML^2 = \frac{1}{24}ML^2$$

Since there are two such halves, the total rotational inertia is twice that:

$$I = I_{\text{half}} + I_{\text{half}} = \frac{1}{12}ML^2$$

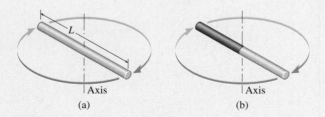

Figure 8.3
(a) A rod rotating about a vertical axis through its center. (b) The same rod, viewed as two rods, each half as long, rotating about an axis through an end.

continued on next page

Example 8.1 *continued*

Discussion The rotational inertia is less than $\frac{1}{3}ML^2$, as expected. That it is $\frac{1}{4}$ as much is a result of the distances r_i being *squared* in the definition of rotational inertia. The various particles that compose the rod are at distances that range from 0 to $\frac{1}{2}L$ from the rotation axis, instead of from 0 to L. Think of it as if the rod were compressed to half its length, still pivoting about the endpoint. All the distances r_i are half as much as before; since each r_i is squared in the sum, the rotational inertia is $\left(\frac{1}{2}\right)^2 = \frac{1}{4}$ its former value.

See the Physics at Home that follows the Practice Problem.

Practice Problem 8.1 Playground merry-go-round

A playground merry-go-round is essentially a uniform disk that rotates about a vertical axis through its center (Fig. 8.4).

Suppose the disk has a radius of 2.0 m and a mass of 160 kg; a child weighing 180 N sits at the edge of the merry-go-round. What is the merry-go-round's rotational inertia, including the contribution due to the child? [*Hint:* Treat the child as a point mass at the edge of the disk.]

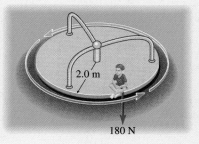

2.0 m

180 N

Figure 8.4
Child on merry-go-round.

Physics at Home

The change in rotational inertia of a rod as the rotation axis changes can be easily felt. Hold a baseball bat in the usual way, with your hands gripping the bottom of the bat. Swing the bat a few times. Now "choke up" on the bat—move your hands up the bat—and swing a few times. The bat is easier to swing because it now has a smaller rotational inertia. Children often choke up on a bat that is too massive for them. Even major league baseball players occasionally choke up on the bat when they want more control over their swing to place a hit in a certain spot (Fig. 8.5). On the other hand, choking up on the bat makes it impossible to hit a home run. To hit a long fly ball, you want the pitched baseball to encounter a bat that is swinging with a lot of rotational inertia.

Figure 8.5 Hank Aaron choking up on the bat.

Example 8.2

Atwood's Machine

Atwood's machine consists of a cord around a pulley of rotational inertia I, radius R, and mass M, with two blocks (masses m_1 and m_2) hanging from the ends of the cord as in Fig. 8.6. (Note that in Example 3.7 we analyzed Atwood's machine for the special case of a massless pulley; for a massless pulley $I = 0$.) Assume that the pulley is free to turn without friction and that the cord does not slip. Ignore air resistance. If the masses are released from rest, find how fast they are moving after they have moved a distance h (one up, the other down).

Strategy Neglecting both air resistance and friction means that no nonconservative forces act on the system; therefore its mechanical energy is conserved. Gravitational potential energy is converted into the translational kinetic

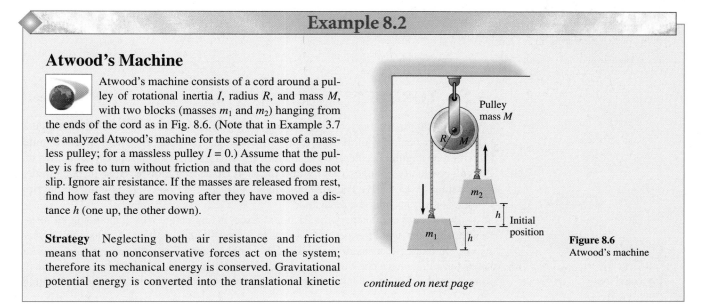

Pulley mass M

R M

m_2

h Initial position

m_1 h

Figure 8.6
Atwood's machine

continued on next page

Example 8.2 *continued*

energies of the two blocks and the rotational kinetic energy of the pulley.

Solution For our convenience, we assume that $m_1 > m_2$. Mass m_1 therefore moves down and m_2 moves up. After the masses have each moved a distance h, the changes in gravitational potential energy are

$$\Delta U_1 = -m_1 gh$$

$$\Delta U_2 = +m_2 gh$$

Since mechanical energy is conserved,

$$\Delta U + \Delta K = 0$$

The mechanical energy of the system includes the kinetic energies of three objects: the two masses and the pulley. All start with zero kinetic energy, so

$$\Delta K = \tfrac{1}{2}(m_1 + m_2)v^2 + \tfrac{1}{2}I\omega^2$$

The speed v of the masses is the same since the cord's length is fixed. The speed v and the angular speed of the pulley ω are related if the cord does not slip: the tangential speed of the pulley must equal the speed at which the cord moves. The tangential speed of the pulley is its angular speed times its radius:

$$v = \omega R$$

After substituting v/R for ω, the energy conservation equation becomes

$$\Delta U + \Delta K = [-m_1 gh + m_2 gh] + \left[\tfrac{1}{2}(m_1 + m_2)v^2 + \tfrac{1}{2}I\left(\frac{v}{R}\right)^2\right] = 0$$

or

$$\left[\tfrac{1}{2}(m_1 + m_2) + \tfrac{1}{2}\frac{I}{R^2}\right]v^2 = (m_1 - m_2)gh$$

Solving this equation for v yields

$$v = \sqrt{\frac{2(m_1 - m_2)gh}{m_1 + m_2 + I/R^2}}$$

Discussion This answer is rich in information, in the sense that we can ask many "What if?" questions. Not only do these questions provide checks as to whether the answer is reasonable, they also enable us to perform thought experiments,

which could then be checked by constructing an Atwood's machine and comparing the results.

For instance: What if m_1 is only slightly greater than m_2? Then the final speed v is small—as m_2 approaches m_1, v approaches 0. This makes intuitive sense: a small imbalance in weights produces a small acceleration. You should practice this kind of reasoning by making other such checks.

It is also enlightening to look at terms in an algebraic solution and connect them with physical interpretations. The quantity $(m_1 - m_2)g$ is the imbalance in the gravitational forces pulling on the two sides. The denominator $(m_1 + m_2 + I/R^2)$ is a measure of the total inertia of the system—the sum of the two masses plus an inertial contribution due to the pulley. The pulley's contribution is *not* simply equal to its mass. If, for example, the pulley is a uniform disk with $I = \tfrac{1}{2}MR^2$, the term I/R^2 would be equal to *half* the mass of the pulley.

The same principles used to analyze Atwood's machine have many applications in the real world. One such application is in elevators, where one of the hanging masses is the elevator and the other is the counterweight. Of course, the elevator and counterweight are not allowed to hang freely from a pulley—we must also consider the energy supplied by the motor.

Practice Problem 8.2 Modified Atwood's machine

Figure 8.7 shows a modified form of Atwood's machine where one of the blocks slides on a table instead of hanging from the pulley. The coefficient of kinetic friction between the sliding mass and the table is μ_k. The blocks are released from rest. Find the speed of the blocks after they have moved a distance h in terms of μ_k, m_1, m_2, I, R, and h.

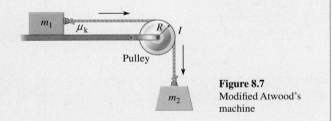

Figure 8.7
Modified Atwood's machine

8.2   TORQUE

Suppose you place a bicycle upside down, as if you're going to repair it. First, you give one of the wheels a spin. If everything is working as it should, the wheel spins for quite a while; its angular acceleration is small. If the wheel doesn't spin for very long, then its angular velocity changes rapidly and the angular acceleration is large in magnitude; there must be excessive friction somewhere. Perhaps the brakes are rubbing on the rim or the bearings need to be repacked.

If we could eliminate *all* the frictional forces acting on the wheel, including air resistance, then we would expect the wheel to keep spinning without diminishing angular speed. In that case, its angular acceleration would be zero. The situation is reminiscent of Newton's first law: a body with no external interactions, or at least no net force acting on it, moves with constant velocity. We can state a "Newton's first law for rotation": a rotating body with no external interactions, and whose rotational inertia doesn't change, keeps rotating at constant angular velocity.

Of course, the hypothetical frictionless bicycle wheel does have external interactions. The Earth's gravitational field exerts a downward force and the axle exerts an upward force to keep the wheel from falling. Then is it true that, as long as there is no net external force, the angular acceleration is zero? No; it is easy to give the wheel an angular acceleration while keeping the net force zero. Imagine bringing the wheel to rest by pressing two hands against the tire on opposite sides. On one side, the motion of the rim of the tire is downward and the kinetic frictional force is upward (see Fig. 8.8). On the other side, the tire moves upward and the frictional force is downward. In a similar way, we could apply equal and opposite forces to the opposite sides of a wheel at rest to make it start spinning. In either case, we exert equal magnitude forces, so that the net force is zero, and still give the wheel an angular acceleration.

A quantity related to force, called **torque**, plays the role in rotational dynamics that force itself played in translational dynamics. A torque is not separate from a force; it is impossible to exert a torque without exerting a force. Torque is a measure of how effective a given force is at twisting or turning something. For something rotating about a fixed axis such as the bicycle wheel, a torque can *change* the rotational motion either by making it rotate faster or by slowing it down.

When stopping the bicycle wheel with two equal and opposite forces, as in Fig. 8.8, the net applied force is zero and thus the wheel is in translational equilibrium; but the net torque is not zero so it is not in rotational equilibrium. Both forces tend to give the wheel the same sign of angular acceleration; they are both making the wheel slow down. The two torques are in fact equal, with the same sign.

What determines the torque produced by a particular force? Imagine trying to push open a massive bank vault door. Certainly you would push as hard as you can; the torque is proportional to the magnitude of the force. It also matters where and in what direction the force is applied. For maximum effectiveness, you would push perpendicular to the door (Fig. 8.9a). If you pushed radially, straight in toward the axis of rotation that passes through the hinges, the door wouldn't rotate, no matter how hard you push (Fig. 8.9b). A force acting in any other direction could be decomposed into radial and perpendicular components, with the radial component contributing nothing to the torque. Only the perpendicular component of the force produces a torque.

Furthermore, *where* you apply the force is critical (Fig. 8.10). Instinctively, you would push at the outer edge, as far from the rotation axis as possible. If you pushed close to the axis, it would be difficult to open the door. Torque is proportional to the distance between the rotation axis and the **point of application** of the force (the point at which the force is applied).

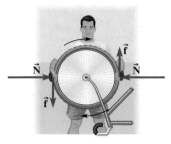

Figure 8.8 A spinning bicycle wheel slowed to a stop by friction. Each hand exerts a normal force $\vec{N}$ and a frictional force $\vec{f}$ on the tire.

The *radial* direction is directly toward or away from the axis of rotation. The *perpendicular* or *tangential* direction is perpendicular to both the radial direction and the axis of rotation; it is tangent to the circular path followed by a point as the object rotates.

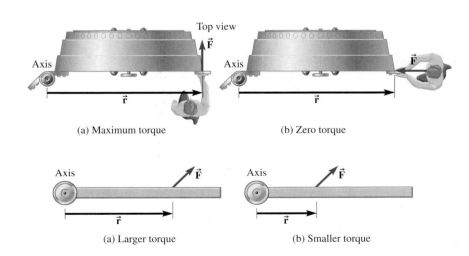

Top view

(a) Maximum torque

(b) Zero torque

(a) Larger torque

(b) Smaller torque

Figure 8.9 Torque on a bank vault door depends on the direction of the applied force. (a) Pushing perpendicularly gives the maximum torque. (b) Pushing radially inward with the same magnitude force gives zero torque.

Figure 8.10 Torques; the same force at different distances from the axis.

The symbol ⊥ stands for *perpendicular*; ∥ stands for *parallel*.

To satisfy the requirements of the previous paragraphs we define the magnitude of the torque as the product of the distance between the rotation axis and the point of application of the force (r) with the perpendicular component of the force ($F_\perp$):

Definition of torque:

$$\tau = \pm r F_\perp \qquad (8\text{-}3)$$

---

where r is the shortest distance between the rotation axis and the point of application of the force and $F_\perp$ is the perpendicular component of the force.

The symbol for torque is τ, the Greek letter tau. The SI unit of torque is the N·m. The SI unit of *energy*, the joule, is equivalent to N·m, but we do not write torque in joules. Even though both energy and torque can be written using the same SI base units, the two quantities have different meanings; torque is not a form of energy. To help maintain the distinction, the joule is used for energy but *not* for torque.

The sign of the torque indicates the direction of the angular acceleration that torque would cause *by itself*. Recall from Section 5.1 that by convention a positive angular velocity ω means counterclockwise (CCW) rotation and negative angular velocity ω means clockwise (CW) rotation. A positive angular acceleration α either increases the rate of CCW rotation (increases the magnitude of a positive ω) or decreases the rate of CW rotation (decreases the magnitude of a negative ω).

Figure 8.11 (a) When the cyclist climbs a hill, the top half of the chain exerts a large force $\vec{F}$ on one of the gears attached to the rear wheel. As viewed here, the torque about the axis of rotation (the axle) due to this force is clockwise. By convention, we call this a negative torque. (b) When the brakes are applied, the brake pads are pressed onto the rim, giving rise to frictional forces on the rim. As viewed here, the frictional force $\vec{f}$ causes a counterclockwise (positive) torque on the wheel about the axle.

We use the same sign convention for torque. A force whose perpendicular component tends to cause rotation in the CCW direction gives rise to a positive torque; if it is the only torque acting, it would cause a positive angular acceleration α (see Fig. 8.11). A force whose perpendicular component tends to cause rotation in the CW direction gives rise to a negative torque. The symbol $\pm$ in Eq. (8-3) reminds us to assign the appropriate algebraic sign each time we calculate a torque.

The sign of the torque is *not* determined by the sign of the angular velocity (in other words, whether the wheel is spinning CCW or CW); rather, it is determined by the sign of the angular *acceleration* the torque would cause if acting alone. To determine the sign of a torque, imagine which way the torque would make the object begin to spin if it is initially not rotating.

Example 8.3

A Spinning Bicycle Wheel

To stop a spinning bicycle wheel, suppose you push radially inward on opposite sides of the wheel, as shown in Fig. 8.8, with equal forces of magnitude 10.0 N. The radius of the wheel is 32 cm and the coefficient of kinetic friction between the tire and your hand is 0.75. The wheel is spinning in the clockwise sense. What is the net torque on the wheel?

Strategy The 10.0-N forces are directed radially toward the rotation axis, so they produce no torques themselves; only perpendicular components of forces give rise to torques. The forces of kinetic friction between the hands and the tire are tangent to the tire, so they do produce torques. The normal force applied to the tire is 10.0 N on each side; using the coefficient of friction, we can find the frictional forces.

Solution The frictional force exerted by each hand on the tire has magnitude

$$f = \mu_k N = (0.75) \times (10.0 \text{ N}) = 7.5 \text{ N}$$

The frictional force is tangent to the wheel, so $f_\perp = f$. Then the magnitude of each torque is

$$|\tau| = r f_\perp = (0.32 \text{ m}) \times (7.5 \text{ N}) = 2.4 \text{ N·m}$$

The two torques have the same sign, since they are both tending to slow down the rotation of the wheel. Is the torque positive or negative? The angular velocity of the wheel is negative since it rotates clockwise. The angular acceleration has the opposite sign because the angular speed is decreasing. Since $\alpha > 0$, the net torque is also positive. Therefore,

$$\tau_{net} = +4.8 \text{ N·m}$$

Discussion The trickiest part of calculating torques is determining the sign. To check, look at the frictional forces in Fig. 8.8. Imagine which way the forces would make the wheel begin to rotate if the wheel were not originally rotating. The frictional forces point in a direction that would tend to cause a counterclockwise rotation, so the torques are positive.

Practice Problem 8.3   Disk brakes

In the disk brakes that slow down a car, the brake pads squeeze a spinning rotor; friction between the pads and the rotor provides the torque that slows down the car. If the normal force that a pad exerts on a rotor is 85 N and the coefficient of friction is 0.62, what is the frictional force on the rotor? If this force acts 8.0 cm from the rotation axis, what is the magnitude of the torque on the rotor?

Lever Arms

There is another, completely equivalent, way to calculate torques that is often more convenient than finding the perpendicular component of the force. Figure 8.12 shows a force $\vec{F}$ acting at a distance r from an axis. The distance r is the length of a line perpendicular to the axis that runs from the axis to the force's point of application. The force makes an angle θ with that line. The torque is then

$$\tau = \pm r F_\perp = \pm r(F \sin \theta)$$

The factor $\sin \theta$ could be grouped with r instead of with F. Then $\tau = \pm (r \sin \theta)F$, or

$$\tau = \pm r_\perp F \tag{8-4}$$

The distance $r_\perp$ is called the **lever arm** (or **moment arm**). The magnitude of the torque is therefore the magnitude of the force times the lever arm.

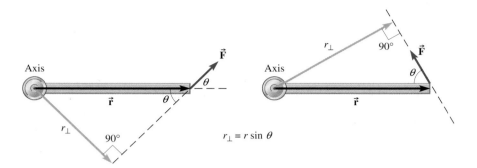

Figure 8.12 Finding the
magnitude of a torque using lever arms.

$$r_\perp = r \sin \theta$$

Finding Torques Using the Lever Arm

1. Draw a line parallel to the force through the force's point of application; this line (dashed in Fig. 8.12) is called the force's **line of action**.

2. Draw a line from the rotation axis to the line of action. This line must be perpendicular to both the axis and the line of action. The distance from the axis to the line of action along this perpendicular line is the lever arm ($r_\perp$). If the line of action of the force goes through the rotation axis, the lever arm and the torque are both zero (Fig. 8.9b).

3. The magnitude of the torque is the magnitude of the force times the lever arm:

$$\tau = \pm r_\perp F$$

4. Determine the algebraic sign of the torque as before.

Example 8.4

Screen Door Closer

An automatic screen door closer attaches to a door 47 cm away from the hinges and pulls on the door with a force of 25 N, making an angle of 15° with the door (Fig. 8.13). Find the magnitude of the torque exerted on the door due to this force about the rotation axis through the hinges using (a) the perpendicular component of the force and (b) the lever arm. (c) What is the sign of this torque as viewed from above?

Strategy For method (a), we must find the component of the 25-N force perpendicular to the radial direction. Then this component is multiplied by the length of the radial line. For method (b), we draw in the line of action of the force. Then the lever arm is the perpendicular distance from the line of action to the rotation axis. The torque is the magnitude of the force times the lever arm. We must be careful not to combine the two methods: the torque is *not* equal to the perpendicular force component times the lever arm. For (c), we determine whether this torque would tend to make the door rotate CCW or CW.

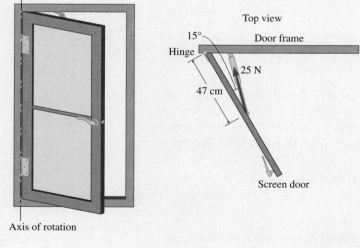

Axis of rotation

Figure 8.13
Screen door with automatic closing mechanism.

continued on next page

Example 8.4 *continued*

Solution   (a) As shown in Fig. 8.14a, the radial component of the force ($F_\parallel$) passes through the rotation axis. The angle labeled 15° would actually be a bit larger than 15°, but since the thickness of the door is much less than 47 cm, we approximate it as 15°. The perpendicular component is

$$F_\perp = F \sin 15°$$

The magnitude of the torque is

$$|\tau| = rF_\perp = 0.47 \text{ m} \times 25 \text{ N} \times \sin 15° = 3.0 \text{ N·m}$$

(b) Figure 8.14b shows the line of action of the force, drawn parallel to the force and passing through the point of application. The lever arm is the perpendicular distance between the rotation axis and the line of action. The distance r is approximately 47 cm (again neglecting the thickness of the door). Then the lever arm is

$$r_\perp = r \sin 15°$$

and the magnitude of the torque is

$$|\tau| = r_\perp F = 0.47 \text{ m} \times \sin 15° \times 25 \text{ N} = 3.0 \text{ N·m}$$

(c) Using the top view of Fig. 8.13, the torque tends to close the door by making it rotate counterclockwise (assuming the door is initially at rest and no other torques act). The torque is therefore positive as viewed from above.

Discussion   The most common mistake to make in either solution method would be to use cosine instead of sine (or, equivalently, to use the complementary angle 75° instead of 15°). A check is a good idea. If the automatic closer were more nearly parallel to the door, the angle would be less than 15°. The torque would be smaller because the force is more nearly pulling straight in toward the axis. Since the sine function gets smaller for angles closer to zero, the expression checks out correctly.

It might seem silly for a door closer to pull at such an angle that the perpendicular component is relatively small. The reason it's done that way is so the door closer does not get in the way. A closer that pulled in a perpendicular direction would stick straight out from the door. As discussed in Section 8.5, the situation is much the same in our bodies. In order to not inhibit the motion of our limbs, our tendons and muscles are nearly parallel to the bones. As a result, the forces they exert must be much larger than we might expect.

Practice Problem 8.4   Exercise is good for you

A person is lying on an exercise mat and lifts one leg at an angle of 30.0° from the horizontal with an 89-N (20-lb) weight attached to the ankle (Fig. 8.15). The distance between the ankle weight and the hip joint (which is the rotation axis for the leg) is 84 cm. What is the torque due to the ankle weight on the leg?

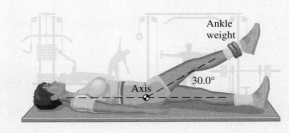

Figure 8.15
Exercise leg lifts.

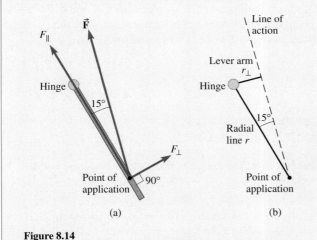

Figure 8.14
(a) Finding the perpendicular component of the force. (b) Finding the lever arm.

Center of Gravity

We have seen that the torque produced by a force depends on the point of application of the force. What about gravity? The gravitational force on a body is not exerted at a single point, but is distributed throughout the volume of the body. When we talk of "the" force of gravity on something, we really mean the total force of gravity acting on each particle making up the system.

Fortunately, when we need to find the total torque due to the forces of gravity acting on an object, the total force of gravity can be considered to act at a single point. This point is called the **center of gravity**. The torque found this way is the same as finding all the torques due to the forces of gravity acting at every point in the body and adding

Tips   When calculating the torque due to gravity, consider the entire gravitational force to act at the center of gravity.

them together. As you can verify in Problem 71, if the gravitational field is uniform in magnitude and direction, then the center of gravity of an object is located at the object's center of mass.

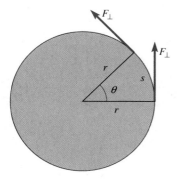

Figure 8.16 Work done by torque.

8.3   WORK DONE BY A TORQUE

Torques can do work, as anyone who has started a lawnmower with a pull cord can verify. Actually it is the force that does the work, but in rotational problems it is often simpler to calculate the work done from the torque. Just as the work done by a constant force is the product of force and the parallel component of displacement, work done by a constant torque can also be calculated as the torque times the *angular* displacement.

Imagine a torque acting on a wheel that spins through an angular displacement θ while the torque is applied. The work done by the force that gives rise to the torque is the product of the perpendicular component of the force ($F_\perp$) with the arc length s through which the point of application of the force moves (see Fig. 8.16). We use the perpendicular force component because that is the component parallel to the *displacement*, which is instantaneously tangent to the arc of the circle. Thus,

$$W = F_\perp s \tag{8-5}$$

To write the work in terms of torque, note that $\tau = rF_\perp$ and $s = r\theta$; then

$$W = F_\perp s = \frac{\tau}{r} \times r\theta = \tau\theta$$

$$W = \tau\theta \quad (\theta \text{ in radians}) \tag{8-6}$$

Work is indeed the product of torque and the angular displacement. If τ and θ have the same sign, the work done is positive; if they have opposite signs, the work done is negative. The *power* due to a constant torque—the rate at which the torque does work—is

$$P = \tau\omega \tag{8-7}$$

Example 8.5

Work Done on Potter's Wheel

A potter's wheel is a heavy stone disk upon which the pottery is shaped. Potter's wheels were once driven by the potter pushing on a foot treadle; today most potter's wheels are driven by electric motors. (a) If the potter's wheel is a uniform disk of mass 40.0 kg and diameter 0.50 m, how much work must be done by the motor to bring the wheel from rest to 80.0 rpm? (b) If the motor delivers a constant torque of 8.2 N·m during this time, through how many revolutions does the wheel turn in coming up to speed?

Strategy   Work is an energy transfer. In this case, the motor is increasing the rotational kinetic energy of the potter's wheel. Thus, the work done by the motor is equal to the change in rotational kinetic energy of the wheel, ignoring frictional losses. In the expression for rotational kinetic energy, we must express ω in rad/s; we cannot substitute 80.0 rpm for ω. Once we know the work done, we use the torque to find the angular displacement.

Solution   (a) The change in rotational kinetic energy of the wheel is

$$\Delta K = \tfrac{1}{2}I\,(\omega_f^2 - \omega_i^2) = \tfrac{1}{2}I\,\omega_f^2$$

Initially the wheel is at rest, so the initial angular velocity ω_i is zero. From Table 8.1, the rotational inertia of a uniform disk is

$$I = \tfrac{1}{2}MR^2$$

Substituting this for I,

$$\Delta K = \tfrac{1}{4}MR^2\omega_f^2$$

Before substituting numerical values, we convert 80.0 rpm to rad/s:

$$\omega_f = 80.0\,\frac{\text{rev}}{\text{min}} \times 2\pi\frac{\text{rad}}{\text{rev}} \times \frac{1}{60}\frac{\text{min}}{\text{s}} = 8.38 \text{ rad/s}$$

Substituting the known values for mass and radius,

$$\Delta K = \tfrac{1}{4} \times 40.0 \text{ kg} \times (\tfrac{0.50}{2}\text{ m})^2 \times (8.38 \text{ rad/s})^2 = 43.9 \text{ J}$$

Therefore, the work done by the motor, rounded to two significant figures, is 44 J.

continued on next page

Example 8.5 continued

(b) The work done by a constant torque is

$$W = \tau\theta$$

Solving for the angular displacement θ gives

$$\theta = \frac{W}{\tau} = \frac{43.9 \text{ J}}{8.2 \text{ N·m}} = 5.35 \text{ rad} \qquad (1)$$

Since 2π rad = 1 revolution,

$$\theta = 5.35 \text{ rad} \times \frac{1 \text{ rev}}{2\pi \text{ rad}} = 0.85 \text{ rev}$$

Discussion   As always, work is an energy transfer. In this problem, the work done by the motor is the means by which the potter's wheel acquires its rotational kinetic energy. But work done by a torque does not *always* appear as a change in rotational kinetic energy. For instance, when you wind up a mechanical clock or a windup toy, the work done by the torque you apply is stored as elastic potential energy in some sort of spring.

Practice Problem 8.5   Work done on air conditioner

A belt wraps around a pulley of radius 7.3 cm that drives the compressor of an automobile air conditioner. The tension in the belt on one side of the pulley is 45 N and on the other side of the pulley it is 27 N (Fig. 8.17). How much work is done by the belt on the compressor during one revolution of the pulley?

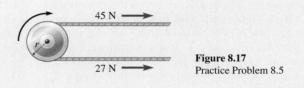

Figure 8.17
Practice Problem 8.5

8.4   EQUILIBRIUM REVISITED

In Chapter 2, we said that an object is in equilibrium when the net force acting on it is zero. That statement is true but incomplete. It is quite possible for the net force acting to be zero, while the net torque is nonzero; the object would then have a nonzero angular acceleration. When designing a bridge or a new house, it would be unacceptable for any of the parts to have nonzero angular acceleration! Zero net force is sufficient to ensure *translational* equilibrium; if an object is also in *rotational* equilibrium, then the net torque acting on it must also be zero.

Conditions for equilibrium:

$$\vec{\mathbf{F}}_{\text{net}} = 0 \ \text{ and } \ \tau_{\text{net}} = 0 \qquad (8\text{-}8)$$

Before tackling equilibrium problems, we must resolve a conundrum: if something is not rotating, then where is the axis of rotation? How can we calculate torques without knowing where the axis of rotation is? In some cases, perhaps involving axles or hinges, there may be a clear axis about which the object would rotate if the balance of forces and torques is disturbed. In many cases, though, it is not clear what the rotation axis would be, and in general it depends on exactly how the equilibrium is upset. Fortunately, the axis can be chosen *arbitrarily* when calculating torques *in equilibrium problems*.

In equilibrium, the net torque about *any* rotation axis must be zero. Does that mean that we have to write down an infinite number of torque equations, one for each possible axis of rotation? Fortunately, no. Although the proof is complicated, it can be shown that if the net force acting on an object is zero and the net torque about one rotation axis is zero, then the net torque about every other axis parallel to that axis must also be zero. Therefore, one torque equation is all we need.

Since the torque can be calculated about any desired axis, a judicious choice can greatly simplify the solution of the problem. The best place to choose the axis is usually at the point of application of an unknown force so that the unknown force does not appear in the torque equation.

Example 8.6

Carrying a 6 × 6 Beam

Two carpenters are carrying a uniform 6 × 6 beam. The beam is 2.44 m (8.00 ft) long and weighs 425 N (95.5 lb). One of the carpenters, being a bit stronger than the other, agrees to carry the beam 1.00 m in from the end; the other carries the beam at its opposite end. What is the upward force exerted on the beam by each carpenter?

Strategy The conditions for equilibrium are that the net external force equal zero and the net external torque equal zero. Should we start with forces or with torques? In this problem, it is easiest to start with torques. If we choose the axis of rotation where one of the unknown forces acts, then that force has a lever arm of zero and its torque is zero. The torque equation can be solved for the other unknown force. Then with only one force still unknown, we set the sum of the y-components of the forces equal to zero.

Solution The first step is to draw a force diagram (Fig. 8.18). Each force is drawn at the point where it acts. Known distances are labeled.

We choose a rotation axis perpendicular to the xy-plane and passing through the point of application of $\vec{F}_2$. The simplest way to find the torques for this example is to multiply each force by its lever arm. The lever arm for $\vec{F}_1$ is

$$2.44 \text{ m} - 1.00 \text{ m} = 1.44 \text{ m}$$

and the magnitude of the torque due to this force is

$$|\tau| = Fr_\perp = F_1 \times 1.44 \text{ m}$$

Since the beam is uniform, its center of gravity is at its midpoint. We imagine the entire gravitational force to act at this point. Then the lever arm for the gravitational force is

$$\tfrac{1}{2} \times 2.44 \text{ m} = 1.22 \text{ m}$$

and the torque due to gravity has magnitude

$$|\tau| = Fr_\perp = 425 \text{ N} \times 1.22 \text{ m} = 518.5 \text{ N·m}$$

The torque due to $\vec{F}_1$ is negative since, if it were the only torque, it would make the beam start to rotate clockwise about our chosen axis of rotation. The torque due to gravity is positive since, if it were the only torque, it would make the beam start to rotate counterclockwise. Therefore,

$$\tau_{\text{net}} = -F_1 \times 1.44 \text{ m} + 518.5 \text{ N·m} = 0$$

Solving for F_1,

$$F_1 = \frac{518.5 \text{ N·m}}{1.44 \text{ m}} = 360 \text{ N}$$

Since another condition for equilibrium is that the net force be zero,

$$\Sigma F_y = F_1 + F_2 - mg = 0$$

Solving for F_2,

$$F_2 = 425 \text{ N} - 360 \text{ N} = 65 \text{ N}$$

Discussion A good way to check this result is to make sure that the net torque about a *different axis* is zero—for an object in equilibrium, the net torque about any axis must be zero. Suppose we choose an axis through the point of application of $\vec{F}_1$. Then the lever arm for $m\vec{g}$ is 1.22 m − 1.00 m = 0.22 m and the lever arm for $\vec{F}_2$ is 2.44 m − 1.00 m = 1.44 m. Setting the net torque equal to zero:

$$\tau_{\text{net}} = -425 \text{ N} \times 0.22 \text{ m} + F_2 \times 1.44 \text{ m} = 0$$

Solving for F_2 gives

$$F_2 = \frac{425 \text{ N} \times 0.22 \text{ m}}{1.44 \text{ m}} = 65 \text{ N}$$

which agrees with the value calculated before. We could have used this second torque equation to find F_2 instead of setting ΣF_y equal to zero.

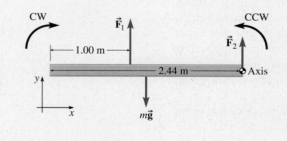

Figure 8.18
Diagram of the beam with rotation axis, forces, and distances shown.

continued on next page

Example 8.6 *continued*

Practice Problem 8.6   A diving board

A uniform diving board of length 5.0 m is supported at two points; one support is located 3.4 m from the end of the board and the second is at 4.6 m from the end (Fig. 8.19). The supports exert vertical forces on the diving board. A diver stands at the end of the board over the water. Determine the directions of the support forces. [*Hint:* In this problem, consider torques about different rotation axes.]

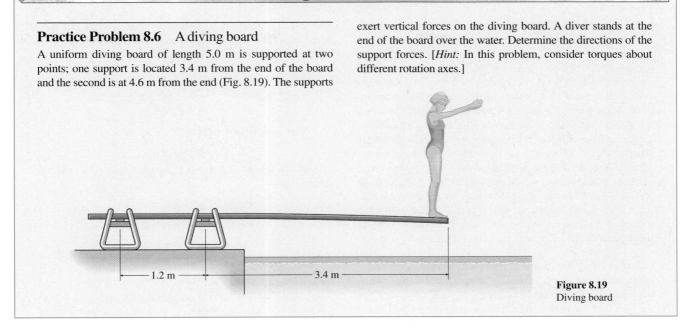

Figure 8.19
Diving board

A diving board is an example of a cantilever—a beam or pole that extends beyond its support. The forces exerted by the supports on a diving board are considerably larger than if the same board were supported at both ends (see Problem 28). The advantage is that the far end of the board is left free to vibrate; as it does, the support forces adjust themselves to keep the board from tipping over. The architect Frank Lloyd Wright was fond of using cantilever construction to open up the sides and corners of a building, allowing corner windows that give buildings a lighter and more spacious feel (Fig. 8.20).

Making The Connection:
cantilever building
construction

Figure 8.20   The cantilevered master bedroom in the north wing of Wingspread by Frank Lloyd Wright juts well out over its brick foundation. The cypress trellis extending even farther beyond the bedroom balcony filters the natural light and serves to emphasize the free-floating nature of the structure with views of the landscape below.

Example 8.7

The Slipping Ladder

A 15.0-kg uniform ladder leans against a wall in the atrium of a large hotel (Fig. 8.21a). The ladder is 8.00 m long; it makes an angle $\theta = 60.0°$ with the floor. The coefficient of static friction between the floor and the ladder is $\mu_s = 0.45$. How far along the ladder can a

continued on next page

Example 8.7 *continued*

60.0-kg person climb before the ladder starts to slip? Assume that the wall is frictionless.

Strategy Normal forces act on the ladder due to the wall ($\vec{N}_w$) and the floor ($\vec{N}_f$). A frictional force acts on the base of the ladder due to the floor ($\vec{f}$), but no frictional force acts on the top of the ladder since the wall is frictionless. Gravitational forces act on the ladder and on the person climbing it. Consider the ladder and the climber as a single system. Until the ladder starts to slip, this system is in equilibrium. Therefore, the net external force and the net external torque acting on the system are both equal to zero. As the person ascends the ladder, the frictional force $\vec{f}$ has to increase to keep the ladder in equilibrium. The ladder begins to slip when the frictional force required to maintain equilibrium is larger than its maximum possible value $\mu_s N_f$. The ladder is about to slip when $f = \mu_s N_f$.

Solution The first step is to make a careful drawing of the ladder and label all distances and forces (Fig. 8.21b). Instead of cluttering the diagram with numerical values, we use L for the length of the ladder, d for the unknown distance from the bottom of the ladder to the point where the person stands, and M and m for the masses of the person and ladder, respectively. The weight of the ladder acts at the ladder's center of gravity, which is the ladder's midpoint since it is uniform.

The conditions for equilibrium are

$$\Sigma F_x = 0, \Sigma F_y = 0, \text{ and } \Sigma \tau = 0$$

Starting with $\Sigma F_x = 0$, we find

$$N_w - f = 0$$

where, if the climber is at the highest point possible, the frictional force must have its maximum possible magnitude:

$$f = \mu_s N_f$$

Combining these two equations, we obtain a relationship between the magnitudes of the two normal forces:

$$N_w = \mu_s N_f$$

Next we use the condition $\Sigma F_y = 0$, which gives

$$N_f - Mg - mg = 0$$

The only unknown quantity in this equation is N_f, so we can solve for it:

$$Mg = 60.0 \text{ kg} \times 9.8 \text{ m/s}^2 = 588 \text{ N}$$

$$mg = 15.0 \text{ kg} \times 9.8 \text{ m/s}^2 = 147 \text{ N}$$

$$N_f = Mg + mg = 588 \text{ N} + 147 \text{ N} = 735 \text{ N}$$

Now we can find the other normal force, N_w:

$$N_w = \mu_s N_f = 0.45 \times 735 \text{ N} = 331 \text{ N}$$

At this point, we know the magnitudes of all the forces. We do not know the distance d, which is the goal of the problem. To find d we must set the net torque equal to zero.

First we choose a rotation axis. The most convenient choice is an axis perpendicular to the plane of Fig. 8.21 and passing through the bottom of the ladder. Since two of the five forces ($\vec{N}_f$ and $\vec{f}$) act at the bottom of the ladder, these two forces have zero lever arms and thus produce zero torque. Another reason why this is a convenient choice of axis is that the distance d is measured from the bottom of the ladder.

continued on next page

Figure 8.21   (a) A ladder and (b) forces acting on the ladder.

Example 8.7 *continued*

To find torques we must either find the perpendicular component of each force or find the lever arm for each force. In this situation, with the forces either vertical or horizontal, it is probably easiest to find the lever arms. In three diagrams (Fig. 8.22), we first draw the line of action for each force; then the lever arm is the perpendicular distance between the axis and the line of action.

Using the usual convention that counterclockwise torques are positive, the torque due to $\vec{N}_w$ is negative while the torques due to gravity are positive. The magnitude of each torque is the magnitude of the force times its lever arm:

$$\tau = Fr_\perp$$

Setting the net torque equal to zero yields:

$$-N_w L \sin\theta + mg\left(\tfrac{1}{2}L \cos\theta\right) + Mgd\cos\theta = 0$$

Substituting known values,

$$-331\,\text{N} \times 8.00\,\text{m} \times \sin 60.0° + 147\,\text{N} \times 4.00\,\text{m} \times \cos 60.0° +$$
$$588\,\text{N} \times d\cos 60.0° = 0$$

$$-2293\,\text{N·m} + 294\,\text{N·m} + 294\,\text{N} \times d = 0$$

$$d = \frac{2293\,\text{N·m} - 294\,\text{N·m}}{294\,\text{N}} = 6.8\,\text{m}$$

The person can climb 6.8 m up the ladder without having it slip. This is the distance *along the ladder*, not the height above the ground, which is

$$h = 6.8\,\text{m} \times \sin 60.0° = 5.9\,\text{m}$$

Discussion If the person goes any higher, then his weight produces a larger CCW torque about our chosen rotation axis. To stay in equilibrium, the total CW torque would have to get larger. The only force providing a CW torque is the normal force due to the wall, which pushes to the right. However, if this force were to get larger, the frictional force would have to get larger to keep the net horizontal force equal to zero. Since friction already has its maximum magnitude, there is no way for the ladder to be in equilibrium if the person climbs any higher.

Practice Problem 8.7 Another ladder leaning on a wall

A uniform ladder of mass 10.0 kg and length 3.2 m leans against a frictionless wall with its base located 1.5 m from the wall. If the ladder is not to slip, what must be the minimum coefficient of static friction between the bottom of the ladder and the ground? Assume the wall is frictionless.

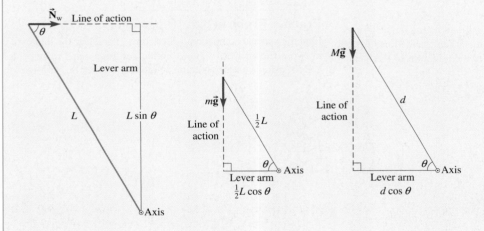

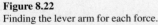

Figure 8.22
Finding the lever arm for each force.

Physics at Home

Take a dumbbell and wrap some string around the center of its axle. (An alternative: slide two spools of thread onto a pencil near its center with a small gap between the spools. Wrap some thread around the pencil between the two spools.) Place the dumbbell on a table (or on the floor). Unwind a short length of string and try pulling perpendicular to the axle at different angles to the horizontal (see Fig. 8.23). Depending on the direction of your pull, the dumbbell can roll in either direction. Try to find the angle at which the rolling changes direction; at this angle the dumbbell does not roll at all. (If using the pencil and spools of thread, pull gently and try to find the angle at which the whole thing *slides* along the table without any rotation.)

What is special about this angle? Since the dumbbell is in equilibrium when pulling at this angle, we can analyze the torques using any rotation axis we choose. A convenient choice is the axis that passes through point *P*, the point of contact with the table. Then the contact force between the table and the dumbbell acts at the rotation axis and its torque is zero. The torque due to gravity is also zero since the line of action passes through point *P*. The dumbbell can only be in equilibrium if the torque due to the remaining force (the tension in the string) is zero. This torque is zero if the lever arm is zero, which means the line of action passes through point *P*.

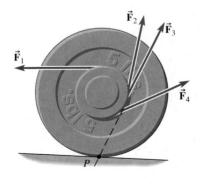

Figure 8.23 Forces $\vec{F}_1$ and $\vec{F}_2$ make the dumbbell roll to the left; $\vec{F}_4$ makes it roll to the right; $\vec{F}_3$ does not make it roll.

Example 8.8

The Sign and the Breaking Cord

A uniform beam of weight 196 N and of length 1.00 m is attached to a hinge on the outside wall of a restaurant. A cord is attached at the center of the beam and is attached to the wall, making an angle of 30.0° with the beam (Fig. 8.24a). The cord keeps the beam perpendicular to the wall. If the breaking tension of the cord is 620 N, how large can the mass of the sign be without breaking the cord?

Strategy   The beam is in equilibrium; both the net force and the net torque acting on it must be zero. To find the maximum weight of the sign, we let the tension in the cord have its maximum value of 620 N. We do not know the force exerted by the hinge on the beam, so we choose an axis of rotation through the hinge. Then the force exerted by the hinge on the beam has a zero lever arm and does not enter the torque equation.

Before doing anything else, we draw a diagram showing each force acting on the beam and the chosen rotation axis. The free-body diagrams in previous chapters often placed all the force vectors starting from a single point. Now we draw each force vector starting at its point of application so that we can find the torque—either by finding the lever arm or by finding the perpendicular force component and the distance from the axis to the point of application.

Solution   Figure 8.24b shows the forces acting on the beam; three of these contribute to the torque. The gravitational force on the beam can be taken to act at the midpoint of the beam

since it is uniform. The force due to the cord has a perpendicular component (Fig. 8.24c) of

$$F_\perp = 620 \text{ N sin } 30.0° = 310 \text{ N}$$

The two gravitational forces tend to rotate the beam clockwise, while the tension in the cord tends to rotate it counterclockwise. As an alternative to setting the net torque equal to zero, we can set the total magnitude of the CW torques equal to the total magnitude of the CCW torques:

$$0.50 \text{ m} \times 196 \text{ N} + 1.00 \text{ m} \times Mg = 0.50 \text{ m} \times 310 \text{ N}$$

or

$$1.00 \text{ m} \times Mg = 0.50 \text{ m} \times (310 \text{ N} - 196 \text{ N})$$

Now we solve for the unknown mass M:

$$M = \frac{0.50 \text{ m} \times (310 \text{ N} - 196 \text{ N})}{1.00 \text{ m} \times 9.8 \text{ N/kg}} = 5.8 \text{ kg}$$

Discussion   In this problem we did not have to set the net force equal to zero. By placing the axis of rotation at the hinge we eliminated two of the three unknowns from the torque equation: the horizontal and vertical components of the hinge force (or, equivalently, its magnitude and direction). If we wanted to find the hinge force as well, setting the net force equal to zero would be necessary.

Practice Problem 8.8   Hinge forces
Find the vertical component of the force exerted by the hinge in two different ways: (a) setting the net force equal to zero and (b) using a torque equation about a different axis.

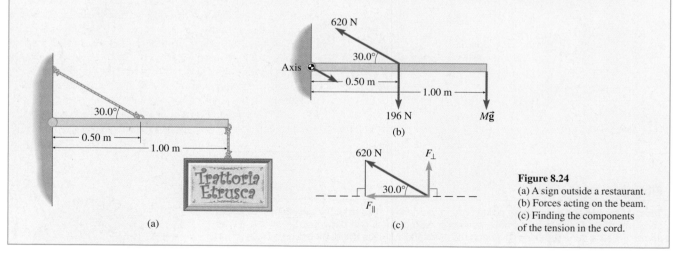

Figure 8.24
(a) A sign outside a restaurant.
(b) Forces acting on the beam.
(c) Finding the components of the tension in the cord.

Distributed Forces

Gravity is not the only force that is distributed rather than acting at a point. Contact forces, including both the normal component and friction, are spread over the contact surface. Just as for gravity, we can consider the contact force to act at a single point, but the location of that point is often not at all obvious. For a book sitting on a horizontal table, it seems reasonable that the normal force effectively acts at the geometric center of the book cover that touches the table. It is less clear where that effective point is if the book is on an incline or is sliding. As Example 8.9 shows, when something is about to topple over, contact is about to be lost everywhere except at the corner around which the toppling object is about to rotate. That corner then must be the location of the contact forces.

Example 8.9

The Toppling File Cabinet

 A file cabinet of height a and width b is on a ramp at angle θ (Fig. 8.25a). The file cabinet's center of gravity is at its geometric center. Find the largest θ for which the file cabinet does not tip over. Assume the coefficient of static friction is large enough to prevent sliding.

Strategy Until the file cabinet begins to tip over, it is in equilibrium; the net force acting on it must be zero and the total torque about any axis must also be zero. We first draw a force diagram showing the three forces (gravity, normal, friction) acting on the file cabinet. The point of application of the two contact forces (normal, friction) must be at the lower edge of the file cabinet if it is on the steepest possible incline, just about to tip over. In that case, contact has been lost over the rest of the bottom surface of the file cabinet so that only the lower edge makes good contact with the ramp.

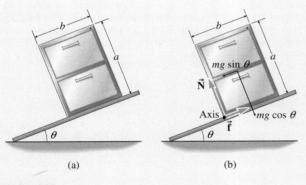

Figure 8.25
(a) File cabinet on an incline. (b) Forces acting on the file cabinet.

As in all equilibrium problems, a good choice of rotation axis makes the problem easier to solve. We know that, at the maximum angle, the contact forces act at the bottom edge of the file cabinet. A good choice of rotation axis is along the bottom edge of the file cabinet, because then the normal and frictional forces have zero lever arm.

Solution Figure 8.25b shows the forces acting on the file cabinet at the maximum angle θ. The gravitational force is drawn at the center of gravity. Instead of drawing a single vector arrow for the gravitational force, we represent the gravitational force by its components parallel and perpendicular to the ramp. Then we find the lever arm for each of the components. The lever arm for the parallel component of the weight ($mg \sin \theta$) is $\frac{1}{2}a$ and the lever arm for the perpendicular component ($mg \cos \theta$) is $\frac{1}{2}b$. Setting the net torque equal to zero:

$$\tau_{net} = -mg \cos \theta \times \tfrac{1}{2}b + mg \sin \theta \times \tfrac{1}{2}a = 0$$

After dividing out the common factors of $\frac{1}{2}mg$,

$$\cos \theta \times b = \sin \theta \times a$$

Solving for θ,

$$\theta = \tan^{-1} \frac{b}{a}$$

Discussion As a check, we can regard the normal and friction forces as two components of a single contact force. We can think of that contact force as acting at a single point—a "center of contact" analogous to the center of gravity. As the file cabinet is put on steeper and steeper surfaces, the effective point of application of the contact force moves toward the lower edge of the file cabinet (see Fig. 8.26). If we take the rotation axis through the center of gravity so there is no gravitational torque, then the torque due to the contact force must be zero. The only way that can happen is if its lever arm is zero, which means that the contact force must point directly toward the center of gravity. If the angle θ has its maximum

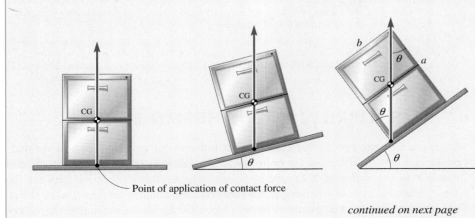

Point of application of contact force

Figure 8.26
Contact force for various incline angles.

continued on next page

Example 8.9 *continued*

value, the contact force acts at the lower edge and tan $\theta = b/a$. The file cabinet is about to tip when its center of gravity is directly above the lower edge. Any object supported only by contact forces can be in equilibrium only if the point of application of the total contact force is directly below the object's center of gravity.

Conceptual Practice Problem 8.9   Gymnast holding a pike position

Figure 8.27 shows a gymnast holding a pike position. What can you say about the location of the gymnast's center of gravity?

Figure 8.27
Yuri Chechi of Italy holds the pike position on the rings at the World Gymnastic Championships in Sabae, Japan.

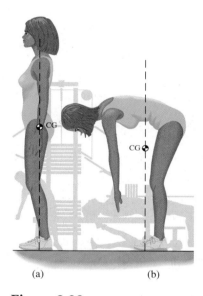

Figure 8.28 Location of the center of gravity when (a) standing and (b) touching toes.

Physics at Home

When a person stands up straight, the body's center of gravity lies directly above a point between the feet, about 3 cm in front of the ankle joint (see Fig. 8.28a). When a person bends over to touch her toes, the center of gravity lies outside the body (Fig. 8.28b). Note that the lower half of the body must move backward to keep the center of gravity from moving out in front of the toes, which would cause the person to fall over.

An interesting experiment can be done that illustrates what happens to your balance when you shift your center of gravity. Stand against a wall with the heels of your feet touching the wall and your back pressed against the wall. Then carefully try to bend over as if to touch your toes, without bending your knees. Can you do this without falling over? Explain.

Problem-Solving Steps in Equilibrium Problems

- Identify an object or system in equilibrium. Draw a diagram showing all the forces acting on that object, each drawn at its point of application. Use the center of gravity as the point of application of any gravitational forces.
- To apply the force condition $\Sigma \vec{F} = 0$, choose a convenient coordinate system and resolve each force into its x- and y-components.
- To apply the torque condition $\Sigma \tau = 0$, choose a convenient rotation axis—generally one that passes through the point of application of an unknown force. Then find the torque due to each force. Use whichever method is easier: either the lever arm times the magnitude of the force or the distance times the perpendicular component of the force. Determine the direction of each torque; then either set the sum of all the torques (with their correct signs) equal to zero or set the magnitude of the CW torques equal to the magnitude of the CCW torques.
- Not all problems require all three equations (two force component equations and one torque equation). Sometimes it is easier to use more than one torque equation, with a different axis. Before diving in and writing down all the equations, think about which approach is the easiest and most direct.

Making The Connection:
flexor versus extensor muscles

8.5   EQUILIBRIUM IN THE HUMAN BODY

We can use the concepts of torque and equilibrium to understand some of how the musculoskeletal system of the human body works. A muscle has tendons at each end that connect it to two different bones across a joint (the flexible connection between the bones). When the muscle contracts, it pulls the tendons, which in turn pull on the bones. Thus, the muscle produces a pair of forces of equal magnitude, one acting on each of the two bones. The biceps muscle (Fig. 8.29) in the upper arm attaches the scapula to the forearm (radius) across the inside of the elbow joint. When the biceps contracts, the forearm is pulled toward the upper arm. The biceps is a *flexor* muscle; it moves one bone closer to another.

A muscle can pull but not push, so a flexor muscle such as the biceps cannot reverse its action to push the forearm away from the upper arm. The *extensor* muscles make bones move apart from each other. In the upper arm an extensor muscle, the triceps (Fig. 8.29), connects the humerus to the ulna (another bone in the forearm parallel to the radius) across the outside of the elbow. Since the biceps and triceps connect to the forearm on opposite sides of the elbow joint, they tend to cause rotation about the joint in opposite directions. When the triceps contracts it pulls the forearm away from the upper arm. Using flexor and extensor muscles on opposite sides of the joint, the body can produce both positive and negative torques, even though both muscles pull in the same direction.

Suppose the arm is held in a horizontal position. The deltoid muscle exerts a force $\vec{\mathbf{F}}$ on the humerus at an angle of about 15° above the horizontal. This force has to do two things. The vertical component (magnitude $F \sin 15° \approx 0.26F$) supports the weight of the arm, while the horizontal component (magnitude $F \cos 15° \approx 0.97F$) stabilizes the joint by pulling the humerus in against the shoulder (scapula). In Example 8.10 we estimate the magnitude of $\vec{\mathbf{F}}$.

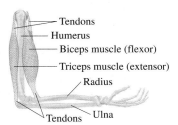
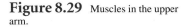

Figure 8.29   Muscles in the upper arm.

Example 8.10

Force to Hold Arm Horizontal

A person is standing with his arm outstretched in a horizontal position. The weight of the arm is 30.0 N and its center of gravity is at the elbow joint, 27.5 cm from the shoulder joint (Fig. 8.30). The deltoid pulls on the upper arm at an angle of 15° above the horizontal and at a distance of 12 cm from the joint. What is the magnitude of the force exerted by the deltoid muscle on the arm?

Strategy   The arm is in equilibrium, so we can apply the conditions for equilibrium: $\vec{\mathbf{F}}_{net} = 0$ and $\tau_{net} = 0$. When calculating torques, we choose the rotation axis at the shoulder joint because then the unknown force $\vec{\mathbf{F}}_s$, which acts on the arm at the joint, has a zero lever arm and produces zero torque. With only one unknown in the torque equation, we can solve immediately for F_m. We do not need to apply the condition $\vec{\mathbf{F}}_{net} = 0$ unless we want to find $\vec{\mathbf{F}}_s$.

Solution   The gravitational force is perpendicular to the line between its point of application and the rotation axis. Gravity produces a clockwise torque of magnitude

$$|\tau| = Fr = 30.0 \text{ N} \times 0.275 \text{ m} = 8.25 \text{ N·m}$$

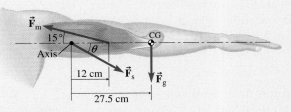

Figure 8.30
Forces exerted on an outstretched arm by the deltoid muscle ($\vec{\mathbf{F}}_m$), the scapula ($\vec{\mathbf{F}}_s$), and gravity ($\vec{\mathbf{F}}_g$)

For the torque due to $\vec{\mathbf{F}}_m$, we find the component of $\vec{\mathbf{F}}_m$ that is perpendicular to the line between its point of application and the rotation axis. Since this line is horizontal, we need the vertical component of $\vec{\mathbf{F}}_m$, which is $F_m \sin 15°$. Then the magnitude of the counterclockwise torque due to $\vec{\mathbf{F}}_m$ is

$$|\tau| = F_\perp r = F_m \sin 15° \times 0.12 \text{ m}$$

These torques must be equal in magnitude:

magnitude of CCW torque = magnitude of CW torque

$$F_m \sin 15° \times 0.12 \text{ m} = 8.25 \text{ N·m}$$

Solving for F_m,

$$F_m = \frac{8.25 \text{ N·m}}{\sin 15° \times 0.12 \text{ m}} = 270 \text{ N}$$

Discussion   The force exerted by the muscle is much larger than the 30.0-N weight of the arm. The muscle must exert a larger force because the lever arm is small; the point of application is less than half as far from the joint as the center of gravity (0.12 m/0.275 m ≈ 4/9). Also, the muscle cannot pull straight up on the arm; the vertical component of the muscle force is only about $\frac{1}{4}$ of the magnitude of the force. These two factors together make the weight supported (30.0 N) only $\frac{4}{9} \times \frac{1}{4} = \frac{1}{9}$ as large as the force exerted by the muscle.

Practice Problem 8.10   Holding a juice carton

Find the force exerted by the same person's deltoid muscle when holding a 1-L juice carton (weight 9.9 N) with the arm outstretched and parallel to the floor (as in Fig. 8.30). Assume that the juice carton is 60.0 cm from the shoulder.

Making The Connection:
muscle forces for
the iron cross
(gymnastics)

The Iron Cross

When a gymnast does the iron cross (Fig. 8.31a), the primary muscles involved are the latissimus dorsi ("lats") and pectoralis major ("pecs"). Since the rings are supporting the gymnast's weight, they exert an upward force on the gymnast's arms. Thus, the task for the muscles is not to hold the arm up, but to pull it down. The lats pull on the humerus about 3.5 cm from the shoulder joint (Fig. 8.31b). The pecs pull on the humerus about 5.5 cm from the joint (Fig. 8.31c). The other ends of these two muscles connect to bone in many places, widely distributed over the back (lats) and chest (pecs). As a reasonable simplification we can assume that these muscles pull at a 45° angle below the horizontal in the iron cross maneuver. We also assume that the two muscles exert equal forces, so we can replace the two with a single force acting at 4.5 cm from the joint.

To determine the force exerted, we look at the entire arm as a system in equilibrium. This time we can ignore the weight of the arm itself since the force exerted on the arm by the ring is much larger—half the gymnast's weight is supported by each ring. The ring exerts an upward force that acts on the hand about 60 cm from the shoulder joint (see Fig. 8.31d). Taking torques about the shoulder, in equilibrium we have

$$|\text{CW torque}| = |\text{CCW torque}|$$

$$F_m \times 0.045 \text{ m} \times \sin 45° = \tfrac{1}{2}W \times 0.60 \text{ m}$$

$$F_m = \frac{\tfrac{1}{2}W \times 0.60 \text{ m}}{0.045 \text{ m} \times \sin 45°}$$

$$= 9.4W$$

Thus, the force exerted by the lats and pecs *on one side* of the gymnast's body is more than nine times his weight.

The design of the human body makes large muscular forces necessary. Are there advantages to the design? Due to the small lever arms, the muscle forces are much larger than they would otherwise be, but the human body has traded this for a wide range of movement of the bones. The biceps and triceps muscles can move the lower arms through almost 180° while they change their lengths by only a few centimeters. The muscles also remain nearly parallel to the bones. If the biceps and triceps muscles were attached to the lower arm much farther from the elbow, there would have to be a large flap of skin to allow them to move so far away from the bones. The arrangement of our bones and muscles favors a wide range of movement.

Another advantage of the design is that it tends to minimize the rotational inertia of our limbs. For example, the muscles that control the motion of the lower arm are contained mostly within the *upper* arm. This keeps the rotational inertia of the lower arms about the elbow smaller. It also keeps the rotational inertia of the entire arm about the shoulder smaller. Smaller rotational inertia means that the energy we have to expend to move our limbs around is smaller.

The biceps muscle with its tendons is almost parallel to the humerus. One interesting observation is that the tendon connects to the radius at different points in different people. In one person this point may be 5.0 cm from the elbow joint, while in another person whose arm is the same length it may be 5.5 cm from the elbow. Thus, some people are naturally stronger than others because of their internal structure. Chimpanzees have an advantage over humans because their biceps muscle has a longer lever arm. Do not make the mistake of arm wrestling with an adult chimp; challenge the chimp to a game of chess instead.

Making The Connection:
forces on the
human spine during
heavy lifting

Heavy Lifting

When lifting an object from the floor, our first instinct is to bend over and pick it up. This is not a good way to lift something heavy. The spine is an ineffective lever and is susceptible to damage when a heavy object is lifted with bent waist. It is much better to squat down and use the powerful leg muscles to do the lifting instead of using our back muscles. Analyzing torques in a simplified model of the back can illustrate why.

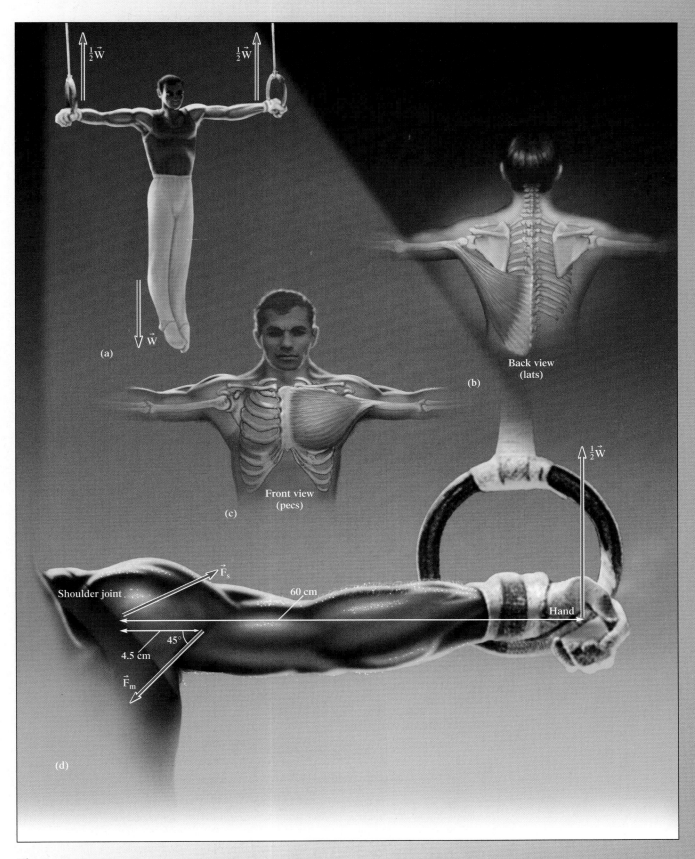

Figure 8.31   (a) Gymnast doing the iron cross. The principal muscles involved are (b) the "lats" and (c) the "pecs." (d) Simplified model of the forces acting on the arm of the gymnast.

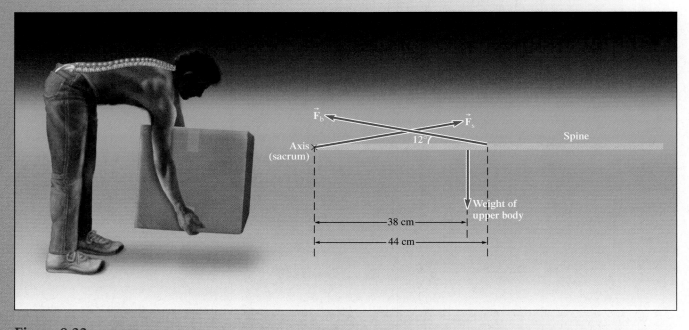

Figure 8.32 A simplified model of the human back when bent over.

The spine can be modeled as a rod with an axis at the tailbone (the sacrum). The sacrum exerts a force, marked $\vec{\mathbf{F}}_s$ in Fig. 8.32, when a person bends at the waist with the back horizontal. The forces due to the complicated set of back muscles can be replaced with a single equivalent force $\vec{\mathbf{F}}_b$ as shown. This equivalent force makes an angle of 12° with the spine and acts about 44 cm from the sacrum. The weight of the upper body, $m\vec{\mathbf{g}}$ in Fig. 8.32, is about 65% of total body weight; its center of gravity is about 38 cm from the sacrum. By placing an axis at the sacrum we can ignore the force $\vec{\mathbf{F}}_s$ in our torque equation. Since the vertical component of $\vec{\mathbf{F}}_b$ is $F_b \sin 12° \approx 0.21 F_b$, only about $\frac{1}{5}$ the magnitude of the forces exerted by the back muscles is supporting the body weight. The rest, the horizontal component, is pressing the rod representing the spine into the sacrum.

If we put some numbers into this example we can get an idea of the forces required for just supporting the upper body in this position. If the person's total weight is 710 N (160 lb), then the upper body weight is

$$mg = 0.65 \times 710 \text{ N}$$

Now we set the magnitude of the CCW torques about the axis equal to the magnitude of the CW torques:

$$F_b \times 0.44 \text{ m} \times \sin 12° = mg \times 0.38 \text{ m}$$

Substituting and solving,

$$F_b = \frac{0.65 \times 710 \text{ N} \times 0.38 \text{ m}}{0.44 \text{ m} \times \sin 12°} = 1920 \text{ N}$$

The muscular force that compresses the spine is the horizontal component of $\vec{\mathbf{F}}_b$:

$$F_b \cos 12° = 1900 \text{ N}$$

or about 430 lb. This is over four times the weight of the upper body.

Now if the person tries to lift something with his arms in this position, the lever arm for the weight of the load is even longer than for the weight of the upper body. The back muscles must supply a much larger force. The spine is now compressed with a dangerously large force. A cushioning disk called the lumbosacral disk, at the bottom of the spine, separates the last vertebra from the sacrum. This disk can be ruptured or deformed, causing great pain when the back is misused in such a fashion.

If, instead of bending over, we bend our knees and lower our body, keeping it vertically aligned as much as possible while lifting a load, the centers of gravity of the body and load are positioned more closely in a line above the sacrum, as in Fig. 8.33. Then the lever arms of these forces with respect to an axis through the sacrum are relatively small and the force on the lumbosacral disk is roughly equal to the upper body weight plus the weight being lifted.

Figure 8.33   A safer way to lift a heavy object.

8.6   ROTATIONAL FORM OF NEWTON'S SECOND LAW

The concepts of torque and rotational inertia can be used to formulate a "Newton's second law for rotation"—a law that fills the role of $\Sigma\vec{F} = m\vec{a}$ for rotation about a fixed axis. What is that role? Newton's second law determines the translational acceleration of an object if its mass and the net force acting on it are known. For rotation, we want to determine the angular acceleration, so α takes the place of $\vec{a}$. Net torque takes the place of net force. If the net torque is zero, then there is zero angular acceleration; the greater the net torque, the greater the angular acceleration. In place of mass, rotational inertia measures how difficult it is to change the angular velocity. The law that enables us to calculate the angular acceleration of a rigid object given the applied torques takes a form analogous to $\Sigma\vec{F} = m\vec{a}$:

> **Rotational form of Newton's second law**
>
> $$\Sigma\tau = I\alpha \qquad (8\text{-}9)$$

Thus, the angular acceleration of a rigid body is proportional to the net torque (more torque causes a larger α) and is inversely proportional to the rotational inertia (more inertia causes a smaller α). In equilibrium, the angular acceleration must be zero; Eq. (8-9) then requires that the net torque be zero. We used $\Sigma\tau = 0$ as one of the conditions of equilibrium in Sections 8.4 and 8.5.

Equation (8-9) is proved in Problem 48. It is subject to an important restriction. Just as $\Sigma\vec{F} = m\vec{a}$ is valid only if the mass of the object is constant, $\Sigma\tau = I\alpha$ is valid only if the rotational inertia of the object is constant. For a rigid object rotating about a fixed axis, I cannot change, so Eq. (8-9) is always applicable.

Tips   When calculating the net torque, remember to assign the correct algebraic sign to each torque before adding them.

Tips   The sum of the torques due to internal forces acting on a rigid object is always zero. Therefore, only *external* torques need be included in Eq. (8-9)

Example 8.11

The Grinding Wheel

A grinding wheel is a solid, uniform disk of mass 2.50 kg and radius 9.00 cm. Starting from rest, what constant torque must a motor supply so that the wheel attains a rotational speed of 126 rev/s in a time of 6.00 s?

Strategy   Since the grinding wheel is a uniform disk, we can find its rotational inertia using Table 8.1. After converting the revolutions per second to radians per second, we can find the angular acceleration from the change in angular velocity over the given time interval. Once we have I and α, we can find the net torque from Newton's second law for rotation.

Solution   The grinding wheel is a uniform disk, so its rotational inertia is

$$I = \tfrac{1}{2}mr^2$$

$$\tfrac{1}{2} \times 2.50\ \text{kg} \times (0.0900\ \text{m})^2 = 0.010125\ \text{kg·m}^2$$

A single rotation of the wheel is equivalent to 2π radians, so

$$\omega = 126\,\frac{\text{rev}}{\text{s}} \times 2\pi\,\frac{\text{rad}}{\text{rev}}$$

The angular acceleration is

$$\alpha = \frac{\Delta\omega}{\Delta t}$$

continued on next page

Example 8.11 *continued*

Then the torque required is

$$\tau_{net} = I\alpha = I\frac{\Delta\omega}{\Delta t} = (0.010125 \text{ kg·m}^2) \times \frac{(126 \text{ rev/s} \times 2\pi \text{ rad/rev})}{6.00 \text{ s}}$$

$$= 1.34 \text{ N·m}$$

If there are no other torques on the wheel, the motor must supply a constant torque of 1.34 N·m.

Discussion   We assumed that no other torques are exerted on the wheel. There is certain to be at least a small frictional

torque on the wheel with a sign opposite to the sign of the motor's torque. Then the motor would have to supply a torque larger than 1.34 N·m. The *net* torque would still be 1.34 N·m.

Practice Problem 8.11   Another approach

Verify the answer to Example 8.11 by: (a) finding the angular displacement of the wheel using kinematic equations for constant α; (b) finding the change in rotational kinetic energy of the wheel; and (c) finding the torque from $W = \tau\theta$.

Newton's second law for rotation explains why a tightrope walker carries a long pole to help maintain balance. If the acrobat is about to topple over sideways, the pole would have to go with him, rotating up and over the rope in a large arc. The pole has a large rotational inertia due to its length, so a large torque is required to make the pole start to rotate. The angular acceleration of the system (acrobat plus pole) due to a small gravitational torque is much smaller than it would be without the pole. The pole greatly increases the stability of the acrobat.

8.7   THE DYNAMICS OF ROLLING OBJECTS

A rolling object combines translational motion of the center of mass with rotation about an axis that passes through the center of mass (Section 5.1). For an object that is rolling without slipping, $v_{CM} = \omega R$. As a result, there is a specific relationship between the rolling object's translational and rotational kinetic energies. The total kinetic energy of a rolling object is the sum of its translational and rotational kinetic energies.

A wheel with mass M and radius R has a rotational inertia that is some pure number times MR^2; it couldn't be anything else and still have the right units. We can write the rotational inertia about an axis through the center of mass as $I_{CM} = \beta MR^2$ where β is a pure number that measures how far from the axis of rotation the mass is distributed. Larger β means the mass is, on average, farther from the axis. From Table 8.1, a hoop has $\beta = 1$; a disk, $\beta = \frac{1}{2}$; and a solid sphere, $\beta = \frac{2}{5}$.

Using $I_{CM} = \beta MR^2$ and $v_{CM} = \omega R$, the rotational kinetic energy for a rolling object can be written

$$K_{rot} = \frac{1}{2}I_{CM}\omega^2 = \frac{1}{2} \times \beta MR^2 \times \left(\frac{v_{CM}}{R}\right)^2 = \beta \times \frac{1}{2}Mv_{CM}^2$$

Since $\frac{1}{2}Mv_{CM}^2$ is the translational kinetic energy,

$$K_{rot} = \beta K_{tr} \tag{8-10}$$

This is convenient since β depends only on the shape, not on the mass or radius of the object. For a given shape rolling without slipping, the ratio of its rotational to translational kinetic energy is always the same (β).

The total kinetic energy can be written

$$K = K_{tr} + K_{rot}$$

$$K = \frac{1}{2}Mv_{CM}^2 + \frac{1}{2}I_{CM}\omega^2 \tag{8-11}$$

or in terms of β,

$$K = (1 + \beta) K_{tr}$$

$$K = (1 + \beta) \frac{1}{2}Mv_{CM}^2 \tag{8-12}$$

Thus, two objects of the same mass rolling at the same translational speed do *not* have the same kinetic energy. The object with the larger value of β has more rotational kinetic energy.

Conceptual Example 8.12

Hollow and Solid Rolling Balls

Starting from rest, two balls are rolled down a hill as in Fig. 8.34. One is solid, the other hollow. Which one is moving faster when it reaches the bottom of the hill?

Strategy and Solution   Energy conservation is the best way to approach this problem. As a ball rolls down the hill, its gravitational potential energy decreases as its kinetic energy increases by the same amount. The total kinetic energy is the sum of the translational and rotational contributions.

We do not know the mass or the radius of either ball and we cannot assume they are the same. Since both kinetic and potential energies are proportional to mass, mass does not affect the final speed. Also, the total kinetic energy does not depend on the radius of the ball [see Eq. (8-12)]. The final speeds of the two balls differ because a different *fraction* of their total kinetic energies is translational.

One ball is a solid sphere and the other is approximately a spherical shell. The mass of a spherical shell is all concentrated on the surface of a sphere, while a solid sphere has its mass distributed throughout the sphere's volume. Therefore, the shell has a larger β than the solid sphere. When the shell rolls, it converts a bigger fraction of the lost potential energy into rotational kinetic energy; therefore, a smaller fraction becomes translational kinetic energy. The final speed of the solid sphere is larger since it puts a larger fraction of its kinetic energy into translational motion.

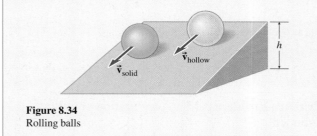

Figure 8.34
Rolling balls

Discussion   We can make this conceptual question into a quantitative one: what is the ratio of the speeds of the two balls at the bottom of the hill?

Let the height of the hill be h. Then for a ball of mass M, the loss of gravitational potential energy is Mgh. This amount of gravitational potential energy is converted into translational and rotational kinetic energy:

$$Mgh = K_{tr} + K_{rot} = (1 + \beta)K_{tr} = (1 + \beta)\frac{Mv^2}{2}$$

Mass cancels out, as expected. We can solve for the final speed in terms of g, h, and β. The final speed is independent of the ball's mass and radius.

$$v_{CM} = \sqrt{\frac{2gh}{1 + \beta}}$$

The ratio of the final speeds for two balls rolling down the same hill is therefore

$$\frac{v_1}{v_2} = \sqrt{\frac{1 + \beta_2}{1 + \beta_1}}$$

To evaluate the ratio, we look up the rotational inertias in Table 8.1. The solid sphere has $\beta = \frac{2}{5}$ and the spherical shell has $\beta = \frac{2}{3}$. Then

$$\frac{v_{solid}}{v_{hollow}} = \sqrt{\frac{1 + \frac{2}{3}}{1 + \frac{2}{5}}} \approx 1.091$$

The solid ball's final speed is therefore 9.1% faster than that of the hollow ball. This ratio depends neither on the masses of the balls, the radii of the balls, the height of the hill, nor the slope of the hill.

Practice Problem 8.12   Fraction of kinetic energy that is rotational energy

What fraction of a rolling ball's kinetic energy is rotational kinetic energy? Answer both for a solid ball and a hollow one.

What is the acceleration of a ball rolling down an incline? Figure 8.35 includes an FBD for such a ball. Static friction is the force that makes the ball rotate; if there were no friction, instead of rolling the ball would just *slide* down the incline. This is true because friction is the only force acting that yields a nonzero torque about the rotation axis, which is through the ball's center of mass. Gravity gives zero torque because it acts at the axis, so the lever arm is zero. The normal force points directly at the axis, so its lever arm is also zero.

The frictional force $\vec{f}$ provides a torque

$$\tau = rf$$

where r is the ball's radius. An analysis of the forces and torques combined with Newton's second law in both forms enables us to calculate the acceleration of the ball in Example 8.13.

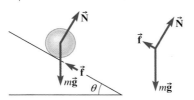

Figure 8.35   Forces acting on a ball rolling downhill.

Example 8.13

Acceleration of a Rolling Ball

Calculate the acceleration of a solid ball rolling down a slope inclined at an angle θ to the horizontal (Fig. 8.36a).

Strategy   The axis of rotation is through the ball's center of mass. As already discussed, neither gravity nor the normal force produce a torque about this axis; the net torque is $\tau_{net} = rf$, where f is the magnitude of the frictional force. The net torque is related to the angular acceleration by $\tau_{net} = I\alpha$, Newton's second law for rotation. Similarly, the net force acting on the ball gives the acceleration of the center of mass: $\vec{F}_{net} = m\vec{a}_{CM}$. One problem is that the force of friction is unknown. We must resist the temptation to assume that $f = \mu_s N$; there is no reason to assume that static friction has its maximum possible magnitude. We do know that the two accelerations, translational and rotational, are related. We know that v_{CM} and ω are proportional since r is constant. To stay proportional they must change in lock step; their rates of change, a_{CM} and α, are proportional to each other by the same factor of r. Thus, $a_{CM} = \alpha r$. This connection should enable us to eliminate f and solve for the acceleration. Since the speed of a ball after rolling a certain distance was found to be independent of the mass and radius of the ball in Example 8.12, we expect the same to be true of the acceleration.

Solution   Since the net torque is

$$\tau_{net} = rf$$

the angular acceleration is

$$\alpha = \frac{\tau}{I} = \frac{rf}{I} \qquad (1)$$

where I is the ball's rotational inertia about its center of mass.

Figure 8.36b shows the forces along the incline acting on the ball. The acceleration of the center of mass is found from Newton's second law. The component of the net force acting along the incline (in the direction of the acceleration) is

$$F = mg \sin \theta - f = ma_{CM} \qquad (2)$$

Because the ball is rolling without slipping, the acceleration of the center of mass and the angular acceleration are related by

$$a_{CM} = \alpha r$$

Now we try to eliminate the unknown frictional force f from the equations above. Solving Eq. (1) for f gives

$$f = \frac{I\alpha}{r}$$

Substituting this into Eq. (2), we get

$$mg \sin \theta - \frac{I\alpha}{r} = ma_{CM}$$

Now to eliminate α, we can substitute $\alpha = a_{CM}/r$:

$$mg \sin \theta - \frac{Ia_{CM}}{r^2} = ma_{CM}$$

Solving for a_{CM},

$$a_{CM} = \frac{g \sin \theta}{1 + (I/mr^2)}$$

For a solid sphere, $I = \frac{2}{5} mr^2$, so

$$a_{CM} = \frac{g \sin \theta}{1 + \frac{2}{5}} = \frac{5}{7} g \sin \theta$$

Discussion   The acceleration of an object *sliding* down an incline without friction is $a = g \sin \theta$. The acceleration of the rolling ball is smaller than $g \sin \theta$ due to the frictional force directed up the incline.

We can check the answer using the result of Example 8.12. Since the ball's acceleration is constant, the kinematic quantities are related by familiar equations. If the ball starts from rest as in Fig. 8.36a, after it has rolled a distance d, its speed v is

$$v = \sqrt{2ad} = \sqrt{2\left(\frac{g \sin \theta}{1 + \beta}\right)d}$$

where $\beta = \frac{2}{5}$. The vertical drop during this time is $h = d \sin \theta$, so

$$v = \sqrt{\frac{2gh}{1 + \beta}}$$

Practice Problem 8.13   Acceleration of hollow cylinder

Calculate the acceleration of a hollow cylinder rolling down a slope inclined at an angle θ to the horizontal.

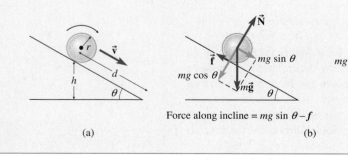

Force along incline = $mg \sin \theta - f$

(a)   (b)

Figure 8.36
(a) A ball accelerating downhill and (b) forces along the incline with the gravitational force resolved into components perpendicular and parallel to the incline.

8.8 ANGULAR MOMENTUM

Newton's second law for translational motion can be written in two ways:

$$\vec{F}_{net} = \lim_{\Delta t \to 0} \frac{\Delta \vec{p}}{\Delta t} \text{ (general form)} \quad \text{or} \quad \vec{F}_{net} = m\vec{a} \text{ (constant mass)}$$

In Eq. (8-9) we wrote Newton's second law for rotation as $\tau_{net} = I\alpha$, which applies only when I is constant—that is, for a rigid body rotating about a fixed axis. A more general form of Newton's second law for rotation uses the concept of **angular momentum** (symbol L).

The net external torque acting on a system is equal to
the rate of change of the angular momentum of the system.

$$\tau_{net} = \lim_{\Delta t \to 0} \frac{\Delta L}{\Delta t} \tag{8-13}$$

The angular momentum of a rigid body rotating about a fixed axis is the rotational inertia times the angular velocity, which is analogous to the definition of linear momentum (mass times velocity):

Angular momentum

$$L = I\omega \tag{8-14}$$

(rigid body, fixed axis)

Either Eq. (8-13) or Eq. (8-14) can be used to show that the SI units of angular momentum are $kg \cdot m^2/s$.

For a rigid body rotating around a fixed axis, angular momentum doesn't tell us anything new. The rotational inertia is constant for such a body since the distance r_i between every point on the object and the axis stays the same. Then any change in angular momentum must be due to a change in angular velocity ω:

$$\tau_{net} = \lim_{\Delta t \to 0} \frac{\Delta L}{\Delta t} = \lim_{\Delta t \to 0} \frac{I\Delta\omega}{\Delta t} = I \lim_{\Delta t \to 0} \frac{\Delta\omega}{\Delta t} = I\alpha$$

However, Eq. (8-13) is *not* restricted to rigid objects or to fixed rotation axes. In particular, if the net external torque acting on a system is zero, then the angular momentum of the system cannot change. This is the **law of conservation of angular momentum**:

Conservation of angular momentum

If $\tau_{net} = 0$,

$$L_i = L_f \tag{8-15}$$

Here L_i and L_f represent the angular momentum of the system at two different times. Conservation of angular momentum is one of the most basic and fundamental laws of physics, along with the two other conservation laws we have studied so far (energy and linear momentum). For an isolated system, the total energy, total linear momentum, and total angular momentum of the system are each conserved. None of these quantities can change unless some external agent causes the change.

In this section, we restrict our consideration to cases where the axis of rotation is fixed but where the rotational inertia is not necessarily constant. One familiar example of a changing rotational inertia occurs when a figure skater spins (Fig. 8.37). To start the spin, the skater glides along with her arms outstretched and then begins to rotate her body about a vertical axis by pushing against the ice with a skate. The push of the ice against the skate provides the external torque that gives the skater her initial angular momentum. Initially the skater's arms and the leg not in contact with the ice are extended away from her body. The mass of the arms and leg when extended contribute

Tips Conservation of angular momentum can be applied to any system if the net external torque on the system is zero (or negligibly small).

Making The Connection:
rotational inertia of
a figure skater

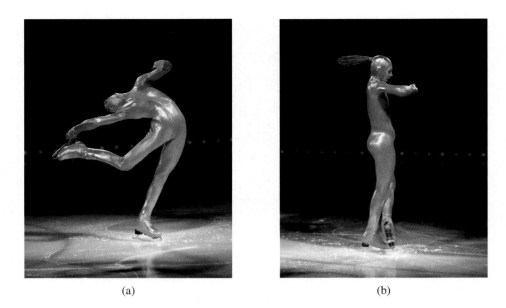

Figure 8.37 Figure skater Lucinda Ruh at the (a) beginning and (b) end of a spin. Her angular velocity is much higher in (b) than in (a).

(a)   (b)

more to her rotational inertia than they do when held close to the body. As the skater spins, she pulls her arms and leg close to her body to decrease her rotational inertia. As she does, her angular velocity increases dramatically. The concept of angular momentum helps explain why she spins faster as she pulls in her arms and leg.

As the skater spins, there is a negligibly small external torque due to friction between her skates and the ice. With no significant external torque acting, her angular momentum must be conserved:

$$L_i = L_f$$

Even though her rotational inertia changes, at the beginning and again at the end of the spin we can think of her as a rigid body. Then

$$I_i \omega_i = I_f \omega_f$$

where the subscripts "i" and "f" refer to before and after the skater pulls her arms and leg close to her body. The final rotational inertia is less than the initial rotational inertia because the arms and leg are much closer to the axis of rotation. The skater decreases her inertia by moving her arms and leg in toward her body:

$$I_f < I_i$$

In order for her angular momentum to stay the same,

$$\omega_f = \frac{I_i}{I_f} \omega_i > \omega_i$$

which means that she is spinning faster at the end than she was initially.

Many natural phenomena can be understood in terms of angular momentum. In a hurricane, circulating air is sucked inward by a low pressure region at the center of the storm (the *eye*). As the air moves closer and closer to the axis of rotation, it circulates faster and faster. An even more dramatic example is the formation of a pulsar. Under certain conditions, a star can implode under its own gravity, forming a neutron star (a collection of tightly packed neutrons). If the Sun were to collapse into a neutron star, its radius would be only about 13 km. If a star is rotating before its collapse, then as its rotational inertia decreases dramatically, its angular velocity must increase to keep its angular momentum constant. Such rapidly rotating neutron stars are called pulsars because they emit regular pulses of x-rays, at the same frequency as their rotation, that can be detected when they reach Earth. Some pulsars rotate in only a few thousandths of a second per revolution.

Example 8.14

Mouse on a Wheel

A 0.10-kg mouse is perched at point B on the rim of a 2.00-kg bicycle wheel that rotates freely in a horizontal plane at 1.00 rev/s (Fig. 8.38). The mouse crawls to point A at the center. Assume the mass of the wheel is concentrated at the rim. What is the frequency of rotation in rev/s when the mouse arrives at point A?

Strategy Assuming that frictional torques are negligibly small, there is no external torque acting on the mouse/wheel system. Then the angular momentum of the mouse/wheel system must be conserved; it takes an external torque to change angular momentum. The mouse and wheel exert torques on each other, but these *internal* torques only transfer some angular momentum between the wheel and the mouse without changing the total angular momentum. We can think of the system as initially being a rigid body with rotational inertia I_i. When the mouse reaches the center, we think of the system as a rigid body with a different rotational inertia I_f. The mouse changes the rotational inertia of the mouse/wheel system by moving from the outer rim, where its mass makes the maximum possible contribution to the rotational inertia, to the rotation axis, where its mass makes no contribution to the rotational inertia.

Solution Initially, all of the mass of the system is at a distance R from the rotation axis, where R is the radius of the wheel. Therefore,

$$I_i = (M + m)R^2$$

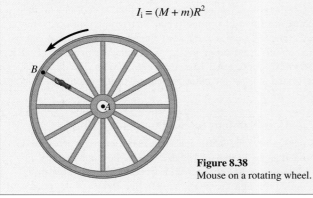

Figure 8.38
Mouse on a rotating wheel.

where M is the mass of the wheel and m is the mass of the mouse. After the mouse moves to the center of the wheel, its mass contributes nothing to the rotational inertia of the system:

$$I_f = MR^2$$

From conservation of angular momentum,

$$I_i \omega_i = I_f \omega_f$$

Substituting the rotational inertias and $\omega = 2\pi f$,

$$(M + m)R^2 \times 2\pi f_i = MR^2 \times 2\pi f_f$$

Factors of $2\pi R^2$ cancel from each side, leaving

$$(M + m)f_i = Mf_f$$

Solving for f_f,

$$f_f = \frac{M + m}{M} f_i = \frac{2.10 \text{ kg}}{2.00 \text{ kg}} (1.00 \text{ rev/s}) = 1.05 \text{ rev/s}$$

Discussion Conservation laws are powerful tools. We do not need to know the details of what happens as the mouse crawls along the spoke from the outer edge of the wheel; we need only look at the initial and final conditions.

A common mistake in this sort of problem is to say that the initial rotational kinetic energy is equal to the final rotational kinetic energy. This is not true because the mouse crawling in toward the center must expend energy to do so. In other words, the mouse does work, converting some internal energy into rotational kinetic energy.

Practice Problem 8.14 Change in rotational kinetic energy

What is the percent change in the rotational kinetic energy of the mouse/wheel system?

Angular Momentum in Planetary Orbits

Conservation of angular momentum applies to planets orbiting the Sun in elliptical orbits. Kepler's second law says that the orbital speed varies in such a way that the planet sweeps out area at a constant rate (Fig. 8.39a). In Problem 83 you can show that Kepler's second law is a direct result of conservation of angular momentum, where the angular momentum of the planet is calculated using an axis of rotation perpendicular to the plane of the orbit and passing through the Sun. When the planet is closer to the Sun, it moves faster; when it is farther away, it moves more slowly. Conservation of angular momentum can be used to relate the orbital speeds and radii at two different points in the orbit. The same applies to satellites and moons orbiting planets.

Making The Connection:
Kepler's laws of planetary motion

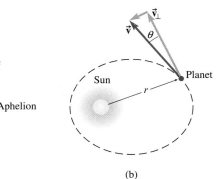

Figure 8.39 The planet's speed varies such that it sweeps out equal areas in equal time intervals. The eccentricity of the planetary orbit is exaggerated for clarity.

(a)   (b)

Example 8.15

Earth's Orbital Speed

At perihelion (closest approach to the Sun), Earth is 1.47×10^8 km from the Sun and its orbital speed is 30.3 km/s. What is Earth's orbital speed at aphelion (greatest distance from the Sun), when it is 1.52×10^8 km from the Sun? Note that at these two points Earth's velocity is perpendicular to a radial line from the Sun (see Fig. 8.39a).

Strategy We take the axis of rotation through the Sun. Then the gravitational force on Earth points directly toward the axis; with zero lever arm, the torque is zero. With no other external forces acting on the Earth, the net external torque is zero. Earth's angular momentum about the rotation axis through the Sun must therefore be conserved. To find Earth's rotational inertia, we treat it as a point particle since its radius is much less than its distance from the axis of rotation.

Solution The rotational inertia of the Earth is

$$I = mr^2$$

where m is Earth's mass and r is its distance from the Sun. The angular velocity is

$$\omega = \frac{v_\perp}{r}$$

where $v_\perp$ is the component of the velocity perpendicular to a radial line from the Sun. At the two points under consideration, $v_\perp = v$. As the distance from the Sun r varies, its speed v must vary to conserve angular momentum:

$$I_i \omega_i = I_f \omega_f$$

By substitution,

$$mr_i^2 \times \frac{v_i}{r_i} = mr_f^2 \times \frac{v_f}{r_f}$$

or

$$r_i v_i = r_f v_f \qquad (1)$$

Solving for v_f,

$$v_f = \frac{r_i}{r_f} v_i = \frac{1.47 \times 10^8 \text{ km}}{1.52 \times 10^8 \text{ km}} \times 30.3 \text{ km/s} = 29.3 \text{ km/s}$$

Discussion Earth moves slower at a point farther from the Sun. This is what we expect from energy conservation. The potential energy is greater at aphelion than at perihelion. Since the mechanical energy of the orbit is constant, the kinetic energy must be smaller at aphelion.

Equation (1) implies that the orbital speed and orbital radius are inversely proportional, but strictly speaking this equation only applies to the perihelion and aphelion. At a general point in the orbit, the *perpendicular component* $v_\perp$ is inversely proportional to r (see Fig. 8.39b). The orbits of Earth and most of the other planets are nearly circular so that $\theta \approx 0°$ and $v_\perp \approx v$.

Practice Problem 8.15   Puck on a string

A puck on a frictionless, horizontal air table is attached to a string that passes down through a hole in the table. Initially the puck moves at 12 cm/s in a circle of radius 24 cm. If the string is pulled through the hole, reducing the radius of the puck's circular motion to 18 cm, what is the new speed of the puck?

8.9   THE VECTOR NATURE OF ANGULAR MOMENTUM

Until now we have treated torque and angular momentum as scalar quantities. Such a treatment is adequate in the cases we have considered so far. However, the law of conservation of angular momentum applies to *all* systems, including rotating objects whose axis of rotation changes direction. Torque and angular momentum are actually vector quantities. Angular momentum is conserved in *both magnitude and direction* in the absence of external torques.

An important special case is that of a symmetric object rotating about an axis of symmetry, such as the spinning disk in Fig. 8.40. The magnitude of the angular

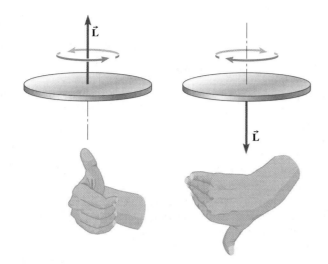

Figure 8.40   Right-hand rule for finding the direction of the angular momentum of a spinning disk.

momentum of such an object is $L = I\omega$. The direction of the angular momentum vector points along the axis of rotation. To find which of the two directions along the axis is correct, a **right-hand rule** is used. Align your right hand so that, as you curl your fingers in toward your palm, your fingertips follow the object's rotation; then your thumb points in the direction of $\vec{L}$.

Making The Connection:
angular momentum
of a gyroscope

A disk with a large rotational inertia can be used as a *gyroscope*. When the gyroscope spins at a large angular velocity, it has a large angular momentum. It is then difficult to change the orientation of the gyroscope's rotation axis, because to do so requires changing its angular momentum. To change the direction of a large angular momentum requires a correspondingly large torque. Thus a gyroscope can be used to maintain stability. Gyroscopes are used in guidance systems in airplanes, submarines, and space vehicles to maintain a constant direction in space.

The same principle explains the great stability of rifle bullets and spinning tops. A rifle bullet is made to spin as it passes through the rifle's barrel. The spinning bullet then keeps its correct orientation—nose first—as it travels through the air. Otherwise a small torque due to air resistance could make the bullet turn around randomly, greatly increasing air resistance and undermining accuracy. A properly thrown football is made to spin for the same reasons. A spinning top can stay balanced for a long time, while the same top falls over immediately when it is not spinning.

The Earth's rotation gives it a large angular momentum. As the Earth orbits the Sun, the axis of rotation stays in a fixed direction in space. The axis points nearly at Polaris (the North Star), so even as the Earth rotates around its axis, Polaris maintains its position in the northern sky. The fixed direction of the rotation axis also gives us the regular progression of the seasons (see Fig. 8.41).

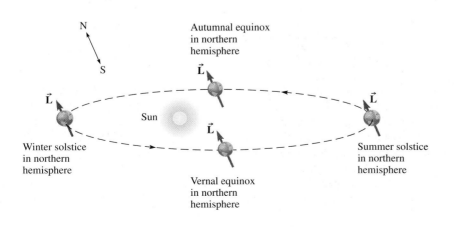

Figure 8.41   Spinning like a top, the Earth maintains the direction of its angular momentum due to rotation as it revolves around the Sun.

A Classic Demonstration

A demonstration often done in physics classes is for a student to hold a spinning bicycle wheel while standing on a platform that is free to rotate. The wheel's rotation axis is initially horizontal (Fig. 8.42a). Then the student repositions the wheel so that its axis of rotation is vertical (Fig. 8.42b). As he repositions the wheel, the platform begins to rotate opposite to the wheel's rotation. If we assume *no* friction acts to resist rotation of the platform, then the platform continues to rotate as long as the wheel is held with its axis vertical. If the student returns the wheel to its original orientation, the rotation of the platform stops.

The platform is free to rotate about a vertical axis. As a result, once the student steps onto the platform, *the vertical component L_y of the angular momentum of the system* (student + platform + wheel) is conserved. The horizontal components of $\vec{L}$ are *not* conserved. The platform is not free to rotate about any horizontal axis since the floor can exert external torques to keep it from doing so. In vector language, we would say that only the vertical component of the external torque is zero, so only the vertical component of angular momentum is conserved.

Initially $L_y = 0$ since the student and the platform have zero angular momentum and the wheel's angular momentum is horizontal. When the wheel is repositioned so that it spins with an upward angular momentum ($L_y > 0$), the rest of the system (the student and the platform) must acquire an equal magnitude of downward angular momentum ($L_y < 0$) so that the vertical component of the total angular momentum is still zero. Thus, the platform and student rotate in the opposite sense from the rotation of the wheel. Since the platform and student have more rotational inertia than the wheel, they do not spin as fast as the wheel, but their vertical angular momentum is just as large.

The student and the wheel apply torques to each other to transfer angular momentum from one part of the system to the other. These torques are equal and opposite and they have both vertical and horizontal components. As the student lifts the wheel, he feels a strange twisting force that tends to rotate him about a horizontal axis. The platform prevents the horizontal rotation by exerting unequal normal forces on the student's feet. The horizontal component of the torque is so counterintuitive that, if the student is not expecting it, he can easily be thrown from the platform!

Figure 8.42 A demonstration of angular momentum conservation.

(a)   (b)

MASTER THE CONCEPTS

Summary

- The rotational kinetic energy of a rigid object with rotational inertia I and angular velocity ω is

$$K_{rot} = \tfrac{1}{2}I\omega^2 \qquad (8\text{-}1)$$

In this expression, ω must be measured in *radians* per unit time.

- Rotational inertia is a measure of how difficult it is to change an object's angular velocity. It is defined as:

$$I = \sum_{i=1}^{N} m_i r_i^2 \qquad (8\text{-}2)$$

where r_i is the perpendicular distance between a particle of mass m_i and the rotation axis. The rotational inertia depends on the location of the rotation axis.

- Torque measures the effectiveness of a force for twisting or turning an object. It can be calculated in two equivalent ways: either as the product of the perpendicular component of the force with the shortest distance between the rotation axis and the point of application of the force

$$\tau = \pm r F_{\perp} \qquad (8\text{-}3)$$

or as the product of the magnitude of the force with its lever arm (the perpendicular distance between the line of action of the force and the axis of rotation)

$$\tau = \pm r_{\perp} F \qquad (8\text{-}4)$$

- A force whose perpendicular component tends to cause rotation in the CCW direction gives rise to a positive torque; a force whose perpendicular component tends to cause rotation in the CW direction gives rise to a negative torque.

- The work done by a constant torque is the product of the torque and the angular displacement:

$$W = \tau\theta \quad (\theta \text{ in radians}) \qquad (8\text{-}6)$$

- The conditions for equilibrium are

$$\vec{F}_{net} = 0 \text{ and } \tau_{net} = 0 \qquad (8\text{-}8)$$

The rotation axis can be chosen *arbitrarily* when calculating torques in equilibrium problems. Generally, the best place to choose the axis is at the point of application of an unknown force so that the unknown force does not appear in the torque equation.

- Newton's second law for rotation is

$$\Sigma\tau = I\alpha \qquad (8\text{-}9)$$

where radian measure must be used for α. A more general form is

$$\Sigma\tau = \lim_{\Delta t \to 0}\frac{\Delta L}{\Delta t} \qquad (8\text{-}13)$$

where L is the angular momentum of the system.

- The total kinetic energy of a body that is rolling without slipping is the sum of the rotational kinetic energy about an axis through the center of mass and the translational kinetic energy:

$$K = \tfrac{1}{2}Mv_{CM}^2 + \tfrac{1}{2}I_{CM}\omega^2 \qquad (8\text{-}11)$$

- The angular momentum of a rigid body rotating about a fixed axis is the rotational inertia times the angular velocity:

$$L = I\omega \qquad (8\text{-}14)$$

- The law of conservation of angular momentum: if the net external torque acting on a system is zero, then the angular momentum of the system cannot change.

$$\text{If } \tau_{net} = 0, L_i = L_f \qquad (8\text{-}15)$$

- This table summarizes the analogous quantities in translational and rotational motion.

Translation	Rotation
m	I
$\vec{F}$	τ
$\vec{a}$	α
$\Sigma\vec{F} = m\vec{a}$	$\Sigma\tau = I\alpha$
Δx	θ
$W = F_x \Delta x$	$W = \tau\theta$
$\vec{v}$	ω
$K = \tfrac{1}{2}mv^2$	$K = \tfrac{1}{2}I\omega^2$
$\vec{p} = m\vec{v}$	$L = I\omega$
$\Sigma\vec{F} = \lim\limits_{\Delta t \to 0}\dfrac{\Delta\vec{p}}{\Delta t}$	$\Sigma\tau = \lim\limits_{\Delta t \to 0}\dfrac{\Delta L}{\Delta t}$
If $\Sigma\vec{F} = 0$, $\vec{p}$ is conserved	If $\Sigma\tau = 0$, L is conserved

Highlighted Figures and Tables

F8.1 and **F8.2** Rotational inertia depends on the rotational axis (pp. 253, 254)

T8.1 Rotational inertia for uniform objects with various geometrical shapes (p. 253)

F8.8 Nonzero net torque with zero net force (p. 257)

F8.9 Torque as a product of the perpendicular component of the force (p. 257)

F8.11 Positive and negative torque (p. 258)

F8.12 Using lever arms to find torques (p. 260)

F8.16 Work done by torque (p. 262)

F8.18 Rotational axis, forces, and distances in equilibrium problems (p. 264)

F8.22 Finding the lever arm for each force (p. 267)

F8.26 Contact force for various incline angles (p. 269)

F8.35 Forces acting on an object rolling downhill (p. 277)

F8.39 Angular momentum in planetary orbits (p. 282)

F8.42 Angular momentum conservation on a platform that is free to rotate (p. 284)

CONCEPTUAL QUESTIONS

1. Explain why it is easier to drive a wood screw using a screwdriver with a large diameter handle rather than one with a thin handle.

2. One way to find the center of gravity of an irregular flat object is to suspend it from various points so that it is free to rotate (Fig. 8.43). When the object hangs in equilibrium, a vertical line is drawn downward from the support point. After drawing lines from several different support points, the center of gravity is the point where the lines all intersect. Explain how this works.

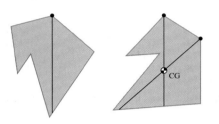
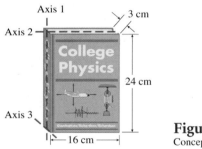

Figure 8.43
Conceptual Question 2

3. One of the effects of significant global warming would be the melting of part or all of the polar ice caps. This, in turn, would change the length of the day (the period of the Earth's rotation). Explain why. Would the day get longer or shorter?

4. A book measures 3 cm by 16 cm by 24 cm. About which of the axes shown in Fig. 8.44 is its rotational inertia smallest?

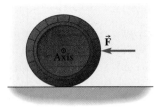

Figure 8.44
Conceptual Question 4

5. A body in static equilibrium has only two forces acting on it. We found in Section 2.4 that the forces must be equal in magnitude and opposite in direction in order to give a net force of zero. What else must be true of the two forces? [*Hint:* Consider the lines of action of the forces.]

6. Why do many helicopters have a small propeller attached to the tail that rotates in a vertical plane? Why is this attached at the tail rather than somewhere else? [*Hint:* Most of the helicopter mass is forward, in the cab.]

7. In the "Pinewood Derby," Cub Scouts construct cars and then race them down an incline. Some say that, everything else being equal (friction, drag coefficient, same wheels, etc.), a heavier car will win; others maintain that the weight of the car does not matter. Who is right? Explain. [*Hint:* Think about the fraction of the car's kinetic energy that is rotational.]

8. A large barrel lies on its side. In order to roll it across the floor, you apply a horizontal force, as shown in Fig. 8.45. If the applied force points toward the axis of rotation, which runs down the center of the barrel through the center of mass, it produces zero torque about that axis. How then can this applied force make the barrel start to roll?

Figure 8.45
Conceptual Question 8

9. Animals that can run fast always have thin legs. Their leg muscles are concentrated close to the hip joint; only tendons extend into the lower leg. Using the concept of rotational inertia, explain how this helps them run fast.

10. Figure 8.46a shows a simplified model of how the triceps muscle connects to the forearm. As the angle θ is changed, the tendon wraps around a nearly circular arc. Explain how this is much more effective than if the tendon is connected as in Fig. 8.46b. [*Hint:* Look at the lever arm as θ changes.]

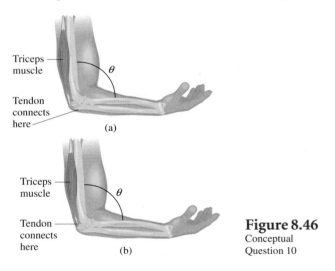

Figure 8.46
Conceptual
Question 10

11. Figure 8.47a shows a simplified model of how the biceps muscle enables the forearm to support a load. What are the

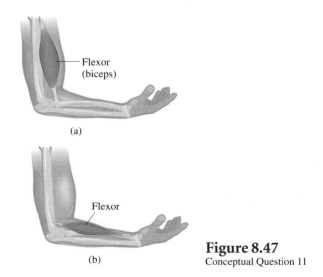

Figure 8.47
Conceptual Question 11

advantages of this arrangement as opposed to the alternative shown in Fig. 8.47b, where the flexor muscle is in the forearm instead of in the upper arm? Are the two equally effective when the forearm is horizontal? What about for other angles between the upper arm and the forearm? Consider also the rotational inertia of the forearm about the elbow and of the entire arm about the shoulder.

12. In Section 8.6, it was asserted that the sum of all the internal torques (that is, the torques due to internal forces) acting on a rigid object is zero. Figure 8.48 shows two particles in a rigid object. The particles exert forces $\vec{F}_{12}$ and $\vec{F}_{21}$ on each other. These forces are directed along a line that joins the two particles. Explain why the torques due to these two forces must be equal and opposite even though the forces are applied at different points (and therefore possibly different distances from the axis).

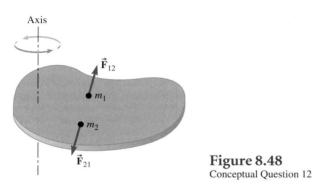

Figure 8.48
Conceptual Question 12

13. A playground merry-go-round (Fig. 8.4) spins with negligible friction. A child moves from the center out to the rim of the merry-go-round platform. Let the system be the merry-go-round plus the child. Which of these quantities change: angular velocity of the system, rotational kinetic energy of the system, angular momentum of the system? Explain your answer.

14. Figure 8.49 shows a common balancing toy with weights extending on either side. The toy is extremely stable. It can be pushed quite far off center one way or the other but it does not fall over. Explain why it is so stable.

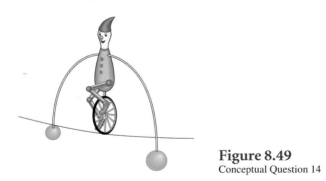

Figure 8.49
Conceptual Question 14

15. Explain why the posture taken by defensive football linemen (Fig. 8.50) makes them more difficult to push out of the way. Consider both the height of the center of gravity and the size

Figure 8.50
Conceptual Question 15

of the support base (the area on the ground bounded by the hands and feet touching the ground). In order to knock a person over, what has to happen to the center of gravity? Which do you think needs a more complex neurological system for maintaining balance: four legged animals or humans?

16. The CG of the upper body of a bird is located below the hips; in a human, the CG of the upper body is located well above the hips (see Fig. 8.51). Since the upper body is supported by the hips, are birds or humans more stable? Consider what happens if the upper body is displaced a little so that its CG is not directly above or below the hips. In what direction does the torque due to gravity tend to make the upper body rotate about an axis through the hips?

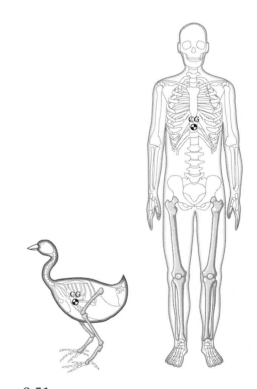

Figure 8.51   Conceptual Question 16

MULTIPLE CHOICE QUESTIONS

1. A heavy box is resting on the floor (Fig. 8.52). You would like to push the box to tip it over on its side, using the minimum force possible. Which of the force vectors in the diagram shows the correct location and direction of the force? Assume enough friction so that the box does not slide; instead it rotates about point P.

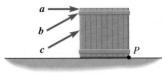

Figure 8.52 Multiple Choice Question 1

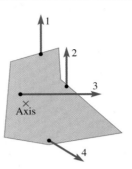

Figure 8.53 Multiple Choice Questions 6–8

2. When both are expressed in terms of SI *base* units, torque has the same units as
   (a) angular acceleration   (b) angular momentum
   (c) force   (d) energy
   (e) rotational inertia   (f) angular velocity

Questions 3–4. A uniform solid cylinder rolls without slipping down an incline. At the bottom of the incline, the speed, v, of the cylinder is measured and the translational and rotational kinetic energies (K_{tr}, K_{rot}) are calculated. A hole is drilled through the cylinder along its axis and the experiment is repeated; at the bottom of the incline the cylinder now has speed v' and translational and rotational kinetic energies K'_{tr} and K'_{rot}.

3. How does the speed of the cylinder compare to its original value?
   (a) $v' < v$ (b) $v' = v$ (c) $v' > v$
   (d) answer depends on the radius of the hole drilled

4. How does the ratio of rotational to translational kinetic energy of the cylinder compare to its original value?
   (a) $\dfrac{K'_{rot}}{K'_{tr}} < \dfrac{K_{rot}}{K_{tr}}$ (b) $\dfrac{K'_{rot}}{K'_{tr}} = \dfrac{K_{rot}}{K_{tr}}$ (c) $\dfrac{K'_{rot}}{K'_{tr}} > \dfrac{K_{rot}}{K_{tr}}$
   (d) answer depends on the radius of the hole drilled

5. The SI units of angular momentum are
   (a) $\dfrac{\text{rad}}{\text{s}}$ (b) $\dfrac{\text{rad}}{\text{s}^2}$ (c) $\dfrac{\text{kg·m}}{\text{s}^2}$
   (d) $\dfrac{\text{kg·m}^2}{\text{s}^2}$ (e) $\dfrac{\text{kg·m}^2}{\text{s}}$ (f) $\dfrac{\text{kg·m}}{\text{s}}$

6. Which of the forces in Fig. 8.53 produces the largest magnitude torque about the rotation axis indicated?
   (a) 1   (b) 2   (c) 3   (d) 4

7. Which of the forces in Fig. 8.53 produces a clockwise torque about the rotation axis indicated?
   (a) 3 only   (b) 4 only   (c) 1 and 2
   (d) 1, 2, and 3   (e) 1, 2, and 4

8. Which pair of forces in Fig. 8.53 might produce equal magnitude torques with opposite signs?
   (a) 2 and 3   (b) 2 and 4   (c) 1 and 2
   (d) 1 and 3   (e) 1 and 4   (f) 3 and 4

9. A high diver in midair pulls her legs inward toward her chest in order to rotate faster. Doing so changes which of these quantities: her angular momentum L, her rotational inertia I, and her rotational kinetic energy K_{rot}?
   (a) L only   (b) I only   (c) K_{rot} only
   (d) L and I only   (e) I and K_{rot} only   (f) all three

10. A student, moving with angular velocity ω on a platform that is free to rotate, is holding a 3-kg dumbbell in each hand at arm's length from his axis of rotation. Without changing the positions of his arms, he drops the two dumbbells to the ground. What effect does this have on the system of student and platform?
   (a) The angular velocity increases.
   (b) The angular velocity decreases.
   (c) The angular velocity is unchanged.
   (d) Both the angular velocity and the rotational kinetic energy increase.
   (e) Both the angular velocity and the rotational kinetic energy decrease.

PROBLEMS

Note: Ⓒ indicates a combination conceptual/quantitative problem. Gold diamonds ✦, ✦✦ are used to indicate the increasing level of difficulty of each problem. Problem numbers appearing in blue, 9., denote problems that have a detailed solution available in the Student Solutions Manual. Some problems are *paired* by concept; their numbers are connected by a ruled box.

8.1 Rotational Kinetic Energy and Rotational Inertia

1. Verify that $\frac{1}{2}I\omega^2$ has dimensions of energy.
2. Four point masses of 3.0 kg each are arranged in a square on massless rods (Fig. 8.54). The length of a side of the square is

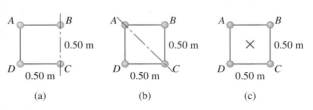

Figure 8.54 Problem 2

0.50 m. What is the rotational inertia for rotation about an axis (a) passing through masses B and C? (b) passing through masses A and C? (c) passing through the center of the square and perpendicular to the plane of the square?

3. What is the rotational inertia of a solid iron disk of mass 49 kg, with a thickness of 5.00 cm and radius of 20.0 cm, about an axis through its center and perpendicular to it?

4. A bowling ball made for a child has half the radius of an adult bowling ball. They are made of the same material (and therefore have the same mass *per unit volume*). By what factor is the (a) mass and (b) rotational inertia of the child's ball reduced compared to the adult ball?

5. How much work is done by the motor in a CD player to make a CD spin, starting from rest? The CD has a diameter of 12.0 cm and a mass of 15.8 g. The laser scans at a constant tangential velocity of 1.20 m/s. Assume that the music is first detected at a radius of 20.0 mm from the center of the disk. Ignore the small circular hole at the CD's center.

6. Find the ratio of the rotational inertia of the Earth for rotation about its own axis to its rotational inertia for rotation about the Sun.

7. A bicycle has wheels of radius 0.32 m. Each wheel has a rotational inertia of 0.080 kg·m² about its axle. The total mass of the bicycle including the wheels and the rider is 79 kg. When coasting at constant speed, what fraction of the total kinetic energy of the bicycle (including rider) is the rotational kinetic energy of the wheels?

8. In many problems in previous chapters (especially Chapter 6), cars and other objects that roll on wheels were considered to act as if they were sliding without friction. (a) Can the same assumption be made for a wheel rolling *by itself*? Explain your answer. (b) If a moving car of total mass 1300 kg has four wheels, each with rotational inertia of 0.705 kg·m² and radius of 35 cm, what fraction of the total kinetic energy is rotational?

9. A centrifuge has a rotational inertia of 6.5×10^{-3} kg·m². How much energy must be supplied to bring it from rest to 420 rad/s (4000 rpm)?

8.2 Torque

10. A mechanic turns a wrench using a force of 25 N at a distance of 16 cm from the rotation axis. The force is perpendicular to the wrench handle. What magnitude torque does she apply to the wrench?

11. The pull cord of a lawnmower engine is wound around a drum of radius 6.00 cm. While the cord is pulled with a force of 75 N to start the engine, what magnitude torque does the cord apply to the drum?

12. Any pair of equal and opposite forces acting on the same object is called a *couple*. Consider the couple in Fig. 8.55a. The rotation axis is perpendicular to the page and passes through point P. (a) Show that the net torque due to this couple is equal to Fd, where d is the distance between the lines of action of the two forces. Because the distance d is independent of the location of the rotation axis, this shows that the torque is the same for any rotation axis. (b) Repeat for the couple in Fig. 8.55b. Show that the torque is still Fd if d is the *perpendicular* distance between the lines of action of the forces.

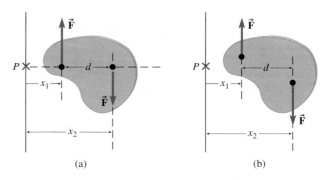

(a)   (b)

Figure 8.55 (a) A couple acting along a radius from the rotation axis located at P and perpendicular to the page; (b) a couple offset from the radial line containing the rotation axis. (Problem 12)

13. A uniform door weighs 50.0 N and is 1.0 m wide and 2.6 m high. What is the magnitude of the torque due to the door's own weight about a horizontal axis perpendicular to the door and passing through a corner?

14. A child of mass 40.0 kg is sitting on a horizontal seesaw at a distance of 2.0 m from the supporting axis. What is the magnitude of the torque about the axis due to the weight of the child?

15. A 124-g mass is placed on one pan of a balance, at a point 25 cm from the support of the balance. What is the magnitude of the torque about the support exerted by the mass?

16. A tower outside the Houses of Parliament in London has a famous clock commonly referred to as Big Ben, the name of its 13-ton chiming bell. The hour hand of each clock face is 2.7 m long and has a mass of 60.0 kg. Assume the hour hand to be a uniform rod attached at one end. (a) What is the torque on the clock mechanism due to the weight of one of the four hour hands when the clock strikes noon? The axis of rotation is perpendicular to a clock face and through the center of the clock. (b) What is the torque due to the weight of one hour hand about the same axis when the clock tolls 9:00 A.M.?

17. A weightless rod, 10.0 m long, supports three weights (Fig. 8.56). Where is its center of gravity?

5.0 kg   15.0 kg   10.0 kg

0.0   5.0 m   10.0 m   **Figure 8.56** Problem 17

18. A door weighing 300.0 N measures 2.00 m × 3.00 m and is of uniform density; that is, the mass is uniformly distributed throughout the volume. A doorknob is attached to the door as

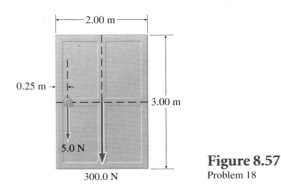

Figure 8.57 Problem 18

shown in Fig. 8.57. Where is the center of gravity if the door-knob weighs 5.0 N and is located 0.25 m from the edge?

19. A plate of uniform thickness is shaped as shown in Fig. 8.58. Where is the center of gravity? Assume the origin (0, 0) is located at the lower left corner of the plate; the upper left corner is at (0, s) and upper right corner is at (s, s).

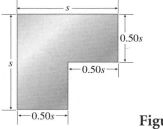

Figure 8.58   Problem 19

8.3 Work Done by a Torque

20. The radius of a wheel is 0.500 m. A rope is wound around the outer rim of the wheel. The rope is pulled with a force of magnitude 5.00 N, unwinding the rope and making the wheel spin counterclockwise about its central axis. Ignore the mass of the rope. (a) How much rope unwinds while the wheel makes 1.00 revolution? (b) How much work is done by the rope on the wheel during this time? (c) What is the torque on the wheel due to the rope? (d) What is the angular displacement θ, in radians, of the wheel during 1.00 revolution? (e) Show that the numerical value of the work done is equal to the product $\tau\theta$.

21. A stone used to grind wheat into flour is turned through 12 revolutions by a constant force of 20.0 N applied to the rim of a 10.0-cm-radius shaft connected to the wheel. How much work is done on the stone during the 12 revolutions?

22. A flywheel of mass 182 kg has an effective radius of 0.62 m (assume the mass is concentrated along a circumference located at the effective radius of the flywheel). (a) What torque is required to bring this wheel from rest to a speed of 120 rpm in a time interval of 30.0 s? (b) How much work is done during the 30.0 s?

23. A Ferris wheel rotates because a motor exerts a torque on the wheel. The radius of the London Eye, a huge observation wheel on the banks of the Thames, is 67.5 m and its mass is 1.90×10^6 kg. The cruising angular speed of the wheel is 3.50×10^{-3} rad/s. (a) How much work does the motor need to do to bring the stationary wheel up to cruising speed? [*Hint:* Treat the wheel as a hoop.] (b) What is the torque (assumed constant) the motor needs to provide to the wheel if it takes 20.0 seconds to reach the cruising angular speed?

8.4 Equilibrium Revisited

24. A sculpture is 4.00 m tall and has its CG located 1.80 m above the center of its base. The base is a square with a side of 1.10 m (Fig. 8.59). To what angle θ can the sculpture be tipped before it falls over?

25. A sign painter is standing on a uniform, horizontal platform that is held in equilibrium at the tenth-story level by two cables attached to supports on the roof of the building (Fig. 8.60). The painter has a mass of 75 kg and the mass of

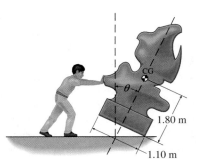

Figure 8.59
Problem 24

Figure 8.60
Problem 25

the platform is 20.0 kg. The distance from the left end of the platform to where the painter is standing is $d = 2.0$ m and the total length of the platform is 5.0 m. (a) How large is the force exerted by the left-hand cable on the platform? (b) How large is the force exerted by the right-hand cable?

26. A rod being used as a lever is shown in Fig. 8.61. The fulcrum is 1.2 m from the load and 2.4 m from the applied force. If the load has a mass of 20.0 kg, what force must be applied to lift the load?

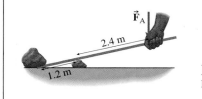

Figure 8.61
Problem 26

27. A weight of 1200 N rests on a lever at a point 0.50 m from a support (Fig. 8.62). On the same side of the support, at a distance of 3.0 m from it, an upward force with magnitude F is applied. Neglect the weight of the board itself. If the system is in equilibrium, what is F?

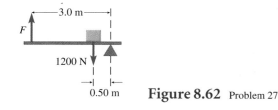

Figure 8.62   Problem 27

28. A uniform diving board, of length 5.0 m and mass 55 kg, is supported at two points; one support is located 3.4 m from the end of the board and the second is at 4.6 m from the end (see Fig. 8.19). What are the forces acting on the board due to the two supports when a diver of mass 65 kg stands at the end

of the board over the water? Assume that these forces are vertical. [*Hint:* In this problem, consider using two different torque equations about different rotation axes. This may help you determine the directions of the two forces.]

29. A house painter stands 3.0 m above the ground on a 5.0-m-long ladder that leans against the wall at a point 4.7 m above the ground (Fig. 8.63.) The painter weighs 680 N and the ladder weighs 120 N. Assuming no friction between the house and the upper end of the ladder, find the force of friction that the driveway exerts on the bottom of the ladder.

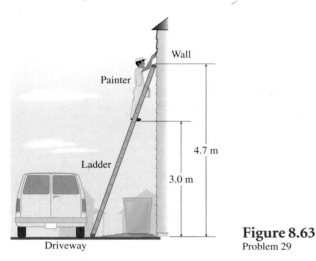

Figure 8.63
Problem 29

30. A mountain climber is rappelling down a vertical wall (Fig. 8.64). The rope attaches to a buckle strapped to the climber's waist 15 cm to the right of his center of gravity. If the climber weighs 770 N, find (a) the tension in the rope and (b) the magnitude and direction of the contact force exerted by the wall on the climber's feet.

Figure 8.64
Problem 30

31. A boom of mass m supports a steel girder of weight W hanging from its end (Fig. 8.65). One end of the boom is hinged at the floor; a cable attaches to the other end of the boom and pulls horizontally on it. The boom makes an angle θ with the horizontal. Find the tension in the cable as a function of m, W, θ, and g. Comment on the tension at $\theta = 0$ and $\theta = 90°$.

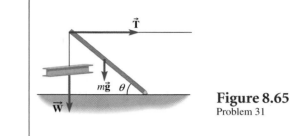

Figure 8.65
Problem 31

32. A sign is supported by a uniform horizontal boom of length 3.00 m and weight 80.0 N. A cable, inclined at an angle of 35° with the boom, is attached at a distance of 2.38 m from the hinge at the wall (Fig. 8.66). The weight of the sign is 120 N. What is the tension in the cable and what are the horizontal and vertical forces F_x and F_y exerted on the boom by the hinge? Comment on the magnitude of F_y.

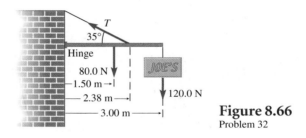

Figure 8.66
Problem 32

33. A man is doing push-ups (Fig. 8.67). He has a mass of 68 kg and his center of gravity is located at a horizontal distance of 0.70 m from his palms and 1.00 m from his feet. Find the forces exerted by the floor on his palms and feet.

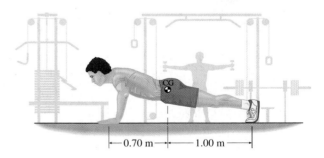

Figure 8.67 Problem 33

8.5 Equilibrium in the Human Body

34. Find the tension in the Achilles tendon and the force that the tibia exerts on the ankle joint when a person who weighs 750 N supports himself on the ball of one foot (Fig. 8.68). The normal force $N = 750$ N pushes up on the ball of the foot on one side of the ankle joint, while the Achilles tendon pulls up on the foot on the other side of the joint.

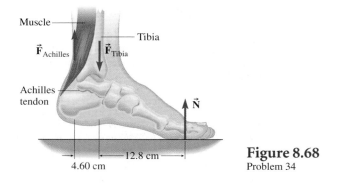

Figure 8.68
Problem 34

35. In the movie *Terminator*, Arnold Schwarzenegger lifts someone up by the neck and, with both arms fully extended and horizontal, holds the person off the ground. If the person being held weighs 700 N, is 60 cm from the shoulder joint, and Arnold has an anatomy analogous to that in Fig. 8.30, what force must *each* of the deltoid muscles exert to perform this task?

36. A person is doing leg lifts with 3.0-kg ankle weights. She is sitting in a chair with her legs bent at a right angle initially. The quadriceps muscles are attached to the patella via a tendon; the patella is connected to the tibia by the patellar tendon, which attaches to bone 10.0 cm below the knee joint (Fig. 8.69). Assume that the tendon pulls at an angle of 20.0° *with respect to the lower leg*, regardless of the position of the lower leg. The lower leg has a mass of 5.0 kg and its center of gravity is 22 cm below the knee. The ankle weight is 41 cm from the knee. If the person lifts one leg, find the force exerted by the patellar tendon to hold the leg at an angle of (a) 30.0° and (b) 90.0° with respect to the vertical.

37. Find the force exerted by the biceps muscle in holding a 1-L milk carton (weight 9.9 N) with the forearm parallel to the floor (Fig. 8.70). Assume that the hand is 35.0 cm from the elbow and that the upper arm is 30.0 cm long. The elbow is bent at a right angle and one tendon of the biceps is attached to the forearm at a position 5.00 cm from the elbow, while the other tendon is attached at 30.0 cm from the elbow. The weight of the forearm and empty hand is 18.0 N and the center of gravity of the forearm is at a distance of 16.5 cm from the elbow.

38. A man is trying to lift 60.0 kg off the floor. (a) First he does so by bending at the waist (see Fig. 8.32). Assume that the man's upper body weighs 455 N and the upper body's center of gravity is 38 cm from the sacrum (tailbone). If, when bent over, the hands are a horizontal distance of 76 cm from the sacrum, what torque must be exerted by the erector spinae muscles to lift 60.0 kg off the floor? (The axis of rotation passes through the sacrum, as shown in Fig. 8.32.) (b) When bent over, the erector spinae muscles are a horizontal distance of 44 cm from the sacrum and act at a 12° angle above the horizontal. What force ($\vec{F}_b$ in Fig. 8.32) do the erector spinae

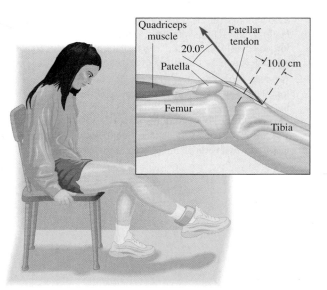

Figure 8.69   Problem 36

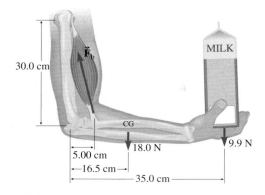

Figure 8.70   Problem 37

muscles need to exert to lift the weight? (c) What is the component of this force that compresses the spinal column?

8.6 Rotational Form of Newton's Second Law

39. Verify that the units of the rotational form of Newton's second law [Eq. (8-15)] are consistent. In other words, show that the product of a rotational inertia expressed in kg·m^2 and an angular acceleration expressed in rad/s^2 is a torque expressed in N·m.

40. A spinning flywheel has rotational inertia $I = 400.0$ kg·m^2. Its angular velocity decreases from 20.0 rad/s to zero in 300.0 s due to friction. What is the frictional torque acting?

41. An LP turntable must spin at 33.3 rpm (3.49 rad/s) to play a record. How much torque must the motor deliver if the turntable is to reach its final angular speed in 2.0 revolutions, starting from rest? The turntable is a uniform disk of diameter 30.5 cm and mass 0.22 kg.

42. Refer to Atwood's machine (Example 8.2). (a) Assuming that the cord does not slip as it passes around the pulley, what is the relationship between the angular acceleration of the pulley (α) and the magnitude of the linear acceleration of the blocks (a)? (b) What is the net torque on the pulley about its

axis of rotation in terms of the tensions T_1 and T_2 in the left and right sides of the cord? (c) Explain why the tensions cannot be equal if $m_1 \neq m_2$. (d) Apply Newton's second law to each of the blocks and Newton's second law for rotation to the pulley. Use these three equations to solve for a, T_1, and T_2. (e) Since the blocks move with constant acceleration, use the result of Example 8.2 along with the constant acceleration equation $v_y^2 - v_{0y}^2 = 2a_y \Delta y$ to check your answer for a.

43. Four masses are arranged as shown in Fig. 8.71. They are connected by rigid, massless rods of lengths 0.75 m and 0.50 m. What torque must be applied to cause an angular acceleration of 0.75 rad/s² about the axis shown?

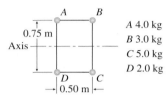

A 4.0 kg
B 3.0 kg
C 5.0 kg
D 2.0 kg

Figure 8.71 Problem 43

44. A grinding wheel, with a mass of 20.0 kg and a radius of 22.4 cm, is a uniform cylindrical disk. (a) Find the rotational inertia of the wheel about its central axis. (b) When the grinding wheel's motor is turned off, friction causes the wheel to slow from 1200 rpm to rest in 60.0 s. What torque must the motor provide to accelerate the wheel from rest to 1200 rpm in 4.00 s? Assume that the frictional torque is the same regardless of whether the motor is on or off.

45. A playground merry-go-round (Fig. 8.4), made in the shape of a solid disk, has a diameter of 2.50 m and a mass of 350.0 kg. Two children, each of mass 30.0 kg, sit on opposite sides at the edge of the platform. Approximate the children as point masses. (a) What torque is required to bring the merry-go-round from rest to 25 rpm in 20.0 s? (b) If two other bigger children are going to push on the merry-go-round rim to produce this acceleration, with what force magnitude must each child push?

46. Two children standing on opposite sides of a merry-go-round (Fig. 8.4) are trying to rotate it. They each push in opposite directions with forces of magnitude 10.0 N. (a) If the merry-go-round has a mass of 180 kg and a radius of 2.0 m, what is the angular acceleration of the merry-go-round? (Assume the merry-go-round is a uniform disk.) (b) How fast is the merry-go-round rotating after 4.0 s?

47. A bicycle wheel, of radius 0.30 m and mass 2 kg (concentrated on the rim), is rotating at 4.00 rev/s. After 50 s the wheel comes to a stop because of friction. What is the magnitude of the average torque due to frictional forces?

48. Derive the rotational form of Newton's second law as follows. Consider a rigid object that consists of a large number N of particles (Fig. 8.72). Let F_i, m_i, and r_i represent the tangential

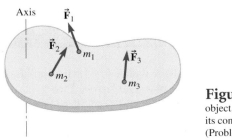

Figure 8.72 A rigid object, showing three of its constituent particles. (Problem 48)

component of the net force acting on the ith particle, the mass of that particle, and the particle's distance from the axis of rotation, respectively. (a) Use Newton's second law to find a_i, the particle's tangential acceleration. (b) Find the torque acting on this particle. (c) Replace a_i with an equivalent expression in terms of the angular acceleration α. (d) Sum the torques due to all the particles and show that $\sum_{i=1}^{N} \tau_i = I\alpha$.

8.7 The Dynamics of Rolling Objects

49. A hollow cylinder, a uniform solid sphere, and a uniform solid cylinder all have the same mass m. The three objects are rolling on a horizontal surface with identical translational speeds v. Find their total kinetic energies in terms of m and v and order them from smallest to largest.

50. A solid sphere of mass 0.600 kg rolls without slipping along a horizontal surface with a translational speed of 5.00 m/s. It comes to an incline that makes an angle of 30° with the horizontal surface. Neglecting energy losses due to friction, (a) what is the total energy of the rolling sphere? and (b) to what vertical height above the horizontal surface does the sphere rise on the incline?

51. A uniform solid cylinder rolls without slipping down an incline. A hole is drilled through the cylinder along its axis. The radius of the hole is 0.50 times the (outer) radius of the cylinder. (a) Does the cylinder take more or less time to roll down the incline now that the hole has been drilled? Explain. (b) By what percentage does drilling the hole change the time for the cylinder to roll down the incline?

52. A solid sphere of radius R and mass M *slides* without friction down a loop-the-loop track (Fig. 8.73). The sphere starts from rest at a height of h above the horizontal. Assume that the radius of the sphere is small compared to the radius r of the loop. (a) Find the minimum value of h in terms of r so that the sphere remains on the track all the way around the loop. (b) Find the minimum value of h if, instead, the sphere rolls without slipping on the track.

53. A hollow cylinder, of radius R and mass M, rolls without slipping down a loop-the-loop track of radius r (Fig. 8.73). The cylinder starts from rest at a height h above the horizontal section of track. What is the minimum value of h so that the cylinder remains on the track all the way around the loop?

Figure 8.73 Problems 52 and 53

54. The string in a yo-yo is wound around an axle of radius 0.500 cm. The yo-yo has both rotational and translational motion, like a rolling object, and has mass 0.200 kg and outer radius 2.00 cm. Starting from rest, it rotates and falls a distance of 1.00 m (the length of the string). Assume for simplicity that the yo-yo is a uniform circular disk and that the string is thin compared to the radius of the axle. (a) What is the speed of the yo-yo when it reaches the distance of 1.00 m? (b) How long does it take to fall? [*Hint:* The translational and rotational kinetic energies are related, but the yo-yo is *not* rolling on its outer radius.]

8.8 Angular Momentum

55. A turntable of mass 5.00 kg has a radius of 0.100 m and spins with a frequency of 0.550 rev/s. What is its angular momentum? Assume the turntable is a uniform disk.

56. Assume the Earth is a uniform solid sphere with radius of 6.37×10^6 m and mass of 5.98×10^{24} kg. Find the magnitude of the angular momentum of the Earth due to rotation about its axis.

57. The mass of a flywheel is 5.6×10^4 kg. This particular flywheel has its mass concentrated at the rim of the wheel. If the radius of the wheel is 2.6 m and it is rotating at 350 rpm, what is the magnitude of its angular momentum?

58. The angular momentum of a spinning wheel is 240 kg·m²/s. After application of a constant braking torque for 2.5 s, it slows and has a new angular momentum of 115 kg·m²/s. What is the torque applied?

59. How long would a braking torque of 4.00 N·m have to act to just stop a spinning wheel that has an initial angular momentum of 6.40 kg·m²/s?

60. A rotating star collapses under the influence of gravitational forces to form a pulsar. The radius of the star after collapse is 1.0×10^{-4} times the radius before collapse. There is no change in mass. In both cases, the mass of the star is uniformly distributed in a spherical shape. Find the ratios of the (a) angular momentum, (b) angular velocity, and (c) rotational kinetic energy of the star after collapse to the value before collapse. (d) If the period of the star's rotation before collapse is 1.0×10^7 s, what is its period after collapse?

61. A figure skater is spinning at a rate of 1.0 rev/s with her arms outstretched. She then draws her arms in to her chest, reducing her rotational inertia to 67% of its original value. What is her new rate of rotation?

62. A skater is initially spinning at a rate of 10.0 rad/s with a rotational inertia of 2.50 kg·m² when her arms are extended. What is her angular velocity after she pulls her arms in and reduces her rotational inertia to 1.60 kg·m²?

63. A diver can change his rotational inertia by drawing his arms and legs close to his body in the tuck position. After he leaves the diving board (with some unknown angular velocity), he pulls himself into a ball as closely as possible and makes 2.00 complete rotations in 1.33 s. If his rotational inertia decreases by a factor of 3.00 when he goes from the straight to the tuck position, what was his angular velocity when he left the diving board?

64. The rotational inertia for a diver in a pike position is about 15.5 kg·m²; it is only 8.0 kg·m² in a tucked position (Fig. 8.74). (a) If the diver gives himself an initial angular momentum of 106 kg·m²/s as he jumps off the board, how many turns can he make when jumping off a 10.0-m platform in a

(a)

(b)

Figure 8.74 Problem 64. (a) Gregory Louganis in the pike position. (b) Mark Ruiz in the tuck position.

tuck position? (b) How many in a pike position? [*Hint:* Gravity exerts no torque on the person as he falls; assume he is rotating throughout the 10.0 m dive.]

8.9 The Vector Nature of Angular Momentum

Problems 65 and 66. A solid cylindrical disk is to be used as a stabilizer in a ship. By using a massive disk rotating in the hold of the ship, the captain knows that a large torque is required to tilt its angular momentum vector. The mass of the disk to be used is 1.00×10^5 kg and it has a radius of 2.00 m.

65. If the cylinder rotates at 300.0 rpm, what is the magnitude of the average torque required to tilt its axis by 60.0° in a time of 3.00 s? [*Hint:* Draw a vector diagram of the initial and final angular momenta.]

66. How should the disk be oriented to prevent rocking from side to side and from bow to stern? Does this orientation make it difficult to steer the ship? Explain.

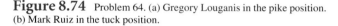

COMPREHENSIVE PROBLEMS

67. The Moon's distance from Earth varies between 3.56×10^5 km at perigee and 4.07×10^5 km at apogee. What is the ratio of its orbital speed around Earth at perigee to that at apogee?

68. A painter (mass 61 kg) is walking along a trestle, consisting of a uniform plank (mass 20.0 kg, length 6.00 m) balanced on

two sawhorses. Each sawhorse is placed 1.40 m from an end of the plank. A paint bucket (mass 4.0 kg, diameter 28 cm) is placed as close as possible to the right-hand edge of the plank while still having the whole bucket in contact with the plank (Fig. 8.75). (a) How close to the right-hand edge of the

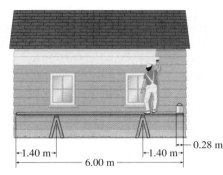

←1.40 m→ ←1.40 m→ →0.28 m←
←────── 6.00 m ──────→

Figure 8.75
Problem 68

plank can the painter walk before tipping the plank and spilling the paint? (b) How close to the left-hand edge can the same painter walk before causing the plank to tip? [*Hint:* As the painter walks toward the right-hand edge of the plank and the plank starts to tip clockwise, what is the force acting upward on the plank from the left-hand sawhorse support?]

✦ 69. An experimental flywheel, used to store energy and replace an automobile engine, is a solid disk of mass 200.0 kg and radius 0.40 m. (a) What is its rotational inertia? (b) When driving at 22.4 m/s (50 mph), the fully energized flywheel is rotating at an angular speed of 3160 rad/s. What is the initial rotational kinetic energy of the flywheel? (c) If the total mass of the car is 1000.0 kg, find the ratio of the initial rotational kinetic energy of the flywheel to the translational kinetic energy of the car. (d) If the force of air resistance on the car is 670.0 N, how far can the car travel at a speed of 22.4 m/s (50 mph) with the initial stored energy? Ignore losses of mechanical energy due to means other than air resistance.

✦✦ 70. (a) Assume the Earth is a uniform solid sphere. Find the kinetic energy of the Earth due to its rotation about its axis. (b) Suppose we could somehow extract 1.0% of the Earth's rotational kinetic energy to use for other purposes. By how much would that change the length of the day? (c) For how many years would 1.0% of the Earth's rotational kinetic energy supply the world's energy usage (assume a constant 1.0×10^{21} J per year)?

✦ 71. A flat object in the xy-plane is free to rotate about the z-axis. The gravitational field is uniform in the $-y$-direction. Think of the object as a large number of particles with masses m_i located at coordinates (x_i, y_i), as in Fig. 8.76. (a) Show that the torques on the particles about the z-axis can be written $\tau_i = -x_i m_i g$. (b) Show that if the center of gravity is located at (x_{CG}, y_{CG}), the total torque due to gravity on the object must be $\Sigma \tau_i = -x_{CG}Mg$, where M is the total mass of the object. (c) Show that $x_{CG} = x_{CM}$. (This same line of reasoning can be applied to objects that are not flat and to other axes of rotation to show that $y_{CG} = y_{CM}$ and $z_{CG} = z_{CM}$.)

72. The operation of the Princeton Tokomak Fusion Test Reactor requires large bursts of energy. The power needed exceeds the amount that can be supplied by the utility company. Prior to pulsing the reactor, energy is stored in a giant flywheel of mass 7.27×10^5 kg and rotational inertia 4.55×10^6 kg·m². The flywheel rotates at a maximum angular speed of 386 rpm. When the stored energy is needed to operate the reactor, the flywheel is connected to an electrical generator, which converts some of the rotational kinetic energy into electrical energy. (a) If the flywheel is a uniform disk, what is its radius? (b) If the flywheel is a hollow cylinder with its mass concentrated at the rim, what is its radius? (c) If the flywheel slows to 252 rpm in 5.00 s, what is the average power supplied by the flywheel during that time?

73. The distance from the center of the breastbone to a man's hand, with the arm outstretched and horizontal to the floor, is 1.0 m. The man is holding a 10.0-kg dumbbell, oriented vertically, in his hand, with the arm horizontal. What is the torque due to this weight about a horizontal axis through the breastbone perpendicular to his chest?

74. A uniform rod of length L is free to pivot around an axis through its upper end. If it is released from rest when horizontal, at what speed is the lower end moving at its lowest point? [*Hint:* The gravitational potential energy change is determined by the change in height of the center of gravity.]

✦ 75. A gymnast is performing a giant swing on the high bar (Fig. 8.77). In a simplified model of the giant swing, assume that the gymnast keeps his arms and body straight as he swings all the way around the upper bar. Assume also that the gymnast does no work during the swing. With what angular speed should he be moving at the bottom of the giant swing in order to make it all the way around? The distance from the bar to his feet is 2.0 m and his center of gravity is 1.0 m from his feet.

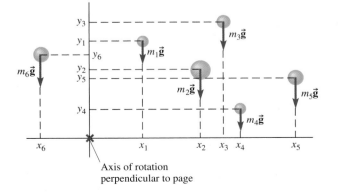

Figure 8.76   Problem 71

Figure 8.77
Problem 75

✦✦ 76. A box of mass 42 kg sits on top of a ladder (Fig. 8.78). Neglecting the weight of the ladder, find the tension in the brace. Assume that the brace exerts horizontal forces on the ladder at each end. [*Hint:* Use a symmetry argument; then analyze the forces and torques on one side of the ladder.]

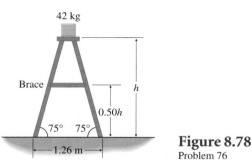

Figure 8.78
Problem 76

✦ 77. A person is trying to lift a ladder of mass 15 kg and length 8.0 m. The person is exerting a vertical force on the ladder at a point of contact 2.0 m from the center of gravity. The opposite end of the ladder rests on the floor. (a) When the ladder makes an angle of 60.0° with the floor, what is this vertical force? (b) A person tries to help by lifting the ladder at the point of contact with the floor. Does this help the person trying to lift the ladder?

78. A crustacean (*Hemisquilla ensigera*) rotates its anterior limb to strike a mollusk, intending to break it open. The limb reaches an angular velocity of 175 rad/s in 1.50 ms. We can approximate the limb as a thin rod rotating about an axis perpendicular to one end (the joint where the limb attaches to the crustacean). (a) If the mass of the limb is 28.0 g and the length is 3.80 cm, what is the rotational inertia of the limb about that axis? (b) If the extensor muscle is 3.00 mm from the joint and acts perpendicular to the limb, what is the muscular force required to achieve the blow?

79. A block of mass m_2 hangs from a string. The string wraps around a pulley of rotational inertia I and then attaches to a Ⓖ second block of mass m_1, which sits on a frictionless table (Fig. 8.79). What is the acceleration of the blocks when they are released?

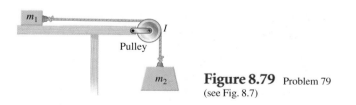

Figure 8.79 Problem 79
(see Fig. 8.7)

80. A 2.0-kg uniform flat disk is thrown into the air with a linear speed of 10.0 m/s. As it travels, the disk spins at 3.0 revolutions per second. If the radius of the disk is 10.0 cm, what is the magnitude of its angular momentum?

Ⓖ ✦ 81. A hoop of 2.00-m circumference is rolling down an inclined plane of length 10.0 m in a time of 10.0 s. It started out from rest. (a) What is its angular velocity when it arrives at the bottom? (b) If the mass of the hoop, concentrated at the rim, is 1.50 kg, what is the angular momentum of the hoop when it reaches the bottom of the incline? (c) What force(s) supplied

the net torque to change the hoop's angular momentum? Explain. [*Hint:* Use a rotation axis through the hoop's center.] (d) What is the magnitude of this force?

82. A large clock has a second hand with a mass of 0.10 kg concentrated at the tip of the pointer. (a) If the length of the second hand is 30.0 cm, what is its angular momentum? (b) The same clock has an hour hand with a mass of 0.20 kg concentrated at the tip of the pointer. If the hour hand has a length of 20.0 cm, what is its angular momentum?

✦ 83. A planet moves around the Sun in an elliptical orbit (see Fig. 8.39). (a) Show that the external torque acting on the planet about an axis through the Sun is zero. (b) Since the torque is zero, the planet's angular momentum is constant. Write an expression for the planet's angular momentum in terms of its mass m, its distance r from the Sun, and its angular velocity ω. (c) Given r and ω, how much area is swept out during a short time Δt? [*Hint:* Think of the area as a fraction of the area of a *circle*, like a slice of pie; if Δt is short enough, the radius of the orbit during that time is nearly constant.] (d) Show that the area swept out per unit time is constant. You have just proved Kepler's second law!

✦ 84. A 68-kg woman stands straight with both feet flat on the floor. Her center of gravity is a horizontal distance of 3.0 cm in front of a line that connects her two ankle joints. The Achilles tendon attaches the calf muscle to the foot a distance of 4.4 cm behind the ankle joint. If the Achilles tendon is inclined at an angle of 81° with respect to the horizontal, find the force that each calf muscle needs to exert while she is standing. [*Hint:* Consider the equilibrium of the part of the body *above* the ankle joint.]

85. A merry-go-round (radius R, rotational inertia I_0) spins with negligible friction. Its initial angular velocity is ω_0. A child (mass m) on the merry-go-round moves from the center out to the rim. (a) Calculate the angular velocity after the child moves out to the rim. (b) Calculate the rotational kinetic energy and angular momentum of the system (merry-go-round + child) before and after.

Ⓖ ✦ 86. In a motor, a flywheel (solid disk of radius R and mass M) is rotating with angular velocity ω_0. When the clutch is released, a second disk (radius r and mass m) initially not rotating is brought into frictional contact with the flywheel. The two disks spin around the same axle with frictionless bearings. After a short time, friction between the two disks brings them to a common angular velocity. (a) Ignoring external influences, what is the final angular velocity? (b) Does the total angular momentum of the two change? If so, account for the change. If not, explain why it does not. (c) Repeat (b) for the rotational kinetic energy.

87. Since humans are generally not symmetrically shaped, the height of our center of gravity is generally not half of our height. One way to determine the location of the center of gravity is shown in Fig. 8.80. A 2.2-m-long uniform plank is supported by two bathroom scales, one at either end. Initially the scales each read 100.0 N. A 1.60-m-tall student then lies on top of the plank, with the soles of his feet directly above scale B. Now scale A reads 394.0 N and scale B reads 541.0 N. (a) What is the student's weight? (b) How far is his center of gravity from the soles of his feet? (c) When standing, how far above the floor is his center of gravity, expressed as a fraction of his height?

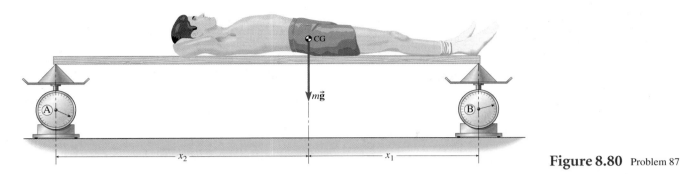

Figure 8.80 Problem 87

88. A spool of thread of mass m rests on a plane inclined at angle θ. The end of the thread is tied as shown in Fig. 8.81. The outer radius of the spool is R and the inner radius (where the thread is wound) is r. The rotational inertia of the spool is I. Give all answers in terms of m, θ, R, r, I, and g. (a) If there is no friction between the spool and the incline, describe the motion of the spool and calculate its acceleration. (b) If the coefficient of friction is large enough to keep the spool from slipping, calculate the magnitude and direction of the frictional force. (c) What is the minimum possible coefficient of friction to keep the spool from slipping in part (b)?

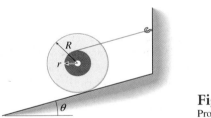

Figure 8.81
Problem 88

89. A bicycle travels up an incline at constant velocity (Fig. 8.82). The magnitude of the frictional force due to the road on the rear wheel is $f = 3.8$ N. The upper section of chain pulls on the sprocket wheel, which is attached to the rear wheel, with a force $\vec{\mathbf{F}}_\mathrm{C}$. The lower section of chain is slack. If the radius of the rear wheel is 6.0 times the radius of the sprocket wheel, what is the magnitude of the force $\vec{\mathbf{F}}_\mathrm{C}$ with which the chain pulls?

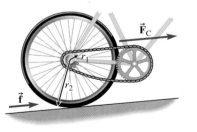

Figure 8.82
Problem 89

90. A circus roustabout is attaching the circus tent to the top of the main support post of length L when the post suddenly breaks at the base (Fig. 8.83). The worker's weight is negligible compared to that of the uniform post. What is the speed with which the roustabout reaches the ground if (a) he jumps at the instant he hears the post crack or (b) if he clings to the

post and rides to the ground with it. (c) Which is the safest course of action for the roustabout?

Figure 8.83 Problem 90

91. The 12.2-m crane (Fig. 8.84) weighs 18 kN and is lifting a 67-kN load. The hoisting cable (tension T_1) passes over a pulley at the top of the crane and attaches to an electric winch in the cab. The pendant cable (tension T_2), which supports the crane, is fixed to the top of the crane. Find the tensions in the two cables and the force $\vec{\mathbf{F}}_\mathrm{p}$ at the pivot.

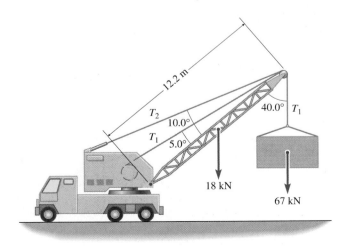

Figure 8.84 Problem 91

92. A collection of objects is set to rolling, without slipping, down a slope inclined at 30°. The objects are a solid sphere, a hollow sphere, a solid cylinder, and a hollow cylinder. A frictionless cube is also allowed to slide down the same incline. Which one gets to the bottom first? List the others in the order they arrive at the finish line.

© ✦93. A student stands on a platform that is free to rotate and holds two dumbbells, each at a distance of 65 cm from his central axis (Fig. 8.85). Another student gives him a push and starts the system of student, dumbbells, and platform rotating at 0.50 rev/s. The student on the platform then pulls the dumbbells in close to his chest so that they are each 22 cm from his central axis. Each dumbbell has a mass of 1.00 kg and the rotational inertia of the student, platform, and dumbbells is initially 2.40 kg·m². Model each arm as a uniform rod of mass 3.00 kg with one end at the central axis; the length of the arm is initially 65 cm and then is reduced to 22 cm. (a) What is his new rate of rotation? (b) The student becomes nervous and drops the two dumbbells so that they fall away from the platform onto the floor. What happens to the student's rate of rotation? Explain.

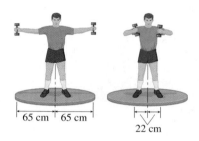

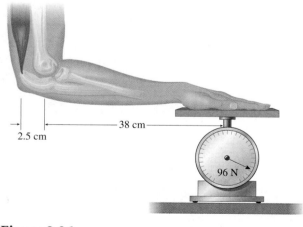

Figure 8.85
Problem 93

65 cm | 65 cm

22 cm

94. A person places his hand palm downward on a scale and pushes down on the scale until it reads 96 N (Fig. 8.86). The triceps muscle is responsible for this arm extension force. Find the force exerted by the triceps muscle. The bottom of the triceps muscle is 2.5 cm to the left of the elbow joint and the palm is pushing at approximately 38 cm to the right of the elbow joint.

38 cm

2.5 cm

96 N

Figure 8.86 Problem 94

95. The posture of small animals may prevent them from being blown over by the wind. For example, with wind blowing from the side, a small insect stands with bent legs (Fig. 8.87); the more bent the legs, the lower the body, and thus the smaller the angle θ. The wind exerts a force on the insect, which causes a torque about the point where the downwind feet touch. The torque due to the weight of the insect must be equal and opposite to keep the insect from being blown over. For example, the drag force on a blowfly due to a sideways wind is $F_{wind} = cAv^2$, where v is the velocity of the wind, A is the cross-sectional area on which the wind is blowing, and $c \approx 1.3$ N·s²·m⁻⁴. (a) If the blowfly has a cross-sectional side area of 0.10 cm², a mass of 0.070 g, and crouches such that $\theta = 30.0°$, what is the maximum wind speed in which the blowfly can stand? (Assume that the drag force acts at the center of gravity.) (b) How about if it stands such that $\theta = 80.0°$? (c) Compare to the maximum wind velocity that a dog can withstand, if the dog stands such that $\theta = 80.0°$, has a cross-sectional area of 0.030 m², and weighs 10.0 kg. (Assume the same value of c.)

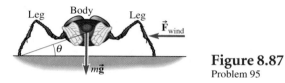

Leg   Body   Leg

$\vec{F}_{wind}$

θ

$m\vec{g}$

Figure 8.87
Problem 95

✦✦ 96. (a) Redo Example 8.7 to find an algebraic solution for d in terms of M, m, μ_s, L, and θ. (b) Use this to show that placing the ladder at a larger angle θ (that is, more nearly vertical) enables the person to climb farther up the ladder without having it slip, all other things being equal. (c) Using the numerical values from Example 8.7, find the minimum angle θ that enables the person to climb all the way to the top of the ladder.

ANSWERS TO PRACTICE PROBLEMS

8.1 390 kg·m^2

8.2 $v = \sqrt{\dfrac{2(m_2 - \mu_k m_1)gh}{m_1 + m_2 + I/R^2}}$

8.3 53 N; 4.2 N·m

8.4 −65 N·m

8.5 8.3 J

8.6 left support, downward; right support, upward

8.7 0.27

8.8 57 N, downward

8.9 It must lie in the same vertical plane as the two ropes holding up the rings. Otherwise the gravitational force would have a nonzero lever arm with respect to a horizontal axis that passes through the contact points between his hands and the rings; thus, gravity would cause a net torque about that axis.

8.10 460 N

8.11 (a) 2380 rad; (b) 3.17 kJ; (c) 1.34 N·m

8.12 solid ball, $\frac{2}{7}$; hollow ball, $\frac{2}{5}$.

8.13 $\frac{1}{2}g \sin \theta$

8.14 5% increase

8.15 16 cm/s

Fluids

9

A hippopotamus in Kruger National Park, South Africa, wants to feed on the vegetation growing on the bottom of a pond. When the hippo wades into the pond, it floats. How does a hippopotamus get its floating body to sink to the bottom of a pond?

Fluids (liquids and gases) are materials that flow.

Gases are much more compressible than liquids.

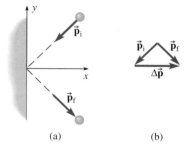

Figure 9.1 A single fluid molecule bouncing off a container wall.

9.1   STATES OF MATTER

Macroscopic matter is usually classified into three states or phases: solids, liquids, and gases. Solids tend to hold their shapes. Many solids are quite rigid; they are not easily deformed by external forces because strong forces due to neighboring atoms hold each atom in a particular position. Although the atoms vibrate around fixed equilibrium positions, they do not have enough energy to break the bonds with their neighbors. To bend an iron bar, for example, the arrangement of the atoms must be altered, which is not easy to do. A blacksmith heats iron in a forge to loosen the bonds between atoms so that he can bend the metal into the required shape.

In contrast to solids, liquids and gases do not hold their shapes. A liquid flows and takes the shape of its container and a gas expands to fill its container. **Fluids**—both liquids and gases—are easily deformed by external forces. This chapter deals mainly with properties that are common to both liquids and gases.

The atoms or molecules in a fluid do not have fixed positions, so a fluid does not have a definite shape. An applied force can easily make a fluid flow; for instance, the squeezing of the heart muscle exerts an applied force that pumps blood through the blood vessels of the body. However, this squeezing does not change the *volume* of the blood by much. In many situations we can think of liquids as **incompressible**—that is, as having a fixed volume that is impossible to change. The shape of the liquid can be changed by pouring it from a container of one shape into a container of a different shape, but the volume of the liquid remains the same.

In most liquids, the atoms or molecules are almost as closely packed as those in the solid phase of the same material. The intermolecular forces in a liquid are almost as strong as those in solids, but the molecules are not locked in fixed positions as they are in solids. That is why the volume of the liquid can remain nearly constant while the shape is easily changed. Water is one of the exceptions: in cold water, the molecules in the liquid phase are actually *more* closely packed than those in the solid phase (ice).

Gases, on the other hand, cannot be characterized by a definite volume nor by a definite shape. A gas expands to fill its container and can easily be compressed. The molecules in a gas are very far apart compared to the molecules in liquids and solids. The molecules are almost free of interactions with each other except when they collide.

9.2   PRESSURE

A **static** fluid does not flow; it is everywhere at rest. In the study of fluid statics (*hydrostatics*), we also assume that any solid object in contact with the fluid—whether a vessel containing the fluid or an object submerged in the fluid—is at rest. The atoms or molecules in a static fluid are not themselves static; they are continually moving. The motion of people bouncing up and down and bumping into each other in a mosh pit gives you a rough idea of the motion of the closely packed atoms or molecules in a liquid; in gases, the atoms or molecules are much farther apart than in liquids so they travel greater distances between collisions.

Fluid pressure is caused by collisions of the fast-moving atoms or molecules of a fluid. When a single molecule hits a container wall and rebounds, its momentum changes due to the force exerted on it by the wall. Figure 9.1a shows a molecule of a fluid within a container making an elastic collision with one of the container walls. In

Physics at Home

Drop a very tiny speck of dust or lint into a container of water and push the speck below the surface. The motion of the speck—called *Brownian motion*—is easily observed as it is pushed and bumped about randomly by collisions with water molecules. The water molecules themselves move about randomly, but at much higher speeds than the speck of dust due to their much smaller mass.

this case, the *y*-component of momentum is unchanged while the *x*-component reverses direction. The momentum change is in the +*x*-direction (Fig. 9.1b), which occurs because the wall exerts a force to the right on the molecule. By Newton's third law, the molecule exerts a force to the left on the wall during the collision. If we consider all the molecules colliding with this wall, *on average* they exert no force on the wall in the ±*y*-direction, but all exert a force in the –*x*-direction. The frequent collisions of fluid molecules with the walls of the container cause a net force pushing outward on the walls.

A static fluid exerts a force on any surface with which it comes in contact; the direction of the force is perpendicular to the surface (Fig. 9.2). A static fluid *cannot* exert a force *parallel* to the surface. If it did, the surface would exert a force on the fluid parallel to the surface, by Newton's third law. This force would make the fluid flow along the surface, contradicting the premise that the fluid is static.

The *average pressure* of a fluid at points on a planar surface is

| The force due to a static fluid on a surface is always perpendicular to the surface.

Definition of average pressure

$$P_{av} = \frac{F}{A} \qquad (9\text{-}1)$$

---

where F is the magnitude of the force acting perpendicular to the surface and A is the area of the surface. By imagining a tiny surface at various points within the fluid and measuring the force that acts on it, we can define the pressure at any point within the fluid. In the limit of a small area A, $P = F/A$ is the **pressure** P of the fluid.

Pressure is a scalar quantity; it does not have a direction. The force acting on an object submerged in a fluid—or on some portion of the fluid itself—is a vector quantity; its direction is perpendicular to the contact surface. Pressure is defined as a scalar because, at a given location in the fluid, the magnitude of the force per unit area is the same for any orientation of the surface. The molecules in a static fluid are moving in random directions; there can be no preferred direction since that would constitute fluid flow. There is no reason that a surface would have a greater number of collisions, or collisions with more energetic molecules, for one particular surface orientation compared with any other orientation.

The SI unit for pressure is the newton per square meter (N/m²), which is named the *pascal* (symbol Pa) after the French scientist Blaise Pascal (1623–1662). Another commonly used unit of pressure is the *atmosphere* (atm). One atmosphere is the average air pressure at sea level. The conversion factor between atmospheres and pascals is

Figure 9.2 Forces due to a static fluid acting on the walls of the container and on a submerged object. Here the weight of the fluid is neglected.

1 atm = 101.3 kPa

Other units of pressure in common use are introduced in Section 9.5.

Example 9.1

The Danger of Stiletto-Heeled Shoes

A young woman weighing 534 N (120 lb) walks to her bedroom while wearing tennis shoes. She then gets dressed for her evening date, putting on her new "spike-heeled" dress shoes. The area of the heel section of her tennis shoe is 60.0 cm² and the area of the heel of her dress shoe is 1.00 cm². For each pair of shoes, find the average pressure caused by the heel making contact with the floor when her entire weight is supported by one heel.

Strategy The average pressure is the force applied to the floor divided by the contact area. The force that the heel exerts on the floor is 534 N. To keep the units straight, we convert the areas from square centimeters to square meters.

Solution To convert the area of the tennis shoe heel and the dress shoe heel from cm² to m², we multiply by the conversion factor $\left(\dfrac{1 \text{ m}}{10^2 \text{ cm}}\right)^2$.

For the tennis shoe

$$A = 60.0 \text{ cm}^2 \times \left(\frac{1 \text{ m}}{10^2 \text{ cm}}\right)^2 = 6.00 \times 10^{-3} \text{ m}^2$$

and for the dress shoe heel

$$A = 1.00 \text{ cm}^2 \times \left(\frac{1 \text{ m}}{10^2 \text{ cm}}\right)^2 = 1.00 \times 10^{-4} \text{ m}^2$$

continued on next page

Example 9.1 *continued*

The average pressure is the woman's weight divided by the area of the heel. For the tennis shoe:

$$P = \frac{F}{A} = \frac{534\ N}{6.00 \times 10^{-3}\ m^2} = 8.90 \times 10^4\ N/m^2 = 89.0\ kPa$$

For the "spike-heeled" shoe:

$$P = \frac{534\ N}{1.00 \times 10^{-4}\ m^2} = 5.34 \times 10^6\ N/m^2 = 5.34\ MPa$$

Discussion   In atmospheres, these pressures are 0.879 atm and 52.7 atm, respectively. The pressure due to the dress shoe is 60 times the pressure due to the tennis shoe since the same force is spread over $\frac{1}{60}$ the area. This high pressure caused a problem for the manufacturers of vinyl floors when such shoes were in fashion. Permanent dents were made in floors due to the stress of tremendous pressure exerted on a tiny area

by such shoes. Finally the manufacturers sent out warnings that their floors were not guaranteed from damage caused by stiletto-heeled shoes. Women had to remove their shoes before stepping onto a vinyl kitchen floor.

Practice Problem 9.1   Pressure from ordinary dress shoe heel

Fortunately for floor manufacturers, and for women's feet, stiletto heels are out of fashion more often than they are in fashion. Suppose that a woman's dress shoes have heels that are each 4.0 cm² in area. Find the pressure on the floor, when the entire weight is on a single heel, for such a shoe worn by the same woman as in Example 9.1. Find the factor by which this pressure exceeds the pressure from the tennis shoe.

Atmospheric Pressure

On the surface of the Earth, we live at the bottom of an ocean of fluid called air. The forces exerted by air on our bodies and on surfaces of other objects may be surprisingly large: 1 atm is approximately 10 N/cm² of surface area, or nearly 15 lb/in². We are not crushed by this pressure because most of the fluids in our bodies are at approximately the same pressure as the air around us. As an analogy, consider a suitcase. Why is an empty suitcase not crushed by the air pushing in on all sides? Because the air inside the suitcase is at the same pressure and pushes out on the sides of the suitcase. The pressure of the fluids inside our cells matches the pressure of the surrounding fluids pushing in on the cell membranes, so the cells do not rupture.

By contrast, the blood pressure in the arteries is as much as 20 kPa higher than atmospheric pressure. The strong, elastic arterial walls are stretched by the pressure of the blood inside; the walls squeeze the arterial blood to keep its higher pressure from being transmitted to other fluids in the body.

Changing weather conditions cause variations of approximately 5% in the actual value of air pressure at sea level; 101.3 kPa is only the average value. Air pressure also decreases with increasing elevation. The average air pressure in Leadville, Colorado, the highest incorporated city in the United States (elevation 3100 m), is 70 kPa. Some Tibetans live at altitudes of over 5000 m, where the average air pressure is only half its value at sea level.

9.3   PASCAL'S PRINCIPLE

If the weight of a static fluid is negligible, then the pressure must be the same every-where in the fluid. Why? In Fig. 9.3, imagine the submerged cube to be composed of the same fluid as its surroundings. Neglecting the fluid's weight, the only forces acting on the cubical piece of fluid are those due to the surrounding fluid pushing inward. The forces pushing on each pair of opposite sides of the cube must be equal in magnitude, since the fluid inside the cube is in equilibrium. Therefore, the pressure must be the same on both sides. Since we can extend this argument to any size and shape piece of fluid, *the fluid pressure must be the same everywhere.*

Even when the weight of the fluid is not negligible, the same reasoning can be used (see Conceptual Question 15) to establish **Pascal's principle**.

Figure 9.3   Forces acting on a cube of fluid

Pascal's principle
A *change in pressure* at any point in a confined fluid is transmitted everywhere throughout the fluid.

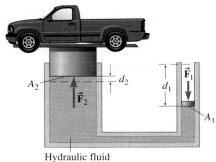

Hydraulic fluid

Figure 9.4 A hydraulic lift

For example, when a truck needs to have its muffler replaced, it is lifted into the air by a mechanism called a hydraulic lift (Fig. 9.4). A force is exerted on a liquid by a piston with a relatively small area; the resulting increase in pressure is transmitted everywhere throughout the liquid. Then the truck is lifted by the fluid pressure on a piston of much larger area. The upward force on the truck is much larger than the force applied to the small piston. Pascal's principle has many other applications, such as the hydraulic brakes in cars and trucks and the hydraulic controls in airplanes.

To analyze the forces in the hydraulic lift, let force F_1 be applied to the small piston of area A_1, causing a pressure increase:

$$\Delta P = \frac{F_1}{A_1}$$

A truck is supported by a piston of much larger area A_2 on the other side of the lift. The increase in pressure due to the small piston is transmitted everywhere in the liquid, so

$$\frac{F_1}{A_1} = \frac{F_2}{A_2}$$

Since A_2 is larger than A_1, the force exerted on the large piston (F_2) is larger than the force applied to the small piston (F_1). We are not getting something for nothing; just as for the pulley systems discussed in Chapter 6, the advantage of the smaller force applied to the small piston is balanced by a greater distance it must be moved. The small piston has to move a long distance d_1 while the large piston moves a short distance d_2. Assuming the liquid to be incompressible, the volume of fluid displaced by each piston is the same, so

$$A_1 d_1 = A_2 d_2$$

The displacements of the pistons are inversely proportional to their areas, while the forces are directly proportional to the areas; then the product of force and displacement is the same:

$$\frac{F_1}{A_1} \times A_1 d_1 = \frac{F_2}{A_2} \times A_2 d_2; \text{ therefore, } F_1 d_1 = F_2 d_2$$

The work (force times displacement) done by moving the small piston equals the work done by the large piston in raising the truck.

Example 9.2

The Hydraulic Lift

In a hydraulic lift, if the diameter of the smaller piston is 4.0 cm and the diameter of the larger piston is 40.0 cm, what weight can the larger piston support when a force of 250 N is applied to the smaller piston?

Strategy According to Pascal's principle, the pressure increases the same amount at every point in the fluid. A natu-

ral way to work is in terms of proportions since the forces are proportional to the areas of the pistons.

Solution Since the pressure on the two pistons increases by the same amount,

$$\frac{F_1}{A_1} = \frac{F_2}{A_2}$$

continued on next page

Example 9.2 *continued*

Equivalently, the forces are proportional to the areas:

$$\frac{F_2}{F_1} = \frac{A_2}{A_1}$$

The ratio of the diameters is $d_2/d_1 = 10$, so the ratio of the areas is $A_2/A_1 = (d_2/d_1)^2 = 100$. Then the weight that can be supported is

$$F_2 = 100F_1 = 25{,}000 \text{ N} = 25 \text{ kN}$$

⚠ **Discussion**   One common error in this sort of problem is to think of the area and the force as a *trade-off*—in other words, that the piston with the *large* area has the *small* force and

vice versa. Since the pressures are the same, the force exerted by the fluid on either piston is proportional to the piston's area. We make the piston that lifts the truck large because we know the force on it will be large, *in direct proportion to* its area.

Practice Problem 9.2   Application of Pascal's principle
Consider the hydraulic lift of the preceding example. (a) What is the increase in pressure caused by the 250-N force on the small piston? (b) If the larger piston moves 5.0 cm, how far does the smaller piston move?

9.4   THE EFFECT OF GRAVITY ON FLUID PRESSURE

On a drive through the mountains or on a trip in a small plane, the feeling of our ears popping is evidence that pressure is not the same everywhere in a static fluid. Gravity makes fluid pressure increase as you move down and decrease as you move up. To understand more about this variation, we must first define the density of a fluid.

The density of a uniform substance is its mass divided by its volume.

The **density** of a substance is its mass per unit volume. The Greek letter ρ (rho) is used to represent density. The density of a uniform substance of mass m and volume V is

$$\rho = \frac{m}{V} \tag{9-2}$$

The SI units of density are kilograms per cubic meter: kg/m^3. For a nonuniform substance, Eq. (9-2) defines the **average density**.

Table 9.1 lists the densities of some common substances. Note that temperatures and pressures are specified in the table. For solids and liquids, density is only weakly dependent on temperature and pressure. On the other hand, gases are highly compressible, so even a relatively small change in temperature or pressure can change the density of a gas significantly.

Now, using the concept of density, we can find how pressure increases with depth due to gravity. Suppose we have a glass beaker containing a static liquid of uniform density ρ. Within this liquid, imagine a cylinder of liquid with cross-sectional area A and height d (Fig. 9.5a). The mass of the liquid in this cylinder is

$$m = \rho V$$

where the volume of the cylinder is

$$V = Ad$$

The weight of the cylinder of liquid is therefore

$$mg = (\rho Ad)g$$

The vertical forces acting on this column of liquid are shown in Fig. 9.5b. The pressure at the top of the cylinder is P_1 and the pressure at the bottom is P_2. Since the liquid in the column is in equilibrium, the net vertical force acting on it must be zero by Newton's second law:

$$\Sigma F_y = P_2 A - P_1 A - \rho Adg = 0$$

Dividing by the common factor A and rearranging yields:

Pressure variation with depth in a static fluid

$$P_2 = P_1 + \rho gd$$

where point 2 is a depth d below point 1   (9-3)

Table 9.1

Densities of Common Substances
(at 0°C and 1 atm unless otherwise indicated)

Gases	Density (kg/m^3)	Liquids	Density (kg/m^3)	Solids	Density (kg/m^3)
Hydrogen	0.090	Gasoline	680	Polystyrene	100
Helium	0.18	Ethanol	790	Cork	240
Steam (100°C)	0.60	Oil	800–900	Wood (pine)	350–550
Nitrogen	1.25	Water (0°C)	999.87	Wood (oak)	600–900
Air (20°C)	1.20	Water (3.98°C)	1000.00	Ice	917
Air (0°C)	1.29	Water (20°C)	1001.80	Wood (ebony)	1000–1300
Oxygen	1.43	Seawater	1025	Bone	1500–2000
Carbon dioxide	1.98	Blood (37°C)	1060	Concrete	2000
		Mercury	13,600	Quartz, granite	2700
				Aluminum	2702
				Iron, steel	7860
				Copper	8920
				Lead	11,300
				Gold	19,300
				Platinum	21,500

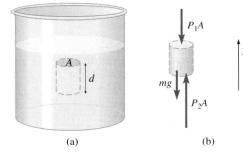

(a)   (b)

Figure 9.5 (a) A cylinder of liquid of height d and area A. (b) Vertical forces on the cylinder of liquid.

Since we can put the top and bottom of the cylinder anywhere we choose within the liquid, Eq. (9-3) relates the pressure at any two points in a static liquid where point 2 is a depth d below point 1. Note that in deriving this result, we assumed a constant density.

For gases, Eq. (9-3) can be applied as long as the depth d is small enough that changes in the density due to gravity are negligible. Since liquids are nearly incompressible, Eq. (9-3) holds to great depths in liquids.

For a liquid that is open to the atmosphere, suppose we take point 1 at the surface and point 2 a depth d below. Then $P_1 = P_{atm}$, so the pressure at a depth d below the surface is

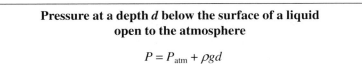

Pressure at a depth d below the surface of a liquid open to the atmosphere

$$P = P_{atm} + \rho g d \qquad (9\text{-}4)$$

Example 9.3

A Diver

A diver swims to a depth of 3.2 m in a freshwater lake. What is the increase in the force pushing in on her eardrum, compared to what it was at the lake surface? The area of the eardrum is 0.60 cm^2.

Strategy   We can find the increase in pressure at a depth of 3.2 m and then find the corresponding increase in force on the eardrum. If the force on the eardrum at the surface is P_1A and the force at a depth of 3.2 m is P_2A, then the increase in the force is $(P_2 - P_1)A$.

Solution   The increase in pressure depends on the depth d and the density of water. From Table 9.1, the density of water is 1000 kg/m^3 to two significant figures for any reasonable temperature.

$$P_2 - P_1 = \rho g d$$
$$\Delta P = 1000 \text{ kg/m}^3 \times 9.8 \text{ m/s}^2 \times 3.2 \text{ m}$$
$$= 31.4 \text{ kPa}$$

The increase in force on the eardrum is

$$\Delta F = \Delta P \times A$$

where $A = 0.60 \text{ cm}^2 = 6.0 \times 10^{-5} \text{ m}^2$. Then

$$\Delta F = (3.14 \times 10^4 \text{ Pa}) \times (6.0 \times 10^{-5} \text{ m}^2)$$
$$= 1.9 \text{ N}$$

Discussion   A force also pushes *outward* on the eardrum due to the pressure inside the ear canal. If the diver descends rapidly so that the pressure inside the ear canal does not change, then a 1.9-N net force due to fluid pressure pushes inward on the eardrum. When the diver's ear "pops," the pressure inside the ear canal increases to equal the fluid pressure outside the eardrum, so that the net force due to fluid pressure on the eardrum is zero.

Practice Problem 9.3   Limits on submarine depth

A submarine is constructed so that it can safely withstand a pressure of 1.6×10^7 Pa. How deep may this submarine descend in the ocean if the average density of seawater is 1025 kg/m^3?

Conceptual Example 9.4

The Hydrostatic Paradox

Three vessels have different shapes, but the same base area and the same weight when empty (Fig. 9.6). The vessels are filled with water to the same level and then the air is pumped out. The top surface of the water is then at a low pressure that, for simplicity, we assume to be zero. (a) Are the water pressures at the bottom of each vessel the same? If not, which is largest and which is smallest? (b) If the three vessels containing water are weighed on a scale, do they give the same reading? If not, which weighs the most and which weighs the least? (c) If the water exerts the same downward force on the bottom of each vessel, is that force equal to the weight of water in the vessel? Is there a paradox here? [*Hint:* Think about the forces due to fluid pressure on the *sides* of the containers; do they have vertical components?]

Solution and Discussion   (a) The water at the bottom of each vessel is the same depth d below the surface. Water at the surface of each vessel is at a pressure $P_{\text{surface}} = 0$. Therefore, the pressures at the bottom must be equal:

$$P = P_{\text{surface}} + \rho g d = \rho g d$$

(b) The weight of each filled vessel is equal to the weight of the vessel itself plus the weight of the water inside. The vessels themselves have equal weights, but vessel A holds more water than C while vessel B holds less water than C. Vessel A weighs the most and vessel B weighs the least.

continued on next page

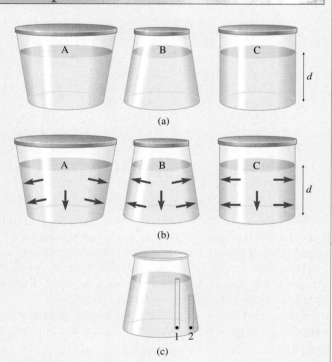

Figure 9.6
(a) Three differently shaped vessels filled with water to same level.
(b) Forces exerted on the containers by the water. (c) Two different points on the bottom of an open vessel.

Conceptual Example 9.4 *continued*

(c) Each container supports the water inside by exerting an upward force equal in magnitude to the weight of the water. By Newton's third law, the water exerts a downward force on the container of the same magnitude. Figure 9.6b shows the forces acting on each container due to the water. In vessel C, the horizontal forces on any two diametrically opposite points on the walls of the container are equal and opposite; thus, the net force on the container walls is zero. The force on the bottom is

$$F = PA = \rho g d \times \pi r^2 = \rho g \times \text{volume of water} = \text{weight of water}$$

as expected. However, the force on the bottom of vessel A is less than the weight of the water in the container, while the force on the bottom of vessel B is greater than the weight of the water. Then how can the water be in equilibrium? In vessel A, the forces on the container walls have downward components as well as horizontal components. The horizontal components of the forces on any two diametrically opposite points are equal and opposite, so the horizontal components add to zero. The sum of the downward components of the forces on the walls and the downward force on the bottom of the container is equal to the weight of the water. Similarly, the forces on the walls of vessel B have upward components. In each case, the *total* force on the bottom *and sides* of the container due to the water is equal to the weight of the water.

Conceptual Practice Problem 9.4   Is pressure determined by column height?

Figure 9.6c shows a vessel with two points marked at the bottom of the water in the vessel. A narrow column of water is drawn above each point. (a) Is the pressure at point 2, P_2, the same as the pressure at point 1, P_1, even though the column of water above point 2 is not as tall? (b) Does $P = P_{atm} + \rho g d$ imply that $P_2 < P_1$? Explain.

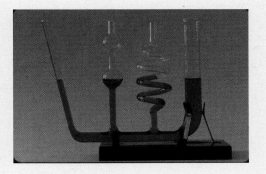

"Water seeks its own level."

9.5   MEASURING PRESSURE

Many other units are used for pressure besides atmospheres and pascals. In the United States, the pressure in an automobile tire is measured in pounds per square inch (lb/in^2). Weather bureaus record atmospheric pressure in bars or millibars. In the United States, television weather reports and home barometers measure pressure in inches of mercury. One atmosphere is equal to approximately 1 bar (1000 millibars), 76 cm of mercury, or 29.9 inches of mercury. Blood pressures are measured in millimeters of mercury (mm Hg), also called the torr. Inches or millimeters of mercury may seem like strange units for pressure: how can a force per unit area be equal to a *distance* (so many mm of Hg)? There is an assumption inherent in using these pressure units that we can understand by studying the mercury manometer.

Units of pressure: 1 atm = 101.3 kPa = 1.013 bar = 14.7 lb/in^2 = 760.0 mm Hg = 760.0 torr = 29.9 in Hg

Manometer

A mercury manometer consists of a vertical U-shaped tube, containing some mercury, with one side typically open to the atmosphere and the other connected to a vessel containing a gas whose pressure we want to measure. Figure 9.7a shows the manometer before it is connected to such a vessel. When both sides of the manometer are open to the atmosphere, the mercury levels are the same.

Now we connect a vessel containing a gas to the left side of the U-tube (Fig. 9.7b). If the gas is at a higher pressure than the atmosphere, the gas pushes the mercury down on the left side and forces it up on the right side. The density of a gas is small compared to the density of mercury, so every point within the gas is assumed to be at the same pressure. At point B, the mercury pushes on the gas with the same magnitude force with which the gas pushes on the mercury, so point B is at the same pressure as the gas. Since point B' is at the same height within the mercury as point B, the pressure at B' is the same as at B. Point C is at atmospheric pressure.

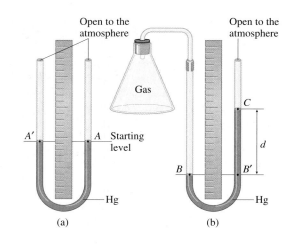

Figure 9.7 (a) A mercury manometer open on both sides. Points A and A' are at atmospheric pressure. (b) The manometer connected on one side to a container of gas at a pressure greater than atmospheric pressure.

The pressure at B is

$$P_B = P_{B'} = P_C + \rho g d$$

where ρ is the density of mercury. The difference in the pressures on the two sides of the manometer is

$$\Delta P = P_B - P_C = \rho g d \tag{9-5}$$

Thus, the difference in mercury levels d is a measure of the pressure *difference*—commonly reported in millimeters of mercury (mm Hg).

The pressure measured when one side of the manometer is open is the *difference* between atmospheric pressure and the gas pressure rather than the absolute pressure of the gas. This difference is called the **gauge pressure**, since it is what most gauges measure:

Gauge pressure

$$P_{\text{gauge}} = P_{\text{abs}} - P_{\text{atm}} \tag{9-6}$$

Since the density of mercury is 13,600 kg/m³, 1.00 mm Hg can be converted to pascals by substituting $d = 1.00$ mm in Eq. (9-5):

$$1.00 \text{ mm Hg} = \rho g d = (13{,}600 \text{ kg/m}^3)(9.80 \text{ m/s}^2)(0.00100 \text{ m}) = 133 \text{ Pa}$$

The liquid in a manometer may be something other than mercury, such as water or oil. Equation (9-5) still applies, as long as we use the correct density ρ of the liquid in the manometer.

Example 9.5

The Mercury Manometer

A manometer is attached to a container of gas to determine its pressure. Before the container is attached, both sides of the manometer are open to the atmosphere. After the container is attached, the mercury on the side attached to the gas container rises 12 cm above its previous level. (a) What is the gauge pressure of the gas in Pa? (b) What is the absolute pressure of the gas in Pa?

Strategy The mercury column is higher on the side connected to the container of gas, so we know that the pressure of the enclosed gas is lower than atmospheric pressure. We need to find the *difference* in levels of the mercury columns on the two sides.

⚠ Careful: it is *not* 12 cm! If one side went up by 12 cm, then the other side has gone down by 12 cm, since the same volume of mercury is contained in the manometer.

Solution (a) The difference d in the mercury levels is 24 cm. Since the mercury on the gas side went *up*, the absolute pressure of the gas is lower than atmospheric pressure. Therefore, the gauge pressure of the gas is *less than zero*. The gauge pressure in Pa is

$$P_{\text{gauge}} = \rho g d$$

continued on next page

Example 9.5 continued

where $d = -24$ cm (the mercury is 24 cm *lower* on the gas side). Then

$$P_{\text{gauge}} = 13{,}600 \text{ kg/m}^3 \times 9.8 \text{ m/s}^2 \times (-0.24 \text{ m}) = -32 \text{ kPa}$$

(b) The absolute pressure of the gas is

$$P = P_{\text{gauge}} + P_{\text{atm}}$$
$$= -32 \text{ kPa} + 101 \text{ kPa} = 69 \text{ kPa}$$

Discussion As a check, the manometer tells us directly that the gauge pressure of the gas is −240 mm Hg. Converting to pascals gives

$$(-240 \text{ mm Hg}) \times (133 \text{ Pa/mm Hg}) = -32 \text{ kPa}$$

Practice Problem 9.5 Column heights in manometer

A mercury manometer is connected to a container of gas. (a) The height of the mercury column on the side connected to the gas is 22.0 cm (measured from the bottom of the manometer). What is the height of the mercury column on the open side if the gauge pressure is measured to be 13.3 kPa? (b) If the gauge pressure of the gas doubles, what are the new heights of the two columns?

Barometer

A manometer can act as a **barometer**—a device to measure atmospheric pressure. Instead of attaching a container with a gas to one end of the manometer, attach a container and a vacuum pump. Pump the air out of the container to get as close to a vacuum—zero pressure—as possible. Then the atmosphere pushes down on one side and pushes the fluid up on the other side toward the evacuated container.

Figure 9.8 shows a barometer in which the vacuum is not created by a vacuum pump. A tube, of length greater than 76 cm and closed at one end, is filled with mercury. The tube is then inverted into an open container of mercury. Some mercury flows down from the tube into the bowl. The space left at the top of the tube is nearly a vacuum because nothing is left but a negligible amount of mercury vapor. Points A and B are at the same level in the mercury and therefore are both at atmospheric pressure since the bowl is open to the air. The distance d from A to the top of the mercury column in the closed tube is a measure of the atmospheric pressure (often called *barometric pressure* because it is measured with a barometer). The barometer was invented by Evangelista Torricelli, an assistant to Galileo, in the 1600s; in his honor, one millimeter of mercury is called one torr.

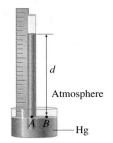

Figure 9.8 A barometer

Physics at Home

When you next have a drink with a straw, insert the straw into the drink and place your finger over the upper opening of the straw so that no more air can enter the straw. Raise the lower end of the straw up out of your drink. Does the liquid in the straw flow back down into your glass? What do you suppose is holding the liquid in place? Make a free-body diagram on your paper napkin.

Some air is trapped between your finger and the top of the liquid in the straw; that air exerts a downward force on the liquid of magnitude P_1A (Fig. 9.9). A downward gravitational force mg also acts on the liquid. The air at the bottom of the straw exerts an upward force on the liquid of magnitude $P_{\text{atm}}A$; this upward force is what holds the liquid in place. Because the liquid does not pour out of the straw, but instead is in equilibrium,

$$P_{\text{atm}}A = P_1A + mg$$

Thus, the pressure P_1 of the air trapped above the liquid must be less than atmospheric pressure.

How did P_1 become less than atmospheric pressure? As you pulled the straw up and out, the liquid in the straw falls a bit, expanding the volume available to the air trapped above the liquid. As we learn in Chapter 15, when a gas expands under conditions like this, its pressure decreases.

When you remove your finger from the top of the straw, air can get in at the top of the straw. Then the pressures above and below the liquid are equal, so the gravitational force pulls the liquid down and out of the straw.

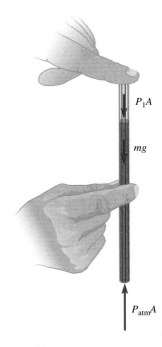

Figure 9.9 Force acting on the liquid inside a straw.

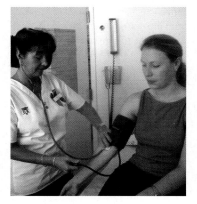

Figure 9.10 A sphygmomanometer

Sphygmomanometer

Blood pressure is measured with a sphygmomanometer (Fig. 9.10). The oldest kind of sphygmomanometer consists of a mercury manometer on one side attached to a closed bag—the cuff. The cuff is wrapped around the upper arm at the level of the heart and is then pumped up with air. The manometer measures the gauge pressure of the air in the cuff.

When the cuff pressure exceeds the highest pressure in the brachial artery, blood flow into the forearm is cut off. A valve on the cuff is then opened to allow the air to slowly escape. The sound of turbulent blood flow past the constriction is heard through a stethoscope when the pressure of the cuff begins to allow some blood to get through the brachial artery. At this point the pressure of the air in the cuff is equal to the maximum or *systolic* pressure—the pressure when the heart contracts.

As air continues to escape from the cuff, the sound of blood flowing through the constriction in the artery continues to be heard. When the pressure in the cuff reaches the *diastolic* pressure in the artery—the minimum pressure that occurs when the heart muscle is relaxed—there is no longer a constriction in the artery, so the pulsing sounds cease. The *gauge* pressures for a healthy heart are nominally around 120 mm Hg (systolic) and 80 mm Hg (diastolic).

9.6 ARCHIMEDES' PRINCIPLE

When an object is immersed in a fluid, the pressure on the lower surface of the object is higher than the pressure on the upper surface. The difference in pressures leads to an upward net force acting on the object due to the fluid pressure. If you try to push a beach ball underwater, you feel the effects of the buoyant force pushing the ball back up. It takes a rather large force to hold such an object completely underwater; the instant you let go, the object pops back up to the surface.

Consider a rectangular solid immersed in a fluid of density ρ (Fig. 9.11a). For each vertical face (left, right, front, and back), there is a face of equal area opposite it. The forces on these two faces due to the fluid are equal in magnitude since the areas and the average pressures are the same. The directions are opposite, so the forces acting on the vertical faces cancel in pairs.

Let the top and bottom surfaces each have area A. The force on the lower face of the block is $F_2 = P_2A$; the force on the upper face is $F_1 = P_1A$. The total force on the block due to the fluid, called the **buoyant force** F_B, is upward since $F_2 > F_1$ (Fig. 9.11b).

$$\vec{F}_B = \vec{F}_1 + \vec{F}_2$$
$$F_B = (P_2 - P_1)A$$

Figure 9.11 (a) Forces due to fluid pressure on the top and bottom of an immersed rectangular solid. (b) The buoyant force is the sum of $\vec{F}_1$ and $\vec{F}_2$.

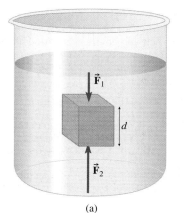

(a)   (b)

Since $P_2 - P_1 = \rho g d$, the magnitude of the buoyant force can be written

---
Buoyant force

$$F_B = \rho g dA = \rho g V \qquad (9\text{-}7)$$
---

where $V = Ad$ is the volume of the block.

Note that ρV is the mass of the volume V of the fluid that the block displaces. Thus, the buoyant force on the submerged block is equal to the weight of an equal volume of fluid, a result called **Archimedes' principle**.

Archimedes' principle

A fluid exerts an upward buoyant force on a submerged object equal in magnitude to the weight of the volume of fluid displaced by the object.

Archimedes' principle applies to a submerged object of any shape even though we derived it for a rectangular block. Why? Imagine replacing an irregular submerged object with enough fluid to fill the object's place. This "piece" of fluid is in equilibrium, so the buoyant force must be equal to its weight. The buoyant force is the net force exerted on the "piece" of fluid by the surrounding fluid, which is identical to the buoyant force on the irregular object since the two have the same shape and surface area.

The same argument can be used to show that if an object is only partly submerged, the buoyant force is still equal to the weight of fluid displaced. Equation (9-7) applies as long as V is the part of the object's volume below the fluid surface rather than the entire volume of the object.

The net force due to gravity and buoyancy acting on an object totally or partially immersed in a fluid (Fig. 9.12) is

$$\vec{F} = \vec{F}_B + \vec{W}$$

The force of gravity for an object of volume V_o and average density ρ_o is

$$W = mg = \rho_o g V_o$$

and the buoyant force is

$$F_B = \rho_f g V_f$$

where V_f and ρ_f are the volume of fluid displaced and the fluid density, respectively. Choosing up to be the $+y$-direction, the net force is

$$F_y = \rho_f g V_f - \rho_o g V_o \qquad (9\text{-}8a)$$

Here F_y can be positive or negative, depending on which density is larger. Imagine releasing a pebble and an air bubble underwater. The pebble's average density is greater than the density of water, so the net force on it is downward; the pebble sinks. An air bubble's average density is less than the density of water, so the net force is upward, causing the bubble to rise toward the surface of the water.

As long as the object is completely submerged, the volumes of the object and the displaced fluid are the same and

$$F_y = (\rho_f - \rho_o)g V \qquad (9\text{-}8b)$$

We might think of F_y as the *apparent weight* of the object when immersed in the fluid. If the object is suspended from a scale, the reading of the scale is the magnitude of F_y in Eq. (9-8b).

If $\rho_o < \rho_f$, the object floats with only part of its volume submerged. In equilibrium, the object displaces a volume of fluid whose weight is equal to the object's weight. At that point the gravitational force equals the buoyant force and the object floats. Setting $F_y = 0$ in Eq. (9-8a) yields

$$\rho_f g V_f = \rho_o g V_o$$

which can be rearranged as:

$$\frac{V_f}{V_o} = \frac{\rho_o}{\rho_f}$$

On the left side of this equation is the fraction of the object's volume that is submerged; it is equal to the ratio of the densities.

Figure 9.12 Forces acting on a floating ice cube

This ratio of densities is called the **specific gravity** of the material when ρ_f is the density of water at 4°C. Specific gravity is without units because it is a ratio of two densities. Water at 4°C is chosen as the reference material because at that temperature, the density of water is a maximum (at atmospheric pressure). The gram was originally defined as the mass of one cubic centimeter of water at 4°C. Thus, water at 4°C has a density of 1 g/cm³ (1000 kg/m³). The specific gravity of seawater is 1.025, which means that seawater has a density of 1.025 g/cm³ (1025 kg/m³).

Figure 9.13 The buoyant force due to the air keeps these balloons aloft.

Making The Connection:
hot air balloons

Specific gravity

$$\text{S.G.} = \frac{\rho}{\rho_{\text{water}}} = \frac{\rho}{1000 \text{ kg/m}^3} \qquad (9\text{-}9)$$

Gases such as air are fluids and exert buoyant forces just as liquids do. The buoyant force due to air is often negligible if an object's average density is much larger than the density of air. To see a significant buoyant force in air, we must use an object with a small average density. A hot air balloon has an opening at the bottom and a burner for heating the air within (Fig. 9.13). Many molecules of the heated air escape through the opening, decreasing the balloon's average density. When the balloon is less dense on average than the surrounding air, it rises; at higher altitudes, the surrounding air becomes less and less dense. At some particular altitude, the buoyant force is equal in magnitude to the weight of the balloon. Then, by Newton's second law, the net force on the balloon is zero. The balloon is in *stable* equilibrium at this altitude: if the balloon rises a bit, it experiences a net force downward, while if the balloon sinks down a bit, it is pushed back upward.

Example 9.6

The Golden (?) Falcon

A small statue in the shape of a falcon is painted black and has a weight of 24.1 N. The owner of the statue claims it is made of solid gold. When the statue is completely submerged in a container brimful of water, the weight of the water that spills over the top and into a bucket is 1.25 N. Find the density and specific gravity of the metal. Is the density consistent with the claim that the falcon is solid gold? Let $g = 9.80$ N/kg.

Strategy   When the solid is totally submerged, it displaces a volume of water equal to its own volume. Thus, the volume of water spilled is equal to the volume of the statue. From the weight and the volume of the object, we can calculate its average density. Dividing by the density of water gives the specific gravity. We can compare the density with that of gold found in Table 9.1.

Solution   The mass of 1.25 N of water is

$$m = \frac{W}{g}$$

$$= \frac{1.25 \text{ N}}{9.80 \text{ N/kg}} = 0.1276 \text{ kg}$$

The volume of the water is then

$$V = \frac{m}{\rho} = \frac{0.1276 \text{ kg}}{1000 \text{ kg/m}^3} = 1.276 \times 10^{-4} \text{ m}^3$$

This is also the volume of the statue. The mass of the metal is

$$\frac{24.1 \text{ N}}{9.80 \text{ N/kg}} = 2.459 \text{ kg}$$

and its average density is

$$\rho = \frac{m}{V} = \frac{2.459 \text{ kg}}{1.276 \times 10^{-4} \text{ m}^3} = 1.93 \times 10^4 \text{ kg/m}^3$$

Its specific gravity is therefore

$$\text{S.G.} = \frac{1.93 \times 10^4 \text{ kg/m}^3}{1000 \text{ kg/m}^3} = 19.3$$

From Table 9.1, we find that gold also has a density of 1.93×10^4 kg/m³. The metal has the correct density; it may *possibly* be gold.

Discussion   An alternative solution recognizes that 24.1 N and 1.25 N are the weights of equal volumes of metal and water. Then the ratio of the weights gives us the ratio of the densities, which is the specific gravity of the statue:

$$\frac{W_{\text{metal}}}{W_{\text{water}}} = \frac{\rho_{\text{metal}} \, Vg}{\rho_{\text{water}} \, Vg} = \frac{\rho_{\text{metal}}}{\rho_{\text{water}}} = \text{S.G.}$$

$$\text{S.G.} = \frac{24.1 \text{ N}}{1.25 \text{ N}} = 19.3$$

This method shows that the value of g does not affect the result.

According to legend, this method to determine the specific gravity of a solid was discovered by Archimedes in the third century B.C.E. King Hieron II asked Archimedes to find a way to check whether his crown was made of pure gold—without melting down the crown, of course! Archimedes came up with his method while he was taking a bath; he noticed the water level rising as he got in and connected the rising water level

continued on next page

Example 9.6 continued

with the volume of water displaced by his body. In his excitement, he jumped out of the bath and ran naked through the streets of Siracusa (a city in Sicily) shouting "Eureka!"

Practice Problem 9.6 Identifying an unknown substance

An unknown solid substance has a weight of 142.0 N. The object is suspended from a scale and hung so that it is completely submerged in the water (but not touching bottom). The scale reads 129.4 N. Find the specific gravity of the object and determine whether the substance could be anything listed in Table 9.1.

Example 9.7

Hidden Depths of an Iceberg

What percentage of a floating iceberg's volume is above water? The specific gravity of ice is 0.917 and the specific gravity of the surrounding seawater is 1.025.

Strategy The ratio of the density of ice to the density of seawater tells us the ratio of the volume of ice that is submerged in the seawater to the total volume of the iceberg. The rest of the ice is above the water.

Solution We could calculate the densities of seawater and of ice in SI units from their specific gravities, but that is unnecessary; the ratio of the specific gravities is equal to the ratio of the densities:

$$\frac{S.G._{ice}}{S.G._{seawater}} = \frac{\rho_{ice}/\rho_{water}}{\rho_{seawater}/\rho_{water}} = \frac{\rho_{ice}}{\rho_{seawater}}$$

The fraction of the iceberg's volume that is submerged is equal to the ratio of the densities of ice and seawater. Thus, the ratio of the volume submerged to the total volume of ice is

$$\frac{V_{submerged}}{V_{ice}} = \frac{\rho_{ice}}{\rho_{seawater}} = \frac{S.G._{ice}}{S.G._{seawater}}$$

$$= \frac{0.917}{1.025} = 0.895$$

89.5% of the ice is below the surface of the water, leaving only 10.5% above the surface.

Discussion An alternative solution does not depend on remembering that the ratio of the volumes is equal to the ratio of the densities. The buoyant force is equal to the weight of a volume $V_{submerged}$ of water:

$$\text{buoyant force} = \rho_{seawater} V_{submerged} g$$

The weight of the iceberg is $mg = \rho_{ice} V_{ice} g$. From Newton's second law, the buoyant force must be equal in magnitude to the weight of the iceberg when it is floating in equilibrium:

$$\rho_{seawater} V_{submerged} g = \rho_{ice} V_{ice} g$$

or

$$\frac{V_{submerged}}{V_{ice}} = \frac{\rho_{ice}}{\rho_{seawater}}$$

The fact that ice floats is of great importance for the balance of nature. If ice were more dense than water, it would gradually fill up the ponds and lakes *from the bottom*. It would not form on top of lakes and remain there. The consequences for fish and other bottom dwellers of solidly frozen lakes would be catastrophic. The water below the surface layer of ice formed in winter remains just above freezing so that the fish are able to survive.

Practice Problem 9.7 Floating in freshwater versus seawater

If the average density of a human being is 980 kg/m³, what fraction of a human body floats above water in a freshwater pond and what fraction floats above seawater in the ocean? The specific gravity of seawater is 1.025.

Blood tests often include determination of the specific gravity of the blood—normally around 1.040 to 1.065. A reading that is too low may indicate anemia, since the presence of red blood cells increases the average density of the blood. Before taking blood from a donor, a drop of the blood is placed in a saline solution. If the drop does not sink, it is not safe for the donor to give blood because the concentration of red blood cells is too low. Urinalysis also includes a specific gravity measurement (normally 1.015 to 1.030); too high a value indicates an abnormally high concentration of dissolved salts, which can signal a medical problem.

Freighters, aircraft carriers, and cruise ships float, although they are made from steel and other materials that are more dense than seawater. When a ship floats, the buoyant force

acting on the ship is equal to the ship's weight. A ship is constructed so that it displaces a volume of seawater larger than the volume of the steel and other construction materials. The *average* density of the ship is its weight divided by its total volume. A large part of a ship's interior is filled with air. All of the "empty" space contributes to the total volume; the resulting average density is less than that of seawater, allowing the ship to float.

Now we can understand how a hippopotamus can sink to the bottom of a pond: it can expel some of the air in its body by exhaling. Exhalation increases the average density of the hippopotamus so that it is just slightly above the density of the water; thus, it sinks. An armadillo does just the opposite: it swallows air, inflating its stomach and intestines, to increase the buoyant force for a swim across a large lake.

Conceptual Example 9.8

A Hovering Fish

How is it that a fish is able to hover almost motionless in one spot—until some attractive food is spotted and, with a flip of the tail, off it swims after the food? Fish have a thin-walled bladder, called a swim bladder, located under the spinal column. The swim bladder contains a mixture of oxygen and nitrogen obtained from the blood of the fish. How does the swim bladder help the fish keep the buoyant and gravitational forces balanced so that it can hover?

Solution and Discussion If the fish's average density is greater than that of the surrounding water, it will sink; if its average density is smaller than that of the water, it will rise. By varying the volume of the swim bladder, the fish is able to vary its own volume and thus its average density. By adjusting its average density to match the density of the surrounding water, the fish can remain suspended in position. The fish can also adjust the volume of the bladder when it wants to rise or sink.

Conceptual Practice Problem 9.8   The diving beetle

A diving beetle traps a bubble of air under its wings. While under the water, the beetle uses the air in the bubble to breathe, gradually exchanging the oxygen for carbon dioxide. (a) What does the beetle do to the air bubble so that it can dive under the water? (b) Once under water, what does the beetle do so that it can rise to the surface? [*Hint:* Treat the beetle and the air bubble as a single system. How can the beetle change the buoyant force acting on the system?]

9.7   FLUID FLOW

Fluid dynamics (or *hydrodynamics*) is the study of *moving* fluids. The blood flowing in our bodies is a hydrodynamic system. The air all around us is another hydrodynamic system. We simplify this wonderfully complex subject—to illustrate some important ideas in less complex situations—by limiting our study at first to fluids flowing under special conditions.

One difference between moving fluids and static fluids is that a moving fluid *does* exert a force *parallel* to any surface over or past which it flows; a static fluid does not. Since the moving fluid exerts a force against a surface, the surface must also exert a force on the fluid. This **viscous force** opposes the flow of the fluid; it is the counterpart to the kinetic frictional force between solids. An external force must act on a viscous fluid (and thereby do work) to keep it flowing. Viscosity is considered in Section 9.9. Until then, we consider only nonviscous fluids—fluid flow where the viscous forces are negligibly small.

Fluid flow can be characterized as steady or unsteady. When the flow is **steady**, the velocity of the fluid *at any point* is constant in time. The velocity is not necessarily the same everywhere, but at any particular point, the velocity of the fluid passing that point remains constant in time. The density and pressure of a steadily flowing fluid are also constant in time.

Steady flow is **laminar**. The fluid flows in neat layers so that each small portion of fluid that passes a particular point follows the same path as every other portion of fluid that passes the same point. The path that the fluid follows, starting from any point, is called a **streamline** (see Fig. 9.14). The streamlines may curve and bend, but they cannot

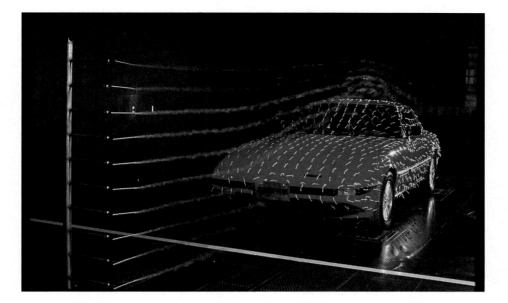

Figure 9.14 A wind tunnel shows the streamlines in the flow of air past a car.

cross each other; if they did, the fluid would have to "decide" which way to go when it gets to such a point. The direction of the fluid velocity at any point must be tangent to the streamline passing through that point. Streamlines are a convenient way to depict fluid flow in a sketch.

The special case that we consider first is the flow of an **ideal fluid**. An ideal fluid is incompressible, undergoes laminar flow, and has no viscosity. Under some conditions real fluids can be modeled as (nearly) ideal, but not under all conditions.

The flow of an ideal fluid is governed by two principles: the continuity equation and Bernoulli's equation. The continuity equation is an expression of conservation of mass: since no fluid is created or destroyed, the total mass of the fluid must be constant. Bernoulli's equation, discussed in Section 9.8, is a form of the energy conservation law applied to fluid flow. Together, these two equations enable us to predict the flow of an ideal fluid.

The Continuity Equation

We start by deriving the continuity equation, which relates the speed of flow to the cross-sectional area of the fluid. Suppose an incompressible fluid flows into a pipe of nonuniform cross-sectional area under conditions of steady flow. In Fig. 9.15, the fluid on the left moves at speed v_1. During a time Δt, the fluid travels a distance

$$x_1 = v_1 \Delta t$$

If A_1 is the cross-sectional area of this section of pipe, then the mass of water moving past point 1 in time Δt is

$$\Delta m_1 = \rho V_1 = \rho A_1 x_1 = \rho A_1 v_1 \Delta t$$

During this same time interval, the mass of fluid moving past point 2 is

$$\Delta m_2 = \rho V_2 = \rho A_2 x_2 = \rho A_2 v_2 \Delta t$$

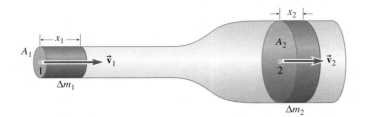

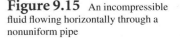

Figure 9.15 An incompressible fluid flowing horizontally through a nonuniform pipe

But, if the flow is steady, the mass passing through one section of pipe in time interval Δt must pass through any other section of the pipe in the same time interval. Therefore,

$$\Delta m_1 = \Delta m_2$$

or

$$\rho A_1 v_1 \Delta t = \rho A_2 v_2 \Delta t \qquad (9\text{-}10)$$

The quantity $\rho A v$ is the *mass flow rate* of the fluid:

Mass flow rate	
$\dfrac{\Delta m}{\Delta t} = \rho A v$ (SI unit: kg/s)	(9-11)

Since the time intervals Δt are the same, Eq. (9-11) says that the mass flow rate past any two points is the same. Since the density of an incompressible fluid is constant, the volume flow rate past any two points must also be the same:

Volume flow rate	
$\dfrac{\Delta V}{\Delta t} = A v$ (SI unit: m³/s)	(9-12)

The **continuity equation** for an ideal fluid equates the volume flow rates past two different points:

Continuity equation for ideal fluid	
$A_1 v_1 = A_2 v_2$	(9-13)

The same volume of fluid that enters the pipe in a given time interval exits the pipe in the same time interval. Where the diameter of the tube is large, the speed of the fluid is small; where the diameter is small, the fluid speed is large. A familiar example is what happens when you use your thumb to partially block the end of a garden hose to make a jet of water. The water moves past your thumb, where the cross-sectional area is smaller, at a greater speed than it moves in the hose. Similarly, water traveling along a river speeds up, forming rapids, when the riverbed narrows or is partially blocked by rocks and boulders.

Streamlines are closer together where the fluid flows faster and farther apart where it flows more slowly (Fig. 9.16). Thus streamlines help us visualize fluid flow. The fluid velocity at any point is tangent to a streamline through that point.

Figure 9.16 Streamlines in a pipe of varying diameter.

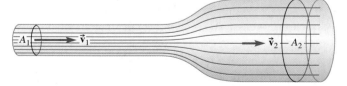

Figure 9.17 Demonstrating the continuity equation at the kitchen sink.

Physics at Home

The continuity equation applies to an ideal fluid even if it is not flowing through a pipe. Turn on a faucet so that the water flows out in a moderate stream (Fig. 9.17). The falling water is in free fall, accelerated by gravity until it hits the sink below. As the water falls, its speed increases. The stream of water gradually narrows as it falls so that the product of speed and cross-sectional area is constant, as predicted by the continuity equation.

Example 9.9

Speed of Blood Flow

The heart pumps blood into the aorta, which has an inner radius of 1.0 cm. The aorta feeds 32 major arteries. If blood in the aorta travels at a speed of 28 cm/s, at approximately what average speed does it travel in the arteries? Assume that blood can be treated as an ideal fluid and that the arteries each have an inner radius of 0.21 cm.

Strategy Since we have assumed blood to be an ideal fluid, we can apply the continuity equation. The main tube (the aorta) is connected to multiple tubes (the major arteries), so this problem seems to be more complicated than a single pipe with a constriction in it. What matters here is the total cross-sectional area into which the blood flows.

Solution We start by finding the cross-sectional area of the aorta

$$A_1 = \pi r_{aorta}^2$$

and then the total cross-sectional area of the arteries

$$A_2 = 32\pi r_{artery}^2$$

Now we apply the continuity equation and solve for the unknown speed.

$$A_1 v_1 = A_2 v_2$$

$$v_2 = v_1 \frac{A_1}{A_2} = 0.28 \text{ m/s} \times \frac{\pi \times (0.010 \text{ m})^2}{32\pi \times (0.0021 \text{ m})^2} = 0.20 \text{ m/s}$$

Discussion The blood flow slows in the arteries because the total cross-sectional area is greater than that of the aorta alone. From the arteries, the blood travels to the many capillaries of the body. Each capillary has a tiny cross-sectional area, but there are so many of them that the blood flow slows greatly once it reaches the capillaries—allowing time for the exchange of oxygen, carbon dioxide, and nutrients between the blood and the body tissues.

Practice Problem 9.9   Hosing down a wastebasket

A garden hose fills a 32-L wastebasket in 120 s. The opening at the end of the hose has a diameter of 1.95 cm. (a) How fast is the water traveling as it leaves the hose? (b) How fast does the water travel if half the exit area is obstructed by placing a finger over the opening?

9.8   BERNOULLI'S EQUATION

The continuity equation relates the flow velocities of an ideal fluid at two different points, based on the change in cross-sectional area of the pipe. However, its application requires that one of the two velocities be known. We haven't yet addressed the question of how the fluid is made to flow in the first place. Now we do so by analyzing the forces that act on the fluid.

According to the continuity equation, the fluid must speed up as it enters a constriction (Fig. 9.18) and then slow down to its original speed when it leaves the constriction. What forces cause the flow speed to change? Consider a small volume of fluid entering the constriction at position A. Its acceleration is to the right since it is speeding up; therefore, the net force must be to the right. The only horizontal forces acting on that volume of fluid are due to the pressure of the surrounding fluid. The pressure ahead of that volume (P_2) must be smaller than the pressure behind it (P_1). At position B, the volume of fluid has an acceleration to the left since it is slowing down. The net force is to the left, so the pressure ahead of it (P_1) is larger than the pressure behind it (P_2). For horizontal flow *the speed is higher where the pressure is lower*. This principle is often called the **Bernoulli effect**.

The Bernoulli effect: Fluid flows faster where the pressure is lower.

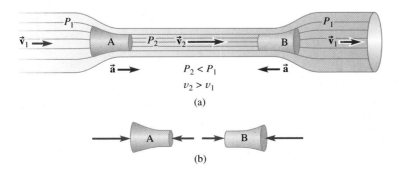

Figure 9.18 (a) A small volume of fluid speeds up as it moves into a constriction (position A) and then slows down as it moves out of the constriction (position B). (b) Unequal forces due to fluid pressure cause acceleration to the right at position A and to the left at position B.

The Bernoulli effect can seem counterintuitive at first; isn't rapidly moving fluid at *high* pressure? For instance, if you were hit with the fast-moving water out of a firehose, you would be knocked over easily. The force that knocks you over is indeed due to fluid pressure; you would justifiably conclude that the pressure was high. However, the pressure is not high *until you slow down the water* by getting in its way. The rapidly moving water in the jet is, in fact, approximately at atmospheric pressure (zero gauge pressure). So your conclusion is correct: when you stop the water, its pressure increases dramatically.

To find the quantitative relationship between pressure changes and flow speed changes for an ideal fluid, we apply conservation of energy. In Fig. 9.19, the shaded volume of fluid flows to the right. If the left end moves a distance Δx_1, then the right end moves a distance Δx_2. Since the fluid is incompressible,

$$A_1 \Delta x_1 = A_2 \Delta x_2 = V$$

Work is done by the neighboring fluid during this flow. Fluid behind (to the left) pushes forward, doing positive work, while fluid ahead pushes backward, doing negative work. The total work done on the shaded volume by neighboring fluid is

$$W = P_1 A_1 \Delta x_1 - P_2 A_2 \Delta x_2 = (P_1 - P_2)V$$

Since no dissipative forces act on an ideal fluid, the work done is equal to the total change in kinetic and gravitational potential energy. The net effect of the displacement is to move a volume V of fluid from height y_1 to height y_2 and to change its speed from v_1 to v_2. The energy change is therefore

$$\Delta E = \Delta K + \Delta U = \tfrac{1}{2}m(v_2^2 - v_1^2) + mg(y_2 - y_1)$$

where the $+y$-direction is up. Substituting $m = \rho V$ and equating the work done on the fluid to the change in its energy yields

$$(P_1 - P_2)V = \tfrac{1}{2}\rho V(v_2^2 - v_1^2) + \rho Vg(y_2 - y_1)$$

Dividing both sides by V and rearranging yields Bernoulli's equation, named after Swiss mathematician Daniel Bernoulli.

Bernoulli's equation

$$P_1 + \rho gy_1 + \tfrac{1}{2}\rho v_1^2 = P_2 + \rho gy_2 + \tfrac{1}{2}\rho v_2^2$$

$$\text{(or } P + \rho gy + \tfrac{1}{2}\rho v^2 = \text{constant)} \tag{9-14}$$

Bernoulli's equation relates the pressure, flow speed, and height at two points in an ideal fluid. Although we derived Bernoulli's equation in a relatively simple situation, it applies to the flow of any ideal fluid as long as points 1 and 2 are on the same streamline.

Each term in Bernoulli's equation has units of pressure, which in the SI system is Pa or N/m². Since a joule is a newton-meter, the pascal is also equal to a joule per cubic meter (J/m³). Each term represents energy per unit volume (energy density). The pressure is the work done by the fluid on the fluid ahead of it per unit volume of flow. The kinetic energy density (kinetic energy per unit volume) is $\tfrac{1}{2}\rho v^2$ and the gravitational potential energy density is ρgy.

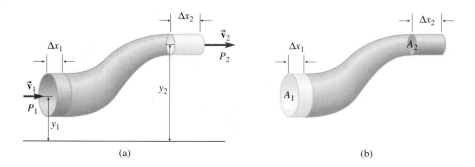

Figure 9.19 Applying conservation of energy to the flow of an ideal fluid. The shaded volume of fluid in (a) is flowing to the right; (b) shows the same volume of fluid a short time later.

(a)   (b)

We can check Bernoulli's equation in some special cases. For horizontal flow, $y_1 = y_2$. Then the pressure is higher where the speed is smaller, as expected. For a static fluid, $v_1 = v_2 = 0$. In this case Bernoulli's equation reduces to

$$P_1 + \rho g y_1 = P_2 + \rho g y_2$$

or, if $d = y_1 - y_2$,

$$P_2 - P_1 = -(\rho g y_2 - \rho g y_1) = \rho g d$$

which is the pressure dependence with depth for a static fluid that we found in Section 9.4.

Example 9.10

The Venturi Meter

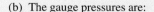 A *Venturi meter* (Fig. 9.20) measures fluid speed in a pipe. A constriction (of cross-sectional area A_2) is put in a pipe of normal cross-sectional area A_1. Two vertical tubes, open to the atmosphere, rise from two points, one of which is in the constriction. The vertical tubes function like manometers, enabling the pressure to be determined. From this information the flow speed in the pipe can be determined.

Suppose that the pipe in question carries water, $A_1 = 2.0A_2$, and the fluid heights in the vertical tubes are $h_1 = 1.20$ m and $h_2 = 0.80$ m. Let $g = 9.80$ N/kg. (a) Find the ratio of the flow speeds v_2/v_1. (b) Find the gauge pressures P_1 and P_2. (c) Find the flow speed v_1 in the pipe.

Strategy   Neither of the two flow speeds is given. We need more than Bernoulli's equation to solve this problem. Since we know the ratio of the areas, the continuity equation gives us the ratio of the speeds. The height of the water in the vertical tubes enables us to find the pressures at points 1 and 2. The fluid pressure at the bottom of each vertical tube is the same as the pressure of the moving fluid just beneath each tube—otherwise water would flow into or out of the vertical tubes until the pressure equalized. The water in the vertical tubes is static, so the gauge pressure at the bottom is $P = \rho g d$. Once we have the ratio of the speeds and the pressures, we apply Bernoulli's equation.

Solution   (a) From the continuity equation, the product of flow speed and area must be the same at points 1 and 2. Therefore,

$$\frac{v_2}{v_1} = \frac{A_1}{A_2} = 2.0$$

The water flows twice as fast in the constriction as in the rest of the pipe.

(b) The gauge pressures are:

$$P_1 = \rho g d_1 = (1000 \text{ kg/m}^3) \times (9.80 \text{ N/kg}) \times (1.20 \text{ m}) = 11.8 \text{ kPa}$$

$$P_2 = \rho g d_2 = (1000 \text{ kg/m}^3) \times (9.80 \text{ N/kg}) \times (0.80 \text{ m}) = 7.8 \text{ kPa}$$

(c) Now we apply Bernoulli's equation. We can use gauge pressures as long as we do so on both sides—in effect we are just subtracting atmospheric pressure from both sides of the equation:

$$P_1 + \rho g y_1 + \tfrac{1}{2}\rho v_1^2 = P_2 + \rho g y_2 + \tfrac{1}{2}\rho v_2^2$$

Since the tube is horizontal, $y_1 \approx y_2$ and we can neglect the small change in gravitational potential energy density $\rho g y$. Then

$$P_1 + \tfrac{1}{2}\rho v_1^2 = P_2 + \tfrac{1}{2}\rho v_2^2$$

We are trying to find v_1, so we can eliminate v_2 by substituting $v_2 = 2.0v_1$:

$$P_1 + \tfrac{1}{2}\rho v_1^2 = P_2 + \tfrac{1}{2}\rho(2v_1)^2$$

Simplifying,

$$P_1 - P_2 = \tfrac{3}{2}\rho v_1^2$$

$$v_1 = \sqrt{\frac{2 \times (11{,}800 \text{ Pa} - 7800 \text{ Pa})}{3 \times 1000 \text{ kg/m}^3}} = 1.6 \text{ m/s}$$

Discussion   The assumption that $y_1 \approx y_2$ is fine as long as the pipe radius is small compared to the difference between the static water heights (40 cm). Otherwise, we would have to account for the different y values in Bernoulli's equation.

One subtle point: recall that we assumed that the fluid pressure at the bottom of the vertical tubes was the same as the pressure of the moving fluid just beneath. Does that contradict Bernoulli's equation? Since there is an abrupt change in fluid speed, shouldn't there be a significant difference in the pressures? No, because these points are not on the same streamline.

Practice Problem 9.10   Garden hose

Water flows horizontally through a garden hose of radius 1.0 cm at a speed of 1.4 m/s. The water shoots horizontally out of a nozzle of radius 0.25 cm. What is the gauge pressure of the water inside the hose?

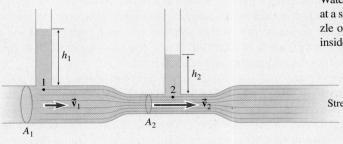

Streamlines

Figure 9.20
Venturi meter

Example 9.11

Torricelli's Theorem

A barrel full of rainwater has a spigot near the bottom, at a depth of 0.80 m beneath the water surface. (a) When the spigot is directed horizontally (Fig. 9.21a) and is opened, how fast does the water come out? (b) If the opening points upward (Fig. 9.21b), how high does the resulting "fountain" go?

Strategy　The water at the surface is at atmospheric pressure. The water emerging from the spigot is *also* at atmospheric pressure since it is in contact with the air. If the pressure of the emerging water were different than that of the air, the stream would expand or contract until the pressures were equal. We apply Bernoulli's equation to two points: point 1 at the water surface and point 2 in the emerging stream of water.

Solution　(a)　Since $P_1 = P_2$, Bernoulli's equation is

$$\rho g y_1 + \tfrac{1}{2}\rho v_1^2 = \rho g y_2 + \tfrac{1}{2}\rho v_2^2$$

Point 1 is 0.80 m above point 2, so

$$y_1 - y_2 = 0.80 \text{ m}$$

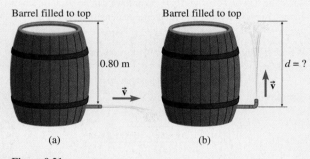

Barrel filled to top　　　　Barrel filled to top

0.80 m　　　　　　　$d = ?$

$\vec{v}$　　　　　　　$\vec{v}$

(a)　　　　　　　(b)

Figure 9.21
Full barrel of rainwater with open spigot (a) horizontal and (b) upward.

The speed of the emerging water is v_2. What is v_1, the speed of the water at the surface? The water at the surface is moving slowly, since the barrel is draining. The continuity equation requires that

$$\frac{v_2}{v_1} = \frac{A_1}{A_2}$$

Since the cross-sectional area of the spigot A_2 is much smaller than the area of the top of the barrel A_1, the speed of the water at the surface v_1 is negligibly small compared with v_2. Setting $v_1 = 0$, Bernoulli's equation reduces to

$$\rho g y_1 = \rho g y_2 + \tfrac{1}{2}\rho v_2^2$$

After dividing through by ρ, we solve for v_2:

$$g(y_1 - y_2) = \tfrac{1}{2}v_2^2$$

$$v_2 = \sqrt{2g(y_1 - y_2)} = 4.0 \text{ m/s}$$

(b)　Now take point 2 to be at the top of the fountain. Then $v_2 = 0$ and Bernoulli's equation reduces to

$$\rho g y_1 = \rho g y_2$$

The "fountain" goes right back up to the top of the water in the barrel!

Discussion　The result of part (b) is called Torricelli's theorem. In reality, the fountain does not reach as high as the original water level; some energy is dissipated due to viscosity and air resistance.

Practice Problem 9.11　Fluid in free fall

Verify that the speed found in part (a) is the same as if the water just fell 0.80 m straight down. That shouldn't be too surprising since Bernoulli's equation is an expression of energy conservation.

Figure 9.22　A perfume atomizer

Making The Connection:
air circulation in
underground burrows

Making The Connection:
household plumbing

Bernoulli's equation applies in an approximate way to moving air. Even though air is not incompressible, in many instances the density changes are small enough to be negligible. A perfume atomizer (Fig. 9.22) is an application of the Bernoulli effect. A squeeze bulb forces air at a high speed over the top of an open tube with its lower end in the perfume liquid. The reduced pressure at the top of the tube then pulls the perfume up into the airflow so that it is ejected with the air in a spray of fine droplets.

Prairie dogs, woodchucks, and rabbits need air circulation in their underground burrows (Fig. 9.23). The burrows always have at least two entrances. Due to differences in the size, shape, and location of the holes, the wind blows across them with different speeds. Where the wind blows across an entrance with a higher speed, the pressure is lower, so air flows out of the burrow at that entrance and air is pulled in at the other entrance. The pressure difference is increased if one entrance is higher than another, since the wind speed is generally higher on higher ground.

Household plumbing also must take the Bernoulli effect into account. In the drain pipe under a sink you can see a U-shaped pipe fitting called a trap (Fig. 9.24). Some water remains in the trap at all times to prevent sewer gases from backing up into the house. The drain that leads to the sewer line is vented to the outside air. Without such a vent, a rush of water through the sewer line from a toilet or washing machine would reduce the pressure at one side of the trap. Atmospheric pressure on the other side of the trap could then push the water out of the trap. The vent to the outside air maintains atmospheric pressure at both sides of the trap.

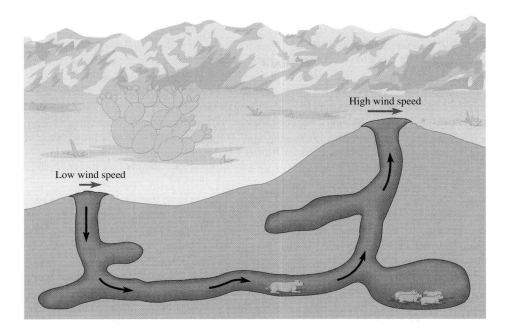

Figure 9.23 Air circulation in a prairie dog burrow.

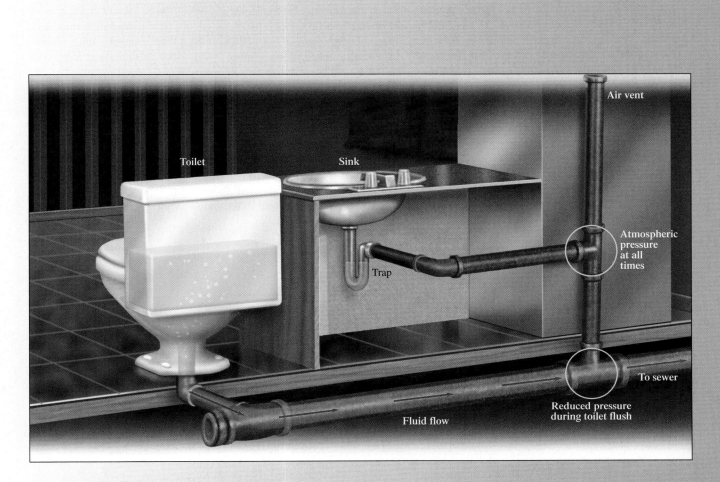

Figure 9.24 Household plumbing must be vented due to the Bernoulli effect.

9.9   VISCOSITY

Bernoulli's equation ignores viscosity (fluid friction). According to Bernoulli's equation, an ideal fluid can continue to flow in a horizontal pipe at constant velocity on its own, just as a hockey puck would slide across frictionless ice at constant velocity without anything pushing it along. However, all real fluids have some viscosity; to maintain flow in a viscous fluid, we have to apply an external force since viscous forces oppose the flow of the fluid (Fig. 9.25). A *pressure difference* between the ends of the pipe must be maintained to keep a real liquid moving through a horizontal pipe. The pressure difference is important—in everything from blood flowing through arteries to oil pumped through a pipeline.

To visualize viscous flow in a tube of circular cross section, imagine the fluid to flow in cylindrical layers, or shells. If there were no viscosity, all the layers would move at the same speed (Fig. 9.26a). In viscous flow, the fluid speed depends on the distance from the tube walls (Figs. 9.26b and c). The fastest flow is at the center of the tube. Layers closer to the wall of the tube move more slowly. The outermost layer of fluid, which is in contact with the tube, does not move. Each layer of fluid exerts viscous forces on the neighboring layers; these forces oppose the relative motion of the layers. The outermost layer exerts a viscous force on the tube. In reality, layers of discrete thickness do not exist; rather, the fluid speed increases continuously from zero (at the wall of the tube) to the maximum (at the center).

A liquid is more viscous if the cohesive forces between molecules are stronger. The viscosity of a liquid decreases with increasing temperature because the molecules become less tightly bound. A decrease in the temperature of the human body is dangerous because the viscosity of the blood increases and the flow of blood through the body is hindered. Gases, on the other hand, have an increase in viscosity for an increase in temperature. At higher temperatures the gas molecules move faster and collide more often with each other.

The coefficient of viscosity (or simply *the viscosity*) of a fluid is written as the Greek letter eta (η) and has units of pascal-seconds (Pa·s) in SI. Other viscosity units in common use are the poise (pronounced *pwäz*, symbol P; 1 P = 0.1 Pa·s) and the centipoise (1 cP = 0.01 P = 0.001 Pa·s). Table 9.2 lists the viscosities of some common fluids.

Poiseuille's Law

The volume flow rate $\Delta V/\Delta t$ for laminar flow of a viscous fluid through a horizontal, cylindrical pipe depends on several factors. First of all, the volume flow rate is proportional to the *pressure drop per unit length* ($\Delta P/L$)—also called the pressure gradient. If a pressure drop ΔP maintains a certain flow rate in a pipe of length L, then a similar pipe of length $2L$ needs twice the pressure drop to maintain the same flow rate (ΔP across the first half and another ΔP across the second half). Thus, the flow rate must be proportional to the pressure drop per unit length.

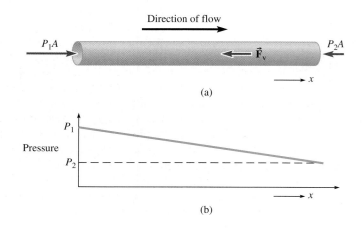

Figure 9.25 (a) To maintain viscous flow, a net force due to fluid pressure $(P_1 - P_2)A$ must be applied in the direction of flow to balance the viscous force F_v due to the pipe, which opposes flow. (b) The pressure in the fluid decreases from P_1 at the left end to P_2 at the right end.

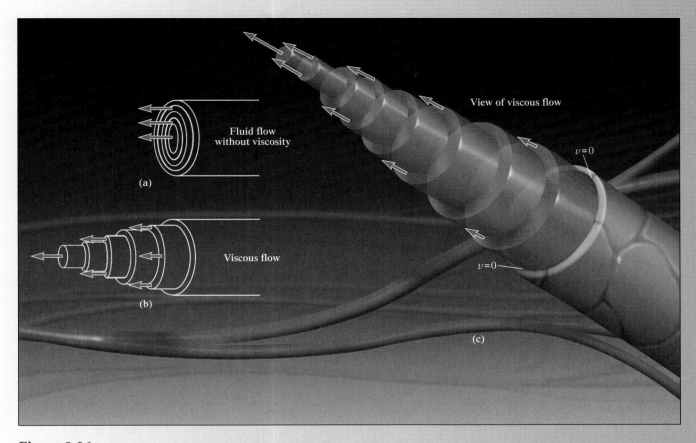

Figure 9.26 (a) Nonviscous flow. (b) Viscous flow. (c) Viscous flow of blood in an artery.

Next, the flow rate is inversely proportional to the viscosity of the fluid. The more viscous the fluid, the smaller the flow rate, if all other factors are equal.

The only other consideration is the radius of the pipe. In the nineteenth century, during a study of flow in blood vessels, French physician Jean-Louis Marie Poiseuille discovered that the flow rate is proportional to the *fourth power* of the pipe radius:

Poiseuille's law

$$\frac{\Delta V}{\Delta t} = \frac{\pi}{8} \frac{\Delta P/L}{\eta} r^4$$

(9-15)

---

where $\Delta V/\Delta t$ is the volume flow rate, ΔP is the pressure difference between the ends of the pipe, r and L are the inner radius and length of the pipe, and η is the viscosity of the fluid. Poiseuille's name is pronounced *pwahzoy*, in a rough English approximation.

It isn't often that we encounter a *fourth-power* dependence. Why such a strong dependence on radius? First of all, if fluids are flowing through two different pipes at the *same speed*, the volume flow rates are proportional to radius squared (flow rate = speed multiplied by cross-sectional area). But, in viscous flow, the average flow speed is larger for wider pipes; fluid farther away from the walls can flow faster. It turns out that the average flow speed for a given pressure gradient is also proportional to radius squared, giving the overall fourth power dependence on the pipe radius of Poiseuille's law.

The strong dependence of flow rate on radius is important in blood flow. A person with cardiovascular disease has arteries narrowed by plaque deposits. To maintain the necessary

Table 9.2

Viscosities of Some Fluids

Substance	Temperature	Viscosity (Pa·s)
Gases		
Water vapor	100°C	1.3×10^{-5}
Air	0°C	1.7×10^{-5}
	20°C	1.8×10^{-5}
	30°C	1.9×10^{-5}
	100°C	2.2×10^{-5}
Liquids		
Acetone	30°C	0.30×10^{-3}
Methanol	30°C	0.51×10^{-3}
Ethanol	30°C	1.0×10^{-3}
Water	0°C	1.8×10^{-3}
	20°C	1.0×10^{-3}
	30°C	0.80×10^{-3}
	40°C	0.66×10^{-3}
	60°C	0.47×10^{-3}
	80°C	0.36×10^{-3}
	100°C	0.28×10^{-3}
Blood plasma	37°C	1.3×10^{-3}
Blood, whole	20°C	3.0×10^{-3}
	37°C	2.1×10^{-3}
SAE #10 oil	30°C	0.20
Glycerin	20°C	0.83
	30°C	0.63

blood flow to keep the body functioning, the blood pressure increases. If the diameter of an artery narrows to $\frac{1}{2}$ of its original value due to plaque deposits, the blood flow rate decreases to $\frac{1}{16}$ of its original value should the pressure drop across it stay the same. To compensate for some of this decrease in blood flow, the heart pumps harder, increasing the blood pressure. High blood pressure is not good either; it introduces its own set of health problems, not least of which is the increased demands placed on the heart muscle.

Example 9.12

Arterial Blockage

A cardiologist reports to her patient that the radius of the left anterior descending artery of the heart has narrowed by 10.0%. What percent increase in the blood pressure drop across the artery is required to maintain the normal blood flow through this artery?

Strategy We assume that the viscosity of the blood has not changed, nor has the length of the artery. To maintain normal blood flow, the volume flow rate $\Delta V/\Delta t$ must stay the same. From Poiseuille's law, then, the product of r^4 and ΔP must be the same before and after the narrowing of the artery.

Solution If r_1 is the normal radius and r_2 is the actual radius, a 10.0% reduction in radius means $r_2 = 0.900 r_1$. Then, from Poiseuille's law,

$$r_1^4 \Delta P_1 = r_2^4 \Delta P_2$$

We solve for the ratio of the pressure drops:

$$\frac{\Delta P_2}{\Delta P_1} = \frac{r_1^4}{r_2^4} = \frac{1}{(0.900)^4} = 1.52$$

Discussion A factor of 1.52 means there is a 52% increase in the blood pressure difference across that artery. The

continued on next page

Example 9.12 *continued*

increased pressure must be provided by the heart. If the normal pressure drop across the artery is 10 mm Hg, then it is now 15.2 mm Hg. The person's blood pressure either must increase by 5.2 mm Hg or there will be a reduction in blood flow through this artery. The heart is under greater strain as it works harder, attempting to maintain an adequate flow of blood.

Practice Problem 9.12   New water pipe

The town water supply is operating at nearly full capacity. The town board decides to replace the water main with a bigger one to increase capacity. If the maximum flow rate is to increase by a factor of 4.0, by what factor should they increase the radius of the water main?

Turbulence

When the fluid velocity at a given point changes, the flow is *unsteady*. **Turbulence** is an extreme example of unsteady flow (Fig. 9.27). In turbulent flow, swirling vortices—little whirlpools of fluid—appear. The vortices are not stationary; they move with the fluid. The flow velocity at any point changes erratically; prediction of the direction or speed of fluid flow under turbulent conditions is difficult.

Figure 9.27   Colored smoke rising from the ground shows the turbulent flow of air past the wingtip of an airplane.

9.10   VISCOUS DRAG

When an object moves through a fluid, the fluid exerts a drag force on it. When the relative velocity between the object and the fluid is low enough for the flow around the object to be laminar, the drag force derives from viscosity and is called **viscous drag**. The viscous drag force is proportional to the speed of the object. For larger relative speeds, the flow becomes turbulent and the drag force is proportional to the square of the object's speed. The air resistance discussed in Chapter 3 was due to the turbulent flow of air.

The viscous drag force depends also on the shape and size of the object. For a spherical object, the viscous drag force is given by Stokes's law:

Stokes's law (viscous drag on a sphere)

$$F_D = 6\pi\eta r v \qquad (9\text{-}16)$$

---

where r is the radius of the sphere, η is the viscosity of the fluid, and v is the speed of the object with respect to the fluid.

An object's **terminal velocity** is the velocity that produces just the right drag force so that the net force is zero. An object falling at its terminal velocity has zero acceleration, so it continues moving at that constant velocity. Using Stokes's law, we can find the terminal velocity of a spherical object falling through a fluid. When the object moves at terminal velocity, the net force acting on it is zero. If $\rho_o > \rho_f$, the object sinks; the terminal velocity is downward and the viscous drag force acts upward to oppose the motion. For an object, such as a helium balloon in air or an air bubble in oil, that rises rather than sinks ($\rho_o < \rho_f$), the terminal velocity is *upward* and the drag force is *downward*.

Example 9.13

Falling Droplet

In an experiment to measure the electric charge of the electron, a fine mist of oil droplets is sprayed into the air and observed through a telescope as they fall. These droplets are so tiny that they soon reach their terminal velocity. If the radius of the droplets is 2.40 μm and the average density of the oil is 862 kg/m³, find the terminal speed of the droplets. The density of air is 1.29 kg/m³ and the viscosity of air is 1.8×10^{-5} Pa·s.

continued on next page

Example 9.13 continued

Strategy   When the droplets fall at their terminal velocity, the net force on them is zero. We set the net force equal to zero and use Stokes' law for the drag force.

Solution   We set the sum of the forces equal to zero when $v = v_t$.

$$\Sigma F_y = +F_D + F_B - W = 0$$

If m_{air} is the mass of displaced air, then

$$6\pi\eta r v_t + m_{air}g - m_{oil}g = 0$$

Solving for v_t,

$$v_t = \frac{g(m_{oil} - m_{air})}{6\pi\eta r}$$

$$= \frac{\frac{4}{3}\pi r^3 g(\rho_{oil} - \rho_{air})}{6\pi\eta r}$$

After dividing the numerator and denominator by πr, we substitute numerical values:

$$v_t = \frac{\frac{4}{3} \times (2.40 \times 10^{-6}\,\text{m})^2 \times 9.8\,\text{N/kg} \times (862\,\text{kg/m}^3 - 1.29\,\text{kg/m}^3)}{6 \times 1.8 \times 10^{-5}\,\text{Pa}\cdot\text{s}}$$

$$= 6.0 \times 10^{-4}\,\text{m/s} = 0.60\,\text{mm/s}$$

Discussion   We should check the units in the final expression:

$$\frac{\text{m}^2 \times \text{N/kg} \times \text{kg/m}^3}{\text{Pa}\cdot\text{s}} = \frac{\text{N/m}}{\text{N/m}^2 \times \text{s}} = \text{m/s}$$

Stokes's law was applied in this way by Robert Millikan in his experiments in 1909–1913 to measure the charge of the electron. Using an atomizer, Millikan produced a fine spray of oil droplets. The droplets picked up electric charge as they were sprayed through the atomizer. Millikan kept a droplet suspended without falling by applying an upward electric force. After removing the electric force, he measured the terminal speed of the droplet as it fell through the air. He calculated the mass of the droplet from the terminal speed and the density of the oil using Stokes's law. By setting the magnitude of the electric force equal to the weight of a suspended droplet, Millikan calculated the electric charge of the droplet. He measured the charges of hundreds of different droplets and found that they were all multiples of the same quantity—the charge of an electron.

Practice Problem 9.13   Rising bubble

Find the terminal velocity of an air bubble of 1.00 mm diameter in a cup of vegetable oil. The specific gravity of the oil is 0.840 and the viscosity is 0.160 Pa·s. Assume the diameter of the bubble does not change as it rises.

Physics at Home

A demonstration of terminal velocity can be done at home. Climb up a small stepladder, or lean over an upstairs balcony, and drop two objects at the same time: a coin and two or three nested cone-shaped paper coffee filters. You will see the effects of viscous drag on the coffee filters as they fall with a constant terminal velocity. Enlist the help of a friend so you can get a side view of the two objects falling. (See also Fig. 3.25.) Why do the coffee filters work so well?

For small particles falling in a liquid, the terminal velocity is also called the *sedimentation velocity*. The sedimentation velocity is often small for two reasons. First, if the particle isn't much more dense than the fluid, then the apparent weight (vector sum of the gravitational and buoyant forces) is small. Second, notice that the terminal velocity is proportional to r^2; viscous drag is most important for small particles. Thus, it can take a long time for the particles to sediment out of solution. Because the sedimentation velocity is proportional to g, it can be increased by the use of a centrifuge, a rotating container that creates artificial gravity of magnitude $g_{eff} = \omega^2 r$ (see Eq. (5-11) and Section 5.7). Ultracentrifuges are capable of rotating at 100,000 rev/min and produce artificial gravity approaching a million times g.

9.11   SURFACE TENSION

The surface of a liquid has special properties not associated with the interior of the liquid. The surface acts like a stretched membrane under tension. The **surface tension** (symbol γ, the Greek letter gamma) of a liquid is the force per unit *length* with which the surface pulls *on its edge*. The direction of the force is tangent to the surface at its edge. Surface tension is caused by the cohesive forces that pull the molecules toward each other.

Physics at Home

Place a needle (or a flat plastic-coated paper clip) gently on the surface of a glass of water. It may take some practice, but you should be able to get it to "float" on top of the water (Fig. 9.28). Now add some detergent to the water and try again. The detergent reduces the surface tension of the water so it is unable to support the needle. Soaps and detergents are *surfactants*—substances that reduce the surface tension of a fluid. The reduced surface tension allows the water to spread out more, wetting more of a surface to be cleaned.

Figure 9.28 Surface tension supports the weight of a needle.

The high surface tension of water enables water striders and other small insects to walk on the surface of a pond. The foot of the insect makes a small indentation in the water surface (Fig. 2.5); the deformation of the surface enables it to push upward on the foot as if the water surface were a thin sheet of rubber. Visually it looks similar to a person walking across the mat of a trampoline. Other small water creatures, such as mosquito larvae and planaria, hang from the surface of water, using surface tension to hold themselves up. In plants, surface tension aids in the transport of water from the roots to the leaves.

The high surface tension of water is a hindrance in the lungs. The exchange of oxygen and carbon dioxide between inspired air and the blood takes place in tiny sacs called *alveoli*, 0.05 to 0.15 mm in radius, at the end of the bronchial tubes.

If the mucus coating the alveoli had the same surface tension as other body fluids, the pressure difference between the inside and outside of the alveoli would not be great enough for them to expand and fill with air. The alveoli secrete a surfactant that decreases the surface tension in their mucous coating.

Bubbles

In an underwater air bubble, the surface tension of the water surface tries to contract the bubble while the pressure of the enclosed air pushes outward on the surface. In equilibrium, the air pressure inside the bubble must be larger than the water pressure outside so that the net outward force due to pressure balances the inward force due to surface tension. The excess pressure $\Delta P = P_{in} - P_{out}$ depends both on the surface tension and the size of the bubble. In Problem 62, you can show that the excess pressure is

$$\Delta P = \frac{2\gamma}{r} \qquad (9\text{-}17)$$

Look closely at a glass of champagne and you can see strings of bubbles rising, originating from the same points in the liquid. Why don't bubbles spring up from random locations? A very small bubble would require an insupportably large excess pressure. The bubbles need some sort of nucleus—a small dust particle, for instance—on which to form so they can start out larger, with excess pressures that aren't so large. The strings of bubbles in the glass of champagne are showing where suitable nuclei have been "found."

Example 9.14

Lung Pressure

During inhalation the gauge pressure in the alveoli is about −400 Pa to allow air to flow in through the bronchial tubes. Suppose the mucous coating on an alveolus of initial radius 0.050 mm had the same surface tension as water (0.070 N/m). What lung pressure outside the alveoli would be required to begin to inflate the alveolus?

Strategy We model an alveolus as a sphere coated with mucus. Due to the surface tension of the mucus, the alveolus must have a lower pressure outside than inside, as for a bubble.

Solution The excess pressure is

$$\Delta P = \frac{2\gamma}{r} = \frac{2 \times 0.070 \text{ N/m}}{0.050 \times 10^{-3} \text{ m}} = 2.8 \text{ kPa}$$

Thus, the pressure inside the alveolus would be 2.8 kPa higher than the pressure outside. The gauge pressure inside is −400 Pa, so the gauge pressure outside would be

$$P_{out} = -0.4 \text{ kPa} - 2.8 \text{ kPa} = -3.2 \text{ kPa}$$

Discussion The *actual* gauge pressure outside the alveoli is about −0.5 kPa rather than −3.2 kPa; then $\Delta P = P_{in} - P_{out} =$

continued on next page

Example 9.14 *continued*

−0.4 kPa − (−0.5 kPa) = 0.1 kPa rather than 2.8 kPa. Here the surfactant comes to the rescue; by decreasing the surface tension in the mucus, it decreases ΔP to about 0.1 kPa and allows the expansion of the alveoli to take place. For a newborn baby, the alveoli are initially collapsed, making the required pressure difference about 4 kPa. That first breath is as difficult an event as it is significant.

Practice Problem 9.14   Champagne bubbles

A bubble in a glass of champagne is filled with CO_2. When it is 2.0 cm below the surface of the champagne, its radius is 0.50 mm. What is the gauge pressure inside the bubble? Assume that champagne has the same average density as water and a surface tension of 0.070 N/m.

MASTER THE CONCEPTS

Summary

- Fluids are materials that flow and include both liquids and gases. A liquid is nearly incompressible, whereas a gas expands to fill its container.

- Pressure is the perpendicular force per unit area that a fluid exerts on any surface with which it comes in contact. The SI unit of pressure is the pascal (1 Pa = 1 N/m^2).

- The average air pressure at sea level is 1 atm = 101.3 kPa.

- Pascal's principle: A change in pressure at any point in a confined fluid is transmitted everywhere throughout the fluid.

- The average density of a substance is the ratio of its mass to its volume

$$\rho = \frac{m}{V} \qquad (9\text{-}2)$$

- The specific gravity of a material is the ratio of its density to that of water at 4°C.

- Pressure variation with depth in a static fluid:

$$P_2 = P_1 + \rho g d \qquad (9\text{-}3)$$

where point 2 is a depth d below point 1.

- Instruments to measure pressure include the manometer and the barometer. The barometer measures the pressure of the atmosphere. The manometer measures a pressure difference.

- Gauge pressure is the amount by which the absolute pressure exceeds atmospheric pressure:

$$P_{gauge} = P_{abs} - P_{atm} \qquad (9\text{-}6)$$

- Archimedes' principle: a fluid exerts an upward buoyant force on a completely or partially submerged object equal in magnitude to the weight of the volume of fluid displaced by the object.

- In steady flow, the velocity of the fluid *at any point* is constant in time. In laminar flow, the fluid flows in neat layers so that each small portion of fluid that passes a particular point follows the same path as every other portion of fluid that passes the same point. The path that the fluid follows, starting from any point, is called a streamline. Laminar flow is steady. Turbulent flow is chaotic and unsteady. The viscous force opposes the flow of the fluid; it is the counterpart to the frictional force for solids.

- An ideal fluid exhibits laminar flow, has no viscosity, and is incompressible. The flow of an ideal fluid is governed by two principles: the continuity equation and Bernoulli's equation.

- The continuity equation states that the volume flow rate for an ideal fluid is constant:

$$\frac{\Delta V}{\Delta t} = A_1 v_1 = A_2 v_2 \qquad (9\text{-}12, 9\text{-}13)$$

- Bernoulli's equation relates pressure changes to changes in flow speed and height:

$$P_1 + \rho g y_1 + \tfrac{1}{2}\rho v_1^2 = P_2 + \rho g y_2 + \tfrac{1}{2}\rho v_2^2 \qquad (9\text{-}14)$$

- Poiseuille's law gives the volume flow rate $\Delta V/\Delta t$ for viscous flow in a horizontal pipe:

$$\frac{\Delta V}{\Delta t} = \frac{\pi}{8}\frac{\Delta P/L}{\eta} r^4 \qquad (9\text{-}15)$$

where ΔP is the pressure difference between the ends of the pipe, r and L are the inner radius and length of the pipe, and η is the viscosity of the fluid.

- Stokes's law gives the viscous drag force on a spherical object moving in a fluid:

$$F_D = 6\pi \eta r v \qquad (9\text{-}16)$$

- The surface tension γ (the Greek letter gamma) of a liquid is the force *per unit length* with which the surface pulls on its edge.

Highlighted Figures and Tables

F9.1 Fluid pressure caused by collisions (p. 302)

F9.2 Forces due to a static fluid (p. 303)

F9.4 Pascal's principle and the hydraulic lift (p. 305)

T9.1 Densities of some common substances (p. 307)

F9.7 Mercury manometer and measuring pressure (p. 310)

F9.11 Archimedes' principle (p. 312)

F9.15 Continuity equation (p. 317)

F9.17 Streamlines of fluid (p. 318)

F9.18 Unequal forces due to fluid pressure cause accelerations (p. 319)

F9.19 Applying conservation of energy to the flow of an ideal fluid (p. 320)

CONCEPTUAL QUESTIONS

1. Does a manometer (with one side open) measure absolute pressure or gauge pressure? How about a barometer? A tire pressure gauge? A sphygmomanometer?

2. A volunteer firefighter holds the end of a firehose as a strong jet of water emerges. (a) The hose exerts a large backward force on the firefighter. Explain the origin of this force. (b) If another firefighter steps on the hose, forming a constriction (a place where the area of the hose is smaller), the hose begins to pulsate wildly. Explain.

3. The weight of a boat is listed on specification sheets as its "displacement." Explain.

4. In tall buildings, the water supply system uses multiple pumping stations on different floors. At each station, water pumped up from below collects in a storage tank held at atmospheric pressure before it enters the pump. The storage tank supplies water to the floors below it. What are some of the reasons why these multiple pumping stations are used?

5. Can an astronaut on the Moon use a straw to drink from a normal drinking glass? How about if he pokes a straw through an otherwise sealed juice box? Explain.

6. It is commonly said that wood floats because it is "lighter than water" or that a stone sinks because it is "heavier than water." Are these accurate statements? If not, correct them.

7. Why must a blood pressure cuff be wrapped around the arm at the same vertical level as the heart?

8. A hot air balloon is floating in equilibrium with the surrounding air. (a) How does the pressure inside the balloon compare to the pressure outside? (b) How does the density of the air inside compare to the density outside?

9. When helium weather balloons are released, they are purposely underinflated. Why? [*Hint:* The balloons go to very high altitudes.]

10. Bernoulli's equation applies only to *steady flow*. Yet Bernoulli's equation allows the fluid velocity at one point to be different than the velocity at another point. For the fluid velocity to change, the fluid must be accelerated as it moves from one point to another. In what way is the flow *steady*, then?

11. Before getting an oil change, it is a good idea to drive a few miles to warm up the engine. Why?

12. Your ears "pop" when you change altitude quickly—such as during takeoff or landing in an airplane, or during a drive in the mountains. Curiously, if you are a passenger in a high-speed train, your ears sometimes pop as the speed of the train increases rapidly—even though there is little or no change in altitude. Explain.

13. It is easier to get a good draft in a chimney on a windy day than when the outside air is still, all other things being equal. Why?

14. Two soap bubbles of *different radii* are formed at the ends of a tube with a closed valve in the middle. What happens to the bubbles when the valve is opened? (If the alveoli in the lung did not have a surfactant that reduces surface tension in the smaller alveoli, the same thing would happen in the lung, with disastrous results!)

15. *Pascal's principle: proof by contradiction.* Points A and B are near each other at the same height in a fluid. Suppose $P_A > P_B$. (a) Can both v_A and v_B be zero? Explain. (b) Point C is just above point D in a static fluid. Suppose the pressure at C increases by an amount ΔP. What would happen if the pressure at D did not increase by the same amount?

16. Two pilots are arguing over competing explanations of what holds an airplane up. One says that air flows over the top of the wing faster than the air beneath the wing; therefore, the pressure beneath is higher than the pressure above. The other says that the wing deflects air downward; by Newton's third law (or conservation of momentum), the air pushes up on the wing. Use Fig. 9.29, which shows streamlines of air flowing past an airplane wing, to answer these questions. (a) Does air above the wing move faster than air beneath the wing? (b) Is the pressure beneath greater than the pressure above? (Changes in the air's density are negligible at subsonic speeds.) (c) Does the wing deflect air downward? (d) Which pilot is right—or are they both right?

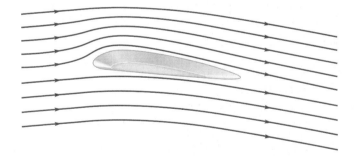

Figure 9.29 Streamlines about an airplane wing

17. What are the advantages of using hydraulic systems rather than mechanical systems to operate automobile brakes or the control surfaces of an airplane?

18. In any hydraulic system, it is important to "bleed" air out of the line. Why?

19. A large gravel truck is loosely covered with a tarpaulin. The edges of the tarp are fastened to the truck. If the truck moves fast enough, the tarp lifts up off the gravel. Why? [*Hint:* The air under the tarp is stationary with respect to the tarp, while the air above is moving with respect to the tarp.]

20. Is it possible for a skin diver to dive to any depth as long as his snorkel tube is sufficiently long? (A snorkel is a face mask with a breathing tube that sticks above the surface of the water.)

MULTIPLE CHOICE QUESTIONS

1. A glass of ice water is filled to the brim with water; the ice cubes stick up above the water surface. After the ice melts, which is true?

   (a) The water level is below the top of the glass.
   (b) The water level is at the top of the glass but no water has spilled.
   (c) Some water has spilled over the sides of the glass.
   (d) Impossible to say without knowing the initial densities of the water and the ice.

2. A dam holding back the water in a reservoir exerts a horizontal force on the water. The magnitude of this force depends on

   (a) The maximum depth of the reservoir.
   (b) The depth of the water at the location of the dam.
   (c) The surface area of the reservoir.
   (d) Both (a) and (b).
   (e) All three—(a), (b), and (c).

3. Bernoulli's equation applies to

   (a) Any fluid.
   (b) An incompressible fluid, whether viscous or not.
   (c) An incompressible, nonviscous fluid, whether the flow is turbulent or not.
   (d) An incompressible, nonviscous, nonturbulent fluid.
   (e) A static fluid only.

Questions 4–5. Two spheres, A and B, fall through the same viscous fluid.

Answer choices for Questions 4 and 5:

   (a) A has the larger terminal velocity
   (b) B has the larger terminal velocity
   (c) A and B have the same terminal velocity
   (d) Insufficient information is given to reach a conclusion

4. A and B have the same radius; A has the larger mass. Which has the larger terminal velocity?

5. A and B have the same density; A has the larger radius. Which has the larger terminal velocity?

6. Bernoulli's equation is an expression of

   (a) Conservation of mass.
   (b) Conservation of energy.
   (c) Conservation of momentum.
   (d) Conservation of angular momentum.

7. The continuity equation is an expression of

   (a) Conservation of mass.
   (b) Conservation of energy.
   (c) Conservation of momentum.
   (d) Conservation of angular momentum.

8. What is the gauge pressure of the gas in the closed tube in Fig. 9.30? (Take the atmospheric pressure to be 76 cm Hg.)

   (a) 20 cm Hg　　(b) –20 cm Hg　　(c) 96 cm Hg
   (d) 56 cm Hg　　(e) –96 cm Hg　　(f) –56 cm Hg

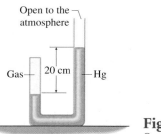

Figure 9.30 Multiple Choice Question 8

9. A manometer contains two different fluids of different densities (Fig. 9.31). Both sides are open to the atmosphere. Which pair(s) of points have equal pressure?

   (a) $P_1 = P_5$　　　(b) $P_2 = P_5$
   (c) $P_3 = P_4$　　　(d) Both (a) and (c)
   (e) Both (b) and (c)

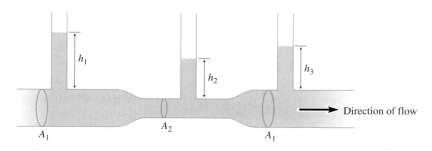

Figure 9.31 Multiple Choice Question 9

10. A Venturi meter is used to measure the flow speed of a *viscous* fluid (Fig. 9.32). Which is true?

    (a) $h_3 = h_1$　　(b) $h_3 > h_1$
    (c) $h_3 < h_1$　　(d) Insufficient information to determine

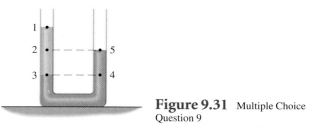

Figure 9.32 Multiple Choice Question 10

PROBLEMS

Note: **C** indicates a combination conceptual/quantitative problem. Gold diamonds ✦, ✦✦ are used to indicate the increasing level of difficulty of each problem. Problem numbers appearing in blue, 9., denote problems that have a detailed solution available in the Student Solutions Manual. Some problems are *paired* by concept; their numbers are connected by a ruled box.

9.2 Pressure

1. Someone steps on your toe, exerting a force of 500 N on an area of 1.0 cm^2. What is the average pressure on that area in atm?

C 2. Atmospheric pressure is about 1.0×10^5 Pa on average. (a) What is the downward force of the air on a desktop with surface area 1.0 m^2? (b) Convert this force to pounds so you really understand how large it is. (c) Why does this huge force not crush the desk?

3. What is the average pressure on the soles of the feet of a standing 90.0-kg person due to the contact force with the floor? Each foot has a surface area of 0.020 m^2.

4. The pressure inside a bottle of champagne is 4.5 atm higher than the air pressure outside. The neck of the bottle has an inner radius of 1.0 cm. What is the frictional force on the cork due to the neck of the bottle?

9.3 Pascal's Principle

5. A container is filled with gas at a pressure of 4.0×10^5 Pa. The container is a cube, 0.10 m on a side, with one side facing south. What is the magnitude and direction of the force on the south side of the container due to the gas inside?

6. A nurse applies a force of 4.40 N to the piston of a syringe. The piston has an area of $5.00 \times 10^{-5} \text{ m}^2$. What is the pressure increase in the fluid within the syringe?

7. In a hydraulic lift, the diameters of the pistons are 5.00 cm and 20.0 cm. A car weighing $W = 10.0$ kN is to be lifted by the force of the large piston. (a) What force F_a must be applied to the small piston? (b) When the small piston is pushed in by 10.0 cm, how far is the car lifted? (c) Find the mechanical advantage of the lift, which is the ratio W/F_a.

8. A hydraulic lift is lifting a car that weighs 12 kN. The area of the piston supporting the car is A, the area of the other piston is a, and the ratio A/a is 100.0. How far must the small piston be pushed down to raise the car a distance of 1.0 cm? [*Hint:* Consider the work to be done.]

✦ 9. Depressing the brake pedal in a car pushes on a piston with cross-sectional area 3.0 cm^2. The piston applies pressure to the brake fluid, which is connected to two pistons, each with area 12.0 cm^2. Each of these pistons presses a brake pad against one side of a rotor attached to one of the rotating wheels (Fig. 9.33). (a) When the force applied by the brake pedal to the small piston is 7.5 N, what is the normal force applied to each side of the rotor? (b) If the coefficient of kinetic friction between a brake pad and the rotor is 0.80 and each pad is (on average) 12 cm from the rotation axis of the rotor, what is the torque on the rotor due to the two pads?

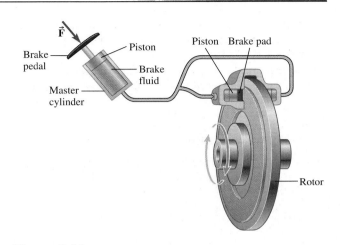

Figure 9.33 Problem 9

9.4 The Effect of Gravity on Fluid Pressure

10. At the surface of a freshwater lake the air pressure is 1.0 atm. At what depth under water in the lake is the water pressure 4.0 atm?

11. What is the pressure on a fish 10 m under the ocean surface?

12. How high can you suck water up a straw? The pressure in the lungs can be reduced to about 10 kPa below atmospheric pressure.

13. At the surface of a freshwater lake the pressure is 105 kPa. (a) What is the pressure increase in going 35.0 m below the surface? (b) What is the approximate pressure decrease in going 35 m above the surface? Air at 20°C has density of 1.20 kg/m^3.

14. The maximum pressure most organisms can survive is about 1000 times atmospheric pressure. Only small, simple organisms such as tadpoles and bacteria can survive such high pressures. What then is the maximum depth at which these organisms can live under the sea (assuming that the density of seawater is 1025 kg/m^3)?

15. In the Netherlands, a dike holds back the sea from a town below sea level. The dike springs a leak 3.0 m below the water surface. If the area of the hole in the dike is 1.0 cm^2, what force must the Dutch boy exert to save the town?

16. A container has a large cylindrical lower part with a long thin cylindrical neck (Fig. 9.34). The lower part of the container holds 12.5 m^3 of water and the surface area of the

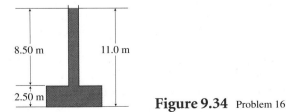

8.50 m   11.0 m

2.50 m

Figure 9.34 Problem 16

bottom of the container is 5.00 m². The height of the lower part of the container is 2.50 m and the neck contains a column of water 8.50 m high. The total volume of the column of water in the neck is 0.200 m³. What is the magnitude of the force exerted by the water on the bottom of the container?

17. The density of platinum is 21,500 kg/m³. Find the volume of 1.00 kg of platinum and compare it to the volume of 1.00 kg of aluminum.

9.5 Measuring Pressure

18. A woman's systolic blood pressure when resting is 160 mm Hg. What is this pressure in (a) Pa, (b) lb/in², (c) atm, (d) torr?

19. The gauge pressure of the air in an automobile tire is 32 lb/in². Convert this to (a) Pa, (b) torr, (c) atm.

20. An IV is connected to a patient's vein. The blood pressure in the vein has a gauge pressure of 12 mm Hg. At least how far above the vein must the IV bag be hung in order for fluid to flow into the vein? Assume the fluid in the IV has the same density as blood.

21. When a mercury manometer is connected to a gas main, the mercury stands 40.0 cm higher in the tube that is open to the air than in the tube connected to the gas main. A barometer at the same location reads 74.0 cm Hg. Determine the absolute pressure of the gas in cm Hg.

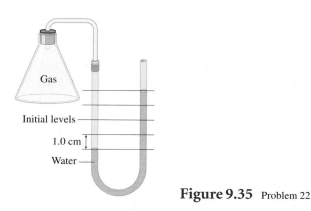

Figure 9.35 Problem 22

22. An experiment to determine the specific heat of a gas makes use of a water manometer attached to a flask (Fig. 9.35). Initially the two columns of water are even. Atmospheric pressure is 1.0×10^5 Pa. After heating the gas, the water levels change to those shown. Find the change in pressure of the gas in Pa.

23. A manometer using oil (density 0.90 g/cm³) as a fluid is connected to an air tank. Suddenly the pressure in the tank increases by 0.74 cm Hg. (a) By how much does the fluid level rise in the side of the manometer that is open to the atmosphere? (b) What would your answer be if the manometer used mercury instead?

24. Estimate the average blood pressure in a person's foot, if the foot is 1.37 m below the aorta, where the average blood pressure is 104 mm Hg. For the purposes of this estimate, assume the blood isn't flowing.

9.6 Archimedes' Principle

25. A Canada goose floats with 25% of its volume below water. What is the average density of the goose?

26. A flat-bottomed barge, loaded with coal, has a mass of 3.0×10^5 kg. The barge is 20.0 m long and 10.0 m wide. It floats in freshwater. What is the depth of the barge below the waterline?

27. (a) What is the buoyant force on 0.90 kg of ice floating freely in liquid water? (b) What is the buoyant force on 0.90 kg of ice that is held completely submerged under water?

28. A block of wood floats in oil with 90.0% of its volume submerged. What is the density of the oil? The density of the block of wood is 0.67 g/cm³.

29. A piece of metal is released under water. The volume of the metal is 50.0 cm³ and its specific gravity is 5.0. What is its initial acceleration? [*Hint:* When $v = 0$, there is no drag force.]

30. (a) A piece of balsa wood with density 0.50 g/cm³ is released under water. What is its initial acceleration? [*Hint:* When $v = 0$, there is no drag force.] (b) Repeat for a piece of pine with density 0.750 g/cm³. (c) Repeat for a ping-pong ball with an average density of 0.125 g/cm³.

31. A cylindrical disk has volume 8.97×10^{-3} m³ and mass 8.16 kg. The disk is floating on the surface of some water with its flat surfaces horizontal. The area of each flat surface is 0.640 m². (a) What is the specific gravity of the disk? (b) How far below the water level is its bottom surface? (c) How far above the water level is its top surface?

32. An aluminum cylinder weighs 1.03 N. When this same cylinder is completely submerged in alcohol, the volume of the displaced alcohol is 3.90×10^{-5} m³. The apparent weight of the cylinder when completely submerged is 0.730 N. What is the specific gravity of the alcohol?

33. A fish uses a swim bladder to change its density so it is equal to that of water, enabling it to remain suspended under water. If a fish has an average density of 1080 kg/m³ and mass 10.0 g with the bladder completely deflated, to what volume must the fish inflate the swim bladder in order to remain suspended in seawater of density 1060 kg/m³?

34. The average density of a fish can be found by first weighing it in air and then finding the apparent weight of the fish as it hangs completely immersed in water. If a fish has weight 200.0 N and apparent weight 15.0 N in water, what is the average density of the fish?

9.7 Fluid Flow; 9.8 Bernoulli's Equation

35. A garden hose of inner diameter 2.0 cm carries water at 2.0 m/s. The nozzle at the end has diameter 0.40 cm. How fast does the water move through the nozzle?

36. If the average volume flow of blood through the aorta is 8.5×10^{-5} m³/s and the cross-sectional area of the aorta is

3.0×10^{-4} m², what is the average speed of blood in the aorta?

37. A nozzle of inner radius 1.00 mm is connected to a hose of inner radius 8.00 mm. The nozzle shoots out water moving at 25.0 m/s. (a) At what speed is the water in the hose moving? (b) What is the volume flow rate? (c) What is the mass flow rate?

38. Water entering a house flows with a speed of 0.20 m/s through a pipe of 2.0 cm inside diameter. What is the speed of the water at a point where the pipe tapers to a diameter of 5.0 mm?

39. Use Bernoulli's equation to estimate the upward force on an airplane's wing if the average flow speed of air is 190 m/s above the wing and 160 m/s below the wing. The density of the air is 1.3 kg/m³ and the area of each wing surface is 28 m².

40. An airplane flies on a level path. There is a pressure difference of 500 Pa between the lower and upper surfaces of the wings. The area of each wing surface is about 100 m². The air moves below the wings at a speed of 80.5 m/s. Estimate (a) the weight of the plane and (b) the air speed above the wings.

◆41. The volume flow rate of the water supplied by a well is 2.0×10^{-4} m³/s. The well is 40.0 m deep. (a) What is the power output of the pump—in other words, at what rate does the well do work on the water? (b) Find the pressure difference the pump must maintain. (c) Can the pump be at the top of the well or must it be at the bottom? Explain.

42. A nozzle is connected to a horizontal hose. The nozzle shoots out water moving at 25 m/s. What is the gauge pressure of the water in the hose? Neglect viscosity and assume that the diameter of the nozzle is much smaller than the inner diameter of the hose.

9.9 Viscosity

43. Using Poiseuille's law [Eq. (9-15)], show that viscosity has SI units of pascal-seconds.

44. A viscous liquid is flowing steadily through a pipe of diameter D. Suppose you replace it by two parallel pipes, each of diameter $D/2$, but the same length as the original pipe. If the pressure difference between the ends of these two pipes is the same as for the original pipe, what is the total rate of flow in the two pipes compared to the original flow rate?

45. A hypodermic syringe is attached to a needle that has an internal radius of 0.300 mm and a length of 3.00 cm. The needle is filled with a solution of viscosity 2.00×10^{-3} Pa·s; it is injected into a vein at a gauge pressure of 16.0 mm Hg. (a) What must the pressure of the fluid in the syringe be in order to inject the solution at a rate of 0.250 mL/s? (b) What force must be applied to the plunger, which has an area of 1.00 cm²?

Problems 46–48. Four identical sections of pipe are connected in various ways to pumps that supply water at the pressures indicated in Fig. 9.36 (in units of 10^5 Pa). The water exits at the right at atmospheric pressure. Assume viscous flow.

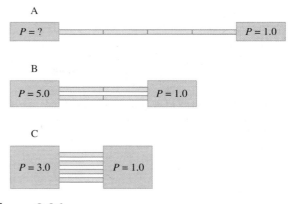

Figure 9.36 Problems 46–48

46. If the *total* volume flow rates in systems A and C are the same and the flow speed in each of the pipes in C is 3.0 m/s, what is the flow speed in system A?

47. If the total volume flow rate in system B is 0.020 m³/s, what is the total volume flow rate in system C?

48. If the total volume flow rates in systems A and B are the same, at what pressure does the pump supply water in system A?

49. (a) What is the pressure difference required to make blood flow through an artery of inner radius 2.0 mm and length 0.20 m at a speed of 6.0 cm/s? (b) What is the pressure difference required to make blood flow at 0.60 mm/s through a capillary of radius 3.0 μm and length 1.0 mm? (c) Compare both answers to your average blood pressure, about 100 torr.

50. (a) Since the flow rate is proportional to the pressure difference, show that Poiseuille's law can be written in the form $\Delta P = IR$, where I is the volume flow rate and R is a constant of proportionality called the fluid flow *resistance*. (Written this way, Poiseuille's law is analogous to *Ohm's law* for electrical current to be studied in Chapter 18: $\Delta V = IR$, where ΔV is the potential drop across a conductor, I is the electrical current flowing through the conductor, and R is the electrical resistance of the conductor.) (b) Find R in terms of the viscosity of the fluid and the length and radius of the pipe.

9.10 Viscous Drag

51. A dinoflagellate takes 5.0 s to travel 1.0 mm. Approximate a dinoflagellate as a sphere of diameter 70.0 μm (ignoring the flagellum). (a) What is the drag force on the dinoflagellate in seawater of viscosity 0.0010 Pa·s? (b) What is the power output of the flagellate?

52. An air bubble of 1.0-mm radius is rising in a container with vegetable oil of specific gravity 0.85 and viscosity 0.12 Pa·s. The container of oil and the air bubble are at 20°C. What is its terminal velocity?

53. Two identical spheres are dropped into two different columns: one column contains a liquid of viscosity 0.5 Pa·s, while the other contains a liquid of the same density but unknown viscosity. The sedimentation velocity in the second tube is 20% higher than the sedimentation velocity in the first tube. What is the viscosity of the second liquid?

54. A sphere of radius 1.0 cm is dropped into a glass cylinder filled with a viscous liquid. The mass of the sphere is 12.0 g

and the density of the liquid is 1200 kg/m³. The sphere reaches a terminal speed of 0.15 m/s. What is the viscosity of the liquid?

◆55. An aluminum sphere (specific gravity = 2.7) falling through water reaches a terminal speed of 5.0 cm/s. What is the terminal speed of an air bubble of the same radius rising through water? Assume viscous drag in both cases and ignore the possibility of changes in size or shape of the air bubble; the temperature is 20°C. [*Hint:* Compare the apparent weights of the two.]

©56. This table gives the terminal speeds of various spheres falling through the same fluid. The spheres all have the same radius.

$m =$	5.0	11.3	20.0	31.3	45.0	80.0	(g)
$v_t =$	1.0	1.5	2.0	2.5	3.0	4.0	(cm/s)

Is the drag force primarily viscous or turbulent? Explain your reasoning.

©57. This table gives the terminal speeds of various spheres falling through the same fluid. The spheres all have the same radius.

$m =$	8	12	16	20	24	28	(g)
$v_t =$	1.0	1.5	2.0	2.5	3.0	3.5	(cm/s)

Is the drag force primarily viscous or turbulent? Explain your reasoning.

58. *What keeps a cloud from falling?* A cumulus (fair-weather) cloud consists of tiny water droplets of average radius 5.0 μm. Find the terminal velocity for these droplets at 20°C, assuming viscous drag. (Besides the viscous drag force, there are also upward air currents called *thermals* that push the droplets upward.)

9.11 Surface Tension

59. An underwater air bubble has an excess inside pressure of 10 Pa. What is the excess pressure inside an air bubble with twice the radius?

60. Assume a water strider has a roughly circular foot of radius 0.02 mm. (a) What is the maximum possible upward force on the foot due to surface tension of the water? (b) What is the maximum mass of this water strider so that it can keep from breaking through the water surface? The strider has six legs.

◆◆61. There is potential energy associated with surface tension much like the elastic potential energy of a stretched spring or a balloon. Suppose we do work on a puddle of liquid, spreading it out through a distance of Δs along a line L perpendicular to the force (see Fig. 9.37). (a) What is the work done on the fluid surface in terms of γ, L, and Δs? (b) The work done is equal to the increase in surface energy of the fluid. Show that the increase in energy is proportional to the increase in area. (c) Show that we can think of γ as the surface energy per unit area. (d) Show that the SI units of surface tension can be expressed either as N/m (force per unit length) or J/m² (energy per unit area).

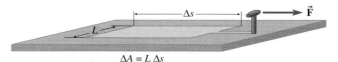

Figure 9.37 Problem 61

◆◆62. A hollow hemispherical object is filled with air (Fig. 9.38a). (a) Show that the magnitude of the force due to fluid pressure on the curved surface of the hemisphere has magnitude $F = \pi r^2 P$, where r is the radius of the hemisphere and P is the pressure of the air. Neglect the weight of the air. [*Hint:* First find the force on the *flat* surface. What is the net force on the hemisphere due to the air?] (b) Consider an underwater air bubble to be divided into two hemispheres along the circumference (Fig. 9.38b). One hemisphere of the water surface exerts a force of magnitude $2\pi r\gamma$ (circumference times force per unit length) on the other due to surface tension. Show that the air pressure inside the bubble must exceed the water pressure outside by $\Delta P = (2\gamma)/r$.

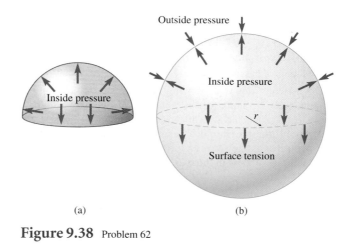

Figure 9.38 Problem 62

COMPREHENSIVE PROBLEMS

63. The deepest place in the ocean is the Mariana Trench in the South Pacific, which is over 11.0 km deep. On January 23, 1960, the research sub *Trieste* went to a depth of 10.915 km, nearly to the bottom of the trench. This still is the deepest dive on record. The density of seawater is 1025 kg/m³. (a) What is the water pressure at that depth? (b) What was the force due to water pressure on a flat section of area 1.0 m² on the top of the sub's hull?

64. The pressure in a water pipe in the basement of an apartment house is 4.10×10^5 Pa, but on the seventh floor it is only 1.85×10^5 Pa. What is the height between the basement and the seventh floor? Assume the water is not flowing; no faucets are opened.

65. The body of a 90.0-kg person contains 0.020 m³ of body fat. If the density of fat is 890 kg/m³, what percentage of the person's body weight is composed of fat?

66. Near sea level, how high a hill must you ascend for the reading of a barometer you are carrying to drop by 1.0 cm Hg? Assume the temperature remains at 20°C as you climb. The reading of a barometer on an average day at sea level is 76.0 cm Hg.

✦ 67. A plastic beach ball has radius 20.0 cm and mass 0.10 kg, not including the air inside. (a) What is the weight of the beach ball including the air inside? Assume the air density is 1.3 kg/m³ both inside and outside. (b) What is the buoyant force on the beach ball in air? The thickness of the plastic is about 2 mm—negligible compared to the radius of the ball. (c) The ball is thrown straight up in the air. At the top of its trajectory, what is its acceleration? [*Hint:* When $v = 0$, there is no drag force.]

68. A stone of weight W has specific gravity 2.50. (a) What is the apparent weight of the stone when under water? (b) What is its apparent weight in oil (specific gravity = 0.90)?

69. A simple hydrometer is an instrument for measuring the specific gravity of a liquid. Markings along a stem are calibrated to indicate the specific gravity for the level at which the hydrometer floats in a liquid (Fig. 9.39). The weighted base ensures that the hydrometer floats vertically. Suppose the hydrometer has a cylindrical stem of cross-sectional area 0.400 cm². The total volume of the bulb and stem is 8.80 cm³ and the mass of the hydrometer is 4.80 g. (a) How far from the top of the cylinder should a mark be placed to indicate a specific gravity of 1.00? (b) When the hydrometer is placed in alcohol, it floats with 7.25 cm of stem above the surface. What is the specific gravity of the alcohol? (c) What is the lowest specific gravity that can be measured with this hydrometer?

✦ 70. Are evenly spaced specific gravity markings on the cylinder of a hydrometer (Fig. 9.39) equal distances apart? In other words, is the depth d to which the cylinder is submerged linearly related to the density ρ of the fluid? To answer this question, assume that the cylinder has radius r and mass m. Find an expression for d in terms of ρ, r, and m and see if d is a linear function of ρ.

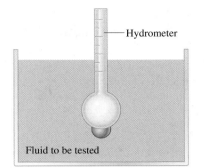

Figure 9.39
Simple hydrometer
(Problems 69 and 70)

71. The average speed of blood in the aorta is 0.3 m/s and the radius of the aorta is 1 cm. There are about 2×10^9 capillaries with an average radius of 6 μm. What is the approximate average speed of the blood flow in the capillaries?

72. If the cardiac output of a small dog is 4.1×10^{-3} m³/s, the diameter of its aorta is 1.0 cm, and the aorta length is 40.0 cm, determine the pressure drop across the aorta of the dog. Assume the viscosity of blood is 4.0×10^{-3} Pa·s.

✦ 73. To measure the airspeed of a plane, a device called a Pitot tube is used. A simplified model of the Pitot tube (Fig. 9.40) is a manometer with one side connected to a tube facing directly into the "wind" (stopping the air that hits it head-on) and the other side connected to a tube so that the "wind" blows across its openings. If the manometer uses mercury and the levels differ by 25 cm, what is the plane's airspeed? The density of air at the plane's altitude is 0.90 kg/m³.

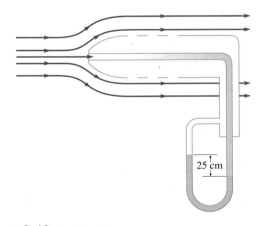

25 cm

Figure 9.40 Problem 73

✦ 74. A house with its own well has a pump in the basement with an output pipe of inner radius 6.3 mm. Assume that the pump can maintain a gauge pressure of 410 kPa in the output pipe. A showerhead on the second floor (6.7 m above the pump's output pipe) has 36 holes, each of radius 0.33 mm. The shower is on "full blast" and no other faucet in the house is open. (a) Ignoring viscosity, with what speed does water leave the showerhead? (b) With what speed does water move through the output pipe of the pump?

75. In an aortic aneurysm, a bulge forms where the walls of the aorta are weakened. If blood flowing through the aorta (radius 1.0 cm) enters an aneurysm with a radius of 3.0 cm, how much on average is the blood pressure higher inside the aneurysm than the pressure in the unenlarged part of the aorta? The average flow rate through the aorta is 120 cm³/s. Assume the blood is nonviscous and the patient is lying down so there is no change in height.

76. The diameter of a certain artery has decreased by 25% due to arteriosclerosis. (a) If the same amount of blood flows through it per unit time as when it was unobstructed, by what percentage has the blood pressure difference between its ends increased? (b) If, instead, the pressure drop across the artery stays the same, by what factor does the blood flow rate through it decrease? (In reality we are likely to see a combination of some pressure increase with some reduction in flow.)

77. Scuba divers are admonished not to rise faster than their air bubbles when rising to the surface. This rule helps them avoid the rapid pressure changes that cause the bends. Air bubbles of 2.0 mm diameter are rising from a scuba diver to the surface of the sea. Assume a water temperature of 20°C. (a) If the viscosity of the water is 1.0×10^{-3} Pa·s, what is the terminal velocity of the bubbles? (b) What is the largest rate of pressure change tolerable for the diver according to this rule?

78. A U-shaped tube is partly filled with water and partly filled with a liquid that does not mix with water (Fig. 9.41). Both sides of the tube are open to the atmosphere. What is the density of the liquid (in g/cm³)?

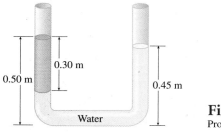

0.30 m
0.50 m
0.45 m
Water

Figure 9.41
Problem 78

79. Atmospheric pressure is equal to the weight of a vertical column of air, extending all the way up through the atmosphere, divided by the cross-sectional area of the column. (a) Explain why that must be true. [*Hint:* Apply Newton's second law to the column of air.] (b) If the air all the way up had a uniform density of 1.29 kg/m³ (the density at sea level at 0°C), how high would the column of air be? (c) In reality, the density of air decreases with increasing altitude. Does that mean that the height found in (b) is a lower limit or an upper limit on the height of the atmosphere?

80. On a nice day when the temperature outside is 20°C, you take the elevator to the top of the Sears Tower in Chicago, which is 440 m tall. (a) How much less is the air pressure at the top than the air pressure at the bottom? Express your answer both in pascals and atm. [*Hint:* The altitude change is small enough to treat the density of air as constant.] (b) How many pascals does the pressure decrease for every meter of altitude? (c) If the pressure gradient—the pressure decrease per meter of altitude—were uniform, at what altitude would the atmospheric pressure reach zero? (d) Atmospheric pressure does *not* decrease with a uniform gradient since the density of air decreases as you go up. Which is true: the pressure reaches zero at a *lower* altitude than your answer to (c), or the pressure is nonzero at that altitude and the atmosphere extends to a higher altitude? Explain.

81. A shallow well usually has the pump at the top of the well. (a) What is the deepest possible well for which a surface pump will work? [*Hint:* A pump maintains a pressure difference, keeping the outflow pressure higher than the intake pressure.] (b) Why is there not the same depth restriction on wells with the pump at the bottom?

82. A bug from South America known as *Rhodnius prolixus* extracts the blood of animals. Suppose *Rhodnius prolixus* extracts 0.30 cm³ of blood in 25 minutes from a human arm through its feeding tube of length 0.20 mm and radius 5.0 μm. What is the absolute pressure at the bug's end of the feeding tube if the absolute pressure at the other end (in the human arm) is 105 kPa? Assume the viscosity of blood is 0.0013 Pa·s. [*Note:* Negative absolute pressures are possible in liquids in very slender tubes.]

ANSWERS TO PRACTICE PROBLEMS

9.1 1.3×10^6 N/m^2 = 1.3 MPa; pressure is a factor of 15 greater than the pressure from the tennis shoe.

9.2 (a) 2.0×10^5 Pa; (b) 5.0 m

9.3 1.6 km

9.4 (a) Yes, $P_2 = P_1$. The column above point 2 is not as tall, but the pressure at the top of that column is *greater than* atmospheric pressure. (b) No, $P = P_{atm} + \rho gd$ gives the pressure at a depth d below a point where the pressure is P_{atm}.

9.5 (a) 32.0 cm; (b) 17.0 cm and 37.0 cm

9.6 S.G. = 11.3; could be lead

9.7 2% and 4%

9.8 (a) The beetle can squeeze the air bubble with its wings, compressing the air to reduce the bubble volume and decreasing the buoyant force. (b) When it is time to rise to the surface, the beetle relaxes the tension on the bubble, allowing it to expand again.

9.9 (a) 0.89 m/s; (b) 1.8 m/s

9.10 250 kPa

9.11 $\sqrt{2gh}$ = 4.0 m/s

9.12 1.4

9.13 2.9 mm/s upward

9.14 480 Pa

Elasticity and Oscillations

In the fairy tale, Jack's beanstalk grew taller and taller until it reached the sky, enabling Jack to climb up to the giant's kingdom. The giant is depicted as a greatly enlarged version of a man with his body having the same relative proportions as a human. If there were giants with bones composed of the same material as human bones, could their proportions be the same as those of humans? Is there a limit to how high a beanstalk might grow and still support its own weight?

Figure 10.1 A tennis ball is flattened by contact with the strings of the tennis racquet.

10.1 ELASTIC DEFORMATIONS OF SOLIDS

If the net force and the net torque on an object are zero, the object is in equilibrium—but that does not mean that the forces and torques have no effect. An object is deformed when contact forces are applied to it (Fig. 10.1). A **deformation** is a change in the size or shape of the object. Many solids are stiff enough that the deformation cannot be seen with the human eye; a microscope or other sensitive device is required to detect the change in size or shape.

When the contact forces are removed, an **elastic** object returns to its original shape and size. Many objects are elastic as long as the deforming forces are not too large. On the other hand, any object may be permanently deformed or even broken if the forces acting are too large. An automobile that collides with a tree at a low speed may not be damaged; but at a higher speed the car suffers a permanent deformation of the body-work and the driver may suffer a broken bone.

10.2 HOOKE'S LAW FOR TENSILE AND COMPRESSIVE FORCES

Suppose we stretch a wire by applying tensile forces of magnitude F to each end. The length of the wire increases from L to $L + \Delta L$. How does the elongation ΔL depend on the original length L? Conceptual Example 10.1 helps answer this question.

Conceptual Example 10.1

Stretching Wires

If a given tensile force stretches a wire an amount ΔL, by how much would the same force stretch a wire twice as long but identical in thickness and composition?

Strategy and Solution Think of the wire of length $2L$ as two wires of length L placed end-to-end (Fig. 10.2). Under the same tension, each of the two imagined wires stretches by an amount ΔL, so the total deformation of the long wire is $2\Delta L$.

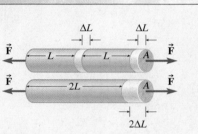

Figure 10.2
Two identical wires are joined end-to-end and stretched by tensile forces. Each wire stretches an amount ΔL. The elongation of a wire is proportional to its original length L.

Practice Problem 10.1   Cutting a spring in half

If a spring (spring constant k) is cut in half, what is the spring constant of each of the two newly formed springs?

When stretched by the same tensile forces, the two wires in Conceptual Example 10.1 get longer by an amount proportional to their original lengths: $\Delta L \propto L$. In other words, the two wires have the same *fractional length change* $\Delta L/L$. The fractional length change is called the **strain**; it is a dimensionless measure of the degree of deformation.

$$\text{strain} = \frac{\Delta L}{L} \tag{10-1}$$

Suppose we had wires of the same composition and length but different thicknesses. It would require larger tensile forces to stretch the thicker wire the same amount as the thinner one; a thick steel cable is harder to stretch than the same length of a thin strand of steel. In Conceptual Question 13, we conclude that the tensile force required is proportional to the cross-sectional area of the wire ($F \propto A$). Thus, the same applied force *per unit area* produces the same deformation on wires of the same length and composition. The force per unit area is called the **stress**:

$$\text{stress} = \frac{F}{A} \tag{10-2}$$

The SI units of stress are the same as those of pressure: N/m^2 or Pa.

Suppose that a solid object of initial length L is subjected to tensile or compressive forces of magnitude F. As a result of the forces, the length of the object is changed by magnitude ΔL. According to Hooke's law, the deformation is proportional to the deforming forces as long as they are not too large:

$$F = k \Delta L \tag{10-3}$$

In Eq. (10-3), k is a constant analogous to the spring constant of a spring. This constant k depends on the length and cross-sectional area of the object. A larger cross-sectional area A makes k larger; a greater length L makes k smaller.

It is more convenient to write Hooke's law in terms of stress (F/A) and strain ($\Delta L/L$):

Hooke's law

stress $\propto$ strain

$$\frac{F}{A} = Y \frac{\Delta L}{L} \tag{10-4}$$

Equation (10-4) still says that the length change (ΔL) is proportional to the magnitude of the deforming forces (F). Stress and strain account for the effects of length and cross-sectional area; the proportionality constant Y depends only on the inherent stiffness of the material from which the object is composed; it is independent of the length and cross-sectional area. Comparing Eqs. (10-3) and (10-4), the "spring constant" k for the object is

$$k = Y \frac{A}{L} \tag{10-5}$$

The constant of proportionality Y in Eqs. (10-4) and (10-5) is called the **elastic modulus** or **Young's modulus**; Y has the same units as those of stress (Pa or N/m^2) since strain is dimensionless. Young's modulus can be thought of as the inherent stiffness of a material; it measures the resistance of the material to elongation or compression. Material that is flexible and stretches easily (for example, rubber) has a *low* Young's modulus. A stiff material (such as steel) has a high Young's modulus; it takes a larger stress to produce the same strain. Table 10.1 gives Young's modulus for a variety of common materials.

Table 10.1

Approximate Values of Young's Modulus for Various Substances

Substance	Young's Modulus (10^9 Pa)	Substance	Young's Modulus (10^9 Pa)
Rubber	0.002–0.008	Wood, along the grain	10–15
Human cartilage	0.024	Brick	14–20
Human vertebra	0.088 (compression); 0.17 (tension)	Concrete	20–30 (compression)
		Marble	50–60
Collagen, in bone	0.6	Aluminum	70
Human tendon	0.6	Cast iron	100–120
Wood, across the grain	1	Copper	120
Nylon	2–6	Wrought iron	190
Spider silk	4	Steel	200
Human femur	9.4 (compression); 16 (tension)	Diamond	1200

Hooke's law holds up to a maximum stress called the *proportional limit*. For many materials, Young's modulus has the same value for tension and compression. Some composite materials, such as bone and concrete, have significantly different Young's moduli for tension and compression. The components of bone include fibers of collagen (a protein found in all connective tissue) that give it strength under tension and hydroxyapatite crystals (composed of calcium and phosphate) that give it strength under compression. The different properties of these two substances lead to different values of Young's modulus for tension and compression.

Example 10.2

Compression of Femur

A man whose weight is 0.80 kN is standing upright (Fig. 10.3). By approximately how much is his femur (thighbone) shortened compared to when he is lying down? Assume that the compressive force on each femur is about half his weight. The average cross-sectional area of the femur is 8.0 cm^2 and the length of the femur when lying down is 43.0 cm.

Strategy A change in length of the femur involves a strain. After finding the stress and looking up the Young's modulus, we can find the strain using Hooke's law. We assume that each femur supports *half* the man's weight.

Solution The strain is proportional to the stress:

$$\frac{F}{A} = Y\frac{\Delta L}{L}$$

Solving this equation for ΔL gives

$$\Delta L = \frac{F/A}{Y}L$$

From Table 10.1, Young's modulus for a femur *in compression* is:

$$Y = 9.4 \times 10^9 \text{ Pa}$$

We need to convert the cross-sectional area to m^2 since 1 Pa = 1 N/m^2:

$$A = 8.0 \text{ cm}^2 \times \left(\frac{1 \text{ m}}{100 \text{ cm}}\right)^2 = 0.00080 \text{ m}^2$$

The force on each leg is 0.40 kN, or 4.0×10^2 N. The length change is then

$$\Delta L = \frac{F/A}{Y}L = \frac{(4.0 \times 10^2 \text{ N})/(0.00080 \text{ m}^2)}{9.4 \times 10^9 \text{ Pa}} \times 43.0 \text{ cm}$$

$$= 5.3 \times 10^{-5} \times 43.0 \text{ cm} = 0.0023 \text{ cm}$$

Discussion The strain—or fractional length change—is 5.3×10^{-5}. Since the strain is much smaller than 1, we are justified in not worrying about whether the length is 43.0 cm with or without the compressive load; we would calculate the same value of ΔL (to two significant figures) either way.

Practice Problem 10.2   Fractional length change of a cable

A steel cable of diameter 3.0 cm supports a load of 2.0 kN. What is the fractional length increase of the cable compared to the length when there is no load if $Y = 2.0 \times 10^{11}$ Pa?

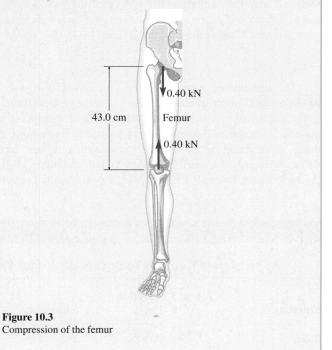

43.0 cm

0.40 kN

Femur

0.40 kN

Figure 10.3
Compression of the femur

10.3   BEYOND HOOKE'S LAW

If the tensile or compressive stress exceeds the proportional limit, the strain is no longer proportional to the stress (Fig. 10.4). The solid still returns to its original length when the stress is removed as long as the stress does not exceed the *elastic limit*. If the stress exceeds the elastic limit, the material is permanently deformed. For still larger stresses, the solid fractures when the stress reaches the *breaking point*. The maximum stress that can be withstood without breaking is called the *ultimate strength*. The ultimate strength

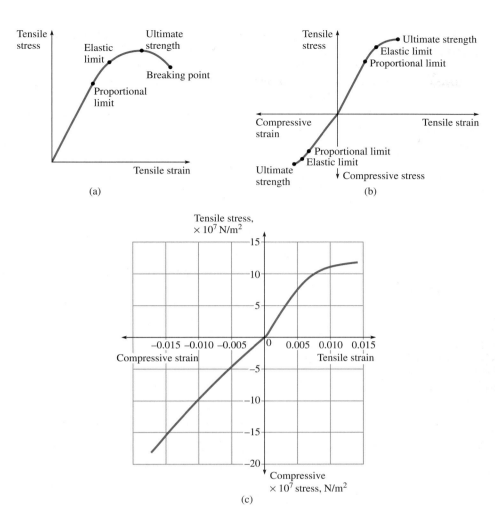

(a)

(b)

(c)

Figure 10.4 A stress-strain curve showing limits for (a) a ductile material, (b) a brittle material, and (c) compact bone. The elastic limit, ultimate strength, and breaking point are well separated for ductile materials, but close together for a brittle material. The breaking point is so close to the ultimate strength for the brittle material that it is not shown as a separate point on the stress-strain curve.

Figure 10.5 This machine, used to make thin wire, stretches metal beyond its elastic limit.

Figure 10.6 Pouring concrete over stretched steel rods

Making The Connection:
strength of
building materials

can be different for compression and tension; then we refer to the compressive strength or the tensile strength of the material.

A *ductile* material continues to stretch beyond its ultimate tensile strength without breaking; the stress then *decreases* from the ultimate strength (Fig. 10.4a). Examples of ductile solids are the relatively soft metals, such as gold, silver, copper, and lead. These metals can be pulled like taffy, becoming thinner and thinner until finally reaching the breaking point (Fig. 10.5).

For a *brittle* substance, the ultimate strength and the breaking point are close together (Fig. 10.4b). Bone is an example of a brittle material; it fractures abruptly if the stress becomes too large. Under either tension or compression, its elastic limit, breaking point, and ultimate strength are approximately the same. For compression, the stress and strain are roughly proportional right up to the breaking point, whereas for tension there is a marked deviation from Hooke's law. Babies have more flexible bones than adults because they have built up less of the calcium compound hydroxyapatite. As people age, their bones become more brittle as the collagen fibers lose flexibility and they also become weaker as calcium gets reabsorbed.

Like bone, concrete has one component for tensile strength and another for compressive strength. Reinforced concrete contains steel rods that provide tensile strength that concrete itself lacks. In prestressed concrete, the steel rods are stretched when the concrete is poured (Fig. 10.6). After the concrete hardens, the frame holding the rods under tension is removed. The rods contract, compressing the concrete. Then, when the prestressed concrete is subject to a tensile force, the compression of the concrete is lessened but not eliminated so that the concrete itself is never subjected to a tensile stress.

Example 10.3

Crane with Steel Cable

A crane is required to lift loads of up to 1.0×10^5 N (11 tons). (a) What is the minimum diameter of the steel cable that must be used? (b) If a cable of twice the minimum diameter is used and it is 8.0 m long when no load is present, how much longer is it when supporting 1.0×10^5 N? (Data for steel: $Y = 2.0 \times 10^{11}$ Pa; proportional limit = 2.0×10^8 Pa; elastic limit = 3.0×10^8 Pa; tensile strength = 5.0×10^8 Pa.)

Strategy   The data given for steel consists of four quantities that all have the same units. It would be easy to mix them up if we didn't understand what each one means. Young's modulus is the proportionality constant between stress and strain. That will be useful in part (b) where we find the elongation of the cable; the elongation is the strain times the original length. However, we should first check that the stress is less than the proportional limit before using Young's modulus to find the strain.

The elastic limit is the maximum stress so that no permanent deformation occurs; the tensile strength is the maximum stress so that the cable does not break. We certainly don't want the cable to break, but it would be prudent to keep the stress under the elastic limit to give the cable a long useful life. Therefore, we choose a minimum diameter in (a) to keep the stress below the elastic limit.

Solution   (a) We choose the minimum diameter to keep the stress less than the elastic limit:

$$\frac{F}{A} < \text{elastic limit} = 3.0 \times 10^8 \text{ Pa}$$

for $F = 1.0 \times 10^5$ N. Then

$$A > \frac{F}{\text{elastic limit}} = \frac{1.0 \times 10^5 \text{ N}}{3.0 \times 10^8 \text{ Pa}} = 3.33 \times 10^{-4} \text{ m}^2$$

The minimum cross-sectional area corresponds to the minimum diameter. The cross-sectional area of the cable is πr^2 or $\pi d^2/4$, so

$$d = \sqrt{\frac{4A}{\pi}} = \sqrt{\frac{4 \times 3.33 \times 10^{-4} \text{ m}^2}{\pi}} = 2.1 \text{ cm}$$

The minimum *diameter* is therefore 2.1 cm.

(b) If we double the diameter and keep the same load, the stress is reduced by a factor of four since the cross-sectional area is proportional to the square of the diameter. Therefore, the stress is

$$\frac{F}{A} = \frac{3.0 \times 10^8 \text{ Pa}}{4} = 7.5 \times 10^7 \text{ Pa}$$

The strain is then

$$\frac{\Delta L}{L} = \frac{F/A}{Y} = \frac{7.5 \times 10^7 \text{ Pa}}{2.0 \times 10^{11} \text{ Pa}} = 0.000375$$

The strain is the fractional length change. Then the length change is

$$\Delta L = 0.000375L = 0.000375 \times 8.0 \text{ m} = 0.0030 \text{ m} = 3.0 \text{ mm}$$

Discussion   By using a cable twice as thick as the minimum, we build in a safety factor. We don't want to be right at the edge of disaster! Since doubling the diameter of the cable increases the cross-sectional area of the cable by a factor of four, the maximum stress on the cable is $\frac{1}{4}$ of the elastic limit.

Practice Problem 10.3   Tuning a harpsichord string

A harpsichord string is made of yellow brass (Young's modulus 9.0×10^{10} Pa, tensile strength 6.3×10^8 Pa). When tuned correctly, the tension in the string is 59.4 N, which is 93% of the maximum tension that the string can endure without breaking. What is the radius of the string?

Human anatomy has special features for adapting to the compressive stress associated with standing upright. For example, the vertebrae in the spinal column gradually increase in size from the neck to the tailbone. Such an arrangement places the stronger vertebrae in the lower positions, where they must support more weight. The vertebrae are separated by fluid-filled disks, which have a cushioning effect by spreading out the compressive forces.

Height Limits

What about the question of the height of Jack's beanstalk and the proportions of the giant? For simplicity, imagine the beanstalk to be a single cylindrical column. If the beanstalk grows too tall, it could be crushed under its own weight. The maximum height of a cylindrical column is limited since the compressive stress at the bottom cannot exceed the compressive strength of the material (see Problem 78). However, the beanstalk would *buckle* long before being crushed; the maximum height at which a vertical column buckles is generally less than the height at which it would be crushed.

The bones of our limbs are hollow; the inside of the structural material is filled with marrow, which is structurally weak. A hollow bone is better able to resist fracture from bending and twisting forces than a solid bone with the same amount of structural material, although the hollow bone would buckle more easily under a compressive force along the central axis.

Physics at Home

Challenge a friend to use a single sheet of 8.5" × 11" paper and two paper clips (or tape) to support a book at least 8 in. above a table. If your friend has no idea what to do, roll the sheet of paper into a narrow cylinder about 2.5 cm (an inch) in diameter; then fasten the cylinder at the top and bottom with paper clips (or with tape). Carefully place the book so that it is balanced on top of the cylinder. If you have difficulty, try using thicker paper or a lighter book.

Use the same "apparatus" to get some insight into the buckling of columns. Try making the diameter of the paper cylinder twice as large. The walls of this column are thinner because there are fewer layers of the paper in the cylinder wall, although the same cross-sectional area *of paper* supports the book. If nothing happens, try again with a heavier book. You will likely see the walls crumple in on themselves as the cylinder buckles and the book falls to the table.

Why would the design of a giant's bones have to be different from a human's? If the giant's average density is the same as a human's, then his weight is larger by the same factor that his *volume* is larger. If the giant is five times as tall as a human, for instance, and has the same relative proportions, then his volume is $5^3 = 125$ times as large, since each of the three dimensions of any body part has increased by a factor of five. On the other hand, the cross-sectional area of a bone is proportional to the *square* of its radius. So while the leg bones must support 125 times as much weight, the maximum compressive force they can withstand has only increased by a factor of 25. The giant would need much thicker legs (in relation to their length) to support his increased weight. Similar analysis can be applied to the twisting and bending forces that are more likely to break bones than are compressive forces. The result is the same: the bones of a giant could not have human proportions.

Some science fiction or horror movies portray giant insects as greatly magnified versions of a normal insect. Such a giant insect's legs would collapse under the weight of the insect.

Making The Connection:
size limitations
on organisms

10.4 SHEAR AND VOLUME DEFORMATIONS

In this section we consider two other kinds of deformation. In each case we define a stress (force per unit area), a strain (dimensionless), and a modulus (the constant of proportionality between stress and strain).

Shear Deformation

Unlike tensile and compressive forces, which act perpendicular to two opposite surfaces of an object, a shear deformation is the result of a pair of equal and opposite forces that act *parallel* to two opposite surfaces. Consider a book placed on a desk. If we push horizontally on the top cover of the book while pushing in the opposite direction on the bottom cover to hold it in place, the book is deformed as shown in Fig. 10.7. Such a deformation is called a **shear deformation**.

Shear forces produce the same kind of deformation in a solid block; the amount of the deformation is just smaller. The **shear stress** is the magnitude of the shear force divided by the area of the surface on which the force acts:

$$\text{shear stress} = \frac{\text{shear force}}{\text{area of surface}} = \frac{F}{A} \qquad (10\text{-}6)$$

Shear strain is the ratio of the relative displacement Δx to the separation L of the two surfaces:

$$\text{shear strain} = \frac{\text{displacement of surfaces}}{\text{separation of surfaces}} = \frac{\Delta x}{L} \qquad (10\text{-}7)$$

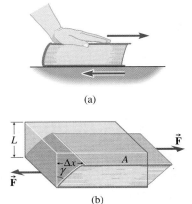

Figure 10.7 A book under shear stress

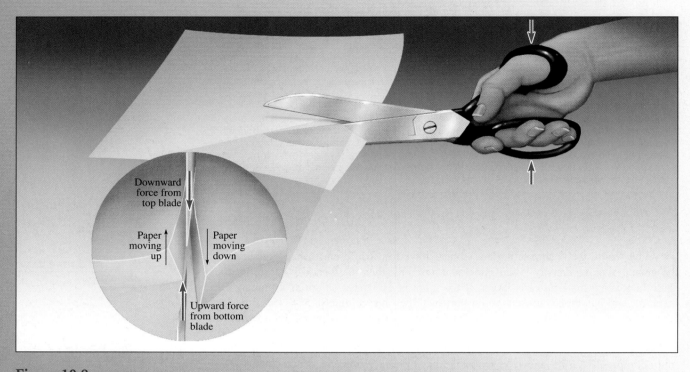

Figure 10.8 Scissors apply shear stress to a sheet of paper. The shear stress is the force exerted by a blade divided by the cross-sectional area of the paper—the thickness of the paper times the length of blade that is in contact with the paper.

The shear strain is proportional to the shear stress as long as the stress is not too large. The constant of proportionality is the **shear modulus** S.

Hooke's law for shear deformations:

shear stress $\propto$ shear strain

$$\frac{F}{A} = S\frac{\Delta x}{L}$$ (10-8)

The units of shear stress and the shear modulus are the same as for tensile or compressive stress and Young's modulus: Pa or N/m^2. The strain is once again dimensionless. Table 10.2 lists shear moduli for various substances.

An example of shear stress is the cutting action of a pair of scissors (or "shears") on a piece of paper. The forces acting on the paper from above and below are offset from each other and act parallel to the cross-sectional surfaces of the paper (see Fig. 10.8).

Example 10.4

Cutting Paper

A sheet of paper of thickness 0.20 mm is cut with scissors that have blades of length 10.0 cm and width 0.20 cm. While cutting, the scissors blades each exert a force of 3.0 N on the paper; the length of each blade

that makes contact with the paper is approximately 0.5 mm. What is the shear stress on the paper?

Strategy   Shear stress is a force divided by an area. In this problem, identifying the correct area is tricky. The blades

continued on next page

Example 10.4 *continued*

push two *cross-sectional* paper surfaces in opposite directions so they are displaced with respect to each other. The shear stress is the force exerted by each blade divided by this cross-sectional area—the thickness of the paper times the length of blade *in contact with the paper*. The total length and the width of the blades are irrelevant.

Solution The cross-sectional area is

$$A = \text{thickness} \times \text{contact length}$$
$$= 2.0 \times 10^{-4}\,\text{m} \times 5 \times 10^{-4}\,\text{m} = 1 \times 10^{-7}\,\text{m}^2$$

The shear stress is

$$\frac{F}{A} = \frac{3.0\,\text{N}}{1 \times 10^{-7}\,\text{m}^2} = 3 \times 10^7\,\text{N/m}^2$$

Discussion To identify the correct area, remember that shear forces act *in the plane of* the surfaces that are displaced with respect to each other. By contrast, tensile and compressive forces act perpendicular to the area used to find tensile and compressive stresses.

Practice Problem 10.4   Shear stress due to hole punch

A hole punch has a diameter of 8.0 mm and presses onto 10 sheets of paper with a force of 6.7 kN. If each sheet of paper is of thickness 0.20 mm, find the shear stress. [*Hint:* Be careful in deciding what area to use. Remember that a shear force acts *parallel* to the surface whose area is relevant.]

Table 10.2

Shear and Bulk Moduli for Various Materials

Material	Shear Modulus S (10^9 Pa)	Bulk Modulus B (10^9 Pa)
Gases		
Air (1)		0.00010
Air (2)		0.00014
Liquids		
Ethanol		0.9
Water		2.2
Mercury		25
Solids		
Cast iron	40–50	60–90
Marble		70
Aluminum	25–30	70
Copper	40–50	120–140
Steel	80–90	140–160
Diamond		620

(1) At 0°C and 1 atm; constant temperature expansion or compression

(2) At 0°C and 1 atm; no heat flow during expansion or compression

When a bone is twisted, it is subjected to a shear stress. Shear stress is a more common cause of fracture than a compressive or tensile stress along the length of the bone. The twisting of a bone can result in a spiral fracture (Fig. 10.9).

Volume Deformation

As discussed in Chapter 9, a fluid exerts inward forces on an immersed solid object. These forces are perpendicular to the surfaces of the object. Since the fluid presses inward on all sides of the object (Fig. 10.10), the solid is compressed—its volume is reduced. The fluid pressure P is the force per unit surface area; it can be thought of as the **volume stress** on the solid object. Pressure has the same units as the other kinds of stress: N/m^2 or Pa.

$$\text{volume stress} = \text{pressure} = \frac{F}{A} = P$$

(a)

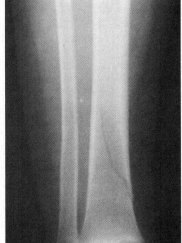

(b)

Figure 10.9 (a) An Olympic skier falls and his leg is subjected to a shear stress. (b) X-ray of a spiral fracture of the tibia

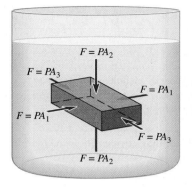

Figure 10.10 Forces on an object when submerged in a fluid

The resulting deformation of the object is characterized by the **volume strain**, which is the fractional change in volume:

$$\text{volume strain} = \frac{\text{change in volume}}{\text{original volume}} = \frac{\Delta V}{V} \qquad (10\text{-}9)$$

Unless the stress is too large, the stress and strain are proportional within a constant of proportionality called the **bulk modulus** B. A substance with a large bulk modulus is more difficult to compress than a substance with a small bulk modulus.

An object at atmospheric pressure is already under volume stress: the air pressure already compresses the object slightly compared to what its volume would be in vacuum. For solids and liquids the volume strain due to atmospheric pressure is, for most purposes, negligibly small (5×10^{-5} for water). Since we are usually concerned with the deformation due to a *change* in pressure from atmospheric pressure, we can write Hooke's law as:

Hooke's law for volume deformations

$$\Delta P = -B\frac{\Delta V}{V} \qquad (10\text{-}10)$$

---

where V is the volume at atmospheric pressure. The negative sign in Eq. (10-10) allows the bulk modulus to be positive—an increase in the volume stress causes a *decrease* in volume, so ΔV is negative. Table 10.2 lists bulk moduli for various substances.

Unlike the stresses and strains discussed previously, volume stress may be applied to fluids (liquids and gases) as well as solids. The bulk moduli of liquids are generally not much less than those of solids, since the atoms in liquids are nearly as close together as those in solids. In Chapter 9 we assume that liquids are incompressible, which is often a good approximation since the bulk moduli of liquids are generally large. In gases, the atoms are much farther apart on average than in solids or liquids. Gases are much easier to compress than solids or liquids, so their bulk moduli are much smaller.

Example 10.5

Marble Statue Under Water

A marble statue of volume 1.5 m³ is being transported by ship from Athens to Cyprus. The statue topples into the ocean when an earthquake-caused tidal wave sinks the ship; the statue ends up on the ocean floor, 1.0 km below the surface. Find the change in volume of the statue in cm³ due to the pressure of the water. The density of seawater is 1025 kg/m³.

Strategy The water pressure is the volume stress; it is the force per unit area pressing inward and perpendicular to all the surfaces of the statue. The water pressure at a depth d is greater than the pressure at the water surface; we can find the pressure using the given density of seawater. Then, using the bulk modulus of marble given in Table 10.2, we find the change in volume from Hooke's law.

Solution The pressure at a depth d = 1.0 km is larger than atmospheric pressure by

$$\Delta P = \rho g d$$
$$= 1025 \text{ kg/m}^3 \times 9.8 \text{ N/kg} \times 1000 \text{ m}$$
$$= 1.005 \times 10^7 \text{ Pa}$$

According to Table 10.2, the bulk modulus for marble is 70×10^9 Pa. This is the constant of proportionality between the volume stress (pressure increase) and the strain (fractional change in volume).

$$\Delta P = -B\frac{\Delta V}{V}$$

Solving for ΔV,

$$\Delta V = -\frac{\Delta P}{B}V = -\frac{1.005 \times 10^7 \text{ Pa}}{70 \times 10^9 \text{ Pa}} \times 1.5 \text{ m}^3$$
$$= -2.15 \times 10^{-4} \text{ m}^3 \times \left(\frac{100 \text{ cm}}{1 \text{ m}}\right)^3 = -200 \text{ cm}^3$$

The statue's volume decreases approximately 200 cm³.

Discussion The decrease in volume is 1.005×10^7 Pa/ 70×10^9 Pa $\approx 1/7000$ of the initial volume; a reduction of 0.014%.

In calculating the pressure increase, we assumed that the density of seawater is constant—the equation $\Delta P = \rho g d$ is derived for a constant fluid density ρ. Should we worry that

continued on next page

Example 10.5 *continued*

our calculation of ΔP is wrong? The result of Practice Problem 10.5 shows that density of seawater at a depth of 1.0 km is only about 0.43% greater than its density at the surface. The calculation of ΔP is inaccurate by less than 0.5%—negligible here since we only know the depth to two significant figures.

Practice Problem 10.5   Compression of water

Show that a pressure increase of 1.0×10^7 Pa (100 atm) on one cubic meter of seawater causes a 0.43% decrease in volume. The bulk modulus of seawater is 2.3×10^9 Pa.

10.5   SIMPLE HARMONIC MOTION

Vibration, one of the most common kinds of motion, is repeated motion back and forth along the same path. A common and important example of vibration is sound. At a rock concert, vibrations of the singer's vocal cords cause the air to vibrate. Vibrations in the air make the diaphragm of a microphone vibrate; the electrical signal generated by the microphone and amplifier makes the speaker cones vibrate. The speakers produce more vibrations in the air; we hear the music when the air makes our eardrums vibrate.

Vibrations occur in the vicinity of a point of **stable equilibrium**. An equilibrium point is *stable* if the net force on an object when it is displaced from equilibrium points back toward the equilibrium point (Fig. 10.11). Such a force is called a **restoring force** since it tends to restore equilibrium.

A special kind of vibratory motion—called **simple harmonic motion** (or **SHM**)—occurs whenever the restoring force is proportional to the displacement from equilibrium. The ideal spring is a favorite model of physicists because the restoring force it provides is proportional to the displacement from equilibrium ($F = kx$). Hooke's law applies to small deformations of many kinds of objects, not just springs. Thus, simple harmonic motion occurs in many situations as long as the vibrations are not too large.

Figure 10.12a shows an $F(x)$ curve for some restoring force. Since the curve is not linear, the resulting oscillations are not SHM—unless the amplitude is small. For small amplitudes, we can approximate the $F(x)$ curve near equilibrium by a straight line tangent to the curve at the equilibrium point (Fig. 10.12b). For small amplitude oscillations, the restoring force is approximately linear, so the resulting oscillations are (approximately) SHM.

Consider a relaxed ideal spring with spring constant k and zero mass. The spring is fixed at one end and attached at the other to an object of mass m (Fig. 10.13) that slides without friction. Since the normal force is equal and opposite to the weight of the object, the net force on the object is that due to the spring. When the spring is relaxed, the net force is zero; the object is in equilibrium.

If the object is now pulled to the right to the position $x = A$ and then released, the net force on the object is

$$F_x = -kx \qquad (10\text{-}11)$$

where the negative sign tells us that the spring force is opposite in direction to the displacement from equilibrium. At first the object is to the right of the equilibrium position and the spring pulls to the left. Notice that the force exerted by the spring is in the correct direction to restore the object to the equilibrium position; it always pushes or pulls back toward the equilibrium point.

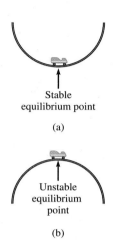

Stable
equilibrium point

(a)

Unstable
equilibrium
point

(b)

Figure 10.11   (a) A point of *stable* equilibrium for a roller-coaster car. If the car is displaced slightly from its position at the bottom of the track, gravity pulls the car back toward the equilibrium point. (b) A point of *unstable* equilibrium for a roller-coaster car. If the car is displaced slightly from the very top of the track, gravity pulls the car *away from* the equilibrium point.

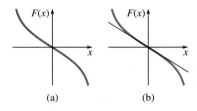

(a)   (b)

Figure 10.12   (a) A nonlinear restoring force. (b) For small vibrations we can approximate the restoring force as linear.

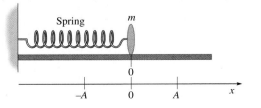

Figure 10.13   Spring in relaxed position

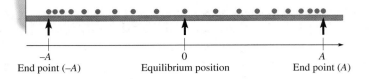

Figure 10.14 Positions of an oscillating body at equal time intervals over half a period. The spring is omitted for clarity.

Imagine taking a series of photos at equal time intervals as the object oscillates back and forth. In Fig. 10.14 the blue dots are the positions of the object at equal time intervals over one-half of a full cycle, from one endpoint to the other. (A full cycle would include the return trip.)

Figure 10.14 suggests that the speed is greatest as the object passes through the equilibrium position. The object slows as it approaches the endpoints and gains speed as it approaches the equilibrium point. At the endpoints ($x = \pm A$), the body is instantaneously at rest before heading back in the other direction. Conservation of energy supports these observations. The total mechanical energy of the mass and spring is constant.

$$E = K + U = \text{constant}$$

where K is the kinetic energy and U is the elastic potential energy stored in the spring. As the object oscillates back and forth, energy is converted from potential to kinetic and back to potential in the half-cycle shown in Fig. 10.14. From Chapter 6, the elastic potential energy of the spring is

$$U = \tfrac{1}{2}kx^2 \tag{6-11}$$

The speed at any point x can be found from the energy equation

$$E = \tfrac{1}{2}kx^2 + \tfrac{1}{2}mv_x^2 \tag{10-12}$$

The maximum displacement of the body is the **amplitude** A. At the maximum displacement, where the motion changes direction, the velocity is zero. Since the kinetic energy is zero at $x = \pm A$, all the energy is elastic potential energy at the endpoints. Therefore, the total energy E at the endpoints is

$$E_{\text{total}} = \tfrac{1}{2}kA^2 \tag{10-13}$$

and, since energy is conserved, this must be the total energy at any point in the object's motion. The maximum speed v_{m} occurs at $x = 0$ where all the energy is kinetic. Thus, at $x = 0$, the total energy equals the kinetic energy

$$E_{\text{total}} = \tfrac{1}{2}mv_{\text{m}}^2$$

and, from Eq. (10-13),

$$\tfrac{1}{2}mv_{\text{m}}^2 = \tfrac{1}{2}kA^2$$

Solving for v_{m} yields

$$v_{\text{m}} = \sqrt{\frac{k}{m}}\,A \tag{10-14}$$

The maximum speed is proportional to the amplitude; larger-amplitude oscillations have a larger maximum speed.

The force on the object at any point x is given by Hooke's law; Newton's second law then gives the acceleration:

$$F_x = -kx = ma_x$$

Solving for the acceleration,

$$a_x(t) = -\frac{k}{m}x(t) \tag{10-15}$$

Thus, the acceleration is a negative constant ($-k/m$) times the displacement; the acceleration and displacement are always in opposite directions. Whenever the acceleration is a negative constant times the displacement, the motion is SHM.

The acceleration has its maximum magnitude a_m, where the force is largest, which is at the maximum displacement $x = \pm A$:

$$a_m = \frac{k}{m} A \qquad (10\text{-}16)$$

Example 10.6

Oscillating Model Rocket

A model rocket of 1.0-kg mass is attached to a horizontal spring with a spring constant of 6.0 N/cm. The spring is compressed by 18.0 cm and then released. The intent is to shoot the rocket horizontally, but the release mechanism fails to disengage, so the rocket starts to oscillate horizontally. Ignore friction and assume the spring to be ideal. (a) What is the amplitude of the oscillation? (b) What is the maximum speed? (c) What are the rocket's speed and acceleration when it is 12.0 cm from the equilibrium point?

Strategy   Initially all of the energy is elastic potential energy and the kinetic energy is zero. The initial displacement must be the maximum displacement—or amplitude—of the oscillations since to get farther from equilibrium would require more elastic energy than the total energy available. The speed at any position can be found using energy conservation ($\frac{1}{2}kx^2 + \frac{1}{2}mv_x^2 = \frac{1}{2}kA^2$). The maximum speed occurs when all of the energy is kinetic. The acceleration can be found from Newton's second law.

Solution   (a) The amplitude of the oscillation is the maximum displacement, so $A = 18.0$ cm.

(b) From energy conservation, the maximum kinetic energy is equal to the maximum elastic potential energy:

$$K_m = \tfrac{1}{2}mv_m^2 = E = \tfrac{1}{2}kA^2$$

Solving for v_m,

$$v_m = \sqrt{\frac{k}{m}}\,A = \sqrt{\frac{6.0 \times 10^2\ \text{N/m}}{1.0\ \text{kg}}} \times 0.180\ \text{m} = 4.4\ \text{m/s}$$

(c) For the speed at a displacement of 0.120 m, we again use energy conservation.

$$\tfrac{1}{2}kx^2 + \tfrac{1}{2}mv^2 = \tfrac{1}{2}kA^2$$

Solving for v,

$$v = \sqrt{\frac{kA^2 - kx^2}{m}} = \sqrt{\frac{k}{m}(A^2 - x^2)}$$

$$= \sqrt{\frac{6.0 \times 10^2\ \text{N/m}}{1.0\ \text{kg}\,[(0.180\ \text{m})^2 - (0.120\ \text{m})^2]}} = 3.3\ \text{m/s}$$

From Newton's second law,

$$F_x = -kx = ma_x$$

At $x = \pm 0.120$ m,

$$a_x = -\frac{k}{m}x = \frac{6.0 \times 10^2\ \text{N/m}}{1.0\ \text{kg}} \times (\pm 0.120\ \text{m}) = \pm 72\ \text{m/s}^2$$

The magnitude of the acceleration is 72 m/s^2; the direction is toward the equilibrium point.

Discussion   Note that at a given position (say $x = +0.120$ m), we can find the *speed* of the rocket, but the direction of the velocity can be either left or right; the rocket passes through each point (other than the endpoints) both on its way to the left and on its way to the right. By contrast, the *acceleration* at $x = +0.120$ m is always in the $-x$-direction, regardless of whether the rocket is moving to the left or to the right. If the rocket is moving to the right, then it is slowing down as it approaches $x = +A$; if it is moving to the left, then it is speeding up as it approaches $x = 0$.

Practice Problem 10.6   Maximum acceleration of rocket

What is the maximum acceleration of the rocket and at what position(s) does it occur?

10.6   THE PERIOD AND FREQUENCY FOR SHM

SHM is *periodic* motion because the same motion repeats over and over—a particle goes back and forth over the same path in exactly the same way. Each time the particle repeats its original motion, we say that it has completed another cycle. To complete one cycle of motion, the particle must be at the same point *and heading in the same direction* as it was at the start of the cycle. The period and frequency are defined exactly as for uniform circular motion, which is another kind of periodic motion. The **period** T is the time taken by one complete cycle. The **frequency** f is the number of cycles per unit time:

$$f = \frac{1}{T} \qquad \text{(SI unit: Hz = cycles per second)} \qquad (5\text{-}7)$$

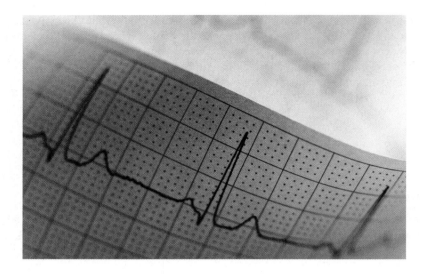

Figure 10.15 An electrocardiogram

SHM is a special kind of periodic motion in which the restoring force is proportional to the displacement from equilibrium. Not all periodic vibrations are examples of simple harmonic motion since not all restoring forces are proportional to the displacement. Any restoring force can cause oscillatory motion. An electrocardiogram (Fig. 10.15) traces the periodic pattern of a beating heart, but the motion of the recorder needle is not simple harmonic motion. As we are about to show, in SHM the position is a sinusoidal function of time.

To learn more about SHM, imagine setting up an experiment (Fig. 10.16). We attach an object to an ideal spring, move the object away from the equilibrium position, and then release it. The object vibrates back and forth in simple harmonic motion with amplitude A. At the same time a horizontal circular disk, of radius $r = A$ and with a pin projecting vertically up from its outer edge, is set into rotation with uniform circular motion. Both the pin and the object attached to the spring are illuminated so that shadows of the vibrating object and of the pin on the rotating disk are seen on a screen. The speed of the disk is adjusted until the shadows oscillate with the same period. We will show that the motion of the two shadows is identical, so the mathematical description of one can be used for the other.

To find the mathematical description of SHM, we analyze the uniform circular motion of the pin as done in Chapter 5. Figure 10.16b shows the pin P moving counterclockwise around a circle of radius A at a constant angular velocity ω. For simplicity, let the pin start at $\theta = 0$ at time $t = 0$. The location of the pin at any time is then given by the angle θ:

$$\theta(t) = \omega t$$

The motion of the pin's shadow has the same x-component as the pin itself. Using a right triangle (Fig. 10.16c), we find that

$$x(t) = A \cos \theta = A \cos \omega t \qquad (10\text{-}17)$$

Since the pin moves in uniform circular motion, its acceleration is constant in *magnitude* but not in direction; the acceleration is toward the center of the circle. In Chapter 5, the magnitude of the centripetal acceleration is shown to be

$$a = \omega^2 r = \omega^2 A \qquad (5\text{-}11)$$

At any instant the direction of the acceleration vector is opposite to the direction of the displacement vector in Fig. 10.16b—that is, toward the center of the circle. Therefore,

$$a_x = -a \cos \theta = -\omega^2 A \cos \omega t \qquad (10\text{-}18)$$

Comparing Eqs. (10-17) and (10-18), we see that, at any time t,

$$a_x(t) = -\omega^2 x(t) \qquad (10\text{-}19)$$

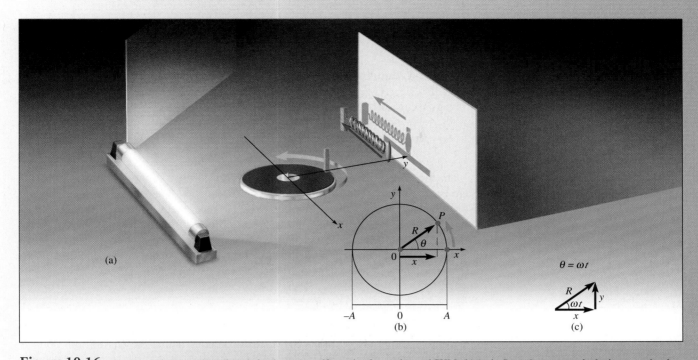

Figure 10.16 (a) An experiment to show the relation between uniform circular motion and SHM. (b) A pin P moving counterclockwise around a circle as a disk rotates with constant angular velocity ω. (c) Finding the x-component of the displacement.

In Eq. (10-15) we showed that in SHM the acceleration is proportional to the displacement:

$$a_x = -\frac{k}{m}x \qquad (10\text{-}15)$$

The motions of the two shadows are identical as long as

$$\omega = \sqrt{\frac{k}{m}} \qquad (10\text{-}20a)$$

The position and acceleration of an object in SHM are sinusoidal functions of time [Eqs. (10-17) and (10-18)]. The sinusoidal functions are sine and cosine. In Problem 46, you can show that v_x is also a sinusoidal function of time.

Since the object in SHM and the pin in circular motion have the same frequency and period, the relationships between ω, f, and T still apply. Therefore, the frequency and period of a mass-spring system are

$$f = \frac{\omega}{2\pi} = \frac{1}{2\pi}\sqrt{\frac{k}{m}} \qquad (10\text{-}20b)$$

and

$$T = \frac{1}{f} = 2\pi\sqrt{\frac{m}{k}} \qquad (10\text{-}20c)$$

In the context of SHM the quantity ω is called the **angular frequency**. Note that the angular frequency is determined by the mass and the spring constant but is independent of the amplitude.

With the identification of ω for a mass-spring system, we can write the maximum speed and acceleration from Eqs. (10-14) and (10-16):

$$v_{\mathrm{m}} = \omega A \qquad (10\text{-}21)$$

$$a_{\mathrm{m}} = \omega^2 A \qquad (10\text{-}22)$$

These expressions are more general than Eqs. (10-14) and (10-16)—they apply to any system in SHM, not just a mass-spring system.

To Find the Angular Frequency for Any Object in SHM

- Write down the restoring force as a function of the displacement from equilibrium. Since the restoring force is linear, it always takes the form $F = -kx$, where k is a constant.
- Use Newton's second law to relate the restoring force to the acceleration.
- Solve for ω using $a_x = -\omega^2 x$ [Eq. (10-19)].

The term *harmonic* in *simple harmonic motion* refers to a sinusoidal vibration; this usage is related to similar usage in music and acoustics. The sinusoidal functions are also called harmonic functions. In Chapter 12 we show that a complex vibration can be formed by combining harmonic vibrations at different frequencies, which is why the study of SHM is the basis for understanding more complex vibrations. The term *simple* in SHM means that the amplitude of the vibration is constant; we assume there is no energy dissipation to cause the vibration to die out.

A Vertical Mass and Spring

The mass and spring systems discussed so far oscillate horizontally. An oscillating mass on a vertical spring also exhibits SHM; the difference is that the equilibrium point is shifted downward by gravity. In our discussions, we assume ideal springs that obey Hooke's law and have a negligibly small mass of their own.

Suppose that an object of weight mg is hung from an ideal spring of spring constant k (Fig. 10.17). In equilibrium, the spring is stretched downward a distance d from its relaxed length so that the spring pulls up with a force equal to mg. Taking the $+y$-axis in the upward direction, the condition for equilibrium is

$$\Sigma F_y = +kd - mg = 0 \qquad (10\text{-}23)$$

Let us take the origin ($y = 0$) at the equilibrium point. If the object is displaced vertically from the equilibrium point to a position y, the spring force becomes

$$F = k(d - y)$$

If y is positive, the object is displaced upward and the spring force is less than kd. The y-component of the net force is then

$$\Sigma F_y = k(d - y) - mg = kd - ky - mg$$

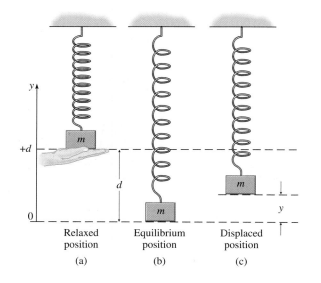

Figure 10.17 (a) A relaxed spring, of spring constant k, with mass m attached. (b) The same spring is extended to its equilibrium position, a distance d below the relaxed position, after mass m is allowed to hang freely. (c) The spring is displaced a distance y from the equilibrium position.

From Eq. (10-23), we know that $kd = mg$; therefore,

$$\Sigma F_y = -ky$$

The restoring force provided by the spring and gravity together is $-k$ times the displacement from equilibrium. Therefore, the vertical mass-spring exhibits SHM with the same period and frequency as if it were horizontal.

Example 10.7

A Vertical Spring

A spring with spring constant k is suspended vertically. A model goose of mass m is attached to the unstretched spring and then released so that the bird oscillates up and down. (Ignore friction and air resistance; assume an ideal massless spring.) Calculate the kinetic energy, the elastic potential energy, the gravitational potential energy, and the total mechanical energy at (a) the point of release and (b) the equilibrium point. Take the gravitational potential energy to be zero at the equilibrium point.

Strategy   The bird oscillates in SHM about its equilibrium point $y = 0$ between two extreme positions $y = +A$ and $y = -A$ (see Fig. 10.18). The amplitude A is equal to the distance the spring is stretched at the equilibrium point; it can be found by setting the net force on the bird equal to zero. The total mechanical energy is the sum of the kinetic energy, the elastic potential energy, and the gravitational potential energy. We expect the total energy to be the same at the two points; since no dissipative forces act, mechanical energy is conserved.

Solution   The equilibrium point is where the net force on the bird is zero:

$$\Sigma F_y = +kd - mg = 0 \qquad (10\text{-}23)$$

In this equation, d is the extension of the spring at equilibrium. Since the bird is released where the spring is relaxed, d is also the amplitude of the oscillations:

$$A = d = \frac{mg}{k}$$

(a) At the point of release, $v = 0$ and the kinetic energy is zero. The elastic energy is also zero—the spring is unstretched. The gravitational potential energy is

$$U_g = mgy = mgA = \frac{(mg)^2}{k}$$

The total mechanical energy is the sum of the kinetic and potential (elastic + gravitational) energies,

$$E = K + U_e + U_g = \frac{(mg)^2}{k}$$

(b) At the equilibrium point, the bird moves with its maximum speed $v_m = \omega A$. The angular frequency is the same as for a horizontal spring: $\omega = \sqrt{k/m}$. Then the kinetic energy is

$$K = \tfrac{1}{2}mv_m^2 = \tfrac{1}{2}m\omega^2 A^2$$

Substituting $A = mg/k$ and $\omega^2 = k/m$,

$$K = \frac{1}{2}m\frac{k}{m}\frac{(mg)^2}{k^2} = \frac{1}{2}\frac{(mg)^2}{k}$$

The spring is stretched a distance A, so the elastic energy is

$$U_e = \frac{1}{2}kA^2 = \frac{1}{2}k\frac{(mg)^2}{k^2} = \frac{1}{2}\frac{(mg)^2}{k}$$

The gravitational potential energy is zero at $y = 0$. Therefore, the total mechanical energy is

$$E = K + U_e + U_g = \frac{1}{2}\frac{(mg)^2}{k} + \frac{1}{2}\frac{(mg)^2}{k} + 0 = \frac{(mg)^2}{k}$$

which is the same as at $y = +A$.

continued on next page

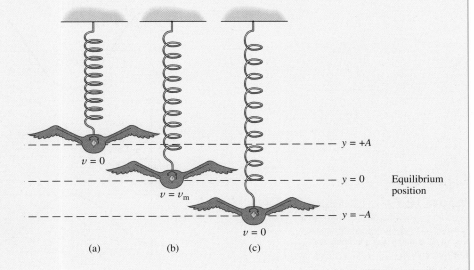

Figure 10.18
(a) The spring is unstretched before the model bird is released at position $y = +A$;
(b) the model bird passes through the equilibrium position $y = 0$ with maximum speed;
(c) the spring's maximum extension occurs when the bird is at $y = -A$.

Example 10.7 *continued*

Discussion   As the bird moves down from the release point toward the equilibrium point, gravitational potential energy is converted into elastic energy and kinetic energy. After the bird passes the equilibrium point, both kinetic and gravitational energy are converted into elastic energy. At the lowest point in the motion, the gravitational potential energy has its lowest value while the elastic potential energy has its greatest value. The *total* potential energy (gravitational plus elastic) has its

minimum value at the equilibrium point since the kinetic energy is maximum there.

Practice Problem 10.7   Energy at maximum extension

Calculate the energies at the lowest point in the oscillations in Example 10.7.

10.7   GRAPHICAL ANALYSIS OF SHM

We have shown that the position of a particle vibrating along the *x*-axis is

$$x(t) = A \cos \omega t \qquad (10\text{-}17)$$

Since the cosine function goes from −1 to +1, multiplying it by *A* gives us a displacement from −*A* to +*A*. Figure 10.19a is a graph of the position as a function of time.

The velocity at any time is the slope of the *x*(*t*) graph. Note that the maximum slope in Fig. 10.19a occurs when *x* = 0, which confirms what we already know from energy conservation: the velocity is maximum at the equilibrium point. Note also that the velocity is zero when the displacement is a maximum (+*A* or −*A*). Figure 10.19b shows a graph of $v_x(t)$. The equation describing this graph is (see Problem 46):

$$v_x(t) = -v_m \sin \omega t = -\omega A \sin \omega t \qquad (10\text{-}24)$$

The acceleration is the slope of the $v_x(t)$ graph. Figure 10.19c is a graph of $a_x(t)$, which is described by the equation

$$a_x(t) = -a_m \cos \omega t = -\omega^2 A \cos \omega t \qquad (10\text{-}18)$$

Observe the interrelationships between the three graphs. The velocity graph is one quarter cycle ahead of the position graph; that is, $v_x(t)$ reaches its positive maximum one quarter period before *x*(*t*) reaches its positive maximum. Likewise, the acceleration is one quarter cycle ahead of the velocity and one half cycle ahead of the position.

Figure 10.19 (a) Position of a particle in simple harmonic motion as a function of time, (b) velocity as a function of time, and (c) acceleration as a function of time

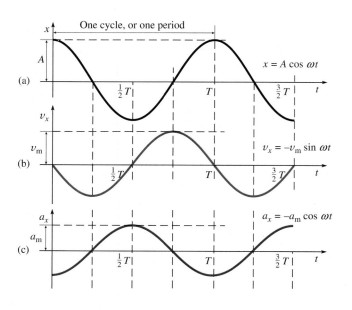

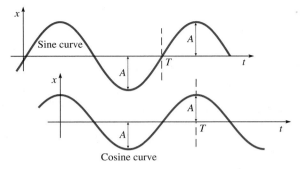

Figure 10.20 Sine and cosine curves of displacement versus time

We have written the position as a function of time in terms of the cosine function, but we can just as correctly use the sine function. The difference between the two is the initial position at time $t = 0$ (see Fig. 10.20). If the position is at a maximum ($x = A$) at $t = 0$, $x(t)$ is a cosine function. If the position is at the equilibrium point ($x = 0$) at $t = 0$, $x(t)$ is a sine function. By analyzing the slopes of the graphs, you can show (Problem 43) that if the position as a function of time is

$$x(t) = A \sin \omega t \qquad (10\text{-}25\text{a})$$

then the velocity is

$$v_x(t) = v_m \cos \omega t \qquad (10\text{-}25\text{b})$$

and the acceleration is

$$a_x(t) = -a_m \sin \omega t \qquad (10\text{-}25\text{c})$$

Example 10.8

A Vibrating Loudspeaker Cone

A loudspeaker has a movable diaphragm (the *cone*) that vibrates back and forth to produce sound waves. The displacement of a loudspeaker cone playing a sinusoidal test tone is graphed in Fig. 10.21. Find (a) the amplitude of the motion, (b) the period of the motion, and (c) the frequency of the motion. (d) Write equations for $x(t)$ and $v_x(t)$.

Strategy   The amplitude and period can be read directly from the graph. The frequency is the inverse of the period. Since $x(t)$ begins at the maximum displacement, it is described by a cosine function. By looking at the slope of $x(t)$ we can tell whether the velocity is a positive or negative sine function.

Solution   (a) The amplitude is the maximum displacement shown on the graph: $A = 0.015$ m.

(b) The period is the time for one complete cycle. From the graph: $T = 0.040$ s.

(c) The frequency is the inverse of the period.

$$f = \frac{1}{T} = \frac{1}{0.040 \text{ s}} = 25 \text{ Hz}$$

(d) Since $x = +A$ at $t = 0$, we write $x(t)$ as a cosine function:

$$x(t) = A \cos \omega t$$

where $A = 0.015$ m and

$$\omega = 2\pi f = 160 \text{ rad/s}$$

The slope of $x(t)$ is initially zero and then goes negative. Therefore, $v_x(t)$ is a negative sine function:

$$v_x(t) = -v_m \sin \omega t$$

where $\omega = 2\pi f = 160$ rad/s and

$$v_m = \omega A = 160 \text{ rad/s} \times 0.015 \text{ m} = 2.4 \text{ m/s}$$

Discussion   As a check, the velocity should be $\frac{1}{4}$ of a cycle ahead of the position. If we imagine shifting the vertical axis to the right (ahead) by 0.01 s, the graph would have the shape of a negative sine function.

Practice Problem 10.8   Acceleration of speaker cone

Sketch a graph and write an equation for $a_x(t)$.

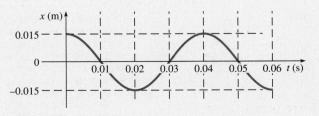

Figure 10.21
Horizontal displacement of a vibrating cone as a function of time

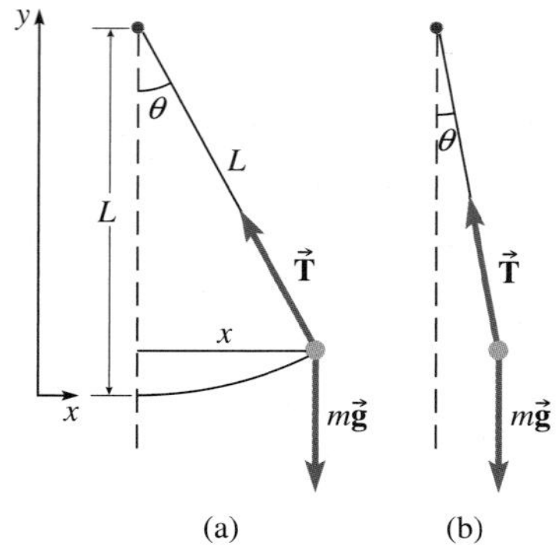

(a)   (b)

Figure 10.22 Forces on a pendulum bob (a) far from equilibrium and (b) near equilibrium.

10.8 THE PENDULUM

Simple Pendulum

When a pendulum swings back and forth, a string or thin rod constrains the bob to move along a circular arc (Fig. 10.22). However, *for oscillations with small amplitude*, we assume that the bob moves *back and forth along the x-axis*; the vertical motion of the bob is negligible.

Since the weight of the bob has no x-component, the restoring force is the x-component of the force due to the string. We expect the restoring force to be proportional to the displacement for small oscillations. From Fig. 10.22,

$$\Sigma F_x = -T \sin \theta \approx -\frac{Tx}{L}$$

where L is the length of the string and $\sin \theta = x/L$. The y-component of the acceleration is negligibly small, so

$$\Sigma F_y = T \cos \theta - mg = ma_y \approx 0$$

Since $\cos \theta \approx 1$ for small θ, $T \approx mg$. Then

$$\Sigma F_x \approx -\frac{mgx}{L} = ma_x$$

Solving for a_x:

$$a_x = -\frac{g}{L}x$$

To identify the angular frequency, we recall that

$$a_x = -\omega^2 x \tag{10-19}$$

Therefore, the angular frequency is

$$\omega = \sqrt{\frac{g}{L}} \tag{10-26a}$$

and the period is

$$T = \frac{2\pi}{\omega} = 2\pi\sqrt{\frac{L}{g}} \tag{10-26b}$$

Note that the period depends on L and g but not on the mass of the pendulum.

Be careful not to confuse the *angular frequency* of the pendulum with its *angular velocity*. Even though the two have the same units (rad/s in SI) and are written with the same symbol (ω), for a pendulum they are *not* the same. When dealing with the pendulum, we use the symbol ω to stand for the *angular frequency* only. The angular frequency $\omega = 2\pi f$ of a given pendulum is constant, while the angular velocity (the rate of change of θ) changes with time between zero (at the extremes) and its maximum magnitude (at the equilibrium point).

Physics at Home

The relation between the period and the length of the pendulum is easily tested at home. Make a simple pendulum by tying a thin string to one end of a paper clip and sliding the clip over a coin. Some tape can be used to help hold the coin if it slips out of place. Holding the end of the string, let the coin swing through a small arc and note the time for the coin to make ten complete oscillations, starting from one extreme position and returning to the same position ten times. Divide the time by ten to get the period. (This gives a more accurate value than timing a single period.) Measure the length of the pendulum and test Eq. (10-26b).

Repeat the experiment by holding the string at a position closer to the coin, effectively shortening the length of the pendulum. What do you find? Is the period for the shorter pendulum longer, shorter, or the same as that measured for the longer pendulum?

The effect of a different mass on the period can also be tested by using two or three coins taped together, with the same length pendulum as used for the first measurement. Does a heavier coin affect the result?

Example 10.9

Grandfather Clock

A grandfather clock uses a pendulum with period 2.0 s to keep time. In one such clock, the pendulum bob has mass 150 g; the pendulum is set into oscillation by displacing it 33 mm to one side. (a) What is the length of the pendulum? (b) Does the initial displacement satisfy the small angle approximation?

Strategy The period depends on the length of the pendulum and on the gravitational field strength g. It does not depend on the mass of the bob. It also does not depend on the initial displacement, as long as it is small compared to the length.

Solution (a) Assuming small amplitudes, the period is

$$T = 2\pi\sqrt{\frac{L}{g}}$$

Solving for L,

$$L = \frac{T^2 g}{(2\pi)^2}$$

$$= \frac{(2.0 \text{ s})^2 \times 9.8 \text{ m/s}^2}{(2\pi)^2} = 0.99 \text{ m}$$

(b) The small angle approximation is valid if the maximum displacement is small compared to the length of the pendulum.

$$\frac{x}{L} = \frac{33 \text{ mm}}{990 \text{ mm}} = 0.033$$

Is that small enough? If $\sin\theta = x/L = 0.033$, then

$$\theta = \sin^{-1} 0.033 = 0.033006$$

Sin θ and θ differ by less than 0.02%. Since we only know T and g to two significant figures, the approximation is good.

Discussion We should check that we didn't write the expression for the period "upside down," which is the most likely error we could make. Besides checking that the units work out, we know that a longer pendulum has a longer period, so L must go in the numerator. On the other hand, if g were larger, the restoring force would be larger and we would expect the period to shorten; thus, g belongs in the denominator.

Practice Problem 10.9 Pendulum on the Moon

A pendulum of length 0.99 m is taken to the Moon by an astronaut. The period of the pendulum is 4.9 s. What is the gravitational field strength on the surface of the Moon?

The period of a pendulum as just determined is valid only for small amplitudes. For larger amplitudes, the pendulum's motion is still periodic (though not SHM). Why would the period be any different for large amplitudes? Remember that we assumed the bob was moving horizontally back and forth along the x-axis. This simplification breaks down for large amplitudes. For instance, if we pull the pendulum out horizontally ($\theta = 90°$), the tangential component of the weight is mg, but $F_x = 0$! Since we have overestimated F_x, we have underestimated the time for the bob to return to $x = 0$; thus, the period for large amplitudes is greater than $2\pi\sqrt{L/g}$. Another way of looking at it is in terms of the tangential force. The expression for the tangential component of the weight is correct even for large amplitudes. However, the distance the bob must move to return to equilibrium is larger than x. For instance, starting at $\theta = 90°$, the bob must move one-quarter of the circumference, a distance $\frac{1}{4}(2\pi L) \approx 1.6L$, to return to equilibrium. Assuming linear motion along the x-axis would make the distance only L. With a longer distance to travel, the time is longer.

Physical Pendulum

Imagine that you have a simple pendulum of length L. Beside it you have a uniform metal bar of the same length, which is free to swing about an axis at one end (see Fig. 10.23). Would the two have the same period if they are set into oscillation?

For the simple pendulum, the bob is assumed to be a point mass; all the mass of the pendulum is at a distance L from the rotation axis. For the metal bar, however, the mass is uniformly distributed from the axis to a *maximum* distance L away from the axis. The center of mass is located at the midpoint, a distance $d = \frac{1}{2}L$ from the axis (Fig. 10.23). Since the mass is on average closer to the axis, the period is shorter than that of the simple pendulum.

Would this bar have a period equal to that of a simple pendulum of length $\frac{1}{2}L$? That is a good guess, since the center of mass of the bar is a distance $\frac{1}{2}L$ away from the rotation axis. Unfortunately, it isn't quite that easy. The gravitational force acts at the center of mass, but we *cannot* think of all the mass as being concentrated at that point—that would give the wrong rotational inertia. When set into oscillation, the bar, or any other rigid object free to rotate about a fixed axis, is called a **physical pendulum**.

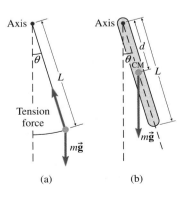

Figure 10.23 (a) A simple pendulum and (b) a physical pendulum

To find the period of a physical pendulum, we first find the net torque acting on the physical pendulum and then use the rotational form of Newton's second law. Taking torques about the rotation axis, only gravity gives a nonzero torque. If the pendulum has mass m and the distance from the axis to the center of mass is d, then the torque is

$$\tau = F_\perp r = -(mg \sin\theta)d$$

where θ is the angle indicated in Fig. 10.23. The other component of the gravitational force, $mg \cos\theta$, passes through the axis of rotation so it does not contribute to the torque. In the equation above, both τ and θ are positive if they are counterclockwise; the minus sign says that they always have opposite sign, since the torque always acts to bring θ closer to zero. Assuming small amplitudes, $\sin\theta \approx \theta$ (in radians) and the torque is

$$\tau = -mgd\theta$$

Thus, the restoring torque is proportional to the displacement angle θ, just as the restoring force was proportional to the displacement for the simple pendulum and the mass on a spring.

The net torque is equal to the rotational inertia times the angular acceleration:

$$\tau = -mgd\theta = I\alpha$$

and the angular acceleration is

$$\alpha = -\frac{mgd}{I}\theta \qquad (10\text{-}27)$$

Since the angular acceleration is a negative constant times the angular displacement from equilibrium, we indeed have SHM. Equation (10-27) is analogous to the equation for the linear acceleration of the oscillating spring

$$a_x = -\omega^2 x$$

where

$$\frac{mgd}{I} = \omega^2$$

Therefore, the angular frequency of the physical pendulum is

$$\omega = 2\pi f = \sqrt{\frac{mgd}{I}} \qquad (10\text{-}28\text{a})$$

and the period is

$$T = \frac{2\pi}{\omega} = 2\pi\sqrt{\frac{I}{mgd}} \qquad (10\text{-}28\text{b})$$

where d is the distance from the axis to the CM and I is the rotational inertia.

For a uniform bar of length L, the center of mass is halfway down the bar:

$$d = \tfrac{1}{2}L$$

From Table 8.1, the rotational inertia of a uniform bar rotating about an axis through an endpoint is $I = \tfrac{1}{3}mL^2$. The period of oscillation is

$$T = 2\pi\sqrt{\frac{I}{mgd}}$$

Substituting for I and d,

$$T = 2\pi\sqrt{\frac{\tfrac{1}{3}mL^2}{(mg)\tfrac{1}{2}L}} = 2\pi\sqrt{\frac{2L}{3g}}$$

The bar has the same period as a simple pendulum of length $\tfrac{2}{3}L$.

Example 10.10

Comparison of Walking Frequencies and Speeds for Various Creatures

During a relaxed walking pace, an animal's leg can be thought of as a physical pendulum of length L that pivots about the hip. (a) What is the relaxed walking frequency for a cat ($L = 30$ cm), dog (60 cm), human (1 m), giraffe (2 m), and a mythological titan (10 m)? (b) Derive an equation that gives the walking speed (amount of ground covered per unit time) for a given walking frequency f.

continued on next page

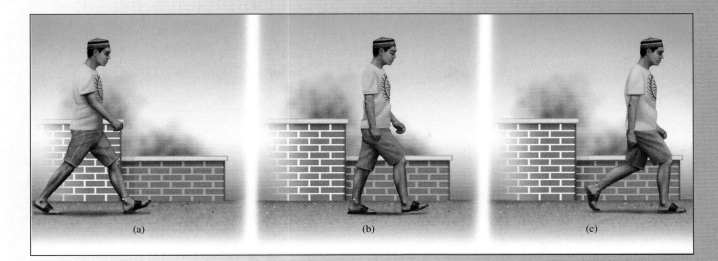

Figure 10.24   The forward motion of a leg during walking is similar to the swing of a physical pendulum. From (a) to (b), the right leg swings forward like a pendulum. In (c), the right foot is on the ground and the left leg is about to swing forward.

Example 10.10 *continued*

[*Hint:* Start by drawing a picture of the leg position at the start of the swing (leg back) and the end of the swing (leg forward) and assume a comfortable angle of about 30° between these two positions. To how many steps does a complete period of the pendulum correspond?] (c) Find the walking speed for each of the animals listed in part (a).

Strategy   We have to use an idealized model of the leg, since we don't know the exact location of the center of mass or the rotational inertia. The simple pendulum is not a good model, since it would assume all the mass of the leg at the foot! A much better model is to think of the leg as a uniform cylinder pivoting about one end.

Solution   (a) For a uniform cylinder, the center of mass is a distance $d = \frac{1}{2}L$ from the pivot and the rotational inertia about an axis at one end is $I = \frac{1}{3}mL^2$. Then the period is

$$T = 2\pi\sqrt{\frac{I}{mgd}} = 2\pi\sqrt{\frac{\frac{1}{3}mL^2}{(mg)\frac{1}{2}L}} = 2\pi\sqrt{\frac{2L}{3g}}$$

and the frequency f is

$$f = \frac{1}{T} = \frac{1}{2\pi}\sqrt{\frac{3g}{2L}} \approx 0.2\sqrt{\frac{g}{L}}$$

Substituting the numerical values of L for each animal, we find the frequencies to be 1 Hz (cat), 0.8 Hz (dog), 0.6 Hz (human), 0.4 Hz (giraffe), and 0.2 Hz (titan).

(b) One period of the "pendulum" corresponds to two steps. In Fig. 10.24a, the right leg is about to step forward. The step occurs as the pendulum swings forward through half a cycle. In Fig. 10.24b, the right foot is about to touch the ground; in

Fig. 10.24c the right foot touches the ground and now the left leg is about to step forward. During this step, the right foot stays in place on the ground, but the right leg is swinging backward relative to the hip joint. During each step, the distance covered is approximately the length of a 30° arc of radius L, which is $\frac{1}{12}$ the circumference of a circle of radius L. So during one period, the distance walked is

$$D = 2 \times \frac{1}{12} \times 2\pi L = \frac{\pi}{3}L \approx L$$

and the walking speed is

$$v = \frac{D}{T} = Lf = 0.2\sqrt{gL}$$

(c) The speeds are 0.3 m/s (cat), 0.5 m/s (dog), 0.6 m/s (human), 0.9 m/s (giraffe), and 2 m/s (titan).

Discussion   You may be more familiar with walking speeds in mi/h. Converting the units, 0.6 m/s ≈ 1.3 mi/h, which is just about right for a leisurely walk. A brisk walk is about 3 mi/h for most people; to go much faster than that, you need to jog or run.

The solution says that longer legs walk faster, but the frequency of the steps is lower. You can verify that by walking beside a friend who is much taller or much shorter than you, or by taking your dog for a walk.

Practice Problem 10.10   Walking speed for human

A more realistic model of a human leg of length 1.0 m has the center of mass 0.45 m from the hip and a rotational inertia of $\frac{1}{6}mL^2$. What is the walking speed predicted by this model?

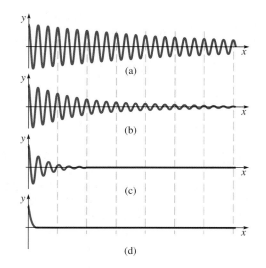

Figure 10.25 Graphs of $x(t)$ for a mass-spring system with increasing amounts of damping (a), (b), (c), (d). In (d) the damping is sufficient to prevent oscillations from occurring.

10.9   DAMPED OSCILLATIONS

In simple harmonic motion, we assume that no dissipative forces such as friction or viscous drag exist. Since the mechanical energy is constant, the oscillations continue forever with constant amplitude. SHM is a simplified model. The oscillations of a swinging pendulum or a vibrating tuning fork gradually die out as energy is dissipated. The amplitude of each cycle is a little smaller than that of the previous cycle (see Figs. 10.25a and b). This kind of motion is called **damped oscillation**, where the word *damped* is used in the sense of *extinguished* or *restrained*. For a small amount of damping, oscillations occur at approximately the same frequency as if there were no damping. A greater degree of damping lowers the frequency slightly (Fig. 10.25c). Even more damping prevents oscillations from occurring at all (Fig. 10.25d).

Damping is not always a disadvantage. The suspension system of a car includes shock absorbers that cause the vibration of the body—a mass connected to the chassis by springs—to be quickly damped. The shock absorbers reduce the discomfort that passengers would otherwise experience due to the bouncing of an automobile as it travels along a bumpy road. Figure 10.26 shows how a shock absorber works. In order to compress or expand the shock absorber, a viscous oil must flow through the holes in the piston. The viscous force dissipates energy regardless of which direction the piston moves. The shock absorber enables the spring to smoothly return to its equilibrium length without oscillating up and down (see Fig. 10.25d). When the oil leaks out of the shock absorber, the damping is insufficient to prevent oscillations. After hitting a bump, the body of the car oscillates up and down (Fig. 10.25c).

Making The Connection:
shock absorbers
in a car

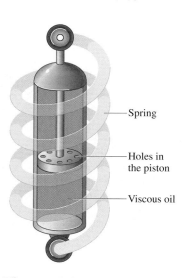

Figure 10.26   A shock absorber

10.10 FORCED OSCILLATIONS AND RESONANCE

When damping forces are present, the only way to keep the amplitude of oscillations from diminishing is to replace the dissipated energy from some other source. When a child is being pushed on a swing, the parent replaces the energy dissipated with a small push. In order to keep the amplitude of the motion constant, the parent gives a little push once per cycle, adding just enough energy each time to compensate for the energy dissipated in one cycle. The frequency of the *driving force* (the parent's push) matches the *natural frequency* of the system (the frequency at which it would oscillate on its own).

Forced oscillations (or driven oscillations) occur when a periodic external driving force acts on a system that can oscillate. The frequency of the driving force does not have to match the natural frequency of the system. Ultimately, the system oscillates at the driving frequency, even if it is far from the natural frequency. However, the amplitude of the oscillations is generally quite small unless the driving frequency is close to the natural frequency (see Fig. 10.27). When the driving frequency is equal to the natural frequency of the system, the amplitude of the motion is a maximum. This condition is called **resonance**.

At resonance, the driving force is always in the same direction as the object's velocity. Since the driving force is always doing positive work, the energy of the oscillator builds up until the energy dissipated balances the energy added by the driving force. For an oscillator with little damping, this requires a large amplitude. When the driving and natural frequencies differ, the driving force and velocity are no longer synchronized; sometimes they are in the same direction and sometimes in opposite directions. The driving force is not at resonance, so it sometimes does negative work. The net work done by the driving force decreases as the driving frequency moves away from resonance. Therefore, the oscillator's energy and amplitude are smaller than at resonance.

Large-amplitude vibrations due to resonance can be dangerous in some situations. In 1940, a strong wind set the Tacoma Narrows Bridge in Washington state into vibration with increasing amplitude. Turbulence in the air as it flowed across the bridge caused the air pressure to fluctuate with a frequency matching one of the bridge's resonant frequencies. As the amplitude of the oscillations grew, the bridge was closed; soon after, the bridge collapsed (Fig. 10.28). Engineers now design bridges with much higher resonant frequencies so the wind cannot cause resonant vibrations.

In the nineteenth century, bridges were sometimes set into resonant vibration when the cadence of marching soldiers matched a resonant frequency of the bridge. After the collapse of several bridges due to resonance, soldiers were told to break step when crossing a bridge to eliminate the danger of their cadence setting the bridge into resonance.

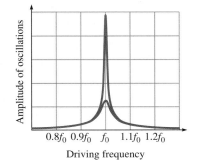

Figure 10.27 Two resonance curves for an oscillator with natural frequency f_0. The amplitude of the driving force is constant. In the red graph, the oscillator has one-fourth as much damping as in the blue graph.

Making The Connection:
vibration of a bridge

Figure 10.28 The collapse of the Tacoma Narrows Bridge

MASTER THE CONCEPTS

Summary

- A deformation is a change in the size or shape of an object.

- When deforming forces are removed, an *elastic* object returns to its original shape and size.

- Hooke's law, in a generalized form, says that the deformation of a material (measured by the strain) is proportional to the magnitude of the forces causing the deformation (measured by the stress). The definitions of stress and strain are as given in the table.

Type of Deformation

	Tensile or Compressive	Shear	Volume
Stress	Force per unit cross-sectional area F/A	Shear force divided by the parallel area of the surface on which it acts F/A	Pressure P
Strain	Fractional length change $\Delta L/L$	Ratio of the relative displacement Δx to the separation L of the two parallel surfaces $\Delta x/L$	Fractional volume change $\Delta V/V$
Constant of Proportionality	Young's modulus Y	Shear modulus S	Bulk modulus B

- If the tensile or compressive stress exceeds the *proportional limit,* the strain is no longer proportional to the stress. The solid still returns to its original length when the stress is removed as long as the stress does not exceed the *elastic limit.* If the stress exceeds the elastic limit, the material is permanently deformed. For larger stresses yet, the solid fractures when the stress reaches the *breaking point.* The maximum stress that can be withstood without breaking is called the *ultimate strength.*

- Vibrations occur in the vicinity of a point of stable equilibrium. An equilibrium point is *stable* if the net force on an object when it is displaced from equilibrium points back toward the equilibrium point. Such a force is called a restoring force since it tends to restore equilibrium.

- Simple harmonic motion is periodic motion that occurs whenever the restoring force is proportional to the displacement from equilibrium. In SHM, the position, velocity, and acceleration as functions of time are sinusoidal (i.e., sine or cosine functions). Any oscillatory motion is approximately SHM if the amplitude is small, because for small oscillations the restoring force is approximately linear.

- The maximum velocity and acceleration in SHM are

$$v_{\mathrm{m}} = \omega A \quad \text{and} \quad a_{\mathrm{m}} = \omega^2 A \quad (10\text{-}21,\ 10\text{-}22)$$

where ω is the angular frequency. The acceleration is proportional to and in the opposite direction from the displacement:

$$a_x(t) = -\omega^2 x(t) \quad (10\text{-}19)$$

- The equations of motion for SHM are

If $x = A$ at $t = 0$,   If $x = 0$ at $t = 0$,

$$x = A \cos \omega t \qquad x = A \sin \omega t$$

$$v_x = -v_{\mathrm{m}} \sin \omega t \qquad v_x = v_{\mathrm{m}} \cos \omega t$$

$$a_x = -a_{\mathrm{m}} \cos \omega t \qquad a_x = -a_{\mathrm{m}} \sin \omega t$$

In either case, the velocity is $\frac{1}{4}$ of a cycle ahead of the position and the acceleration is $\frac{1}{4}$ of a cycle ahead of the velocity.

- The angular frequency for a mass-spring system is

$$\omega = \sqrt{\frac{k}{m}} \quad (10\text{-}20a)$$

For a simple pendulum it is

$$\omega = \sqrt{\frac{g}{L}} \quad (10\text{-}26a)$$

and for a physical pendulum it is

$$\omega = \sqrt{\frac{mgd}{I}} \quad (10\text{-}28a)$$

- In the absence of dissipative forces, the total mechanical energy of a simple harmonic oscillator is constant and proportional to the square of the amplitude:

$$E = \tfrac{1}{2}kA^2 \quad (10\text{-}13)$$

where the potential energy has been chosen to be zero at the equilibrium point. At any point, the sum of the kinetic and potential energies is constant:

$$E = \tfrac{1}{2}kx^2 + \tfrac{1}{2}mv_x^2 = \tfrac{1}{2}kA^2 \quad (10\text{-}12)$$

Highlighted Figures and Tables

T10.1 Approximate values of Young's modulus for various substances (p. 343)

F10.4 Stress-strain curve showing limits for ductile material, brittle material, and compact bone (p. 345)

F10.7 Shear stress (p. 347)

T10.2 Shear and bulk moduli for various materials (p. 349)

F10.10 Forces on an object when submerged in fluid (p. 350)

F10.11 Points of stable and unstable equilibrium (p. 351)

F10.14 Positions of an oscillating body at equal time intervals (p. 352)

F10.16 Relation between uniform circular motion and SHM (p. 355)

F10.17 Relaxed, equilibrium, and displaced positions for a relaxed spring of constant k with mass m attached (p. 356)

F10.19 Position, velocity, and acceleration of a particle in SHM as a function of time (p. 358)

F10.22 Forces on a pendulum bob far from equilibrium and near equilibrium (p. 360)

F10.23 A simple and a physical pendulum (p. 361)

F10.25 Oscillations with increasing amounts of damping (p. 364)

CONCEPTUAL QUESTIONS

1. Young's modulus for diamond is about 20 times as large as that of glass. Does that tell you which is stronger? If not, what does it tell you?

2. A grandfather clock is running too fast. To fix it, should the pendulum be lengthened or shortened? Explain.

3. A karate student hits downward on a stack of concrete blocks supported at both ends (Fig. 10.29). A block breaks. Explain where it starts to break first, at the bottom or at the top. (The block experiences shear, compressive, and tensile stresses. Recall that concrete has much less tensile strength than compressive strength. Which part of the block is stretched and which is compressed when the block bends in the middle?)

Figure 10.29
Conceptual Question 3

4. A cylindrical steel bar is compressed by the application of forces of magnitude F at each end. What magnitude forces would be required to compress by the same amount (a) a steel bar of the same cross-sectional area but one half the length? (b) a steel bar of the same length but one half the radius?

5. The columns built by the ancient Greeks and Romans to support temples and other structures are tapered; they are thicker at the bottom than at the top (Fig. 10.30). This certainly has

Figure 10.30
Conceptual Question 5

an aesthetic purpose, but is there an engineering purpose as well? What might it be?

6. Explain how the period of a mass-spring system can be independent of amplitude, even though the distance traveled during each cycle is proportional to the amplitude.

7. In a saber saw, a *Scotch yoke* converts the rotation of the motor into the back-and-forth motion of the blade. The Scotch yoke is a mechanical device used to convert oscillatory motion to circular motion or *vice versa* (Fig. 10.31). A wheel with a fixed knob rotates at constant angular velocity; the knob is constrained within a vertical slot causing the saw blade to move left and right without moving up and down. Is the motion of the saw blade SHM? Explain.

Figure 10.31
Conceptual Question 7

8. A mass hanging vertically from a spring and a simple pendulum both have a period of oscillation of 1 s on Earth. An astronaut takes the two devices to another planet where the gravitational field is stronger than that of Earth. For each of the two systems, state whether the period is now longer than 1 s, shorter than 1 s, or equal to 1 s. Explain your reasoning.

9. A bungee jumper leaps from a bridge and comes to a stop a few centimeters above the surface of the water below. At that lowest point, is the tension in the bungee cord equal to the jumper's weight? Explain why or why not.

10. Does it take more force to break a longer rope or a shorter rope? Assume the ropes are identical except for their lengths and are ideal—there are no weak points. Does it take more *energy* to break the long rope or the short rope? Explain.

11. A pilot is performing vertical loop-the-loops over the ocean at noon. The plane speeds up as it approaches the bottom of the circular loop and slows as it approaches the top of the loop. An observer in a helicopter is watching the shadow of the plane on the surface of the water. Does the shadow exhibit SHM? Explain.

12. Are you more likely to find steel rods in a horizontal concrete beam or in a vertical concrete column? Is concrete more in need of reinforcement under tensile or compressive stress?

13. Suppose that it takes tensile forces of magnitude F to produce a given strain $\Delta L/L$ in a steel wire of cross-sectional area A. If you had two such wires side by side and stretched them simultaneously, what magnitude tensile forces would be required to produce the same strain? By thinking of a thick wire as two (or more) thinner wires side by side, explain why the force to produce a given strain must be proportional to the cross-sectional area. Thus, the strain depends on the stress—the force per unit area.

14. Think of a crystalline solid as a set of atoms connected by ideal springs (Fig. 10.32). When a wire is stretched, how is the elongation of the wire related to the elongation of each of the interatomic springs? Use your answer to explain why a given tensile stress produces an elongation of the wire proportional to the wire's initial length—or, equivalently, that a given stress produces the same strain in wires of different lengths.

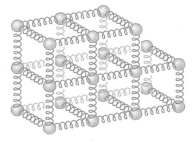

Figure 10.32 Conceptual Question 14

15. What are the advantages of using the concepts of stress and strain to describe deformations?

16. An old highway is built out of concrete blocks of equal length. A car traveling on this highway feels a little bump at the joint between blocks. The passengers in the car feel that the ride is uncomfortable at a speed of 45 mi/h, but much smoother at speeds either lower or higher than that. Explain.

17. The period of oscillation of a simple pendulum does not depend on the mass of the bob. By contrast, the period of a mass-spring system does depend on mass. Explain the apparent contradiction. [*Hint:* What provides the restoring force in each case? How does the restoring force depend on mass?]

18. A mass connected to an ideal spring is oscillating without friction on a horizontal surface. Sketch graphs of the kinetic energy, potential energy, and total energy as functions of time for one complete cycle.

MULTIPLE CHOICE QUESTIONS

Questions 1–4. A body is suspended vertically from an ideal spring. The spring is initially in its relaxed position. The body is then released and oscillates about the equilibrium position. Answer choices for Questions 1–4:
  (a) The spring is relaxed.
  (b) The body is at the equilibrium point.
  (c) The spring is at its maximum extension.
  (d) The spring is somewhere between the equilibrium point and maximum extension.

1. The acceleration is greatest in magnitude and is directed upward when:

2. The speed of the body is greatest when:

3. The acceleration of the body is zero when:

4. The acceleration is greatest in magnitude and is directed downward when:

5. Two simple pendulums, A and B, have the same length, but the mass of A is twice the mass of B. Their vibrational amplitudes are equal. Their periods are T_A and T_B, respectively, and their energies are E_A and E_B. Choose the correct statement.
  (a) $T_A = T_B$ and $E_A > E_B$ (b) $T_A < T_B$ and $E_A > E_B$
  (c) $T_A > T_B$ and $E_A < E_B$ (d) $T_A = T_B$ and $E_A < E_B$

6. A force F applied to each end of a steel wire (length L, diameter d) stretches it by 1.0 mm. How much does F stretch another steel wire, of length $2L$ and diameter $2d$?
  (a) 0.50 mm   (b) 1.0 mm   (c) 2.0 mm
  (d) 4.0 mm   (e) 0.25 mm

7. A stiff material is characterized by
  (a) High ultimate strength.   (b) High breaking strength.
  (c) High Young's modulus.   (d) High proportional limit.

8. A brittle material is characterized by
  (a) High breaking strength and low Young's modulus.
  (b) Low breaking strength and high Young's modulus.
  (c) High breaking strength and high Young's modulus.
  (d) Low breaking strength and low Young's modulus.

9. Which pair of quantities can be expressed in the same units?
  (a) Stress and strain   (b) Young's modulus and strain
  (c) Young's modulus and stress
  (d) Ultimate strength and strain

10. Two wires have the same diameter and length. One is made of copper, the other brass. The wires are connected together end to end. When the free ends are pulled in opposite directions, the two wires *must* have the same
  (a) Stress.   (b) Strain.   (c) Ultimate strength.
  (d) Elongation.   (e) Young's modulus.

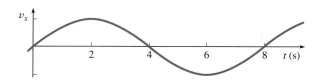

Figure 10.33 Multiple Choice Questions 11–20

Questions 11–20. Figure 10.33 is the graph of $v_x(t)$ for an object in SHM. Answer choices for each question:
  (a) 1 s, 2 s, 3 s   (b) 5 s, 6 s, 7 s   (c) 0 s, 1 s, 7 s, 8 s
  (d) 3 s, 4 s, 5 s   (e) 0 s, 4 s, 8 s   (f) 2 s, 6 s
  (g) 3 s, 5 s   (h) 1 s, 3 s   (i) 5 s, 7 s
  (j) 3 s, 7 s   (k) 1 s, 5 s

11. When is the kinetic energy maximum?

12. When is the kinetic energy zero?

13. When is the potential energy maximum?

14. When is the potential energy minimum?

15. When is the object at the equilibrium point?

16. When does the acceleration have its maximum magnitude?

17. Which answer specifies times when the net force is in the $+x$-direction?

18. Which answer specifies times when the object is on the $-x$-side of the equilibrium point ($x < 0$)?

19. Which answer specifies times when the object is moving away from the equilibrium point?

20. Which answer specifies times when the potential energy is decreasing?

PROBLEMS

Note: **C** indicates a combination conceptual/quantitative problem. Gold diamonds ✦, ✦✦ are used to indicate the increasing level of difficulty of each problem. Problem numbers appearing in blue, 9., denote problems that have a detailed solution available in the Student Solutions Manual. Some problems are *paired* by concept; their numbers are connected by a ruled box.

10.2 Hooke's Law for Tensile and Compressive Forces

1. A steel beam is placed vertically in the basement of a building to keep the floor above from sagging. The load on the beam is 5.8×10^4 N, the length of the beam is 2.5 m, and the cross-sectional area of the beam is 7.5×10^{-3} m^2. Find the vertical compression of the beam.

2. A 91-kg man's thighbone has a relaxed length of 0.50 m, a cross-sectional area of 7.0×10^{-4} m^2, and a Young's modulus of 1.1×10^{10} N/m^2. By how much does the thighbone compress when the man is standing on both feet?

3. A brass wire with Young's modulus of 9.2×10^{10} Pa is 2.0 m long and has a cross-sectional area of 5.0 mm^2. If a weight of 5.0 kN is hung from the wire, by how much does it stretch?

4. A wire of length 5.00 m with a cross-sectional area of 0.100 cm^2 stretches by 6.50 mm when a load of 1.00 kN is hung from it. What is the Young's modulus for this wire?

✦ 5. It takes a flea 1.0×10^{-3} s to reach a peak velocity of 0.74 m/s. (a) If the mass of the flea is 0.45×10^{-6} kg, what is the average power required? (b) Insect muscle has a maximum output of 60 W/kg. If 20% of the flea's weight is muscle, can the muscle provide the power needed? (c) The flea has a resilin pad at the base of the hind leg that compresses when the flea bends its leg to jump. If we assume the pad is a cube with a side of 6.0×10^{-5} m, and the pad compresses fully, what is the energy stored in the compression of the pads of the two hind legs? Assume the Young's modulus for resilin is 1.7×10^6 N/m^2. (d) Does this provide enough power for the jump?

6. A 0.50-m-long guitar string, of cross-sectional area 1.0×10^{-6} m^2, has Young's modulus $Y = 2.0 \times 10^9$ N/m^2. By how much must you stretch the string to obtain a tension of 20 N?

7. Two steel wires (of the same length and different radii) are connected together, end to end, and tied to a wall (Fig. 10.34). An applied force stretches the combination by 1.0 mm. How far does the *midpoint* move?

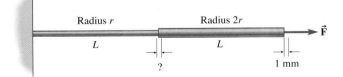

Figure 10.34 Problem 7

8. Abductin is an elastic protein found in scallops, with a Young's modulus of 4.0×10^6 N/m^2. It is used as an inner hinge ligament, with a cross-sectional area of 0.78 mm^2 and a relaxed length of 1.0 mm. When the muscles in the shell relax, the shell opens. This increases efficiency as the muscles do not need to exert any force to open the shell, only to close it. If the muscles must exert a force of 1.5 N to keep the shell closed, by how much is the abductin ligament compressed?

10.3 Beyond Hooke's Law

C 9. Using the stress-strain graph for bone (Fig. 10.4c), calculate Young's moduli for tension and for compression. Consider only small stresses.

10. An acrobat of mass 55 kg is going to hang by her teeth from a steel wire and she does not want the wire to stretch beyond its elastic limit. The elastic limit for the wire is 2.5×10^8 Pa. What is the minimum diameter the wire should have to support her?

11. A hair breaks under a tension of 1.2 N. What is the diameter of the hair? The tensile strength is 2.0×10^8 Pa.

12. The ratio of the tensile (or compressive) strength to the density of a material is a measure of how strong the material is "pound for pound." (a) Compare tendon (tensile strength 80.0 MPa, density 1100 kg/m^3) to steel (tensile strength 0.50 GPa, density 7700 kg/m^3): which is stronger "pound for pound" under tension? (b) Compare bone (compressive strength 160 MPa, density 1600 kg/m^3) to concrete (compressive strength 0.40 GPa, density 2700 kg/m^3): which is stronger "pound for pound" under compression?

13. What is the maximum load that could be suspended from a copper wire of length 1.0 m and radius 1.0 mm without permanently deforming the wire? Copper has an elastic limit of 2.0×10^8 Pa and a tensile strength of 4.0×10^8 Pa.

14. What is the maximum load that could be suspended from a copper wire of length 1.0 m and radius 1.0 mm without breaking the wire? Copper has an elastic limit of 2.0×10^8 Pa and a tensile strength of 4.0×10^8 Pa.

15. A marble column with a cross-sectional area of 25 cm^2 supports a load of 7.0×10^4 N. (Marble: $Y = 6.0 \times 10^{10}$ Pa; compressive strength $= 2.0 \times 10^8$ Pa.) (a) What is the stress in the column? (b) What is the strain in the column? (c) If the column is 2.0 m high, how much is its length changed by supporting the load? (d) What is the maximum weight the column can support?

16. A copper wire of length 3.0 m is observed to stretch by 2.1 mm when a weight of 120 N is hung from one end. (a) What is the diameter of the wire and what is the tensile stress in the wire? (b) If the tensile strength of copper is 4.0×10^8 N/m^2, what is the maximum weight that may be hung from this wire?

17. The leg bone (femur) breaks under a compressive force of about 5×10^4 N for a human and 10×10^4 N for a horse. The human femur has a compressive strength of 1.6×10^8 Pa, while the horse femur has a compressive strength of 1.4×10^8 Pa. What is the effective cross-sectional area of the femur in a human and in a horse? (*Note:* Since the center of the femur contains bone marrow, which has essentially no compressive strength, the effective cross-sectional area is about 80% of the total cross-sectional area.)

18. The maximum strain of a steel wire ($Y = 2.0 \times 10^{11}$ N/m^2), just before breaking, is 0.20%. What is the stress at its breaking point, assuming that strain is proportional to stress up to the breaking point?

10.4 Shear and Volume Deformations

19. A sphere of copper is subjected to 1.0×10^8 N/m^2 of pressure. The copper has a bulk modulus of 130 GPa. By what fraction does the volume of the sphere change? By what fraction does the radius of the sphere change?

20. By what percentage does the density of water increase at a depth of 1.0 km below the surface?

21. Two steel plates are fastened together using four bolts (Fig. 10.35). The bolts each have a shear modulus of 8.0×10^{10} Pa and a shear strength of 6.0×10^8 Pa. The radius of each bolt is 1.0 cm. What is the maximum shearing force F on the plates that the four bolts can withstand?

Figure 10.35 Problem 21

22. An anchor, made of cast iron of bulk modulus 60.0×10^9 Pa and of volume 0.230 m^3, is lowered over the side of the ship to the bottom of the harbor where the pressure is greater than sea level pressure by 1.75×10^6 Pa. Find the change in the volume of the anchor.

23. The upper surface of a cube of gelatin, 5.0 cm on a side, is displaced 0.64 cm by a tangential force. If the shear modulus of the gelatin is 940 Pa, what is the magnitude of the tangential force?

24. A large sponge has forces of magnitude 12 N applied in opposite directions to two opposite faces of area 42 cm^2 (see Fig. 10.7). The thickness of the sponge (L) is 2.0 cm. The deformation angle (γ) is 8.0°. (a) What is Δx? (b) What is the shear modulus of the sponge?

10.5 Simple Harmonic Motion; 10.6 The Period and Frequency for SHM

25. The period of oscillation of an object in an ideal spring-and-mass system is 0.50 s and the amplitude is 5.0 cm. What is the speed at the equilibrium point?

26. The period of oscillation of a spring-and-mass system is 0.50 s and the amplitude is 5.0 cm. What is the magnitude of the acceleration at the point of maximum extension of the spring?

27. A sewing machine needle moves with a rapid vibratory motion, rather like SHM, as it sews a seam. Suppose the needle moves 8.4 mm from its highest to its lowest position and it makes 24 stitches in 9.0 s. What is the maximum needle speed?

28. The prong of a tuning fork moves back and forth when it is set into vibration. The distance the prong moves between its extreme positions is 2.24 mm. If the frequency of the tuning fork is 440.0 Hz, what are the maximum velocity and the maximum acceleration of the prong? Assume SHM.

29. Show that the equation $a = -\omega^2 x$ is consistent for units, and that $\sqrt{k/m}$ has the same units as ω.

30. A small bird's wings can undergo a maximum displacement amplitude of 5.0 cm (distance from the tip of the wing to the horizontal). If the maximum acceleration of the wings is 12 m/s^2, and we assume the wings are undergoing simple harmonic motion when beating, what is the oscillation frequency of the wing tips?

Figure 10.36 Problem 31

31. An empty cart, tied between two ideal springs, oscillates with $\omega = 10.0$ rad/s (Fig. 10.36). A load is placed in the cart, making the total mass 4.0 times what it was before. What is the new value of ω?

32. The air pressure variations in a sound wave cause the eardrum to vibrate. (a) For a given vibration amplitude, are the maximum velocity and acceleration of the eardrum greatest for high-frequency sounds or low-frequency sounds? (b) Find the maximum velocity and acceleration of the eardrum for vibrations of amplitude 1.0×10^{-8} m at a frequency of 20.0 Hz. (c) Repeat (b) for the same amplitude but a frequency of 20.0 kHz.

33. Show that, for SHM, the maximum displacement, velocity, and acceleration are related by $v_m^2 = a_m A$.

34. Equipment to be used in airplanes or spacecraft is often subjected to a shake test to be sure it can withstand the vibrations that may be encountered during flight. A radio receiver of mass 5.24 kg is set on a platform that vibrates in SHM at 120 Hz and with a maximum acceleration of 98 m/s^2 ($= 10g$). Find the radio's (a) maximum displacement, (b) maximum speed, and (c) the maximum net force exerted on it.

35. In an aviation test lab, pilots are subjected to vertical oscillations on a shaking rig to see how well they can recognize objects in times of severe airplane vibration. The frequency can be varied from 0.02 to 40.0 Hz and the amplitude can be set as high as 2 m for low frequencies. What are the maximum velocity and acceleration to which the pilot is subjected if the frequency is set at 25.0 Hz and the amplitude at 1.00 mm?

36. The diaphragm of a speaker has a mass of 50.0 g and responds to a signal of frequency 2.0 kHz by moving back and forth with an amplitude of 1.8×10^{-4} m at that frequency. (a) What is the maximum force acting on the diaphragm? (b) What is the mechanical energy of the diaphragm?

37. An ideal spring has a spring constant $k = 25$ N/m. The spring is suspended vertically. A 1.0-kg body is attached to the unstretched spring and released. It then performs oscillations. (a) What is the magnitude of the acceleration of the body when the extension of the spring is a maximum? (b) What is the maximum extension of the spring?

38. An ideal spring with a spring constant of 15 N/m is suspended vertically. A body of mass 0.60 kg is attached to the unstretched spring and released. (a) What is the extension of the spring when the speed is a maximum? (b) What is the maximum speed?

39. A 0.50-kg object, suspended from an ideal spring of spring constant 25 N/m, is oscillating vertically. How much change of kinetic energy occurs while the object moves from the equilibrium position to a point 5.0 cm lower?

40. A baby jumper consists of a cloth seat suspended by an elastic cord from the lintel of an open doorway. The unstretched length of the cord is 1.2 m and the cord stretches by 0.20 m

when a baby of mass 6.8 kg is placed into the seat. The mother then pulls the seat down by 8.0 cm and releases it. (a) What is the period of the motion? (b) What is the maximum speed of the baby?

10.7 Graphical Analysis of SHM

41. The displacement of an object in SHM is given by $y(t) =$ (8.0 cm) sin [(1.57 rad/s)t]. What is the frequency of the oscillations?

42. A body is suspended vertically from an ideal spring of spring constant 2.5 N/m. The spring is initially in its relaxed position. The body is then released and oscillates about its equilibrium position. The motion is described by

$$y = (4.0 \text{ cm}) \sin [(0.70 \text{ rad/s})t]$$

What is the maximum kinetic energy of the body?

43. (a) Sketch a graph of $x(t) = A \sin \omega t$ (the position of an object in SHM that is at the equilibrium point at $t = 0$). (b) By analyzing the slope of the graph of $x(t)$, sketch a graph of $v_x(t)$. Is $v_x(t)$ a sine or cosine function? (c) By analyzing the slope of the graph of $v_x(t)$, sketch $a_x(t)$. (d) Verify that $v_x(t)$ is $\frac{1}{4}$ cycle ahead of $x(t)$ and that $a_x(t)$ is $\frac{1}{4}$ cycle ahead of $v_x(t)$.

44. A mass-and-spring system oscillates with amplitude A and angular frequency ω. (a) What is the *average* speed during one complete cycle of oscillation? (b) What is the maximum speed? (c) Find the ratio of the average speed to the maximum speed. (d) Sketch a graph of $v_x(t)$, and refer to it to explain why this ratio is greater than $\frac{1}{2}$.

45. A ball is dropped from a height h onto the floor and keeps bouncing. No energy is dissipated, so the ball regains the original height h after each bounce. Sketch the graph for $y(t)$ and list several features of the graph that indicate that this motion is *not* SHM.

46. (a) Given that $x(t) = A \cos \omega t$, show that $v_x(t) = -\omega A \sin \omega t$.[*Hint:* Draw the velocity vector for point P in Fig. 10.16b and then find its x-component.] (b) Verify that the expressions for $x(t)$ and $v_x(t)$ are consistent with energy conservation. [*Hint:* Use the trigonometric identity $\sin^2 \omega t + \cos^2 \omega t = 1$.]

10.8 The Pendulum

47. What is the period of a pendulum consisting of a 6.0-kg mass oscillating on a 4.0-m-long string?

48. A pendulum of length 75 cm and mass 2.5 kg swings with a mechanical energy of 0.015 J. What is the amplitude?

49. A 0.50-kg mass is suspended from a string, forming a pendulum. The period of this pendulum is 1.5 s when the amplitude is 1.0 cm. The mass of the pendulum is now reduced to 0.25 kg. What is the period of oscillation now, when the amplitude is 2.0 cm?

50. A bob of mass m is suspended from a string of length L, forming a pendulum. The period of this pendulum is 2.0 s. If the pendulum bob is replaced with one of mass $\frac{1}{3}m$ and the length of the pendulum is increased to $2L$, what is the period of oscillation?

51. A pendulum (mass m, unknown length) moves according to $x = A \sin \omega t$. (a) Write the equation for $v_x(t)$ and sketch one cycle of the $v_x(t)$ graph. (b) What is the maximum kinetic energy?

52. A clock has a pendulum that performs one full swing every 1.0 s (back *and* forth). The object at the end of the pendulum weighs 10.0 N. What is the length of the pendulum?

53. A pendulum of length L_1 has a period $T_1 = 0.950$ s. The length of the pendulum is adjusted to a new value L_2 such that $T_2 = 1.00$ s. What is the ratio L_2/L_1?

54. A pendulum of length 120 cm swings with an amplitude of 2.0 cm. Its mechanical energy is 5.0 mJ. What is the mechanical energy of the same pendulum when it swings with an amplitude of 3.0 cm?

55. A grandfather clock is constructed so that it has a simple pendulum that swings from one side to the other, a distance of 20.0 mm, in 1.00 s. What is the maximum speed of the pendulum bob? Use two different methods. First, assume SHM and use the relationship between amplitude and maximum speed. Second, use energy conservation. Assume $g = 9.81$ m/s^2.

10.9 Damped Oscillations

56. (a) What is the energy of a pendulum ($L = 1.0$ m, $m = 0.50$ kg) oscillating with an amplitude of 5.0 cm? (b) The pendulum's energy loss (due to damping) is replaced in a clock by allowing a 2.0-kg mass to drop 1.0 m in 1 week. What percentage of the pendulum's energy is lost during one cycle?

57. Because of dissipative forces, the amplitude of an oscillator decreases 5.0% in 10 cycles. How many cycles does it take for the *energy* to decrease 5.0%? [*Hint:* Assume the same amount of loss per cycle.]

58. The amplitude of oscillation of a pendulum decays by a factor of 20.0 in 120 s. By what factor has its energy decayed in that time?

COMPREHENSIVE PROBLEMS

59. A pendulum passes $x = 0$ with a speed of 0.50 m/s; it swings out to $A = 0.20$ m. What is the period T of the pendulum? (Assume the amplitude is small.)

60. A person drops a cylindrical steel bar ($Y = 2.0 \times 10^{11}$ Pa) from a height of 1.0 m (distance between the floor and the bottom of the vertically oriented bar). The bar, of length 0.50 m, radius 0.75 cm, and mass 0.70 kg, hits the floor and bounces up, maintaining its vertical orientation. Assuming the collision with the floor is elastic, and that no rotation occurs, what is the maximum

compression of the bar? [*Hint:* At maximum compression, the kinetic energy is zero and the elastic potential energy in the bar is maximum. Start by finding the "spring constant" of the bar.]

61. What is the period of a pendulum formed by placing a horizontal axis (a) through the end of a meterstick (100-cm mark)? (b) through the 75-cm mark? (c) through the 60-cm mark? Assume $g = 9.80$ m/s^2.

62. A naval aviator had to eject from her plane before it crashed at sea. She is rescued from the water by helicopter and dangles

from a cable that is 45 m long while being carried back to the aircraft carrier. What is the period of her vibration as she swings back and forth while the helicopter hovers over her ship?

63. A spider's web can undergo SHM when a fly lands on it and displaces the web. For simplicity, assume that a web obeys Hooke's law (which it does not really as it deforms permanently when displaced). If the web is initially horizontal, and a fly landing on the web is in equilibrium when it displaces the web by 0.030 mm, what is the frequency of oscillation when the fly lands?

64. Spider silk has a Young's modulus of 4.0×10^9 N/m^2 and can withstand stresses up to 1.4×10^9 N/m^2. A single web strand has a cross-sectional area of 1.0×10^{-11} m^2, and a web is made up of 50 radial strands. A bug lands in the center of a horizontal web so that the web stretches downward (Fig. 10.37). (a) If the maximum stress is exerted on each strand, what angle θ does the web make with the horizontal? (b) What does the mass of a bug have to be in order to exert this maximum stress on the web? (c) If the web is 0.10 m in radius, how far down does the web extend?

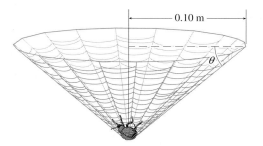

Figure 10.37 Problem 64

65. A pendulum is made from a uniform rod of mass m_1 and a small block of mass m_2 attached at the lower end. (a) If the length of the pendulum is L and the oscillations are small, find the period of the oscillations in terms of m_1, m_2, L, and g. (b) Check your answer to part (a) in the two special cases $m_1 \gg m_2$ and $m_1 \ll m_2$.

66. A thin circular hoop is suspended from a knife edge (Fig. 10.38). Its rotational inertia about the rotation axis (along the knife) is $I = 2mr^2$. Show that it oscillates with the same frequency as a simple pendulum of length equal to the diameter of the hoop.

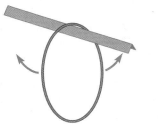

Figure 10.38 Problem 66

67. A hedge trimmer has a blade that moves back and forth with a frequency of 28 Hz. The blade motion is converted from the rotation provided by the electric motor to an oscillatory motion by means of a Scotch yoke (see Conceptual Question 7). The blade moves 2.4 cm during each stroke. Assuming that the blade moves with SHM, what are the maximum speed and maximum acceleration of the blade?

68. The motion of a simple pendulum is approximately SHM only if the amplitude is small. Consider a simple pendulum that is released from a horizontal position ($\theta_i = 90°$ in Fig. 10.22). (a) Using conservation of energy, find the speed of the pendulum bob at the bottom of its swing? Express your answer in terms of the mass m and the length L of the pendulum. Do *not* assume SHM. (b) Assuming (incorrectly, for such a large amplitude) that the motion *is* SHM, determine the maximum speed of the pendulum. Based on your answers, is the period of a pendulum for large amplitudes larger or smaller than that given by Eq. (10-26b)?

69. The gravitational potential energy of a pendulum is $U = mgy$. (a) Taking $y = 0$ at the lowest point, show that $y = L(1 - \cos \theta)$, where θ is the angle the string makes with the vertical. (b) If θ is small, $(1 - \cos \theta) \approx \frac{1}{2}\theta^2$ and $\theta \approx x/L$ (Appendix A.7). Show that the potential energy can be written $U \approx \frac{1}{2}kx^2$ and find the value of k (the equivalent of the spring constant for the pendulum).

70. The simple pendulum can be thought of as a special case of the physical pendulum where all of the mass is at a distance L from the rotation axis. For a simple pendulum of mass m and length L, show that the expression for the angular frequency of a physical pendulum (Eq. 10-28a) reduces to the expression for the angular frequency of a simple pendulum (Eq. 10-26a).

71. A 4.0-N body is suspended vertically from an ideal spring of spring constant 250 N/m. The spring is initially in its relaxed position. Write an equation to describe the motion of the body if it is released at $t = 0$. [*Hint:* Let $y = 0$ at the equilibrium point and take $+y =$ up.]

72. Show using dimensional analysis that the frequency f at which a mass-spring system oscillates is independent of the amplitude A and proportional to $\sqrt{k/m}$. [*Hint:* Start by assuming that f does depend on A (to some power).]

73. A steel piano wire ($Y = 2.0 \times 10^{11}$ Pa) has a diameter of 0.80 mm. At one end it is wrapped around a tuning pin of diameter 8.0 mm (Fig. 10.39). The length of the wire (not including the wire wrapped around the tuning pin) is 66 cm. Initially, the tension in the wire is 381 N. To tune the wire, the tension must be increased to 402 N. Through what angle must the tuning pin be turned?

Figure 10.39 Problem 73

74. When the tension is 402 N, what is the tensile stress in the piano wire in Problem 73? How does that compare to the elastic limit of steel piano wire (8.26×10^8 Pa)?

75. An ice cube slides back and forth without friction in a shallow bowl of radius R (Fig. 10.40). If the amplitude of oscillation is small, what is the period of oscillation?

Figure 10.40 Problem 75

76. A gibbon, hanging onto a horizontal tree branch with one arm, swings with a small amplitude. The gibbon's center of mass is

0.40 m from the branch and its rotational inertia divided by its mass is $I/m = 0.25$ m^2. Estimate the frequency of oscillation.

77. A tightrope walker who weighs 640 N walks along a steel cable. When he is halfway across, the cable makes an angle of 0.040 rad below the horizontal (Fig. 10.41). (a) What is the strain in the cable? Assume the cable is horizontal with a tension of 80 N before he steps onto it. Neglect the weight of the cable itself. (b) What is the tension in the cable when the tightrope walker is standing at the midpoint? (c) What is the cross-sectional area of the cable? (d) Has the cable been stretched beyond its elastic limit (2.5×10^8 Pa)?

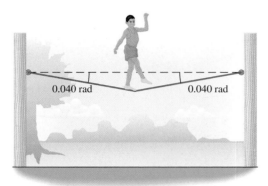

Figure 10.41 Problem 77 (the 0.040-rad angles are greatly exaggerated)

🄯 78. The maximum height of a cylindrical column is limited by the compressive strength of the material; if the compressive stress at the bottom were to exceed the compressive strength of the material, the column would be crushed under its own weight. (a) For a cylindrical column of height h, radius r, made of material of density ρ, calculate the compressive stress at the bottom of the column. (b) Since the answer to part (a) is independent of the radius r, there is an absolute limit to the height of a cylindrical column, regardless of how wide it is. For marble, which has a density of 2.7×10^3 kg/m^3 and a compressive strength of 2.0×10^8 Pa, find the maximum height of a cylindrical column. (c) Is this limit a practical concern in the construction of marble columns? Might it limit the height of a beanstalk?

79. In Problem 8.34 we found that the force of the tibia (shinbone) on the ankle joint for a person (of weight 750 N) standing on the ball of one foot was 2800 N (Fig. 8.68). The ankle joint therefore pushes upward on the bottom of the tibia with a force of 2800 N, while the top end of the tibia must feel a net downward force of approximately 2800 N (ignoring the weight of the tibia itself). The tibia (Fig. 10.42) has a length of 0.40 m, an

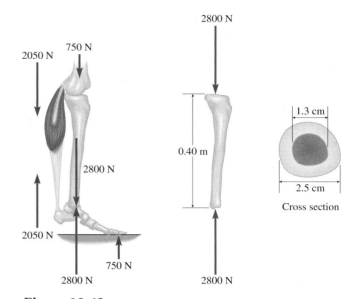

Figure 10.42 Problem 79

average inner diameter of 1.3 cm, and an average outer diameter of 2.5 cm. (The central core of the bone contains marrow that has negligible compressive strength.) (a) Find the average cross-sectional area of the tibia. (b) Find the compressive stress in the tibia. (c) Find the change in length for the tibia due to the compressive forces.

🄯 80. A bungee jumper leaps from a bridge and undergoes a series of oscillations. Assume $g = 9.78$ m/s^2. (a) If a 60.0-kg jumper uses a bungee cord that has an unstretched length of 33.0 m and she jumps from a height of 50.0 m above a river, coming to rest just a few centimeters above the water surface on the first downward descent, what is the period of the oscillations? Assume the bungee cord follows Hooke's law. (b) The next jumper in line has a mass of 80.0 kg. Should he jump using the same cord? Explain.

✦ 81. Resilin is a rubber-like protein that helps insects to fly more efficiently. The resilin, attached from the wing to the body, is relaxed when the wing is down and is extended when the wing is up. As the wing is brought up, some elastic energy is stored in the resilin. The wing is then brought back down with little muscular energy, since the potential energy in the resilin is converted back into kinetic energy. Resilin has a Young's modulus of 1.7×10^6 N/m^2. (a) If an insect wing has resilin with a relaxed length of 1.0 cm and with a cross-sectional area of 1.0 mm^2, how much force must the wings exert to extend the resilin to 4.0 cm? (b) How much energy is stored in the resilin?

ANSWERS TO PRACTICE PROBLEMS

10.1 $2k$ (When the original spring is stretched an amount L, each of the half-springs stretches only $\frac{1}{2}L$. Each of the newly formed springs stretches half as far as the original spring for a given applied force.) **10.2** 1.4×10^{-5} **10.3** 0.18 mm **10.4** 1.3×10^8 Pa

10.5 $-\dfrac{\Delta P}{B} = -\dfrac{1.0 \times 10^7 \text{ Pa}}{2.3 \times 10^9 \text{ Pa}} = -0.0043 = \dfrac{\Delta V}{V}$ and $\Delta V = -0.43\%$ V

10.6 110 m/s^2 at $x = \pm A$ **10.7** $K = 0$, $U_e = 2(mg)^2/k$, $U_g = -(mg)^2/k$, $E = (mg)^2/k$

10.8

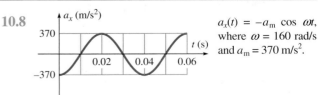

$a_x(t) = -a_m \cos \omega t$, where $\omega = 160$ rad/s and $a_m = 370$ m/s^2.

10.9 1.6 m/s^2 (about 1/6 that of the Earth) **10.10** 0.82 m/s or 1.8 mi/h

Waves

On January 17, 1995, a terrible earthquake struck the Hanshin region of Japan, killing over 6400 people and injuring about 40,000 others. Some 200,000 homes and buildings were damaged, causing the evacuation to shelters of 320,000 people. The heaviest damage occurred in the city of Kobe, including the buckling and collapse of an elevated highway. However, geologists found that the point of origin of the earthquake was 15–20 km below the northern tip of Awaji Island, about 20 km southwest of Kobe. How did the energy released by the earthquake travel that far with enough energy to cause such great devastation?

11.1 WAVES AND ENERGY TRANSPORT

Physicists use only a few basic models to describe the physical world. One such model is the particle: a point-like object with no inner structure and with certain characteristics such as mass and electrical charge. Another basic model is the **wave**. Water waves are familiar examples. When a pebble is dropped into a pond, it disturbs the surface of the water. Ripples on the surface of the pond travel away from the spot where the pebble landed. This traveling disturbance of the water from its equilibrium position is a wave on the surface of the water.

Any wave is characterized as some sort of "disturbance" that travels away from its source. In Chapters 11 and 12, we concentrate on mechanical waves traveling through a material medium, such as water waves, sound waves, and the seismic waves caused by earthquakes. Particles in the medium are disturbed from their equilibrium positions as the wave passes, returning to their equilibrium positions after the wave has passed. In Chapter 22, we discuss electromagnetic waves such as radio waves and light waves, in which the disturbance consists of oscillating electromagnetic fields. Two of our five human senses are wave detectors: the ear is sensitive to the tiny fluctuations in air pressure caused by compressional waves in air (audible sound) and the eye is sensitive to electromagnetic waves in a certain frequency range (visible light).

Energy Transport

Suppose we drop a pebble into a still pond. The kinetic energy of the pebble just before it hits the pond is partly converted into the energy carried off by the water wave. That waves carry energy is clear to anyone who has been surfing or swimming in the ocean. Speaking of surfing, information on the Internet is carried by waves of various sorts: electrical waves in wires, microwaves between Earth and communications satellites, light waves in optical fibers. Microwaves in ovens carry energy from their source to the food; the electromagnetic energy of the microwaves is absorbed by water molecules in the food and appears as thermal energy. Electromagnetic waves from the Sun bring the energy that makes life possible. Seismic waves carry energy released by an earthquake to other parts of the Earth, sometimes with devastating results.

When a baseball pitcher throws a ball to the catcher, the ball carries energy with it (Fig. 11.1a). The pitcher gives the ball kinetic energy; the catcher receives the energy when the ball hits his hand and his hand recoils. Suppose instead that they hold a rope stretched between them. If the pitcher suddenly moves his hand up and down quickly, a wave pulse travels along the rope until it reaches the catcher's hand (Fig. 11.1b). Once again, the

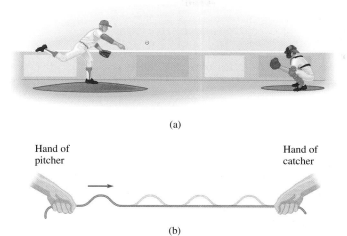

(a)

Hand of
pitcher

Hand of
catcher

(b)

Figure 11.1 Two different ways to transfer energy. (a) The baseball carries energy from the pitcher to the catcher. (b) A wave pulse also carries energy from the pitcher to the catcher, but the piece of rope held by the pitcher's hand does not move to the catcher's hand.

pitcher sends the energy and the catcher receives it when the rope makes his hand recoil. However, in this case the pitcher is still holding his end of the rope; it never leaves his hand. Energy is transferred *without any matter moving from the pitcher to the catcher*.

Physics at Home

Stretch a heavy rope between yourself and a friend and test out the transfer of energy from one to the other. Does your hand recoil when the pulse arrives?

Similarly, seismic waves travel away from the *focus* of an earthquake (the point of origin) both through the Earth (*body waves*) and along the Earth's crust (*surface waves*), transporting vibrations and energy. However, the material through which the waves travel is *not* transported. Most earthquake damage is caused by seismic waves rather than caused by fault movement. In the Hanshin earthquake, damage to buildings was caused by seismic waves at distances over 100 km from the *epicenter* (the point on the surface directly above the focus) but the motion of the vibrating particles in the ground never moved more than about 1.5 m.

The sound of thunder travels for miles in all directions, but none of the air molecules zapped by lightning travel more than a meter or so during the few seconds that it takes the sound to reach our ears. A wave can transmit energy from one point to another without transporting any matter between the two points.

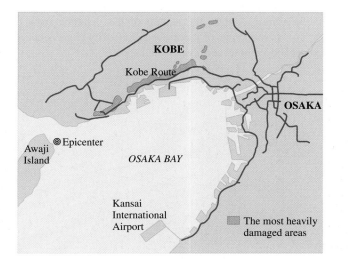

Physics at Home

Observe carefully what happens when you snap your fingers. You start by pressing your thumb against your fingers and then sideways, the thumb in one direction and the fingers in the opposite direction. Initially friction keeps them from moving sideways, but suddenly they slip, releasing the built-up energy partly as a sound wave.

Similarly, the rock on two sides of a fault line is pressed together and sideways. Friction keeps them from moving sideways as elastic (or strain) energy builds up. Then suddenly they slip, releasing a tremendous amount of energy largely in the form of seismic waves that carry vibrations far from the focus of the earthquake.

Intensity

For a wave that travels in a three-dimensional medium (such as sound waves or seismic waves traveling through the Earth), the **intensity** (symbol I, SI unit W/m^2) is a measure of the *average power per unit area* carried by the wave past a surface perpendicular to the wave's direction of propagation. For example, if a sound wave's intensity is a fairly loud 10^{-5} W/m^2 when it reaches the eardrum and the area of the eardrum is 10^{-4} m^2, then the power delivered to the eardrum is 10^{-9} W (assuming that all the energy incident on the eardrum is absorbed). The energy absorbed by the eardrum at this rate in *one hour* would be

$$10^{-9}\,W \times 3600\,s \approx 4\,\mu J$$

The human ear is a very sensitive detector indeed.

For most waves, the intensity decreases as the distance from the source increases. Some of the energy can be absorbed (dissipated) by the wave medium. The amount of energy absorbed depends on the medium. Air absorbs relatively little sound energy, which is why we can hear sounds generated far away.

Another reason that intensity decreases with distance is that, as the wave spreads out, the energy gets spread over a larger and larger area. Consider a point source emitting a wave uniformly in all directions—an *isotropic* source (Fig. 11.2). The average power (energy per unit time) emitted is constant. Imagine a sphere surrounding the source; the rate at which energy passes through the surface of the sphere is the same no matter what the radius. The surface area of a sphere is $4\pi r^2$, so as the wave moves farther from the source, the energy spreads out over a larger and larger area. Thus, the power per unit area (intensity) decreases with distance. Assuming that no energy is absorbed by the medium and there are no obstacles to reflect or absorb sound,

$$I = \frac{power}{area} = \frac{P}{4\pi r^2} \qquad (11\text{-}1)$$

(point source emitting uniformly in all directions; no reflection or absorption)

Therefore, if energy absorption by the medium can be neglected, the intensity of the sound is inversely proportional to the square of the distance from the source. This "inverse square law" is the result of a conserved quantity (here, energy) radiating uniformly from a point source in three-dimensional space.

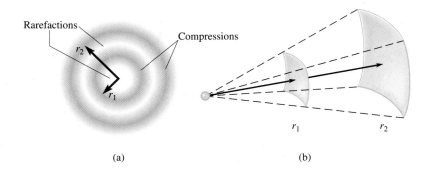

Figure 11.2 (a) A point source of sound radiating energy uniformly in all directions. (b) The intensity at a distance r_2 is smaller than the intensity at a distance r_1 since the same power is spread out over a greater area.

Rarefactions

Compressions

(a)

(b)

11.2   TRANSVERSE AND LONGITUDINAL WAVES

A Slinky toy can be used to demonstrate two different kinds of wave. In a **transverse** wave, the motion of particles in the medium is perpendicular to the direction of propagation of the wave. To send a transverse wave down a Slinky, wiggle the end of the Slinky back and forth in a direction perpendicular to the length of the Slinky (Fig. 11.3a). In a **longitudinal** wave, the motion of particles in the medium is along the same line as the direction of propagation of the wave. To send a longitudinal wave down the Slinky, jiggle the end in and out along its length to alternately stretch and compress the coils (Fig. 11.3b). A red dot painted on one coil of the Slinky helps illustrate the difference. In a transverse wave, the dot moves back and forth about a fixed position with its motion perpendicular to the direction of propagation of the wave; in a longitudinal wave, the dot also moves back and forth about a fixed position but along the direction of propagation of the wave. In both cases, the wave itself moves from one end of the Slinky to the other while the dot is moving about its fixed position.

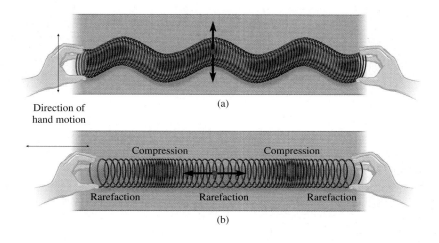

Direction of
hand motion

(a)

Compression   Compression

Rarefaction   Rarefaction   Rarefaction

(b)

Figure 11.3 (a) Transverse and (b) longitudinal waves on a Slinky

Physics at Home

Ask a friend to sit at a table and hold one end of a long, loose spring (or a Slinky). With the spring supported by the table surface, grasp the other end and stretch the spring. Figure out how to move your hand to send transverse and longitudinal waves down the spring.

The Slinky—or any long spring—is a better approximation to solid materials than the stretched rope. In solids both types of waves can exist; a transverse wave results from a shear disturbance and a longitudinal wave from a compressional disturbance. An earthquake generates both longitudinal and transverse seismic waves. The two travel at different speeds so the time lapse between the arrival of the two wave types at a detector gives a clue to the distance to the earthquake's focus.

Fluids can be compressed but, because they flow, they do not sustain shear stresses. Therefore, longitudinal waves travel through fluids but transverse waves do not. However, gravity or surface tension can provide the transverse restoring force that allows a transverse wave to travel *along the surface* of a liquid.

A sound wave is longitudinal; each small volume of air vibrates back and forth along the direction of travel of the wave. Air molecules are compressed together in some places and more thinly spaced (*rarefied*) in others; the air has regions of higher and lower density called **compressions** and **rarefactions** (see Fig. 11.3b).

Seismic body waves can be either longitudinal or transverse. Longitudinal body waves (*P waves*) are the fastest seismic waves (typically 4–8 km/s). They are similar to sound waves in air: particles in the Earth's interior are pushed together and pulled apart in the same direction that the wave propagates (Fig. 11.4a). Transverse body waves

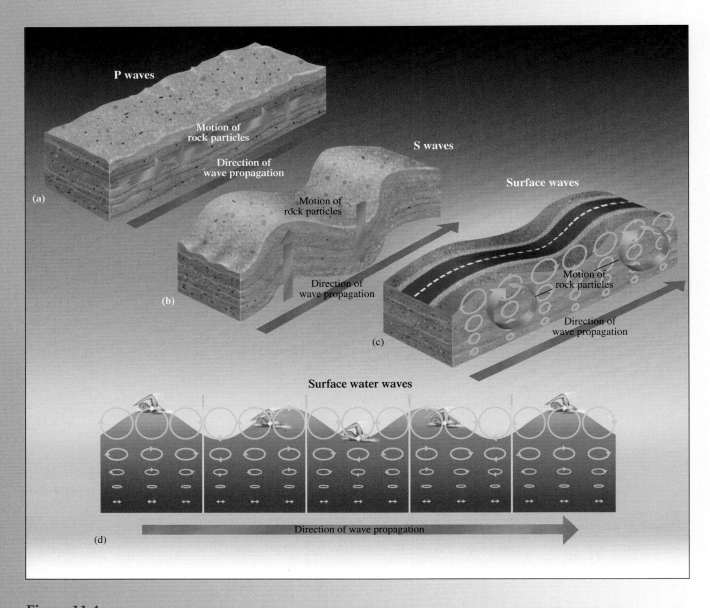

Figure 11.4 The motion of ground particles in (a) P waves, (b) S waves, and (c) one kind of surface wave. (d) Motion of a swimmer as a water wave passes by.

(*S waves*) travel more slowly—typically 2–5 km/s. In an S wave, particles in the Earth's interior vibrate at right angles to the direction that the wave travels (Fig. 11.4b). By measuring the time between the first arrivals of the two types of waves at different detection stations, geologists are able to determine the origin of the earthquake.

Waves That Combine Transverse and Longitudinal Motion

Not all seismic waves are purely transverse or purely longitudinal. In a surface wave, the ground near the surface rolls approximately in a circle. Thus, the motion of the ground has components both parallel and perpendicular to the direction of propagation. The transverse component can either be up and down (as shown in Fig. 11.4c) or side to side. The motion of the ground is greatest at the surface. Surface waves are generally slower than body waves.

Ocean waves are similar to the surface seismic wave shown in Fig. 11.4c. Deep underwater, the wave is mostly longitudinal (Fig. 11.4d); as the wave passes, water moves back and forth along the direction of propagation of the wave. Higher up, the wave has both transverse and longitudinal components; water moves in an oval as the wave passes. A swimmer floating at the surface moves approximately in a circle. The air above the surface presents little resistance, so water swells upward more easily there and then is pulled back downward by gravity (or, for small amplitudes, by surface tension). If the wave were purely transverse the swimmer would move only up and down, but the longitudinal component of the wave superimposed on the transverse component results in an approximately circular motion. When the wave gets close to shore, the crest often collapses or *breaks*; the motion of the water particles is then much more complex.

When a guitar string is plucked *gently*, the wave on the string is almost purely transverse; stretching of the string is negligible. When it is plucked more forcefully, the resulting wave is a combination of transverse and longitudinal waves. At any instant, the string is stretched more in some places than in others; a point on the string has longitudinal motion as well as transverse motion.

11.3 SPEED OF TRANSVERSE WAVES ON A STRING

The speed of a mechanical wave depends on properties of the wave medium. What properties of a string determine the speed of a transverse wave moving along it? Suppose that a string of length L and mass m is under tension F. In Problem 41, you can show that $\sqrt{FL/m}$ is the only combination of those three quantities with the correct units for speed. There could be a dimensionless constant multiplier, but a derivation using more advanced mathematics shows that the constant is 1; the speed of a transverse wave on a string is

$$v = \sqrt{\frac{FL}{m}} \qquad (11\text{-}2)$$

We can rewrite Eq. (11-2) in another form. Length and mass are not independent; for a given string composition and diameter (say, a yellow brass string of 0.030 in. diameter), the mass of the string is proportional to its length. By defining the **linear mass density** (mass per unit length) of the string to be

$$\mu = m/L \qquad (11\text{-}3)$$

the speed of a transverse wave on a string can be written

$$v = \sqrt{\frac{F}{\mu}} \qquad (11\text{-}4)$$

An advantage of Eq. (11-4) is that it shows clearly that the wave speed depends on *local* properties of the medium; it does not depend on how much of the medium there is. The wave speed in the vicinity of some point P, for instance, does not depend on how long the string is; only properties of the string in the immediate vicinity of point P can determine how fast the wave travels past that point.

Note that as tension increases, wave speed increases; as mass density increases, wave speed decreases. A somewhat more general way to think about it, applicable to other waves as well, is *more restoring force makes faster waves; more inertia makes slower waves.*

The speed at which a wave propagates is not the same as the speed at which a particle in the medium moves. Suppose a horizontal string is stretched along the *x*-axis and a transverse pulse in the *y*-direction is sent down the string. The speed of propagation of the wave v is the speed at which the *pattern* or disturbance moves along the string (in the *x*-direction); for a uniform string, the wave speed is constant. A point on the string vibrates up and down in the ±*y*-direction with a *different* speed that is *not* constant.

The wave speed is not the same as the speed of a particle in the wave medium.

Example 11.1

A Piñata

A string of length 0.50 m has a mass of 125 mg. The string is attached to the ceiling and a piñata of mass 0.500 kg hangs from the other end. A tall child grabs the piñata and shakes it to see if there is candy inside; as a result, a transverse pulse travels up the string toward the ceiling. At what speed does the pulse travel?

Strategy We start with a diagram of the situation (Fig. 11.5). The piñata puts the string under tension. The tension in the string is equal to the weight of the piñata. The mass and length of the string are given, so the linear mass density can be found. Then we can find the wave speed.

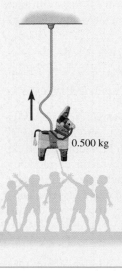

Figure 11.5
A transverse pulse travels up the string supporting a piñata.

0.500 kg

Solution The speed of a transverse wave on a string is given by Eq. (11-4):

$$v = \sqrt{F/\mu}$$

where F is the tension in the string and μ is the linear mass density of the string. The tension is equal to the weight hanging on the string:

$$F = Mg$$

The linear mass density of the string is mass per unit length ($\mu = m/L$). Substituting the tension and mass density, we have

$$v = \sqrt{\frac{F}{m/L}} = \sqrt{\frac{(Mg)L}{m}}$$

$$= \sqrt{\frac{0.500 \text{ kg} \times 9.8 \text{ m/s}^2 \times 0.50 \text{ m}}{125 \times 10^{-6} \text{ kg}}} = 140 \text{ m/s}$$

Discussion The *weight* of the string (mg) is negligible in comparison with the weight hanging from the end of the string (Mg). That is not always the case, as can be seen in Practice Problem 11.1.

Practice Problem 11.1   Initial velocity of another wave pulse traveling on a string

A string of length 10.0 m has a linear mass density of 25 g/m. The string is fixed at the top and has an object of mass 0.200 kg hanging from the bottom. (a) What is the *initial* wave speed of a pulse sent up the string from the bottom? (b) What is the speed of the pulse as it approaches the top of the string? [*Hint*: Does the weight of the string itself affect the tension in either case?]

11.4   PERIODIC WAVES

A **periodic** wave repeats the same pattern over and over, each repeating section transporting the energy that was used to generate it. A periodic water wave can be produced by steadily dropping a series of pebbles into the water; a periodic wave on a cord can be produced by taking one end of the cord and moving it up and down, over and over, in a repeating pattern. As the wave propagates along the cord, every point on the cord oscillates with the same up and down pattern, though with a time delay that depends on the wave speed. While musical sounds are often periodic waves, noise is *aperiodic*. The human voice makes a periodic sound wave when a vowel is sung at a steady pitch (constant frequency); most of the consonant sounds are aperiodic (Fig. 11.6).

The terminology for periodic waves is similar to that used for uniform circular motion (Chapter 5) and for simple harmonic motion (SHM, Chapter 10). At any given point, the wave repeats itself after a time T called the **period**. The inverse of the period is the **frequency** f.

$$f = \frac{1}{T} \quad (\text{SI unit Hz} = \text{s}^{-1}) \tag{5-7}$$

The frequency tells how often the pattern of motion repeats itself at any single point. For instance, if the frequency is 20 Hz, then there are 20 repetitions, or cycles, per second. Each cycle takes a time $T = 1/f = 0.05$ s. The angular frequency is $\omega = 2\pi f$ and is measured in rad/s.

During one period T, a periodic wave traveling at speed v moves a distance vT. In Fig. 11.7, note that, at any instant, points separated by a distance vT along the direction of propagation of a wave move "in sync" with each other. Thus, vT is the *repetition distance* of the wave, just as the period is the *repetition time*. This distance is called the **wavelength** (symbol λ, the Greek letter lambda).

$$\lambda = vT \tag{11-5}$$

Combining this relation and the expression for frequency, we obtain

$$v = \frac{\lambda}{T} = f\lambda \tag{11-6}$$

Equations (11-5) and (11-6) are true for all periodic waves, no matter how the wave is produced or what the shape of the wave.

The maximum displacement of any particle from its equilibrium position is the **amplitude** A of the wave. For a sinusoidal wave traveling along a stretched string in the x-direction, the amplitude A is the maximum displacement in the positive or negative y-direction.

$$y_m = \pm A$$

For surface water waves, the amplitude is the height of a crest (a high point) above or the depth of a trough (a low point) below the undisturbed water level.

Harmonic waves are a special kind of periodic wave in which the disturbance is sinusoidal (either a sine or cosine function). In a harmonic transverse wave on a string, for instance, every point on the string moves in SHM with the same amplitude and angular frequency, although different points reach their maximum displacements at different times. The maximum speed and maximum acceleration of a point on the string depend on both the angular frequency and the amplitude of the wave:

$$v_m = \omega A \tag{10-21}$$

$$a_m = \omega^2 A \tag{10-22}$$

Since the total energy of an object moving in SHM is proportional to the amplitude squared (Section 10.5), the total energy of a harmonic wave is proportional to the square of its amplitude. That turns out to be a general result not limited to harmonic waves:

The intensity of a wave is proportional to the square of its amplitude.

11.5 MATHEMATICAL DESCRIPTION OF A WAVE

A wave is represented mathematically by a variation in some quantity (such as pressure or displacement) that is described as a function of both position and time. For a transverse wave on a guitar string, the function specifies the displacement of each point on the string from its equilibrium position. If the string is oriented along the x-axis and the displacement of any point on the string is in the $\pm y$-direction, then the wave is described by a function of two variables: $y(x, t)$.

Consider a long stretched string along the x-axis. One end of the string (at $x = 0$) is moved by an external agent according to some function $y = h(t)$; as a result, a transverse

(a)

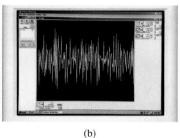
(b)

Figure 11.6 Sound wave pattern produced by a singer (a) singing the vowel "ah" and (b) hissing the consonant "s."

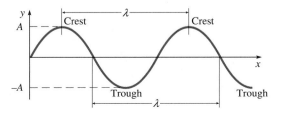

Figure 11.7 Sinusoidal wave moving with speed v in the x-direction. The amplitude A and the wavelength λ are shown.

wave is produced that travels in the $+x$-direction with wave speed v. If the wave retains the same shape as it moves down the string, then the motion of any point x on the string copies the motion of the left end after a time delay x/v (the time it takes the wave to travel a distance x at speed v—see Fig. 11.8). Thus, $y(x, t) = h(t - x/v)$. Even though the function that describes the wave has two variables (x and t), these variables must occur in the particular combination $(t - x/v)$ in order to describe a wave that retains its form as it propagates in the $+x$-direction. For a wave moving in the $-x$-direction, the variables would occur in the combination $(t + x/v)$. A wave that retains its shape as it moves in a single direction is called a *traveling wave*.

+x-direction: $y(x, t) = h(t - x/v)$
−x-direction: $y(x, t) = h(t + x/v)$

A Harmonic Traveling Wave

Suppose the motion of the left end of the string is described by $y = A \cos \omega t$. By substituting $(t - x/v)$ for t, we obtain the function that describes the motion of *any* point $x > 0$:

$$y(x, t) = A \cos \omega(t - x/v)$$

To simplify the writing, we introduce a constant called the **wavenumber** (symbol k, SI unit rad/m):

$$k = \frac{\omega}{v} = \frac{2\pi f}{v} = \frac{2\pi}{\lambda} \qquad (11\text{-}7)$$

Then the equation for the harmonic wave can be written

$$y(x, t) = A \cos (\omega t - kx) \qquad (11\text{-}8)$$

The argument of the sine or cosine function, $(\omega t \pm kx)$, is called the **phase** of the wave at x and t. Phase is measured in units of angle (usually radians). The phase of a wave at a given point and instant of time tells us how far along that point is in the repeating pattern of its motion. Since a sine or cosine function repeats every 2π radians, the motions of two different points x_1 and x_2 that differ in phase by an integer times 2π are exactly the same; the points move "in sync" or *in phase* with each other. The distance between the two points is an integral number of wavelengths:

If

$$k(x_2 - x_1) = 2\pi n \quad \text{(where } n \text{ is any integer)}$$

then

$$x_2 - x_1 = \frac{2\pi n}{k} = \frac{2\pi n}{2\pi/\lambda} = n\lambda$$

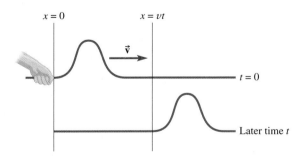

$x = 0$

$x = vt$

$\vec{v}$

$t = 0$

Later time t

Figure 11.8 A wave pulse, with the same shape, at successive times. The motion of the point x repeats the motion of the point $x = 0$ with a time delay $\Delta t = x/v$.

Example 11.2

A Traveling Wave on a String

A wave on a string is described by $y(x, t) = a \sin (bt + cx)$, where a, b, and c are constants. (a) Does this wave retain its shape as it travels? (b) In what direction does the wave travel? (c) What is the wave speed?

Strategy   We try to manipulate the function to see if it can be written as a function of either $(t - x/v)$ or $(t + x/v)$ as in the general harmonic wave equation $y(x, t) = A \cos \omega(t - x/v)$. The wave speed v does not appear explicitly in the function as

continued on next page

Example 11.2 *continued*

written, but it may be some combination of the other constants in the function. We are starting from an equation similar to that of Eq. (11-8), $y(x, t) = A \cos(\omega t - kx)$.

Solution The coefficient of t in our equation should be the constant that represents ω. Factoring out that constant, we have

$$y(x, t) = a \sin b\left(t + \frac{cx}{b}\right) = a \sin b\left(t + \frac{x}{b/c}\right)$$

Now we see that $y(x, t)$ is a function of $t + x/v$, where $v = b/c$,

$$y(x, t) = a \sin b\left(t + \frac{x}{v}\right) \text{ where } v = \frac{b}{c}$$

Therefore: (a) yes, the wave retains its shape since it is a function of $(t + x/v)$; (b) it travels in the $-x$-direction since the t and x/v terms have the same sign; and (c) the wave speed is b/c.

Discussion Before being completely satisfied with this solution, it is a good idea to check that b/c has the right units for wave speed. The two terms bt and cx that are added together must have the same units. In SI, the argument of a sine function is measured in radians. Then b is measured in rad/s and c is measured in rad/m. Then the units of b/c are (rad/s)/(rad/m) = m/s, which is correct for wave speed.

Practice Problem 11.2   Another traveling wave on a string

A wave on a string is described by

$$y(x,t) = 0.0050 \sin(4.0t - 0.50x)$$

where x and y are in meters and t is in seconds. (a) Does this wave retain its shape as it travels? (b) In what direction does the wave travel? (c) What is the wave speed?

11.6   GRAPHING WAVES

To graph a one-dimensional wave $y(x, t)$, only one of the two independent variables (x, t) can be plotted. The other must be "frozen"; it is treated as a constant. If x is held constant, then one particular point (determined by the value of x) is singled out; the graph shows the motion of *that point* as a function of time (Fig. 11.9a). If instead t is held constant and y is plotted as a function of x, then the graph is like a snapshot—an instantaneous picture of what the wave looks like *at that particular instant* (Fig. 11.9b).

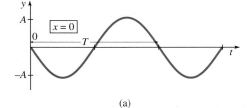

(a)

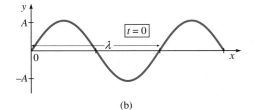

(b)

Figure 11.9 Two graphs of a harmonic wave $y(x, t) = A \sin(\omega t - kx)$ on a string. (a) The vertical displacement of a particular point on the string ($x = 0$) as a function of time. (b) The vertical displacement as a function of horizontal position at a single instant of time ($t = 0$).

Example 11.3

A Transverse Harmonic Wave

A transverse harmonic wave travels in the $+x$-direction on a string at a speed of 5.0 m/s. Figure 11.10 shows a graph of $y(t)$ for the point $x = 0$. (a) What is the period of the wave? (b) What is the wavelength? (c) What is the amplitude? (d) Write the

function $y(x, t)$ that describes the wave. (e) Sketch a graph of $y(x)$ at $t = 0$.

Strategy Since the graph uses time as the independent variable, the period can be read from the graph as the time for one

continued on next page

Example 11.3 *continued*

cycle. The wavelength is the distance traveled by the wave during one period. The amplitude can be read from the graph as the maximum displacement. These are all the constants needed to write the function $y(x, t)$. We do have to think about the direction of travel and whether to write sine or cosine.

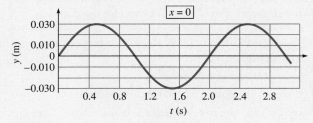

Figure 11.10
Graph of a transverse harmonic wave

Solution (a) The period T is the time for one cycle. From the graph, $T = 2.0$ s.

(b) The wavelength λ is the distance traveled by the wave at speed $v = 5.0$ m/s during one period:

$$\lambda = vT = 5.0 \text{ m/s} \times 2.0 \text{ s} = 1.0 \times 10^1 \text{ m}$$

(c) The amplitude A is the maximum displacement from equilibrium. From the graph, $A = 0.030$ m.

(d) Figure 11.10 is a sine function. The motion of the point $x = 0$ is

$$y(t) = A \sin 2\pi \frac{t}{T}$$

Since the wave moves in the $+x$-direction, a point at $x > 0$ duplicates the motion of $x = 0$ with a time delay of $\Delta t = x/v$ (the time for the wave to travel a distance x). Then

$$y(x, t) = A \sin 2\pi \frac{t - x/v}{T}$$

where $v = 5.0$ m/s and $T = 2.0$ s.

(e) Substituting $t = 0$,

$$y(x) = A \sin \left(-2\pi \frac{x}{vT} \right)$$

Substituting $vT = \lambda$ and using the identity $\sin(-\theta) = -\sin \theta$ (Appendix A.7), we have

$$y(x, t = 0) = -A \sin 2\pi \frac{x}{\lambda}$$

A graph of this function is an inverted sine function with amplitude $A = 0.030$ m and wavelength $\lambda = vT = 5.0$ m/s $\times$ 2.0 s $= 1.0 \times 10^1$ m.

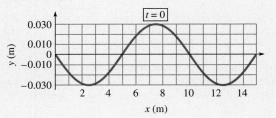

Discussion Figure 11.10 shows that the point $x = 0$ is initially at $y = 0$ and then moves up (in the $+y$-direction) until it reaches the crest (maximum y) at $t = 0.50$ s. Imagine the graph in (e) to represent the first frame (at $t = 0$) of a movie of the wave. Since the wave moves to the right, the sinusoidal patterns shifts a little to the right in each successive frame. The point $x = 0$ moves up until it reaches the crest when the wave has traveled 2.5 m to the right. Since the wave speed is 5.0 m/s, the point $x = 0$ reaches the crest at $t = \dfrac{2.5 \text{ m}}{5.0 \text{ m/s}} = 0.50$ s.

Practice Problem 11.3   Another harmonic transverse wave

A wave is described by $y(x, t) = (1.2 \text{ cm}) \sin(10.0\pi t + 2.5\pi x)$, where x is in meters and t is in seconds. (a) Sketch a graph of $y(t)$ at $x = 0$. (b) Sketch a graph of $y(x)$ at $t = 0$. (c) What is the period of the wave? (d) What is the wavelength? (e) What is the amplitude? (f) What is the speed of the wave? (g) In what direction does the wave move?

11.7   PRINCIPLE OF SUPERPOSITION

Suppose two waves of the same type pass through the same region of space. Do the waves affect each other? If the amplitudes of the waves are large enough, then particles in the medium are displaced far enough from their equilibrium positions that Hooke's law (restoring force $\propto$ displacement) no longer holds; in that case the waves *do* affect each other. However, for small amplitudes, the waves can pass through each other and emerge *unchanged*. More generally, when the amplitudes are not too large, the principle of superposition applies:

> **Principle of superposition**
> When two or more waves overlap, the net disturbance at any point is the sum of the individual disturbances due to each wave.

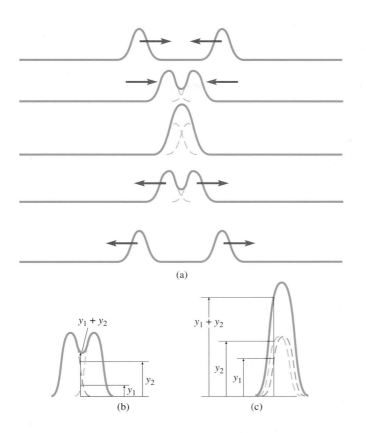

(a)

(b)   (c)

Figure 11.11 (a) Two identical wave pulses traveling toward and through each other. (b), (c) Details of the wave pulse summation; dashed lines are the separate wave pulses and solid line is the sum.

Suppose two wave pulses are traveling toward each other on a string (Fig. 11.11a). If one of the pulses (acting alone) would produce a displacement y_1 at a certain point and the other would produce a displacement y_2 at the same point, the result when the two overlap is a displacement of $y_1 + y_2$. Figures 11.11b, c show in greater detail how y_1 and y_2 add together to produce the net displacement when the pulses overlap. The dashed curves represent the individual pulses; the solid line represents the superposition of the pulses. In Fig. 11.11b the pulses are starting to overlap and in Fig. 11.11c they are just about to coincide.

The wave pulses pass right through each other without affecting each other; once they have separated, their shapes and heights are the same as before the overlap (Fig. 11.11a). The principle of superposition enables us to distinguish two voices speaking in the same room at the same time; the sound waves pass through each other unaffected.

Example 11.4

Superposition of Two Wave Pulses

Two identical wave pulses travel at 0.5 m/s toward each other on a long cord (Fig. 11.12). Sketch the shape of the cord at $t = 1.0$, 1.5, and 2.0 s.

Strategy   We start by sketching the two pulses in their new positions at each time given. Wherever they overlap, we

apply superposition by adding the individual displacements at each point to find the net displacement of the cord at that point.

Solution   Using graph paper, we draw the wave pulses at $t = 0$ (Fig 11.13a). At $t = 1.0$ s, each pulse has moved 0.5 m toward the other. The leading edges of the pulses are just starting to overlap (Fig. 11.13b). At $t = 1.5$ s, each pulse has moved another 0.25 m; the crests overlap exactly. By adding the displacements point by point, we see that the string has the shape of a single pulse twice as high as either of the individual pulses (Fig. 11.13c). At $t = 2.0$ s, the pulses have each moved another 0.25 m (Fig. 11.13d).

Discussion   When the two pulses exactly overlap, the displacement of points on the string is larger than for corresponding points on a single pulse because we add displacements *in the same direction* ($y > 0$ for both). However, superposition does not *always* produce larger displacements (see Practice Problem 11.4).

Practice Problem 11.4   Superposition of two opposite wave pulses

Repeat Example 11.4, except now let the pulse on the right be inverted (Fig. 11.14). [*Hint:* Points on the string below the x-axis have negative displacements ($y < 0$).]

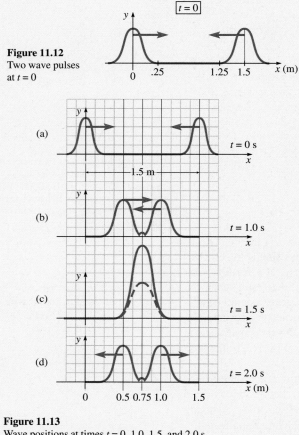

Figure 11.12
Two wave pulses at $t = 0$

Figure 11.13
Wave positions at times $t = 0$, 1.0, 1.5, and 2.0 s

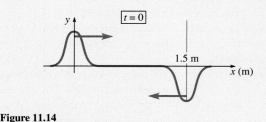

Figure 11.14
Wave pulses for Practice Problem 11.4

11.8   REFLECTION AND REFRACTION

Reflection

A traveling wave sends energy in the direction of propagation without sending any backward, as long as the wave medium is uniform. At an abrupt boundary between one medium and another, **reflection** occurs; a reflected wave carrying some of the energy of the incident wave travels backward from the boundary. A sound wave in air, for instance, does not produce a reflection as long as it travels through air; it *does* reflect when it reaches a wall. Reflection occurs whenever the wave speed changes abruptly. The amount of energy reflected depends on the ratio of the wave speeds in the two media; the farther the ratio is from 1, the more energy is reflected. There is no reason why sound waves cannot travel through a wall; in fact, they do. However, the speed of sound in plaster or wallboard is considerably faster than the speed in air; consequently, much of the sound wave's energy is reflected rather than transmitted through the wall.

The reflected wave can be inverted or not, depending on which side of the boundary has the faster wave speed. Let's look at an extreme example: a string tied to a wall. If you send a wave pulse down the string, the reflected pulse is inverted (Fig. 11.15). One way to understand why it is inverted is that, by the principle of superposition, the shape of the string at any point is the sum of the incident and reflected waves. This is true even at the fixed point at the end. The only way the end can stay in place is if the reflected wave is an upside down version of the incident wave. Another way to understand the inversion is by considering the force exerted on the string by the wall. When an upward pulse reaches the fixed end, the force exerted by the string on the wall has an upward component. By Newton's third law, the wall exerts a force on the string with a downward component. This downward force produces a downward reflected pulse.

Now, instead of tying the string to the wall, tie it to another string with an enormous linear mass density—so large that its motion is too small to measure. The original string doesn't know the difference; it just knows that one end is fixed in place. The second string with the huge density has a much slower wave speed than the first string. Now make the mass density of the second string not huge, but still greater than the first string. The greater inertia inhibits the motion of the boundary point and causes the reflected wave to be inverted. In general, when a wave reflects from a boundary with a region of slower wave speed, the reflected wave is inverted. On the other hand, when a wave reflects from a boundary with a region of *faster* wave speed, the reflected wave is *not* inverted.

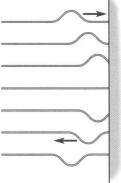
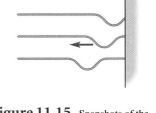

Figure 11.15 Snapshots of the reflection of a wave pulse from a fixed end. The reflected pulse is upside down.

A wave that reflects from a boundary with a *slower* medium is inverted.

Refraction

When there is an abrupt change in wave speed, an incident wave splits up at the boundary; part is reflected and part is transmitted past the boundary into the other medium. The frequencies of both the reflected and transmitted waves are the same as the frequency of the incident wave. To understand why, think of a wave incident on the knot between two different strings. Both the reflected and the transmitted waves are generated by the up-and-down motion of the knot; the knot vibrates at the frequency dictated by the incident wave. However, since the wave speed changes at the boundary, *the wavelength of the transmitted wave is not the same* as the wavelength of the incident and reflected waves. Since $v = \lambda f$ and the frequencies are the same,

$$f = \frac{v_1}{\lambda_1} = \frac{v_2}{\lambda_2} \tag{11-9}$$

Equation (11-9) applies to any kind of wave and is of particular importance in the study of optics.

When a wave passes from one medium into another, the frequency of the transmitted wave is the same as that of the incident wave.

Example 11.5

Wavelength in Air and Under Water

A horn near the beach emits a 440-Hz sound wave. (a) What is the wavelength of the sound wave in air? The speed of sound in air is 340 m/s. (b) What is the wavelength of the sound wave in seawater? The speed of sound in seawater is 1520 m/s.

Strategy The *frequency* of the sound wave in water is the same as in air. The wavelengths depend on both the frequency and the speed of sound in the medium. Sound travels faster in solids and liquids than in gases; during one period, the wave travels farther in water than it does in air, so the wavelength is longer in water.

Solution The wavelength in air is related to the speed of sound in air and the frequency:

$$\lambda_{air} = v_{air}T = \frac{v_{air}}{f}$$

Substituting numerical values,

$$\lambda_{air} = \frac{340 \text{ m/s}}{440 \text{ Hz}} = 0.77 \text{ m}$$

continued on next page

Example 11.5 *continued*

The wave in the water has the *same frequency*, but the speed of sound is different:

$$\lambda_{water} = \frac{v_{water}}{f} = \frac{1520 \text{ m/s}}{440 \text{ Hz}} = 3.5 \text{ m}$$

Discussion   The wavelength in water is longer, as expected. As a quick check, the ratio of the wavelengths should be equal to the ratio of the wave speeds:

$$\frac{0.77 \text{ m}}{3.5 \text{ m}} = 0.22; \quad \frac{340 \text{ m/s}}{1520 \text{ m/s}} = 0.22$$

Practice Problem 11.5   Working on the railroad

A railroad worker, driving in spikes, misses the spike and hits the iron rail; a sound wave travels through the air and through the rail. (We ignore the *transverse* wave that also travels in the rail.) The wavelength of the sound in air is 0.548 m. The speed of sound in air is 340 m/s; the speed of sound in iron is 5300 m/s. A coworker is standing 980 m away. (a) What is the frequency of the wave? (b) What is the wavelength of the sound wave in the rail?

A transmitted wave not only has a different wavelength than the incident wave; it also travels in a different direction unless the incident wave's direction of propagation is along the *normal* (the direction perpendicular to the boundary). This change in propagation direction is called **refraction**. If the change in wave speed is gradual, then the change in direction is gradual as well. The speed of ocean waves depends on the depth of the water; the waves are slower in shallower water. As waves approach the shore, they gradually slow down; as a result, they gradually bend until they reach shore nearly head-on. A sudden change in wave speed causes a sudden refraction. When a beam of light passes from air into water, in which the speed of light is slower, it bends toward the normal (Fig. 11.16a).

The law of refraction can be derived from the geometry of Fig. 11.16b. In a time interval Δt, the wave moving with speed v_1 through air travels a distance $v_1 \Delta t$. During that same time t, the wave in the water moving with speed v_2 travels a distance $v_2 \Delta t$. Note in Fig. 11.16b the two right triangles sharing a common side, s, along the boundary between air and water. From the definition of the sine function,

$$\sin \theta_1 = \frac{v_1 \Delta t}{s} \quad \text{and} \quad \sin \theta_2 = \frac{v_2 \Delta t}{s}$$

Dividing one equation by the other, we obtain the law of refraction.

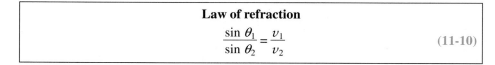

Law of refraction

$$\frac{\sin \theta_1}{\sin \theta_2} = \frac{v_1}{v_2} \tag{11-10}$$

The angles in Eq. (11-10) are called the *angle of incidence* and the *angle of refraction*. Note that these angles are measured between the propagation direction of the wave and the *normal*.

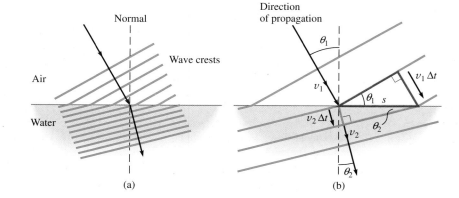

Figure 11.16   (a) A broad beam of light refracts when it passes from air into water. The reflected wave is omitted for clarity. The normal is the direction perpendicular to the boundary. The wave's propagation direction is closer to the normal in the slower medium (water, in this case). (b) Detailed view for derivation of the law of refraction

Although Fig. 11.16 shows a beam entering water from air, the same diagram holds for the reverse situation. If a flashlight under water is pointed toward the surface of the water at an angle of incidence θ_2 with the normal, the light beam is bent away from the normal as it enters the air, so that the angle of refraction becomes θ_1. Where the speed of the waves is greater, the angle with respect to the normal is greater.

Example 11.6

Refraction of an S Wave During an Earthquake

An earthquake emits seismic waves of two types, called P waves (longitudinal waves) and S waves (transverse waves), which travel at different speeds through the Earth. A typical P-wave speed is around 9.0 km/s and an S-wave speed is around 5.0 km/s. Suppose the S wave passes through a boundary in rock where its speed decreases from 5.0 km/s to 4.0 km/s. If the angle of incidence at the boundary is 40.0°, what is the angle of refraction?

Strategy   We know the speed of the S wave in the two materials and we know the angle of incidence. From the law of refraction, we can find the angle of refraction. We let the subscript 1 represent the first material and 2 the second. Then $\theta_1 = 40.0°$, $v_1 = 5.0$ km/s, and $v_2 = 4.0$ km/s; we want to find θ_2.

Solution   The law of refraction is

$$\frac{\sin \theta_1}{\sin \theta_2} = \frac{v_1}{v_2}$$

Solving for $\sin \theta_2$ yields

$$\sin \theta_2 = \frac{v_2}{v_1} \sin \theta_1$$

Now we substitute numerical values:

$$\sin \theta_2 = \frac{4.0 \text{ km/s}}{5.0 \text{ km/s}} \sin 40.0° = 0.514$$

Solving for θ_2,

$$\theta_2 = \sin^{-1} 0.514 = 31°$$

Discussion   The wave speed is slower in the second material so we expect the angle of refraction to be smaller than the angle of incidence; the wave is bent toward the normal.

Practice Problem 11.6   Refraction of a P wave

A P wave from the earthquake is traveling at 9.0 km/s. It is incident on a boundary between materials at an angle of incidence of 45°; the angle of refraction is 34°. What is the speed of the P wave in the other material?

Understanding the propagation of seismic waves, including reflection and refraction due to boundaries between geological features, is an essential part of the effort to reduce damage from future earthquakes. Scientists create small seismic waves with a large vibrator, then use seismographs to record ground vibrations at various locations. The goal is to produce a seismic hazard map so that preventative measures can be targeted to areas with the highest risk of earthquake damage.

11.9   INTERFERENCE AND DIFFRACTION

Interference

The principle of superposition leads to dramatic effects when applied to coherent waves. Two waves are **coherent** if they have the *same frequency* and they maintain a *fixed phase relationship* with each other. One way to obtain coherent waves is to get them from the same source. Such is the case, for example, if one *monophonic* amplifier sends the same signal to two speakers. Should some fluctuation occur in the amplifier driving the speakers, the same fluctuation occurs in both speakers at the same time and they maintain their coherence. Waves are **incoherent** if the phase relationship between them varies randomly. Whenever two waves come from two different sources they are incoherent.

Suppose coherent waves with amplitudes A_1 and A_2 pass through the same point in space. If the waves are *in phase* at that point—that is, the phase difference is any *even*

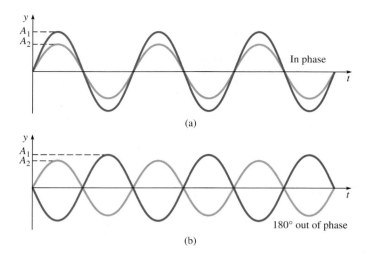

Figure 11.17 Coherent waves (a) in phase and (b) 180° out of phase. (One wave is drawn with a lighter line to distinguish it from the other.)

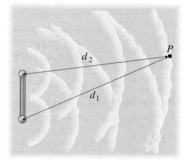

Figure 11.18 Overhead view of coherent surface water waves. The two waves travel different distances d_1 and d_2 to reach a point P. The phase difference between the waves at point P is $k(d_1 - d_2)$.

When two coherent waves are interfering and have a phase difference = $n\pi$, the interference is constructive for $n = 2, 4, 6, 8$, and so forth, and the interference is destructive for $n = 1, 3, 5, 7$, and so forth.

integral multiple of π rad—then the two waves consistently reach their maxima at exactly the same instants of time (Fig. 11.17a). The superposition of the waves that are in phase with each other is called **constructive interference**; the amplitude of the combined waves is the sum of the amplitudes of the two individual waves ($A_1 + A_2$).

Two waves that are *180° out of phase* at a given point have a phase difference of π rad, 3π rad, 5π rad, and so on. The waves are half a cycle apart; when one reaches its maximum, the other reaches its minimum (Fig. 11.17b). The superposition of waves that are 180° out of phase is called **destructive interference**—the amplitude of the combined waves is the *difference* of the amplitudes of the two individual waves ($|A_1 - A_2|$). For any other fixed phase relationship between the two waves, the superposition has an amplitude between $A_1 + A_2$ and $|A_1 - A_2|$.

Suppose two coherent waves start out in phase with each other. In Fig. 11.18, two rods vibrate up and down in step with each other to generate circular waves on the surface of the water. If the two waves travel the same distance to reach a point on the water surface, they arrive *in phase* and interfere constructively. At points where the distances are different, we calculate the phase difference as follows. One wavelength of path difference corresponds to a phase difference of 2π radians (one full cycle). Then, working by proportions,

$$\frac{d_1 - d_2}{\lambda} = \frac{\text{phase difference}}{2\pi \text{ rad}}$$

Thus, the phase difference is

$$\text{phase difference} = \frac{2\pi \text{ rad}}{\lambda} \times (d_1 - d_2) = k(d_1 - d_2) \qquad (11\text{-}11)$$

If the phase difference is an even integral multiple of π rad, then constructive interference occurs at point P; if the phase difference is an odd integral multiple of π rad, then destructive interference occurs at point P.

Intensity Effects for Interfering Waves

When coherent waves interfere, the amplitudes add (for constructive interference) or subtract (for destructive interference). However, since intensity is proportional to amplitude *squared*, we cannot simply add or subtract the *intensities* of coherent waves when they interfere. *Incoherent* waves, on the other hand, have no fixed phase relationship; interference effects are averaged out due to the rapidly varying phase difference. In the superposition of incoherent waves, the total intensity is the sum of the individual intensities.

Why don't we see and hear interference effects all the time? Light from ordinary sources—incandescent bulbs, fluorescent bulbs, or the Sun—is incoherent because it is

Example 11.7

Intensity of Interfering Waves

Two coherent waves interfere. The intensity of one of them (alone) is 9.0 times the intensity of the other. What is the ratio of the maximum possible intensity to the minimum possible intensity of the resulting wave?

Strategy   The intensity is *not* the sum or difference of the individual intensities because the waves are coherent. Since the waves maintain a fixed phase relationship, the principle of superposition tells us that the maximum and minimum *amplitudes* of the interfering waves are the sum and difference of the *individual amplitudes*. Intensity is proportional to amplitude squared, so we find the ratio of the amplitudes and then add or subtract them.

Solution   The intensities of the two individual waves are related by $I_1 = 9.0 I_2$ or $I_1/I_2 = 9.0$. Since intensity is proportional to amplitude squared,

$$\frac{A_1}{A_2} = \sqrt{\frac{I_1}{I_2}} = 3.0$$

Thus, $A_1 = 3.0 A_2$. The maximum possible amplitude for the superposition occurs if the waves are in phase:

$$A_{max} = A_1 + A_2 = 4.0 A_2$$

The minimum possible amplitude for the superposition occurs if the waves are 180° out of phase:

$$A_{min} = |A_1 - A_2| = 2.0 A_2$$

The ratio of the maximum to minimum intensity is

$$\frac{I_{max}}{I_{min}} = \left(\frac{A_{max}}{A_{min}}\right)^2 = \left(\frac{4.0}{2.0}\right)^2 = 4.0$$

Discussion   Had we added and subtracted the *intensities* instead of the amplitudes, we would have found a ratio of $10/8 = 1.25$ between the maximum and minimum intensities. We must be careful to add or subtract the *amplitudes* of the interfering waves instead of the intensities themselves when the waves are coherent.

Practice Problem 11.7   Two more coherent waves

Repeat Example 11.7, but change the ratio of the individual intensities to 4.0 (instead of 9.0).

---

generated by large numbers of independent atomic sources. A single source of sound normally contains many different frequencies, so a point of constructive interference for one frequency is not a point of constructive interference for other frequencies. Furthermore, in most situations there are many different sound waves that reach our ears after traveling different paths due to the reflection of sound from walls, ceilings, chairs, and so forth.

Diffraction

Diffraction is the bending of a wave around an obstacle in its path. The amount of diffraction depends on the relative size of the obstacle and the wavelength of the waves. Figure 11.19 shows water waves encountering a boat at anchor. If the distance

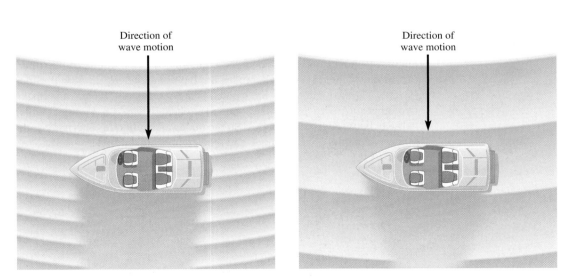

(a)   (b)

Direction of wave motion   Direction of wave motion

Figure 11.19   Diffraction of a wave when it encounters an obstacle. (a) Short wavelength waves are blocked by the boat without much bending. (b) Longer wavelength waves bend easily around the boat.

between crests (the wavelength) is small compared to the length of the boat, the waves do not bend significantly; the "shadow" of the boat is quite clear. If the distance between crests (the wavelength) is comparable to the size of the boat, the waves bend significantly so that the boat has a "shadow" up close but not farther away. Diffraction enables you to hear around a corner but not to see around a corner—the wavelengths of visible light are less than 1 mm while the wavelengths of sound waves in air are typically around 1 m. Higher frequencies have shorter wavelengths and do not diffract as much, so you hear mostly lower frequencies when a sound wave bends around a corner.

11.10 STANDING WAVES

Standing waves occur when a wave is reflected at a boundary and the reflected wave interferes with the incident wave so that the wave appears to stand still. Suppose that a harmonic wave on a string, coming from the right, hits a boundary where the string is fixed. The equation of the incident wave is

$$y(x, t) = A \sin (\omega t + kx)$$

The + sign is chosen in the phase because the wave travels to the left.

The reflected wave travels to the right, so $+kx$ is replaced with $-kx$; and the reflected wave is inverted, so $+A$ is replaced with $-A$. Then the reflected wave is described by

$$y(x, t) = -A \sin (\omega t - kx)$$

Applying the principle of superposition, the motion of the string is described by

$$y(x, t) = A [\sin (\omega t + kx) - \sin (\omega t - kx)]$$

This can be rewritten in a form that shows the motion of the string more clearly. Using the trigonometric identity (Appendix A.7),

$$\sin \alpha - \sin \beta = 2 \cos [\tfrac{1}{2}(\alpha + \beta)] \sin [\tfrac{1}{2}(\alpha - \beta)]$$

where

$$\alpha = \omega t + kx \quad \text{and} \quad \beta = \omega t - kx$$

we have

$$y(x, t) = 2A \cos \omega t \sin kx$$

Notice that t and x are separated. Every point moves in SHM with the same frequency. However, in contrast to a *traveling* harmonic wave, every point reaches its maximum distance from equilibrium simultaneously. In addition, different points move with different amplitudes; the amplitude at any point x is $2A \sin kx$.

Figure 11.20 shows the string at time intervals of $\tfrac{1}{8}T$, where T is the period. What you actually see when looking at a standing wave is a blur of moving string, with points that never move (**nodes**, labeled "N") halfway between points of maximum amplitude (**antinodes**, labeled "A"). The nodes are the points where $\sin kx = 0$. Since $\sin n\pi = 0$ ($n = 0, 1, 2, \ldots$), the nodes are located at $x = n\pi/k = n\lambda/2$. Thus, the distance between two adjacent nodes is $\tfrac{1}{2}\lambda$. The antinodes occur where $\sin kx = \pm 1$, which is exactly halfway between a pair of nodes. So the nodes and antinodes alternate, with one-quarter of a wavelength between a node and the neighboring antinode.

Node–node distance is $\tfrac{1}{2}\lambda$.
Node–antinode distance is $\tfrac{1}{4}\lambda$.

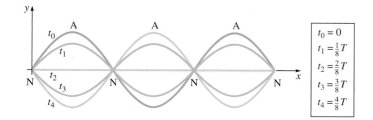

Figure 11.20 A standing wave at various times: $t = 0, \tfrac{1}{8}T, \tfrac{2}{8}T, \tfrac{3}{8}T,$ and $\tfrac{4}{8}T$, where T is the period.

$t_0 = 0$	
$t_1 = \tfrac{1}{8}T$	
$t_2 = \tfrac{2}{8}T$	
$t_3 = \tfrac{3}{8}T$	
$t_4 = \tfrac{4}{8}T$	

So far we have ignored what happens at the other end of the string. If the other end is fixed as well, then it too is a node. The string thus has two or more nodes, with one at each end. The distance between each pair of nodes is $\frac{1}{2}\lambda$, so

$$\frac{n\lambda}{2} = L$$

where L is the length of the string and $n = 1, 2, 3, \ldots$. The possible wavelengths for standing waves on a string are

$$\lambda_n = \frac{2L}{n} \quad (n = 1, 2, 3, \ldots) \tag{11-12}$$

The frequencies are

$$f_n = \frac{v}{\lambda_n} = \frac{nv}{2L} \quad (n = 1, 2, 3, \ldots) \tag{11-13}$$

The lowest frequency standing wave ($n = 1$) is called the **fundamental**. Notice that the higher frequency standing waves are all integral multiples of the fundamental; the set of standing wave frequencies makes an evenly spaced set:

$$f_1, 2f_1, 3f_1, 4f_1, \ldots, nf_1, \ldots$$

These frequencies are called the *natural frequencies* or *resonant frequencies* of the string. *Resonance* occurs when a system is driven at one of its natural frequencies; the resulting vibrations are large in amplitude compared to when the driving frequency is not close to any of the natural frequencies.

Resonance is responsible for much of the structural damage caused by seismic waves. If the frequency at which the ground vibrates is close to a resonant frequency of a structure, the vibration of the structure builds up to a large amplitude. Thus, to construct a building that can survive an earthquake, it is not enough to make it stronger. Either the building must be designed so it is isolated from ground vibrations, or a damping mechanism—something like a shock absorber—must be incorporated to dissipate energy and reduce the amplitude of the vibrations. Damping is becoming increasingly common in large buildings since it is just as effective and much less expensive than isolation.

Large sections of the Hanshin expressway vibrated in a twisting motion due to ground vibrations near a resonant frequency. The road now has rubber base isolators instead of steel bearings connecting the roadway to the concrete piers. Part of their function is to act like shock absorbers.

Figure 11.21 shows the first four standing wave patterns on a string. The two ends are always nodes since they are fixed in place. Notice that each successive pattern has one more node and one more antinode than the previous one. The fundamental has the fewest possible number of nodes (2) and antinodes (1).

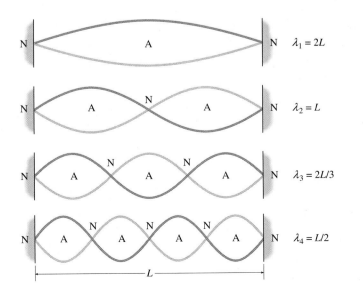

Figure 11.21 Standing waves on a string fixed at both ends. "N" marks the locations of the nodes and "A" marks the locations of the antinodes.

Example 11.8

Wavelength of a Standing Wave

A string is attached to a vibrator driven at 1.20×10^2 Hz. A weight hangs from the other end of the string; the weight is adjusted until a standing wave is formed (Fig. 11.22). What is the wavelength of the standing wave on the string?

Strategy   The measured distance of 42 cm encompasses six "loops"—that is, six segments of string between one node and the next. Each of the loops represents a length of $\frac{1}{2}\lambda$.

Solution   The length of one loop is

$$42 \text{ cm} \times \frac{1}{6} = 7.0 \text{ cm}$$

Since the length of one loop is $\frac{1}{2}\lambda$, the wavelength is 14 cm.

Discussion   This string is *not* fixed at both ends. The left end is connected to a moving vibrator, so it is not a node. The right end wraps around a pulley; it may not be easy to determine precisely where the 'end' is. For this case, it is more accurate to measure the distance between two actual nodes rather than to assume that the ends are nodes.

Practice Problem 11.8   Standing wave with seven loops

The vibrator frequency is increased until there are seven loops within the 42-cm length. What is the new standing wave frequency for this string (assuming the same tension)?

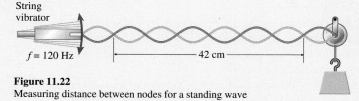

$f = 120$ Hz   42 cm

Figure 11.22
Measuring distance between nodes for a standing wave

MASTER THE CONCEPTS

Summary

- An isotropic source radiates sound uniformly in all directions. Assuming that no energy is absorbed by the medium and there are no obstacles to reflect or absorb sound,

$$I = \frac{\text{power}}{\text{area}} = \frac{P}{4\pi r^2} \qquad (11\text{-}1)$$

- In a transverse wave, the motion of particles in the medium is perpendicular to the direction of propagation of the wave. In a longitudinal wave, the motion of particles in the medium is along the same line as the direction of propagation of the wave.

- The speed of a transverse wave on a string is

$$v = \sqrt{\frac{F}{\mu}} \qquad (11\text{-}4)$$

where

$$\mu = m/L \qquad (11\text{-}3)$$

- A periodic wave repeats the same pattern over and over. Harmonic waves are a special kind of periodic wave characterized by a sinusoidal function (either a sine or cosine function).

- If a periodic wave has period T and travels at speed v, the repetition distance of the wave is the wavelength:

$$\lambda = vT \qquad (11\text{-}5)$$

- The principle of superposition: When two or more waves overlap, the net disturbance at any point is the sum of the individual disturbances due to each wave.

- A harmonic traveling wave can be described by

$$y(x, t) = A \cos (\omega t - kx) \qquad (11\text{-}8)$$

- The argument of the sinusoidal function, $(\omega t \pm kx)$, is called the phase of the wave at x and t. The constant k is the wave number

$$k = \frac{\omega}{v} = \frac{2\pi f}{v} = \frac{2\pi}{\lambda} \qquad (11\text{-}7)$$

- Reflection occurs at a boundary between media with different wave speeds. Some energy may be transmitted into the new medium and the rest is reflected. The wave transmitted past the boundary is refracted:

$$\frac{\sin \theta_1}{\sin \theta_2} = \frac{v_1}{v_2} \qquad (11\text{-}10)$$

- Coherent waves have the *same frequency* and maintain a *fixed phase relationship* with each other. Coherent waves that are in phase with each other interfere constructively; those that are 180° out of phase interfere destructively.

- Diffraction occurs when a wave bends around an obstacle in its path.

- In a standing wave on a string, every point moves in SHM with the same frequency. Nodes are points of zero amplitude; antinodes are points of maximum amplitude. The distance between two adjacent nodes is $\frac{1}{2}\lambda$.

Highlighted Figures and Tables

F11.2 A point source of sound radiating energy uniformly in all directions (p. 378)

F11.3 Transverse and longitudinal waves (p. 379)

F11.4 Motion of ground particles in P waves and S waves (p. 380)

MASTER THE CONCEPTS *continued*

F11.7 Sinusoidal wave moving with speed v in the x-direction (p. 383)

F11.8 A wave pulse at successive times (p. 384)

F11.9 Two graphs of a harmonic wave $y(x, t) = A \sin(\omega t - kx)$ on a string (p. 385)

F11.11 Two identical wave pulses traveling toward and through each other (p. 387)

F11.15 Reflection of a wave pulse from a fixed end (p. 389)

F11.16 Change in wave speed causing sudden refraction and geometric derivation (p. 390)

F11.17 Coherent waves in phase and 180° out of phase (p. 392)

F11.19 Diffraction of a wave when it encounters an obstacle (p. 393)

F11.21 Standing waves on a string fixed at both ends (p. 395)

CONCEPTUAL QUESTIONS

1. The piano strings that vibrate with the lowest frequencies consist of a steel wire around which a thick coil of copper wire is wrapped. Only the inner steel wire is under tension. What is the purpose of the copper coil?

2. Is the vibration of a string in a piano, guitar, or violin a *sound* wave? Explain.

3. The wavelength of the fundamental standing wave on a cello string depends on which of these quantities: length of the string, mass per unit length of the string, or tension? The wavelength of the *sound wave* resulting from the string's vibration depends on which of the same three quantities?

4. If the length of a guitar string is decreased while the tension remains constant, what happens to each of these quantities? (a) the wavelength of the fundamental, (b) the frequency of the fundamental, (c) the time for a pulse to travel the length of the string, (d) the maximum velocity of a point on the string (assuming the amplitude is the same both times), (e) the maximum acceleration of a point on the string (assuming the amplitude is the same both times).

5. Why is it possible to understand the words spoken by two people at the same time?

6. A cello player can change the frequency of the sound produced by her instrument by (a) increasing the tension in the string, (b) pressing her finger on the string at different places along the fingerboard, or (c) bowing a different string. Explain how each of these methods affects the frequency.

7. Why is a transverse wave sometimes called a shear wave?

8. Figure 11.23 shows a complex wave moving to the right along a cord. Draw the shape of the cord an instant later and determine which parts of the cord are moving upward and which are moving downward. Indicate the directions on your drawing with arrows.

9. When an earthquake occurs, the S waves (transverse waves) are not detected on the opposite side of the Earth while the P waves (longitudinal waves) are. How does this provide evidence that the Earth's solid core is surrounded by liquid?

10. Water waves can travel in any direction when far offshore, but they always arrive at the beach nearly head on. Explain.

11. Simple ear-protection devices use materials that reflect or absorb sound before it reaches the ears. A newer technology, sometimes called *noise cancellation*, uses a microphone to produce an electrical signal that mimics the noise. The signal is modified electronically, then fed to the speakers in a pair of headphones. The speakers emit sound waves that *cancel* the noise. On what principle is this technology based? What kind of modification is made to the electrical signal?

12. When connecting speakers to a stereo, it is important to connect them with the correct polarity so that, if the same electrical signal is sent, they both move in the same direction. If the wires going to one speaker are reversed, the listener hears a noticeably weaker bass (low frequencies). Explain what causes this and why low frequencies are affected more than high frequencies.

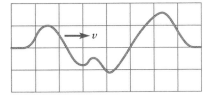

Figure 11.23   Conceptual Question 11.8

MULTIPLE CHOICE QUESTIONS

1. Standing waves are produced by the superposition of two waves with
   - (a) the same amplitude, frequency, and direction of propagation.
   - (b) the same amplitude and frequency, and opposite propagation directions.
   - (c) the same amplitude and direction of propagation, but different frequencies.
   - (d) the same amplitude, different frequencies, and opposite directions of propagation.

2. A transverse wave travels on a string of mass m, length L, and tension F. Which statement below is correct?
   - (a) The energy of the wave is proportional to the square root of the wave amplitude.
   - (b) The speed of a moving point on the string is the same as the wave speed.
   - (c) The wave speed is determined by the values of m, L, and F.
   - (d) The wavelength of the wave is proportional to L.

3. A transverse wave on a string is described by $y(x, t) = A \cos(\omega t + kx)$. It arrives at the point $x = 0$ where the string is fixed in place. Which function describes the reflected wave?

   (a) $A \cos(\omega t + kx)$ (b) $A \cos(\omega t - kx)$ (c) $-A \sin(\omega t + kx)$
   (d) $-A \cos(\omega t - kx)$ (e) $A \sin(\omega t + kx)$

4. A violin string of length L is fixed at both ends. Which one of these is *not* a wavelength of a standing wave on the string?

   (a) L (b) $2L$ (c) $L/2$ (d) $L/3$ (e) $2L/3$ (f) $3L/2$

5. The speed of waves in a stretched string depends upon which one of the following?

   (a) The tension in the string   (b) The amplitude of the waves
   (c) The wavelength of the waves
   (d) The gravitational field strength

6. The higher the frequency of a wave

   (a) the smaller its speed.   (b) the shorter its wavelength.
   (c) the greater its amplitude.   (d) the longer its period.

7. In a transverse wave, the individual particles of the medium

   (a) move in circles.   (b) move in ellipses.
   (c) move parallel to the direction of the wave's travel.
   (d) move perpendicular to the direction of the wave's travel.

8. Of these properties of a wave, the one that is independent of the others is its

   (a) amplitude   (b) wavelength
   (c) speed   (d) frequency

9. Two successive transverse pulses, one caused by a brief displacement to the right and the other by a brief displacement to the left, are sent down a Slinky that is fastened at the far end. At the point where the first reflected pulse meets the second advancing pulse, the deflection (compared with that of a single pulse) is

   (a) quadrupled.   (b) doubled.
   (c) canceled.   (d) halved.

10. The intensity of an isotropic sound wave is

    (a) directly proportional to the distance from the source.
    (b) inversely proportional to the distance from the source.
    (c) directly proportional to the square of the distance from the source.
    (d) inversely proportional to the square of the distance from the source.
    (e) none of the above.

PROBLEMS

Note: **ⓒ** indicates a combination conceptual/quantitative problem. Gold diamonds ◆, ◆◆ are used to indicate the increasing level of difficulty of each problem. Problem numbers appearing in blue, 9., denote problems that have a detailed solution available in the Student Solutions Manual. Some problems are *paired* by concept; their numbers are connected by a ruled box.

11.1 Waves and Energy Transport

1. The intensity of sunlight that reaches the Earth's atmosphere is 1400 W/m². What is the intensity of the sunlight that reaches Jupiter? Jupiter is 5.2 times as far from the sun as Earth. [*Hint:* Treat the Sun as an isotropic source of light waves.]

2. The intensity of the sound wave from a jet airplane is 9.2 W/m² at a distance of 120 m. What is the intensity of the sound wave that reaches the ground when the airplane flies overhead at an altitude of 8.6 km? Assume that the sound wave radiates from the airplane equally in all directions.

3. At what rate does the jet airplane in Problem 2 radiate energy in the form of sound waves?

4. The Sun emits electromagnetic waves (including light) equally in all directions. The intensity of the waves at the Earth's upper atmosphere is 1.4 kW/m². At what rate does the Sun emit electromagnetic waves? (In other words, what is the power output?)

11.3 Speed of Transverse Waves on a String

5. (a) What is the speed of propagation of the pulse shown in Fig. 11.24? (b) At what average speed does the point at $x = 2.0$ m move during this time interval?

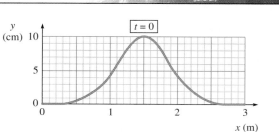

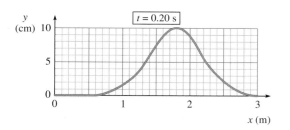

Figure 11.24   Problems 5, 6, 32, 33, 58, and 59. A pulse on a string at times $t = 0$ and $t = 0.20$ s. (Note that the horizontal and vertical scales are different.)

6. (a) What is the position of the peak of the pulse shown in Fig. 11.24 at $t = 3.00$ s? (b) When does the peak of the pulse arrive at $x = 4.00$ m?

7. A metal guitar string has a linear mass density of $\mu = 3.20$ g/m. What is the speed of transverse waves on this string when its tension is 90.0 N?

8. When the tension in a cord is 75 N, the wave speed is 140 m/s. What is the linear mass density of the cord?

9. A uniform string of length 10.0 m and weight 0.25 N is attached to the ceiling. A weight of 1.00 kN hangs from its lower end. The lower end of the string is suddenly displaced horizontally. How long does it take the resulting wave pulse to travel to the upper end? [*Hint*: Is the weight of the string negligible in comparison with that of the hanging mass?]

11.4 Periodic Waves

10. What is the speed of a wave whose frequency and wavelength are 500.0 Hz and 0.500 m, respectively?

11. What is the wavelength of a wave whose speed and period are 75.0 m/s and 5.00 ms, respectively?

12. What is the frequency of a wave whose speed and wavelength are 120 m/s and 30.0 cm, respectively?

13. The speed of sound in air at room temperature is 340 m/s. (a) What is the frequency of a sound wave in air with wavelength 1.0 m? (b) What is the frequency of a radio wave with the same wavelength? (Radio waves are electromagnetic waves that travel at 3.0×10^8 m/s in air or in vacuum.)

14. Light visible to humans consists of electromagnetic waves with wavelengths (in air) in the range 400–700 nm (4.0×10^{-7} m – 7.0×10^{-7} m). The speed of light in air is 3.0×10^8 m/s. What are the frequencies of electromagnetic waves that are visible?

15. A fisherman notices a buoy bobbing up and down in the water in ripples produced by waves from a passing speedboat. These waves travel at 2.5 m/s and have a wavelength of 7.5 m. At what frequency does the buoy bob up and down?

11.5 Mathematical Description of a Wave

16. You are swimming in the ocean as water waves with wavelength 9.6 m pass by. What is the closest distance that another swimmer could be so that his motion is exactly opposite yours (he goes up when you go down)?

17. The equation of a wave is

$$y(x, t) = 3.5 \sin \frac{\pi}{3.0}(x - 66t) \text{ cm}$$

where t is in seconds and x and y are both in cm. Find (a) the amplitude and (b) the wavelength of this wave.

18. Write an equation for a sine wave with amplitude 0.120 m, wavelength 0.300 m, and wave speed 6.40 m/s traveling in the $-x$-direction.

19. A wave on a string has equation $y(x, t) = (4.0 \text{ mm}) \sin(\omega t - kx)$, where $\omega = 6.0 \times 10^2$ rad/s and $k = 6.0$ rad/m. (a) What is the amplitude of the wave? (b) What is the wavelength? (c) What is the period? (d) What is the wave speed? (e) In which direction does the wave travel?

11.6 Graphing Waves

20. Figure 11.25 shows a snapshot of a transverse wave traveling along a string at 10.0 m/s. The equation for the wave is $y(x, t) = A \cos(\omega t + kx)$. (a) Is the wave moving to the right or to the left? (b) What are the numerical values of A, ω, and k? (c) At

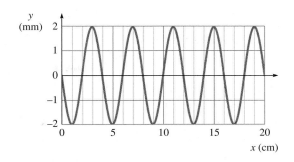

Figure 11.25   Problem 20

what times could this snapshot have been taken? (Give the three smallest nonnegative possibilities.)

21. (a) Plot a graph for $y(x, t) = 4.0 \sin(378t - 314x)$, where t is in s and x and y are in cm, versus x at $t = 0$ and at $t = \frac{1}{480}$ s. From the plots determine the amplitude, wavelength, and speed of the wave. (b) For the same function, plot a graph of $y(x, t)$ versus t at $x = 0$ and find the period of the vibration. Show that $\lambda = vT$.

22. Figure 11.26 shows a sine wave traveling to the right on a cord. The lighter line represents the shape of the cord at time $t = 0$; the darker line represents the shape of the cord at time $t = 0.10$ s. (Note that the horizontal and vertical scales are different.) What are (a) the amplitude and (b) the wavelength of the wave? (c) What is the speed of the wave? What are (d) the frequency and (e) the period of the wave?

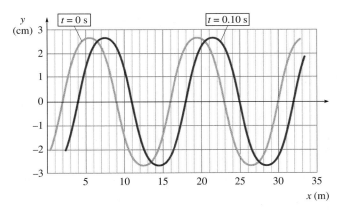

Figure 11.26   Problem 22

23. For $y(x, t) = 0.0050 \cos(4.0\pi t - 1.0\pi x)$ m, where all measurements are in meters and seconds, find the maximum speed and the maximum acceleration of a point on the string. Plot graphs for one cycle of displacement y versus t, velocity v_y versus t, and acceleration a_y versus t.

24. A transverse wave on a string is described by $y(x, t) = (1.2 \text{ mm}) \sin(2.0\pi t - 0.50\pi x)$, where x is in meters and t is in seconds. Plot the displacement y and the velocity v_y versus t for one complete cycle of the point $x = 0$ on the string.

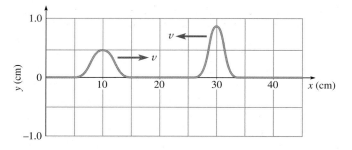

Figure 11.27 Problem 25

11.7 Principle of Superposition

25. Figure 11.27 shows two pulses on a cord at time $t = 0$. The pulses are moving toward each other; the speed of each pulse is 40 cm/s. Sketch the shape of the cord at 0.15, 0.25, and 0.30 s.

26. Figure 11.28 shows two pulses on a cord at time $t = 0$. The pulses are moving toward each other; the speed of each pulse is 2.5 m/s. Sketch the shape of the cord at 0.60, 0.80, and 0.90 s.

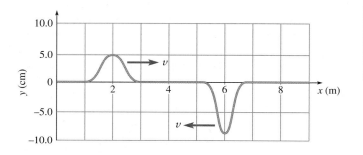

Figure 11.28 Problem 26

✦✦ 27. A traveling sine wave is the result of the superposition of two other sine waves with equal amplitudes, wavelengths, and frequencies. The two component waves each have amplitude 5.00 cm. If the resultant wave has amplitude 6.69 cm, what is the phase difference ϕ between the component waves? [*Hint:* Let $y_1 = A \sin(\omega t + kx)$ and $y_2 = A \sin(\omega t + kx - \phi)$. Make use of the trigonometric identity (Appendix A.7) for $\sin \alpha + \sin \beta$ when finding $y = y_1 + y_2$ and identify the new amplitude in terms of the original amplitude.]

✦ 28. Two traveling sine waves, identical except for a phase difference ϕ, add so that their superposition produces another traveling wave with the same amplitude as the two component waves. What is the phase difference between the two waves?

29. Using graph paper, sketch two identical sine waves of amplitude 4.0 cm that differ in phase by (a) 60.0° and (b) 90.0°. Use superposition to find the amplitude of the resultant wave.

11.8 Reflection and Refraction

30. Light of wavelength 0.500 μm (in air) enters the water in a swimming pool. The speed of light in water is 0.750 times the speed in air. What is the wavelength of the light in water?

31. An S wave crosses the boundary between two different types of rock. The angle of incidence is 22° and the angle of refraction is 34°. If the speed of the S wave before the boundary is 3.2 m/s, what is the speed past the boundary?

Problems 32–33: The pulse of Fig. 11.24 travels to the right on a string whose ends at $x = 0$ and $x = 4.0$ m are both fixed in place. Imagine a reflected pulse that begins to move onto the string at an endpoint at the same time the incident pulse reaches that endpoint. The superposition of the incident and reflected pulses gives the shape of the string.

32. When is the next time t that the string referred to in Fig. 11.24 looks exactly as it does at $t = 0$?

✦ 33. The pulse of Fig. 11.24 travels on a string whose ends at $x = 0$ and $x = 4.0$ m are both fixed in place. When does the string look completely flat? Find the first time for $t > 0$.

11.9 Interference and Diffraction

34. Two waves with identical frequency but different amplitudes $A_1 = 5.0$ cm and $A_2 = 3.0$ cm, occupy the same region of space (are superimposed). (a) At what phase difference does the resulting wave have the largest amplitude? What is the amplitude of the resulting wave in that case? (b) At what phase difference does the resulting wave have the smallest amplitude and what is its amplitude? (c) What is the ratio of the largest and smallest amplitudes?

35. Two waves with identical frequency but different amplitudes $A_1 = 6.0$ cm and $A_2 = 3.0$ cm, occupy the same region of space (i.e., are superimposed). (a) At what phase difference will the resulting wave have the highest intensity? What is the amplitude of the resulting wave in that case? (b) At what phase difference will the resulting wave have the lowest intensity and what will its amplitude be? (c) What is the ratio of the two intensities?

36. A sound wave with intensity 25 mW/m^2 interferes constructively with a sound wave with intensity 15 mW/m^2. What is the intensity of the superposition of the two?

37. A sound wave with intensity 25 mW/m^2 interferes destructively with a sound wave with intensity 28 mW/m^2. What is the intensity of the superposition of the two?

38. Two coherent sound waves have intensities of 0.040 and 0.090 W/m^2 where you are listening. (a) If the waves interfere constructively, what is the resultant intensity that you hear? (b) What if they interfere destructively? (c) If they were incoherent, what would be the resultant intensity? [*Hint:* If your answers are correct, then (c) is the average of (a) and (b).]

11.10 Standing Waves

39. In order to decrease the fundamental frequency of a guitar string by 4.0%, by what percentage should you reduce the tension?

✦ 40. The longest "string" (a thick metal wire) on a particular piano is 2.0 m long and has a tension of 300.0 N. It vibrates with a fundamental frequency of 27.5 Hz. What is the total mass of the wire?

✦ 41. Suppose that a string of length L and mass m is under tension F. (a) Show that $\sqrt{FL/m}$ has units of speed. (b) Show that there is no other combination of L, m, and F with units of speed. [*Hint:* Of the dimensions of the three quantities L, m, and F, only F includes time.] Thus, the speed of transverse waves on the string can only be some dimensionless constant times $\sqrt{FL/m}$.

42. A standing wave has wavenumber 2.0×10^2 rad/m. What is the distance between two adjacent nodes?

43. A piano string of length 1.50 m and mass density 25.0 mg/m vibrates at a (fundamental) frequency of 450.0 Hz. (a) What is the speed of the transverse string waves? (b) What is the tension? (c) What are the wavelength and frequency of the sound wave in air produced by vibration of the string? The speed of sound in air at room temperature is 340 m/s.

44. A cord of length 1.5 m is fixed at both ends. Its mass per unit length is 1.2 g/m and the tension is 12 N. (a) What is the frequency of the fundamental oscillation? (b) What tension is required if the $n = 3$ mode has a frequency of 0.50 kHz?

45. A guitar's E-string has length 65 cm and is stretched to a tension of 82 N. It vibrates at a fundamental frequency of 329.63 Hz. Determine the mass per unit length of the string.

46. A string 2.0 m long is held fixed at both ends. If a sharp blow is applied to the string at its center, it takes 0.050 s for the pulse to travel to the ends of the string and return to the middle. What is the fundamental frequency of oscillation for this string?

COMPREHENSIVE PROBLEMS

47. The speed of waves on a lake depends on frequency. For waves of frequency 1.0 Hz, the wave speed is 1.56 m/s; for 2.0 Hz waves, the speed is 0.78 m/s. The 2.0-Hz waves from a speedboat's wake reach you 120 s after the 1.0-Hz waves generated by the same boat. How far away is the boat?

© 48. The formula for the speed of transverse waves on a spring is the same as for a string. (a) A spring is stretched to a length much greater than its relaxed length. Explain why the tension in the spring is approximately proportional to the length. (b) A wave takes 4.00 s to travel from one end of such a spring to the other. Then the length is increased 10.0%. Now how long does a wave take to travel the length of the spring? [*Hint:* Is the mass per unit length constant?]

49. What is the wavelength of the radio waves transmitted by an FM station at 90 MHz? (Radio waves travel at 3.0×10^8 m/s.)

✦ 50. Figure 11.29 shows the graph of ground vibrations recorded by a seismograph 180 km from the focus of a small earthquake. It took the waves 30.0 s to travel from their source to the seismograph. Estimate the wavelength.

51. (a) Write an equation for a surface seismic wave moving along the $-x$-axis with amplitude 2.0 cm, period 4.0 s, and wavelength 4.0 km. Assume the wave is harmonic, x is measured in m, and t is measured in s. (b) What is the maximum speed of the ground as the wave moves by? (c) What is the wave speed?

52. A transverse wave on a string is described by $y(x, t) = (1.2 \text{ cm}) \sin (0.50\pi t - \pi x)$, where x is in meters and t is in seconds. Find the maximum velocity and the maximum acceleration of a point on the string. Plot graphs for displacement y versus t, velocity v_y versus t, and acceleration a_y versus t at $x = 0$.

53. Deep-water waves are *dispersive* (their wave speed depends on the wavelength). The restoring force is provided by gravity. Using dimensional analysis, find out how the speed of deep-water waves depends on wavelength λ, assuming that λ and g are the only relevant quantities. (For the curious: Mass density does not enter into the expression because the restoring force, arising from the weight of the water, is itself proportional to the mass density.)

54. In contrast to deep-water waves, shallow ripples on the surface of a pond are due to surface tension. The surface tension γ of water characterizes the restoring force; the mass density ρ of water characterizes the water's inertia. Use dimensional analysis to determine whether the surface waves are *dispersive* (the wave speed depends on the wavelength) or *nondispersive* (their wave speed is independent of wavelength). [*Hint:* Start by assuming that the wave speed is determined by γ, ρ, and the wavelength λ.]

55. An underground explosion sends out both transverse (S waves) and longitudinal (P waves) mechanical wave pulses (seismic waves) through the crust of the Earth. Suppose the speed of transverse waves is 8.0 km/s and that of longitudinal waves is 10.0 km/s. On one occasion, both waves follow the same path from a source to a detector (a seismograph); the longitudinal pulse arrives 2.0 s before the transverse pulse. What is the distance between the source and the detector?

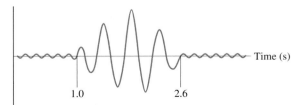

Time (s)

1.0   2.6

Figure 11.29   Problem 50

Figure 11.30 Problem 56

56. Figure 11.30 shows a snapshot of a transverse wave moving to the left on a string. The wave speed is 10.0 m/s. At the instant the snapshot is taken, (a) in what direction is point A moving? (b) In what direction is point B moving? (c) At which of these points is the speed of the string segment (not the wave speed) larger? Explain.

57. When the string of a guitar is pressed against a fret, the shortened string vibrates at a frequency 5.95% higher than when the previous fret is pressed. If the length of the part of the string that is free to vibrate is 64.8 cm, how far from one end of the string are the first three frets located?

Problems 58–59: The pulse of Fig. 11.24 travels to the right on a string whose ends at $x = 0$ and $x = 4.0$ m are both fixed in place. Imagine a reflected pulse that begins to move onto the string at an endpoint at the same time the incident pulse reaches that endpoint. The superposition of the incident and reflected pulses gives the shape of the string.

58. The pulse of Fig. 11.24 travels on a string whose ends at $x = 0$ and $x = 4.0$ m are both fixed in place. Sketch the shape of the string at $t = 1.6$ s.

59. The pulse of Fig. 11.24 travels on a string whose ends at $x = 0$ and $x = 4.0$ m are both fixed in place. Sketch the shape of the string at $t = 2.2$ s.

60. A guitar string has a fundamental frequency of 300.0 Hz. (a) What are the next three lowest standing wave frequencies? (b) If you press a finger *lightly* against the string at its midpoint so that both sides of the string can still vibrate, you force there to be a node at the midpoint. What are the lowest four standing wave frequencies now? (c) If you press *hard* at the same point, only one side of the string can vibrate. What are the lowest four standing wave frequencies?

61. A harpsichord string is made of yellow brass (Young's modulus 9.0×10^{10} Pa, tensile strength 6.3×10^8 Pa, mass density 8500 kg/m^3). When tuned correctly, the tension in the string is 59.4 N, which is 93% of the maximum tension that the string can endure without breaking. The length of the string that is free to vibrate is 9.4 cm. What is the fundamental frequency?

62. Two speakers spaced a distance 1.5 m apart emit coherent sound waves at a frequency of 680 Hz in all directions. The waves start out in phase with each other. A listener walks in a circle of radius greater than one meter centered on the midpoint of the two speakers. At how many points does the listener observe destructive interference? The listener and the speakers are all in the same horizontal plane and the speed of sound is 340 m/s. [*Hint:* Start with a diagram; then determine the *maximum* path difference between the two waves at points on the circle.] Experiments like this must be done in a special room so that reflections are negligible.

ANSWERS TO PRACTICE PROBLEMS

11.1 (a) 8.9 m/s; (b) 13 m/s

11.2 (a) Yes, the traveling wave retains its shape; (b) it travels in the $+x$-direction because the t and x/v terms have *opposite* signs; (c) the wave speed is 8.0 m/s.

11.3

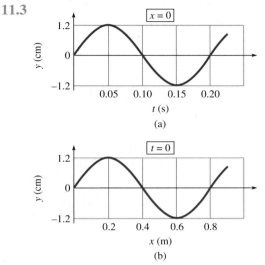

(a)

(b)

(c) $T = 0.20$ s; (d) $\lambda = 0.80$ m; (e) $A = 1.2$ cm; (f) $v = 4.0$ m/s; (g) the wave travels in the $-x$-direction because the signs of the terms containing x and t are the same.

11.4

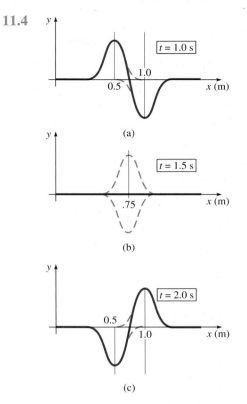

(a)

(b)

(c)

11.5 (a) 620 Hz; (b) 8.5 m

11.6 7.1 km/s

11.7 9.0

11.8 140 Hz

Chapter 12

Sound

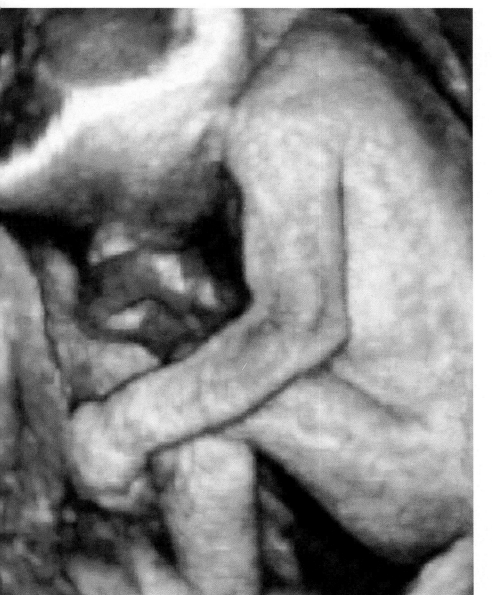

Ultrasonic imaging of the fetus is an increasingly important part of prenatal care. Could an image of the fetus be produced just as well using sound in the audible range rather than ultrasound? Why is ultrasound used rather than some other imaging technology, such as x-rays? Are there other medical applications of ultrasound?

Making The Connection:
sound from a guitar

Making The Connection:
sound from a loudspeaker

12.1 SOUND WAVES

When a guitar string is plucked, a transverse wave travels along the string. The wave on the string is not what we hear, of course, since the string has no direct connection to our eardrums. The vibration of the string is transmitted through the bridge to the body of the guitar, which in turn transmits the vibration to the air—a sound wave. A transverse wave on a guitar string is not a sound wave, though it does *cause* a sound wave.

In the absence of a sound wave, air molecules dart around in random directions. On average, they are uniformly distributed and the pressure is the same everywhere (neglecting the insignificant variation of pressure due to small changes in altitude). In a sound wave, the uniform distribution of air molecules is disturbed. A loudspeaker produces pressure fluctuations that travel through the air in all directions (Fig. 12.1). In some regions (*compressions*), the air molecules are bunched together and the pressure is higher than the average pressure. In other regions (*rarefactions*), the molecules are spread out and the pressure is lower than average. The sound wave can be described mathematically by the gauge pressure p (the difference between the pressure at a given point and the average pressure in the surroundings) as a function of position and time (Fig. 12.2a).

The speaker cone produces these pressure variations by displacing air molecules from their uniform distribution (Fig. 12.2b). When the cone moves to the left of its equilibrium position, air molecules spread out into a region of lower pressure (rarefaction). When the cone moves to the right, air molecules are shoved together into a region of higher pressure (compression).

Thus, the regions of higher and lower pressure are formed when air molecules are displaced from a uniform distribution. A sound wave can be described equally well by the displacement s of an *element* of the air—a region of air that can be considered to move together as a unit (Fig. 12.2c). An element is much smaller than the wavelength of the wave but still large enough to contain many molecules. Near a compression, where the air molecules are bunched together, elements on the left have been displaced to the right, and those on the right have been displaced to the left, as shown by the displacement arrows in Fig. 12.2c. Elements at points of maximum or minimum pressure have zero displacement; they stay put while the neighboring elements move in toward them (a compression) or away from them (a rarefaction). Conversely, where the gauge pressure is zero, the displacement has its maximum magnitude. In a standing wave, a pressure node is a displacement antinode; a pressure antinode is a displacement node.

If the pressure is higher on one side than on the other, the net force pushes air toward the side with lower pressure. The uneven distribution of pressure results in air molecules being pushed toward rarefactions and away from compressions, as shown by the force arrows in Fig. 12.2b. Note that the directions of these force arrows, pointing

Figure 12.1 The vibrating speaker cone in this boombox creates alternating regions of high and low pressure in the air. Air nearby is affected by a net force due to the nonuniform air pressure; as a result, variations in pressure travel in all directions away from the speaker. This traveling disturbance is a sound wave.

opposite to the displacement arrows in a corresponding region, are such that where there is a compression at a given instant, there will later be a rarefaction, and *vice versa;* the pressure at a given point fluctuates above and below the average pressure.

What makes the sound wave propagate away from the speaker cone? At the instant shown in Fig. 12.2, the sound wave has propagated out a certain distance from the speaker. Beyond the leading edge of the wave, the air molecules are not yet disturbed from their uniform distribution. The higher pressure at the leading edge pushes molecules to the right, so that the compression moves outward and the leading edge of the wave moves away from the speaker. When the compressions and rarefactions reach our ear, the changing force due to the air pressure on the eardrum causes the eardrum to vibrate in and out.

Frequencies of Sound Waves

The human ear responds to sound waves within a limited range of frequencies. We generally consider the **audible range** to extend from 20 Hz to 20 kHz. Very few people can actually hear sounds over that entire range. Even for a person with excellent hearing, the sensitivity of the ear declines rapidly below 100 Hz and above 10 kHz. The terms **infrasound** and **ultrasound** are used to describe sound waves with frequencies below 20 Hz and above 20 kHz, respectively.

The audible ranges for animals can be quite different. Dogs can hear frequencies as high as 50 kHz, which is why we can make a dog whistle that is inaudible to humans.

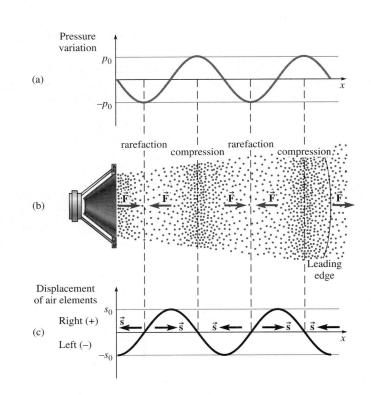

Figure 12.2 A sound wave generated by a loudspeaker. (a) Graph of the pressure variation of the air as a function of position. Pressure is high where air is squeezed together and low where it is more spread out. (b) Elements of the air are displaced from their equilibrium positions. Since the pressure is not uniform, air elements experience a net force due to air pressure; the force arrows indicate the direction of this net force. The force is always directed away from a compression (higher pressure) and toward a rarefaction. (c) Graph of the displacement of air elements from their equilibrium positions as a function of position; the arrows indicate the directions of the displacements in each region. Air elements are displaced towards compressions and away from rarefactions.

Dolphins make use of frequencies as high as 250 kHz. Elephants communicate over long distances (up to 4 km) using sounds with fundamental frequencies as low as 14 Hz. A rhinoceros uses frequencies down to 10 Hz. Such low-frequency sounds cannot be heard by humans, but the vibrations can be felt and the sounds can be recorded using special equipment.

12.2   THE SPEED OF SOUND WAVES

Just as for transverse waves on a string, the speed of sound waves is determined by a balance between two characteristics of the wave medium: the restoring force and the inertia. For string waves, the restoring force is characterized by the tension in the string F and the inertia is characterized by the linear mass density μ (mass per unit length). The speed of transverse waves on a string is

$$v = \sqrt{\frac{F}{\mu}} \tag{11-4}$$

For sound waves in a fluid, the restoring force is characterized by the bulk modulus B, defined in Section 10.4 as the increase in pressure needed to compress the fluid a certain fraction of its volume:

$$\Delta P = -B\frac{\Delta V}{V} \tag{10-10}$$

The inertia of the fluid is characterized by its mass density ρ. Following our dictum "more restoring force makes faster waves; more inertia makes slower waves," we expect the speed of sound to be faster in a medium with a larger bulk modulus (harder to compress means more restoring force) and slower in a medium with a larger density. By analogy with Eq. (11-4), we might *guess* that

$$v = \sqrt{\frac{\text{a measure of the restoring force}}{\text{a measure of the inertia}}} = \sqrt{\frac{B}{\rho}} \quad \text{(in fluids)} \tag{12-1}$$

In Problem 4, you can show that no other combination of B and ρ can have dimensions of speed, so that Eq. (12-1) *must* be correct except for the possibility of multiplication

More restoring force ⇒ faster waves; more inertia ⇒ slower waves.

by a dimensionless constant. The dimensionless constant turns out to be 1; Eq. (12-1) is the correct expression for the speed of sound in fluids.

The bulk modulus B of an ideal gas turns out to be directly proportional to the density ρ and to T, the *absolute temperature* ($B \propto \rho T$). As a result, the speed of sound in an ideal gas is proportional to the absolute temperature but is independent of pressure and density (at a fixed temperature):

$$v = \sqrt{\frac{B}{\rho}} \propto \sqrt{\frac{\rho T}{\rho}} \propto \sqrt{T} \quad \text{(ideal gas)}$$

The SI unit of absolute temperature is the kelvin (symbol K). To find absolute temperature in kelvins, add 273.15 to the temperature in degrees Celsius:

$$T \text{ (in K)} = T_C \text{ (in °C)} + 273.15 \tag{12-2}$$

Since $v \propto \sqrt{T}$, the speed of sound in an ideal gas at any absolute temperature T can be found if it is known at one temperature:

$$v = v_0 \sqrt{\frac{T}{T_0}} \tag{12-3}$$

where the speed of sound is v_0 at absolute temperature T_0. For example, the speed of sound in air at 0°C (or 273 K) is 331 m/s. At room temperature (20°C or 293 K), the speed of sound in air is

$$v = 331 \text{ m/s} \times \sqrt{\frac{293 \text{ K}}{273 \text{ K}}} = 343 \text{ m/s}$$

An *approximate* formula that can be used for the speed of sound in air is

$$v = (331 + 0.606 T_C) \text{ m/s} \tag{12-4}$$

where T_C is air temperature *in degrees Celsius* (see Problem 5 at the end of this chapter). The speed of sound in air increases 0.606 m/s for each degree Celsius increase in temperature. Equation (12-4) gives speeds accurate to better than 1% all the way from –66°C to +89°C.

The speed of sound in a *solid* depends on the Young's modulus Y and the shear modulus S. For sound waves traveling along the length of a thin solid rod, the speed is approximately

$$v = \sqrt{\frac{Y}{\rho}} \quad \text{(thin solid rod)} \tag{12-5}$$

Table 12.1 gives the speed of sound in various materials.

Table 12.1

Speed of Sound in Various Materials (at 0°C and 1 atm unless otherwise noted)

Medium	Speed (m/s)
Carbon dioxide	259
Air	331
Nitrogen	334
Air (20°C)	343
Helium	972
Hydrogen	1284
Mercury (25°C)	1450
Fat (37°C)	1450
Water (25°C)	1493
Seawater (25°C)	1533
Blood (37°C)	1570
Muscle (37°C)	1580
Lead	1322
Concrete	3100
Copper	3560
Bone (37°C)	4000
Pyrex glass	5640
Aluminum	5100
Steel	5790
Granite	6500

Conceptual Example 12.1

Speed of Sound in Hydrogen and Mercury

From Table 12.1, the speed of sound in hydrogen gas at 0°C is almost as large as the speed of sound in mercury, even though the density of mercury is 150,000 times larger than the density of hydrogen. How is that possible? Shouldn't the speed in mercury be much smaller, since it has so much more inertia?

Solution and Discussion   The speed of sound depends on *two* characteristics of the medium: the restoring force (measured by the bulk modulus) and the inertia (measured by the

density). The bulk modulus of mercury is much larger than the bulk modulus of hydrogen. The bulk modulus is a measure of how hard it is to compress a material. Liquids (such as mercury) are much more difficult to compress than are gases. Thus, the restoring forces in mercury are much larger than those in hydrogen; this allows sound to travel a bit faster in mercury than it does in hydrogen gas.

Conceptual Practice Problem 12.1   Speed of sound in solids versus liquids

Why does sound generally travel faster in a solid than in a liquid?

12.3   AMPLITUDE AND INTENSITY OF SOUND WAVES

Since there are two ways to describe a sound wave—pressure and displacement—the amplitude of a sound wave can take one of two forms: the pressure amplitude p_0 or the displacement amplitude s_0. The pressure amplitude p_0 is the maximum pressure fluctuation above or below the pressure in the absence of a sound wave; the displacement amplitude s_0 is the maximum displacement of an element of the medium from its equilibrium position. The pressure amplitude is proportional to the displacement amplitude. For a harmonic sound wave at angular frequency ω, an advanced analysis shows that

$$p_0 = \omega v \rho s_0 \tag{12-6}$$

where v is the speed of sound and ρ is the mass density of the medium.

Is a larger amplitude sound wave perceived as *louder*? Yes, all other things being equal. However, the relationship between our perception of loudness and the amplitude of a sound wave is complex. Loudness is a subjective aspect of how sound is perceived; it has to do with how the ear responds to sound and how the brain interprets signals from the ear. Perceived loudness turns out to be *roughly* proportional to the logarithm of the amplitude. If the amplitude of a sound wave doubles repeatedly, the perceived loudness does not double; it increases by a series of roughly equal steps.

Discussions of loudness are more often phrased in terms of intensity rather than amplitude since we are interested in how much energy the sound wave carries. The intensity of a sound wave is

$$I = \frac{p_0^2}{2\rho v} \tag{12-7}$$

where ρ is the mass density of the medium and v is the speed of sound in that medium. The most important thing to remember is that *intensity is proportional to amplitude squared*, which is true for all waves, not just sound. It is closely related to the fact that energy in SHM is proportional to amplitude squared [see Section 10.5, Eq. (10-13)].

Table 12.2 gives the pressure amplitudes and intensities for a wide range of sounds. Notice that the pressure amplitudes are given in *atmospheres;* thus, even for sounds that

Intensity $\propto$ (Amplitude)^2

Table 12.2

Pressure Amplitudes, Intensities, and Intensity Levels of a Wide Range of Sounds in Air at 20°C

Sound	Pressure Amplitude (atm)	Pressure Amplitude (Pa)	Intensity (W/m²)	Intensity Level (dB)
Threshold of hearing	3×10^{-10}	3×10^{-5}	10^{-12}	0
Leaves rustling	1×10^{-9}	1×10^{-4}	10^{-11}	10
Whisper (1 m away)	3×10^{-9}	3×10^{-4}	10^{-10}	20
Library background noise	1×10^{-8}	0.001	10^{-9}	30
Living room background noise	3×10^{-8}	0.003	10^{-8}	40
Office or classroom	1×10^{-7}	0.01	10^{-7}	50
Normal conversation at 1 m	3×10^{-7}	0.03	10^{-6}	60
Inside a moving car, light traffic	1×10^{-6}	0.1	10^{-5}	70
City street (heavy traffic)	3×10^{-6}	0.3	10^{-4}	80
Shout (at 1 m); or inside a subway train; risk of hearing damage if exposure lasts several hours	1×10^{-5}	1	10^{-3}	90
Car without muffler at 1 m	3×10^{-5}	3	10^{-2}	100
Construction site	1×10^{-4}	10	10^{-1}	110
Indoor rock concert; threshold of pain; hearing damage occurs rapidly	3×10^{-4}	30	1	120
Jet engine at 30 m	1×10^{-3}	100	10	130

are quite loud, the pressure fluctuations due to sound waves are small compared to the "background" atmospheric pressure.

Example 12.2

The Brown Creeper

The song of the Brown Creeper (*Certhia americana*) is very high in frequency—as high as 8 kHz. Many people who have lost some of their high-frequency hearing can't hear it at all. Suppose that you are out in the woods and hear the song. If the intensity of the song at your position is 1.4×10^{-8} W/m² and the frequency is 6.0 kHz, what are the pressure and displacement amplitudes? (Assume the temperature is 20°C.)

Strategy   The displacement and pressure amplitudes are related through Eq. (12-6); the pressure amplitude is related to the intensity through Eq. (12-7). These relationships can be used to solve for both pressure amplitude, p_0, and displacement amplitude, s_0. The density of air at 20°C is $\rho = 1.20$ kg/m³ (Table 9.1). The speed of sound in air at 20°C is $v = 343$ m/s. We need to multiply the frequency by 2π to get the angular frequency ω.

Solution   Intensity and pressure amplitude are related by

$$I = \frac{p_0^2}{2\rho v} \qquad (12\text{-}7)$$

Solving for p_0,

$$p_0 = \sqrt{2I\rho v} = \sqrt{2 \times 1.4 \times 10^{-8} \text{ W/m}^2 \times 1.20 \text{ kg/m}^3 \times 343 \text{ m/s}}$$
$$= 3.4 \times 10^{-3} \text{ Pa}$$

The pressure and displacement amplitudes are related by

$$p_0 = \omega v \rho s_0 \qquad (12\text{-}6)$$

Substituting in Eq. (12-7) yields

$$I = \frac{(\omega v \rho s_0)^2}{2\rho v}$$

Solving for s_0,

$$s_0 = \sqrt{\frac{2I}{\rho \omega^2 v}} = \sqrt{\frac{2 \times 1.4 \times 10^{-8} \text{ W/m}^2}{1.20 \text{ kg/m}^3 \times (2\pi \times 6000 \text{ Hz})^2 \times 343 \text{ m/s}}}$$
$$= 2.2 \times 10^{-10} \text{ m}$$

Discussion   This problem illustrates how sensitive the human ear is. The pressure amplitude is a fluctuation of one part in 30 million in the air pressure. Since the pressure amplitude is 3.4×10^{-3} Pa, the maximum force on the eardrum would be about

$$F_{\text{max}} = 3.4 \times 10^{-3} \text{ N/m}^2 \times 10^{-4} \text{ m}^2 \approx 3 \times 10^{-7} \text{ N}$$

which is about the weight of a large amoeba. The displacement amplitude is about the size of an atom.

Practice Problem 12.2   Pressure and intensity at an outdoor concert

At a distance of 5.0 m from the stage at an outdoor rock concert, the sound intensity is 1.0×10^{-4} W/m². Estimate the intensity and pressure amplitude at a distance of 25 m if there were no speakers other than those on stage. Explain the assumptions you make.

Decibels

Since the perception of loudness by the human ear is roughly proportional to the logarithm of the intensity, it is also roughly proportional to the logarithm of the amplitude (since $\log x^2 = 2 \log x$). An intensity of $I_0 = 10^{-12}$ W/m² is about the lowest intensity sound wave that can be heard under ideal conditions by a person with excellent hearing; it is therefore called the **threshold of hearing**. The threshold of hearing is used as a reference intensity in the definition of the intensity level.

A sound intensity I is compared to the reference level I_0 by taking the ratio of the two intensities. Suppose a sound has an intensity of 10^{-5} W/m²; the ratio is

$$\frac{I}{I_0} = \frac{10^{-5} \text{ W/m}^2}{10^{-12} \text{ W/m}^2} = 10^7$$

so the intensity is 10^7 times that of the hearing threshold level. The power to which 10 is raised is the **sound intensity level** β in units of bels (after Alexander Graham Bell). A ratio of 10^7 indicates a sound intensity of 7 bels or, as it is more commonly stated, 70 decibels (dB). Since $\log_{10}(10^x) = x$, the sound intensity level in decibels is

$$\beta = (10 \text{ dB}) \log_{10} \frac{I}{I_0} \qquad (12\text{-}8)$$

An intensity level of 0 dB corresponds to the threshold of hearing ($I = 10^{-12}$ W/m^2). Although the intensity level is really a pure number, the "units" (dB) remind us what the number means.

Example 12.3

Decibels from a Jackhammer

The sound intensity 5 m from a jackhammer is 4.20×10^{-2} W/m^2. What is the sound intensity level in decibels? (Use a reference level of $I_0 = 1.00 \times 10^{-12}$ W/m^2.)

Strategy   We are given the intensity in W/m^2 and asked for the intensity level in dB. First we find the ratio of the given intensity to the reference level. Then we take the logarithm of the result (to get the level in bels) and multiply by 10 (to convert from bels to dB).

Solution   The ratio of the intensity to the reference level is

$$\frac{I}{I_0} = \frac{4.20 \times 10^{-2} \text{ W/m}^2}{10^{-12} \text{ W/m}^2} = 4.20 \times 10^{10}$$

The intensity level in bels is

$$\log_{10} \frac{I}{I_0} = \log_{10} 4.20 \times 10^{10} = 10.6 \text{ bels}$$

The intensity level in decibels is

$$\beta = 10.6 \text{ bels} \times (10 \text{ dB/bel}) = 106 \text{ dB}$$

Discussion   As a quick check, 100 dB corresponds to $I = 10^{-2}$ W/m^2 and 110 dB corresponds to $I = 10^{-1}$ W/m^2; since the intensity is between 10^{-2} W/m^2 and 10^{-1} W/m^2, the intensity level must be between 100 dB and 110 dB.

Practice Problem 12.3   Consequences of a hole in the muffler

When rust creates a hole in the muffler of a car, the sound intensity level inside the car is 26 dB higher than when the muffler was intact. By what factor does the intensity increase?

As we saw in Section 11.9, when two sounds are coming from different sources, the waves are incoherent. If we know the intensity of each wave alone at a certain point, then the intensity due to the two waves together at that point is the sum of the two intensities:

$$I = I_1 + I_2 \quad \text{(incoherent waves)}$$

This is *not* true for two coherent waves, where the total intensity depends on the phase relationship between the waves. Since there is no fixed phase relationship between two incoherent waves, on average there is neither constructive nor destructive interference. The total power per unit area is the sum of the power per unit area of each wave.

Example 12.4

The Sound Intensity of Two Lathes Compared to That of One Lathe

A metal lathe in a workshop produces a 90.0-dB sound intensity level at a distance of 1 m. What is the intensity level when a second identical lathe starts operating? Assume the listener is at the same distance from both lathes.

Strategy   The noise is coming from two different machines and thus they are incoherent sources. We *cannot* add 90.0 dB to 90.0 dB to get 180.0 dB, which would be a senseless result—two lathes are not going to drown out a jet

engine at close range (see Table 12.1). Instead, what doubles is the *intensity*. We must work in terms of intensity rather than intensity level.

Solution   First find the intensity due to one lathe:

$$\beta = 90.0 \text{ dB} = (10 \text{ dB}) \log_{10} \frac{I}{I_0}$$

$$\log_{10} \frac{I}{I_0} = 9.00, \quad \text{so} \frac{I}{I_0} = 1.00 \times 10^9$$

continued on next page

Example 12.4 *continued*

We could solve for I numerically but it is not necessary. With two machines operating, the intensity doubles, so

$$\frac{I'}{I_0} = 2.00 \times 10^9$$

and the new intensity level is

$$\beta' = (10 \text{ dB}) \log_{10} \frac{I'}{I_0} = (10 \text{ dB}) \log_{10} (2.00 \times 10^9) = 93.0 \text{ dB}$$

 Discussion The new intensity level is just 3 dB higher than the original one, even though the intensity is twice as big. This turns out to be a general result: a 3-dB increase represents a doubling of the intensity.

Practice Problem 12.4   Intensity change for an increment of 5 dB

Maximum recommended exposure time to a sound level of 90 dB is 8 hours. For every increase of 5.0 dB in sound level up to 120 dB, the exposure time should be reduced by a factor of 2. (At 120 dB, damage occurs almost immediately; there is no safe exposure time.) What factor of intensity change does an intensity level increment of 5.0 dB represent?

Sound intensity level is useful because it roughly approximates the way we perceive loudness (since it is a logarithmic function of intensity). Equal increments in intensity level roughly correspond to equal increases in loudness. Two useful rules of thumb: every time the intensity increases by a *factor* of 10, the intensity level *adds* 10 dB; since $\log_{10} 2 = 0.30$, adding 3.0 dB to the intensity level *doubles* the intensity (see Problem 14). In Example 12.4, when both lathes are running at the same time, the intensity is twice as big as for one lathe, but the two do not sound twice as loud as one. Intensity *level* is a better guide to loudness; two lathes produce a level 3 dB higher than one lathe.

Decibels can also be used in a relative sense; instead of comparing an intensity to I_0, we can compare two intensities directly. Suppose we have two intensities I_1 and I_2 and two corresponding intensity levels β_1 and β_2. Then

$$\beta_2 - \beta_1 = 10 \text{ dB} \left(\log_{10} \frac{I_2}{I_0} - \log_{10} \frac{I_1}{I_0} \right)$$

Since $\log x - \log y = \log \frac{x}{y}$ [see Appendix A.3, Eq. (A-21)],

$$\beta_2 - \beta_1 = (10 \text{ dB}) \log_{10} \frac{I_2/I_0}{I_1/I_0} = (10 \text{ dB}) \log_{10} \frac{I_2}{I_1} \qquad (12\text{-}9)$$

Example 12.5

Variation of Intensity Level with Distance

At a distance of 30 m from a jet engine the sound intensity level is 130 dB. Serious, permanent hearing damage occurs rapidly at intensity levels this high, which is why you see air-

port personnel using hearing protection out on the runway. Assume the engine is an isotropic source of sound and ignore reflections and absorption. At what distance is the intensity level 110 dB—still quite loud but below the threshold of pain?

Strategy The intensity level drops 20 dB. According to the rule of thumb, each 10-dB change represents a factor of 10 in intensity. Therefore, we must find the distance at which the intensity is 2 factors of 10 smaller—that is, $\frac{1}{100}$ the original intensity. The intensity is proportional to $1/r^2$ since we assume an isotropic source.

Solution We set up a ratio between the intensities and the inverse square of the distances:

$$\frac{I_1}{I_2} = \left(\frac{r_2}{r_1} \right)^2$$

continued on next page

Example 12.5 *continued*

From the rule of thumb, we know that $I_2 = \frac{1}{100} I_1$. Then

$$\frac{r_2}{r_1} = \sqrt{\frac{I_1}{I_2}} = \sqrt{100} = 10$$

$$r_2 = 10 r_1 = 300 \text{ m}$$

Discussion   It is not necessary to use the rule of thumb. Let $\beta_1 = 130$ dB and $\beta_2 = 110$ dB. Then

$$\beta_1 - \beta_2 = 20 \text{ dB} = (10 \text{ dB}) \log_{10} \frac{I_1}{I_2}$$

From this, we find that

$$\log_{10} \frac{I_1}{I_2} = 2 \quad \text{or} \quad \frac{I_1}{I_2} = 100$$

We can only consider 300 m an estimate. The jet engine may not radiate sound equally in all directions; it might be louder in front than on the side. Sound is partly absorbed and partly reflected by the runway, by the plane, and by any nearby objects. The air itself absorbs some of the sound energy—that is, some of the energy of the wave is dissipated.

Practice Problem 12.5   A plane as quiet as a library

At what distance from the jet engine would the intensity level be comparable to the background noise level of a library (30 dB)? Is your answer realistic?

Making The Connection:
sources of
musical sound

12.4 STANDING SOUND WAVES

Pipe Open at Both Ends

Recall (Section 11.8) that a transverse wave on a string is reflected from a fixed end. A string fixed at both ends reflects the wave at each end. Such a string supports standing waves, caused by the superposition of two waves traveling in opposite directions on the string, only at certain frequencies, since there must be a node at each end. Standing sound waves are also caused by reflections at boundaries. Standing wave patterns for sound waves are in general more complex, since sound is a three-dimensional wave. However, the air inside a pipe open at both ends gives rise to standing waves closely analogous to those on a string, as long as the pipe's diameter is small compared to its length. Such a pipe is an excellent model of some organ pipes and flutes.

If the pipe is open at both ends, then the pipe has the same boundary condition at each end. At each open end, the column of air inside the pipe communicates with the outside air, so the pressure at the ends can't deviate much from atmospheric pressure. The open ends are therefore *pressure nodes* (Fig. 12.3). They are also *displacement antinodes*—elements of air vibrate back and forth with maximum amplitude at the ends. Since nodes and antinodes alternate with equal spacing ($\lambda/4$), the wavelengths of standing sound waves in a pipe open at both ends are the same as for a string fixed at both ends (compare Fig. 12.3 with Fig. 11.21), regardless of whether you consider the pressure or the displacement description.

Standing sound waves (thin pipe open at both ends)	
$$\lambda_n = \frac{2L}{n}$$	(11-12)
$$f_n = \frac{v}{\lambda_n} = n \frac{v}{2L} = n f_1$$	(11-13)
where $n = 1, 2, 3, \ldots$	

Pipe Closed at One End

Some organ pipes are *closed at one end* and open at the other (Fig. 12.4). The closed end is a pressure *antinode;* the air at the closed end meets a rigid surface, so there is no restriction on how far the pressure can deviate from atmospheric pressure. The closed end is also a *displacement node* since the air near it cannot move beyond that rigid surface. Some wind instruments are effectively pipes closed at one end. The reed of a clarinet admits only brief puffs of air into the instrument; the rest of the time the reed closes

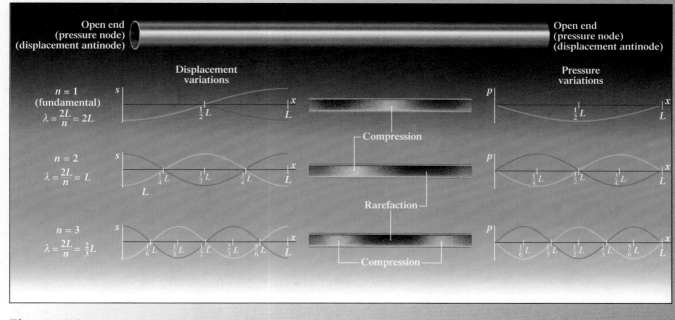

Figure 12.3 Standing waves in a pipe open at both ends

off that end of the pipe. The pressure at the reed end fluctuates above and below atmospheric pressure. The reed end is a pressure antinode and a displacement node.

The wavelengths and frequencies of the standing waves can be found using either the pressure or displacement descriptions of the wave. Using displacement, the fundamental has a node at the closed end, an antinode at the open end, and no other nodes or antinodes (see Fig. 12.5). The distance from a node to the nearest antinode is always $\frac{1}{4}\lambda$, so for the fundamental

$$L = \tfrac{1}{4}\lambda \text{ or } \lambda = 4L$$

which is twice as large as the wavelength ($2L$) of the fundamental in a pipe of the same length open at both ends. Two thin organ pipes of the same length, one open at both

Figure 12.4 Some organ pipes are open at the top; others are closed.

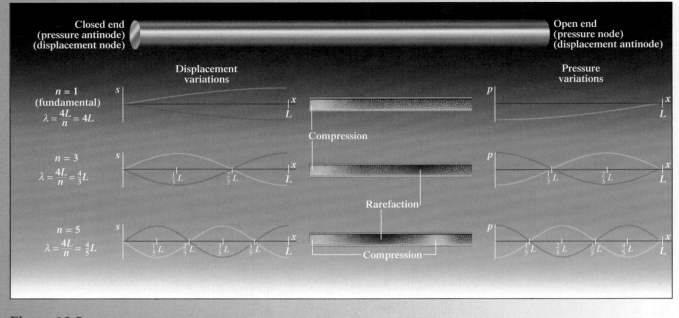

Figure 12.5   Standing waves in a pipe closed at one end

ends and one closed at one end, do not have the same wavelength fundamental. The pipe closed at one end has a wavelength twice as large and therefore a frequency half as large, assuming the pipes are thin. (For musicians: the pitch of the pipe closed at one end sounds an octave lower than the other, since the interval of an octave corresponds to a factor of two in frequency.)

What are the other standing wave frequencies? The next standing wave mode is found by adding one node and one antinode. Then the length of the pipe is 3 quarter-cycles: $L = \frac{3}{4}\lambda$ or $\lambda = \frac{4}{3}L$. This is $\frac{1}{3}$ the wavelength of the fundamental and the frequency is 3 times that of the fundamental. Adding one more node and one more antinode, the wavelength is $\frac{4}{5}L$. Continuing the pattern, we find that the wavelengths and frequencies for standing waves are

---

Standing sound waves (thin pipe closed at one end)

$$\lambda_n = \frac{4L}{n} \qquad (12\text{-}10a)$$

$$f_n = \frac{v}{\lambda_n} = n\frac{v}{4L} = nf_1 \qquad (12\text{-}10b)$$

where $n = 1, 3, 5, 7, \ldots$

 Note that the standing wave frequencies for a pipe closed at one end are only *odd* multiples of the fundamental.

The "missing" standing wave patterns for even values of n require a clarinet to have many more keys and levers than a flute (Fig. 12.6). What the keys do is effectively shorten the length of the pipe, making the standing wave frequencies higher. If a flute's fundamental frequency is f_1 with no keys pressed, the next highest frequency possible without using any keys is $2f_1$—the flutist overblows, exciting the next highest standing wave frequency rather than the fundamental. The flute needs enough keys to fill in all the notes with frequencies between f_1 and $2f_1$. For a clarinet, if the fundamental frequency is

Figure 12.6 A clarinet (a pipe open at one end) needs many more keys than a flute (a pipe open at both ends).

f_1 with no keys pressed, the next highest frequency possible without using any keys is $3f_1$. The clarinet must have more keys because it has to accommodate all the notes with frequencies between f_1 and $3f_1$.

Problem-Solving Strategy for Standing Waves

There is no need to memorize equations for standing wave frequencies and wavelengths. Just sketch the standing wave patterns as in Figs. 12.3 and 12.5. Make sure that nodes and antinodes alternate and that the boundary conditions at the ends are correct. Then determine the wavelengths by setting the distance between a node and antinode equal to $\frac{1}{4}\lambda$. Once the wavelengths are known, the frequencies are found from $v = f\lambda$.

Example 12.6

A Demonstration of Resonance

A thin hollow tube of length 1.00 m is inserted vertically into a tall container of water (Fig. 12.7a). A tuning fork ($f = 520.0$ Hz) is struck and held near the top of the tube as the tube is slowly pulled up and out of the water. At certain distances (L) between the top of the tube and the water surface, the otherwise faint sound of the tuning fork is greatly amplified. At what values of L does this occur? The temperature of the air in the tube is 18°C.

Strategy   Sound waves in the air inside the tube reflect from the water surface. Thus, we have an air column of variable length L, closed at one end by the water surface and open at the other end. The sound is amplified due to resonance; when the frequency of the tuning fork matches one of the natural frequencies of the air column, a large-amplitude standing wave builds up in the column. For standing waves in a column of air, the wavelength and frequency are related by the speed of sound in air. We start by finding the speed of sound in air from the temperature given. Then we can find the wavelength of the sound waves emanating from the tuning fork. Last, we

find the column lengths that support standing waves of that wavelength.

Solution   The speed of sound in air at 18°C is

$$v = (331 + 0.606 \times 18) \text{ m/s} = 342 \text{ m/s}$$

With the speed of sound and the frequency known, we can find the wavelength. The wavelength is the distance traveled by a wave during one period:

$$\lambda = vT = \frac{v}{f}$$

$$\lambda = \frac{342 \text{ m/s}}{520.0 \text{ Hz}} = 0.6577 \text{ m} = 65.77 \text{ cm}$$

The first possible resonance for a tube closed on one end occurs when there is a pressure node at the open end, a pressure antinode at the closed end, and no other pressure nodes or antinodes. Therefore,

$$L_1 = \tfrac{1}{4}\lambda = \tfrac{1}{4} \times 65.77 \text{ cm} = 16.4 \text{ cm}$$

continued on next page

Example 12.6 *continued*

To reach other resonances, the tube must be pulled out to accommodate additional pressure nodes and antinodes. To add one node and one antinode, the additional distance is $\frac{1}{2}\lambda = 32.9$ cm. The resonances occur at intervals of 32.9 cm:

$$L_2 = 16.4 \text{ cm} + 32.9 \text{ cm} = 49.3 \text{ cm}$$
$$L_3 = 49.3 \text{ cm} + 32.9 \text{ cm} = 82.2 \text{ cm}$$

The next one would require a tube longer than 1.00 m, so there are three values of L that produce resonance in this tube.

Discussion   As a check, we can sketch the standing wave pattern for the third resonance (Figs. 12.7b,c). There are 5 quarter-wavelengths in the length of the column so

$$L_3 = \tfrac{5}{4}\lambda = \tfrac{5}{4} \times 65.77 \text{ cm} = 82.2 \text{ cm}$$

At the open end of the tube, the node for pressure and the antinode for maximum displacement is actually a little *above* the opening. For this reason it is best to measure the distance between two successive resonances to find an accurate value for a half-wavelength rather than measuring the distance for the first possible resonance, the shortest distance between the opening and the water surface, and setting it equal to a quarter-wavelength.

Practice Problem 12.6   A roundabout way to measure temperature

A tuning fork of frequency 440.0 Hz is held above the hollow tube in Example 12.6. If the distance ΔL that the tube is moved between resonances is 39.3 cm, what is the temperature of the air inside the tube?

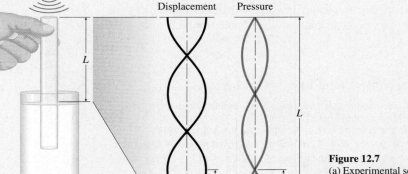

Figure 12.7
(a) Experimental setup for Example 12.6.
(b) Standing wave pattern, showing *pressure* nodes and antinodes, for the third resonance. (c) Standing wave pattern, showing *displacement* nodes and antinodes, for the third resonance.

Physics at Home

You can set up a resonance in an empty water bottle by blowing horizontally across the top of the bottle. Once you have heard one resonance, add varying amounts of water to raise the level within and listen for other resonances. The resonant sound is noticeably louder than the nonresonant sounds. Notice that the longer the air column within the bottle, the lower the pitch heard.

Making The Connection:
human ear

12.5   THE HUMAN EAR

Figure 12.8 shows the structure of the human ear. The human ear has an external part or *pinna* that acts something like a funnel, collecting sound waves and concentrating them at the opening of the auditory canal. The pinna is better at collecting sound coming from in front than from behind, which helps with localization. Resonance in the *auditory canal* (see Problem 44) boosts the ear's sensitivity in the 2- to 5-kHz frequency range—a crucial range for understanding speech.

At the end of the auditory canal, the eardrum (*tympanum*) vibrates in response to the incident sound wave. The region just beyond the eardrum is called the middle ear.

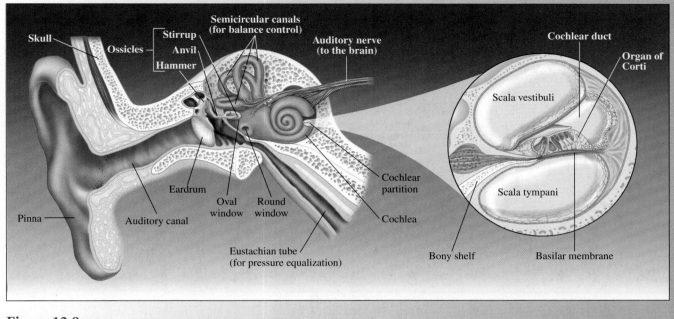

Figure 12.8   Structure of the human ear with a cross section of the cochlea

The vibrations of the eardrum are transmitted through three tiny bones of the middle ear (the *auditory ossicles*) to the *oval window* of the *cochlea*, a tapered spiral-shaped organ filled with fluid. The oval window is a membrane that is in contact with the fluid in the cochlea. The ossicles act as levers; the force exerted by the "stirrup" on the oval window is 1.5 to 2.0 times the force the eardrum exerts on the "hammer." The area of the oval window is one-twentieth that of the eardrum, so there is an overall amplification in pressure by a factor of 30 to 40. The ossicles protect the ear from damage: in response to a loud sound, a muscle pulls the stirrup away from the oval window. At the same time, another muscle increases the eardrum tension. These two changes make the ear temporarily less sensitive. It takes a few milliseconds for the muscles to respond in this way, so they provide no protection against *sudden* loud sounds.

The *cochlear partition* (Fig. 12.8) runs most of the length of the cochlea, separating it into two chambers (the *scala vestibuli* and the *scala tympani*). Vibration of the oval window sends a compressional wave down the fluid in the scala vestibuli, around the end of the partition, and back up the scala tympani to the *round window*. This wave sets the *basilar membrane*, located on the cochlear partition, into vibration. The basilar membrane is thinnest and under greatest tension near the oval and round windows; it gradually increases in thickness and decreases in tension toward its other end. High-frequency waves cause the membrane to vibrate with maximum amplitude near its thin, high-tension end; low-frequency waves cause maximum amplitude vibrations near its thicker, lower-tension end. The location of the maximum amplitude vibrations is one way the ear determines frequency; for low-frequency sounds (up to about 1 kHz), the ear sends periodic nerve signals to the brain at the frequency of the sound wave. For complex sounds, which consist of the superposition of many different frequencies (see Section 12.6), the ear performs a spectral analysis—it decomposes the complex sound into its constituent frequencies.

Located on the basilar membrane is the sensory organ (the *organ of Corti*). Rows of hair cells on the basilar membrane excite neurons when they bend in response to vibration. These neurons send electrical signals to the brain.

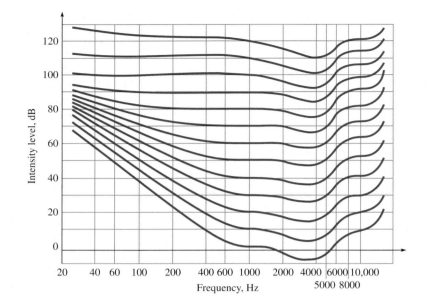

Figure 12.9 Curves of equal loudness

Loudness

While loudness is most closely correlated to intensity level, it also depends on frequency (as well as other factors). In other words, the sensitivity of the ear is frequency-dependent. Figure 12.9 shows a set of *curves of equal loudness* for a typical person. Each curve shows the intensity levels required so that sounds of different frequencies are equally loud. The curves show that the ear is most sensitive to frequencies between 3 kHz and 4 kHz, partly due to resonance in the auditory canal (see Problem 44). The ear's sensitivity falls off rapidly below 800 Hz and above 10 kHz. At any given frequency between 800 Hz and 10 kHz, the curves are approximately evenly spaced: equal steps in intensity level produce equal steps in loudness, which is why intensity level is often used as an approximate measure of loudness. In this frequency range, 1 dB is about the smallest change in intensity level that is perceptible as a change in loudness.

The threshold of hearing is shown by the lowest curve in the set; a person with excellent hearing cannot hear sounds with intensity levels below this curve. The threshold of hearing is at an intensity level of 0 dB only in the vicinity of 1 kHz.

Pitch

Pitch is the perception of frequency. If you sing or play up and down a scale, it is the pitch that is rising and falling. Although pitch is the aspect of sound perception most closely tied to a single physical quantity, frequency, our sense of pitch is affected to a small extent by other factors such as intensity and timbre (Section 12.6).

Our sense of pitch is a *logarithmic* function of frequency, just as loudness is approximately a logarithmic function of intensity. If you start at the lowest note on the piano (which has a fundamental frequency of 27.5 Hz) and play a chromatic scale—every white and black key in turn—all the way to the highest note (4190 Hz), you hear a series of equal steps in pitch. The frequencies do *not* increase in equal steps; the fundamental frequency of each note is 5.95% higher than the previous note. Under ideal conditions, most people can sense frequency changes as small as 0.3%. A trained musician can sense a frequency change of 0.1% or so.

Localization

How can you tell where a sound comes from? The ear has several different tools it uses to localize sounds:

- The principal method for high-frequency sounds (> 4 kHz) is the difference in intensity sensed by the two ears. The head casts a "sound shadow," so a sound coming from the right has a larger intensity at the right ear than at the left ear.

- The shape of the pinna makes it slightly preferential to sounds coming from the front. This helps with front-back localization for high-frequency sounds.

- For lower-frequency sounds, both the difference in arrival time and the phase difference between the waves arriving at the two ears is used for localization.

12.6 TIMBRE

The sound produced by the vibration of a tuning fork is nearly a pure sinusoid at a single frequency. In contrast, most musical instruments produce complex sounds that are the superposition of many different frequencies. The standing wave on a string or in a column of air is almost always the superposition of many standing wave patterns at different frequencies. The lowest frequency in a complex sound wave is called the fundamental; the rest of the frequencies are called **overtones**. All the overtones of a periodic sound wave have frequencies that are integral multiples of the fundamental; the fundamental and the overtones are then called **harmonics**.

Middle C played on an oboe does not sound the same as middle C played on a trumpet, even though the fundamental frequency is the same, largely because the two instruments produce overtones with different relative amplitudes. What is different about the two sounds is the **tone quality** or **timbre** (pronounced either "tamber" or "timber").

Any periodic wave, no matter how complicated, can be decomposed into a set of harmonics, each of which is a simple sinusoid. The characteristic wave form for a note played on a clarinet, for example, can be decomposed into its harmonic series (Fig. 12.10). This process is called harmonic analysis, or Fourier analysis, in honor of the French mathematician, Jean Baptiste Joseph Fourier (1768–1830), who developed mathematical methods for analyzing periodic functions.

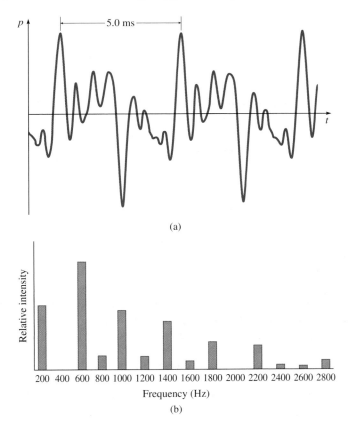

(a)

(b)

Figure 12.10 (a) A graph of the sound wave produced by a clarinet. (b) A bar graph showing the relative intensities of the harmonics, often called the *spectrum*. Notice that *odd* multiples of the fundamental dominate the spectrum. (Data courtesy of P. D. Krasicky, Cornell University.)

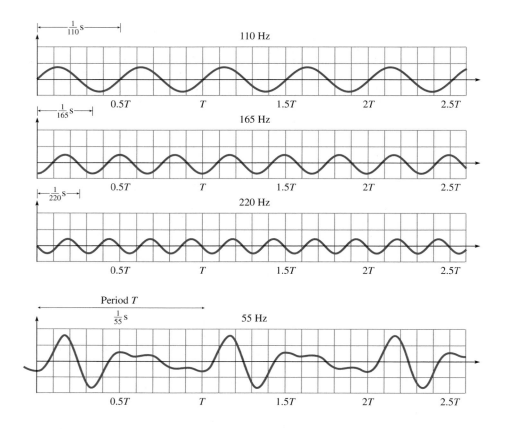

Figure 12.11 Complex wave form (bottom wave) composed by super-position of three sinusoidal waves (three upper waves). Note that the fundamental of the sum is not present in any of the constituent waves.

The opposite of harmonic analysis is harmonic synthesis: combining various harmonics to produce a complex wave. Electronic synthesizers can mimic the sounds of various instruments. Realistic-sounding synthesizers must also allow the adjustment of other parameters such as the attack and decay of the sound.

Although the spectrum of a periodic wave consists only of members of a harmonic sequence, not all members of the sequence need be present, not even the fundamental. A wave with three harmonic components having frequencies of 110, 165, and 220 Hz repeats at a frequency of 55 Hz because each of these three frequencies is an integral multiple of 55 Hz (Fig. 12.11). Even though the fundamental is missing—there is no harmonic component at 55 Hz—the ear is clever enough to "reconstruct" a 55-Hz tone. That's why you can listen to and recognize music on an inexpensive radio whose speaker may reproduce only a small range of frequencies.

12.7 BEATS

When two sound waves are close in frequency (within about 15 Hz of each other), the superposition of the two produces a pulsation that we call **beats**. Beats can be produced by any kind of wave; they are a general result of the principle of superposition when applied to two waves of nearly the same frequency.

Beats are caused by the slow change in the phase difference between the two waves. Suppose that at one instant ($t = 0$ in Fig. 12.12), the two waves are in phase with each other and interfere constructively. According to the superposition principle, the resultant amplitude is maximum—the sum of the amplitudes of the two waves shown in Fig. 12.12a, b. However, since the frequencies are different, the waves do not stay in phase. The higher-frequency wave has a shorter cycle, so it gets ahead of the other one. The phase difference between the two steadily increases; as it does, the resultant amplitude decreases. At a later time ($t = 5T_0$), the phase difference reaches 180°; now the waves are half a cycle out of phase and interfere destructively (Fig. 12.12c). Now the

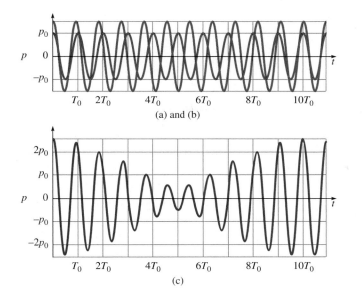

Figure 12.12 (a) Graph (red) of a sound wave with frequency $f_1 = 1/T_0$ and amplitude p_0. (b) Graph (blue) of a second sound wave with frequency $f_2 = 1.1f_1$ and amplitude $1.5p_0$. (c) The superposition of the two has maximum amplitude $2.5p_0$ and minimum amplitude $0.5p_0$.

resultant amplitude is minimum—the difference between the amplitudes of the two waves. As the phase difference continues to increase, the amplitude increases until constructive interference occurs again ($t = 10T_0$). The ear perceives the amplitude (and intensity) cycling from large to small to large to small as a pulsation in loudness.

At what frequency do the beats occur? It depends on how far apart the frequencies of the two waves are. We can measure the time between beats T_{beat} as the time to go from constructive interference to the next occurrence of constructive interference. During that time, each wave must go through a whole number of cycles, with one of them going through one more cycle than the other. Since frequency (f) is the number of cycles per second, the number of cycles a wave goes through during a time T_{beat} is fT_{beat}. (To illustrate: in Fig. 12.12, $T_{beat} = 10T_0$. During that time, wave 1 goes through $f_1T_{beat} = 10$ cycles, while wave 2 goes through $f_2T_{beat} = 1.1/T_0 \times 10T_0 = 11$ cycles.) If $f_2 > f_1$, then wave 2 goes through one cycle more than wave 1:

$$f_2T_{beat} - f_1T_{beat} = (\Delta f)T_{beat} = 1$$

The beat frequency f_{beat} is $1/T_{beat}$:

$$f_{beat} = 1/T_{beat} = \Delta f \qquad (12\text{-}11)$$

Thus, we obtain the remarkably simple result that the beat frequency is the difference between the frequencies of the two waves. If the difference in frequencies exceeds roughly 15 Hz, then the ear no longer perceives a pulsating intensity; instead, we hear two tones at different pitches.

Piano tuners listen for beats as they tune. The tuner sounds two strings and listens for the beats. The beat frequency indicates whether the interval is correct or not. If the two strings are played by the same key, they are tuned to the same fundamental frequency, so the beat frequency should be (nearly) zero. If the two strings belong to two different notes, the beat frequency is nonzero. Actually, in this case the tuner listens to beats between two *overtones* that are close in frequency, not the fundamentals. The fundamental frequencies are too far apart for the beats to be perceptible.

Example 12.7

The Piano Tuner

A piano tuner strikes his tuning fork ($f = 523.3$ Hz) and strikes a key on the piano at the same time. The two have nearly the same frequency; he hears 3.0 beats per second. As he tightens the piano string, he hears the beat frequency gradually decrease to 2.0 beats per second when the two sound together. (a) What was the frequency of the piano string before it was tightened? (b) By what percentage did the tension increase?

Strategy The beat frequency is the difference between the two frequencies; we only have to determine which is higher. The wavelength of the string is determined by its length, which does not change. The increase in tension increases the speed of waves on the string, which in turn increases the frequency.

Solution (a) Since the piano tuner heard 3.0 beats per second, the difference in the two frequencies was 3.0 Hz:

$$\Delta f = 3.0 \text{ Hz}$$

Is the piano string's frequency 3.0 Hz higher or 3.0 Hz lower than the tuning fork's frequency? As the tension increases gradually, the beat frequency decreases, which means that the frequency of the piano string is getting *closer* to the frequency of the tuning fork. Therefore, the string frequency must be 3.0 Hz *lower* than the tuning fork frequency:

$$f_{\text{string}} = 523.3 \text{ Hz} - 3.0 \text{ Hz} = 520.3 \text{ Hz}$$

(b) The tension (F) is related to the speed of the wave on the string (v) and the mass per unit length (μ) by

$$v = \sqrt{\frac{F}{\mu}} \qquad (11\text{-}4)$$

The mass per unit length does not change, so $v \propto \sqrt{F}$. The speed of the wave on the string is related to its wavelength and frequency by

$$v = \lambda f$$

⚠ The wavelength λ in this expression is the wavelength of the transverse wave on the string, *not* the wavelength of the sound wave in air. Since λ does not change, $v \propto f$. Therefore, $f \propto \sqrt{F}$ or

$$F \propto f^2$$

This means that the ratio of the tension F to the original tension F_0 is equal to the ratio of the frequencies squared:

$$\frac{F}{F_0} = \left(\frac{f}{f_0}\right)^2 = \left(\frac{521.3 \text{ Hz}}{520.3 \text{ Hz}}\right)^2 = 1.004$$

The tension was increased 0.4%.

Discussion We needed to find whether the original frequency was too high or too low. As the beat frequency decreases, the frequency of the string is getting closer to the frequency of the tuning fork. Tightening the string makes the string's frequency increase; since increasing the string's frequency brings it closer to the tuning fork's frequency, we know that the original frequency of the string was lower than the frequency of the tuning fork. Had an increase in tension *increased* the beat frequency instead, we would know that the original frequency was already too high; the tension would have to be relaxed to tune the string.

Practice Problem 12.7   Tuning a violin

A tuning fork with a frequency of 440.0 Hz produces 4.0 beats per second when sounded together with a violin string of nearly the same frequency. What is the frequency of the string if a slight increase in tension increases the beat frequency?

12.8   THE DOPPLER EFFECT

A police car races by, its sirens screaming. As it passes, we hear the pitch change from higher to lower. The frequency change is called the **Doppler effect**, after the Austrian physicist Christian Andreas Doppler (1803–1853). The observed frequency is different from the frequency transmitted by the source when one or both are in motion relative to the wave medium.

We consider only the motion of the source and observer directly toward or away from each other in the reference frame in which the wave medium is at rest. Velocities of the source and observer are expressed as components along the direction of propagation of the sound wave (from source to observer). A positive component means the velocity is in the direction of propagation of the wave, while a negative component means the velocity is opposite the direction of propagation.

Moving Source

First we consider a moving source. A source emits a sound wave at frequency f_s, which means that wave crests (regions of maximum amplitude, indicated by circles in Fig. 12.13) leave the source spaced by a time interval $T_s = 1/f_s$. If the source is moving at velocity v_s

Figure 12.13 (a) A steamboat is moving to the right at speed v_s while it blows its whistle. The whistle emits wave crests at positions 1, 2, 3, 4, and 5; each wave crest moves outward in all directions, from the point at which it was emitted, at speed v. (b) The observed wavelength λ_o is the distance between wave crests.

toward a stationary observer on the right, Fig. 12.13a shows that the wavelength—the distance between crests—is smaller in front of the source and larger behind the source. In Fig. 12.13b, at the instant that crest 6 is emitted, crest 5 has traveled outward a distance vT_s from point 5, where v is the speed of sound. During the same time interval, the source has advanced a distance v_sT_s. The wavelength λ_o as measured by the observer on the right is the distance between crests 5 and 6:

$$\lambda_o = vT_s - v_sT_s$$

The frequency at which the crests arrive at the observer is the *observed* wave frequency f_o. The observed period T_o between the arrival of two crests is the time it takes sound to travel a distance $(v - v_s)T_s$:

$$T_o = \frac{(v - v_s)T_s}{v}$$

The observed frequency is

$$f_o = \frac{1}{T_o} = \frac{v}{(v - v_s)} \times \frac{1}{T_s}$$

Dividing numerator and denominator by v and substituting $f_s = 1/T_s$ yields

Doppler effect (moving source)

$$f_o = \left(\frac{1}{1 - v_s/v} \right) f_s \qquad (12\text{-}12)$$

Since the denominator $1 - v_s/v$ is less than 1, the observed frequency is higher than the source frequency when the source moves in the same direction as the wave (toward the observer). If the source instead moves *away* from the observer, the correct observed frequency is given by Eq. (12-12) as long as we make v_s negative (the source moves opposite the direction of the wave). With v_s negative, $1 - v_s/v$ is *greater* than 1, so the observed frequency is *less* than the source frequency.

Moving Observer

Now we consider motion of the observer. A stationary source emits a sound wave at frequency f_s and wavelength $\lambda_s = v/f_s$, where v is the speed of sound. A stationary observer would measure the arrival of wave crests spaced by a time interval $T_s = 1/f_s$. An observer moving away from the source at velocity v_o would observe a longer time interval between crests. Just as crest 1 reaches the observer, the next (crest 2) is a distance λ_s away. Crest 2 catches up with the observer at a time T_o later when the distance the wave crest travels toward the observer is equal to the distance the observer travels away from the wave crest plus the wavelength (Fig. 12.14):

$$vT_o = v_oT_o + \lambda_s \quad \text{or} \quad (v - v_o)T_o = \lambda_s = v/f_s$$

Solving for T_o,

$$T_o = \frac{v/f_s}{v - v_o}$$

The observed frequency is

$$f_o = \frac{1}{T_o} = \frac{v - v_o}{v}f_s$$

Dividing numerator and denominator by v yields

Doppler effect (moving observer)

$$f_o = \left(1 - \frac{v_o}{v}\right)f_s \tag{12-13}$$

An observer moving away from the source measures a frequency lower than f_s. An observer moving *toward* the source moves opposite to the direction of the wave; in that case, v_o is negative and the observed frequency is *higher* than f_s.

Example 12.8

Train Whistle and Doppler Shift

A monorail train approaches a platform at a speed of 10.0 m/s while it blows its whistle. A musician with perfect pitch standing on the platform hears the whistle as "middle C," a frequency of 261 Hz. There is no wind and the temperature is a chilly 0°C. What is the observed frequency of the whistle when the train is at rest?

Strategy In this case, the source—the whistle—is moving and the observer is stationary. The source is moving *toward* the observer, so v_s is *positive*. With the source approaching the observer, the observed frequency is higher than the source frequency. When the train is at rest, there is no Doppler shift; the observed frequency then is equal to the source frequency.

Solution For a moving source, the source (f_s) and observed (f_o) frequencies are related by

$$f_o = \left(\frac{1}{1 - v_s/v}\right)f_s$$

where $v = 331$ m/s (the speed of sound in air at 0°C), $v_s = +10.0$ m/s, and $f_o = 261$ Hz. Solving for f_s,

$$f_s = (1 - v_s/v)f_o$$

$$= \left(1 - \frac{10.0 \text{ m/s}}{331 \text{ m/s}}\right) \times 261 \text{ Hz}$$

$$= 253 \text{ Hz}$$

The source frequency is less than the observed frequency, as expected. The observed frequency when the train is at rest is equal to the source frequency: 253 Hz.

continued on next page

Example 12.8 *continued*

Discussion   When the train is moving toward the platform, the distance between source and observer is decreasing. Wave crests emitted later take *less time* to reach the observer than if the train were at rest, so the time between arrival of wave crests is smaller than if the train were stationary. When the distance between source and observer is decreasing, the observed frequency is higher than the source frequency; when the distance is increasing, the observed frequency is lower than the source frequency.

Practice Problem 12.8   A sports car racing by Justine is gardening in her front yard when a Mazda Miata races by at 32.0 m/s (71.6 mi/h). If she hears the sound of the Miata's engine at 220.0 Hz as it approaches her, what frequency does she hear after it passes? Assume the temperature is 20°C and there is no wind.

Motion of Both Source and Observer

If both source and observer are moving, we combine the two Doppler shifts (see Conceptual Question 11) to obtain

$$f_o = \left(\frac{1 - v_o/v}{1 - v_s/v}\right) f_s \tag{12-14}$$

Remember that the signs of v_o and v_s are positive for motion in the direction of propagation of the wave and negative for motion opposite the direction of propagation.

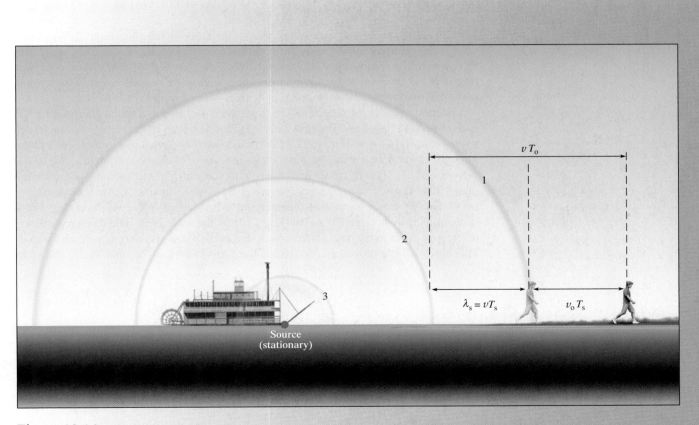

Figure 12.14   An observer moving away from a stationary source. The observed frequency is lower than the source frequency.

Example 12.9

Determining Speed from Horn Frequency

Two cars, with equal ground speeds, are moving in opposite directions away from each other on a straight highway. One driver blows a horn with a frequency of 111 Hz; the other measures the frequency as 105 Hz. If the speed of sound is 338 m/s and there is no wind, what is the ground speed of each car?

Strategy　A sound wave travels from source to observer. The source moves opposite the direction of the wave, so v_s is negative. The observer moves in the direction of the wave, so v_o is positive. The speeds are the same, so $v_s = -v_o$.

Solution　With both the source and observer moving, the frequencies are related by

$$f_o = \left(\frac{1 - v_o/v}{1 - v_s/v}\right) f_s$$

To simplify the algebra, we let $\alpha = v_o/v = -v_s/v$. Then

$$f_o = \left(\frac{1 - \alpha}{1 + \alpha}\right) f_s$$

Now we solve for α:

$$(1 + \alpha)\frac{f_o}{f_s} = 1 - \alpha$$

$$\alpha = \frac{1 - f_o/f_s}{1 + f_o/f_s} = \frac{1 - (105\ \text{Hz}/111\ \text{Hz})}{1 + (105\ \text{Hz}/111\ \text{Hz})} = 0.02777$$

Now we can find v_o:

$$v_o = \alpha v = 0.02777 \times 338\ \text{m/s} = 9.4\ \text{m/s}$$

The speed of each car is 9.4 m/s.

Discussion　Quick check on the algebra: substituting $v = 338$ m/s, $f_s = 111$ Hz, $v_o = 9.4$ m/s, and $v_s = -9.4$ m/s directly into Eq. (12-14),

$$f_o = \frac{1 - (9.4\ \text{m/s})/(338\ \text{m/s})}{1 - (-9.4\ \text{m/s})/(338\ \text{m/s})} \times 111\ \text{Hz} = 105\ \text{Hz}$$

Practice Problem 12.9　Finding speed from the Doppler shift

A car is driving due west at 15 m/s and sounds its horn with a frequency of 260.0 Hz. A passenger in a car heading east away from the first car hears the horn at a frequency of 230.0 Hz. How fast is the second car traveling? The speed of sound is 350 m/s.

12.9　SHOCK WAVES

Let's examine two interesting special cases of the Doppler formula [Eq. (12-14)]. First, what if the observer moves away from the source at the speed of sound ($v_o = v$)? The Doppler-shifted frequency would be zero according to Eq. (12-14). What does that mean? If the observer moves away from the source with a speed equal to (or greater than) the wave speed, the wave crests *never reach the observer*.

Second, what if the source moves toward the observer at a speed approaching the speed of sound ($v_s \to v$)? Then Eq. (12-14) gives an observed frequency that increases without bound ($f_o \to \infty$). Figure 12.15 helps us understand what that means. For a plane moving slower than sound, the wave crests in front of it are closer together due to the plane's motion (Fig. 12.15a). An observer to the right would measure a frequency higher than the source frequency. As the plane's speed increases, the wave crests in front of it get closer and closer together and the observed frequency increases. For a plane moving at the speed of sound (Fig. 12.15b), the wave crests pile up on top of each other; they move to the right at the same speed as the plane, so they can't get ahead of it. This wall of high-pressure air is called the *sound barrier*. An observer to the right

Making The Connection:
shock wave of a
supersonic plane

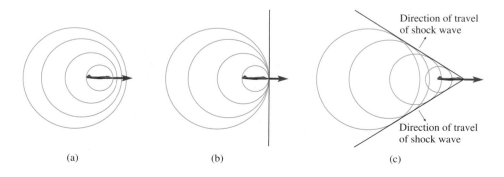

Figure 12.15 (a) Wave crests for a plane moving slower than sound. (b) A plane moving at the speed of sound; the wave crests pile up on each other since the plane moves to the right as fast as the wave crests. (c) Shock wave for a supersonic plane.

(a)　　　　(b)　　　　(c)

Direction of travel of shock wave

Direction of travel of shock wave

time taken for the echo to return. How deep is the ocean at a point where the echo time (down and back) is 7.07 s? The temperature of the seawater is 25°C.

✦38. A bat emits chirping sounds of frequency 82.0 kHz while hunting for moths to eat. If the bat is flying toward the moth at a speed of 4.40 m/s and the moth is flying away from the bat at 1.20 m/s, what is the frequency of the sound wave reflected from the moth as observed by the bat? Assume it is a cool night with a temperature of 10.0°C. [*Hint:* There are two Doppler shifts. Think of the moth as a receiver, which then becomes a source as it "retransmits" the reflected wave.]

39. The bat of Problem 38 emits a chirp that lasts for 2.0 ms and then is silent while it listens for the echo. If the beginning of the echo returns just after the outgoing chirp is finished, how close to the moth is the bat? [*Hint:* Is the change in distance between the two significant during a 2.0-ms time interval?]

✦40. Doppler ultrasound is used to measure the speed of blood flow (see Problem 31). The reflected sound interferes with the emitted sound, producing beats. If the speed of red blood cells is 0.10 m/s, the ultrasound frequency used is 5.0 MHz, and the velocity of sound in blood is 1570 m/s, what is the beat frequency?

✦41. (a) In Problem 31, find the beat frequency between the outgoing and reflected sound waves. (b) Show that the beat frequency is proportional to the speed of the blood cell if $v \ll c$. [*Hint:* Use the binomial approximation from Appendix A.5.]

42. A ship is lost in a dense fog in a Norwegian fjord that is 1.80 km wide. The air temperature is 5.0°C. The captain fires a pistol and hears the first echo after 4.0 s. (a) How far from one side of the fjord is the ship? (b) How long after the first echo does the captain hear the second echo?

COMPREHENSIVE PROBLEMS

43. What are the four lowest standing-wave frequencies for an organ pipe that is 4.80 m long and closed at one end?

44. The length of the auditory canal in humans averages about 2.5 cm. What are the lowest three standing-wave frequencies for a pipe of this length open at one end? What effect might resonance have on the sensitivity of the ear at various frequencies? (Refer to Fig. 12.9. Note that frequencies critical to speech recognition are in the range 2 to 5 kHz.)

45. The Vespertilionidae family of bats detect the distance to an object by timing how long it takes for an emitted signal to reflect off the object and return. Typically they emit sound pulses 3 ms long and 70 ms apart while cruising. (a) If an echo is heard 60 ms later ($v_{sound} = 331$ m/s), how far away is the object? (b) When an object is only 30 cm away, how long will it be before the echo is heard? (c) Will the bat be able to detect this echo?

46. Some bats determine their distance to an object by detecting the difference in intensity between echoes. (a) If intensity falls off at a rate that is inversely proportional to the distance squared, show that the echo intensity is inversely proportional to the fourth power of distance. (b) The bat was originally 0.60 m from one object and 1.10 m from another. After flying closer, it is now 0.50 m from the first and at 1.00 m from the second object. What is the percentage increase in the intensity of the echo from each object?

47. Horseshoe bats use the Doppler effect to determine their location. A Horseshoe bat flies toward a wall at a speed of 15 m/s while emitting a sound of frequency 35 kHz. What is the beat frequency between the emission frequency and the echo?

48. At what frequency f does a sound wave in air have a wavelength of 15 cm, about half the diameter of the human head? Some methods of localization work well only for frequencies below f while others work well only above f. (See Conceptual Questions 4 and 5.)

✦✦49. A periodic wave is composed of the superposition of three sine waves whose frequencies are 36, 60, and 84 Hz. The speed of the wave is 180 m/s. What is the wavelength of the wave?

✦✦50. Analysis of the periodic sound wave produced by a violin's G string includes three frequencies: 392, 588, and 980 Hz. What is the fundamental frequency? [*Hint:* The wave on the string is the superposition of several different standing wave patterns.]

51. According to a treasure map, a treasure lies at a depth of 40.0 fathoms on the ocean floor due east from the lighthouse. The treasure hunters use sonar to find where the depth is 40.0 fathoms as they head east from the lighthouse. What is the elapsed time between an emitted pulse and the return of its echo at the correct depth if the water temperature is 25°C? [*Hint:* One fathom is 1.83 m.]

✦52. An airplane is flying 1 km directly over your position on the ground at Mach 2. How far from that overhead position will the airplane have moved along its horizontal flight path when you hear the sonic boom? [*Hint:* See Problem 34 and Fig. 12.20.]

53. A wind tunnel is used to simulate the flight of a plane. Air at 20°C is blown past the model plane at very high speeds. If a shock cone angle of $\theta = 40.0°$ develops, how fast is the air moving? [*Hint:* See Problem 34 and Fig. 12.20.]

54. In this problem, you will estimate the smallest kinetic energy of vibration that the human ear can detect. Suppose that a harmonic sound wave at the threshold of hearing ($I = 1.0 \times 10^{-12}$ W/m^2) is incident on the eardrum. The speed of sound is 340 m/s and the density of air is 1.3 kg/m^3. (a) What is the maximum speed of an element of air in the sound wave? [*Hint:* See Eq. (10-21).] (b) Assume the eardrum vibrates with displacement s_0 at angular frequency ω; its maximum speed is then equal to the maximum speed of an air element. The mass of the eardrum is approximately 0.1 g.

What is the *average* kinetic energy of the eardrum? (c) The average kinetic energy of the eardrum due to collisions with air molecules *in the absence of a sound wave* is about 10^{-20} J. Compare your answer to (b) and discuss.

✦55. When playing *fortissimo* (very loudly), a trumpet emits sound energy at a rate of 0.800 W out of a bell (opening) of diameter 12.7 cm. (a) What is the sound intensity level right in front of the trumpet? (b) If the trumpet radiates sound waves uniformly in all directions, what is the sound intensity level at a distance of 10.0 m?

✦✦56. During a rehearsal, all eight members of the first violin section of an orchestra play a very soft passage. The sound intensity level at a certain point in the concert hall is 38.0 dB. What is the sound intensity level at the same point if only one of the violinists plays the same passage? [*Hint:* When playing together, the violins are *incoherent* sources of sound.]

ANSWERS TO PRACTICE PROBLEMS

12.1 Although solids usually have somewhat higher densities than liquids, they have *much* higher bulk moduli—they are much stiffer. The greater restoring forces in solids cause sound waves to travel faster.

12.2 Assumptions: Treat the stage as a point source; neglect reflection and absorption of waves. 4.0×10^{-6} W/m^2, 0.057 Pa.

12.3 400

12.4 a factor of 3.2

12.5 3000 km. No; it is not realistic to ignore absorption and reflection over such a great distance.

12.6 24°C

12.7 444.0 Hz

12.8 182 Hz

12.9 27 m/s

Temperature and the Ideal Gas

I n warm-blooded or homeothermic (constant temperature) animals, body temperature is carefully regulated. The hypothalamus, located in the brain, acts as the master thermostat to keep body temperature constant to within a fraction of a degree Celsius in a healthy animal. If the body temperature starts to deviate much from the desired constant level, the hypothalamus causes changes in blood flow and initiates other processes, such as shivering or perspiration, to bring the temperature back to normal. Why is this careful temperature regulation necessary? What would be the problem if body temperature fluctuated with environmental conditions? What evolutionary advantage does a constant body temperature give the warm-blooded animals (birds, mammals) over the cold-blooded (such as reptiles and insects)? What are the disadvantages?

13.1   TEMPERATURE

The measurement of **temperature** is part of everyday life. We measure the temperature of the air outdoors to decide how to dress when going outside; a thermostat measures the air temperature indoors to control heating and cooling systems to keep our homes and offices comfortable. Regulation of oven temperature is important in baking. When we feel ill, we measure our body temperature to see if we have a fever. Despite how matter-of-fact it may seem, temperature is a subtle concept. Although our subjective sensations of hot and cold are related to temperature, they can easily mislead, as the next Physics at Home demonstrates.

The definition of temperature is based on the concept of **thermal equilibrium**. Suppose two objects or systems are allowed to exchange energy. The net flow of energy is always from the object at the higher temperature to the object at the lower temperature. As energy flows, the temperatures of the two objects approach each other. When the temperatures are the same, there is no longer any net flow of energy; the objects are now said to be in thermal equilibrium. Thus, *temperature is a quantity that determines when objects are in thermal equilibrium.* (The objects do *not* necessarily have the same *energy* when in thermal equilibrium.) The flow of energy that occurs between two objects or systems due to a temperature difference between them is called **heat flow**. In Chapter 14 we discuss heat flow in detail. If heat can flow between two objects or systems, the objects or systems are said to be in **thermal contact**.

Temperature measurement relies on the **zeroth law of thermodynamics**.

Heat is the flow of energy due to a temperature difference. Heat always flows from the hotter object to the colder object.

Zeroth law of thermodynamics:

If two objects are each in thermal equilibrium with a third object, then the two are in thermal equilibrium with each other.

The rather odd name *zeroth* law of thermodynamics came about because this law was formulated historically *after* the first, second, and third laws of thermodynamics and yet it is so fundamental that it should come *before* the others. **Thermodynamics**, the subject of Chapters 13 to 15, concerns temperature, heat flow, and the internal energy of systems.

Physics at Home

Try an experiment described by the philosopher John Locke in 1690. Fill one container with water that is hot (but not too hot to touch); fill a second container with lukewarm water; and fill a third container with cold water. Put one hand in the hot water and one in the cold water (Fig. 13.1) for about 10 to 20 seconds. Then plunge both hands into the container of lukewarm water. Although both hands are now immersed in water that is at a single temperature, the hand that had been in the hot water feels cool while the hand that had been in the cold water feels warm. This demonstration shows that we cannot trust our subjective senses to measure temperature.

Cold   Lukewarm   Hot

Figure 13.1 It is easy to trick our sense of temperature.

To measure the temperature of an object, we put a thermometer into thermal contact with the object. If the zeroth law were not true, it would be impossible to define temperature, since different thermometers could give different results. The zeroth law may *seem* obvious, but it is possible to imagine a universe in which there is no net flow of energy when A is in thermal contact with B, nor when B is in thermal contact with C, and yet energy does flow when A is in contact with C.

13.2   TEMPERATURE SCALES

Thermometers measure temperature by exploiting some property of matter that is temperature-dependent. The familiar liquid-in-glass thermometer relies on thermal expansion: the mercury or alcohol expands as its temperature rises (or contracts as its temperature drops) and we read the temperature on a calibrated scale. Since some materials expand more than others, these thermometers must be calibrated on a scale using some easily reproducible phenomenon, such as the melting point of ice or the boiling point of water. The assignment of temperatures to these phenomena is arbitrary.

The most commonly used temperature scale in the world is the Celsius scale. On the Celsius scale, 0°C is the freezing temperature of water at $P = 1$ atm (the *ice point*) and 100°C is the boiling temperature of water at $P = 1$ atm (the *steam point*).

In the United States, the Fahrenheit scale is still commonly used. The ice point is 32°F and the steam point is 212°F, so that the difference between the steam and ice points is 180°F. The size of the Fahrenheit degree interval is therefore smaller than the Celsius degree interval: a temperature difference of 1°C is equivalent to a difference of 1.8°F:

$$\Delta T_F = \Delta T_C \times 1.8 \frac{°F}{°C} \tag{13-1}$$

Since the two scales also have an offset—0°C is not the same temperature as 0°F—conversion between the two is:

$$T_F = 1.8 T_C + 32°F \tag{13-2a}$$

$$T_C = \frac{T_F - 32°F}{1.8} \tag{13-2b}$$

The SI unit of temperature is the **kelvin** (symbol K, *without* a degree sign). The kelvin has the same degree size as the Celsius scale; that is, a temperature *difference* of 1°C is the same as a difference of 1 K. However, 0 K represents *absolute zero*—there are no temperatures below 0 K. The ice point is 273.15 K, so temperature in °C (T_C) and temperature in kelvins (T) are related.

$$T_C = T - 273.15 \tag{13-3}$$

Equation (13-3) is the definition of the Celsius scale in terms of the kelvin. Table 13.1 shows some temperatures in kelvins, °C and °F.

Table 13.1

Some Reference Temperatures in K, °C, and °F

	K	°C	°F		K	°C	°F
Absolute zero	0	−273.15	−459.67	Human body temperature	310	37	98.6
Lowest transient temperature				Water boils	373.15	100.00	212.0
achieved (laser cooling)	10^{-9}			Campfire	1,000	700	1,300
Intergalactic space	3	−270	−454	Gold melts	1,337	1,064	1,947
Helium boils	4.2	−269	−452	Lightbulb filament	2,500	2,200	4,000
Nitrogen boils	77	−196	−321	Surface of Sun; iron welding arc	6,300	6,000	11,000
Carbon dioxide freezes ("dry ice")	195	−78	−108	Center of Earth	16,000	15,700	28,300
Mercury freezes	234	−39	−38	Center of Sun	10^7	10^7	10^7
Ice melts/water freezes	273.15	0	32.0	Interior of neutron star	10^9	10^9	10^9

Example 13.1

A Sick Friend

A friend suffering from the flu has a fever; her body temperature is 38.6°C. What is her temperature in (a) K and (b) °F?

Strategy   (a) Kelvins and °C differ only by a shift of the zero point. Converting from °C to K requires only the addition of 273.15 K since 0°C (the ice point) corresponds to 273.15 K. (b) The °F is a different size than the °C, as well as having a different zero. In the Celsius scale, the zero is at the ice point. First multiply by 1.8°F/°C to find how many °F above the ice point. Then add 32°F (the Fahrenheit temperature of the ice point).

Solution   (a) The temperature is 38.6 K *above* the ice point of 273.15 K. Therefore, the kelvin temperature is

$$T = 38.6\text{ K} + 273.15\text{ K} = 311.8\text{ K}$$

(b) First find how many °F above the ice point:

$$\Delta T_F = 38.6\text{°C} \times (1.8\text{°F/°C}) = 69.5\text{°F}$$

The ice point is 32°F, so

$$T_F = 32.0\text{°F} + 69.5\text{°F} = 101.5\text{°F}$$

Discussion   The answer is reasonable since 98.6°F is normal body temperature.

Practice Problem 13.1   Normal body temperatures with two scales

Convert the normal human body temperature (98.6°F) to degrees Celsius and kelvins.

Figure 13.2   The blacksmith heats rods of iron in his hot forge and, before the iron cools, works quickly to shape it at his anvil into hoops for barrels, rims for wagon wheels, and shoes for horses.

13.3   THERMAL EXPANSION OF SOLIDS AND LIQUIDS

Most objects expand as their temperature increases. Long before the cause of thermal expansion was understood, the phenomenon was put to practical use. The cooper (barrel maker) was an important person in days gone by. Barrels made from wooden staves were often held together by bands of iron that fit tightly around the middle (Fig. 13.2). The iron band was heated red hot to make it expand. Then it was fit around the barrel and allowed to cool so that it would shrink and tighten in place. A circular iron rim was fit around a wooden wagon wheel in the same way.

Tensile stresses and strains were discussed in Chapter 10. An object subject to a tensile stress gets slightly longer. An object's length can also change due to a *change in temperature* rather than an applied force. For tensile stresses, Hooke's law says that the fractional change in length is proportional to the stress (up to some maximum stress). Similarly, as long as the temperature change is not too great, the fractional change in length of a solid is proportional to the temperature change. If the length of a wire, rod, or pipe is L_0 at temperature T_0 (Fig. 13.3), then

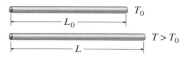

Figure 13.3 Expansion of a solid rod with increasing temperature.

$$\frac{\Delta L}{L_0} = \alpha \Delta T \qquad (13\text{-}4)$$

where $\Delta L = L - L_0$ and $\Delta T = T - T_0$. The length at temperature T is

$$L = L_0 + \Delta L = (1 + \alpha \Delta T) L_0 \qquad (13\text{-}5)$$

The constant of proportionality, α, is called the **coefficient of thermal expansion** of the substance. It plays a role in thermal expansion similar to that of the elastic modulus in tensile stress. If T is measured in kelvins or in degrees Celsius, then α has units of K^{-1} or $°C^{-1}$. Since only the *change* in temperature is involved in Eq. (13-4), either Celsius or Kelvin temperatures can be used to find ΔT; a temperature change of 1 K is the same as a temperature change of 1°C.

As is true for the elastic modulus, the coefficient of thermal expansion has different values for different solids and also depends to some extent on the starting temperature of the object. Table 13.2 lists the coefficients of thermal expansion for various solids at room temperature (20°C).

Figure 13.4 is a graph of the length of a steel girder as a function of temperature over a *wide* range of temperatures. The curvature of this graph shows that the thermal expansion of the girder is in general *not* proportional to the temperature change. However, over a *limited* temperature range, the curve can be approximated by a straight line; the slope of the tangent line is the coefficient α at the temperature T_0. For small temperature changes near T_0, the change in length of the girder can be treated as being proportional to the temperature change with only a small error.

Allowances must be made in building sidewalks, roads, bridges, and buildings to leave space for expansion in hot weather. Old subway tracks have small spaces left between rail sections to prevent the rails from pushing into each other and causing the

Making The Connection:
expansion joints in
bridges and buildings

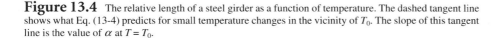

Figure 13.4 The relative length of a steel girder as a function of temperature. The dashed tangent line shows what Eq. (13-4) predicts for small temperature changes in the vicinity of T_0. The slope of this tangent line is the value of α at $T = T_0$.

Table 13.2

Coefficients of Linear Expansion α and Volume Expansion β (at $T_0 = 20°C$ except for ice)

Material	$\alpha(10^{-6}\ K^{-1})$	$\beta(10^{-6}\ K^{-1})$
Solids		
Glass (Vycor)	0.75	2.25
Brick	1.0	3.0
Glass (Pyrex)	3.25	9.75
Granite	8	24
Glass, most types	9.4	28.2
Cement or concrete	12	36
Iron or steel	12	36
Copper	16	48
Silver	18	54
Brass	19	57
Aluminum	22.5	69
Lead	29	87
Ice (at 0°C)	51	153
Liquids		
Mercury		182
Water		207
Gasoline		950
Ethyl alcohol		1120
Benzene		1240
Gases		
Air (and most other gases) at 1 atm		3340

Example 13.2

Expanding Rods

Two metal rods, one aluminum and one brass, are clamped down on one end (Fig. 13.5). At 0.0°C, the rods are each 50.0 cm long and are separated by 0.024 cm at their unfastened ends. At what temperature will the rods just come into contact? (Assume that the base to which the rods are clamped undergoes a negligibly small thermal expansion.)

Strategy Two rods of different materials expand by different amounts. The sum of the two expansions ($\Delta L_{br} + \Delta L_{Al}$) must equal the space between the rods. After finding ΔT, we add it to $T_0 = 0°C$ to obtain the temperature at which the two rods touch.

Known: $L_0 = 50.0$ cm, $T_0 = 0.0°C$ for both
Look up: $\alpha_{br} = 19 \times 10^{-6}\ K^{-1}$; $\alpha_{Al} = 22.5 \times 10^{-6}\ K^{-1}$
Requirement: $\Delta L_{br} + \Delta L_{Al} = 0.024$ cm
Find: $T_f = T_0 + \Delta T$

Solution The brass rod expands by
$$\Delta L_{br} = (\alpha_{br}\, \Delta T) L_0$$
and the aluminum rod by
$$\Delta L_{Al} = (\alpha_{Al}\, \Delta T) L_0$$
The sum of the two expansions is known:
$$\Delta L_{br} + \Delta L_{Al} = 0.024\ \text{cm}$$

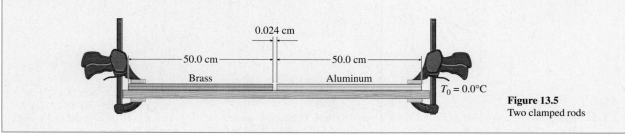

0.024 cm

50.0 cm — Brass   50.0 cm — Aluminum $T_0 = 0.0°C$

Figure 13.5
Two clamped rods

Example 13.2 *continued*

Since both the initial lengths and the temperature changes are the same,

$$(\alpha_{br} + \alpha_{Al}) \Delta T \times L_0 = 0.024 \text{ cm}$$

Solving for ΔT,

$$\Delta T = \frac{0.024 \text{ cm}}{(\alpha_{br} + \alpha_{Al}) L_0}$$

$$= \frac{0.024 \text{ cm}}{(19 \times 10^{-6} \text{ K}^{-1} + 22.5 \times 10^{-6} \text{ K}^{-1}) \times 50.0 \text{ cm}}$$

$$= 11.6°C$$

The temperature at which the two touch is

$$T_f = T_0 + \Delta T = 0.0°C + 11.6°C = 11.6°C$$

Discussion   As a check on the solution, we can find how much each individual rod expands and then add the two amounts.

$$\Delta L_{Al} = \alpha_{Al} \Delta T L_0$$

$$= 22.5 \times 10^{-6} \text{ K}^{-1} \times 11.6 \text{ K} \times 50.0 \text{ cm} = 0.013 \text{ cm}$$

$$\Delta L_{br} = \alpha_{br} \Delta T L_0$$

$$= 19 \times 10^{-6} \text{ K}^{-1} \times 11.6 \text{ K} \times 50.0 \text{ cm} = 0.011 \text{ cm}$$

total expansion = 0.013 cm + 0.011 cm = 0.024 cm

which is correct.

Practice Problem 13.2   Expansion of a wall

The outer wall of a building is constructed from concrete blocks. If the wall is 5.00 m long at 20.0°C, how much longer is the wall on a hot day (30.0°C)? How much shorter is it on a cold day (−5.0°C)?

track to bow. A train riding on such tracks is subject to a noticeable amount of "clickety-clack" as it goes over these small expansion breaks in the tracks. Expansion joints are easily observed in bridges (Fig. 13.6). Concrete roads and sidewalks have tar joints between sections. In hot weather the tar softens, allowing the expanding concrete to push into the joint. Homeowners sometimes build their own sidewalks without realizing the necessity for such joints; these sidewalks begin to crack almost immediately!

Allowances must also be made for contraction in cold weather. If an object is not free to expand or contract, then as the temperature changes it is subjected to *thermal stress* as its environment exerts forces on it to prevent the thermal expansion or contraction that would otherwise occur.

Differential Expansion

When two strips made of different metals are joined together and then heated, one expands more than the other (unless they have the same coefficient of expansion). This differential expansion can be put to practical use: the joined strips bend into a curve, allowing one strip to expand more than the other.

The bimetallic strip (Fig. 13.7) is made by joining a material with a lower coefficient of expansion, such as steel, and one of a higher coefficient of expansion, such as brass. Unequal expansions or contractions of the two materials force the bimetallic strip

Figure 13.6 Expansion joints permit the roadbed of a bridge to expand and contract as the temperature changes.

 Making The Connection: bimetallic strip in a thermostat

Bimetallic strip

Brass

Steel

Cold   Room temperature   Hot

Figure 13.7 A bimetallic strip bends when its temperature changes; brass expands and contracts more than steel for the same temperature change.

to bend. In Fig. 13.7 the brass expands more than the steel when the bimetallic strip is heated. As the strip is cooled, the brass contracts more than the steel.

The bimetallic strip is used in many wall thermostats. The bending of the bimetallic strip closes or opens an electrical switch in the thermostat that turns the furnace or air conditioner on or off. Inexpensive oven thermometers also use a bimetallic strip wound into a spiral coil; the coil winds tighter or unwinds as the temperature changes.

Area Expansion

As you might suspect, *each dimension* of an object expands when the object's temperature increases. For instance, a pipe expands not only in length, but also in radius. An isotropic substance expands uniformly in all directions; any two points move farther apart as the temperature rises. Thus, there are changes in area and volume caused by expansion in all directions. In Problem 14 you can show that, for small temperature changes, the area of any flat surface of a solid changes in proportion to the temperature change:

$$\frac{\Delta A}{A_0} = 2\alpha\,\Delta T \tag{13-6}$$

The factor of two in Eq. (13-6) arises because the surface expands in two perpendicular directions.

Volume Expansion

The fractional change in volume of a solid or liquid is also proportional to the temperature change as long as the temperature change is not too large:

$$\frac{\Delta V}{V_0} = \beta\,\Delta T \tag{13-7}$$

The coefficient of volume expansion, β, is the fractional change in volume per unit temperature change. For solids, the coefficient of volume expansion is three times the coefficient of linear expansion (as shown in Problem 15):

$$\beta = 3\alpha \tag{13-8}$$

The factor of three in Eq. (13-8) arises because the object expands in three-dimensional space. For liquids, the volume expansion coefficient is the only one given in tables. Since liquids do not necessarily retain the same shape as they expand, the quantity that is uniquely defined is the change in volume. Table 13.2 provides values of β for some common solids and liquids.

 A hollow cavity in a solid expands exactly as if it were filled—the interior of a steel gasoline can expands just as if it were a solid steel block. The steel wall of the can does *not* expand inward to make the cavity smaller; that would require pairs of points in the steel to get *closer together* instead of farther apart.

In an ordinary alcohol-in-glass or mercury-in-glass thermometer, it is not just the liquid that expands as temperature rises. The reading of the thermometer is determined by the difference in the volume expansion of the liquid and that of the interior of the glass. The calibration of an accurate thermometer must account for the expansion of the glass. A glance at Table 13.2 shows that, as is usually the case, the liquid expands much more than the glass for a given temperature change.

Example 13.3

Hollow Cylinder Full of Water

A hollow copper cylinder is filled to the brim with water at 20.0°C. If the water and the container are heated to a temperature of 91°C, what percentage of the water spills over the top of the container?

continued on next page

Example 13.3 continued

Strategy The volume expansion coefficient for water is greater than that for copper, so the water expands more than the interior of the cylinder. The cavity expands just as if it were solid copper. Since the problem does not specify the initial volume, we call it V_0. We need to find out how much a volume V_0 of water expands and how much a volume V_0 of copper expands; the difference is the water volume that spills over the top of the container.

Known: Initial copper cylinder interior volume = initial water
volume = V_0
Initial temperature = $T_0 = 20.0°C$
Final temperature = $91°C$; $\Delta T = 71°C$
Look up: $\beta_{Cu} = 48 \times 10^{-6} °C^{-1}$; $\beta_{H_2O} = 207 \times 10^{-6} °C^{-1}$
Find: $\Delta V_{H_2O} - \Delta V_{Cu}$ as a percentage of V_0

Solution The volume expansion of the interior of the copper cylinder is

$$\Delta V_{Cu} = \beta_{Cu} \Delta T\, V_0$$

The volume expansion of the water is

$$\Delta V_{H_2O} = \beta_{H_2O} \Delta T\, V_0$$

The amount of water that spills is

$$\Delta V_{H_2O} - \Delta V_{Cu} = \beta_{H_2O} \Delta T\, V_0 - \beta_{Cu} \Delta T\, V_0$$
$$= (\beta_{H_2O} - \beta_{Cu}) \Delta T\, V_0$$
$$= (207 \times 10^{-6}\,°C^{-1} - 48 \times 10^{-6}\,°C^{-1}) \times 71\,°C \times V_0$$
$$= 0.011 V_0$$

The percentage of water that spills is therefore 1.1%.

Discussion As a check we can find the change in volume of the copper container and of the water and find the difference.

$$\Delta V_{Cu} = \beta_{Cu} \Delta T\, V_0 = 48 \times 10^{-6}\,°C^{-1} \times 71\,°C \times V_0 = 0.0034 V_0$$

$$\Delta V_{H_2O} = \beta_{H_2O} \Delta T\, V_0 = 207 \times 10^{-6}\,°C^{-1} \times 71\,°C \times V_0 = 0.0147 V_0$$

volume of water that spills = $0.0147 V_0 - 0.0034 V_0 = 0.0113 V_0$

which again shows that 1.1% spills.

Practice Problem 13.3   Overflowing gas can

A driver fills an 18.9-L steel gasoline can with gasoline at 15.0°C right up to the top. He forgets to replace the cap and leaves the can in the back of his truck. The temperature climbs to 30.0°C by 1 P.M. How much gasoline spills out of the can?

13.4 MOLECULAR PICTURE OF A GAS

As we saw in Chapter 9, the densities of liquids are generally not much less than the densities of solids. Gases are *much* less dense than liquids and solids because the molecules are, on average, much farther apart. The mass density—mass per unit volume—of a substance depends on two microscopic quantities: the mass m of a single molecule and the number of molecules packed into a given volume of space (see Fig. 13.8). This second quantity, the number of molecules per unit volume, is called the **number density** to distinguish it from mass density. In SI units, number density is written as the number of molecules per cubic meter, usually written simply as m^{-3} (read "per cubic meter"). If a gas has a total mass M, occupies a volume V, and each molecule has a mass m, then the number of gas molecules is

$$N = \frac{M}{m} \tag{13-9}$$

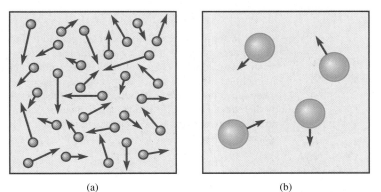

(a)   (b)

Figure 13.8 These two gases have the same mass per unit volume but different number densities. In (a), there are a larger number of molecules in a given volume, but the mass of each molecule in (b) is greater.

and the average number density is

$$\frac{N}{V} = \frac{M}{mV} = \frac{\rho}{m}$$

(13-10)

where $\rho = M/V$ is the mass density.

It is common to express the amount of a substance in units of **moles** (abbreviated mol). The mole is an SI base unit and is defined as follows: one mole of anything contains the same number of units as there are atoms in 12 *grams* (not kilograms) of carbon-12. This number is called **Avogadro's number** and has the value

$$N_A = 6.022 \times 10^{23} \text{ mol}^{-1} \quad \text{(Avogadro's number)}$$

Avogadro's number is written with units, mol^{-1}, to show that this is the number *per mole*. The number of moles, n, is therefore given by

$$\text{number of moles} = \frac{\text{number of molecules}}{\text{number of molecules per mole}}$$

$$n = \frac{N}{N_A}$$

(13-11)

The mass of a molecule is often expressed in units other than kg. The most common is the **atomic mass unit** (symbol u). By definition, one atom of carbon-12 has a mass of 12 u (exactly). Using Avogadro's number, the relationship between atomic mass units and kilograms can be calculated (see Problem 19):

$$1 \text{ u} = 1.66 \times 10^{-27} \text{ kg}$$

(13-12)

The proton, neutron, and hydrogen atom all have masses within 1% of 1 u—which is why the atomic mass unit is so convenient. More precise values are 1.007 u for the proton, 1.009 u for the neutron, and 1.008 u for the hydrogen atom. The mass of an atom is *approximately* equal to the number of nucleons (neutrons plus protons)—the "atomic mass number"—times 1 u.

Tables commonly express the mass of a substance as the **molar mass**—the mass of *one mole* of the substance *in grams*. For an element with several isotopes (such as carbon-12, carbon-13, and carbon-14), the molar mass is averaged according to the naturally occurring abundance of each isotope. The atomic mass unit is chosen so that the mass of a molecule in "u" is numerically the same as the molar mass in g/mol. For example, the molecular mass of O_2 is 32.0 g/mol and the mass of one molecule is 32.0 u. Probably the most important thing to remember from all of this is: *be careful* with units!

The mass of a molecule is the sum of the masses of its constituent atoms. The molecular mass is therefore equal to the sum of the atomic masses. For example, the atomic mass of carbon is 12.0 g/mol and the atomic mass of (atomic) oxygen is 16.0 g/mol; therefore, the molecular mass of carbon dioxide (CO_2) is $(12.0 + 2 \times 16.0)$ g/mol = 44.0 g/mol.

Example 13.4

A Helium Balloon

A helium balloon of volume 0.010 m^3 contains 0.40 mol of He gas. (a) Find the number of atoms, the number density, and the mass density. (b) Estimate the average distance between He atoms.

Strategy   The number of moles tells us the number of atoms as a fraction of Avogadro's number. Once we have the number of atoms, N, the next quantity we are asked to find is N/V. To find the mass density, we can look up the atomic mass of

helium in the Periodic Table. The mass per atom times the number density (atoms per m^3) equals the mass density (mass per m^3). To find the average distance between atoms, imagine a simplified picture in which each atom is at the center of a spherical volume equal to the total volume of the gas divided by the number of atoms. In this approximation, the average distance between atoms is equal to the diameter of each sphere.

continued on next page

Example 13.4 *continued*

Solution   (a) The number of atoms is

$$N = n N_A$$

$$= 0.40 \text{ mol} \times 6.022 \times 10^{23} \text{ atoms/mol} = 2.4 \times 10^{23} \text{ atoms}$$

The number density is

$$\frac{N}{V} = \frac{2.4 \times 10^{23} \text{ atoms}}{0.010 \text{ m}^3} = 2.4 \times 10^{25} \text{ atoms/m}^3$$

The atomic mass of helium is 4.00 u. Then the mass in kg of a helium atom is

$$m = 4.00 \text{ u} \times 1.66 \times 10^{-27} \text{ kg/u} = 6.64 \times 10^{-27} \text{ kg}$$

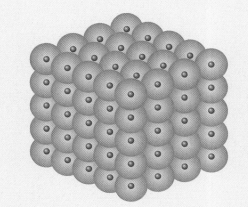

and the mass density of the gas is

$$\rho = \frac{M}{V} = m \times \frac{N}{V}$$

$$= 6.64 \times 10^{-27} \text{ kg} \times 2.4 \times 10^{25} \text{ m}^{-3} = 0.16 \text{ kg/m}^3$$

(b) We assume that each atom is at the center of a sphere of radius r (Fig. 13.9). The volume of the sphere is

$$\frac{V}{N} = \frac{1}{N/V} = \frac{1}{2.4 \times 10^{25} \text{ atoms/m}^3} = 4.2 \times 10^{-26} \text{ m}^3 \text{ per atom}$$

Then

$$\frac{V}{N} = \frac{4}{3} \pi r^3 \approx 4 r^3 \quad (\text{since } \pi \approx 3)$$

Solving for r,

$$r \approx \left(\frac{V}{4N} \right)^{1/3} = 2.2 \times 10^{-9} \text{ m} = 2.2 \text{ nm}$$

The average distance between atoms is $d = 2r \approx 4$ nm (since this is an estimate).

Discussion   For comparison, in *liquid* helium the average distance between atoms is about 0.4 nm, so in the gas the average separation is about ten times larger.

Practice Problem 13.4   Number density for water
The mass density of liquid water is 1000.0 kg/m³. Find the number density.

Figure 13.9
Simplified model in which equally spaced gas molecules sit at the center of a spherical volume of space

13.5   ABSOLUTE TEMPERATURE AND THE IDEAL GAS LAW

We have examined the thermal expansion of solids and liquids. What about gases? Is the volume expansion of a gas proportional to the temperature change? We must be careful; since gases are easily compressed, we must also specify what happens to the pressure. The French scientist Jacques Charles (1746–1823) found experimentally that, if the pressure of a gas is held constant, the change in temperature is indeed proportional to the change in volume (Fig. 13.10a).

Charles's law

$$\Delta V \propto \Delta T \quad (\text{for constant } P)$$

According to Charles's law, a graph of V versus T for a gas held at constant pressure is a straight line, but the line does not necessarily pass through the origin (Fig. 13.10b). Temperature scales such as the Celsius and Fahrenheit scales assign $T = 0$ to some arbitrary phenomenon. We would *not* expect a gas at 1 atm pressure and 0°C to occupy a volume of zero!

However, if we graph V versus T (at constant P) for various gases, something interesting happens. If we extrapolate the straight line to where it reaches $V = 0$, the temperature at that point is the *same* regardless of what gas we use, how many moles of gas are present, or what the pressure of the gas is (Fig. 13.10c). (The reason that we have to extrapolate is that all gases become liquids or solids before they reach $V = 0$.) This temperature, −273.15°C or −459.67°F, is called **absolute zero**—the lower limit of attainable

Absolute zero is the lower limit of attainable temperatures.

Figure 13.10 (a) Apparatus to verify Charles's law. The pressure of the enclosed gas is held constant by the fixed quantity of mercury resting on top of it. If the temperature of the gas is changed, it expands or contracts, moving the mercury column above it. (b) Charles's law: for a gas held at constant pressure, changes in temperature are proportional to changes in volume. (c) Volume versus temperature graphs for various gases, each at a constant pressure, are extrapolated to $V = 0$. The graphs intersect the temperature axis at the same temperature, T_{limit}. (d) An absolute temperature scale sets $T_{limit} = 0$.

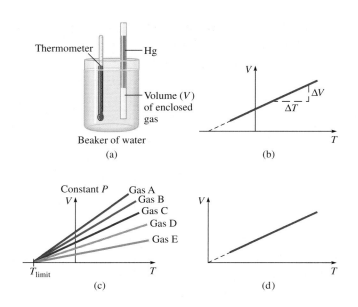

temperatures. In kelvins—an *absolute* temperature scale—absolute zero is defined as 0 K (Fig. 13.10d). As long as it is understood that an absolute temperature scale is to be used, then Charles's law can be written

$$V \propto T \quad \text{(for constant } P)$$

Physics at Home

Take an empty 2-L soda bottle, cap it tightly, and put it in the freezer. Check it an hour later; what has happened? Estimate the percentage change in the volume of the air inside and compare to the percentage change in absolute temperature (if you don't have a thermometer handy, a typical freezer temperature is about –10°C).

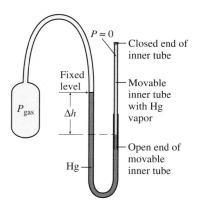

Figure 13.11 A constant volume gas thermometer. A dilute gas is contained in the vessel on the left, which is connected to a mercury manometer. The right side has a tube that can be moved up or down to keep the mercury level on the left at a fixed level, so that the volume of gas is kept constant. Then the manometer is used to measure the absolute pressure of the gas.

Thermal expansion of a gas can be used to measure temperature. Gas thermometers have advantages as well as disadvantages. One huge advantage is that a gas thermometer is universal: it does not matter what gas is used or how many moles of gas are present, as long as the number density is sufficiently low. Gas thermometers give absolute temperature in a natural way and they are extremely accurate and reproducible. The main disadvantage is that they are much less convenient to use than most other thermometers. Gas thermometers are mainly used to calibrate other, more convenient thermometers.

A thermometer based on Charles's law would be called a *constant pressure gas thermometer*. More common is the *constant volume gas thermometer* (Fig. 13.11), which is based on Gay-Lussac's law:

$$P \propto T \quad \text{(for constant } V)$$

Here we keep the volume of the gas constant, measure the pressure and use that to indicate the temperature. (It is much easier to keep the volume constant and measure the pressure than to do the opposite.)

Both Charles's law and Gay-Lussac's law are valid only for a **dilute** gas—a gas where the number density is low enough, or equivalently the average distance between gas molecules is large enough, that interactions between the molecules are negligible except when they collide. Two other experimentally discovered laws that apply to dilute gases are Boyle's law and Avogadro's law. Boyle's law states that the pressure of a gas is inversely proportional to its volume at constant temperature:

$$P \propto \frac{1}{V} \quad \text{(for constant } T)$$

Avogadro's law states that the volume occupied by a gas at a given temperature and pressure is proportional to the number of gas molecules N:

$$V \propto N \quad \text{(constant } P, T\text{)}$$

(A constant number of gas molecules was assumed in the statements of Boyle's, Gay-Lussac's, and Charles's laws.)

One equation combines all four of these gas laws—the **ideal gas law**:

Ideal gas law (microscopic form)

$$PV = NkT \quad (N = \text{number of molecules}) \qquad (13\text{-}13)$$

The constant of proportionality is a universal quantity known as **Boltzmann's constant** (symbol k); its value is

> In the ideal gas law, T stands for *absolute* temperature (in K).

$$k = 1.38 \times 10^{-23} \text{ J/K} \qquad (13\text{-}14)$$

The macroscopic form of the ideal gas law is written in terms of n, the number of *moles* of the gas, in place of N, the number of molecules. Substituting

$$N = nN_A$$

into the microscopic form yields

$$PV = nN_A kT$$

The product of N_A and k is called the **universal gas constant**:

$$R = N_A k = 8.31 \; \frac{\text{J/K}}{\text{mol}} \qquad (13\text{-}15)$$

Then the ideal gas law in macroscopic form is written

Ideal gas law (macroscopic form)

$$PV = nRT \quad (n = \text{number of moles}) \qquad (13\text{-}16)$$

Many problems deal with the changing pressure, volume, and temperature in a gas with a constant number of molecules (and a constant number of moles). In such problems, it is often easiest to write the ideal gas law as follows:

$$\frac{P_1 V_1}{T_1} = \frac{P_2 V_2}{T_2}$$

Example 13.5

Temperature of the Air in a Tire

 Before starting out on a long drive, you check the air in your tires to make sure they are properly inflated. The pressure gauge reads 31.0 lb/in^2 (214 kPa), and the temperature is 15°C. After a few hours of highway driving, you stop and check the pressure again. Now the gauge reads 35.0 lb/in^2 (241 kPa). What is the temperature of the air in the tires now?

Strategy   We treat the air in the tire as an ideal gas. We must work with absolute temperatures and absolute pressures when using the ideal gas law. The pressure gauge reads *gauge* pressure; to get absolute pressure we add 1 atm = 101 kPa. We don't know the number of molecules inside the tire or the volume, but we can reasonably assume that neither changes. The number is

constant as long as the tire does not leak. The volume may actually change a bit as the tire warms up and expands, but this change is small. Since N and V are constant, we can rewrite the ideal gas law as a proportionality between P and T.

Solution   First convert the initial and final gauge pressures to absolute pressures:

$$P_i = 214 \text{ kPa} + 101 \text{ kPa} = 315 \text{ kPa}$$

$$P_f = 241 \text{ kPa} + 101 \text{ kPa} = 342 \text{ kPa}$$

Now convert the initial temperature to an absolute temperature:

$$T_i = 15°C + 273 \text{ K} = 288 \text{ K}$$

continued on next page

Example 13.5 *continued*

According to the ideal gas law, pressure is proportional to temperature, so

$$\frac{T_f}{T_i} = \frac{P_f}{P_i} = \frac{342 \text{ kPa}}{315 \text{ kPa}}$$

Then

$$T_f = \frac{P_f}{P_i} T_i = \frac{342}{315} \times 288 \text{ K} = 313 \text{ K}$$

Now convert back to °C:

$$313 \text{ K} - 273 \text{ K} = 40°\text{C}$$

Discussion The final answer of 40°C seems reasonable since, after a long drive, the tires are noticeably warm, but not hot enough to burn your hand.

It is often most convenient to work with the ideal gas law by setting up a proportion. In this problem, we did not know the volume or the number of molecules, so

we had no choice. In essence, what we used was Gay-Lussac's law. Starting with the ideal gas law, we can "rederive" Gay-Lussac's law or Charles's law or any other proportionality inherent in the ideal gas law.

Practice Problem 13.5
Air pressure in tire after temperature decreases

Suppose you now (unwisely) decide to bleed air from the tires, since the manufacturer suggests keeping the pressure between 28 lb/in² and 32 lb/in². (The manufacturer's specification refers to when the tires are "cold.") If you let out enough air so that the pressure returns to 31 lb/in², what percentage of the air molecules did you let out of the tires? What is the gauge pressure after the tires cool back down to 15°C?

Physics at Home

The next time you take a car trip, check the tire pressure with a gauge just before the trip and then again after an hour or more of highway driving. Calculate the temperature of the air in the tires from the two pressure readings and the initial temperature. Feel the tire with your hand to see if your calculation is reasonable.

Example 13.6

Scuba Diver

A scuba diver needs air delivered at a pressure equal to the pressure of the surrounding water—the pressure in the lungs must match the water pressure on the diver's body to prevent the lungs from collapsing. Since the pressure in the air tank is much higher, a regulator delivers air to the diver at the appropriate pressure. The compressed air in a diver's tank lasts 90 minutes at the water surface. About how long does the same tank last at a depth of 20 meters underwater? (Assume that the volume of air breathed per minute does not change.)

Strategy The compressed air in the *tank* is at a pressure much higher than the pressure at which the diver breathes, whether at the surface or at 20 m depth. The constant quantity is *N*, the number of gas molecules in the tank. We also assume that the temperature of the gas remains the same; it may change slightly but much less than the pressure or volume.

Solution Since *N* and *T* are constant,

$$PV = \text{constant}$$

or

$$P \propto 1/V$$

The pressure at the surface is (approximately) 1 atm, while the pressure at 20 m underwater is

$$P = 1 \text{ atm} + \rho g h$$

$$\rho g h = (1000 \text{ kg/m}^3) \times (9.8 \text{ m/s}^2) \times (20 \text{ m}) = 196 \text{ kPa} \approx 2 \text{ atm}$$

Therefore, at a depth of 20 m,

$$P = 3 \text{ atm}$$

To match the pressure of the surrounding water, the pressure of the compressed air is three times larger at a depth of 20 m; then the volume of air is one-third what it was at the surface. The diver breathes the same volume per minute, so the tank will last one-third as long—30 minutes.

continued on next page

Example 13.6 *continued*

Discussion   To do the same thing a bit more formally, we could write:

$$P_i V_i = P_f V_f$$

After setting $P_i = 1$ atm and $P_f = 3$ atm, we find that $V_f/V_i = \frac{1}{3}$.

In this problem, the only numerical values given (indirectly) were the initial and final pressures. Assuming that N and T remain constant, we then can find the ratio of the final and

 Tips

initial volumes. Whenever there *seems* to be insufficient numerical information given in a problem, think in terms of ratios and look for constants that cancel out.

Practice Problem 13.6   Pressure in air tank after temperature increase

A tank of compressed air is at an absolute pressure of 580 kPa at a temperature of 300.0 K. The temperature increases to 330.0 K. What is the pressure in the tank now?

Problem-Solving Tips for the Ideal Gas Law

- In most problems, some change occurs; decide which of the four quantities (P, V, N or n, and T) remain constant during the change.
- Use the microscopic form if the problem deals with the number of molecules and the macroscopic form if the problem deals with the number of moles.
- Use subscripts (i and f) to distinguish initial and final values.
- Work in terms of ratios so that constant factors cancel out.
- Write out the units when doing calculations.
- Remember that P stands for *absolute* pressure (not gauge pressure) and T stands for *absolute* temperature (in kelvins, not °C or °F).

13.6   KINETIC THEORY OF THE IDEAL GAS

In a gas, the interaction between two molecules gets weak rapidly as the distance between the molecules increases. In a dilute gas, the average distance between gas molecules is large enough that we can ignore interactions between the molecules except when they collide. In addition, the volume of space occupied by the molecules themselves is a small fraction of the total volume of the gas—the gas is mostly "empty space." The **ideal gas** is a simplified model of a dilute gas in which we think of the molecules as pointlike particles that move *independently* in free space with no interactions except for elastic collisions.

This simplified model is a good approximation for many gases under ordinary conditions. Many properties of gases can be understood from this model; the microscopic theory based on it is called the **kinetic theory** of the ideal gas.

Microscopic Basis of Pressure

The force that a gas exerts on a surface is due to collisions that the gas molecules make with that surface. For instance, think of the air inside an automobile tire. Whenever an air molecule collides with the inner tire surface, the tire exerts an inward force to turn the air molecule around and return it to the bulk of the gas. By Newton's third law, the gas molecule exerts an outward force on the tire surface. The net force per unit area on the inside of the tire due to all the collisions of the many air molecules is equal to the air pressure in the tire. The pressure depends on three things: how many molecules there are, how often each one collides with the wall, and the momentum transfer due to each collision.

We want to find out how the pressure of an ideal gas is determined by the motions of the gas molecules. To simplify the discussion, consider a gas contained in a box of length L and side area A (Fig. 13.12a)—the result does not depend on the shape of the container. Figure 13.12b shows a gas molecule about to collide with the rightmost wall of the container. For simplicity we assume that the collision is elastic; a more advanced analysis shows that the result is correct even though not all collisions are elastic.

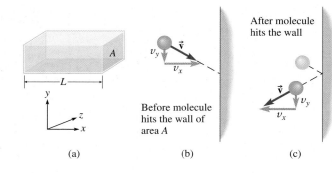

Figure 13.12 (a) Gas molecules confined to a container of length L and area A. (b) A molecule is about to collide with the wall of area A. (c) After an elastic collision, v_x has changed sign, while v_y and v_z are unchanged.

For an elastic collision, the x-component of the molecule's momentum is reversed in direction since the wall is much more massive than the molecule. Since the gas exerts only an outward force on the wall (a static fluid exerts no tangential force on a boundary), the y- and z-components of the molecule's momentum are unchanged. Thus, the molecule's momentum change is $\Delta p_x = 2m|v_x|$.

When does this molecule next collide with the same wall? Ignoring for now collisions with other molecules, its x-component of velocity never changes magnitude—only the sign of v_x changes when it reverses direction (Fig. 13.12c). The time it takes the molecule to travel the length L of the container and hit the other wall is $L/|v_x|$. Then the round-trip time is

$$\Delta t = 2\frac{L}{|v_x|}$$

The average force exerted by the molecule on the wall during one complete round-trip is

$$F_{av,x} = \frac{\Delta p_x}{\Delta t} = \frac{2m|v_x|}{2L/|v_x|} = \frac{m|v_x|^2}{L} = \frac{mv_x^2}{L}$$

The total force on the wall is the sum of the forces due to each molecule in the gas. If there are N molecules in the gas, we can simply multiply N by the *average* force due to one molecule to get the total force on the wall. To represent such an average, we use angle brackets $\langle \rangle$; the quantity inside the brackets is averaged over all the molecules in the gas.

$$F = N\langle F_{av}\rangle = \frac{Nm}{L}\langle v_x^2\rangle$$

The pressure is then

$$P = \frac{F}{A} = \frac{Nm}{AL}\langle v_x^2\rangle$$

The volume of the box is $V = AL$, so

$$P = \frac{Nm}{V}\langle v_x^2\rangle \tag{13-17}$$

which is true regardless of the shape of the container enclosing the gas. Since we end up averaging over all the molecules in the gas, the simplifying assumption about no collisions with other molecules does not affect the result.

The product $m\langle v_x^2\rangle$ suggests kinetic energy. It certainly makes sense that if the average kinetic energy of the gas molecules is larger, the pressure is higher. The average translational kinetic energy of a molecule in the gas is $\langle K_{tr}\rangle = \frac{1}{2}m\langle v^2\rangle$. For any gas molecule, $v^2 = v_x^2 + v_y^2 + v_z^2$, since velocity is a vector quantity. The gas as a whole is at rest, so there is no preferred direction of motion. Then the average value of v_x^2 must be the same as the averages of v_y^2 and v_z^2, so

$$\langle v_x^2\rangle = \tfrac{1}{3}\langle v^2\rangle$$

Therefore,

$$m\langle v_x^2\rangle = \tfrac{1}{3}m\langle v^2\rangle = \tfrac{2}{3}\langle K_{tr}\rangle$$

Substituting this into Eq. (13-17), the pressure is

$$P = \frac{2}{3}\frac{N\langle K_{tr}\rangle}{V} = \frac{2}{3}\frac{N}{V}\langle K_{tr}\rangle \qquad (13\text{-}18)$$

This expression is written with the variables grouped in two different ways to give two different insights. Since $N\langle K_{tr}\rangle$ is the total internal kinetic energy of the gas, the expression with the first grouping says that pressure is proportional to the kinetic energy density (the kinetic energy per unit volume). The second says that pressure is proportional to the product of the number density N/V and the average molecular kinetic energy. The pressure of a gas increases if either the gas molecules are packed closer together or if the molecules have more kinetic energy.

Note that $\langle K_{tr}\rangle$ is the average *translational* kinetic energy of a gas molecule and v is the center-of-mass speed of a molecule. If the gas molecule has more than one atom (such as in N_2), the atoms also have vibrational and rotational kinetic energies, but these are *not* included in $\langle K_{tr}\rangle$.

What about the assumption that the gas molecules never collide with each other? It certainly is *not* true that the same molecule returns to collide with the same wall at a fixed time interval and has the same v_x each time it returns! However, the derivation really only relies on average quantities. In a gas at equilibrium, an average quantity like $\langle v_x^2\rangle$ remains unchanged even though any one particular molecule changes its velocity components as a result of each collision.

Temperature and Translational Kinetic Energy

The temperature of an ideal gas has a direct physical interpretation that we can now bring to light. We found that in an ideal gas, the pressure, volume, and number of molecules are related to the average translational kinetic energy of the gas molecules:

$$P = \frac{2}{3}\frac{N}{V}\langle K_{tr}\rangle \qquad (13\text{-}18)$$

Solving for the average kinetic energy,

$$\langle K_{tr}\rangle = \frac{3}{2}\frac{PV}{N} \qquad (13\text{-}19)$$

The ideal gas law relates P, V, and N to the temperature:

$$PV = NkT \qquad (13\text{-}13)$$

By rearranging the ideal gas law, we find that P, V, and N occur in the same combination as in Eq. (13-19):

$$\frac{PV}{N} = kT$$

Then by substituting kT for $(PV)/N$ in Eq. (13-19), we find that

$$\langle K_{tr}\rangle = \tfrac{3}{2}kT \qquad (13\text{-}20)$$

Therefore, *the absolute temperature of an ideal gas is proportional to the average translational kinetic energy of the gas molecules.* Temperature then is a way to describe the average translational kinetic energy of the gas molecules. At higher temperatures, the gas molecules have (on average) greater kinetic energy.

The speed of a gas molecule that has the average kinetic energy is called the **rms** (root mean square) **speed**. The rms speed is *not* the same as the average speed. Instead, the rms speed is the square *root* of the *mean* (average) of the speed *squared*. Since

$$\langle K_{tr}\rangle = \tfrac{1}{2}m\langle v^2\rangle = \tfrac{1}{2}mv_{rms}^2 \qquad (13\text{-}21)$$

the rms speed is

$$v_{rms} = \sqrt{\langle v^2\rangle}$$

Squaring before averaging emphasizes the effect of the faster-moving molecules, so the rms speed is a bit higher than the average speed—about 9% higher as it turns out.

Since the average kinetic energy of molecules in an ideal gas depends only on temperature, Eq. (13-21) implies that more massive molecules move more slowly on average than lighter ones at the same temperature. If two different gases are placed in a single chamber so that they reach equilibrium and are at the same temperature, their molecules must have the same average translational kinetic energies. If one gas has molecules of larger mass, its molecules must move with a slower average velocity than those of the gas with the lighter mass molecules. In Problem 62 you can show that

$$v_{rms} = \sqrt{\frac{3kT}{m}} \qquad (13\text{-}22)$$

where k is Boltzmann's constant and m is the mass of a molecule. Therefore, at a given temperature, the rms speed is inversely proportional to the square root of the mass of the molecule.

Example 13.7

O_2 Molecules at Room Temperature

Find the average translational kinetic energy and the rms speed of the O_2 molecules in air at room temperature (20°C).

Strategy The average translational kinetic energy depends only on temperature. We must remember to use absolute temperature. The rms speed is the speed of a molecule that has the average kinetic energy.

Solution   The absolute temperature is

$$20°C + 273\,K = 293\,K$$

Therefore, the average translational kinetic energy is

$$\langle K_{tr}\rangle = \tfrac{3}{2}kT = 1.50 \times 1.38 \times 10^{-23}\,J/K \times 293\,K = 6.07 \times 10^{-21}\,J$$

From the Periodic Table, we find the atomic mass of oxygen to be 16.0 u; the molecular mass of O_2 is twice that (32.0 u). First we convert that to kg:

$$32.0\,u \times 1.66 \times 10^{-27}\,kg/u = 5.31 \times 10^{-26}\,kg$$

The rms speed is the speed of a molecule with the average kinetic energy:

$$\langle K_{tr}\rangle = \tfrac{1}{2}mv_{rms}^2$$

$$v_{rms} = \sqrt{\frac{2\langle K_{tr}\rangle}{m}} = \sqrt{\frac{2 \times 6.07 \times 10^{-21}\,J}{5.31 \times 10^{-26}\,kg}} = 478\,m/s$$

Discussion   How can we decide if the result is reasonable, since we have no first-hand experience watching molecules bounce around? Recall from Chapter 12 that the speed of sound in air at room temperature is 343 m/s. Since sound waves in air propagate by the collisions that occur between air molecules, the speed of sound must be of the same order of magnitude as the average speeds of the molecules.

Practice Problem 13.7 CO_2 molecules at room temperature

Find the average translational kinetic energy and the rms speed of the CO_2 molecules in air at room temperature (20°C).

Maxwell-Boltzmann Distribution

So far we have considered only the *average* kinetic energy and *rms* speed of a molecule. Sometimes we may want to know more: how many molecules have speeds in a certain range? The distribution of speeds is called the **Maxwell-Boltzmann distribution**. The distribution for oxygen at two different temperatures is shown in Fig. 13.13. The interpretation of the graphs is that the number of gas molecules having speeds between any two values v_1 and v_2 is proportional to the area under the curve between v_1 and v_2. In Fig. 13.13, the shaded areas represent the number of oxygen molecules having speeds above 800 m/s at the two selected temperatures. A relatively small temperature change has a significant effect on the number of gas molecules with high speeds.

Any given molecule changes its kinetic energy often—at each collision, which means billions of times per second. However, the total number of gas molecules in a given kinetic energy range in the gas stays the same, as long as the temperature is constant. In fact, it is the frequent collisions that maintain the stability of the Maxwell-Boltzmann distribution. The collisions keep the kinetic energy distributed among the gas molecules *in the most disordered way possible*, which is the Maxwell-Boltzmann distribution.

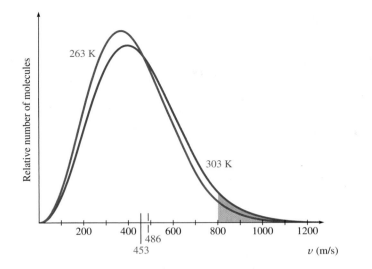

Figure 13.13 The probability distribution of kinetic energies in oxygen at two temperatures: –10°C (263 K) and +30°C (303 K). The area under either curve for any range of speeds is proportional to the number of molecules whose speeds lie in that range. Despite the relatively small difference in rms speeds (453 m/s at 263 K and 486 m/s at 303 K), the number of molecules in the high-speed tail is quite different.

The Maxwell-Boltzmann distribution helps us understand planetary atmospheres. Why does the Earth's atmosphere contain nitrogen, oxygen, and water vapor, among other gases, but not hydrogen or helium? Gas molecules in the upper atmosphere that are moving faster than the *escape speed* (Section 6.6) have enough kinetic energy to escape from the planetary atmosphere to outer space. Those that are heading away from the planet's surface will escape if they avoid colliding with another molecule. The high-energy tail of the Maxwell-Boltzmann distribution does not get depleted by molecules that escape. Other molecules will get boosted to those high kinetic energies as a result of collisions; these replacements will in turn also escape. Thus, the atmosphere gradually leaks away.

How fast the atmosphere leaks away depends on how far the rms speed is from the escape speed. If the rms speed is too small compared to the escape speed, the time for all the gas molecules to escape is so long that the gas is present in the atmosphere indefinitely. This is the case for nitrogen, oxygen, and water vapor in the Earth's atmosphere. On the other hand, since hydrogen and helium are much less massive, their rms speeds are higher. Though only a tiny fraction of the molecules are above the escape speed, the fraction is sufficient for these gases to escape quickly from the Earth's atmosphere (see Fig. 13.14). The Moon is often said to lack an atmosphere. The Moon's low escape speed (2400 m/s) allows most gases to escape, but it does have an atmosphere about 1 cm tall composed of krypton (a gas with molecular mass 83.8 u, about 2.6 times that of oxygen).

Making The Connection:
composition of
planetary
atmospheres

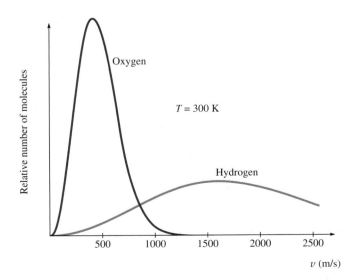

Figure 13.14 Maxwell-Boltzmann distributions for oxygen and hydrogen at *T* = 300 K. Escape speed from the Earth is 11,200 m/s (not shown on the graph).

13.7   TEMPERATURE AND REACTION RATES

What we have learned about the distribution of kinetic energies and its relationship to temperature has a great relevance to the dependence of chemical reaction rates on temperature. Imagine a mixture of two gases, N_2 and O_2, which can react to form nitric oxide (NO):

$$N_2 + O_2 \rightarrow 2NO$$

In order for the reaction to occur, a molecule of nitrogen must collide with a molecule of oxygen. But the reaction does not occur every time such a collision takes place. The reactant molecules must possess enough kinetic energy to initiate the reaction, because the reaction involves the rearrangement of chemical bonds between atoms. Therefore, some chemical bonds must be broken before new ones form; the energy to break these bonds must come from the energy of the reactants. The minimum kinetic energy of the reactant molecules that allows the reaction to proceed is called the **activation energy** (E_a).

If a molecule of N_2 collides with one of O_2, but their total kinetic energy is less than the activation energy, then the two just bounce off each other. Some energy may be transferred from one molecule to the other, or converted between translational, rotational, and vibrational energy, but we are still left with one molecule of N_2 and one of O_2. If the molecules do have enough energy to react, then the reaction may take place.

Now we begin to see why, with few exceptions, rates of reaction increase with temperature. At higher temperatures, the average kinetic energy of the reactants is higher and therefore a greater fraction of the collisions have total kinetic energies exceeding the activation energy. Thus, at higher temperatures the reaction rate is higher. In many situations, only a small fraction of the reactants have enough kinetic energy to react. This situation occurs when the activation energy is much greater than the average kinetic energy of the reactants:

$$E_a \gg \tfrac{3}{2}kT \tag{13-23}$$

Thus, the only candidates for reaction are molecules far off in the high-energy tail of the Maxwell-Boltzmann distribution. In this situation, a small increase in temperature can have an unexpectedly large effect on the reaction rate. Because of this dramatic effect, it turns out that the reaction rate depends *exponentially* on temperature:

$$\text{reaction rate} \propto e^{-E_a/kT} \tag{13-24}$$

Example 13.8

Increase in Reaction Rate with Temperature Increase

 The activation energy for the reaction $N_2O \rightarrow N_2 + O$ is 4.0×10^{-19} J. By what percentage does the reaction rate increase if the temperature is increased from 700.0 K to 707.0 K (a 1% increase in absolute temperature)?

Strategy   We should first check that $E_a \gg \tfrac{3}{2}kT$; otherwise, Eq. (13-24) does not apply. Assuming that checks out, we can set up a ratio of the reaction rates at the two temperatures.

Solution   Start by calculating $E_a/(kT_1)$, where $T_1 = 700.0$ K:

$$\frac{E_a}{kT_1} = \frac{4.0 \times 10^{-19}\ \text{J}}{1.38 \times 10^{-23}\ \text{J/K} \times 700.0\ \text{K}} = 41.41$$

So E_a is about 41 times kT, or about 28 times $\tfrac{3}{2}kT$. The activation energy is much greater than the average kinetic energy;

thus, only a small fraction of the collisions might cause a reaction to occur.

At $T_2 = 707.0$ K,

$$\frac{E_a}{kT_2} = \frac{4.0 \times 10^{-19}\ \text{J}}{1.38 \times 10^{-23}\ \text{J/K} \times 707.0\ \text{K}} = \frac{41.41}{1.01} = 41.00$$

The ratio of the reaction rates is

$$\frac{\text{new rate}}{\text{old rate}} = \frac{e^{-41.00}}{e^{-41.41}} = e^{-(41.00-41.41)} = e^{0.41} = 1.5$$

The reaction rate at 707.0 K is 1.5 times the rate at 700.0 K—a 50% increase in reaction rate for a 1% increase in temperature!

Discussion   Normally we might suspect an error when a 1% change in one quantity causes a 50% change in another! However, this illustrates the dramatic effect of an *exponential* dependence. Reaction rates can be *extremely* sensitive to small temperature changes.

Practice Problem 13.8   Decrease in reaction rate for lower temperature

What is the percentage decrease in the rate of the same reaction if the temperature is lowered from 700.0 K to 699.0 K?

Figure 13.15 Warm-blooded animals use different strategies to maintain a constant body temperature. (a) The fur of an Arctic fox serves as a layer of insulation to help it stay warm. (b) Dogs pant and (c) people sweat when their bodies are in danger of overheating. In cases (b) and (c) the evaporation of water has a cooling effect on the body.

Although we have discussed reactions in terms of gases, the same principles apply to reactions in liquid solutions. In order for two molecules to react, they must collide with sufficient kinetic energy to overcome the "barrier" presented by the activation energy. The temperature in turn determines what fraction of the collisions have enough energy to react, and therefore the reaction rates show the same kind of temperature dependence whether the reaction occurs in a gas mixture or a liquid solution.

At the beginning of this chapter we asked about the necessity for temperature regulation in warm-blooded animals (birds and mammals) (see Fig. 13.15). The temperature dependence of chemical reaction rates has a profound effect on biological functions. In warm-blooded animals, vital internal organs and muscles must be kept at their optimal temperatures so they can perform their functions. If our internal temperatures varied we would have a varying metabolic rate, becoming sluggish in cold weather.

Making The Connection:
evolutionary advantages of warm-blooded versus cold-blooded animals

By being able to regulate internal temperatures and maintain a constant body temperature higher than that of the environment, warm-blooded animals are able to tolerate a wider range of environmental temperatures than cold-blooded animals (such as reptiles and insects). Temperature fluctuation in the environment is much more severe on land than in water; thus, land animals are more likely to be homeothermic than aquatic animals. Keeping muscles at their optimal temperatures contributes to the much larger effort required to move around on land or in the air as opposed to moving through water. Keeping the muscles and vital organs warm allows the high level of aerobic metabolism needed to sustain intense physical activity.

Cold-blooded animals are dependent on the environment for temperature regulation; thus, we see a snake lying on a rock heated by the Sun in an attempt to keep warm. The temperature of the snake's blood depends on the temperature of the environment; as its blood temperature goes down in cold weather, the snake becomes inactive and lethargic. Most insects are inactive below 10°C and many cannot survive the cold of winter.

However, if environmental conditions become too extreme, it may be difficult to maintain ideal body temperature. Hypothermia occurs when the central core of the body becomes too cold; bodily processes slow and eventually cease. People caught outside in blizzards are urged to stay awake and to keep moving; the energy produced by exercise may be up to 20 times that produced by the resting body and can compensate for heat loss in extreme cold.

Warm-blooded animals must consume much more food than cold-blooded animals of a similar size; metabolic processes in warm-blooded animals act like a furnace to keep the body warm. A human must consume about 1500 kcal of food energy per day just to keep warm when resting at 20°C; an alligator of similar body weight needs only 60 kcal/day at rest at 20°C.

13.8   COLLISIONS BETWEEN GAS MOLECULES

How far does a gas molecule move, on average, between collisions? The mean (average) length of the path traveled by a gas molecule as a free particle (no interactions with other particles) is called the **mean free path** (Λ, the Greek capital lambda). The mean free path depends on two things: how large the molecules are and how many of them occupy a given volume. In a simplified model, imagine a molecule of diameter d. Its cross-sectional area is

$$A = \pi r^2 = \tfrac{1}{4}\pi d^2$$

As the molecule moves in a straight line, it sweeps out a cylindrical volume of space (Fig. 13.16). Once the volume it has swept out equals V/N, the total volume divided by the number of molecules, it has "used up" its own space and begins to infringe on the space "belonging" to another molecule—which means that a collision is imminent. Thus, a collision occurs when

$$\text{volume} = A\,\Lambda \approx V/N$$

or

$$\Lambda \approx \frac{1}{\tfrac{1}{4}\pi d^2\,(N/V)}$$

This simplified calculation is correct except for the dimensionless constant of proportionality; a detailed calculation yields

Mean free path
$$\Lambda = \frac{1}{\sqrt{2}\,\pi d^2\,(N/V)}$$

(13-25)

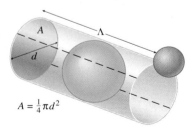

$$A = \tfrac{1}{4}\pi d^2$$

Figure 13.16 A molecule of diameter d moves an average distance Λ before colliding. The volume of space swept out while moving this distance is the volume of a cylinder of length Λ and cross-sectional area $A = \tfrac{1}{4}\pi d^2$.

Example 13.9

Collisions per Second for N_2 at 20°C and 1 atm

Estimate the average number of collisions per second that each N_2 molecule undergoes in air at room temperature and atmospheric pressure. The diameter of an N_2 molecule is 0.3 nm.

Strategy   First find the mean free path. The average time between collisions is the time it takes to travel that distance at the average molecular speed. For the purposes of this estimate, we can use the rms speed instead of the average speed—the rms speed is only 9% higher.

Solution   Use the ideal gas law to find the number density:

$$PV = NkT$$

$$\frac{N}{V} = \frac{P}{kT} = \frac{1.01 \times 10^5 \text{ Pa}}{1.38 \times 10^{-23} \text{ J/K} \times 293 \text{ K}} = 2.50 \times 10^{25} \text{ m}^{-3}$$

The mean free path is then

$$\Lambda = \frac{1}{\sqrt{2}\pi d^2 \, (N/V)}$$

$$\Lambda = \frac{1}{\sqrt{2}\pi \times (3 \times 10^{-10} \text{ m})^2 \times 2.50 \times 10^{25} \text{ m}^{-3}}$$

$$\Lambda = 1 \times 10^{-7} \text{ m} = 0.1 \text{ μm}$$

An estimate of the average time between collisions is the time it takes to travel a distance Λ at speed v_{rms}. In Example 13.7, we found the rms speed of O_2 molecules at 293 K to be 478 m/s. The rms speed of N_2 molecules at that temperature is somewhat higher since the molecular mass of N_2 is smaller (28 u, as opposed to 32 u for O_2). Since average speed is inversely

proportional to the square root of molecular mass, the difference is insignificant for our estimate. Therefore, taking $v_{rms} \approx$ 500 m/s, the average time between collisions is

$$\langle t \rangle = \frac{\Lambda}{v_{rms}} \approx \frac{1 \times 10^{-7} \text{ m}}{500 \text{ m/s}} = 2 \times 10^{-10} \text{ s} = 0.2 \text{ ns}$$

This is the average time per collision, so the number of collisions per second is

$$\frac{1}{\langle t \rangle} = 5 \times 10^9 \text{ s}^{-1}$$

Each molecule collides about 5×10^9 times per second.

Discussion   Note that the mean free path is larger than the average distance between a molecule and its nearest neighbor. At room temperature and atmospheric pressure, the latter distance is about 4 nm; the mean free path is 25 times larger. We should suspect an error if we found the mean free path to be comparable to or smaller than the distance between a molecule and its *nearest* neighbor—since only occasionally does a molecule collide with a nearest neighbor.

Practice Problem 13.9   Mean free path of a hydrogen atom in space

Intergalactic space is nearly a vacuum: there is on average approximately one hydrogen atom per cubic centimeter. The diameter of a hydrogen atom is about 0.1 nm. (a) Estimate the mean free path of a hydrogen atom under these conditions. (b) Find the rms speed of the hydrogen atoms at temperature 2.7 K. (c) Use (a) and (b) to estimate the average time between collisions in years.

Diffusion

A gas molecule moves in a straight line between collisions—the effect of gravity on the velocity of the molecule is negligible during a time interval of only 0.2 ns. At each collision, both the speed and direction of the molecule's motion change. The mean free path tells us the *average* length of the molecule's straight line paths between collisions. The result is that a given molecule follows a *random walk* trajectory (Fig. 13.17).

After an elapsed time *t*, how far on average has a molecule moved from its initial position? The answer to this question is relevant when we consider **diffusion**. Someone across the room opens a bottle of perfume: how long until the scent reaches you? As gas molecules diffuse into the air, the frequent collisions are what determine how long it takes the scent to travel across the room (assuming, as we do here, that there are no air currents). When there is a difference in concentrations between different points in a gas, the random thermal motion of the molecules tends to even out the concentrations (other things being equal). The net flow from regions of high concentration (near the perfume bottle) to regions of lower concentration (across the room) is diffusion (Fig. 13.18).

Consider a molecule of perfume in the air. It has a mean free path Λ. After a large number of collisions *N*, it has traveled a total *distance* $N\Lambda$. However, its displacement from its original position is much less than that, since at each collision it changes direction. It can be shown using statistical analysis of the random walk that the rms magnitude of its displacement after *N* collisions is proportional to $\sqrt{N}$. Since the number of collisions is proportional to the elapsed time, the rms displacement is proportional to $\sqrt{t}$.

Figure 13.17   Successive straight-line paths traveled by a molecule between collisions

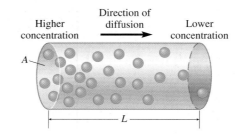

Figure 13.18 Molecules diffuse from regions of higher concentration to regions of lower concentration.

The root mean squared displacement in one direction is

$$x_{rms} = \sqrt{2Dt} \tag{13-26}$$

where D is a diffusion constant such as those given in Table 13.3. The diffusion constant D depends on the molecule or atom that is diffusing and the medium through which it is moving.

Diffusion is crucial in biological processes such as the transport of oxygen. Oxygen molecules diffuse from the air in the lungs through the walls of the alveoli and then through the walls of the capillaries to oxygenate the blood. The oxygen is then carried by hemoglobin in the blood to various parts of the body, where it again diffuses through capillary walls into intercellular fluids and then through cell membranes into cells. Diffusion is a slow process over long distances but can be quite effective over short distances—which is why cell membranes must be thin and capillaries must have small diameters. Evolution has seen to it that the capillaries of animals of widely different sizes are all about the same size—as small as possible while still allowing blood cells to flow through them.

Table 13.3

Diffusion Constants at 1 atm and 20°C

Diffusing Molecule	Medium	D (m²/s)
DNA	Water	1.3×10^{-12}
Oxygen	Tissue (cell membrane)	1.8×10^{-11}
Hemoglobin	Water	6.9×10^{-11}
Sucrose ($C_{12}H_{22}O_{11}$)	Water	5.0×10^{-10}
Glucose ($C_6H_{12}O_6$)	Water	6.7×10^{-10}
Oxygen	Water	1.0×10^{-9}
Oxygen	Air	1.8×10^{-5}
Hydrogen	Air	6.4×10^{-5}

Example 13.10

Diffusion Time for Oxygen into Capillaries

How long on average does it take an oxygen molecule in an alveolus to diffuse into the blood? Assume for simplicity that the diffusion constant for oxygen passing through the two membranes (alveolus and capillary walls) is the same: 1.8×10^{-11} m²/s. The total thickness of the two membranes is 1.2×10^{-8} m.

Strategy Take the x-direction to be through the membranes. Then we want to know how much time elapses until $x_{rms} = 1.2 \times 10^{-8}$ m.

continued on next page

Example 13.10 *continued*

Solution Solving Eq. (13-26) for t yields

$$t = \frac{x_{rms}^2}{2D}$$

Now substitute $x_{rms} = 1.2 \times 10^{-8}$ m and $D = 1.8 \times 10^{-11}$ m²/s:

$$t = \frac{(1.2 \times 10^{-8} \text{ m})^2}{2 \times 1.8 \times 10^{-11} \text{ m}^2/\text{s}} = 4.0 \times 10^{-6} \text{ s}$$

Discussion The time is proportional to the *square* of the membrane thickness. It would take four times as long for an oxygen molecule to diffuse through a membrane twice as thick. The rapid increase of diffusion time with distance is a principal reason why evolution has favored thin membranes over thicker ones.

Practice Problem 13.10   Time for oxygen to get halfway through the membrane

How long on average does it take an oxygen molecule to get *halfway* through the alveolus and capillary wall?

MASTER THE CONCEPTS

Summary

- Temperature is a quantity that determines when objects are in thermal equilibrium. The flow of energy that occurs between two objects or systems due to a temperature difference between them is called heat flow. If heat can flow between two objects or systems, the objects or systems are said to be in thermal contact. When two systems in thermal contact have the same temperature, there is no net flow of heat between them; the objects are said to be in thermal equilibrium.

- Zeroth law of thermodynamics: if two objects are each in thermal equilibrium with a third object, then the two are in thermal equilibrium with each other.

- The SI unit of temperature is the kelvin (symbol K, *without* a degree sign). The kelvin scale is an absolute temperature scale, which means that $T = 0$ is set to absolute zero.

- Temperature in °C (T_C) and temperature in kelvins (T) are related as

$$T_C = T - 273.15 \qquad (13\text{-}3)$$

- As long as the temperature change is not too great, the fractional length change of a solid is proportional to the temperature change:

$$\frac{\Delta L}{L_0} = \alpha \, \Delta T \qquad (13\text{-}4)$$

The constant of proportionality, α, is called the coefficient of thermal expansion of the substance.

- The fractional change in volume of a solid or liquid is also proportional to the temperature change as long as the temperature change is not too large:

$$\frac{\Delta V}{V_0} = \beta \, \Delta T \qquad (13\text{-}7)$$

For solids, the coefficient of volume expansion is three times the coefficient of linear expansion: $\beta = 3\alpha$.

- The mole is an SI base unit and is defined as: one mole of anything contains the same number of units as there are atoms in 12 *grams* (not kilograms) of carbon-12. This number is called Avogadro's number and has the value

$$N_A = 6.02 \times 10^{23} \text{ mol}^{-1}$$

- The mass of a molecule is often expressed in the atomic mass unit (symbol u). By definition, one atom of carbon-12 has a mass of 12 u (exactly).

$$1 \text{ u} = 1.66 \times 10^{-27} \text{ kg} \qquad (13\text{-}12)$$

The atomic mass unit is chosen so that the mass of a molecule in "u" is numerically the same as the molar mass in g/mol.

- In an ideal gas, the molecules move independently in free space with no interactions except when two molecules collide. The ideal gas is a useful model for many real gases, provided that the gas is sufficiently dilute. The ideal gas law:

$$\text{microscopic form: } PV = NkT \qquad (13\text{-}13)$$

$$\text{macroscopic form: } PV = nRT \qquad (13\text{-}16)$$

where Boltzmann's constant and the universal gas constant are

$$k = 1.38 \times 10^{-23} \text{ J/K} \qquad (13\text{-}14)$$

$$R = N_A k = 8.31 \frac{\text{J/K}}{\text{mol}} \qquad (13\text{-}15)$$

- According to kinetic theory, the pressure of a gas is proportional to the average translational kinetic energy of the molecules:

$$P = \frac{2}{3} \frac{N}{V} \langle K_{tr} \rangle \qquad (13\text{-}18)$$

- The average translational kinetic energy of the molecules is proportional to the absolute temperature:

$$\langle K_{tr} \rangle = \tfrac{3}{2} kT \qquad (13\text{-}20)$$

- The speed of a gas molecule that has the average kinetic energy is called the rms speed:

$$\langle K_{tr} \rangle = \tfrac{1}{2} m v_{rms}^2 \qquad (13\text{-}21)$$

- The distribution of molecular speeds in an ideal gas is called the Maxwell-Boltzmann distribution.

- If the activation energy for a chemical reaction is much greater than the average kinetic energy of the reactants, the reaction rate depends *exponentially* on temperature:

$$\text{reaction rate} \propto e^{-E_a/kT} \qquad (13\text{-}24)$$

MASTER THE CONCEPTS continued

- The mean free path (Λ) is the average length of the path traveled by a gas molecule as a free particle (no interactions with other particles) between collisions:

$$\Lambda = \frac{1}{\sqrt{2}\pi d^2 (N/V)} \qquad (13\text{-}25)$$

- The root mean square displacement of a diffusing molecule along the x-axis is

$$x_{rms} = \sqrt{2Dt} \qquad (13\text{-}26)$$

where D is a diffusion constant.

Highlighted Figures and Tables

T13.1 Reference temperatures in K, °C, and °F (p. 443)

T13.2 Coefficients of linear expansion α and volume expansion β (p. 447)

F13.10 Charles's law: If the pressure of a gas is held constant, the change in temperature is proportional to the change in volume (p. 452)

F13.12 How the pressure of an ideal gas is determined by the motions of the gas molecules (p. 456)

F13.13 The probability distribution of kinetic energies in oxygen at two temperatures (p. 459)

F13.14 Maxwell-Boltzmann distributions for oxygen and hydrogen at $T = 300$ K (p. 459)

F13.15 Warm-blooded animals use different strategies to maintain a constant body temperature (p. 461)

CONCEPTUAL QUESTIONS

1. Explain why it would be impossible to uniquely define the temperature of an object if the zeroth law of thermodynamics were violated?

2. Why do we call the temperature 0 K "absolute zero"? How is 0 K fundamentally different from 0°C or 0°F?

3. Under what special circumstances can kelvins or Celsius degrees be used interchangeably?

4. What happens to a hole in a flat metal plate when the plate expands on being heated? Does the hole get larger or smaller?

5. Why would silver and brass probably not be a good choice of metals for a bimetallic strip (leaving aside the question of the cost of silver)? (See Table 13.2.)

6. One way to loosen the lid on a glass jar is to run it under hot water. How does that work?

7. Why must we use absolute temperature (temperature in kelvins) in the ideal gas law ($PV = NkT$)? Explain how using the Celsius scale would give nonsensical results.

8. Natural gas is sold by volume. In the United States, the price charged is usually per cubic foot. Given the price per cubic foot, what other information would you need in order to calculate the price per mole?

9. What are the SI units of mass density and number density? If two different gases have the same number density, do they have the same mass density?

10. Suppose we have two tanks, one containing helium gas and the other nitrogen gas. The two gases are at the same temperature and pressure. Which has the higher number density (or are they equal)? Which has the higher mass density (or are they equal)?

11. The mass of an aluminum atom is 27.0 u. What is the mass of *one mole* of aluminum atoms? (No calculation required!)

12. A ping-pong ball that has been dented during hard play can often be restored by placing it in hot water. Explain why this works.

13. Why does a helium weather balloon expand as it rises into the air? Assume the temperature remains constant.

14. Explain why there is almost no hydrogen (H_2) or helium (He) in the Earth's atmosphere, while both are present in Jupiter's atmosphere. [*Hint:* Escape velocity from the Earth is 11.2 km/s and escape velocity from Jupiter is 60 km/s.]

15. Explain how it is possible that more than half of the molecules in an ideal gas have kinetic energies less than the average kinetic energy. Shouldn't half have less and half have more?

16. In air under ordinary conditions (room temperature and atmospheric pressure), the average intermolecular distance is about 4 nm and the mean free path is about 0.1 μm. The diameter of a nitrogen molecule is about 0.3 nm. Explain how the mean free path can be so much larger than the average distance between molecules.

17. In air under ordinary conditions (room temperature and atmospheric pressure), the average intermolecular distance is about 4 nm and the mean free path is about 0.1 μm. The diameter of a nitrogen molecule is about 0.3 nm. Which two distances should we compare to decide that air is dilute and can be treated as an ideal gas? Explain.

18. In air under ordinary conditions (room temperature and atmospheric pressure), the average intermolecular distance is about 4 nm and the mean free path is about 0.1 μm. The diameter of a nitrogen molecule is about 0.3 nm. What would it mean if the intermolecular distance and the molecular diameter were about the same? In that case, would it make sense to speak of a mean free path? Explain.

19. Explain how an automobile airbag protects the passenger from injury. Why would the airbag be ineffective if the gas pressure inside is too low when the passenger comes into contact with it? What about if it is too high?

20. It takes longer to hard-boil an egg in Mexico City (2200 m above sea level) than it does in Amsterdam (parts of which are below sea level). Why? [*Hint:* At higher altitudes, water boils at less than 100°C.]

MULTIPLE CHOICE QUESTIONS

1. In a mixed gas such as air, the rms speeds of different molecules are

   (a) independent of molecular mass.
   (b) proportional to molecular mass.
   (c) inversely proportional to molecular mass.
   (d) proportional to $\sqrt{\text{molecular mass}}$.
   (e) inversely proportional to $\sqrt{\text{molecular mass}}$.

2. The average kinetic energy of the molecules in an ideal gas increases with the volume remaining constant. Which of these statements *must* be true?

   (a) The pressure increases and the temperature stays the same.
   (b) The number density decreases.
   (c) The temperature increases and the pressure stays the same.
   (d) Both the pressure and the temperature increase.

3. Which of these will increase the average kinetic energy of the molecules in an ideal gas?

   (a) reduce the volume, keeping P and N constant
   (b) increase the volume, keeping P and N constant
   (c) reduce the volume, keeping T and N constant
   (d) increase the pressure, keeping T and V constant
   (e) increase N, keeping V and T constant

4. The absolute temperature of an ideal gas is directly proportional to

   (a) the number of molecules in the sample.
   (b) the average momentum of a molecule of the gas.
   (c) the average translational kinetic energy of the gas.
   (d) the diffusion constant of the gas.

5. The rms speed is

   (a) the speed at which all the gas molecules move.
   (b) the speed of a molecule with the average kinetic energy.
   (c) the average speed of the gas molecules.
   (d) the maximum speed of the gas molecules.

6. What are the most favorable conditions for real gases to approach ideal behavior?

   (a) high temperature and high pressure
   (b) low temperature and high pressure
   (c) low temperature and low pressure
   (d) high temperature and low pressure

7. An ideal gas has the volume V_0. If the temperature and the pressure are each tripled during a process, the new volume is

   (a) V_0.
   (b) $9V_0$.
   (c) $3V_0$.
   (d) $0.33V_0$.

8. The average kinetic energy of a gas molecule can be found from which of these quantities?

   (a) pressure only
   (b) number of molecules only
   (c) temperature only
   (d) pressure and temperature are both required

9. If the temperature of an ideal gas is doubled and the pressure is held constant, the rms speed of the molecules

   (a) remains unchanged.
   (b) is 2 times the original speed.
   (c) is $\sqrt{2}$ times the original speed.
   (d) is 4 times the original speed.

10. A metal box is heated until each of its sides has expanded by 0.1%. By what percent has the *volume* of the box changed?

    (a) –0.3%   (b) –0.2%   (c) +0.1%   (d) +0.2%   (e) +0.3%

PROBLEMS

Note: **C** indicates a combination conceptual/quantitative problem. Gold diamonds ✦, ✦✦ are used to indicate the increasing level of difficulty of each problem. Problem numbers appearing in blue, 9., denote problems that have a detailed solution available in the Student Solutions Manual. Some problems are *paired* by concept; their numbers are connected by a ruled box.

13.2 Temperature Scales

1. On a warm summer day, the air temperature is 84°F. Express this temperature in (a) °C and (b) kelvins.

2. The temperature at which liquid nitrogen boils (at atmospheric pressure) is 77 K. Express this temperature in (a) °C and (b) °F.

3. A room air conditioner causes a temperature change of –6.0°C. (a) What is the temperature change in kelvins? (b) What is the temperature change in °F?

13.3 Thermal Expansion of Solids and Liquids

4. Steel railroad tracks of length 18.30 m are laid at 10.0°C (Fig. 13.19). How much space should be left between the track sections if they are to just touch when the temperature is 50.0°C?

5. A highway is made of concrete slabs that are 15 m long at 20.0°C. (a) If the temperature range at the location of the highway is from –20.0°C to +40.0°C, what size expansion gap should be left (at 20.0°C) to prevent buckling of the highway? (b) How large are the gaps at –20.0°C?

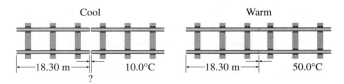

Figure 13.19 Problem 4

6. A lead rod and a common glass rod both have the same length when at 20.0°C. The lead rod is heated to 50.0°C. To what temperature must the glass rod be heated so that they are again at the same length?

7. The coefficient of linear expansion of brass is 1.9×10^{-5} °C^{-1}. At 20.0°C, a hole in a sheet of brass has an area of 1.00 mm^2. How much larger is the area of the hole at 30.0°C?

8. Aluminum rivets used in airplane construction are made slightly too large for the rivet holes to be sure of a tight fit. The rivets are cooled with dry ice (−78.5°C) before they are driven into the holes. If the holes have a diameter of 0.6350 cm at 20.5°C, what should be the diameter of the rivets at 20.5°C if they are to just fit when cooled to the temperature of dry ice?

9. A temperature change ΔT causes a volume change ΔV but has no effect on the mass of an object. (a) Show that the change in density $\Delta \rho$ is given by $\Delta \rho = -\beta \rho \, \Delta T$. (b) Find the fractional change in density ($\Delta \rho / \rho$) of a brass sphere when the temperature changes from 32°C to −10.0°C.

10. An ordinary drinking glass is filled to the brim with water (268.4 mL) at 2.0°C and placed on the sunny pool deck for a swimmer to enjoy. If the temperature of the water rises to 32.0°C before the swimmer reaches for the glass, how much water will have spilled over the top of the glass?

✦11. A steel ring of inner diameter 7.00000 cm at 20.0°C is to be heated and placed over a brass shaft of outer diameter 7.00200 cm at 20.0°C. (a) To what temperature must the ring be heated to fit over the shaft? The shaft remains at 20.0°C. (b) Once the ring is on the shaft and has cooled to 20.0°C, to what temperature must the ring plus shaft combination be cooled to allow the ring to slide off the shaft again?

12. A long, narrow steel rod of length 2.5000 m at 25°C is oscillating as a pendulum about a horizontal axis through one end. If the temperature changes to 0°C, what will be the fractional change in its period?

13. The George Washington Bridge crosses the Hudson River in New York. The span of the steel bridge is about 1.6 km. If the temperature in New York can vary from a low of −15°F in winter to a high of 105°F in summer, by how much might the length of the span change over an entire year?

14. A flat square of side s_0 at temperature T_0 expands by Δs in both length and width when the temperature increases by ΔT. The original area is $s_0^2 = A_0$ and the final area is $(s_0 + \Delta s)^2 = A$. Show that if $\Delta s \ll s_0$,

$$\frac{\Delta A}{A_0} = 2\alpha \, \Delta T \qquad (13\text{-}6)$$

(Although we derive this relation for a square plate, it applies to a flat area of any shape.)

15. The volume of a solid cube with side s_0 at temperature T_0 is $V_0 = s_0^3$. Show that if $\Delta s \ll s_0$, the change in volume ΔV due to a change in temperature ΔT is given by

$$\frac{\Delta V}{V_0} = 3\alpha \, \Delta T$$

and therefore that $\beta = 3\alpha$. (Although we derive this relation for a cube, it applies to a solid of any shape.)

16. A square brass plate, 8.00 cm on a side, has a hole cut into its center of area 4.90874 cm^2 (at 20.0°C). The hole in the plate is to slide over a cylindrical steel shaft of cross-sectional area 4.91000 cm^2 (also at 20.0°C) as shown in Fig. 13.20. To what

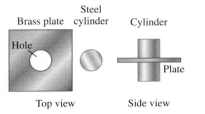

Figure 13.20
Problem 16

temperature must the brass plate be heated so that it can just slide over the steel cylinder (which remains at 20.0°C)? [*Hint:* The steel cylinder is not heated so it does not expand; only the brass plate is heated.]

17. A copper washer is to be fit in place over a steel bolt (Fig. 13.21). Both pieces of metal are at 20.0°C. If the diameter of the bolt is 1.0000 cm and the inner diameter of the washer is 0.9980 cm, at what temperature will the washer fit over the bolt?

Figure 13.21
Problem 17

✦18. A steel rule is calibrated for measuring lengths at 20.00°C. The rule is used to measure the length of a Vycor glass brick; when both are at 20.00°C, the brick is found to be 25.00 cm long. If the rule and the brick are both at 80.00°C, what would be the length of the brick as measured by the rule?

13.4 Molecular Picture of a Gas

19. Use the definition that one mol of ^{12}C (carbon-12) atoms has a mass of exactly 12 g, along with Avogadro's number, to derive the conversion between atomic mass units and kg.

20. Find the molecular mass of ammonia (NH_3).

21. Find the mass (in kg) of one molecule of CO_2.

22. The mass of one mole of ^{13}C (carbon-13) is 13.003 g. (a) What is the mass in u of one ^{13}C atom? (b) What is the mass in kg of one ^{13}C atom?

23. Estimate the number of H_2O molecules in a human body of mass 80.2 kg. Assume that, on average, water makes up about 62% of the mass of a human body.

24. At 0.0°C and 1.00 atm, 1.00 mol of a gas occupies a volume of 0.0224 m^3. (a) What is the number density? (b) Estimate the average distance between the gas molecules. (c) If the gas is nitrogen (N_2), the principal component of air, what is the total mass and mass density?

25. Sand is composed of SiO_2. Find the order of magnitude of the number of silicon (Si) atoms in a grain of sand. Approximate the sand grain as a sphere of diameter 0.5 mm and an SiO_2 molecule as a sphere of diameter 0.5 nm.

26. The mass density of diamond (a crystalline form of carbon) is 3500 kg/m^3. How many carbon atoms per cm^3 are there?

27. How many hydrogen atoms are present in 684.6 g of sucrose ($C_{12}H_{22}O_{11}$)?

28. How many moles of He are in 13 g of He?

29. The principal component of natural gas is methane (CH_4). How many moles of CH_4 are present in 144.36 g of methane?

30. What is the mass of one gold atom in kg?

31. Air at room temperature and atmospheric pressure has a mass density of 1.2 kg/m^3. The average molecular mass of air is 29.0 u. How many air molecules are there in 1.0 cm^3 of air?

13.5 Absolute Temperature and the Ideal Gas Law

32. Verify, using the ideal gas law, the assertion in Problem 24 that 1.00 mol of a gas at 0.0°C and 1.00 atm occupies a volume of 0.0224 m^3.

33. Verify that the SI units of PV (pressure times volume) are joules.

34. Incandescent lightbulbs are filled with an inert gas to lengthen the filament life. With the current off (at $T = 20.0°C$), the gas inside a lightbulb has a pressure of 115 kPa. When the bulb is burning, the temperature rises to 70.0°C. What is the pressure at the higher temperature?

35. What fraction of the air molecules in a house must be pushed outside while the furnace raises the inside temperature from 16.0°C to 20.0°C? The pressure does not change since the house is not 100% airtight.

36. What is the mass density of air at $P = 1.0$ atm and $T =$ (a) –10°C and (b) 30°C? The average molecular mass of air is approximately 29 u.

37. A constant volume gas thermometer containing helium is immersed in boiling ammonia (–33°C) and the pressure is read once equilibrium is reached. The thermometer is then moved to a bath of boiling water (100.0°C). After the manometer was adjusted to keep the volume of helium constant, by what factor was the pressure multiplied?

38. A bubble rises from the bottom of a lake of depth 80.0 m, where the temperature is 4°C. The water temperature at the surface is 18°C. If the bubble's initial diameter is 1.00 mm, what is its diameter when it reaches the surface? (Ignore the surface tension of water. Assume the bubble warms as it rises to the same temperature as the water and retains a spherical shape. Assume $P_{atm} = 1.0$ atm.)

39. A bubble with a volume of 1.00 cm^3 forms at the bottom of a lake that is 20.0 m deep. The temperature at the bottom of the lake is 10.0°C. The bubble rises to the surface where the water temperature is 25.0°C. Assume that the bubble is small enough that its temperature always matches that of its surroundings. What is the volume of the bubble just before it breaks the surface of the water? Ignore surface tension.

40. A hydrogen balloon at the Earth's surface has a volume of 5.0 m^3 on a day when the temperature is 27°C and the pressure is 1.00×10^5 N/m^2. The balloon rises and expands as the pressure drops. What would the volume of the same number of moles of hydrogen be at an altitude of 40 km where the pressure is 0.33×10^3 N/m^2 and the temperature is –13°C?

41. An ideal gas that occupies 1.2 m^3 at a pressure of 1.0×10^5 Pa and a temperature of 27°C is compressed to a volume of 0.60 m^3 and heated to a temperature of 227°C. What is the new pressure?

42. A diver rises quickly to the surface from a 5.0-m depth. If she did not exhale the gas from her lungs before rising, by what factor would her lungs expand? Assume the temperature to be constant

and the pressure in the lungs to match the pressure outside the diver's body. The density of seawater is 1.03×10^3 kg/m^3.

43. An emphysema patient is breathing pure O_2 through a face mask. The cylinder of O_2 contains 0.60 ft^3 of O_2 gas at a pressure of 2200 lb/in^2. (a) What volume would the oxygen occupy at atmospheric pressure (and the same temperature)? (b) If the patient takes in 8 L/min of O_2 at atmospheric pressure, how long will the cylinder last?

44. A scuba diver has an air tank with a volume of 0.010 m^3. The air in the tank is initially at a pressure of 1.0×10^7 Pa. Assuming that the diver breathes 0.500 L/s of air, find how long the tank will last at depths of (a) 2.0 m and (b) 20.0 m.

45. In intergalactic space there is an average of about one hydrogen atom per cm^3 and the temperature is 3 K. What is the absolute pressure?

46. A tank of compressed air of volume 1.0 m^3 is pressurized to 20.0 atm at $T = 273$ K. A valve is opened and air is released until the pressure in the tank is 15.0 atm. How many air molecules were released?

47. A mass of 0.532 kg of molecular oxygen is contained in a cylinder at a pressure of 1.0×10^5 Pa and a temperature of 0.0°C. What volume does the gas occupy?

48. Consider the expansion of an ideal gas at constant pressure. The initial temperature is T_0 and the initial volume is V_0. (a) Show that $\Delta V/V_0 = \beta \Delta T$, where $\beta = 1/T_0$. (b) Compare the coefficient of volume expansion β for an ideal gas at 20°C to the values for liquids and gases listed in Table 13.2.

13.6 Kinetic Theory of the Ideal Gas

49. What is the temperature of an ideal gas whose molecules have an average translational kinetic energy of 3.20×10^{-20} J?

50. What is the total translational kinetic energy of the gas molecules of 0.420 mol of air at atmospheric pressure that occupies a volume of 1.00 L (0.0010 m^3)?

51. What is the kinetic energy per unit volume in an ideal gas at (a) $P = 1.00$ atm and (b) $P = 300.0$ atm?

52. Show that, for an ideal gas,

$$P = \tfrac{1}{3}\rho v_{rms}^2$$

where P is the pressure, ρ is the mass density, and v_{rms} is the rms speed of the gas molecules.

53. Estimate the percentage of the O_2 molecules in air at 0.0°C and 1.00 atm that are moving faster than the speed of sound (see Fig. 13.13).

54. What is the total internal kinetic energy of 1.0 mol of an ideal gas at 0.0°C and 1.00 atm?

55. There are two identical containers of gas at the same temperature and pressure, one containing argon and the other neon. What is the ratio of the rms speed of the argon atoms to that of the neon atoms? The atomic mass of argon is twice that of neon.

56. A smoke particle has a mass of 1.38×10^{-17} kg and it is randomly moving about in thermal equilibrium with room temperature air at 27°C. What is the rms speed of the particle?

57. Find the rms speed in air at 0.0°C and 1.00 atm of (a) the N_2 molecules, (b) the O_2 molecules, and (c) the CO_2 molecules.

58. What are the rms speeds of helium atoms, and nitrogen, hydrogen, and oxygen molecules at 25°C?

59. If the upper atmosphere of Jupiter has a temperature of 160 K and the escape speed is 60 km/s, would an astronaut expect to find much hydrogen there?

60. A sealed cylinder contains a sample of ideal gas at a pressure of 2.0 atm. The rms speed of the molecules is v_0. If the rms speed is then reduced to $0.90v_0$, what is the pressure of the gas?

61. What is the temperature of an ideal gas whose molecules in random motion have an average translational kinetic energy of 3.20×10^{-20} J?

62. Show that the rms speed of a molecule in an ideal gas at absolute temperature T is given by

$$v_{\mathrm{rms}} = \sqrt{\frac{3kT}{m}}$$

where k is Boltzmann's constant and m is the mass of a molecule.

63. Show that the rms speed of a molecule in an ideal gas at absolute temperature T is given by

$$v_{\mathrm{rms}} = \sqrt{\frac{3RT}{M}}$$

where M is the *molar mass*—the mass of the gas per mole.

13.7 Temperature and Reaction Rates

64. The reaction rate for the hydrolysis of benzoyl-l-arginine amide by trypsin at 10.0°C is 1.878 times faster than that at 5.0°C. Assuming that the reaction rate is exponential as in Eq. (13-24), what is the activation energy?

65. The reaction rate for the prepupal development of male *Drosophila* is temperature-dependent. Assuming that the reaction rate is exponential as in Eq. (13-24), the activation energy for this development is then 2.81×10^{-19} J. A *Drosophila* is originally at 10.00°C and its temperature is increasing. If the rate of development has increased 3.5%, how much has its temperature increased?

66. At high altitudes, water boils at a temperature lower than 100.0°C due to the lower air pressure. A rule of thumb states that the time to hard-boil an egg doubles for every 10.0°C drop in temperature. What activation energy does this rule imply for the chemical reactions that occur when the egg is cooked?

13.8 Collisions between Gas Molecules

67. Estimate the mean free path of a N_2 molecule in air at (a) sea level ($P \approx 100$ kPa and $T \approx 290$ K), (b) the top of Mt. Everest (altitude = 8.8 km, $P \approx 50$ kPa, and $T \approx 230$ K), and (c) an altitude of 100 km ($P \approx 0.03$ Pa and $T \approx 1300$ K). For simplicity, assume that air is pure nitrogen gas. The diameter of a N_2 molecule is approximately 0.3 nm.

68. About how long will it take a perfume molecule to diffuse a distance of 5.00 m in one direction in a room if the diffusion constant is 1.00×10^{-5} m²/s? Assume that the air is perfectly still—there are no air currents.

69. Estimate the time it takes a sucrose molecule to move 5.00 mm in one direction by diffusion in water. Assume there are no currents in the water.

COMPREHENSIVE PROBLEMS

70. The automobile driver from Practice Problem 13.3 fills his 18.9-L steel gasoline tank in the cool of the morning when the temperature of the tank and the gasoline is 15.0°C. The temperature climbs to 30.0°C by 1 P.M., but the steel tank is reinforced so that it cannot expand and there is a tight-fitting gas cap with no allowance for an air space or pressure release. What is the pressure on the tank from the heated gasoline, which cannot expand in volume? The bulk modulus for gasoline is 1.00×10^9 N/m².

71. An iron bridge girder ($Y = 2.0 \times 10^{11}$ N/m²) is constrained between two rock faces whose spacing doesn't change. At 20.0°C the girder is relaxed. How large a stress develops in the iron if the sun heats the girder to 40.0°C?

72. A wine barrel has a diameter at its widest point of 134.460 cm at a temperature of 20.0°C. A circular iron band, of diameter 134.448 cm, is to be placed around the barrel at the widest spot. The iron band is 5.00 cm wide and 0.500 cm thick. (a) To what temperature must the band be heated to be able to fit it over the barrel? (b) Once the band is in place and cools to 20.0°C, what will be the tension in the band?

73. The inner tube of a Pyrex glass mercury thermometer has a diameter of 0.120 mm. The bulb at the bottom of the thermometer contains 0.200 cm³ of mercury. How far will the thread of mercury move for a change of 1.00°C? Remember to take into account the expansion of the glass.

74. An iron cannonball of radius 0.08 m has a cavity of radius 0.05 m that is to be filled with gunpowder. If the measurements

were made at a temperature of 22°C, how much extra volume of gunpowder, if any, will be required to fill 500 cannonballs when the temperature is 30°C?

75. A bimetallic strip (Fig. 13.22) is made from metals with expansion coefficients α_1 and α_2 (with $\alpha_2 > \alpha_1$). The thickness of each layer is s. At some temperature T_0 the bimetallic strip is relaxed and straight. (a) Show that, at temperature $T_0 + \Delta T$, the radius of curvature of the strip is

$$R \approx \frac{s}{(\alpha_2 - \alpha_1)\,\Delta T}$$

[*Hint:* At T_0 the lengths of the two layers are the same. At temperature $T_0 + \Delta T$, the layers form circular arcs of radii R and $R + s$, which subtend the same angle θ. Assume a small ΔT so that $\alpha \Delta T \ll 1$ (for either α).] (b) If the layers are made of iron and brass, with $s = 0.1$ mm, what is R for $\Delta T = 20.0$°C?

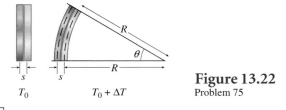

Figure 13.22
Problem 75

76. Estimate the average distance between air molecules at 0.0°C and 1.00 atm.

77. If you wanted to make a scale model of air at 0.0°C and 1.00 atm, using ping-pong balls (diameter, 3.75 cm) to

represent the N_2 molecules (diameter, 0.30 nm), (a) how far apart on average should the ping-pong balls be at any instant? (b) How far would a ping-pong ball travel on average before colliding with another?

78. For divers going to great depths, the composition of the air in the tank must be modified. The ideal composition is to have approximately the same number of O_2 molecules per unit volume as in surface air (to avoid oxygen poisoning), and to use helium instead of nitrogen for the remainder of the gas (to avoid nitrogen narcosis, which results from nitrogen dissolving in the bloodstream). Of the molecules in dry surface air, 78% are N_2, 21% are O_2, and 1% are Ar. (a) How many O_2 molecules per m^3 are there in surface air at 20.0°C and 1.00 atm? (b) For a diver going to a depth of 100.0 m, what percentage of the gas molecules in the tank should be O_2? (Assume that the density of seawater is 1025 kg/m^3 and the temperature is 20.0°C.)

79. Show that, in two gases at the same temperature, the rms speeds are inversely proportional to the square root of the molecular masses:

$$\frac{(v_{rms})_1}{(v_{rms})_2} = \sqrt{\frac{m_2}{m_1}}$$

80. Agnes Pockels, a housewife, was able to determine Avogadro's number using only a few household chemicals, in particular oleic acid, whose formula is $C_{18}H_{34}O_2$. (a) What is the molecular mass of this acid? (b) The mass of one drop of oleic acid is 2.3×10^{-5} g and the volume is 2.6×10^{-5} cm^3. How many moles of oleic acid are there in one drop? (c) Now all Agnes Pockels needed was to find the number of molecules of oleic acid. Luckily, when oleic acid is spread out on water, it lines up in a layer one molecule thick. If the base of the molecule of oleic acid is a square of side d, the height of the molecule is known to be $7d$. Agnes Pockels spread out one drop of oleic acid on some water, and measured the area to be 70.0 cm^2. Using the volume and the area of oleic acid, what is d? (d) If we assume that this film is one molecule thick, how many molecules of oleic acid are there in the drop? (e) What value does this give you then for Avogadro's number?

81. A certain acid has a molecular mass of 63 u. By weight, it consists of 1.6% hydrogen, 22.2% nitrogen, and 76.2% oxygen. What is the chemical formula for this acid?

82. These data are from a constant-volume gas thermometer experiment. The volume of the gas was kept constant, while the temperature was changed. The resulting pressure was measured. Plot the data on a pressure versus temperature diagram. Based on these data, estimate the value of absolute zero in Celsius.

T (°C)	P (atm)
0	1.00
20	1.07
100	1.37
−33	0.88
−196	0.28

83. Given that our body temperature is 98.6°F, (a) what is the average kinetic energy of the air molecules in our lungs? (b) If our temperature has increased to 100.0°F, by what percentage has the kinetic energy of the molecules increased?

84. The volume of air taken in by a warm-blooded vertebrate in the Andes mountains is 210 L/day at standard temperature and pressure (i.e., 0°C and 1 atm). If the air in the lungs is at 39°C, under a pressure of 450 mm Hg, and we assume that the vertebrate takes in an average volume of 100 cm^3 per breath at the temperature and pressure of its lungs, how many breaths does this vertebrate take per day?

85. The alveoli (see Section 13.8) have an average radius of 0.125 mm and are approximately spherical. If the pressure in the sacs is 1.00×10^5 Pa, and the temperature is 310 K (average body temperature), how many air molecules are in an alveolus?

86. In plants, water diffuses out through small openings known as stomatal pores. If $D = 2.4 \times 10^{-5}$ m^2/s for water vapor in air, and the length of the pores is 2.5×10^{-5} m, how long does it take for a water molecule to diffuse out through the pore?

87. During hibernation, an animal's metabolism slows down, and its body temperature lowers. For example, a California ground squirrel's body temperature lowers from 40.0°C to 10.0°C during hibernation. If we assume that the air in the squirrel's lungs is 75.0% N_2 and 25.0% O_2, by how much will the rms speed of the air molecules in the lungs have decreased during hibernation?

C 88. A 10.0-L vessel contains 12 g of N_2 gas at 20°C. (a) Estimate the nearest-neighbor distance. (b) Is the gas dilute? [*Hint:* Compare the nearest-neighbor distance to the diameter of an N_2 molecule, about 0.3 nm.]

ANSWERS TO PRACTICE PROBLEMS

13.1 37.0°C; 310.2 K

13.2 0.60 mm longer; 1.5 mm shorter

13.3 0.26 L

13.4 3.34×10^{28} molecules/m^3

13.5 7.9% of the air molecules; 189 kPa (27 lb/in^2)

13.6 640 kPa

13.7 v_{rms} = 408 m/s—lower than that of oxygen since the molecule is more massive

13.8 6% decrease

13.9 (a) 2×10^{10} km ≈ 100 × the Earth-Sun distance!; (b) 260 m/s; (c) 2000 yr

13.10 7.0×10^{-6} s

Heat

The weather forecast predicts a late spring hard freeze one night; the temperature is to fall several degrees below 0°C and the apple crop is in danger of being ruined. To protect the tender buds, farmers rush out and spray the trees with water. How does that protect the buds?

Concepts & Skills to Review

- energy conservation
  (Section 6.7)
- absolute temperature and ideal
  gas law (Section 13.5)
- kinetic theory of the ideal gas
  (Section 13.6)

14.1   INTERNAL ENERGY

From Section 13.6, the average translational kinetic energy $\langle K_{tr} \rangle$ of the molecules of an ideal gas is proportional to the absolute temperature of the gas:

$$\langle K_{tr} \rangle = \tfrac{3}{2} kT \tag{13-20}$$

The molecules move about in random directions even though, on a macroscopic scale, the gas is neither moving nor rotating. Equation (13-20) also gives the average translational kinetic energy of the random motion of molecules in liquids, solids, and nonideal gases except at very low temperatures. This random microscopic kinetic energy is *part* of what we call the *internal energy* of the system:

> The **internal energy** of a system is the total energy of all of the molecules in the system *except* for the macroscopic kinetic energy (kinetic energy associated with macroscopic translation or rotation) and the external potential energy (energy due to external interactions).

A **system** is whatever we define it to be: one object or a group of objects. Everything that is not part of the system is considered to be external to the system, or in other words, in the surroundings of the system.

Internal energy includes

- Translational and rotational kinetic energy of molecules *due to their individual random motions.*

- Vibrational energy—both kinetic and potential—of molecules and of atoms within molecules due to random vibrations about their equilibrium points.

- Potential energy due to interactions between the atoms and molecules of the system.

- Chemical and nuclear energy—the kinetic and potential energy associated with the binding of atoms to form molecules, the binding of electrons to nuclei to form atoms, and the binding of protons and neutrons to form nuclei.

Internal energy does *not* include

- The kinetic energy of the molecules due to translation, rotation, or vibration of the whole system or of a macroscopic part of the system.

- Potential energy due to interactions of the molecules of the system with something outside of the system (such as a gravitational field due to something outside of the system).

Example 14.1

Dissipation of Energy by Friction

A block of mass 10.0 kg starts at point A at a height of 2.0 m above the horizontal and slides down a frictionless incline (Fig. 14.1). It then continues sliding along the horizontal surface of a table that has friction. The block comes to rest at point C, a distance of 1.0 m along the table surface. How much has the internal energy of the system (block + table) increased?

Strategy Gravitational potential energy is converted to macroscopic translational kinetic energy as the block's speed increases. Friction then converts this macroscopic kinetic energy into internal energy—some of it in the block and some in the table. Since total energy is conserved, the increase in

continued on next page

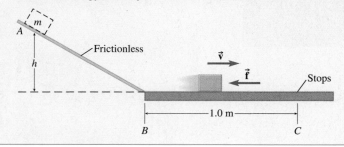

Figure 14.1
An object sliding down a frictionless incline and then across a horizontal surface with friction.

Example 14.1 *continued*

internal energy is equal to the decrease in gravitational potential energy:

decrease in PE from A to B = increase in KE from A to B

= decrease in KE from B to C

= increase in internal energy from B to C

Solution   The initial potential energy (taking $U_g = 0$ at the horizontal surface) is

$$U_g = mgh = 10.0 \text{ kg} \times 9.8 \text{ m/s}^2 \times 2.0 \text{ m} = 200 \text{ J}$$

The final potential energy is zero. The initial and final translational kinetic energies of the block are both zero. Therefore, the increase in the internal energy of the block and table is 200 J.

Discussion   We do not know how much of this internal energy increase appears in the object and how much in the table; we can only find the total. We call friction a *nonconservative* force, but that only means that *macroscopic mechanical* energy is not conserved; total energy is always conserved. Friction merely converts some macroscopic mechanical energy into internal energy of the block and the table. This internal energy increase manifests itself as a slight temperature increase. We often say that mechanical energy is *dissipated* by friction or other nonconservative forces; in other words, energy in an ordered form (translational motion of the block) has been changed into disordered energy (random motion of molecules within the block and table).

Practice Problem 14.1   On the rebound

If a rubber ball of mass 1.0 kg is dropped from a height of 2.0 m and rebounds on the first bounce to 0.75 of the height from which it was dropped, how much energy is dissipated during the collision with the floor?

Physics at Home

Take an elastic band and hold a portion to your lip; it should feel cool to the touch. Then grasp a section of the elastic between your thumbs and forefingers and stretch it back and forth rapidly several times. Hold the stretched region to your lip. For contrast, hold an unstretched portion of the same elastic band to your lip. The unstretched portion feels relatively cool and the stretched portion feels warm. The work you did in stretching the elastic increased its internal energy.

A change in the internal energy of a system does not always cause a temperature change. As we explore further in Section 14.5, the internal energy of a system can change while the temperature of the system remains constant—for instance, when ice melts.

Conceptual Example 14.2

Internal Energy of a Bowling Ball

A bowling ball at rest has a temperature of 18°C. The ball is then rolled down a bowling alley. Neglecting the dissipation of energy by friction and drag forces, is the internal energy of the ball higher, lower, or the same as when the ball was at rest? Is the temperature of the ball higher, lower, or the same as when the ball was at rest?

Strategy, Solution, and Discussion   The only change is that the ball is now rolling—the ball has macroscopic translational and rotational kinetic energy. However, the definition of *internal* energy does *not* include the kinetic energy of the molecules due to translation, rotation, or vibration *of the system as a whole*. Therefore, the *internal* energy of the ball is the same. Temperature is associated with the average translational kinetic energy due to the *individual random* motions of molecules; the temperature is still 18°C.

Conceptual Practice Problem 14.2   Total translational KE

Is the *total* translational kinetic energy of the molecules in the ball higher, lower, or the same as when the ball was at rest?

14.2   HEAT

We defined heat in Chapter 13:

Definition of heat:
Heat is a *flow* of energy between two objects or systems due to a temperature difference between them.

Many eighteenth-century scientists thought that heat was a fluid called "caloric." The flow of heat into an object was thought to cause the object to expand in volume in order to accommodate the additional fluid; why no mass increase occurred was a mystery. Now we know that heat is not a substance but is a flow of energy. One experiment that led to this conclusion was carried out by Count Rumford (Benjamin Thompson, 1753–1814). While supervising the boring of cannon barrels, he noted that the drill doing the boring became quite hot. At the time it was thought that the grinding up of the cannon metal into little pieces caused caloric to be released because the tiny bits of metal could not hold as much caloric as the large piece from which they came. But Rumford noticed that the drill got hot even when it became so dull that metal was no longer being bored out of the cannon and that he could create a limitless amount of what we now call internal energy. He decided that "heat" must be a form of microscopic motion instead of a material substance.

It was not until later experiments were done by James Prescott Joule (1818–1889) that Rumford's ideas were finally accepted. In his most famous experiment (Fig. 14.2), Joule showed that a temperature increase can be caused by mechanical means. In a series of such experiments, Joule determined the "mechanical equivalent of heat," or the amount of mechanical work required to produce the same effect on a system as a given amount of heat. In those days heat was measured in calories, where one calorie was defined as the heat required to change the temperature of 1 g of water by 1°C (specifically from 14.5 to 15.5°C). Joule's experimental results were within 1% of the currently accepted value, which is

$$1 \text{ cal} = 4.186 \text{ J} \qquad (14\text{-}1)$$

Equation (14-1) is now the *definition* of the calorie. The Calorie (with an uppercase letter C) used by dietitians and nutritionists is actually a kilocalorie:

$$1 \text{ Calorie} = 1 \text{ kcal} = 10^3 \text{ cal}$$

Although the calorie is still used, the SI unit for internal energy and for heat is the same as that used for all forms of energy and all forms of energy transfer: the joule.

Heat and work are similar in that both describe a particular kind of energy *transfer*. Work is an energy transfer due to a force acting through a displacement. Heat is a microscopic form of energy transfer involving large numbers of particles; the exchange of energy occurs due to the individual interactions of the particles. No macroscopic displacement occurs when heat flows and no macroscopic force is exerted by one object on the other.

It does not make sense to say that a system *has* 15 kJ of heat, just as it does not make sense to say that a system *has* 15 kJ of work. Similarly, we cannot say that the heat of a system has changed (nor that the work of a system has changed). A system can possess

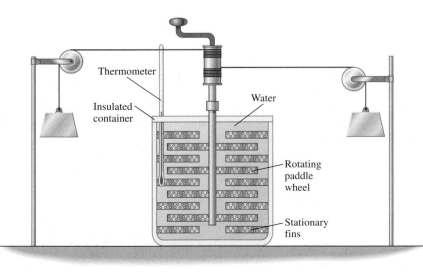

Figure 14.2 Joule's experiment. As the two masses fall under the effects of gravity, they cause a paddle wheel to rotate within an insulated container filled with water. The paddles agitate the water and cause its temperature to rise. By measuring the distance through which the masses fell and the temperature change of the known quantity of water, Joule determined the mechanical work done and the internal energy increase of the water.

energy in various forms (including internal energy); heat and work are two ways that energy can be transferred from one system to another. Joule's experiments showed that a quantity of work done on a system or the same quantity of heat flowing into the system causes the same increase in the system's internal energy. If the internal energy increase comes from mechanical work, as from Joule's paddle wheel, no heat flow occurs.

Heat flows from a system at higher temperature to one at lower temperature. Temperature is associated with the microscopic translational kinetic energy of the molecules; thus the flow of heat tends to equalize the average microscopic translational kinetic energy of the molecules. When two systems are in thermal contact and no net heat flow occurs, the systems are in thermal equilibrium and have the same temperature. In Chapter 13, we saw that heat always flows in the direction from a hotter object to a colder one; we will see in Chapter 15 that this is one way of stating the second law of thermodynamics.

Example 14.3

A Joule Experiment

In an experiment similar to that done by Joule, an object of mass 12.0 kg descends a distance of 1.25 m at constant speed while causing the rotation of a paddle wheel in an insulated container of water. If the descent is repeated 20.0 times, what is the internal energy increase of the water in joules? How many calories of heat flowing into the water would cause the same change in internal energy?

Strategy   Each time the mass descends, it converts gravitational potential energy into kinetic energy of the paddle wheel, which in turn agitates the water and converts kinetic energy into internal energy.

Solution   The change in gravitational potential energy during 20.0 downward trips is

$$\Delta U_g = mg\Delta h = 12.0\ \text{kg} \times 9.80\ \text{N/kg} \times 20.0\ \text{descents} \times 1.25\ \text{m/descent}$$

$$= 2.94\ \text{kJ}$$

If all of this energy goes into the water, the internal energy increase of the water is 2.94 kJ. From the definition of the

calorie (1 cal = 4.186 J), the number of calories of heat that would cause an internal energy change of 2.94 kJ is

$$\text{number of calories} = \frac{2.94 \times 10^3\ \text{J}}{4.186\ \text{J/cal}} = 702\ \text{cal}$$

Discussion   To perform an experiment like Joule's, we can vary the amount of energy delivered to the water. One way is to change the number of times the object is allowed to descend. Other possibilities include varying the mass of the descending object or raising the apparatus so that the object can descend a greater distance. All of these variations allow a change in the amount of gravitational potential energy converted into internal energy without requiring any changes in the complicated mechanism involving the paddle wheel.

Practice Problem 14.3   Temperature change of the water

If the water temperature in the insulated container is found to have increased 2.0°C after 20.0 descents of the falling object, what mass of water is in the container? Assume all of the internal energy increase appears in the water (neglect any internal energy change of the paddle wheel itself). [*Hint*: Recall that the calorie was defined as the heat required to change the temperature of 1 g of water by 1°C.]

The Cause of Thermal Expansion

If not to accommodate additional "caloric," then why do objects generally expand when their temperatures increase? (See Section 13.3.) An object expands when the *average* distance between the atoms (or molecules) increases. The atoms are not at rest; even in a solid, where each atom has a fixed equilibrium position, they *vibrate* to and fro about their equilibrium positions. The energy of vibration is part of the internal energy of the object. When heat flows into the object, raising its temperature, the internal energy increases. Some of the increase goes into vibration, so the average vibrational energy of an atom increases with increasing temperature.

The forces between atoms are highly asymmetric. Two atoms separated by *less* than their equilibrium distance repel each other *strongly*, while two atoms separated by *more* than their equilibrium distance attract each other much less strongly. Therefore, as vibrational energy increases, the maximum distance between the atoms increases more than the minimum distance decreases; the *average* distance between the atoms increases.

The coefficient of expansion varies from material to material because the strength of the interatomic (or intermolecular) bonds varies. As a general rule, the stronger the atomic bond, the smaller the coefficient of expansion. A stronger atomic bond has a potential energy curve with steeper sides; as temperature increases, the maximum displacements of the atoms from their equilibrium positions change relatively little. Liquids have much greater coefficients of volume expansion than do solids because the molecules are more loosely bound in a liquid than in a solid.

14.3   HEAT CAPACITY AND SPECIFIC HEAT

Suppose we have a system on which no mechanical work is done, but we allow heat to flow into the system by placing it in thermal contact with another system at higher temperature. As the internal energy of the system increases, its temperature increases (provided that no part of the system undergoes a change of phase, such as from solid to liquid). Thus, if heat flows into the system, without work being done or a phase change occurring, the temperature increase can be used to measure the heat—the more heat that flows, the larger the temperature increase. If Q represents the heat into the system and U represents the internal energy of the system, then the change in internal energy (ΔU) of the system is

$$\Delta U = Q \qquad (14\text{-}2)$$

(no work done; Q = heat *into* the system; U = internal energy of the system)

Remember that this is a *system on which no work is being done*; otherwise, conservation of energy would require that the change in internal energy be the sum of the heat into the system and the mechanical work done on the system. If heat flows *out* of the system rather than into the system, the internal energy of the system decreases ($\Delta U < 0$). We account for that possibility in Eq. (14-2) by making Q negative if heat flows out of the system; since Q is defined as the heat *into* the system, a negative heat represents heat flow *out of* the system.

Q is defined as the heat into the system.

Heat Capacity

For a large number of substances, under normal conditions, the temperature change ΔT is approximately proportional to the heat Q. The constant of proportionality is called the **heat capacity**:

$$\text{heat capacity} = \frac{Q}{\Delta T} \qquad (14\text{-}3)$$

The heat capacity depends both on the substance and on how much of it is present: 1 cal of heat into 1 g of water causes a temperature increase of 1°C, but 1 cal of heat into 2 g of water causes a temperature increase of 0.5°C. The SI unit of heat capacity is J/K. We can write J/K or J/°C interchangeably since only temperature *changes* are involved; a temperature change of 1 K is equivalent to a temperature change of 1°C.

    The term *heat capacity* is unfortunate since it has nothing to do with a capacity to *hold* heat, or a limited ability to absorb heat, as the name seems to imply. Instead, it relates the heat into a system to the temperature increase. It is better to think of heat capacity as a measure of how much heat must flow into or out of the system to produce a given temperature change.

Specific Heat

Since the heat capacity of a system is proportional to the mass of the system, the **specific heat capacity** (symbol c) of a substance is defined as the heat capacity per unit mass:

Specific heat capacity

$$c = \frac{Q}{m\,\Delta T} \qquad\qquad (14\text{-}4)$$

The SI units of specific heat capacity are J/(kg·K). In SI units, the specific heat is the number of joules of heat required to produce a 1 K temperature change in 1 kg of the substance. Again, since only temperature changes are involved, we can equivalently write J/(kg·°C). Specific heat capacity is often abbreviated to *specific heat*.

specific heat = heat capacity per unit mass

Table 14.1 lists specific heats for some common substances at 1 atm and 20°C (unless otherwise specified). For the range of temperatures in our examples and problems, assume these specific heat values to be valid. Note that water has a relatively large specific heat compared with most other substances. The relatively large specific heat of water causes the oceans to warm slowly in the spring and to cool slowly as winter approaches, moderating the temperature along the coast.

Rearrangement of Eq. (14-4) leads to an expression for the heat required to produce a known temperature change in a system:

$$Q = mc\,\Delta T \qquad\qquad (14\text{-}5)$$

Note that in Eqs. (14-4) and (14-5), the sign convention for Q is consistent: a temperature increase ($\Delta T > 0$) is caused by heat flowing *into* the system ($Q > 0$) while a temperature decrease ($\Delta T < 0$) is caused by heat flowing *out* of the system ($Q < 0$).

Equations (14-3) and (14-4) can be combined to represent the heat capacity:

$$\text{heat capacity} = \frac{Q}{\Delta T} = mc \qquad\qquad (14\text{-}6)$$

Equation (14-6) shows that the heat capacity of a system is proportional to its mass. The specific heat of the water in a drinking glass is the same as that for water in Lake Superior, but the glass of water and the lake have vastly different heat capacities.

Table 14.1

Specific Heats of Common Substances at 1 atm and 20°C

Substance	Specific Heat $\left(\dfrac{\text{kJ}}{\text{kg·K}}\right)$	Specific Heat $\left(\dfrac{\text{kcal}}{\text{kg·K}}\right)$
Gold	0.128	0.0306
Lead	0.13	0.031
Mercury	0.139	0.0332
Silver	0.235	0.0561
Brass	0.384	0.0917
Copper	0.385	0.0920
Steel	0.45	0.11
Iron	0.44	0.11
Granite	0.80	0.19
Marble	0.86	0.21
Aluminum	0.900	0.215
Air (50°C)	1.05	0.25
Wood (average)	1.68	0.40
Steam (110°C)	2.01	0.480
Ice (0°C)	2.1	0.50
Alcohol (ethyl)	2.4	0.57
Human tissue (average)	3.5	0.84
Water (15°C)	4.186	1.000

Example 14.4

Heating Water in a Saucepan

A saucepan containing 5.00 kg of water initially at 20.0°C is heated over a gas burner for 10.0 min. The final temperature of the water is 30.0°C. (a) What is the internal energy increase of the water? (b) What is the expected final temperature if the water were heated for an additional 5.0 min? (c) Is it possible to estimate the flow of heat from the burner during the first 10.0 min?

Strategy We are interested in the internal energy and the temperature *of the water,* so we define a system that consists of the water in the saucepan. Although the pan is also heated, it is not part of this system. The pan, the burner, and the room are all outside the system.

Since no work is done on the water, the internal energy increase is equal to the heat flowing into the water. The heat can be found from the mass of the water, the specific heat of water, and the temperature change. As long as the burner delivers heat at a constant rate, we can find the additional heat delivered in the additional time. Since the temperature change is proportional to the heat delivered, the temperature changes at a constant rate (a constant number of °C per minute). So, in half the time, half as much energy is delivered and the temperature change is half as much.

Solution (a) First find the temperature change:

$$\Delta T = T_f - T_i = 30.0°C - 20.0°C = 10.0 \text{ K}$$

(A change of 10.0°C is equivalent to a change of 10.0 K.) The increase in the internal energy of the water is

$$\Delta U = Q = mc\,\Delta T$$
$$= 5.00 \text{ kg} \times 4.186 \text{ kJ/(kg·K)} \times 10.0 \text{ K} = 209 \text{ kJ}$$

(b) We assume that the heat delivered is proportional to the elapsed time. The temperature change is proportional to the energy delivered, so if the temperature changes 10.0°C in 10.0 min, it changes an additional 5.0°C in an additional 5.0 min. The final temperature is

$$T = 20.0°C + 15.0°C = 35.0°C$$

(c) Not all of the heat flows into the water. Heat also flows from the burner into the saucepan and into the room. All we can say is that *more than* 209 kJ of heat flows from the burner during the 10.0 min.

Discussion As a check, the heat capacity of the water is 5.00 kg × 4.186 kJ/(kg·K) = 20.9 kJ/K; 20.9 kJ of heat must flow for each 1.0 K change in temperature. Since the temperature change is 10.0 K, the heat required is 20.9 kJ/K × 10.0 K = 209 kJ.

Practice Problem 14.4 Price of a bubble bath

If the cost of electricity is $0.080 per kW·h, what does it cost to heat 160 L of water for a bubble bath from 10.0°C (the temperature of the well water entering the house) to 70.0°C? [*Hint:* 1 L of water has a mass of 1 kg. 1 kW·h = 1000 J/s × 3600 s.]

What happens if more than two substances exchange heat? Suppose some water is heated in a large iron pot by dropping a hot piece of copper into the pot. We can define the system to be the water, the copper, and the iron pot; the environment is the room containing the system. Heat continues to flow between the three substances (iron pot, water, copper) until thermal equilibrium is reached—that is, until all three substances are at the same temperature. From energy conservation, if losses to the environment are negligible, all the heat that flows out of the copper flows into either the iron or the water:

$$Q_{Cu} + Q_{Fe} + Q_{H_2O} = 0$$

In this case Q_{Cu} is negative since heat flows *out* of the copper; Q_{Fe} and Q_{H_2O} are positive since heat flows into both the iron and the water.

Calorimetry

A calorimeter is an insulated container that enables the careful measurement of heat (Fig. 14.3). The calorimeter is designed to minimize the heat flow to or from the surroundings. A typical constant volume calorimeter, called a *bomb calorimeter*, consists of a hollow aluminum cylinder of known mass containing a known quantity of water; the cylinder is inside a larger aluminum cylinder with insulated walls. An insulating air space separates the two cylinders. An insulated lid fits over the opening of the cylinders; often there are two small holes in the lid, one for a thermometer to be inserted into the contents of the inner cylinder and one for a stirring device to help the contents reach equilibrium faster.

Suppose an object at one temperature is placed in a calorimeter with the water and aluminum cylinder at another temperature. By conservation of energy, all the heat that flows

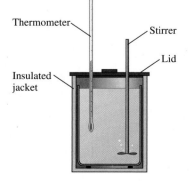

Figure 14.3 A calorimeter

out of one substance ($Q < 0$) flows into some other substance ($Q > 0$). If no heat flows to or from the environment, the total heat into the object, water, and aluminum must equal zero:

$$Q_o + Q_w + Q_a = 0$$

Example 14.5 illustrates the use of a calorimeter to measure the specific heat of an unknown substance. The measured specific heat can be compared with a table of known values to help identify the substance.

Example 14.5

Specific Heat of an Unknown Metal

A sample of unknown metal of mass 0.550 kg is heated in a pan of hot water until it is in equilibrium with the water at a temperature of 75.0°C. The metal is then carefully removed from the heat bath and placed into the inner cylinder of an aluminum calorimeter that contains 0.500 kg of water at 15.5°C. The mass of the inner cylinder is 0.100 kg. When the contents of the calorimeter reach equilibrium, the temperature inside is 18.8°C. Find the specific heat of the metal sample and determine whether it could be any of the metals listed in Table 14.1.

Strategy   Heat flows from the sample to the water and to the aluminum until thermal equilibrium is reached, at which time all three have the same temperature. We use subscripts to keep track of the three heat flows and three temperature changes. Let T_f be the final temperature of all three. Initially, the water and aluminum are both at 15.5°C while the sample is at 75.0°C. When thermal equilibrium is reached, all three are at 18.8°C. We assume negligible heat flow to the environment—in other words, that no heat flows into or out of the system of aluminum + water + sample.

Solution   Heat flows out of the sample ($Q_s < 0$) and into the water and aluminum cylinder ($Q_w > 0$ and $Q_a > 0$). Assuming no heat into or out of the surroundings,

$$Q_s + Q_w + Q_a = 0$$

For each substance, the heat is related to the temperature change. Substituting $Q = mc \, \Delta T$ for each gives

$$m_s c_s \, \Delta T_s + m_w c_w \, \Delta T_w + m_a c_a \, \Delta T_a = 0 \qquad (1)$$

A table helps organize the given information:

Substituting known values into Eq. (1) yields

$$0.550 \text{ kg} \times c_s \times (-56.2°C) + (0.500 \text{ kcal/°C} + 0.0215 \text{ kcal/°C}) \times 3.3°C = 0$$

Now we solve for c_s.

$$0.550 \text{ kg} \times c_s \times 56.2°C = 1.721 \text{ kcal}$$

$$c_s = \frac{1.721 \text{ kcal}}{0.550 \text{ kg} \times 56.2°C} = 0.0557 \, \frac{\text{kcal}}{\text{kg·°C}}$$

By comparing this result with the values in Table 14.1, it appears that the unknown sample could be silver.

Discussion   As a quick check, the heat capacity of the sample is approximately $\frac{1}{17}$ that of the water since its temperature change is 56.2°C/3.3°C ≈ 17 times as much—ignoring the small heat capacity of the aluminum. Since the masses of the water and sample are about equal, the specific heat of the sample is roughly $\frac{1}{17}$ that of the water:

$$\frac{1}{17} \times 1.00 \, \frac{\text{kcal}}{\text{kg·°C}} = 0.059 \, \frac{\text{kcal}}{\text{kg·°C}}$$

That is quite close to our answer.

___

Practice Problem 14.5   Final temperature

If 0.25 kg of water at 90.0°C is added to 0.35 kg of water at 20.0°C in an aluminum calorimeter with an inner cylinder of mass 0.100 kg, find the final temperature of the mixture.

Substance	Mass (m)	Specific Heat (c)	Heat Capacity (mc)	ΔT
Sample	0.550 kg	c_s (unknown)	0.550 kg $\times c_s$	18.8°C – 75.0°C = –56.2°C
H_2O	0.500 kg	1.00 $\frac{\text{kcal}}{\text{kg·K}}$	0.500 kcal/°C	18.8°C – 15.5°C = 3.3°C
Al	0.100 kg	0.215 $\frac{\text{kcal}}{\text{kg·K}}$	0.0215 kcal/°C	18.8°C – 15.5°C = 3.3°C

14.4   SPECIFIC HEAT OF IDEAL GASES

Since the average translational kinetic energy of a molecule in an ideal gas is

$$\langle K_{tr} \rangle = \tfrac{3}{2}kT \qquad \qquad \text{(13-20)}$$

the *total* translational kinetic energy of a gas containing N molecules (n moles) is

$$K_{tr} = \tfrac{3}{2}NkT = \tfrac{3}{2}nRT$$

Table 14.2

Molar Specific Heats at Constant Volume of Gases at 25°C

Gas		$C_V \left(\dfrac{\text{J/K}}{\text{mol}}\right)$
Monatomic	He	12.5
	Ne	12.7
	Ar	12.5
Diatomic	H_2	20.4
	N_2	20.8
	O_2	21.0
Polyatomic	CO_2	28.2
	N_2O	28.4

Suppose we allow heat to flow into a *monatomic* ideal gas—one where the gas molecules consist of single atoms—while keeping the volume of the gas constant. Since the volume is constant, no work is done on the gas, so the change in the internal energy is equal to the heat. If we think of the atoms as point particles, the only way for the internal energy to change when heat flows into the gas is for the translational kinetic energy of the atoms to change. The rest of the internal energy is "locked up" in the atoms and does not change unless something else happens, such as a phase transition or a chemical reaction—neither of which can happen in an ideal gas. Then

$$Q = \Delta U = \Delta K_{\text{tr}} = \tfrac{3}{2}nR\,\Delta T \qquad (14\text{-}7)$$

From Eq. (14-7) we can find the specific heat of the monatomic ideal gas. However, with gases it is more convenient to define the **molar specific heat** at constant volume (C_V) as

$$Q = nC_V\,\Delta T \qquad (14\text{-}8)$$

The subscript "V" is a reminder that the volume of the gas is held constant during the heat flow. The molar specific heat is the heat capacity *per mole* rather than *per unit mass*. In essence, specific heat and molar specific heat can be thought of as the same quantity—heat capacity per amount of substance—expressed in different units. In one case we measure the amount of substance by the mass; in the other case by the number of moles.

From Eqs. (14-7) and (14-8) we can find the molar specific heat of a monatomic ideal gas:

$$Q = \tfrac{3}{2}nR\,\Delta T = nC_V\,\Delta T$$

$$C_V = \tfrac{3}{2}R = 12.5\ \frac{\text{J/K}}{\text{mol}} \quad (\textit{monatomic ideal gas}) \qquad (14\text{-}9)$$

A glance at Table 14.2 shows that this calculation is remarkably accurate at room temperature for monatomic gases.

Diatomic gases have larger molar specific heats than monatomic gases. Why? We cannot model the diatomic molecule as a point mass; the two atoms in the molecule are separated, giving the molecule a much larger rotational inertia about two perpendicular axes (see Fig. 14.4). The molar specific heat is *larger* because not all of the internal energy increase goes into the translational kinetic energy of the molecules; some goes into rotational kinetic energy.

It turns out that the molar specific heat of a diatomic ideal gas at room temperature is approximately

$$C_V = \tfrac{5}{2}R = 20.8\ \frac{\text{J/K}}{\text{mol}} \quad (\textit{diatomic ideal gas}) \qquad (14\text{-}10)$$

Why $\tfrac{5}{2}R$ instead of $\tfrac{3}{2}R$? The diatomic molecule has rotational kinetic energy about two perpendicular axes (Fig. 14.4b and c) in addition to translational kinetic energy associated with motion in three independent directions. Thus, the diatomic molecule has five ways to "store" internal energy while the monatomic molecule has only three. The theorem of **equipartition of energy**—which we cannot prove here—says that internal energy is

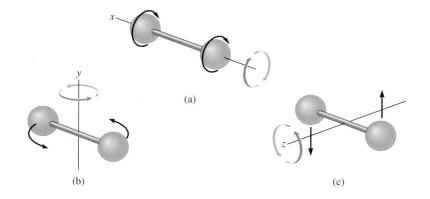

Figure 14.4 Rotation of a model diatomic molecule about three perpendicular axes. The rotational inertia about the *x*-axis (a) is negligible. The rotational inertias about the *y*- and *z*-axes (b) and (c) are much larger than for a single atom of the same mass because of the larger distance between the atoms and the axis of rotation.

distributed equally among all the possible ways in which it can be stored (as long as the temperature is sufficiently high). Each independent form of energy has an average of $\frac{1}{2}kT$ of energy per molecule and contributes $\frac{1}{2}R$ to the molar specific heat at constant volume.

Example 14.6

Heating Some Xenon Gas

A cylinder contains 250 L of xenon gas (Xe) at 20.0°C and a pressure of 5.0 atm. How much heat is required to raise the temperature of this gas to 50.0°C, holding the volume constant? Treat the xenon as an ideal gas.

Strategy   The molar heat capacity is the heat required per degree per mole. The number of moles of xenon (n) can be found from the ideal gas law, $PV = nRT$. Xenon is a monatomic gas, so we expect $C_V = \frac{3}{2}R$.

Solution   First we convert the known quantities into SI units.

$$P = 5.0 \text{ atm} = 5 \times 1.01 \times 10^5 \text{ Pa} = 5.05 \times 10^5 \text{ Pa}$$
$$V = 250 \text{ L} = 250 \times 10^{-3} \text{ m}^3$$
$$T = 20.0°C = 293.15 \text{ K}$$

From the ideal gas law, we find the number of moles,

$$n = \frac{PV}{RT} = \frac{5.05 \times 10^5 \text{ Pa} \times 250 \times 10^{-3} \text{ m}^3}{8.31 \text{ J/(mol·K)} \times 293.15 \text{ K}} = 51.8 \text{ mol}$$

We should check the units. Since $Pa = N/m^2$,

$$\frac{Pa \times m^3}{J/(mol·K) \times K} = \frac{N/m^2 \times m^3}{J/mol} = \frac{N·m}{J} \times mol = mol$$

For a monatomic gas at constant volume, the energy all goes into increasing the translational kinetic energy of the gas molecules. The molar specific heat is defined by $Q = nC_V \Delta T$, where, for a monatomic gas, $C_V = \frac{3}{2}R$. Then,

$$Q = \frac{3}{2}nR \, \Delta T$$

where

$$\Delta T = 50.0°C - 20.0°C = 30.0°C$$

Substituting,

$$Q = \frac{3}{2} \times 51.8 \text{ mol} \times 8.31 \text{ J/(mol·°C)} \times 30.0°C = 19 \text{ kJ}$$

Discussion   Constant volume implies that all the heat is used to increase the internal energy of the gas; if the gas were to expand it could transfer energy by doing work. When we find the number of moles from the ideal gas law, we must remember to convert the Celsius temperature to kelvins. Only when an equation involves a *change* in temperature can we use kelvin or Celsius temperatures interchangeably.

Practice Problem 14.6   Heating some helium gas

A storage cylinder of 330 L of helium gas is at 21°C and is subjected to a pressure of 10.0 atm. How much energy must be added to raise the temperature of the helium in this container to 75°C?

You may wonder why we can ignore rotation for the monatomic molecule—which in reality is not a point particle—or why we can ignore rotation about one axis for the diatomic molecule. The answer comes from quantum mechanics. Energy cannot be added to a molecule in arbitrarily small amounts; energy can only be added in discrete amounts or "steps." At room temperature, there is not enough internal energy to excite the rotational modes with small rotational inertias, so they do not participate in the specific heat. We also ignored the possibility of vibration for the diatomic molecule. That is fine at room temperature, but at higher temperatures vibration becomes significant, adding two more energy modes (one kinetic and one potential). Thus, as temperature increases, the molar specific heat of a diatomic gas increases, approaching $\frac{7}{2}R$ at high temperatures.

14.5   PHASE TRANSITIONS

If heat continually flows into the water in a pot, the water eventually begins to boil; liquid water becomes steam. If heat flows into ice cubes, they eventually melt and turn into liquid water. A **phase transition** occurs whenever a material is changed from one phase, such as the solid phase, to another, such as the liquid phase.

When some ice cubes at 0°C are placed into a glass in a room at 20°C, the ice gradually melts. A thermometer in the water that forms as the ice melts reads 0°C until all the ice is melted. At atmospheric pressure, ice and water can only coexist in equilibrium at 0°C. Once all the ice is melted, the water gradually warms up to room temperature. Similarly, water boiling on a stove remains at 100°C until all the water has boiled away.

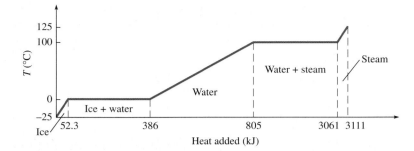

Figure 14.5 Temperature versus heat for 1 kg of ice that starts at a temperature below 0°C. (Horizontal axis not to scale.)

During a phase transition, the temperature of the mixture of two phases does not change.

Suppose we change 1.0 kg of ice at –25°C into steam at 125°C. A graph of the temperature versus heat is shown in Fig. 14.5. If the heat flows at a steady rate, a temperature versus *time* graph would have the same shape. During the two phase transitions, *heat flow continues, but the temperature of the mixture of two phases does not change*. Table 14.3 shows the heat during each step of the process.

The heat required *per unit mass* of substance to produce a phase change is called the **latent heat** (L). The word "latent" is related to the lack of temperature change during a phase transition.

Definition of latent heat

$$|Q| = mL \qquad (14\text{-}11)$$

The heat per unit mass for the solid-liquid phase transition (in either direction) is called the **latent heat of fusion** (L_f). From Table 14.3, it takes 333.7 kJ to change 1 kg of ice to water at 0°C, so for water $L_f = 333.7$ kJ/kg. For the liquid-gas phase transition (in either direction), the heat per unit mass is called the **latent heat of vaporization** (L_v). From Table 14.3, to change 1 kg of water to steam at 100°C takes 2256 kJ, so for water $L_v = 2256$ kJ/kg. Table 14.4 lists latent heats of fusion and vaporization for various materials.

All phase changes can go in either direction. Heat flowing into a substance can cause melting (solid to liquid) or boiling (liquid to gas). Heat flowing out of a substance can cause freezing (liquid to solid) or condensation (gas to liquid). If 2256 kJ must be supplied to turn 1 kg of water into steam, then 2256 kJ of heat is *released* from 1 kg of steam when it condenses to form water. A burn caused by 100°C steam is much more severe than a burn caused by 100°C water because the steam releases a large amount of heat as it condenses into water on the skin; much more energy is transferred to the skin than would be the case for the same amount of water at 100°C. The reverse process, evaporation, facilitates heat flow from a perspiring body.

The large latent heat of fusion of water is partly why spraying fruit trees with water can protect the buds from freezing. Before the buds can freeze, first the water must be cooled to 0°C and then it must freeze. In the process of freezing, the water gives up a large amount of heat and keeps the temperature of the buds from going below 0°C. Even if the water freezes, then the layer of ice over the buds acts like insulation since ice is not a particularly good conductor of heat.

Making The Connection:
using ice to protect buds from freezing

Table 14.3	
Heat to Turn 1 kg of Ice at –25°C to Steam at 125°C	
Phase Transition or Temperature Change	**Q (kJ)**
Ice: –25°C to 0°C	52.3
Melting: ice at 0°C to water at 0°C	333.7
Water: 0°C to 100°C	419
Boiling: water at 100°C to steam at 100°C	2256
Steam: 100°C to 125°C	50

Table 14.4

Latent Heats of Some Common Substances

Substance	Melting Point (°C)	Heat of Fusion		Boiling Point (°C)	Heat of Vaporization	
		(kJ/kg)	(kcal/kg)		(kJ/kg)	(kcal/kg)
Alcohol (ethyl)	−114	104	24.8	78	854	204
Aluminum	660	397	94.8	2,450	11,400	2,720
Copper	1083	205	49	2,340	5,070	1,210
Gold	1063	66.6	15.9	2,660	1,580	377
Lead	327	22.9	5.47	1,620	871	208
Silver	960.8	88.3	21.1	1,950	2,340	558
Water	0.0	333.7	79.7	100	2,256	539

To understand what is happening during a phase change, we must consider the substance on the molecular level. When a substance is in solid form, bonds between the atoms or molecules hold them near fixed equilibrium positions. Energy must be supplied to break the bonds and change the solid into a liquid. When the substance is changed from the liquid to the vapor phase, energy is used to separate the molecules from the loose bonds holding them together and to move the molecules apart as the gas expands. Temperature does not change during these phase transitions because the *kinetic energy* of the molecules is not changing. Instead, the *potential energy* of the molecules changes as work is done against the forces holding them together.

Example 14.7

Making Silver Charms

A jewelry designer plans to make some specially ordered silver charms for a commemorative bracelet. If the melting point of silver is 960.8°C, how much heat (in kcal and in kJ) must the jeweler add to 0.500 kg of silver at 20.0°C to be able to pour silver into her charm molds?

Strategy The solid silver first needs to be heated to its melting point; then more heat has to be added to melt the silver.

Solution The total heat flow into the silver is the sum of the heat to raise the temperature of the solid and the heat that causes the phase transition:

$$Q = mc\,\Delta T + mL_f$$

The temperature change of the solid is

$$\Delta T = 960.8°C - 20.0°C = 940.8°C$$

We look up the specific heat of solid silver and the latent heat of fusion of silver. Substituting numerical values into the equation for Q yields

$$Q = 0.500\ \text{kg} \times 0.0561\ \text{kcal/(kg·°C)} \times 940.8°C + 0.500\ \text{kg} \times 21.1\ \text{kcal/kg}$$

$$= 26.39\ \text{kcal} + 10.55\ \text{kcal} = 36.94\ \text{kcal}$$

Using values of c and L_f in kcal, we get an answer of 36.9 kcal. To convert to kJ, we use the conversion factor 1 kcal = 4.186 kJ:

$$Q = 36.94\ \text{kcal} \times 4.186\ \text{kJ/kcal} = 155\ \text{kJ}$$

Discussion An easy mistake to make is to use the wrong latent heat. Here we were dealing with melting, so we need the latent heat of fusion. Another possible error is to use the specific heat for the wrong phase: here we raised the temperature of *solid* silver, so we need the specific heat of *solid* silver. With water, we must always be careful to use the specific heat of the correct phase; the specific heats of ice, water, and steam have three different values.

Practice Problem 14.7 Making gold medals

Some gold medals are to be made from 750 g of solid gold at 24°C (Fig. 14.6). How much heat is required to melt the gold so that it can be poured into the molds for the medals?

Figure 14.6
A gold medal: the Nobel Prize for physics

Example 14.8

Turning Water into Ice

Ice cube trays are filled with 0.500 kg of water at 20.0°C and placed into the freezer compartment of a refrigerator. How much energy must be removed from the water to turn it into ice cubes at –5.0°C?

Strategy   We can think of this process as three consecutive steps. First, the liquid water is cooled to 0°C. Then the phase change occurs at constant temperature. Now the water is frozen; the ice continues to cool to –5.0°C. The energy that must be removed for the whole process is the sum of the energy removed in each of the three steps.

Solution   For liquid water going from 20.0°C to 0.0°C,

$$Q_1 = mc_{water} \Delta T_1$$

where

$$\Delta T_1 = 0.0°C - 20.0°C = -20.0°C$$

Since ΔT_1 is negative, Q_1 is negative: heat must flow *out of* the water in order for its temperature to decrease. Next the water freezes. The heat is found from the latent heat of fusion:

$$Q_2 = -mL_f$$

Again, heat flows *out* so Q_2 is negative. For phase transitions we supply the correct sign of Q according to the direction of the phase transition (minus sign for freezing, plus sign for melting). Finally, the ice is cooled to –5.0°C:

$$Q_3 = mc_{ice} \Delta T_2$$

where

$$\Delta T_2 = -5.0°C - 0.0°C = -5.0°C$$

We use subscripts on the specific heats to distinguish the specific heat of ice from that of water. The total heat is

$$Q = m(c_{water} \Delta T_1 - L_f + c_{ice} \Delta T_2)$$

Now we look up c_{water}, L_f, and c_{ice} in Tables 14.1 and 14.4 and substitute:

$$Q = 0.500 \text{ kg} \left(1.00 \frac{\text{kcal}}{\text{kg} \cdot °C} \times (-20.0°C) - 79.7 \frac{\text{kcal}}{\text{kg}} + 0.50 \frac{\text{kcal}}{\text{kg} \cdot °C} \times (-5.0°C)\right)$$

$$= -0.500 \text{ kg} \times 102.2 \frac{\text{kcal}}{\text{kg}} = -51.1 \text{ kcal}$$

So 51.1 kcal of heat flows out of the water that becomes ice cubes.

Discussion   We *cannot* consider the entire temperature change from +20°C to –5°C in one step. A phase change occurs, so we must include the flow of heat during the phase change. Also, the specific heat of ice is different from the specific heat of liquid water; we must find the heat to cool water 20°C and then the heat to cool ice 5°C.

Practice Problem 14.8   Frozen popsicles

Nigel pulls a tray of frozen popsicles out of the freezer to share with his friends. If the popsicles are at –4°C and go directly into hungry mouths at 37°C, how much energy is used to bring a popsicle of mass 0.080 kg to body temperature? Assume the frozen popsicles have the same specific heat as ice and the melted popsicle has the specific heat of water.

Example 14.9

Cooling a Drink

Two 50.0-g ice cubes are placed into 0.200 kg of water in a Styrofoam cup. The water is initially at a temperature of 25.0°C and the ice is initially at a temperature of –15.0°C. What is the final temperature of the drink? The average specific heat for ice between –15°C and 0°C is $0.49 \frac{\text{kcal}}{\text{kg} \cdot °C}$.

Strategy   We need to raise the temperature of the ice from –15°C to 0°C before the ice can melt, so we first find how much heat this requires. Then we find how much heat is needed to melt all the ice. Once the ice is melted, the water from the melted ice can be raised to the final temperature of the drink. The heat for all three steps (raising temperature of ice, melting ice, raising temperature of water from melted ice) all comes from the water initially at 25°C. That water cools as heat flows out of it. Assuming no heat flow into or out of the room, the quantity of heat that flows out of the water initially at 25°C flows into the ice or melted ice (before, during, and after melting).

Given: $m_{ice} = 0.1000$ kg at –15.0°C; $m_w = 0.200$ kg at 25.0°C, $c_{ice} = 0.49$ kcal/(kg·°C)

Look up: L_f for water = 79.7 kcal/kg; $c_{water} = 1.00$ kcal/(kg·°C)

Find: T_f

Solution   Since heat flows out of the water and into ice, $Q_{water} < 0$ and $Q_{ice} > 0$. Their sum is zero:

$$Q_{ice} + Q_w = 0$$

The heat flow into the ice is the sum of three terms:

$$Q_{ice} = m_{ice}c_{ice} \Delta T_{ice} + m_{ice}L_f + m_{ice}c_{water}(T_f - 0.0°C)$$

The heat flow out of the water is

$$Q_w = m_w c_{water}(T_f - 25.0°C)$$

The heat required to bring the ice from –15.0°C to 0°C is

$$m_{ice}c_{ice} \Delta T_{ice} = 0.1000 \text{ kg} \times 0.49 \frac{\text{kcal}}{\text{kg} \cdot °C} \times 15.0°C = 0.735 \text{ kcal}$$

The heat required to melt the ice at 0.0°C is

$$m_{ice}L_f = 0.1000 \text{ kg} \times 79.7 \text{ kcal/kg} = 7.97 \text{ kcal}$$

continued on next page

Example 14.9 *continued*

The heat to raise the temperature of the melted ice from 0.0°C to T_f is

$$m_{ice}c_{water}(T_f - 0.0°C) = 0.1000\ kg \times 1.00\frac{kcal}{kg\cdot°C} \times T_f$$

$$= 0.100\frac{kcal}{°C} \times T_f$$

The heat supplied by the water that was initially at 25.0°C is

$$m_w c_{water}(T_f - 25.0°C) = 0.200\ kg \times 1.00\frac{kcal}{kg\cdot°C} \times (T_f - 25.0°C)$$

$$= 0.200\frac{kcal}{°C} \times T_f - 5.00\ kcal$$

Now we substitute these values back into the original equation, $Q_{ice} + Q_w = 0$.

$$0.735\ kcal + 7.97\ kcal + \left(0.100\frac{kcal}{°C} \times T_f\right) + \left(0.200\frac{kcal}{°C} \times T_f - 5.00\ kcal\right) = 0$$

Simplifying yields

$$0.300\frac{kcal}{°C} \times T_f + 3.705\ kcal = 0$$

Solving for T_f, we find

$$T_f = -\frac{3.705\ kcal}{0.300\ kcal/°C} = -12.4°C$$

This result does not make sense: we assumed that all of the ice would melt and that the final mixture would be all liquid, but we cannot have liquid water at –12.4°C. Let's take another look at the solution.

What if the water initially at 25°C cools all the way to 0°C? From cooling the water, how much heat is available to warm the ice and melt it?

$$Q_w = m_w c_{water}(0°C - 25.0°C)$$

$$= 0.200\ kg \times 1.00\frac{kcal}{kg\cdot°C} \times (-25.0°C) = -5.00\ kcal$$

Thus, the water can supply 5.00 kcal when it cools to 0°C. But to warm the ice requires 0.735 kcal and to melt all of the ice requires another 7.97 kcal. The ice can be warmed to 0°C, but there is not enough heat available to melt all of the ice. Only some of the ice melts, so the drink ends up as a mixture of water and ice in equilibrium at 0°C.

Discussion   This example shows the value of checking a result to make sure it is reasonable. We started by assuming incorrectly that all of the ice would melt. When we obtained an answer that was impossible, we went back to see if there was enough heat available to melt all of the ice. Since there was not, the final temperature of the drink can only be 0°C—the only temperature at which ice and water can be in thermal equilibrium at atmospheric pressure.

Practice Problem 14.9   Melting ice

How much of the ice of Example 14.9 melts?

Evaporation

If you leave a cup of water out at room temperature, the water eventually evaporates. Recall that the temperature of the water reflects the average kinetic energy of the water molecules; some have higher than average energies and some have lower. The most energetic molecules have enough energy to break loose from the molecular bonds at the surface of the water. As these highest-energy molecules leave the water, the average energy of those left behind decreases—which is why evaporation is a cooling process. Approximately the same latent heat of vaporization applies to evaporation as to boiling, since the same molecular bonds are being broken. Perspiring basketball players cover up while sitting on the bench for a short time during a game to prevent getting a chill even though the air in the stadium may be warm.

Making The Connection:
chill caused by perspiration

When the humidity is high—meaning there is already a lot of water vapor in the air—evaporation proceeds more slowly. Water molecules in the air can also condense into water; the net evaporation rate is the difference in the rates of evaporation and condensation. A hot, humid day is uncomfortable because our bodies have trouble staying cool when perspiration evaporates slowly.

Physics at Home

The effects of evaporation can easily be felt. Rub some water on the inside of your forearm and then blow on your arm. The motion of the air over your arm removes the newly evaporated molecules from the vicinity of your arm and allows other molecules to evaporate more quickly. You can feel the cooling effect. If you have some rubbing alcohol, repeat the experiment. Since the alcohol evaporates faster, the cooling effect is noticeably greater.

Phase Diagrams

A useful tool in the study of phase transitions is the **phase diagram**—a diagram on which pressure is plotted on the vertical axis and temperature on the horizontal axis.

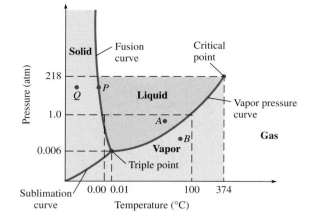

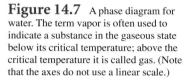

Figure 14.7 A phase diagram for water. The term vapor is often used to indicate a substance in the gaseous state below its critical temperature; above the critical temperature it is called gas. (Note that the axes do not use a linear scale.)

Figure 14.7 is a phase diagram for water. A point on the phase diagram represents water in a state determined by the pressure and the temperature at that point. The curves on the phase diagram are the demarcations between the solid, liquid, and vapor phases. For most temperatures, there is one pressure at which two particular phases can coexist in equilibrium. Since point P lies on the fusion curve, water can exist as liquid, or as solid, or as a mixture of the two at that temperature and pressure. At point Q, water can only be a solid. Similarly, at A water is a liquid; at B it is a vapor.

The one exception is at the **triple point**, where all three phases (solid, liquid, and vapor) can coexist in equilibrium. Triple points are used in precise calibrations of thermometers. The triple point of water is precisely 0.01°C at 0.006 atm.

From the vapor pressure curve, we see that as the pressure is lowered, the temperature at which water boils decreases. It takes longer to cook a hard-boiled egg in the high mountains because the temperature of the boiling water is less than 100°C; the chemical reactions that "cook" the egg proceed more slowly at a lower temperature. It might take as long as half an hour to hard-boil an egg on Pike's Peak, where the pressure is 0.6 atm.

If either the temperature or the pressure or both are changed, the point representing the state of the water moves along some path on the phase diagram. If the path crosses one of the curves, a phase transition occurs and the latent heat for that phase transition is either absorbed or released (depending on direction). Crossing the fusion curve represents freezing or melting; crossing the vapor pressure curve represents condensation or vaporization.

Notice that the vapor pressure curve ends at the **critical point**. Thus, if the path for changing a liquid to a gas goes around the critical point without crossing the vapor pressure curve, no phase transition occurs. At temperatures above the critical temperature or pressures above the critical pressure, it is *impossible* to make a clear distinction between the liquid and gas phases.

The sublimation curve represents another phase change, called **sublimation**, that occurs when a solid becomes vapor (or *vice versa*) without passing through the liquid phase. Sublimation occurs when ice on a car windshield becomes water vapor on a cold dry day. Mothballs and dry ice (solid carbon dioxide) also pass directly from solid to vapor. Sublimation has its own latent heat; the latent heat for sublimation is not the sum of the latent heats for fusion and vaporization.

The phase diagram of water has an unusual feature: the slope of the fusion curve is negative. The fusion curve has a negative slope only for substances (such as water, gallium, and bismuth) that expand on freezing. In these substances the molecules are *closer together* in the liquid than they are in the solid! As liquid water starting at room temperature is cooled, it contracts until it reaches 3.98°C. At this temperature water has its highest density (at a pressure of 1 atm); further cooling makes the water *expand*. When water freezes, it expands even more; ice is less dense than water.

One consequence of the expansion of water upon freezing is that cell walls might rupture when foods are frozen and thawed. The taste of frozen food suffers as a result. Another

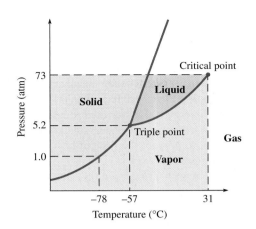

Figure 14.8 Phase diagram for carbon dioxide. (The axes do not use a linear scale.)

consequence is that lakes, rivers, and ponds do not freeze solid in the winter. A layer of ice forms on *top* since ice is less dense than water; underneath the ice, liquid water remains, which permits fish, turtles, and all manner of aquatic life to survive until spring.

The phase diagram for carbon dioxide (Fig. 14.8) is similar to that for water except that the fusion curve has a positive slope, which is the more common situation. At atmospheric pressure, only the solid and vapor phases of CO_2 exist. The liquid phase is not stable below 5.2 atm of pressure, so carbon dioxide does not melt at atmospheric pressure. Instead it sublimes; it goes from the solid directly to the vapor phase. Solid CO_2 is called *dry ice* because it is cold and looks like ice, but does not melt.

14.6   CONDUCTION

Until now we have considered the *effects* of heat flow, but not the mechanism of how heat flow occurs. We now turn our attention to three types of heat flow—*conduction, convection,* and *radiation*.

The **conduction** of heat can take place within solids, liquids, and gases. Conduction is due to collisions between atoms (or molecules) in which energy is exchanged. If the average energy is the same everywhere, there is no net flow of heat. If, on the other hand, the temperature is not uniform, then on average the atoms with more energy transfer some energy to those with less. The net result is that heat flows from the higher temperature region to the lower temperature region.

Conduction also occurs between objects that are in contact. A teakettle on an electric burner receives heat by conduction since the heating coil of the burner is in contact with the bottom of the kettle. The atoms that are vibrating in the object at higher temperature (the coil) collide with atoms in the object at lower temperature (the bottom of the kettle), resulting in a net transfer of energy to the colder object. If conduction is allowed to proceed, with no heat flow to or from the surroundings, then the objects in contact eventually reach thermal equilibrium when the average translational kinetic energies of the atoms are equal.

Suppose we consider a simple geometry such as an object with uniform cross section in which heat flows in a single direction. Examples are a plate of glass, with different temperatures on the inside and outside surfaces, or a cylindrical bar, with its ends at different temperatures (Fig. 14.9). The rate of heat conduction depends on the temperature difference $\Delta T = T_{hot} - T_{cold}$, the length (or thickness) d, the cross-sectional area A through which heat flows, and the nature of the material itself. The greater the temperature difference, the greater the heat flow. The thicker the material, the longer it takes for the heat to travel through—since the energy transfer has to be passed along a longer "chain" of atomic collisions—making the rate of heat flow smaller. A larger cross-sectional area allows more heat to flow.

The nature of the material is the final thing that affects the rate of energy transfer. In metals the electrons associated with the atom are free to move about and they carry the

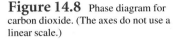

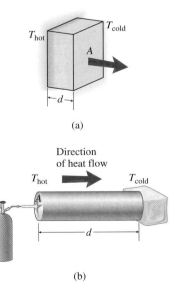

Figure 14.9 (a) Heat conduction through a slab of material of thickness d. (b) Heat conduction along a cylindrical bar of length d.

Table 14.5

Thermal Conductivities at 20°C

Material	$\kappa\left(\dfrac{W}{m \cdot K}\right)$
Air	0.023
Rock wool	0.038
Cork	0.046
Wood	0.13
Soil (dry)	0.14
Asbestos	0.17
Snow	0.25
Sand	0.39
Water	0.6
Glass	0.63
Concrete	1.7
Ice	1.7
Stainless steel	14
Lead	35
Steel	46
Nickel	60
Tin	66.8
Platinum	71.6
Iron	72.8
Brass	122
Zinc	116
Tungsten	173
Aluminum	237
Gold	318
Copper	401
Silver	429

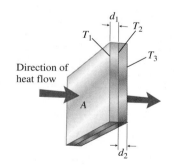

Figure 14.10 Conduction of heat through two different layers ($T_1 > T_2 > T_3$).

heat. When a material has free electrons, the transfer rate is faster; if the electrons are all tightly bound, as in nonmetallic solids, the transfer is slower. Liquids, in turn, conduct heat less readily than solids, because the forces between atoms are weaker. A gas is even less efficient as a conductor of heat than solids or liquids since the atoms of a gas are so much farther apart and have to travel a greater distance before collisions occur. The **thermal conductivity** (symbol κ, the Greek letter kappa) of a substance is directly proportional to the rate at which energy is transferred through the substance. Higher values of κ are associated with good conductors of heat, smaller values with poor conductors or *insulators*. (Insulators tend to prevent the flow of heat.) Table 14.5 lists the thermal conductivities for several common substances.

The dependence of the rate of heat flow within a substance on all the factors mentioned is given by

---
Fourier's law of heat conduction

$$I_{cd} = \frac{Q}{t} = \kappa A \frac{\Delta T}{d} \tag{14-12}$$
---

where $I_{cd} = Q/t$ is the rate of heat flow (or power delivered), κ is the thermal conductivity of the material, A is the cross-sectional area, d is the thickness (or length) of the material, and ΔT is the temperature difference between one side and the other. The quantity $\Delta T/d$ is called the *temperature gradient*; it tells how many °C or K the temperature changes per unit of distance moved along the path of heat flow. Inspection of Eq. (14-12) shows that the SI units of κ are W/(m·K).

In Fig. 14.9a, a slab of material is shown that conducts heat because of a temperature difference between the two sides. By rearranging Eq. (14-12) and solving for ΔT,

$$\Delta T = I_{cd}\frac{d}{\kappa A} = I_{cd}R \tag{14-13}$$

The quantity $d/\kappa A$ is called the **thermal resistance R.**

$$R = \frac{d}{\kappa A} \tag{14-14}$$

Thermal resistance has SI units of K/W (kelvins per watt). Notice that the thermal resistance depends on the nature of the material (through the thermal conductivity κ) and the geometry of the object (d/A). Equation (14-13) is useful for solving problems when heat flows through one material after another.

Suppose we have two layers of material between two temperature extremes as in Fig. 14.10. These layers are in *series* because the heat flows through one and then through the other. Looking at one layer at a time,

$$T_1 - T_2 = I_{cd}R_1 \quad \text{and} \quad T_2 - T_3 = I_{cd}R_2$$

Then, adding the two together

$$(T_1 - T_2) + (T_2 - T_3) = I_{cd}R_1 + I_{cd}R_2$$
$$\Delta T = T_1 - T_3 = I_{cd}(R_1 + R_2)$$

The rate of heat flow through the first layer is the same as the rate through the second layer because otherwise the temperatures would be changing. For n layers,

$$\Delta T = I_{cd}\Sigma R_n \quad n = 1, 2, 3, \ldots \tag{14-15}$$

Equation (14-15) shows that the effective thermal resistance for layers in series is the sum of each layer's thermal resistance.

Equation (14-15) has a familiar form. For a viscous fluid flowing through a pipe, Poiseuille's law can be written in the form $\Delta P = IR$, where I is the volume flow rate and R is called the fluid flow *resistance* (see Problem 9.50). For two or more pipes in series, where the volume flow rate is the same through each, the effective fluid flow resistance is the *sum of the resistances* of the individual pipes. Another equation of this form occurs in the study of electric currents, where the effective electrical resistance of two or more electrical conductors in series is the sum of the resistances of the individual conductors.

Example 14.10

The Rate of Heat Flow through Window Glass

 A windowpane that measures 20.0 cm by 15.0 cm is set into the front door of a house. The glass is 0.32 cm thick. The temperature outdoors is –15°C and inside is 22°C. At what rate does heat leave the house through that one small window?

Strategy   We assume one side of the glass to be at the temperature of the air inside the house and the other to be at the outdoor temperature.

Given: $\Delta T = 22°C - (-15°C) = 37°C$
 thickness of windowpane $d = 0.32 \times 10^{-2}$ m
 area of windowpane $A = 0.200$ m $\times$ 0.150 m = 0.0300 m^2

Look up: thermal conductivity for glass = 0.63 W/(m·K)

Find: rate of heat flow, I_{cd}

Solution   The temperature gradient is

$$\frac{\Delta T}{d} = \frac{37°C}{0.32 \times 10^{-2}\,\text{m}} = 1.16 \times 10^4 \text{ K/m}$$

Now we have all the information we need to find the rate of conductive heat flow:

$$I_{cd} = \kappa A \frac{\Delta T}{d}$$
$$= 0.63 \text{ W/(m·K)} \times 0.0300 \text{ m}^2 \times 1.16 \times 10^4 \text{ K/m}$$
$$= 220 \text{ W}$$

Discussion   A loss of 220 W through one small window is significant. Our assumption about the temperatures of the two glass surfaces exaggerates the temperature difference across the glass. In reality, the inside surface of the glass is colder than the air inside the house, while the outside surface is warmer than the air outside.

Practice Problem 14.10   An igloo

A group of children build an igloo in their garden. The snow walls are 0.30 m thick. If the inside of the igloo is at 10.0°C and the outside is at –10.0°C, what is the rate of heat flow through the snow walls of area 14.0 m^2?

Air has a low thermal conductivity; it is an excellent insulator *when it is still*. An accurate calculation of the energy loss through a single-paned window *must* take into account the thin layer of stagnant air, due to viscosity, on each side of the glass. If the temperature is measured near a window, the temperature of the air just beside the window is intermediate in value between the temperatures of the room air and the outside air (see Fig. 14.11). Thus, the temperature gradient *across the glass* is considerably smaller than the difference between indoor and outdoor temperatures. In fact, much of the thermal resistance of a window is due to the stagnant air layers rather than to the glass.

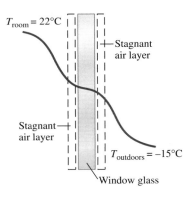

Figure 14.11 Temperature variation on either side of a windowpane. A plot of temperature versus position is superimposed on a cross section of the window glass and the air layers on either side.

Example 14.11

Heat Loss through a Double-Paned Window

 The single-paned window of Example 14.10 is replaced by a double-paned window with an air gap of 0.50 cm between the two panes. For the same temperature difference (22°C to –15°C), what is the new rate of heat loss through the double-paned window?

Strategy   Now there are three layers to consider: two layers of glass and one layer of air. We find the thermal resistance of

each layer and then add them together to find the total thermal resistance. Then we find the temperature difference between the inside of the house and the air outdoors and divide by the total thermal resistance to find the rate at which heat is lost through the replacement window.

Solution   For the first layer of glass,

$$R_1 = \frac{d}{\kappa A} = \frac{0.32 \times 10^{-2}\,\text{m}}{0.63 \text{ W/(m·K)} \times 0.0300 \text{ m}^2} = 0.169 \text{ K/W}$$

continued on next page

Example 14.11 *continued*

For the air gap,

$$R_2 = \frac{d}{\kappa A} = \frac{0.50 \times 10^{-2} \text{ m}}{0.023 \text{ W(m·K)} \times 0.0300 \text{ m}^2} = 7.246 \text{ K/W}$$

The second layer of glass has the same thermal resistance as the first:

$$R_3 = R_1$$

The total thermal resistance is

$$\Sigma R_n = 0.169 + 7.246 + 0.169 = 7.584 \text{ K/W}$$

and the rate of conductive heat flow is

$$I_{cd} = \frac{Q}{t} = \frac{\Delta T}{\Sigma R_n} = \frac{37 \text{ K}}{7.584 \text{ K/W}} = 4.9 \text{ W}$$

Discussion   The reduction in the rate of heat loss by replacing a single-paned window with a double pane is significant. This example, however, overestimates the reduction since we assume that heat can only be conducted through the air layer. In reality, heat can also flow through air by convection and radiation. A more accurate calculation would have to account for the other methods of heat flow.

Practice Problem 14.11   Two panes of glass without the air gap

Repeat the problem of Example 14.11 if the two panes of glass are touching each other, without the intervening layer of air.

The United States building industry rates materials used in construction with *R-factors*. The R-factor is not quite the same as the thermal resistance; thermal resistance cannot be specified without knowing the cross-sectional area. The "R-factor" is the thickness divided by the thermal conductivity:

$$\text{R-factor} = \frac{d}{\kappa} = RA$$

$$\frac{I_{cd}}{A} = \frac{\Delta T}{\text{R-factor}}$$

Unfortunately, SI units are not used. The R-factors quoted in the United States are in units of °F·ft²/(Btu/h)! R-factors are added, just as thermal resistances are, when heat flows through several different layers.

14.7   CONVECTION

Convection involves *fluid currents* that carry heat from one place to another. In conduction, energy flows through a material but the material itself does not move. In convection, *the material itself moves* from one place to another. Thus, convection can occur only in fluids, not in solids. When a wood stove is burning, convection currents in the air carry heat upward to the ceiling. The heated air is less dense than cooler air, so the buoyant force causes it to rise, carrying heat with it. Meanwhile, cooler air that is more dense sinks toward the floor. An example of convection currents at the seashore is shown in Fig. 14.12. Air is a poor *conductor* of heat, but it can easily flow and carry heat by *convection*.

Making The Connection:
offshore and
onshore breezes

The use of sealed, double-paned windows replaces the large air gap of about 6 or 7 cm between a storm window and regular window with a much smaller gap. The smaller air gap minimizes circulating convection currents between the two panes. Down jackets and quilts are good insulators because air is trapped in many little spaces among the feathers, minimizing heat flow due to convection. Materials such as rock wool, glass wool, or fiberglass are used to insulate walls; much of their insulating value is due to the air trapped around and between the fibers.

Convection currents also occur in liquids. As water is heated in a pan on the stove burner, the heat travels from the burner through the bottom surface of the pot by conduction and heats the layer of water in contact with the pot bottom. The heated water then expands, becoming less dense, so buoyant forces help it rise into the cooler regions of water within the pot where it loses heat, coming to equilibrium with cooler water. The slightly cooled water then sinks down to become heated some more and repeat the process (Fig. 14.13).

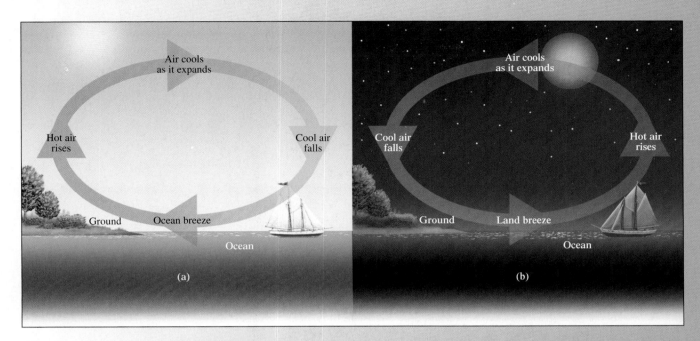

Figure 14.12 (a) During the day, air coming off the ocean is heated as it passes over the warm ground on shore; the heated air rises and expands. The expansion cools the air; it becomes more dense and falls back down. This cycle sets up a convection current that brings cool breezes from the sea to the shore. (b) The reverse circulation occurs at night when the land is cold and the sea is warmer, retaining heat absorbed during the day.

In *natural convection*, the currents are due to gravity. Fluid with a higher density sinks because the buoyant force is smaller than the weight; less dense fluid rises because the buoyant force exceeds the weight (Fig. 14.14). In *forced convection*, fluid is pushed around by mechanical means such as a fan or pump. In forced-hot-air heating, warm air is blown into rooms by a fan; in hot water baseboard heating, hot water is pumped through baseboard radiators (Fig. 14.15). Another example of forced convection is blood circulation in the body. The heart pumps blood around the body. When our body temperature is too high, the blood vessels near the skin dilate so that more blood can be pushed into them by the heart. The blood carries heat from the interior of the body to the skin; heat then flows from the skin into the cooler surroundings. If the surroundings are *hotter* than the skin, such as in a hot tub, this strategy backfires and can lead to dangerous overheating of the body. The hot water delivers heat to the dilated blood vessels; the blood carries the heat back to the central core of the body, *raising* the core temperature.

Figure 14.13 Convection currents in heated water

Physics at Home

In very hot climates, people drink *hot* drinks to cool off. Instead of reaching for a cold drink on a hot day, see if a hot cup of tea or cocoa makes you feel cooler. You are heating your central core so that your blood will carry heat to your skin; then the heat flows to the cooler air surrounding you and you feel cooler as a result.

A Global Warming Worry

Making The Connection:
ocean currents and
global warming

One worry for scientists studying global warming is that northern Europe might be plunged into a deep freeze—a seeming contradiction that results from an interruption of the natural convection cycle. Earth's climate is influenced by convection currents

Figure 14.14 Birds and people flying sailplanes take advantage of thermal updrafts.

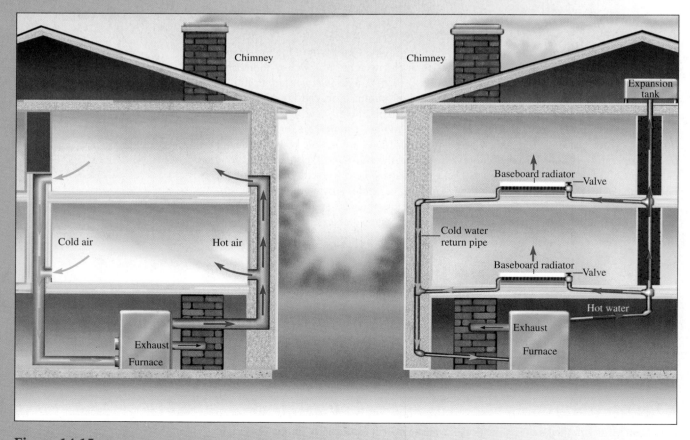

Figure 14.15 Household heating systems rely on forced convection.

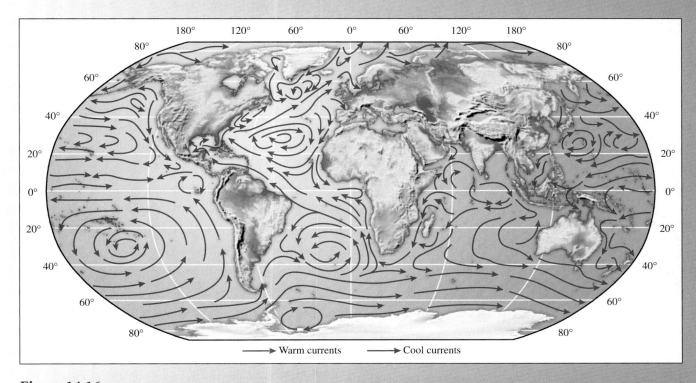

Figure 14.16   Convection currents in the oceans. The Gulf Stream is a current of warm water flowing across the Atlantic.

caused by temperature differences between the poles and the tropics (Fig. 14.16). Massive sea currents travel through the Pacific and Atlantic oceans, carrying about half of the heat from the tropics to the poles, where it is dissipated. Storms moving north from the tropics carry much of the rest of the heat. If the polar regions warm at a faster rate than the tropics, the smaller temperature difference between them changes the patterns of the prevailing winds, the tracks followed by storms, the speed of ocean currents, and the amount of precipitation.

For example, the melting of the ice shelves combined with increased precipitation could lead to a layer of freshwater lying on top of the more dense salt water in the North Atlantic. Normally, the cold ocean water at the surface sinks and starts the process of convection. With the buoyancy of the less dense freshwater layer keeping it from sinking, the convection currents slow down or are stopped. Without the pull of the convection current, the usual northward movement of water from the warm Gulf Stream would slow or cease, causing *colder* temperatures in northern Europe.

Such an effect on climate is not without precedent. At the end of the last Ice Age, freshwater from melting glaciers flowed out the St. Lawrence River and into the North Atlantic. A freshwater layer, buoyed up by the more dense salt water, disrupted the usual ocean currents. The Gulf Stream was effectively shut down and Europe experienced a thousand years of deep freeze.

Mathematical analysis of convection is quite difficult, but a relatively simple approximation is useful in some cases. We can define a **coefficient of convection** (h) for various conditions, such as for dry air moving at various speeds across skin. The rate of

Table 14.6

Coefficients of Convection for Dry Air and Bare Skin

Wind Speed (m/s)	Convective Coefficient W/(m²·°C)
1	15
2	22
3	26
4	28.5
5	32

heat flow due to convection when a fluid moves along a surface is proportional to the surface area and to the temperature difference:

Rate of convective heat flow

$$I_{cv} = \frac{Q}{t} = hA\,\Delta T \qquad (14\text{-}16)$$

Here A is the surface area over which the fluid moves, ΔT is the temperature difference between the surface and the fluid, and h is the coefficient of convection.

If the fluid is at a higher temperature than the surface, the heat flows from the fluid to the surface. If the surface is warmer, then the heat flows from the surface to the fluid. Equation (14-16) can be applied to air moving over skin with the help of Table 14.6, which lists convective coefficients for dry air moving over bare skin.

Example 14.12

Roller Blading in Still Air

A young woman is roller blading in still, dry air at a temperature of 30.0°C. She is moving along at 1.0 m/s and has a body surface area of 1.2 m², of which approximately 75% is exposed to the air. What is the rate of convective heat flow from her skin, at a temperature of 35.0°C, to the outside air?

Strategy   The roller blader is moving at 1.0 m/s along the road, but we can consider her to be still and the air to be moving past her at 1.0 m/s; what matters is the airspeed with respect to the roller blader's skin. The area of skin exposed to the moving air is 75% of the total skin area.

Given: $T_{fluid} = 30.0°C$; $T_{surface} = 35.0°C$;
   surface area of bare skin exposed is $A = 0.75 \times 1.2$ m²

Look up: coefficient of convection h for dry air and bare skin at wind speed of 1.0 m/s

Find: rate of heat flow, I_{cv}

Solution   The surface area of exposed skin is

$$A = 0.75 \times 1.2 \text{ m}^2 = 0.90 \text{ m}^2$$

From Table 14.6, the coefficient of convection h for dry air and bare skin at a wind speed of 1.0 m/s is

$$h = 15\frac{\text{W}}{\text{m}^2 \cdot °\text{C}}$$

To find the rate of heat flow by convection, we use the equation

$$I_{cv} = hA\,\Delta T$$

Substituting values yields

$$I_{cv} = 15\frac{\text{W}}{\text{m}^2 \cdot °\text{C}} \times 0.90 \text{ m}^2 \times (30.0°\text{C} - 35.0°\text{C}) = -68 \text{ W}$$

Heat flows from her skin at a rate of 68 W.

Discussion   The result can only be considered an estimate because the speed at which air moves past the skin is not a uniform 1.0 m/s. Due to the complicated pattern of airflow around the roller blader's body, the air moves faster past some surfaces than others.

In addition to convection, the body loses heat due to radiation (Section 14.8). This heat loss estimate represents only *part* of the heat lost by the body.

Practice Problem 14.12   A sailor standing on the bow of a ship

A sailor on an America's Cup racing yacht is wearing shorts and no shirt; the area of exposed skin is 1.3 m². The apparent wind (speed of the air relative to the sailor's body) is 5.0 m/s. What is the approximate rate of convective heat loss if his skin is at 35.0°C and the air temperature is 29.0°C? Assume that the air is dry.

14.8   RADIATION

Radiation is a third method of heat flow. All bodies emit energy through electromagnetic radiation—due to the oscillation of electric charges in the atoms. Radiation consists of electromagnetic waves that travel at the speed of light. Unlike conduction and convection, radiation need not travel through a material medium; the Sun radiates heat to Earth through the near vacuum of space.

When solar radiation reaches Earth, it is partially absorbed and partially reflected. The absorbed portion increases the Earth's internal energy. If absorption and reflection were the whole story, Earth's internal energy would continuously increase; but the Earth also emits radiation, which carries energy away. Since the temperature of the Earth remains relatively constant, it must emit energy at the same rate, on average, that it absorbs energy from the Sun. Thus, there is an equilibrium between absorption and emission.

Conceptual Example 14.13

An Alligator Lying in the Sun

An alligator crawls out into the Sun to get warm. Solar radiation is incident on the alligator at the rate of 300 W; 70 W of it is reflected. (a) What happens to the other 230 W? (b) If the alligator emits 100 W, does its body temperature rise, fall, or stay the same? Ignore heat flow by conduction and convection.

Solution and Discussion (a) When radiation falls on an object, some can be absorbed, some can be reflected, and—for a transparent or translucent object—some can be transmitted through the object without being absorbed or reflected. Since the alligator is opaque, no radiation is transmitted through it. All the radiation is either absorbed or reflected, so the other 230 W is absorbed. (b) Since 230 W is absorbed while 100 W

is emitted, the alligator absorbs more radiation than it emits. Absorption increases internal energy while emission decreases it, so the alligator's internal energy is increasing at a rate of 130 W. Thus, we expect the alligator's body temperature to rise. (The actual rate of increase of internal energy would be smaller since conduction and convection carry heat away as well.)

Conceptual Practice Problem 14.13
Maintaining constant temperature

After some time elapses, the alligator's body temperature reaches a constant level. The rate of absorption is still 230 W. If the alligator loses heat by conduction and convection at a rate of 90 W, at what rate does it emit radiation?

Stefan's Radiation Law

Some bodies emit more radiation than others per unit surface area even though their temperatures are the same. An idealized body that absorbs all the radiation incident upon it is called a **blackbody**. A *blackbody* absorbs not only all visible light, but infrared, ultraviolet, and all other wavelengths of electromagnetic radiation. Black velvet and a substance called lampblack come close to behaving like blackbodies. (Objects appear black when they absorb most of the visible light incident on them, so that little is reflected. However, an ideal blackbody absorbs all the radiation incident on it, not just the visible radiation.)

What does this have to do with *emission*? It turns out (see Conceptual Question 23) that a good absorber is also a good emitter of radiation. A blackbody emits more radiant power per unit surface area than any real object at the same temperature. The rate at which a blackbody emits radiation per unit surface area is proportional to the fourth power of the *absolute* temperature:

Stefan's law of radiation (blackbody)

$$I_{em} = \frac{Q}{t} = \sigma A T^4 \qquad (14\text{-}17)$$

In Eq. (14-17), A is the surface area and T is the surface temperature of the blackbody *in kelvins*. Since Stefan's law involves the absolute temperature and not a temperature difference, °C *cannot* be substituted. The universal constant σ (Greek letter sigma) is called *Stefan's constant*:

$$\sigma = 5.670 \times 10^{-8} \text{ W/(m}^2 \cdot \text{K}^4) \qquad (14\text{-}18)$$

The fourth-power temperature dependence implies that the power emitted is extremely sensitive to temperature changes. If the absolute temperature of a body doubles, the energy emitted increases by a factor of $2^4 = 16$.

Since real bodies are not perfect absorbers and therefore emit less than a blackbody, we define the **emissivity** (e) as the ratio of the emitted power of the body to that of a blackbody at the same temperature. Then Stefan's law becomes

---

Stefan's law of radiation

$$I_{em} = \frac{Q}{t} = e\sigma A T^4 \qquad (14\text{-}19)$$

The emissivity ranges from 0 to 1; $e = 1$ for a perfect radiator and absorber (a black-body) and $e = 0$ for a perfect reflector. The emissivity for polished aluminum, an excellent reflector, is about 0.05; for soot (carbon black) it is about 0.95. Human skin, no matter what the pigmentation, has an emissivity of about 0.97 in the infrared part of the spectrum. Many objects have high emissivities in the infrared even though they may reflect a fair amount of the visible light incident on them.

Radiation Spectrum

The electromagnetic radiation we are concerned with falls into three wavelength ranges. Infrared radiation includes wavelengths from about 100 μm down to 0.7 μm. The wavelengths of visible light range from about 0.7 μm to about 0.4 μm. Ultraviolet wavelengths are less than 0.4 μm.

The total power radiated is not the only thing that varies with temperature. Figure 14.17 shows the radiation spectrum—a graph of how much radiation occurs as a function of wavelength—for blackbodies at two different temperatures. The wavelength at which the maximum power is emitted decreases as temperature increases. Objects at ordinary temperatures emit primarily in the infrared—around 10 μm in wavelength for 300 K. The Sun, since it is much hotter, radiates primarily at shorter wavelengths. Its radiation peaks in the visible (no surprise there) but includes plenty of infrared and ultraviolet as well. The wavelength of maximum radiation is inversely proportional to the absolute temperature:

Wien's law

$$\lambda_{\max}T = 2.898 \times 10^{-3} \text{ m·K} \qquad (14\text{-}20)$$

---

where the temperature T is the temperature in kelvins and $\lambda_{\max}$ is the wavelength of maximum radiation in meters.

As the temperature of the blackbody rises to 1000 K and above, the peak intensity shifts toward shorter wavelengths until some of the emitted radiation falls in the visible. Since the longest visible wavelengths are for red light, the heated body glows dull red. As the temperature of the blackbody continues to increase, the red glow becomes brighter red, then orange, then yellow-white, and eventually blue-white as the blackbody emits more and more visible light. When the body is emitting all the colors of the visible spectrum, the glow appears white to the eye. When something is red-hot, it is not as hot as something that is white-hot.

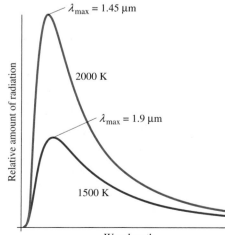

Figure 14.17 Graphs of blackbody radiation as a function of wavelength at two temperatures. At the higher temperature, the wavelength of maximum radiation is shorter, as predicted by Wien's law.

Example 14.14

Temperature of the Sun

 The maximum rate of energy emission from the Sun occurs in the middle of the visible range—at about $\lambda = 0.5 \ \mu m$. Estimate the temperature of the Sun's surface.

Strategy   We assume the Sun to be a blackbody. Then the wavelength of maximum emission and the surface temperature are related by Wien's law.

Solution   Given: $\lambda_{max} = 0.5 \ \mu m = 5 \times 10^{-7}$ m. Then from Wien's law, we know that the product of the wavelength for maximum power emission and the corresponding temperature for the power emission is

$$\lambda_{max} T = 2.898 \times 10^{-3} \ m \cdot K$$

We can solve for the temperature since we know λ_{max}

$$T = \frac{2.898 \times 10^{-3} \ m \cdot K}{5 \times 10^{-7} \ m}$$
$$= 6000 \ K$$

Discussion   Quick check: an object at 300 K has $\lambda_{max} \approx$ 10 μm, which is 20 times the λ_{max} in the radiation from the Sun (0.5 μm). Since λ_{max} and T are inversely proportional, the Sun's surface temperature is 20 times 300 K = 6000 K.

Practice Problem 14.14   Wavelengths of maximum power emission for skin

The temperature of skin varies from 30°C to 35°C depending on the blood flow near the skin surface. What is the range of wavelengths of maximum power emission from skin?

A body emits energy even if it is at the same temperature as its surroundings; it just emits at the same rate that it absorbs. If the surroundings are at a lower temperature than the body, the body emits more than it absorbs. The difference between the power emitted by the body and that absorbed by the body from its surroundings is

$$I_{em} - I_{ab} = e\sigma A(T^4 - T_0^4) \qquad (14\text{-}21)$$

Net rate of energy transfer due to emission and absorption of radiation

In Eq. (14-21), T_0 is the temperature of the surroundings. We have assumed that the emissivity is equal to the fraction of incident energy absorbed. If a body emits and absorbs at very different wavelengths—for example, if T and T_0 differ greatly—this assumption may not be true. *At any particular wavelength*, the fraction absorbed is always equal to the emissivity at that wavelength.

Do not substitute temperature in Celsius degrees into Eq. (14-21). The quantity inside the parentheses might look like a temperature difference, but it is not. The two kelvin temperatures are raised to the fourth power, *then* subtracted—which is not the same as the corresponding two Celsius temperatures subjected to the same mathematical operations. By the same token, do not subtract the temperatures in kelvins and then raise to the fourth power. The difference of the fourth powers is not equal to the difference raised to the fourth power, as can be readily demonstrated:

$$1 = (2 - 1)^4 \neq (2^4 - 1^4) = 15$$

Medical Applications

Thermal radiation from the body is used as a diagnostic tool in medicine. A thermogram shows whether one area is radiating more heat than it should, indicating a higher temperature due to abnormal cellular activity. For example, when a broken bone is healing, heat can be detected at the location of the break just by placing a hand lightly on the area of skin over the break. Infrared detectors, originally developed for military uses (nightscopes, for example), can be used to detect radiation from the skin. The radiation is absorbed and an electrical signal is produced that is then used to produce a visual display (Fig. 14.18).

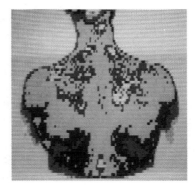

Figure 14.18   Thermography of a backache. The magenta areas are warmer than the surrounding tissue, revealing the location of the source of pain.

Example 14.15

Thermal Radiation from Human Body

 A person of body surface area 2.0 m² is sitting in a doctor's examining room with no clothing on. The temperature of the room is 22°C and the person's average skin temperature is 34°C. Skin emits about 97% as much as a blackbody at the same temperature for wavelengths in the infrared region, where most of the emission occurs. At what rate is energy radiated away from the body?

Strategy Both radiation and absorption occur in the infrared—the absolute temperatures of the skin and the room are not very different. Therefore, we can assume that 97% of the incident radiation from the room is absorbed. Equation (14-21) therefore applies. We must convert the Celsius temperatures to kelvin.

Given: surface area, $A = 2.0$ m²; $T_{room} = 22$°C; skin temperature, $T = 34$°C;
fraction of energy emitted, $e = 0.97$

To find: net rate of energy transfer, $I_{em} - I_{ab}$

Solution The temperature of the skin surface is
$$T = 273 + 34 = 307 \text{ K}$$
and of the room is
$$T_0 = 273 + 22 = 295 \text{ K}$$

The net rate of energy transfer between the room and the body is
$$I_{em} - I_{ab} = e\sigma A(T^4 - T_0^4)$$
Substituting,
$$I_{em} - I_{ab} = 0.97 \times 5.67 \times 10^{-8} \text{ W/(m}^2\cdot\text{K}^4) \times 2.0 \text{ m}^2 \times (307^4 - 295^4) \text{ K}^4$$
$$= 140 \text{ W}$$

Discussion 140 W is a significant heat loss because the body also loses about 10 W by convection and conduction. To stay at a constant body temperature, an inactive person must give off heat at a rate of 90 W to account for basal metabolic activity; if the rate of heat loss exceeds that, the body temperature starts to drop. The patient had better wrap a blanket around his body or start running in place.

We need only the fraction of energy emitted and absorbed by the body; the emissivity of the walls of the room is irrelevant. If the walls are poor emitters, then they also absorb poorly, so they reflect radiation. The amount of radiation incident on the body is the same.

Practice Problem 14.15 The roller blader radiates

Find how much energy per unit time the roller blader in Example 14.12 loses by radiation from her body. Her skin temperature is 35°C and the air temperature is 30°C. Her surface area is 1.2 m² of which 75% is exposed to the air. Assume skin has $e = 0.97$.

Example 14.16

Radiative Equilibrium of Earth

Radiant energy from the Sun reaches Earth at a rate of 1.7×10^{17} W. An average of about 30% is reflected and the rest is absorbed. Energy is also radiated by the atmosphere. Assuming that the atmosphere emits as a blackbody in the infrared ($e = 1$), calculate the temperature of the atmosphere. (The Sun's radiation peaks in the visible part of the spectrum, but the Earth's radiation peaks in the infrared due to its much lower surface temperature.)

Strategy Earth must radiate the same power as it absorbs. We use Stefan's law to find the rate at which energy is radiated as a function of temperature and then equate that to the rate of energy absorption.

Solution Earth absorbs 70% of the incident solar radiation. To have a relatively constant temperature, it must emit radiation at the same rate:
$$I_{em} = 0.7 \times 1.7 \times 10^{17} \text{ W} = 1.2 \times 10^{17} \text{ W}$$
From Stefan's law,
$$I_{em} = e\sigma A T^4$$

where we take $e = 1$ since the atmosphere is assumed to emit as a blackbody. The surface area of the Earth is
$$A = 4\pi R_E^2$$
Solving Stefan's law for T yields
$$T = \left(\frac{I_{em}}{e\sigma A}\right)^{1/4}$$
Now we substitute numerical values:
$$T = \left(\frac{I_{em}}{e\sigma 4\pi R_E^2}\right)^{1/4} = \left(\frac{1.2 \times 10^{17} \text{ W}}{1 \times 5.67 \times 10^{-8} \text{ W/(m}^2\cdot\text{K}^4) \times 4\pi(6.4 \times 10^6 \text{ m})^2}\right)^{1/4}$$
$$= 253 \text{ K} = -20°\text{C}$$

Discussion Remember that –20°C is supposed to be the average temperature of the *atmosphere*, not of the Earth's surface. This relatively simple calculation gives impressively accurate results. To find the temperature of the Earth's surface, we must take the greenhouse effect into account.

Practice Problem 14.16 Reflecting less incident radiation

If the Earth were to reflect 25% of the incident radiation instead of 30%, what would be the average temperature of the atmosphere?

The Greenhouse Effect

Making The Connection:
global warming and
the greenhouse effect

The Earth receives heat by radiation from the Sun. The atmosphere helps trap some of the radiation, acting rather like the glass in a greenhouse. When sunlight falls upon the glass of a greenhouse, most of the visible radiation and short-wavelength infrared (*near-infrared*) travel right on through; the glass is transparent to those wavelengths. The glass absorbs much of the incoming ultraviolet radiation. The radiation that gets through the glass is mostly absorbed inside the greenhouse. Since the inside of the greenhouse is much cooler than the Sun, it emits primarily infrared radiation (IR). The glass is not transparent to this longer-wavelength IR; much of it is absorbed by the glass. The glass itself also emits IR, but in both directions: half of it is emitted back inside the greenhouse. The absorption of IR by the glass keeps the greenhouse warmer than it would otherwise be. (The glass in a greenhouse has a second function not mirrored in the Earth's atmosphere—it prevents heat from being carried away by convection.)

The Earth is something like a greenhouse, where the atmosphere fulfills the role of the glass. Like glass, the atmosphere is largely transparent to visible and near-IR; the ozone layer in the upper atmosphere absorbs some of the ultraviolet. The atmosphere absorbs a great deal of the longer-wavelength IR emitted by Earth's surface. The atmosphere *radiates* IR in two directions: back toward the surface and out toward space (Fig. 14.19). "Greenhouse gases" such as CO_2 are particularly good absorbers of IR. The higher the concentration of greenhouse gases in the atmosphere, the more IR is absorbed and the warmer the Earth's surface becomes. Even small changes in the average surface temperature can have dramatic effects on climate.

In applying Stefan's radiation law to the Earth, there are some complications. One is the effect of the cloud cover. Clouds are quite reflective, but they are sometimes there and sometimes not. The heating of the lakes and oceans causes water to evaporate and form clouds. The clouds then serve as a screen and reflect sunlight away from the Earth, reducing the temperature again.

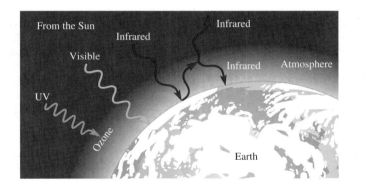

Figure 14.19 The global greenhouse effect. In this *simplified* diagram, all the UV from the Sun is absorbed by the atmosphere, while all the visible and IR from the Sun is transmitted. The Earth absorbs the visible and IR and radiates longer-wavelength IR. The longer-wavelength IR is absorbed by the atmosphere, which itself radiates IR both back toward the surface and out toward space.

MASTER THE CONCEPTS

Summary

- The internal energy of a system is the total energy of all of the molecules in the system except for the macroscopic kinetic energy (kinetic energy associated with macroscopic translation or rotation) and the external potential energy (energy due to external interactions).

- Heat is a *flow* of energy that occurs due to a temperature difference.

- The joule is the SI unit for all forms of energy, for heat, and for work. An alternate unit often used for heat and internal energy is the calorie:

$$1 \text{ cal} = 4.186 \text{ J} \qquad (14\text{-}1)$$

- The ratio of heat flow into a system to the temperature change of the system is the heat capacity of the system:

$$\text{heat capacity} = \frac{Q}{\Delta T} \qquad (14\text{-}3)$$

MASTER THE CONCEPTS *continued*

- The heat capacity per unit mass is the specific heat capacity (or specific heat) of a substance:

$$c = \frac{Q}{m\,\Delta T} \qquad (14\text{-}4)$$

- The *molar specific heat*, C, is the heat capacity per mole:

$$Q = nC_V\,\Delta T \qquad (14\text{-}8)$$

At room temperature, the molar heat capacity at constant volume for a monatomic ideal gas is approximately $C_V = \frac{3}{2}R$; for a diatomic ideal gas it is approximately $C_V = \frac{5}{2}R$.

- Phase transitions occur at constant temperature. The heat *per unit mass* that must flow to melt a solid or to freeze a liquid is the latent heat of fusion L_f. The latent heat of vaporization L_v is the heat *per unit mass* that must flow to change the phase from liquid to gas or from gas to liquid.

- *Sublimation* occurs when a solid changes directly to a gas without going into a liquid form.

- A phase diagram is a graph of pressure versus temperature that indicates solid, liquid, and vapor regions for a substance. The sublimation, fusion, and vapor pressure curves separate the three phases. Crossing one of these curves represents a phase transition.

- Heat flows by three processes: conduction, convection, and radiation.

- Conduction is due to atomic (or molecular) collisions within a substance or from one object to another when they are in contact. The rate of heat flow within a substance is:

$$I_{cd} = \frac{Q}{t} = \kappa A \frac{\Delta T}{d} \qquad (14\text{-}12)$$

where $I_{cd} = Q/t$ is the rate of heat flow (or power delivered), κ is the thermal conductivity of the material, A is the cross-sectional area, d is the thickness (or length) of the material, and ΔT is the temperature difference between one side and the other.

- Convection involves *fluid currents* that carry heat from one place to another. In convection, the material itself moves from one place to another.

- Radiation does not have to travel through a material medium. The energy is carried by electromagnetic waves that travel at the speed of light. All bodies emit energy through electromagnetic radiation. An idealized body that absorbs all the radiation incident on it is called a blackbody. A blackbody emits more radiant power per unit surface area than any real object at the same temperature. Stefan's law of radiation is

$$I_{em} = \frac{Q}{t} = e\sigma AT^4 \qquad (14\text{-}19)$$

where the emissivity e ranges from 0 to 1, A is the surface area, T is the surface temperature of the blackbody *in kelvins*, and Stefan's constant is $\sigma = 5.670 \times 10^{-8}$ W/(m$^2\cdot$K^4). The wavelength of maximum power emission is inversely proportional to the absolute temperature:

$$\lambda_{max}T = 2.898 \times 10^{-3} \text{ m}\cdot\text{K} \qquad (14\text{-}20)$$

The difference between the power emitted by the body and that absorbed by the body from its surroundings is the net power emitted:

$$I_{em} - I_{ab} = e\sigma A(T^4 - T_0^4) \qquad (14\text{-}21)$$

Highlighted Figures and Tables

F14.2 Joule's experiment (p. 476)

T14.1 Specific heats of common substances at 1 atm and 20°C (p. 479)

F14.3 A calorimeter (p. 480)

T14.2 Molar specific heats at constant volume of gases at 25°C (p. 482)

F14.5 Temperature versus heat for 1 kg of ice that starts at a temperature below 0°C (p. 484)

T14.3 Heat to turn 1 kg of ice at −25°C to steam at 125°C (p. 484)

T14.4 Latent heats of some common substances (p. 485)

F14.7 Phase diagram for water (p. 488)

F14.8 Phase diagram for carbon dioxide (p. 489)

F14.9 Conduction through a slab and along a cylindrical bar (p. 489)

T14.5 Thermal conductivities at 20°C (p. 490)

F14.10 Conduction of heat through two different layers ($T_1 > T_2 > T_3$) (p. 490)

F14.12 Example of convection currents at the seashore (p. 493)

F14.14 Thermal updrafts used in flight (p. 494)

F14.15 Household heating systems and forced convection (p. 494)

F14.16 Convection currents in the ocean (p. 495)

T14.6 Coefficients of convection for dry air and bare skin (p. 496)

F14.17 Blackbody radiation spectrum (p. 498)

CONCEPTUAL QUESTIONS

1. What determines the direction of heat flow when two objects at different temperatures are placed in thermal contact?

2. When an old movie has a scene of someone ironing, the person is often shown testing the heat of a hot flat iron with a moistened finger. Why is this safe to do?

3. Why do lakes and rivers freeze first at their surfaces?

4. Why is drinking water in a camp located near the equator often kept in porous jars?

5. Why are several layers of clothing warmer than one coat of equal weight?

6. Why are vineyards planted along lakeshores or riverbanks in cold climates?

7. A metal plant stand on a wooden deck feels colder than the wood around it. Is it necessarily colder? Explain.

8. Near a large lake, in what direction does a breeze passing over the land tend to blow at night?

9. What is the purpose of having fins on an automobile or motorcycle radiator?

10. Why do roadside signs warn that bridges ice before roadways? Explain.

11. Why do cooking directions on packages advise different timing to be followed for some locations?

12. Explain the theory behind the pressure cooker. How does it speed up cooking times?

13. When you eat a pizza that has just come from the oven, why is it that you are apt to burn the roof of your mouth with the first bite although the crust of the pizza feels only warm to your hand?

14. Explain why the molar specific heat of a diatomic gas such as O_2 is larger than that of a monatomic gas such as Ne.

15. At very low temperatures, the molar specific heat of hydrogen (H_2) is $C_V = 1.5R$. At room temperature, $C_V = 2.5R$. Explain.

16. When the temperature as measured in °C of a radiating body is doubled (such as a change from 20°C to 40°C), is the radiation rate necessarily increased by a factor of 16?

17. A cup of hot coffee has been poured, but the coffee drinker has a little more work to do at the computer before she picks up the cup. She intends to add some milk to the coffee. To keep the coffee hot as long as possible, should she add the milk at once, or wait until just before she takes her first sip?

18. Would heat loss be reduced or increased by increasing the usual air gap, 1 to 2 cm, between commercially made double-paned windows? Explain your reasoning. [*Hint:* Consider convection.]

19. A study of food preservation in Britain discovered that the temperature of meat that is kept in transparent plastic packages and stored in open and lighted freezers can be as much as 12°C above the temperature of the freezer. Why is this? How could this be prevented?

20. Which possesses more total internal energy, the water within a large, partially ice-covered lake in winter or a 6-cup teapot filled with hot tea? Explain.

21. A room in which the air temperature is held constant may feel warm in the summer but cool in the winter. Explain. [*Hint:* The walls are not necessarily at the same temperature as the air.]

22. Many homes are heated with "radiators," which are hollow metal devices filled with hot water or steam and located in each room of the house. They are sometimes painted with metallic, high-gloss silver paint so that they look well polished. Does this make them better radiators of heat? If not, what might be a more efficient finish to use?

23. Two objects with the same surface area are inside an evacuated container. The walls of the container are kept at a constant temperature. Suppose one object absorbs a larger fraction of incident radiation than the other. Explain why that object must emit a correspondingly greater amount of radiation than the other. Thus a good absorber must be a good emitter.

MULTIPLE CHOICE QUESTIONS

1. The main loss of heat from the Earth is by
   (a) radiation. (b) convection. (c) conduction.
   (d) All three processes are significant modes of heat loss from the Earth.

2. Assume the average temperature of the Earth's atmosphere to be 253 K. What would be the eventual average temperature of the Earth's atmosphere if the surface temperature of the Sun were to drop by a factor of 2?
   (a) 253 K (b) $\frac{253 \text{ K}}{2} = 127$ K
   (c) $\frac{253 \text{ K}}{4} = 63$ K (d) $\frac{253 \text{ K}}{2^4} = 16$ K

3. In equilibrium, Mars emits as much radiation as it absorbs. If Mars orbits the Sun with an orbital radius that is 1.5 times the orbital radius of the Earth about the Sun, what is the approximate atmospheric temperature of Mars? Assume the atmospheric temperature of Earth to be 253 K.
   (a) $\frac{253 \text{ K}}{1.5} = 170$ K (b) $\frac{253 \text{ K}}{1.5^2} = 112$ K
   (c) $\frac{253 \text{ K}}{1.5^4} = 50$ K (d) $\frac{253 \text{ K}}{\sqrt{1.5}} = 207$ K

4. Which term best represents the relation between a blackbody and radiant energy? A blackbody is an ideal _____ of radiant energy.
   (a) emitter (b) absorber
   (c) reflector (d) emitter and absorber

5. A window conducts power P from a house to the cold outdoors. What power is conducted through a window of *half* the area and *half* the thickness?
   (a) $4P$ (b) $2P$ (c) P (d) $P/2$ (e) $P/4$

6. Iron has a specific heat that is about four times that of gold. A cube of gold and a cube of iron, both of equal mass and at 20°C, are placed in two different Styrofoam cups, each filled with 100 g of water at 40°C. The Styrofoam cups have negligible heat capacities. After equilibrium has been attained,
   (a) the temperature of the gold is lower than that of the iron.
   (b) the temperature of the gold is higher than that of the iron.
   (c) the temperatures of the water in the two cups are the same.
   (d) (a) or (b) depending on how much mass is involved.

7. Sublimation is involved in which of these phase changes?
   (a) liquid to gas (b) solid to liquid
   (c) solid to gas (d) gas to liquid

8. When a vapor condenses to a liquid,
   (a) its internal energy increases.
   (b) its temperature rises.
   (c) its temperature falls.
   (d) it gives off internal energy.

9. When a substance is at its triple point, it
   (a) is in its solid phase. (b) is in its liquid phase.
   (c) is in its vapor phase.
   (d) may be in any or all of these phases.

10. The phase diagram for water is shown in Fig. 14.20. If the temperature of a certain amount of ice is increased by following the path represented by the dashed line from A to B in the phase diagram, which of the graphs of temperature as a function of heat added is correct?

11. Two thin rods are made from the same material and are of lengths L_1 and L_2. The two ends of the rods have the same temperature difference. What should the relation be between their diameters and lengths so that they conduct equal amounts of heat energy in a given time?

(a) $\dfrac{L_1}{L_2} = \dfrac{d_1}{d_2}$ (b) $\dfrac{L_1}{L_2} = \dfrac{d_2}{d_1}$

(c) $\dfrac{L_1}{L_2} = \dfrac{d_1^2}{d_2^2}$ (d) $\dfrac{L_1}{L_2} = \dfrac{d_2^2}{d_1^2}$

12. If you place your hand underneath, but not touching, a kettle of hot water, you *mainly* feel the presence of heat from

(a) conduction.   (b) convection.   (c) radiation.

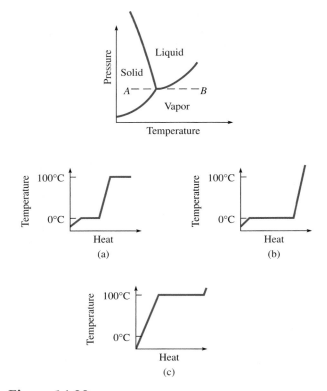

Figure 14.20   Multiple Choice Question 10

PROBLEMS

Note: ⓒ indicates a combination conceptual/quantitative problem. Gold diamonds ✦, ✦✦ are used to indicate the increasing level of difficulty of each problem. Problem numbers appearing in blue, 9., denote problems that have a detailed solution available in the Student Solutions Manual. Some problems are *paired* by concept; their numbers are connected by a ruled box.

14.1 Internal Energy

ⓒ 1. A mass of 1.4 kg of water at 22°C is poured from a height of 2.5 m into a vessel containing 5.0 kg of water at 22°C. (a) How much does the internal energy of the 6.4 kg of water increase? (b) Is it likely that the water temperature increases? Explain.

2. The water passing over Victoria Falls, located along the Zambezi River on the border of Zimbabwe and Zambia, drops about 105 m. How much internal energy is produced per kg as a result of the fall?

3. How much internal energy is generated when a 20.0-g lead bullet, traveling at 7.00×10^2 m/s, comes to a stop as it strikes a metal plate?

4. Nolan threw a baseball, of mass 147.5 g, at a speed of 162 km/h to a catcher. How much internal energy was generated when the ball struck the catcher's mitt?

ⓒ 5. A child of mass 15 kg climbs to the top of a slide that is 1.7 m above a horizontal run that extends for 0.50 m at the base of the slide. After sliding down, the child comes to rest just before reaching the very end of the horizontal portion of the slide. (a) How much internal energy was generated during this process? (b) Where did the generated energy go? (To the slide, to the child, to the air, or to all three?)

6. A 64-kg sky diver jumped out of an airplane at an altitude of 0.90 km. She opened her parachute after a while and eventually landed on the ground with a speed of 5.8 m/s. How much energy was dissipated by air resistance during the jump?

✦ 7. During basketball practice Shane made a jump shot, releasing a 0.60-kg basketball from his hands at a height of 2.0 m above the floor with a speed of 7.6 m/s. The ball swooshes through the net at a height of 3.0 m above the floor and with a speed of 4.5 m/s. How much energy was dissipated by air drag from the time the ball left Shane's hands until it went through the net?

14.2 Heat; 14.3 Heat Capacity and Specific Heat

8. An experiment is conducted with a basic Joule apparatus, where a mass is allowed to descend by 1.25 m and rotate a paddle wheel within an insulated container of water. There are several different sizes of descending masses to choose among. If the investigator wishes to deliver 1.00 kcal to the water within the insulated container after 30.0 descents, what descending mass value should be used? Let $g = 9.80$ m/s^2.

9. Convert 1.00 kcal to kilowatt-hours (kWh).

10. What is the heat capacity of 20.0 kg of silver?

11. What is the heat capacity of a gold ring that has a mass of 5.00 g?

12. If 30.0 kcal of heat are supplied to 5.00×10^2 g of water at 22°C, what is the final temperature of the water?

13. It is a damp, chilly day in a New England seacoast town suffering from a power failure. To warm up the cold, clammy

sheets, Jen decides to fill hot water bottles to tuck between the sheets at the foot of the beds. If she wishes to heat 2.0 L of water on the wood stove from 20.0°C to 80.0°C, how much heat must flow into the water?

14. An 83-kg man eats a banana of energy content 1.00×10^2 Calories (food calories). If all of the energy from the banana is converted into kinetic energy of the man, how fast is he moving, assuming he starts from rest?

15. A high jumper of mass 60.0 kg consumes a meal of 3.00×10^3 Calories (food calories) prior to a jump. If 3.3% of the energy from the food could be converted to gravitational potential energy in a single jump, how high could the athlete jump?

16. What is the heat capacity of a 30.0-kg block of ice?

17. What is the heat capacity of 1.00 m³ of (a) aluminum? (b) iron? See Table 9.1 for density values.

18. What is the heat capacity of a system consisting of (a) a 0.450-kg brass cup filled with 0.050 kg of water? (b) 7.5 kg of water in a 0.75-kg aluminum bucket?

19. A 0.400-kg aluminum teakettle contains 2.00 kg of water at 15.0°C. How much heat is required to raise the temperature of the water (and kettle) to 100.0°C?

20. How much heat is required to raise the body temperature of a 50.0-kg woman from 37.0°C to 38.4°C?

21. It takes 210 cal to raise the temperature of 350 g of lead from 0 to 20.0°C. What is the specific heat of lead?

22. A mass of 1.00 kg of water at temperature T is poured from a height of 0.100 km into a vessel containing water of the same temperature T, and a temperature change of 0.100°C is measured. What mass of water was in the vessel? Neglect heat flow into the vessel, the thermometer, etc.

23. A thermometer containing 0.10 g of mercury is cooled from 15.0°C to 8.5°C. How much energy was lost by the mercury in this process?

24. A bit of space debris penetrates the hull of a spaceship traversing the asteroid belt and comes to rest in a container of water that was at 20.0°C before being hit. The mass of the space rock is 1.0 g and the mass of the water is 1.0 kg. If the space rock traveled at 8.4×10^3 m/s and if all of its kinetic energy is used to heat the water, what is the final temperature of the water?

25. A pot containing 2.00 kg of water is sitting on a hot stove and the water is stirred violently by a mixer that does 6.0 kJ of mechanical work on the water. The temperature of the water rises by 4.00°C. What quantity of heat flowed into the water from the stove during the process?

26. A heating coil inside an electric kettle delivers 2.1 kW of electrical power to the water in the kettle. How long will it take to raise the temperature of 0.50 kg of water from 20.0°C to 100.0°C?

27. A 10.0-g iron bullet with a speed of 4.00×10^2 m/s and a temperature of 20.0°C is stopped in a 0.500-kg block of wood, also at 20.0°C. (a) At first all of the bullet's kinetic energy goes into the internal energy of the bullet. Calculate the temperature increase of the bullet. (b) After a short time the bullet and the block come to the same temperature T. Calculate T, assuming no heat is lost to the environment.

14.4 Specific Heat of Ideal Gases

28. Imagine that 501 people are present in a movie theater of volume 8.00×10^3 m³ that is sealed shut so no air can escape. Each person gives off heat at an average rate of 110 W. By how much will the temperature of the air have increased during a 2.0-h movie? The initial pressure is 1.01×10^5 Pa and the initial temperature is 20.0°C. Assume that all the heat output of the people goes into heating the air (a diatomic gas).

29. A chamber with a fixed volume of 1.0 m³ contains a monatomic gas at 3.00×10^2 K. The chamber is heated to a temperature of 4.00×10^2 K. This operation requires 10.0 J of heat. (Assume all the energy is transferred to the gas.) How many gas molecules are in the chamber?

30. A cylinder contains 250 L of hydrogen gas (H_2) at 0.0°C and a pressure of 10.0 atm. How much energy is required to raise the temperature of this gas to 25.0°C?

14.5 Phase Transitions

31. You are given 2.5×10^2 g of coffee (same specific heat as water) at 80.0°C (too hot to drink). In order to cool this to 60.0°C, how much ice (at 0.0°C) must be added? Neglect heat content of the cup and heat exchanges with the surroundings.

32. Given these data, compute the heat of fusion of water.

Mass of calorimeter = 3.00×10^2 g	Specific heat of calorimeter = 0.090 cal/g·K
Mass of water = 2.00×10^2 g	Initial temperature of water and calorimeter = 20.0°C
Mass of ice = 11.6 g	Initial temperature of ice = −5.0°C
	Final temperature of calorimeter = 15.0°C

33. Given these data, compute the heat of vaporization of water.

Mass of calorimeter = 3.00×10^2 g	Specific heat of calorimeter = 0.090 cal/g·K
Mass of water = 2.00×10^2 g	Initial temperature of water and calorimeter = 15.0°C
Mass of condensed steam = 18.5 g	Initial temperature of steam = 100.0°C
	Final temperature of calorimeter = 62.0°C

34. What mass of water at 25.0°C added to a Styrofoam cup containing two 50.0-g ice cubes from a freezer at −15.0°C will result in a final temperature of 5.0°C for the drink?

35. How much heat in kcal is required to change 1.0 kg of ice, originally at −20.0°C, into steam at 110.0°C? Assume 1.0 atm of pressure.

36. Ice at 0.0°C is mixed with 5.00×10^2 mL of water at 25.0°C. How much ice must melt to lower the water temperature to 0.0°C?

37. A phase diagram is shown in Fig. 14.21. Starting at point A, follow the dashed line to point E and consider what happens to the substance represented by this diagram as its pressure and temperature are changed. (a) Explain what happens for each line segment, AB, BC, CD, and DE. (b) What is the significance of point a and of point b?

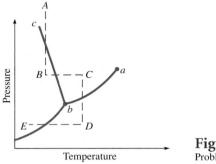

Figure 14.21
Problem 37

38. Compute the heat of fusion of a substance from these data: 11.06 kcal will change 0.500 kg of the solid at 21°C to liquid at 401°C; melting point 325°C; specific heat of solid, 0.030 kcal/kg·K; and specific heat of liquid, 0.10 kcal/kg·K.

39. The graph in Fig. 14.22 shows the change in temperature as heat is supplied to a certain mass of ice at −80.0°C. What is the mass of the ice?

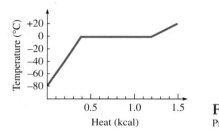

Figure 14.22
Problem 39

40. A 20.0-g lead bullet leaves a rifle at a temperature of 87.0°C and hits a steel plate. If the bullet melts, what is the minimum speed it must have?

41. How many grams of aluminum at 80.0°C would have to be dropped into a hole in a block of ice at 0.0°C to melt 10.0 g of ice?

42. Is it possible to heat the aluminum of Problem 41 to a high enough temperature so that it melts an equal mass of ice? If so, what temperature must the aluminum have?

43. If a leaf is to maintain a temperature of 40°C (reasonable for a leaf), it must lose 250 W/m² by transpiration (evaporative heat loss). Note that the leaf also loses heat by radiation, but we will neglect this. How much water is lost after 1 h through transpiration only? The area of the leaf is 0.005 m².

44. A birch tree loses 618 mg of water per minute through transpiration (evaporation of water through stomatal pores). What is the rate of heat lost through transpiration?

45. A dog loses a lot of heat through panting. The air rushing over the upper respiratory tract causes evaporation and thus heat loss. A dog typically pants at a rate of 670 pants per minute. As a rough calculation, assume that one pant causes 0.010 g of water to be evaporated from the respiratory tract. What is the rate of heat loss for the dog through panting?

14.6 Conduction

46. A window whose glass has $\kappa = 1.0$ W/(m·K) is covered completely with a sheet of foam of the same thickness as the glass, but with $\kappa = 0.025$ W/(m·K). How is the rate at which heat is conducted through the window changed by the addition of the foam?

47. Given a slab of material that is 1.0 m² and 2.0×10^{-2} m thick, (a) what is the thermal resistance if the material is asbestos? (b) What is the thermal resistance if the material is iron? (c) What is the thermal resistance if the material is copper?

48. A copper rod of length 0.50 m and cross-sectional area 6.0×10^{-2} cm² is connected to an iron rod with the same cross section and length 0.25 m (Fig. 14.23). One end of the copper is immersed in boiling water and the other end is at the junction with the iron. If the far end of the iron rod is in an ice bath at 0°C, find the rate of heat transfer passing from the boiling water to the ice bath.

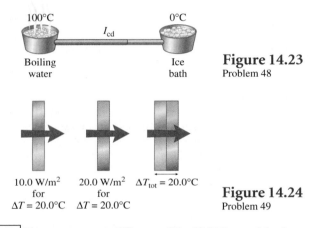

Figure 14.23
Problem 48

Figure 14.24
Problem 49

49. For a temperature difference $\Delta T = 20.0$°C, one slab of material conducts 10.0 W/m²; another of the same shape conducts 20.0 W/m². What is the rate of heat flow per m² of surface area when the slabs are placed side by side, as shown in Fig. 14.24, with $\Delta T_{tot} = 20.0$°C?

50. A wall consists of a layer of wood and a layer of cork insulation of the same thickness. The temperature inside is 20.0°C and the temperature outside is 0.0°C. (a) What is the temperature at the interface between the wood and cork if the cork is on the inside and the wood on the outside? (b) What is the temperature at the interface if the wood is inside and the cork is outside? (c) Does it matter whether the cork is placed on the inside or the outside of the wooden wall? Explain. (See Fig. 14.25.)

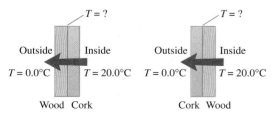

Figure 14.25 Problem 50

51. A copper bar of thermal conductivity 401 W/(m·K) has one end at 104°C and the other end at 24°C. The length of the bar is

0.10 m and the cross-sectional area is 1.0×10^{-6} m². (a) What is the rate of heat conduction, I_{cd}, along the bar? (b) What is the temperature gradient in the bar? (c) If two such bars were placed in series (end to end) between the same temperature baths, what would I_{cd} be? (d) If two such bars were placed in parallel (side by side) with the ends in the same temperature baths, what would I_{cd} be? (e) In the series case, what is the temperature at the junction where the bars meet?

52. A hiker is wearing wool clothing of 0.50-cm thickness to keep warm. Her skin temperature is 35°C and the outside temperature is 4.0°C. Her body surface area is 1.2 m². (a) If the thermal conductivity of wool is 0.040 W/(m·K), what is the rate of heat conduction through her clothing? (b) If the hiker is caught in a rainstorm, the thermal conductivity of the soaked wool increases to 0.60 W/(m·K) (that of water). Now what is the rate of heat conduction?

53. The thermal conductivity of the fur (including the skin) of a male Husky dog is 0.026 W/(m·K). The dog's heat output is measured to be 51 W, its internal temperature is 38°C, its surface area is 1.31 m² and the thickness of the fur is 5.0 cm. How cold can the outside temperature be before the dog must increase its heat output?

54. The thermal resistance of a seal's fur and blubber combined is 0.33 K/W. If the seal's internal temperature is 37°C and the temperature of the sea is about 0°C, what must be the heat output of the seal in order for it to maintain its internal temperature?

55. One cross-country skier is wearing a down jacket that is 2.0 cm thick. The thermal conductivity of goose down is 0.025 W/(m·K). Her companion on the ski outing is wearing a wool jacket that is 0.50 cm thick. The thermal conductivity of wool is 0.040 W/(m·K). (a) If both jackets have the same surface area and the skiers both have the same body temperature, which one will stay warmer longer? (b) How much longer can the person with the warmer jacket stay outside for the same amount of heat loss?

14.7 Convection

56. A marathon runner is moving along the road at 2.0 m/s in still air that is at a temperature of 29.0°C. His surface area is 1.4 m², of which approximately 85% is exposed to the air. What is the rate of convective heat loss from his skin, at a temperature of 35.0°C, to the outside air?

57. A small child is being pulled in a cart behind a bicycle that is traveling at 1.80 m/s in still air at a temperature of 25.0°C. If the exposed skin area of the child is 0.500 m², what is the rate of convective heat loss from the child's skin? The convective coefficient for that wind speed is 21.0 W/(m²·°C) and the child's skin is at 35.0°C.

58. A leaf exposed to the noonday Sun needs to remove energy at the rate of 156 W/m² in order to remain at a temperature of 40.0°C. One method of heat loss is through convection. What is the rate of heat loss per unit area if the temperature difference between the leaf and the air is 5.0°C and the convective coefficient for dry air across a leaf is 84.8 W/(m²·°C) for a wind speed of 4.5 m/s?

14.8 Radiation

59. A black wood stove has a surface area of 1.20 m² and a surface temperature of 175°C. What is the net rate at which heat is radiated into the room? The room temperature is 20°C.

60. An incandescent lightbulb has a tungsten filament that is heated to a temperature of 3.00×10^3 K when an electric current passes through it. If the surface area of the filament is approximately 1.00×10^{-4} m² and it has an emissivity of 0.32, what is the power radiated by the bulb?

61. A tungsten filament in a lamp is heated to a temperature of 2.6×10^3 K by an electric current. The tungsten has an emissivity of 0.32. What is the surface area of the filament if the lamp delivers 40.0 W of power?

62. At a tea party, a coffeepot and a teapot are placed upon the serving table. The coffeepot is a shiny silver-plated pot with emissivity of 0.12; the teapot is ceramic and has an emissivity of 0.65. Both pots hold 1.00 L of liquid at 98°C when the party begins. If the room temperature is at 25°C, what is the rate of radiative heat loss from the two pots? [Hint: To find the surface area, approximate the pots with cubes of similar volume.]

63. A lizard of mass 3.0 g is warming itself in the bright sunlight. It casts a shadow of 1.6 cm² on a piece of paper held perpendicularly to the Sun's rays. The intensity of sunlight at the Earth is 1.4×10^3 W/m² but only half of this energy penetrates the atmosphere and is absorbed by the lizard. (a) If the lizard has a specific heat of 4.2 J/(g·°C), what is the rate of increase of the lizard's temperature? (b) Assuming that there is no heat loss by the lizard (to simplify), how long must the lizard lie in the Sun in order to raise its temperature by 5.0°C?

64. A person of surface area 1.80 m² is risking skin cancer by lying out in the sunlight to get a tan. If the intensity of the incident sunlight is 7.00×10^2 W/m², at what rate must heat be lost by the person in order to maintain a constant body temperature? (Assume the effective area of skin exposed to the Sun is 42% of the total surface area, 57% of the incident radiation is absorbed, and that internal metabolic processes contribute another 90 W for an inactive person.)

65. If the total power per unit area from the Sun incident on a horizontal leaf is 9.00×10^2 W/m², and we assume that 70.0% of this energy goes into heating the leaf, what would be the rate of temperature rise of the leaf? The specific heat of the leaf is 3.70 kJ/(kg·°C), the leaf area is 5.00×10^{-3} m², and its mass is 0.500 g.

66. Consider the leaf of Problem 65. Assume that the top surface of the leaf absorbs 70.0% of 9.00×10^2 W/m² of radiant energy, while the bottom surface absorbs all of the radiant energy incident on it due to its surroundings at 25.0°C. (a) If the only method of heat loss for the leaf were thermal radiation, what would be the temperature of the leaf? (Assume that the leaf radiates like a blackbody.) (b) If the leaf is to remain at a temperature of 25.0°C, how much power per unit area must be lost by other methods such as transpiration (evaporative heat loss)?

COMPREHENSIVE PROBLEMS

✦ 67. A stainless steel saucepan, with a base that is made of 0.350-cm-thick steel (κ = 46.0 W/(m·K)) fused to 0.150 cm thickness of copper (κ = 401 W/(m·K)), sits on a ceramic heating element at 104.00°C. The diameter of the pan is 18.0 cm and it contains boiling water at 100.00°C. (a) If the copper-clad bottom is touching the heat source, what is the temperature at the copper-steel interface? (b) At what rate will the water evaporate from the pan?

68. The inner vessel of a calorimeter contains 2.50×10^2 g of tetrachloromethane, CCl_4, at 40.00°C. The vessel is surrounded by 2.00 kg of water at 18.00°C. After a time, the CCl_4 and the water reach the equilibrium temperature of 18.54°C. What is the specific heat of CCl_4?

69. If the temperature surrounding the sunbather in Problem 64 is greater than normal body temperature of 37°C and the air is still, so that radiation, conduction, and convection play no part in cooling the body, how much water (in liters per hour) from perspiration must be given off to maintain the body temperature? The heat of vaporization of water is 580 cal/g at normal skin temperature.

70. Two 62-g ice cubes are dropped into 186 g of water in a glass. If the water is initially at a temperature of 24°C and the ice is at –15°C, what is the final temperature of the drink? **C** ✦ 81.

71. A 0.500-kg slab of granite is heated so that its temperature increases by 7.40°C. The amount of heat supplied to the granite is 0.700 kcal. Based on this information, what is the specific heat of granite?

72. A spring of force constant $k = 8.4 \times 10^3$ N/m is compressed by 0.10 m. It is placed into a vessel containing 1.0 kg of water and then released. Assuming all the energy from the spring goes into heating the water, find the change in temperature of the water.

✦ 73. A 75-kg block of ice at 0.0°C breaks off from a glacier, slides along the frictionless ice to the ground from a height of 2.43 m, and then slides along a horizontal surface consisting of gravel and dirt. Find how much of the mass of the ice is melted by the friction with the rough surface, assuming 75% of the internal energy generated is used to heat the ice.

74. A blacksmith heats a 0.38-kg piece of iron to 498°C in his forge. After shaping it into a decorative design, he places it into a bucket of water to cool. If the available water is at 20.0°C, what minimum amount of water must be in the bucket to cool the iron to 23.0°C? The water in the bucket should remain in the liquid phase.

75. If 4.0 g of steam at 100.0°C condenses to water on a burn victim's skin and cools to 45.0°C, (a) how much heat is given up by the steam? (b) If the skin was originally at 37.0°C, how much tissue mass was involved in cooling the steam to water? See Table 14.1 for the specific heat of human tissue.

76. If 4.0 g of boiling water at 100.0°C was splashed onto a burn **C** victim's skin, and if it cooled to 45.0°C on the 37.0°C skin, (a) how much heat is given up by the water? (b) How much tissue mass, originally at 37.0°C, was involved in cooling the water? See Table 14.1. Compare the result with that found in Problem 75.

77. The amount of heat generated during the contraction of muscle in an amphibian's leg is given by Q = 0.13 mcal +

(0.35 mcal/cm) $\times \Delta x$, where Δx is the length shortened in cm and Q is in mcal. If a muscle of length 3.0 cm and mass 0.10 g is shortened by 1.5 cm during a contraction, what is the temperature rise? Assume that the specific heat of muscle is 1.0 cal/(g·°C).

78. Many species cool themselves by sweating, because as the sweat evaporates, heat is given up to the surroundings. A human exercising strenuously has an evaporative heat loss rate of about 650 W. If a person exercises strenuously for 30.0 min, how much water must he drink to replenish his fluid loss? The heat of vaporization of water is 580 cal/g at normal skin temperature.

79. Your hot water tank is insulated, but not very well. To reduce heat loss, you wrap some old blankets around it. With the water at 81°C and the room at 21°C, a thermometer inserted between the outside of the original tank and your blanket reads 36°C. By what factor did the blanket reduce the heat loss?

80. A wall consists of a layer of wood outside and a layer of insulation inside. The temperatures inside and outside the wall are +22°C and –18°C; the temperature at the wood/insulation boundary is –8.0°C. By what factor would the heat loss through the wall increase if the insulation were not present?

✦ 81. Small animals eat much more food per kg of body mass than do larger animals. The basal metabolic rate (BMR) is the minimal energy intake necessary to sustain life in a state of complete inactivity. Table 14.7 lists the BMR, mass, and surface area for five animals. (a) Calculate the BMR/kg of body mass for each animal. Is it true that smaller animals must consume much more food per kg of body mass? (b) Calculate the BMR/m² of surface area. (c) Can you explain why the BMR/m² is approximately the same for animals of different sizes? Consider what happens to the food energy metabolized by an animal in a resting state.

Table 14.7

Basal Metabolic Rates for Several Animals

Animal	BMR (kcal/day)	Mass (kg)	Surface Area (m²)
Mouse	3.80	0.018	0.0032
Dog	770	15	0.74
Human	2050	64	2.0
Pig	2400	130	2.3
Horse	4900	440	5.1

C ✦ 82. Imagine a person standing naked in a room at 23.0°C. The walls are well insulated, so they also are at 23.0°C. The person's surface area is 2.20 m² and his basal metabolic rate is 2167 kcal/day. His emissivity is 0.97. (a) If the person's skin temperature were 37.0°C (the same as the internal body temperature), at what net rate would heat be lost through radiation? (Ignore losses by conduction and convection.) (b) Clearly the heat loss in (a) is not sustainable—but skin temperature is less than internal body temperature. Calculate the skin temperature such that the net heat loss due to radiation is equal to the basal metabolic rate. (c) Does wearing clothing slow the loss of heat by radiation, or does it only decrease losses by conduction and convection? Explain.

83. On a very hot summer day Daphne is off to the park for a picnic. She puts 0.10 kg of ice at 0°C in a thermos and then adds a grape-flavored drink, which she has mixed from a powder, using room temperature water (35°C). How much grape-flavored drink will just melt all the ice?

84. It requires 4.085 kcal to melt 1.00×10^2 g of urethane ($CO_2(NH_2)C_2H_5$) at 48.7°C. What is the latent heat of fusion of urethane in kcal/mol?

85. A 20.0-g lead bullet leaves a rifle at a temperature of 47.0°C and travels at a velocity of 5.00×10^2 m/s until it hits a large block of ice at 0°C and comes to rest within it. How much ice will melt?

✦86. Bare, dark-colored basalt has a thermal conductivity of 3.1 W/(m·K), whereas light-colored sandstone's thermal conductivity is only 2.4 W/(m·K). Even though the same amount of radiation is incident on both and their surface temperatures are the same, the temperature gradient within the two materials will differ. For the same patch of area, what is the ratio of the depth in basalt as compared to the depth in sandstone that gives the same temperature difference?

✦87. The power expended by a cheetah is 160 kW while running at 110 km/h, but its body temperature cannot exceed 41.0°C. If 70.0% of the energy expended is dissipated within its body, how far can it run before it overheats? Assume that the initial temperature of the cheetah is 38.0°C, its specific heat is 3.5 kJ/(kg·°C), and its mass is 50.0 kg.

88. A 2.0-kg block of copper at 100.0°C is placed into 1.0 kg of water in a 2.0-kg iron pot. The water and the iron pot are at 25.0°C just before the copper block is placed into the pot. What is the final temperature of the water, assuming negligible heat flow to the environment?

89. A piece of gold of mass 0.250 kg and at a temperature of 75.0°C is placed into a 1.500-kg copper pot containing 0.500 L of water. The pot and water are at 22.0°C before the gold is added. What is the final temperature of the water?

ⓒ ✦✦90. A scientist working late at night in her low-temperature physics laboratory decides to have a cup of hot tea, but discovers the lab hot plate is broken. Not to be deterred, she puts about 8 oz of water, at 12°C, from the tap into a lab Dewar (essentially a large thermos bottle) and begins shaking it up and down. With each shake the water is thrown up and falls back down a distance of 33.3 cm. If she can complete 30 shakes per minute, how long will it take to heat the water to 87°C? Would this really work? If not, why not?

ⓒ ✦91. For a cheetah, 70.0% of the energy expended during exertion is internal work done on the cheetah's system and is dissipated within his body; for a dog only 5.00% of the energy expended is dissipated within the dog's body. Assume that both animals expend the same total amount of energy during exertion, both have the same heat capacity, and the cheetah is 2.00 times as heavy as the dog. (a) How much higher is the temperature change of the cheetah compared to the temperature change of the dog? (b) If they both start out at an initial temperature of 35.0°C, and the cheetah has a temperature of 40.0°C after the exertion, what is the final temperature of the dog? Which animal probably has more endurance? Explain.

ANSWERS TO PRACTICE PROBLEMS

14.1 4.9 J

14.2 Higher. The molecules have the same amount of *random* translational kinetic energy plus the additional kinetic energy associated with the ball's translation and rotation.

14.3 350 g

14.4 at least $0.89

14.5 48°C

14.6 92 kJ

14.7 36 kcal (150 kJ)

14.8 9.5 kcal (40 kJ)

14.9 53.5 g

14.10 230 W

14.11 110 W

14.12 250 W

14.13 To maintain constant temperature, the net heat must be zero. The rate at which energy is emitted is 140 W.

14.14 9.4 μm (at 35°C) to 9.6 μm (at 30°C)

14.15 28 W

14.16 –16°C

Thermodynamics

The gasoline engines in cars are terribly inefficient. Of the chemical energy that is released in the burning of gasoline, typically only about 25% is converted into useful mechanical work done on the car to move it forward. Yet scientists and engineers have been working for decades to make a more efficient gasoline engine. Is there some fundamental limit to the efficiency of a gasoline engine? Is it possible to make an engine that converts all—or nearly all—of the chemical energy in the fuel into useful work?

Concepts & Skills to Review

- internal energy and heat (Sections 14.1–14.2)
- system and environment (Section 14.1)
- area under a graph (Section 3.2)
- work done is the area under a graph of $F(x)$ (Section 6.4)
- specific heat of ideal gases at constant volume (Section 14.4)
- natural logarithm (Appendix A.3)

The choice of a *system* is made in any way convenient for a given problem.

15.1   THE FIRST LAW OF THERMODYNAMICS

Both work and heat can change the internal energy of a system. Work can be done on a rubber ball by squeezing it, stretching it, or slamming it into a wall. Heat can be supplied to the ball by leaving it out in the sun or putting it into a hot oven. These two methods of increasing the internal energy of a system lead to the **first law of thermodynamics**:

> **First law of thermodynamics:**
> The change in internal energy of a system is equal to the heat flow into the system minus the work done *by* the system.

The statement of the first law is actually a statement of energy conservation. The change in internal energy of a system is equal to the energy transferred to it from its surroundings. If the internal energy increases, for example, then the energy in the surroundings decreases by the same amount and the total energy of the universe (system plus surroundings) is unchanged.

We must agree on sign conventions (Table 15.1) for heat and work in order to write the first law as an equation. Q is the heat flow *into* a system; positive Q causes an increase in internal energy ($\Delta U > 0$). If heat flows out of the system, then Q is negative. However, by convention W stands for the work done *by* a system. When a system does positive work, it transfers energy to its surroundings; positive W causes a *decrease* in internal energy ($\Delta U < 0$) while negative W causes an *increase* in internal energy. Therefore,

> **First law of thermodynamics**
> $$\Delta U = Q - W \tag{15-1}$$

These sign conventions for work and heat came from the application of the first law to the steam engine. The usual state of affairs for a steam engine is to put heat into the engine (boiling the water to create steam) and to get work out of the engine. Thus heat flow into the engine and work done by the engine were chosen to be positive quantities. The sign conventions are easily remembered by thinking of the steam engine example.

Table 15.1

Sign Conventions for the First Law of Thermodynamics

Quantity	Definition	Meaning of + Sign	Meaning of – Sign
Q	Heat flow into the system	Heat flow into the system	Heat flow *out of* the system
W	Work done by the system	Work done by the system	Work done *on* the system
ΔU	Internal energy change	Internal energy increase	Internal energy *decrease*

Example 15.1

A Hot Drink at Camp

A pot containing 6.00 kg of water is placed over a campfire (Fig. 15.1). An impatient camper, eager for a hot drink, does 9.2 kJ of mechanical work on the water by agitating it with a spoon. If the internal energy of the water increases by 45.2 kJ, how much heat is delivered to the water by the fire during that time?

Strategy   The first law of thermodynamics relates internal energy changes of a system—in this case, the water—to the heat flow into the system and the work done by the system. The fire is at a higher temperature than the pot of water so heat flows from the fire through the pot and into the water. The spoon does work on the water; this also increases the internal energy of the water.

Solution   The increase in the internal energy of the water is

$$\Delta U = 45.2 \text{ kJ}$$

continued on next page

Example 15.1 *continued*

Work done by camper

Heat flow from fire

Figure 15.1
Energy transfers into a pot heated over a campfire and stirred by a camper.

Mechanical work is done *on* the water so W is negative, according to our sign convention:

$$W = -9.2 \text{ kJ}$$

From the first law,

$$\Delta U = Q - W$$

Solving for Q,

$$Q = \Delta U + W = 45.2 \text{ kJ} + (-9.2 \text{ kJ}) = 36.0 \text{ kJ}$$

Discussion   Fortunately we were not asked to find the *total* heat output of the fire; some went to heating the pot that contains the water, some went to heating the rack over the fire, and some heated the air and campers around the fire. Trying to heat water with a spoon is rather awkward and not very efficient; Joule's experiment requires a lot of patience.

Practice Problem 15.1   Changing internal energy of a gas

If the change in internal energy of a gas is +42 kJ due to a process involving a heat flow into the gas of 14 kJ, was the work done on the gas or by the gas? What is the value of the work, W, using the sign convention already discussed?

15.2   THERMODYNAMIC PROCESSES

A thermodynamic process is the method by which a system is changed from one *state* to another. The state of a system is described by a set of **state variables** such as pressure, temperature, volume, number of moles, and internal energy. State variables describe the state of a system at some instant of time but not how the system got to that state. Thus, heat and work are *not* state variables—they describe *how* a system gets from one state to another.

If a system is changed slowly so that it is always very near equilibrium, the changes in state can be represented by a curve on a plot of pressure versus volume (called a *PV* **diagram**). Each point on the curve represents a state of the system. The *PV* diagram is a useful tool for analyzing thermodynamic processes.

Figure 15.2a shows the expansion of a gas, starting with volume V_i and pressure P_i; Fig. 15.2b is the *PV* diagram for the process. As the gas slowly expands, it does work on the piston. The pressure of the gas pushes against the piston with a force $F = PA$, where A is the cross-sectional area of the piston. This force is not constant since the pressure decreases. As was shown in Section 6.4, the work done by a variable force is the area under a graph of $F(x)$. In thermodynamics, one of the great uses of a *PV* diagram is that the work done is the area under the *PV* curve.

To see how work is related to the area under the curve, first note that the units of $P \times V$ are those of work:

$$[\text{pressure} \times \text{volume}] = [\text{Pa}] \times [\text{m}^3] = \frac{[\text{N}]}{[\text{m}^2]} \times [\text{m}^3] = [\text{N}] \times [\text{m}] = [\text{J}]$$

So far, so good. Imagine that the piston is let out a *small* distance d—small enough that the pressure change is insignificant. The work done by the gas is

$$W = Fd = PAd$$

The volume increase of the gas is

$$\Delta V = Ad$$

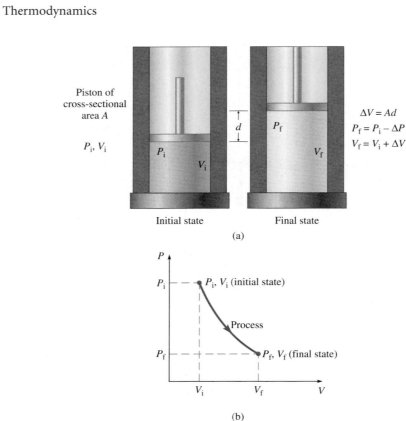

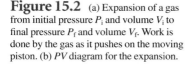

Figure 15.2 (a) Expansion of a gas from initial pressure P_i and volume V_i to final pressure P_f and volume V_f. Work is done by the gas as it pushes on the moving piston. (b) PV diagram for the expansion.

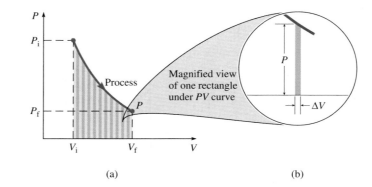

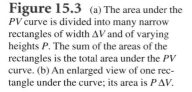

Figure 15.3 (a) The area under the PV curve is divided into many narrow rectangles of width ΔV and of varying heights P. The sum of the areas of the rectangles is the total area under the PV curve. (b) An enlarged view of one rectangle under the curve; its area is $P\,\Delta V$.

So the work done by the gas is

$$W = P\,\Delta V \tag{15-2}$$

To find the *total* work done, we add up the work done during each small volume change. During each small ΔV, the work done is the area of a rectangle of height P and width ΔV under the PV curve (Fig. 15.3). Therefore, the total work done is the total area under the PV curve.

Note that the work done depends not only on the initial and final states of the system, but also on the intermediate states, as depicted by the path taken on the PV curve. Figure 15.4 shows two other possible paths between the same initial and final states as those of Fig. 15.3.

In Fig. 15.4a, the pressure is kept constant at the initial value P_i while the volume is increased from V_i to V_f. Then the volume is kept constant while the pressure is reduced from P_i to P_f. The work done is shown by the shaded area under the PV curve; it is greater than the work done in Fig. 15.3a. Alternatively, in Fig. 15.4b the pressure is first reduced from P_i to P_f while the volume is held fixed; then the volume is allowed to increase from V_i to V_f while the pressure is kept at P_f. We see by the shaded area that the

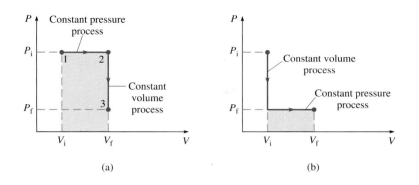

Figure 15.4 Work done (shaded areas) by following different processes between the same initial and final states. (a) Volume is increased from V_i to V_f while pressure is constant at P_i; then pressure is reduced to P_f. (b) Pressure is first decreased from P_i to P_f; then volume is increased from V_i to V_f.

work done this way is less than the work done in Fig. 15.3a. The work done differs from one process to another, even though the initial and final states are the same in each case.

Constant Pressure Processes

A process by which the state of a system is changed while the pressure is held constant is called an *isobaric* process. The word *isobaric* comes from the same Greek root as the word *barometer*. In Fig. 15.4a, the first change of state from V_i to V_f along the line from 1 to 2 occurs at the constant initial pressure P_i. A constant pressure process appears as a horizontal line on the PV diagram. The work done is the area under this horizontal line:

$$W = P_i (V_f - V_i) = P \, \Delta V \quad \text{(constant pressure)} \qquad (15\text{-}3)$$

Constant Volume Processes

A process by which the state of a system is changed while the *volume* remains constant is called an *isochoric* process. Such a process is illustrated in Fig. 15.4a when the system moves along the line from 2 to 3 as the pressure changes from P_i to P_f at the constant volume V_f. No work is done during a constant volume process; without a displacement, work cannot be done. The area under the PV curve—a vertical line—is zero:

$$W = 0 \quad \text{(constant volume)} \qquad (15\text{-}4)$$

If no work is done, then from the first law of thermodynamics, the change in internal energy is equal to the heat flow into the system:

$$\Delta U = Q \quad \text{(constant volume)} \qquad (15\text{-}5)$$

In Section 14.4 we discussed the molar specific heat of an ideal gas *at constant volume*. The first law of thermodynamics enables us to calculate the internal energy change ΔU. Since no work is done during a constant volume process, $\Delta U = Q$. For a constant volume process, $Q = nC_v \Delta T$ and therefore

$$\Delta U = nC_v \Delta T \quad \text{(ideal gas)} \qquad (15\text{-}6)$$

Internal energy is a state variable—its value depends only on the current state of the system, not on the path the system took to get there. Therefore, as long as the number of moles is constant, *the change in internal energy of an ideal gas is determined solely by the temperature change*. Equation (15-6) therefore gives the internal energy change of an ideal gas for *any* thermodynamic process, not just for constant volume processes.

Constant Temperature Processes

A process in which the temperature of the system remains constant is called an **isothermal** process. Such a process must proceed along a path on a PV diagram for which the temperature is constant. Such a path is called an **isotherm** (Fig. 15.5). All the points on an isotherm represent states of the system with the same temperature.

How can we keep the temperature of the system constant? One way is to put the system in thermal contact with a **reservoir**—a system with such a large heat capacity that it can exchange heat in either direction with a negligibly small temperature change.

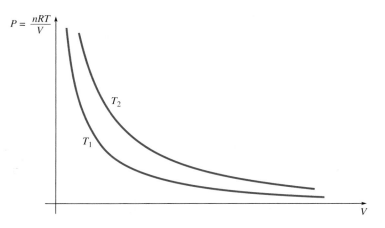

Figure 15.5 Two isotherms for an ideal gas at temperatures $T_1 < T_2$. Each isotherm is a graph of $P = nRT/V$ for a constant temperature.

Then as long as the state of the system does not change too rapidly, heat flows between the system and the reservoir to keep the system's temperature constant.

Isothermal Process for an Ideal Gas

For an ideal gas, we can plot isotherms using the ideal gas law $PV = nRT$. Since the change in internal energy of an ideal gas is determined solely by the temperature change,

$$\Delta U = 0 \quad \text{(ideal gas, isothermal)} \tag{15-7}$$

From the first law of thermodynamics, $\Delta U = 0$ means that $Q = W$; the heat flow into the gas is equal to the work done by the gas. Note that this is only true for an *ideal gas* at constant temperature. Other systems can change internal energy without changing temperature; one example is when the system undergoes a phase change.

It can be shown (using calculus to find the area under the *PV* curve) that the work done by an ideal gas during a constant temperature expansion (or contraction) from volume V_i to volume V_f is

$$W = nRT \ln\left(\frac{V_f}{V_i}\right) \quad \text{(ideal gas, isothermal)} \tag{15-8}$$

In Eq. (15-8), "ln" stands for the natural (or base-*e*) logarithm.

Example 15.2

Constant Temperature Compression of an Ideal Gas

An ideal gas is kept in thermal contact with a heat reservoir at 7°C (280 K) while it is compressed from a volume of 20.0 L to a volume of 10.0 L. During the compression an average force of 33.3 kN is used to move the piston a distance of 0.15 m. How much heat is exchanged between the gas and the reservoir? Does the heat flow into or out of the gas?

Strategy   We can find the work done on the piston from the average force applied and the distance moved. This is equal to the work done on the gas in compressing it. For isothermal compression of an ideal gas, $\Delta U = 0$. Then $W = Q$; the heat is equal to the work done.

Solution   The work done on the gas is

work = Fd = 33.3 kN × 0.15 m = 5.0 kJ

This work adds 5.0 kJ to the internal energy of the gas. Then 5.0 kJ of heat must flow out of the gas if its internal energy does not change. The work done on the gas is positive since the piston is pushed with an inward force as it moves inward. Since positive work is done *on* the gas, negative work is done *by* the gas. Using the sign conventions for the first law,

$$Q = W = -5.0 \text{ kJ}$$

Since Q represents heat flow *into* the gas, the negative sign tells us that heat flows out of the gas into the reservoir.

Discussion   Although the temperature remains constant during the process, it does not mean that no heat

continued on next page

Example 15.2 *continued*

flows. To maintain a constant temperature when work is done on the gas, some heat must flow out of the gas. If the gas were thermally isolated so no heat could flow, then the work done on the gas would increase the internal energy, resulting in an increase in the temperature of the gas.

Practice Problem 15.2 Work done during constant temperature expansion of a gas

Suppose 2.0 mol of an ideal gas are kept in thermal contact with a heat reservoir at 57°C (330 K) while the gas expands from a volume of 20.0 L to a volume of 40.0 L (Fig. 15.6). Does heat flow into or out of the gas? How much heat flows? [*Hint*: Use $W = nRT \ln (V_f/V_i)$, which applies to an ideal gas at constant temperature.]

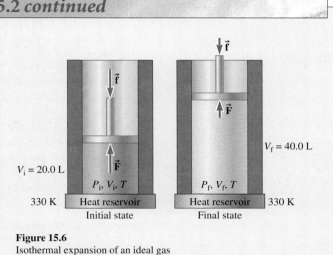

Figure 15.6
Isothermal expansion of an ideal gas

Adiabatic Processes

A process in which no heat is transferred into or out of the system is called an **adiabatic** process. An adiabatic process is *not* the same as a constant temperature (isothermal) process. Heat flows in or out as necessary to maintain a constant temperature in an isothermal process; *no* heat flow occurs in an adiabatic process, so if work is done, the temperature of the system may change. In the adiabatic compression of an ideal gas, work is done on the gas while no heat flows in or out. The internal energy increase due to the work done results in a temperature increase. One way to perform an adiabatic process is to completely insulate the system so that no heat can get in or out; another way is to perform the process so quickly that there is no time for heat to flow in or out.

From the first law of thermodynamics

$$\Delta U = Q - W \qquad (15\text{-}1)$$

With $Q = 0$,

$$\Delta U = -W \quad \text{(adiabatic)}$$

Table 15.2 summarizes all of the thermodynamic processes discussed.

Table 15.2

Summary of Thermodynamic Processes

Process	Name	Condition	Consequences
Constant temperature	Isothermal	T = constant; for an ideal gas, $\Delta U = 0$	For an ideal gas, $Q = W$
Constant pressure	Isobaric	P = constant	$W = P\,\Delta V$
Constant volume	Isochoric	V = constant	$W = 0$; $\Delta U = Q$
No heat flow	Adiabatic	$Q = 0$	$\Delta U = -W$

15.3   CONSTANT PRESSURE EXPANSION OF AN IDEAL GAS

Figure 15.7 is a PV diagram for heat flow into an ideal gas at constant volume. The initial and final states are shown as points on two different isotherms. The area under the vertical line is zero; no work is done when the volume is constant. With $W = 0$, the heat flow increases the internal energy of the gas, so the temperature increases.

Another common situation is when the *pressure* of the gas is constant. In this case, work is done as the gas expands. The first law of thermodynamics enables us to calculate the molar specific heat at constant pressure (C_p).

Figure 15.8 shows a PV diagram for the constant pressure expansion of an ideal gas starting and ending at the same temperatures as for the constant volume process. Applying the first law to the constant pressure process requires that

$$\Delta U = Q - W$$

where the work done is, from the ideal gas law,

$$W = P \Delta V = nR \Delta T$$

The definition of C_p is

$$Q = nC_p \Delta T \tag{15-9}$$

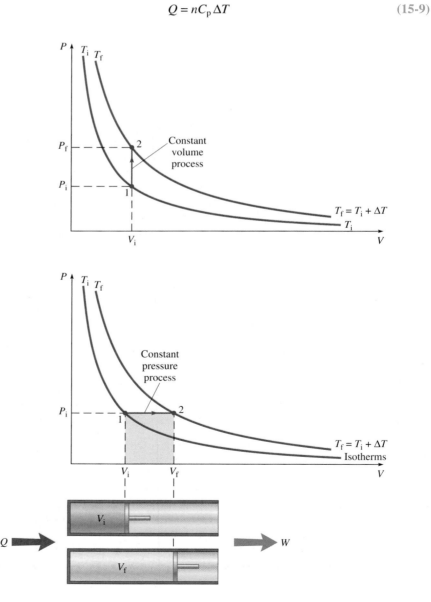

Figure 15.7   A PV diagram for constant volume process for ideal gas.

Figure 15.8   A PV diagram of a constant pressure expansion; heat is added to the ideal gas of volume V_i increasing the internal energy and doing work as the expanding gas moves the frictionless piston and expands to volume V_f. The work is equal to the shaded area under the constant pressure process path from point 1 to point 2.

Substituting into the first law, we obtain

$$\Delta U = nC_p \Delta T - nR\,\Delta T \qquad (15\text{-}10)$$

Since both of the processes discussed ended up at the same temperature and since the internal energy of an ideal gas depends only on the temperature, we can replace ΔU with its equivalent:

$$\Delta U = nC_v \Delta T \qquad (14\text{-}8)$$

Then

$$nC_v \Delta T = nC_p \Delta T - nR\,\Delta T$$

Canceling common factors of n and ΔT, this reduces to

$$C_v = C_p - R \qquad (15\text{-}11)$$

Since R is a positive constant, the molar specific heat of an ideal gas at constant pressure is larger than the molar specific heat at constant volume.

Is this result reasonable? When heat flows into the gas at constant pressure, the gas expands, doing work on the surroundings. Thus, not all of the heat goes into increasing the internal energy of the gas. More heat has to flow into the gas at constant pressure for a given temperature increase than at constant volume.

Example 15.3

Warming a Balloon at Constant Pressure

A weather balloon is filled with helium gas at 20.0°C and 1.0 atm of pressure. The volume of the balloon after filling is measured to be 8.50 m³. The helium is heated until its temperature is 55.0°C. During this process, the balloon expands at constant pressure (1.0 atm). What is the heat flow into the helium?

Strategy We can find how many moles of gas n are present in the balloon by using the ideal gas law. For this problem, we consider the helium to be a system. Helium is a monatomic gas so its molar specific heat at constant *volume* is $C_v = \frac{3}{2}R$. The molar specific heat at constant *pressure* is then $C_p = C_v + R = \frac{5}{2}R$. Then the heat flow into the gas during its expansion is $Q = nC_p \Delta T$.

Solution The ideal gas law is

$$PV = nRT$$

We know the pressure, volume, and temperature: $P = 1$ atm $= 1.01 \times 10^5$ Pa, $V = 8.50$ m³, and $T = 273$ K $+ 20.0$°C $= 293$ K. Solving for the number of moles yields

$$n = \frac{PV}{RT} = \frac{1.01 \times 10^5 \text{ Pa} \times 8.50 \text{ m}^3}{8.31 \text{ J/(mol·K)} \times 293 \text{ K}} = 352.6 \text{ mol}$$

For an ideal gas at constant pressure, the heat required to change the temperature is

$$Q = nC_p \Delta T$$

where $C_p = \frac{5}{2}R$. The temperature change is

$$\Delta T = 55.0\text{°C} - 20.0\text{°C} = 35.0 \text{ K}$$

Now we have everything we need to find Q:

$$Q = nC_p \Delta T = 352.6 \text{ mol} \times \tfrac{5}{2} \times 8.31 \text{ J/(mol·K)} \times 35.0 \text{ K}$$
$$= 256 \text{ kJ}$$

Discussion We do not have to find the work $W = P\,\Delta V$ separately and then add it to the change in internal energy to find Q. The work done is *already* accounted for by the molar specific heat at constant pressure. This simplifies the problem since we use the same method for constant pressure as we use for constant volume; the only change is the choice of C_v or C_p.

Practice Problem 15.3 Air instead of helium

Suppose the balloon were filled with air instead of helium. Find Q for the same temperature change. (Air is mostly N_2 and O_2 so assume an ideal diatomic gas.)

15.4 REVERSIBLE AND IRREVERSIBLE PROCESSES

Have you ever wished you could make time go backward? Perhaps you accidentally broke an irreplaceable treasure in a friend's house, or missed a one-time opportunity to meet your favorite movie star, or said something unforgivable to someone close to you. Why can't the clock be turned around?

Imagine a perfectly elastic collision between two billiard balls. If you were to watch a movie of the collision, you would have a hard time telling whether the movie was being played forward or backward. The laws of physics for an elastic collision are valid even if the direction of time is reversed. Since the total momentum and the total kinetic energy are the same before and after the collision, the reversed collision is physically possible.

The perfectly elastic collision is one example of a **reversible** process. A reversible process is one that does not violate any laws of physics if "played in reverse." Most of the laws of physics do not distinguish forward from backward. A projectile moving in the absence of air resistance (on the Moon, say) is reversible: if we play the movie backward, the total mechanical energy is still conserved and Newton's second law ($\Sigma \vec{\mathbf{F}} = m\vec{\mathbf{a}}$) still holds at every instant in the projectile's trajectory.

Notice the caveats in the examples: "perfectly elastic" and "in the absence of air resistance." If friction or air resistance is present, then the process is **irreversible**. If you played *backward* a movie of a projectile with noticeable air resistance, it would be easy to tell that something is wrong. The force of air resistance on the projectile would act in the wrong direction—in the direction of the velocity instead of opposite it. The same would be true for sliding friction. Slide a book across the table; friction slows it down and brings it to rest. The macroscopic kinetic energy of the book—due to the orderly motion of the book in one direction—has been converted into disordered energy associated with the random motion of molecules; the table and book will be at slightly higher temperatures. The reversed process certainly would never occur, even though it does not violate the first law of thermodynamics (energy conservation). We would not expect a slightly warmed book placed on a slightly warmed table surface to spontaneously begin to slide across the table, gaining speed and cooling off as it goes, even if the total energy is the same before and after. It is easy to convert ordered energy into disordered energy, but not so easy to do the reverse. *The presence of energy dissipation (sliding friction, air resistance) always makes a process irreversible.*

> Irreversible processes do *not* violate energy conservation.

Heat flow is another example of an irreversible process. Put a container of warm lemonade into a cooler with some ice (Fig. 15.9). Some of the ice melts and the lemonade gets cold as heat flows out of the lemonade and into the ice. The reverse would never happen: putting cold lemonade into a cooler with some partially melted ice, we would never find that the lemonade gets warmer as the liquid water freezes. *Heat flow from a hotter body to a colder body is always irreversible.*

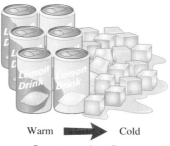
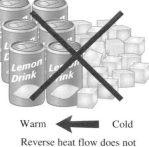

Warm ➡ Cold
Spontaneous heat flow

Warm ⬅ Cold
Reverse heat flow does not happen spontaneously

Figure 15.9 Spontaneous heat flow goes from warm to cool; the reverse does not happen spontaneously.

Conceptual Example 15.4

Eggs for Breakfast

Discuss the scrambling of an egg in terms of irreversible processes and in terms of disorder.

Solution and Discussion Scrambling the egg greatly increases the disorder, since now the yolk and egg-white molecules are mixed together randomly. Once the egg is scrambled, further stirring rearranges molecules but leaves the egg still scrambled—we would never expect the egg to unscramble as we stir. The scrambling is an irreversible process.

The unscrambled egg is in a highly ordered state, with the yolk and the white clearly separated, while the scrambled egg is in a highly disordered state. Scrambling changes the egg from an ordered state to a disordered state; once it is disordered, further stirring cannot return it to the ordered state.

Conceptual Practice Problem 15.4   A campfire

On a camping trip, you gather some twigs and logs and start a fire. Discuss the campfire in terms of irreversible processes and in terms of disorder.

Irreversibility, whether in the dissipation of energy or in heat flow or in scrambling of an egg, is based on statistics. If we looked *microscopically* at any of these processes, any one of the collisions between molecules would be reversible. As we will see in this chapter, the frictional dissipation of energy and the spontaneous flow of heat from a hotter to a colder body also can be thought of in terms of disorder. A system never goes from a disordered state to a more ordered state *spontaneously*. Reversible processes are those that do not change the total amount of disorder in the universe; irreversible processes increase the amount of disorder.

The second law of thermodynamics is the law of physics that determines the direction of time—all the other physical laws we have learned work equally well forward and backward. The second law can be stated in many equivalent forms. One statement is that the total amount of disorder in the universe never decreases. Since we have not yet learned how to measure disorder, we start with an equivalent statement involving heat flow:

Second law of thermodynamics (Clausius statement):
Heat never flows spontaneously from a colder body to a hotter body.

Spontaneous heat flow from a colder body to a hotter body would decrease the total disorder in the universe.

15.5   HEAT ENGINES

We said in Section 15.4 that it is far *easier* to convert ordered energy into disordered energy than to do the reverse. Converting ordered into disordered energy occurs spontaneously, but the reverse does not. A **heat engine** is a device designed to convert disordered energy into ordered energy. We will see that there is a fundamental limitation on how much mechanical work can be produced by a heat engine from a given amount of heat.

The invention of the steam engine—a heat engine that uses steam as its working substance—at the beginning of the eighteenth century was one of the crucial elements in the industrial revolution. The steam engine was the first practical machine with a sustained work output that used an energy source other than muscle, wind, or moving water. Efforts to improve the performance of the steam engine in the nineteenth century gave birth to the branch of physics we call *thermodynamics*. Steam engines are still used in many electrical power plants.

The source of energy in a heat engine is most often the burning of some fuel such as gasoline, coal, oil, natural gas, and the like. A nuclear power plant is a heat engine using energy released by a nuclear reaction instead of a chemical reaction (as in burning). A geothermal engine uses the high temperature found beneath the Earth's crust (which comes to the surface in places such as volcanoes and hot springs).

All practical engines operate in cycles—a "one-shot" engine would be of little use. Each cycle consists of several thermodynamic processes that are repeated the same way during each cycle. In order for these processes to repeat the same way, the engine must end the cycle in the same state in which it started. In particular, the internal energy of the engine must be the same at the end of a cycle as it was in the beginning. Then for one complete cycle,

$$\Delta U = 0$$

From the first law of thermodynamics (energy conservation),

$$Q = W$$

Therefore, for a cyclical heat engine,

The work done by the engine is equal to the *net* heat input.

We stress that it is the *net* heat input since an engine not only takes in heat but exhausts some as well. Figure 15.10 shows the energy transfers involved.

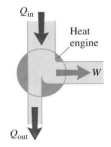

Figure 15.10 A heat engine. The engine is represented by a circle and the arrows indicate the direction of the energy flow. The energy entering the engine equals the energy leaving the engine.

Making The Connection:
operation of
an internal
combustion engine

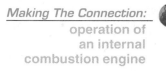

One familiar engine is the internal combustion engine found in automobiles. *Internal combustion* refers to the fact that gasoline is burned inside a cylinder; the resulting hot gases push against a piston and do work. A steam engine is an *external* combustion engine. The coal burned, for example, releases heat that is used to make steam; the steam is the working substance of the engine that drives the turbines.

Most automobile engines work in a cyclic thermodynamic process called the *Otto cycle*. The Otto cycle consists of six steps (Fig. 15.11):

1. Intake stroke: The piston is pulled out, drawing the fuel-air mixture into the cylinder at atmospheric pressure ($a \rightarrow b$).

2. Compression stroke: The piston moves back in, compressing the fuel-air mixture ($b \rightarrow c$). The gas in the cylinder does *negative* work—which means work is done *on* the gas.

3. Ignition: A spark ignites the gases, quickly and dramatically raising the temperature and pressure ($c \rightarrow d$).

4. Power stroke: The high pressure that results from ignition pushes the piston out ($d \rightarrow e$). As the piston is pushed out, the gases do positive work.

5. Cooling: Heat flows out of the cylinder ($e \rightarrow b$).

6. Exhaust stroke: A valve is opened and the exhaust gases are pushed out of the cylinder ($b \rightarrow a$).

An automobile engine is a *four-stroke* engine because each cycle has four strokes (steps 1, 2, 4, and 6) during which the piston moves.

Of the energy released by the burning of the gasoline, only about 25% is turned into mechanical work used to move the car forward and run other systems. The other 75% is discarded. The hot exhaust gases carry energy out of the engine, as does the liquid cooling system.

Making The Connection:
efficiency of a
heat engine

To measure how effectively an engine converts heat into mechanical work, we define the engine's **efficiency** e as what you get (work) divided by what you pay for (heat input):

Efficiency of an engine

$$e = \frac{\text{work output}}{\text{heat input}} = \frac{W}{Q_{in}} \qquad (15\text{-}12)$$

The efficiency is stated as either a fraction or a percentage. It gives the fraction of the heat input that is turned into useful work. Note that the heat input is *not* the same as the net heat Q; rather,

$$Q = Q_{in} + Q_{out} \qquad (15\text{-}13)$$

where Q_{in} is positive and Q_{out} is negative for a heat engine. The efficiency of an engine is less than 100% because some of the heat input, instead of being converted into useful work, is exhausted.

If an engine does work at a constant rate and its efficiency does not change, then it also takes in and exhausts heat at constant rates. The work done, heat input, and heat exhausted during any time interval are all proportional to the elapsed time. Therefore, all the same relationships that are true for the amounts of heat flow and work done apply to the *rates* at which heat flows and work is done. For example, the efficiency is

$$e = \frac{\text{work done}}{\text{heat input}} = \frac{\textit{rate of doing work}}{\textit{rate of taking in heat}} = \frac{W/\Delta t}{Q_{in}/\Delta t}$$

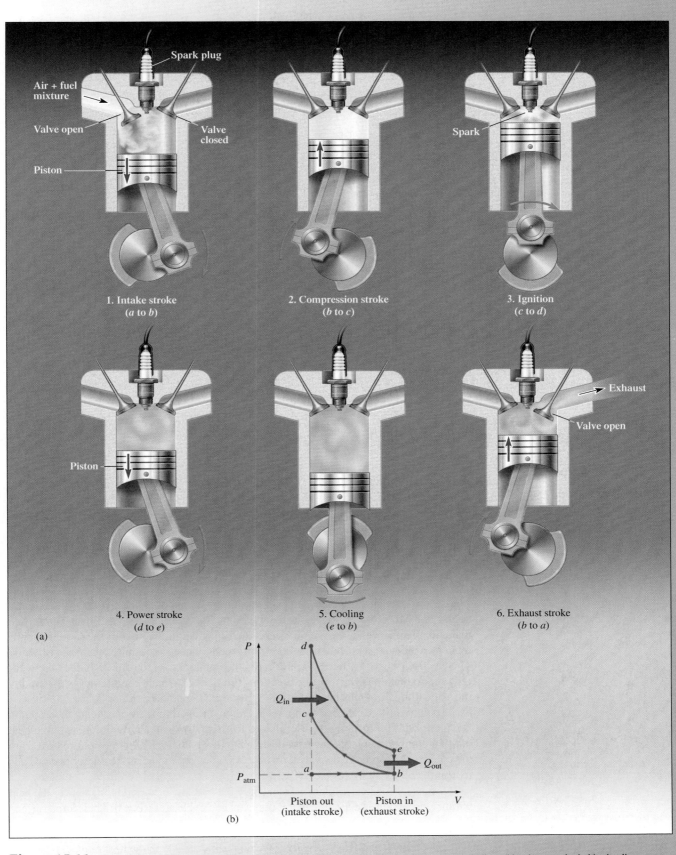

Figure 15.11   (a) The four-stroke Otto cycle. (b) *PV* diagram for the Otto cycle; the net work output is indicated by the area shaded in the diagram.

Example 15.5

Rate at Which Heat Is Exhausted from an Engine

An engine operating at 25% efficiency produces work at a rate of 0.10 MW. At what rate is heat exhausted into the environment?

Strategy   We are given that the engine does work at a constant *rate*. The efficiency is also constant.

Solution   The engine efficiency is the ratio of the rate at which work is done to the rate of heat input:

$$e = \frac{W/\Delta t}{Q_{in}/\Delta t} = \frac{W}{Q_{in}}$$

The *net* rate of heat input Q is

$$\frac{Q}{\Delta t} = \frac{Q_{in}}{\Delta t} + \frac{Q_{out}}{\Delta t}$$

where Q_{out} is negative. Since the internal energy of the engine does not change over a complete cycle, energy conservation (or the first law of thermodynamics) requires that

$$Q - W = 0 \quad \text{or} \quad Q_{in} + Q_{out} = W$$

In terms of the rate at which heat is delivered or exhausted and the rate at which work is done,

$$\frac{Q_{in}}{\Delta t} + \frac{Q_{out}}{\Delta t} = \frac{W}{\Delta t}$$

In other words, the rate at which the engine does work is equal to the *net* rate of heat input. We are asked to find the rate of heat exhausted $Q_{out}/\Delta t$:

$$\frac{Q_{out}}{\Delta t} = \frac{W}{\Delta t} - \frac{Q_{in}}{\Delta t} = \frac{W}{\Delta t} - \frac{W/\Delta t}{e}$$

$$\frac{Q_{out}}{\Delta t} = \frac{W}{\Delta t}\left(1 - \frac{1}{e}\right) = 0.10\ \text{MW} \times \left(1 - \frac{1}{0.25}\right) = -0.30\ \text{MW}$$

Discussion   The negative sign here means that exhausted heat flows *out* of the engine at a rate of 0.30 MW.

As a check: 25% efficiency means that $\frac{1}{4}$ of the heat input does work and $\frac{3}{4}$ of it is exhausted. Therefore, the ratio of work to exhaust is $\dfrac{1/4}{3/4} = \dfrac{1}{3} = \dfrac{0.10\ \text{MW}}{0.30\ \text{MW}}$.

For simplicity, we could have let W, Q_{in}, and Q_{out} refer to *rates* instead of to total amounts—to do this we just cancel common factors of Δt out of the equations for efficiency and energy conservation.

Practice Problem 15.5   Heat engine efficiency

An engine "wastes" 4.0 J of heat for every joule of work done. What is its efficiency?

According to the first law of thermodynamics, the efficiency of a heat engine cannot exceed 100%. An efficiency of 100% would mean that all of the heat input is turned into useful work and no "waste" heat is exhausted. It might seem theoretically possible to make a 100% efficient engine by eliminating all of the imperfections in design such as friction and lack of perfect insulation. However, it is not, as we see in Section 15.7.

15.6   REFRIGERATORS AND HEAT PUMPS

The second law of thermodynamics says that heat cannot *spontaneously* flow from a colder body to a hotter body; but machines such as refrigerators and heat pumps can make that happen. In a refrigerator, heat is pumped "uphill"—out of the food compartment into the warmer room. That doesn't happen by itself; it requires the *input* of work. The electricity used by a refrigerator turns the compressor motor, which does the work required to make the refrigerator function (Fig. 15.12). An air conditioner is essentially the same thing: it pumps heat out of the house into the hotter outdoors.

The only difference between a refrigerator (or an air conditioner) and a heat pump is which end is performing the useful task. Refrigerators and air conditioners pump heat out of a compartment that they are designed to keep cool. Heat pumps pump heat from the colder outdoors into the warmer house. The idea isn't to keep the outdoors cold; it is to heat the house.

Just as for the heat engine, we can apply the first law of thermodynamics to the heat pump. Once again, we assume that it operates in a cycle and $\Delta U = 0$. Therefore,

$$Q = W$$

This time both Q and W are negative. We do positive work on the heat pump, so the heat pump does *negative* work. Heat flows into the heat pump but more flows out since the work input is "converted" to heat.

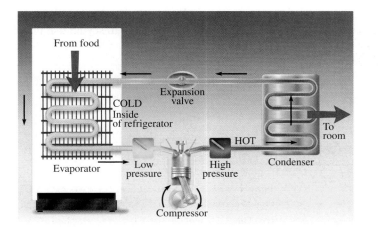

Figure 15.12 In a refrigerator, a fluid is compressed, increasing its temperature. Heat is exhausted as the fluid passes through the condenser. Now the fluid is allowed to expand; its temperature falls. Heat flows from the food compartment into the cold fluid. The fluid returns to the compressor to begin the same cycle again.

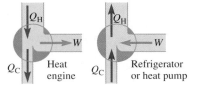

Figure 15.13 A heat engine and a heat pump (or refrigerator)

Notice that the heat pump is a heat engine operating in reverse (Fig. 15.13). In the heat engine, heat flows from hot to cold, with work as the output. In a heat pump, heat flows from cold to hot, with work as the input. It will be most convenient to distinguish the heat transfers not by which is input and which output (since that switches in going from an engine to a heat pump), but rather by the temperature of the environment where the exchange is made, using subscripts "H" and "C" for hot and cold.

Thus, the efficiency of the heat engine is

$$e = \frac{\text{work output}}{\text{heat input}} = \frac{W}{Q_H} \tag{15-12}$$

The efficiency can also be expressed in terms of the heat flows. Since $W = Q_H + Q_C$,

$$e = \frac{Q_H + Q_C}{Q_H} = 1 + \frac{Q_C}{Q_H} \tag{15-14}$$

Since $Q_C < 0$, the efficiency is *less* than 1.

The same quantity e will be useful in our discussions of heat pumps (and refrigerators), but it does not measure *efficiency*. A more efficient heat pump uses a *smaller* work input for a given amount of heat output; therefore, we call e the *inefficiency* of the heat pump.

Inefficiency of a heat pump or refrigerator

$$e = \frac{\text{work input}}{\text{heat output}} = \frac{W}{Q_H} = 1 + \frac{Q_C}{Q_H} \tag{15-15}$$

In a heat pump, the directions of all the energy transfers are reversed from those for the heat engine; the signs of Q_H, Q_C, and W are opposite what they are for an engine. The heat input comes from the *colder* side; the heat exhausted goes to the *hotter* side. Work is not done *by* the pump, it is done *on* the pump. Since the signs are all reversed, e is still positive and less than 1.

Since manufacturers would be loathe to advertise the "inefficiency" of a heat pump or refrigerator, it's common practice to state a **coefficient of performance** K instead. Just as for the efficiency of an engine, the coefficients of performance are ratios of what you get divided by what you put in:

- for a heat pump:

$$K_p = \frac{\text{heat delivered}}{\text{work input}} = \frac{Q_H}{W} = \frac{1}{e} \tag{15-16}$$

- for a refrigerator or air conditioner:

$$K_r = \frac{\text{heat removed}}{\text{work input}} = \left| \frac{Q_C}{W} \right| = \frac{1}{e} - 1 \tag{15-17}$$

A higher coefficient of performance means a better heat pump or refrigerator.

The second law says that heat cannot flow spontaneously from cold to hot—we need to do some work to make that happen. That's equivalent to saying that the coefficient of performance can't be infinite and e can't be zero. So a restatement of the second law is:

Second law of thermodynamics

No refrigerator or heat pump can have $e = 0$.

Example 15.6

A Heat Pump

A heat pump has a performance coefficient of 2.5. (a) How much heat is delivered to the house for every joule of electrical energy consumed? (b) In an electric heater, for each joule of electrical energy consumed, one joule of heat is delivered to the house. Where does the "extra" heat delivered by the heat pump come from? (c) What would its coefficient of performance be when run as an air conditioner instead?

Strategy   There are two slightly different meanings of coefficient of performance. For a heat pump, whose object is to deliver heat to the house, the coefficient of performance is the heat delivered (Q_H) per unit of work done to run the pump. With an air conditioner, the object is to *remove* heat from the house. The coefficient of performance is the heat *removed* (Q_C) per unit of work done.

Solution   (a) As a heat pump,

$$K_p = \frac{\text{heat output}}{\text{work input}} = \frac{Q_H}{W} = 2.5$$

$$Q_H = 2.5\, W$$

For every joule of electrical energy (= work), 2.5 J are delivered to the house.

(b) The 2.5 J include the 1.0 J of work plus 1.5 J that were pumped in from the outside. The electric heater just transforms the joule of work into a joule of heat.

(c) From (b), the coefficient is 1.5:

$$K_r = \frac{\text{heat removed}}{\text{work done}} = \left| \frac{Q_C}{W} \right| = \left| \frac{1.5\,\text{J}}{-1.0\,\text{J}} \right| = 1.5$$

Discussion   One thing that makes a heat pump economical in many situations is that the same machine can function as a heat pump in winter and as an air conditioner in summer. The heat pump delivers heat to the interior of the house, while the air conditioner pumps heat out.

Practice Problem 15.6   Heat exhausted by air conditioner

An air conditioner with a coefficient of performance $K_r = 3$ consumes electricity at an average rate of 1 kW. During 1 h of use, how much heat is exhausted to the outdoors?

15.7   REVERSIBLE ENGINES AND HEAT PUMPS

As we have already seen, a heat pump (or refrigerator) is essentially a reversed heat engine, since all of the energy transfers reverse direction. Many real engines actually can be run in reverse as heat pumps. However, in general the quantity $e = W/Q_H$ is *not* the same when reversed, due to irreversible processes. Recall that a real engine has an efficiency e less than that of an ideal engine operating in the same environment, while a real heat pump has an inefficiency e *greater* than that of an ideal heat pump. You might guess that a real engine's efficiency e_e would therefore be less than the inefficiency e_p of the reversed engine (= heat pump). That guess is correct, as we will see shortly.

An ideal engine or heat pump is one in which no irreversible processes occur: there is no friction and heat only flows across infinitesimal temperature differences. Since all the processes are reversible, we call an ideal engine or heat pump a **reversible engine** or a **reversible heat pump**.

In order to simplify the analysis of reversible engines, let us assume the engine takes heat input from a reservoir at temperature T_H and exhausts heat into another reservoir at temperature T_C (with $T_C < T_H$). Recall that a reservoir is a system with such a large heat capacity that it can exchange heat in either direction with a negligibly small temperature change. Therefore, the cold reservoir stays at temperature T_C and the hot reservoir stays at T_H.

Now imagine connecting an engine to a heat pump, both using the same two reservoirs. The engine has efficiency e_e and the pump has inefficiency e_p. The heat pump uses all the work output of the engine as its work input (Fig. 15.14) so $W_p = -W_e$. The engine takes heat $Q_{H,e}$ from the hot reservoir where $Q_{H,e} = W_e/e_e$. The pump delivers heat $Q_{H,p} = W_p/e_p$ to the hot reservoir. Is there a net flow of heat out of the hot reservoir or into it? If the engine and pump are reversible, then $e_e = e_p$ and $Q_{H,e} = -Q_{H,p}$. There is no net flow of heat into or out of the hot reservoir. The same is true for the cold reservoir. The net effect is: nothing.

What happens if $e_e \neq e_p$? If $e_p > e_e$, then $|Q_{H,p}| < |Q_{H,e}|$. There is a net flow of heat out of the hot reservoir and into the cold. The net effect is heat flowing from a hotter system to a colder system. Nothing unexpected here; that could happen spontaneously. But what if $e_e > e_p$? Then there is a net flow of heat into the hot reservoir and out of the cold reservoir. Without any input of work from the outside, heat is flowing from cold to hot. Since this violates the second law—heat cannot flow spontaneously from cold to hot—we can restate the second law as:

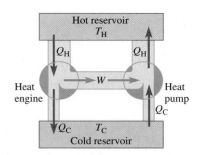

Figure 15.14 Heat engine connected to a heat pump

Second law of thermodynamics:
No engine can have an efficiency greater than the inefficiency of any pump operating between the same hot and cold reservoirs.

If e_r is the efficiency of a reversible engine or the inefficiency of a heat pump operating between the same two reservoirs, then equivalently:

Every engine has $e_e \leq e_r$ and every heat pump has $e_p \geq e_r$.

Since every reversible engine operating between the same temperatures, no matter what the details of its construction, has the same efficiency, the efficiency must depend only on the temperatures of the hot and cold reservoirs. It turns out that e_r is given by the remarkably simple expression:

$$e_r = 1 - \frac{T_C}{T_H} \qquad (15\text{-}18)$$

The temperatures in Eq. (15-18) must be *absolute* temperatures—in SI, kelvin temperatures. [Absolute temperature is also called *thermodynamic temperature* because you can use the efficiency of reversible engines to set a temperature scale. In fact, the definition of the kelvin is based on Eq. (15-18).]

Using Eq. (15-18), the ratio of the heat exhaust to the heat input *for a reversible engine* is

$$\frac{Q_C}{Q_H} = \frac{W - Q_H}{Q_H} = \frac{W}{Q_H} - 1 = e_r - 1 = -\frac{T_C}{T_H} \qquad (15\text{-}19)$$

Since Q_C and Q_H always have opposite signs, while the temperatures are always positive, we can make the following statement:

For a reversible engine, the ratio of the magnitudes of the heats is equal to the ratio of the temperatures.

The efficiency of a reversible engine is always less than 100%, assuming that the cold reservoir is not at absolute zero. Even an ideal, perfectly reversible engine must exhaust some heat, so the efficiency can never be 100% *even in principle*. Efficiencies of real engines are less than those of ideal engines. This provides a way to rephrase the second law of thermodynamics:

Second law of thermodynamics (Kelvin statement):
No engine can have 100% efficiency.

Example 15.7

Efficiency of an Automobile Engine

 In an automobile engine, the combustion of the fuel-air mixture can reach temperatures as high as 3000°C and the exhaust gases leave the cylinder at about 1000°C. (a) Find the efficiency of a *reversible* engine operating between reservoirs at those two temperatures. (b) Theoretically, we might be able to have the exhaust gases leave the engine at the temperature of the outside air (20°C). What would be the efficiency of the hypothetical reversible engine in this case?

Strategy   First we identify the temperatures of the hot and cold reservoirs in each case. We must convert the reservoir temperatures to kelvins in order to find the efficiency of a reversible engine.

Solution   (a) The reservoir temperatures in kelvins are found using

$$T = T_C + 273 \text{ K}$$

Therefore,

$$T_H \approx 3000°C \approx 3273 \text{ K}$$
$$T_C \approx 1000°C \approx 1273 \text{ K}$$

The efficiency of a reversible engine operating between these temperatures is

$$e_r = 1 - \frac{T_C}{T_H} = 1 - \frac{1273 \text{ K}}{3273 \text{ K}} = 0.61 \approx 60\%$$

(b) The high-temperature reservoir is still at 3273 K, while the low-temperature reservoir is now

$$T_C \approx 293 \text{ K}$$

This gives a higher efficiency:

$$e_r = 1 - \frac{T_C}{T_H} = 1 - \frac{293 \text{ K}}{3273 \text{ K}} = 0.91 \approx 90\%$$

Discussion   As mentioned in the chapter opener, real gasoline engines achieve efficiencies of only about 25%. While improvement is possible, the second law of thermodynamics limits the theoretical maximum efficiency to that of a reversible engine operating between the same temperatures. The *theoretical* maximum efficiency can only be increased by using a hotter hot reservoir or a colder cold reservoir. The higher the temperature achieved by the burning gasoline inside the cylinder, the higher the theoretical maximum efficiency.

Practice Problem 15.7   Temperature of hot gases

If the efficiency of a reversible engine is 75% and the temperature of the outdoor world into which the engine sends its exhaust is 27°C, what is the combustion temperature in the engine cylinder?

Example 15.8

Coal-Burning Power Plant

A coal-burning electrical power plant burns coal at 706°C. Heat is exhausted into a river near the power plant; the average river temperature is 19°C. What is the minimum possible rate of thermal pollution (heat exhausted into the river) if the station generates 125 MW of electricity?

Strategy   The minimum discharge of heat into the river would occur if the engine generating the electricity were reversible. As in Example 15.5, we can take all of the rates to be constant.

Solution   First find the absolute temperatures of the reservoirs:

$$T_H = 706°C = 979 \text{ K}$$
$$T_C = 19°C = 292 \text{ K}$$

The efficiency of a reversible engine operating between these temperatures is

$$e_r = 1 - \frac{T_C}{T_H} = 1 - \frac{292 \text{ K}}{979 \text{ K}} = 0.702$$

We want to find the rate at which heat is exhausted, which is $Q_C/\Delta t$. The efficiency is equal to the ratio of the work output to the heat input from the hot reservoir:

$$e = \frac{W}{Q_H}$$

Conservation of energy requires that

$$Q_H + Q_C = W$$

Solving for Q_C,

$$Q_C = W - Q_H = W - \frac{W}{e} = W \times \left(1 - \frac{1}{e}\right)$$

Assuming that all the rates are constant,

$$\frac{Q_C}{\Delta t} = 125 \text{ MW} \times \left(1 - \frac{1}{0.702}\right) = -53 \text{ MW}$$

The rate at which heat enters the river is 53 MW.

Discussion   We expect the actual rate of thermal pollution to be higher. A real, irreversible engine would have a lower efficiency, so more heat is dumped into the river.

Practice Problem 15.8   Generating electricity from coal

What is the minimum possible quantity of heat *input* (from the burning of coal) needed to generate 125 MW of electricity in this same plant?

15.8   CARNOT CYCLE

Sadi Carnot (1796–1832), a French engineer, published a treatise in 1824 that greatly expanded the understanding of how heat engines work. His treatment introduced a hypothetical, ideal engine that uses two heat reservoirs at different temperatures as the source and sink for heat and an ideal gas as the working substance of the engine. We now call this engine a **Carnot engine** and its cycle of operation the **Carnot cycle**. Remember that the Carnot engine is an *ideal* engine, not a real engine.

The Carnot engine is a particular kind of reversible engine. We must assume that all friction has somehow been eliminated—otherwise an irreversible process takes place. We also must avoid heat flow across a finite temperature difference, which would be irreversible. Therefore, whenever the ideal gas takes in or gives off heat, the gas must be at the same temperature as the reservoir with which it exchanges energy.

How can we get heat to flow without a temperature difference? Imagine putting the gas in good thermal contact with a reservoir at the same temperature. Now *slowly* pull a piston so that the gas expands. Since the gas does work, it must lose internal energy—and therefore its temperature drops, since in an ideal gas the internal energy is proportional to absolute temperature. But heat flows in to keep the temperature constant. As long as the expansion occurs slowly, the heat flows in fast enough to keep the temperature constant.

So, to keep everything reversible, we must exchange heat in *isothermal* processes. To take in heat, we expand the gas; to exhaust heat, we compress it. We also need processes to change the gas temperature from T_H to T_C and back to T_H. These processes must be *adiabatic* (no heat flow) since otherwise an irreversible heat flow would occur.

Now that the ground rules are set, here is the four-step Carnot cycle (Fig. 15.15):

$1 \rightarrow 2$: Isothermal expansion. Take in heat Q_H from the hot reservoir, keeping the gas at constant temperature T_H.

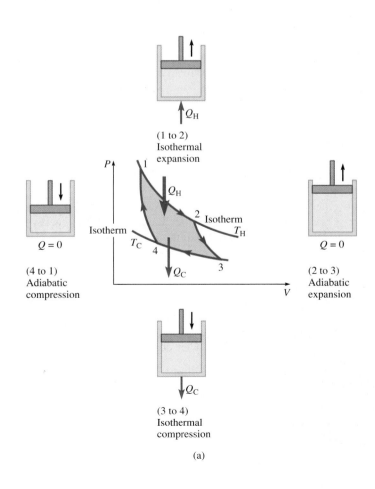

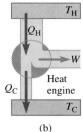

Figure 15.15 (a) The Carnot cycle. (b) Energy transfers in the cycle.

2→3: Adiabatic expansion. The gas does work without any heat flow in, so the temperature decreases. Continue until the gas temperature is T_C.

3→4: Isothermal compression. Heat Q_C is exhausted at constant temperature T_C.

4→1: Adiabatic compression until the temperature is back to T_H.

Carnot was able to calculate the efficiency of an engine operating in this cycle and obtained Eq. (15-18). Since *all* reversible engines operating between the same two reservoirs must have the same efficiency, deriving the efficiency for one particular kind of reversible engine is sufficient to derive it for all of them.

Example 15.9

Carnot Engine

A Carnot engine using 0.020 mol of an ideal gas operates between reservoirs at 1000.0 K and 300.0 K. The engine takes in 25 J of heat from the hot reservoir per cycle. Find the work done by the engine during the two isothermal steps in the cycle.

Strategy During the isothermal processes, the internal energy of the ideal gas stays the same, so the heat input is equal to the work done.

Solution 1→2: During the isothermal expansion, the work done is equal to the heat input—otherwise the temperature of the gas would change.

$$W_{1→2} = +25 \text{ J (per cycle)}$$

3→4: During the isothermal compression, the work done is equal to the heat output. This time the work done is negative: the gas does negative work as it is compressed.

$$W_{3→4} = Q_C < 0$$

The heats are proportional to the temperatures:

$$\frac{Q_C}{Q_H} = -\frac{T_C}{T_H}$$

Therefore,

$$W_{3→4} = Q_C = -\frac{T_C}{T_H} Q_H = -\frac{300.0 \text{ K}}{1000.0 \text{ K}} \times 25 \text{ J} = -7.5 \text{ J (per cycle)}$$

Discussion We will not prove it here, but the total work done during the two adiabatic processes is zero. Then the total work done per cycle is

$$25 \text{ J} + (-7.5 \text{ J}) = 17.5 \text{ J}$$

The efficiency is then

$$e = \frac{17.5 \text{ J}}{25 \text{ J}} = 0.70$$

This should equal

$$e = 1 - \frac{T_C}{T_H} = 1 - \frac{300.0 \text{ K}}{1000.0 \text{ K}} = 0.7000$$

Conceptual Practice Problem 15.9 Adiabatic process in Carnot cycle

Since there is no heat flow during the adiabatic processes and the work done during them adds to zero, why do we need adiabatic processes in the Carnot cycle? Why not just eliminate them?

15.9 ENTROPY

When two systems of different temperatures are in thermal contact, heat flows out of the hotter system and into the colder system. There is no change in the total energy of the two systems; energy just flows out of one and into the other. Why then does heat flow? As we will see, heat flow into a system not only increases the system's internal energy, it also increases the *disorder* of the system. Heat flow out of a system decreases its internal energy and its disorder.

The **entropy** of a system (symbol S) is a quantitative measure of its disorder. Entropy is a state variable (like U, P, V, and T): a system in equilibrium has a unique entropy that does *not* depend on the past history of the system. (Recall that heat and work are *not* state variables.) The word *entropy* was coined by Rudolf Clausius (1822–1888) in 1865; its Greek root means *evolution* or *transformation*.

When heat flows from a hotter system to a colder one, the total energy change is zero, but the total entropy change is *not* zero. The heat entering the colder system increases its entropy *more* than the entropy of the hotter system decreases, so the total entropy of the two—and thus the total entropy of the universe—increases. This makes sense since we expect a spontaneous process to increase the amount of disorder. Heat flow in the opposite direction, from the colder system to the hotter system, would decrease the total entropy of the universe. This would be like the unscrambling of an egg: it *never* happens. The second law of thermodynamics can be stated in terms of entropy:

Second law of thermodynamics (entropy statement):
The entropy of the universe never decreases.

If an amount of heat Q flows into a system at constant absolute temperature T, the entropy change of the system is

$$\Delta S = \frac{Q}{T} \tag{15-20}$$

From this, we can see that the SI unit for entropy is J/K. Heat flowing into a system increases the system's entropy (both ΔS and Q are positive); heat leaving a system decreases the system's entropy (both ΔS and Q are negative). Equation (15-20) is valid as long as the temperature of the system is constant, which is true if

- The heat capacity is large (as for a reservoir);

- The amount of heat flow Q is small; or

- The system does work $W = Q$ so that $\Delta U = 0$.

Note that Eq. (15-20) gives only the *change* in entropy, not the initial and final values of the entropy. As with potential energy, the *change* is what's important in most situations.

If a small amount of heat Q flows between two systems of different temperatures $(T_H > T_C)$, the total entropy change is

$$\Delta S = \frac{-Q}{T_H} + \frac{Q}{T_C}$$

Since $T_H > T_C$,

$$\frac{Q}{T_H} < \frac{Q}{T_C}$$

so the increase in the colder system's entropy is larger than the decrease of the hotter system's entropy:

$$\Delta S > 0 \tag{15-21}$$

A process that increases the entropy of the universe is irreversible, because the reverse process would decrease the universe's entropy. A reversible process must therefore be one that leaves the total entropy unchanged. For example, a reversible engine removes heat $-Q_H$ from a hot reservoir at temperature T_H and exhausts $-Q_C$ to a cold reservoir at T_C. (The minus signs are needed because we consider the heat flow in and out of the *reservoirs* rather than in and out of the engine.) The entropy of the engine itself is left unchanged since it operates in a cycle. The entropy of the hot reservoir decreases by an amount $-Q_H/T_H$ and that of the cold reservoir increases by $-Q_C/T_C$. Since the entropy of the universe must be unchanged by a reversible engine, it must be true that

$$\frac{-Q_H}{T_H} + \frac{-Q_C}{T_C} = 0 \quad \text{or} \quad \frac{Q_C}{Q_H} = -\frac{T_C}{T_H}$$

The efficiency of the engine is therefore

$$e = \frac{W}{Q_H} = \frac{Q_H + Q_C}{Q_H} = 1 + \frac{Q_C}{Q_H} = 1 - \frac{T_C}{T_H}$$

as stated in Section 15.7.

Entropy is *not* a conserved quantity like energy. The entropy of the universe is always increasing. It is possible to decrease the entropy of a *system*, but only at the expense of increasing the entropy of the surroundings by at least as much (usually more).

Example 15.10

Entropy Change of a Freely Expanding Gas

Suppose 1.0 mol of an ideal gas is allowed to freely expand into an evacuated container of equal volume so that the volume of the gas doubles (Fig. 15.16). No work is done by the gas as it expands, since there is nothing for it to push against. No heat flows into or out of the gas. What is the entropy change of the gas?

Strategy   The only way to calculate entropy changes that we've learned so far is for heat flow at a constant temperature. In free expansion, there is no heat flow—but that does not necessarily mean there is no entropy change. Since entropy is a state variable, ΔS depends only on the initial and final states of the gas, not the intermediate states. We can therefore find the entropy change using *any* thermodynamic process with the same initial and final states. The initial and final temperatures of the gas are identical since the internal energy does not change; therefore we find the entropy change for an *isothermal* expansion.

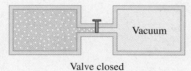

Valve closed

Figure 15.16
Two chambers connected by a valve. One chamber contains a gas and the other has been evacuated. When the valve is opened, the gas expands until it fills both chambers.

Solution   Imagine the gas confined to a cylinder with a moveable piston. In an isothermal expansion, heat flows into the gas at a constant temperature T. As the gas expands, it does work on the piston. If the temperature is to stay constant, the work done must equal the heat flow into the gas:

$$\Delta U = 0 \text{ implies } Q = W$$

In section 15.2 we found that the work done during the isothermal expansion of an ideal gas is

$$W = nRT \ln\left(\frac{V_f}{V_i}\right)$$

The volume of the gas doubles, so $V_f/V_i = 2$:

$$W = nRT \ln 2$$

Now we have an equal amount of heat flow into the gas. The entropy change is therefore

$$\Delta S = \frac{Q}{T} = nR \ln 2 = (1.0 \text{ mol}) \times \left(8.31 \frac{\text{J}}{\text{mol·K}}\right) \times 0.693 = +5.8 \text{ J/K}$$

Discussion   The entropy change is positive, as expected. Free expansion is an irreversible process; the gas molecules do not spontaneously collect back in the original container. The reverse process would cause a decrease in entropy, without a larger increase elsewhere, and so violates the second law.

Practice Problem 15.10   Entropy change of the universe when lump of clay is dropped

A room-temperature lump of clay of mass 400 g is dropped from a height of 2 m and makes a totally inelastic collision with the floor. Approximately what is the entropy change of the universe due to this collision? [*Hint:* The temperature of the clay rises, but only slightly.]

What about the evolution of life on Earth? Some have said that the second law of thermodynamics cannot be true because life has developed from simple life-forms to complex organisms spontaneously. Is this not a spontaneous increase in order—a decrease in entropy?

The second law of thermodynamics does *not* say that the entropy of a system can never decrease, just that the *total* entropy of the universe cannot decrease. When heat flows from a hot body to a cold body, the entropy of the hot body decreases, but the increase in the cold body's entropy is greater, so the entropy of the universe increases. A living organism is not a closed system and neither is the Earth. An adult human, for instance, requires roughly 10 MJ of chemical energy from food per day. What happens to this energy? Some is turned into useful work by the muscles, some more is used to repair body tissues, but most of it is dissipated and leaves the body as heat. The human body therefore is constantly increasing the entropy of its environment. As evolution progresses from simpler to more complicated organisms, the increase in order within the organisms must be accompanied by a larger increase in disorder in the environment.

The Availability of Energy

When people speak of "conserving energy," they usually mean using fuel and electricity sparingly. In the physics sense of the word *conserve*, energy is *always* conserved. Burning natural gas to heat your house does not change the amount of energy around; it just changes it from one form to another.

What we need to be careful not to waste is *high-quality* energy. Our concern is not the total amount of energy but rather whether the energy is in a form that is useful and convenient. The chemical energy stored in fuel is relatively high-quality energy. In a fuel cell, this energy can be turned into useful work with a theoretical maximum efficiency of 100%. When fuel is burned, the energy is degraded into lower-quality energy. The burned fuel could supply the high-temperature reservoir for a heat engine—but now the maximum possible efficiency is no longer 100%. The energy is higher quality if it is burned at a higher temperature, because the theoretical efficiency of a heat engine converting the energy to mechanical work is higher.

A heat engine can only work if a temperature *difference* is available. The smaller the ratio of T_C/T_H, the more work can be extracted.

It may be that, as the entropy of the universe increases inexorably, the universe as a whole comes to thermal equilibrium—the same temperature everywhere. This is called the "heat death of the universe." There would still be just as much energy as ever, but with everything at the same temperature, nothing interesting would happen—there certainly couldn't be any life.

15.10 STATISTICAL INTERPRETATION OF ENTROPY

Suppose that we take four coins, number them, and toss them. We could report the outcome in two different ways: either by specifying the outcome of each coin toss individually (e.g., coin 1 is heads, coin 2 is tails, coin 3 is heads, and coin 4 is heads), or just by reporting the overall outcome as the number of heads. We can use this simple system as an analog of a thermodynamic system. We will find that the second law of thermodynamics is based on statistics, as is the outcome of the four-coin system.

Specifying the outcome of each coin toss individually is analogous to describing the **microstate** of a thermodynamic system. A microstate specifies the state of each constituent particle. For instance, in a monatomic ideal gas with N atoms, a microstate is specified by the position and velocity of each of the N atoms. As the atoms move about and collide, the system changes from one microstate to another. The total number of heads for coin tossing is analogous to a **macrostate** of a thermodynamic system. A macrostate of an ideal gas is determined by the values of the macroscopic state variables (the pressure, volume, and temperature).

In the four-coin model, each of the microstates is equally likely to occur on any toss. Each of the coins has equal probability of landing heads or tails. Since each of 4 coins has 2 possible outcomes, there are $2^4 = 16$ different but equally probable microstates. There are only five macrostates: the number of heads can range from zero to four. The macrostates are *not* equally likely. A good guess would be that 2 heads is much more likely than 4 heads. To find the probability of a macrostate, we count up the number of microstates corresponding to that macrostate and divide by the total number of microstates for all the possible macrostates. From Table 15.3, the probability of the most likely macrostate (2 heads) is $6/16 = 0.375$. The probability of 4 heads is only $1/16 = 0.0625$.

Table 15.3

Possible Results of Tossing Four Coins

Macrostate	Microstates	Number of Microstates
4H (4 heads)	HHHH	1
3H (3 heads)	HHHT HHTH HTHH THHH	4
2H (2 heads)	HHTT HTHT HTTH THHT THTH TTHH	6
1H (1 head)	HTTT THTT TTHT TTTH	4
0H (0 heads)	TTTT	1
	Total number of microstates = 16	

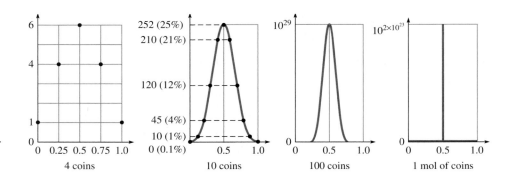

Figure 15.17 Graphs of the number of macrostates versus n/N ($n = 0, 1, \ldots, N$), where n = number of heads for $N = 4$ coins, 10 coins, 100 coins, 1 mole of coins.

Physics at Home

Toss repeatedly a collection of 10 coins. After each toss, count and record the number of heads. After a large number of tosses, are your results similar to the results of a statistical analysis (see Fig. 15.17)? Why are your results not *exactly* the same?

We used 4 coins in our example for convenience, but thermodynamic systems have very large numbers of particles (for instance, there are 6×10^{23} particles in one mole). What happens to the coin-tossing problem if the number of coins gets large? In Fig. 15.17 we have graphed the number of microstates for the various macrostates for systems with $N = 4$ coins, 10 coins, 100 coins, and 1 mole of coins. The horizontal axes for the four graphs specify the macrostate as the fractional number of heads, or (number of heads)/N, which ranges from 0 to 1. The probability of obtaining any macrostate is proportional to the number of microstates since the microstates are equally likely. Notice what happens to the probability peak: as N gets large, the probability of obtaining a macrostate having a number of heads significantly different (say, more than 1%) from $\frac{1}{2}N$ gets smaller and smaller. What if there were 6×10^{23} coins? The probability of getting anything more than 1% away from 3×10^{23} heads is so small that we can call it zero—it is impossible.

This kind of statistical analysis is the basis for the second law of thermodynamics. It turns out, remarkably, that the number of microstates for a given macrostate is related to the entropy of that macrostate in a simple way. Letting W stand for the number of microstates, the relationship is

$$S = k \ln W \tag{15-22}$$

Equation (15-22) is inscribed on the tombstone of Ludwig Boltzmann (1844–1906), the Austrian physicist who made the connection between entropy and statistics in the late nineteenth century. The relationship between S and W has to be logarithmic because entropy is additive: if system 1 has entropy S_1 and system 2 has entropy S_2, then the total entropy is $S_1 + S_2$. However, the number of microstates is *multiplicative*. Think of dice: if die 1 has 6 microstates and die 2 also has 6, the total number of microstates is not 12, but $6 \times 6 = 36$. The entropy is additive since $\ln 6 + \ln 6 = \ln 36$.

Suppose we have a box with 100 coins arranged so that all are showing heads. There is only one way of obtaining this macrostate. This means $W_i = 1$ and

$$S_i = k \ln 1 = k \times 0 = 0$$

Next we shake the box for awhile and we end up with 50 coins showing heads and 50 showing tails.

Statistically, it can be shown that the number of ways to have N coins arranged so that N_1 are in one group (heads) and the remainder, $N - N_1 = N_2$, are in the other group (tails) is

$$\frac{N!}{N_1! \, N_2!} \tag{15-23}$$

where $N!$ is *N-factorial*:

$$N! = N \times (N-1) \times (N-2) \times (N-3) \times \cdots \times 3 \times 2 \times 1$$

Thus, for a collection of 100 coins, with 50 showing heads (and the other 50 showing tails), we have a number of possible arrangements (microstates) given by

$$W_f = \frac{100!}{50! \, 50!}$$

$$= \frac{9.33 \times 10^{157}}{(3.04 \times 10^{64}) \times (3.04 \times 10^{64})}$$

$$= 1.01 \times 10^{29}$$

There are about 10^{29} ways of having 100 coins showing 50 heads. The entropy for this outcome is

$$S_f \approx k \ln 10^{29} = 29k \ln 10 = 66.8k$$

The change in entropy, ΔS, is

$$\Delta S = S_f - S_i = 66.8k - 0 = +66.8k$$

The entropy has increased. No matter what the original arrangement of the coins in the box, we expect to end up with *approximately* 50 heads after we shake the box for awhile.

$$W_H < W_{50} \quad \text{(for H = initial number of heads < 50)}$$

$$\Delta S = k \ln W_{50} - k \ln W_H > 0$$

Thus, the entropy increases no matter what the starting condition.

Entropy never decreases because the macrostate with the highest entropy is the one with the greatest number of microstates, and thus the highest probability. (Recall that since the microstates are equally likely, the probability of a macrostate is proportional to W.) The probability peak is so sharp and narrow in thermodynamic systems that the probability is zero of finding a macrostate not in that peak. The equilibrium macrostate is the one with most microstates. The spontaneous evolution of a system always goes toward higher probability (higher entropy). A process that increases the probability is irreversible—because the reversed process would decrease the probability. Reversible processes leave the probability unchanged.

Example 15.11

Increased Number of Microstates in Free Expansion

Refer to the free expansion of an ideal gas (Example 15.10). How does the number of microstates change when the volume of the gas (containing N molecules) is doubled?

Strategy Since in Example 15.10 we found the entropy change for this process, we can now use the entropy change to find how the number of microstates changes. Since the relationship between S and W is logarithmic, an increase in S will tell us by what *factor* W increases.

Solution The entropy change for n moles was found to be

$$\Delta S = nR \ln 2$$

Since $nR = Nk$, the entropy increase can be written in terms of N:

$$\Delta S = Nk \ln 2$$

If W_i and W_f are the initial and final number of microstates, then

$$\Delta S = k \ln W_f - k \ln W_i = k(\ln W_f - \ln W_i) = k \ln \frac{W_f}{W_i}$$

Equating these last two expressions for ΔS, we find

$$N \ln 2 = \ln \frac{W_f}{W_i}$$

Since $\ln 2^N = N \ln 2$,

$$\frac{W_f}{W_i} = 2^N$$

continued on next page

Example 15.11 *continued*

Discussion   To get an idea of how large the increase in the number of microstates is, let $N = N_A$ (1 mol of gas). To write the number 2^N in ordinary base 10 notation, we would need 2×10^{23} *digits*.

The temperature is the same before and after, so the number of velocity states, rotational states, and vibrational states before and after is the same. But each molecule has twice as much volume in which it can be found, so *for each molecule* there are twice as many possible positions. Therefore, the number of microstates is multiplied by 2 for each molecule, or by 2^N overall.

Practice Problem 15.11   Change in entropy for 100 coins

What is the change in entropy (expressed as a multiple of the Boltzmann constant) if a box of 100 coins starts with 35 heads showing and then is shaken until 50 heads are showing? [*Hint*: Use $\ln A - \ln B = \ln (A/B)$ to simplify the math. $35! = 1.03 \times 10^{40}$ and $65! = 8.25 \times 10^{90}$.]

15.11 THE THIRD LAW OF THERMODYNAMICS

Like the second law, the third law of thermodynamics can be stated in several equivalent ways. We will state just one of them:

> **Third law of thermodynamics**:
> It is impossible to cool a system to absolute zero.

Why does the third law matter? According to one statement of the *second* law, heat can never be 100% converted into work. A reversible engine operating with a cold reservoir at $T = 0$ *could* do just that, since its efficiency would be 100%:

$$e = 1 - \frac{T_C}{T_H} = 1 - \frac{0}{T_H} = 1$$

Since, according to the third law, there can be no cold reservoir at absolute zero, no engine can operate with 100% efficiency.

Suppose we wished to use a refrigerator to cool some object down to absolute zero temperature. The first law tells us

$$W = Q_C + Q_H$$

where Q_H is the heat discharged to the hot reservoir. Assuming an ideal, reversible refrigerator,

$$\frac{Q_H}{T_H} = -\frac{Q_C}{T_C}$$

we can calculate the minimum amount of work that it takes to run the appliance through one cycle,

$$W = Q_H + Q_C = Q_C \times \left(\frac{Q_H}{Q_C} + 1 \right)$$

$$W = Q_C \left(1 - \frac{T_H}{T_C} \right) \tag{15-24}$$

Given a certain amount of heat Q_C to extract in each cycle and a temperature T_H of the hot reservoir, we are going to have to do an infinite amount of work as $T_C \to 0$; so we will find it impossible to reach absolute zero temperature.

While it is impossible to *reach* absolute zero, there is no limit on how *close* we can get. Scientists who study low-temperature physics have attained equilibrium temperatures as low as 1 μK and have sustained temperatures of 2 mK; transient temperatures in the nano- and picokelvin range have been observed.

MASTER THE CONCEPTS

Summary

- The first law of thermodynamics is a statement of energy conservation:

$$\Delta U = Q - W \qquad (15\text{-}1)$$

where Q is the heat flow *into* the system and W is the work done *by* the system.

- Pressure, temperature, volume, number of moles, internal energy, and entropy are state variables; they describe the state of a system at some instant of time but *not* how the system got to that state. Heat and work are *not* state variables—they describe *how* a system gets from one state to another.

- The work done by a system when the pressure is constant—or for a volume change small enough that the pressure change is insignificant—is

$$W = P\,\Delta V \qquad (15\text{-}2)$$

In general, the work done is the total area under the *PV* curve.

- The change in internal energy of an ideal gas is determined solely by the temperature change. Therefore,

$$\Delta U = 0 \quad \text{(ideal gas, isothermal)} \qquad (15\text{-}7)$$

- A process in which no heat is transferred into or out of the system is called an adiabatic process.

- The molar specific heats of an ideal gas at constant volume and constant pressure are related by

$$C_v = C_p - R \qquad (15\text{-}11)$$

- Heat flow from a hotter body to a colder body is always irreversible.

- For a cyclical engine, heat pump, and the like, since $\Delta U = 0$ for a cycle, conservation of energy requires

$$Q = Q_H + Q_C = W$$

- The efficiency of an engine (or the inefficiency of a heat pump, refrigerator, or air conditioner) is defined as

$$e = \frac{W}{Q_H} \qquad (15\text{-}12)$$

- The coefficient of performance for a heat pump is

$$K_p = \frac{\text{heat delivered}}{\text{work input}} = \frac{Q_H}{W} = \frac{1}{e} \qquad (15\text{-}16)$$

- The coefficient of performance for a refrigerator or air conditioner is

$$K_r = \frac{\text{heat removed}}{\text{work input}} = \left| \frac{Q_C}{W} \right| = \frac{1}{e} - 1 \qquad (15\text{-}17)$$

- A reservoir is a system with such a large heat capacity that it can exchange heat in either direction with a negligibly small temperature change.

- The second law of thermodynamics can be stated in various equivalent ways: (1) heat never flows spontaneously from a colder body to a hotter body, (2) no engine can have 100% efficiency, (3) no engine can have an efficiency greater than the inefficiency of any heat pump operating between the same hot and cold reservoirs, (4) no heat pump (or refrigerator) can have an inefficiency of zero, (5) every engine has $e_e \le e_r$ and every heat pump has $e_p \ge e_r$, (6) the entropy of the universe never decreases.

- The efficiency of a *reversible* engine is determined only by the *absolute* temperatures of the hot and cold reservoirs:

$$e_r = 1 - \frac{T_C}{T_H} \qquad (15\text{-}18)$$

- If an amount of heat Q flows into a system at constant absolute temperature T, the entropy change of the system is

$$\Delta S = \frac{Q}{T} \qquad (15\text{-}20)$$

- The number of microstates for a given macrostate is related to the entropy S of that macrostate

$$S = k \ln W \qquad (15\text{-}22)$$

- The third law of thermodynamics: it is impossible to cool a system to absolute zero.

Highlighted Figures and Tables

T15.1 Sign conventions for the first law of thermodynamics (p. 512)

F15.2 Expansion of a gas from initial pressure P_i and volume V_i to final pressure P_f and volume V_f and the *PV* diagram for expansion (p. 514)

F15.3 The *PV* curve shows the work done (p. 514)

F15.4 Work done by following two different processes between the same initial and final states (p. 515)

T15.2 Summary of thermodynamic processes (p. 517)

F15.9 Spontaneous heat flow goes from warm to cool (p. 520)

F15.10 The energy transfers of a heat engine (p. 521)

F15.11 The four-stroke Otto cycle and its *PV* diagram (p. 523)

F15.12 How a refrigerator works (p. 525)

F15.13 A heat engine and a heat pump (p. 525)

F15.15 The Carnot cycle (p. 529)

F15.17 Graphs of the number of macrostates versus n/N (p. 534)

CONCEPTUAL QUESTIONS

1. The entropy of a system increases by 10 J/K. Does this mean that the process is necessarily irreversible? Explain.

2. Is it possible to make a heat pump with an inefficiency equal to 1? Explain.

3. A perfectly elastic collision is reversible. What about an inelastic collision? Justify your answer.

4. An electric baseboard heater can convert 100% of the electrical energy used into heat that flows into the house. Since a gas furnace might be located in a basement and sends exhaust gases up the chimney, the heat flow into the living space is less than 100% of the chemical energy released by burning. Does this mean that electric heating is better? Which heating method consumes less fuel? In your answer, consider how the electricity might have been generated and the efficiency of that process.

5. One whimsical statement of the laws of thermodynamics—probably not one favored by gamblers—goes like this:

   I. You can never win; you can only lose or break even.
   II. You can only break even at absolute zero.
   III. You can never get to absolute zero.
   What do we mean by "win," "lose," and "break even"? [*Hint:* Think about a heat engine.]

6. Why must all reversible engines (operating between the same reservoirs) have the same efficiency? Try an argument by contradiction: imagine that two reversible engines exist with $e_1 > e_2$. Reverse one of them (into a heat pump) and use the work output from the engine to run the heat pump. What happens? (If it seems fine at first, switch the two.)

7. When supplies of fossil fuels such as petroleum and coal dwindle, people might call the situation an "energy crisis." From the standpoint of physics, why is that not an accurate name? Can you think of a better one?

8. If you leave the refrigerator door open and the refrigerator runs continuously, does the kitchen get colder or warmer? Explain.

9. Most heat pumps incorporate an electric heater as a "backup." For relatively mild outdoor temperatures, the electric heater is not used. However, if the outdoor temperature gets very low, the heat pump shuts off and the heater is used instead. Why?

10. Why are heat pumps more often used in mild climates than in areas with severely cold winters?

11. Are entropy changes always caused by the flow of heat? If not, give some other examples of processes that increase entropy.

12. Can a heat engine be made to operate without creating any "thermal pollution," that is, without making its cold reservoir get warmer in the long run? The net work output must be greater than zero.

13. A warm pitcher of lemonade is put into an ice chest. Describe what happens to the entropies of lemonade and ice as heat flows from the lemonade to the ice within the chest.

MULTIPLE CHOICE QUESTIONS

1. A real heat engine, when run between reservoirs at temperatures of 300°C and 30°C, has efficiencies e_e and e_p in the forward and reverse directions (i.e., as an engine or as a heat pump). Which of these pairs of values e_e, e_p are possible?

   (a) 50%, 56%   (b) 56%, 50%   (c) 82%, 95%
   (d) 42%, 57%   (e) 57%, 42%

2. If two different systems are put in thermal contact, so that heat can flow from one to the other, then heat will flow until the systems have the same

   (a) energy.   (b) heat capacity.
   (c) entropy.   (d) temperature.

3. As moisture from the air condenses on the outside of a cold glass of water, the entropy of the condensing moisture

   (a) stays the same.   (b) increases.
   (c) decreases.   (d) not enough information.

4. As a system undergoes a constant volume process

   (a) the pressure does not change.
   (b) the internal energy does not change.
   (c) the work done on or by the system is zero.
   (d) the entropy stays the same.
   (e) the temperature of the system does not change.

5. Which of these statements are implied by the second law of thermodynamics?

   (a) The entropy of an engine (including its fuel and/or heat reservoirs) operating in a cycle never decreases.
   (b) The increase in internal energy of a system in any process is the sum of heat absorbed plus work done on the system.

   (c) A heat engine, operating in a cycle, that rejects no heat to the high temperature reservoir is impossible.
   (d) Both (a) and (c).   (e) All three [(a), (b), and (c)].

6. On a summer day, you keep the air conditioner in your room running. From the list numbered 1 to 4 choose the hot reservoir and the cold reservoir.

   1. the air outside
   2. the compartment inside the air conditioner where the air is compressed
   3. the freon gas that is the working substance (expands and compresses in each cycle)
   4. the air in the room

   (a) 1 is hot, 2 is cold reservoir.   (b) 1 is hot, 3 is cold reservoir.
   (c) 1 is hot, 4 is cold reservoir.   (d) 2 is hot, 3 is cold reservoir.
   (e) 2 is hot, 4 is cold reservoir.   (f) 3 is hot, 4 is cold reservoir.

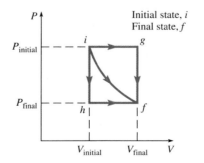

Figure 15.18
Multiple Choice Question 7

7. Figure 15.18 shows a *PV* diagram with several paths to get from an initial to a final state. For which path does the system do the most work?

   (a) path *igf*   (b) path *if*
   (c) path *ihf*   (d) All paths represent equal work.

8. An ideal gas is confined to the left chamber of an insulated container. The right chamber is evacuated. A valve is opened between the chambers, allowing gas to flow into the right chamber. After equilibrium is established, the temperature of the gas ____. [*Hint:* What happens to the internal energy?]

   (a) is lower than the initial temperature
   (b) is higher than the initial temperature
   (c) is the same as the initial temperature
   (d) could be higher than, the same as, or lower than the initial temperature

9. In the first law of thermodynamics ($\Delta U = Q - W$), the variables Q and W stand for

   (a) the heat flow *out of* the system and the work done *on* the system.
   (b) the heat flow *out of* the system and the work done *by* the system.
   (c) the heat flow *into* the system and the work done *by* the system.
   (d) the heat flow *into* the system and the work done *on* the system.

10. As an ideal gas is adiabatically expanding,

    (a) the temperature of the gas does not change.
    (b) the internal energy of the gas does not change.

    (c) work is not done on or by the gas.
    (d) no heat is given off or taken in by the gas.
    (e) both (a) and (d).
    (f) both (a) and (b).

11. As an ideal gas is compressed at constant temperature,

    (a) heat flows out of the gas.
    (b) the internal energy of the gas does not change.
    (c) the work done on the gas is zero.
    (d) none of the above.
    (e) both (a) and (b).
    (f) both (a) and (c).

12. Given 1 mole of an ideal gas, in a state characterized by P_A, V_A, a change occurs so that the final pressure and volume are equal to P_B, V_B, where $V_B > V_A$. Which of these is true?

    (a) The heat supplied to the gas during the process is completely determined by the values P_A, V_A, P_B, and V_B.
    (b) The change in the internal energy of the gas during the process is completely determined by the values P_A, V_A, P_B, and V_B.
    (c) The work done by the gas during the process is completely determined by the values P_A, V_A, P_B, and V_B.
    (d) All three are true.
    (e) None of these is true.

13. Which choice correctly identifies the three processes shown in Fig. 15.19?

    (a) 1 = isobaric; 2 = isochoric; 3 = adiabatic
    (b) 1 = isothermal; 2 = isothermal; 3 = isobaric
    (c) 1 = isochoric; 2 = adiabatic; 3 = isobaric
    (d) 1 = isobaric; 2 = isothermal; 3 = isochoric

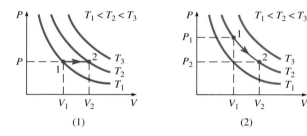

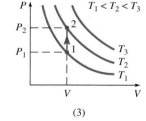

Figure 15.19 Three processes illustrated on *PV* diagrams. (Multiple Choice Question 13)

PROBLEMS

Note: **C** indicates a combination conceptual/quantitative problem. Gold diamonds ◆, ◆◆ are used to indicate the increasing level of difficulty of each problem. Problem numbers appearing in blue, 9., denote problems that have a detailed solution available in the Student Solutions Manual. Some problems are *paired* by concept; their numbers are connected by a ruled box.

15.1 The First Law of Thermodynamics;
15.2 Thermodynamic Processes;
15.3 Constant Pressure Expansion of an Ideal Gas

1. A system takes in 550 J of heat while performing 840 J of work. What is the change in internal energy of the system?

2. The internal energy of a system increases by 400 J while 500 J of work are performed on it. What was the heat flow into or out of the system?

3. A model steam engine of 1.00 kg mass pulls eight cars of 1.00-kg mass each. The cars start at rest and reach a velocity of 3.00 m/s in a time of 3.00 s while moving a distance of 4.50 m. During that time the engine takes in 135 J of heat. What is the change in the internal energy of the engine?

4. A monatomic ideal gas at 27°C undergoes a constant pressure process from *A* to *B* and a constant volume process from *B* to *C*, as shown in Fig. 15.20a. Find the total work done during these two processes.

5. A monatomic ideal gas at 27°C undergoes a constant volume process from *A* to *B* and a constant pressure process from *B* to *C*, as shown in Fig. 15.20b. Find the total work done during these two processes.

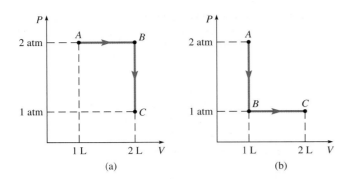

Figure 15.20 Problems 4 and 5

♦ 6. Suppose a monatomic ideal gas is changed from state A to state D by one of the processes shown on the PV diagram of Fig. 15.21. (a) Find the total work done by the gas if it follows the constant volume path A–B followed by the constant pressure path B–C–D. (b) Calculate the total change in internal energy of the gas during the entire process and the total heat flow into the gas.

♦ 7. Repeat Problem 6 for the case when the gas follows the constant temperature path A–C followed by the constant pressure path C–D.

♦ 8. Repeat Problem 6 for the case when the gas follows the constant pressure path A–E followed by the constant temperature path E–D.

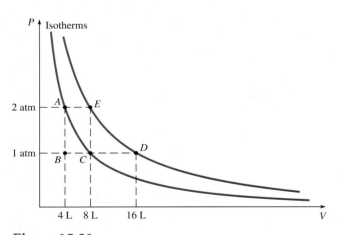

Figure 15.21 Problems 6–8

9. An ideal gas is in contact with a heat reservoir so that it remains at a constant temperature of 300.0 K. The gas is compressed from a volume of 24.0 L to a volume of 14.0 L. During the process the mechanical device pushing the piston to compress the gas is found to expend 5.00 kJ of energy. How much heat flows between the heat reservoir and the gas and in what direction does the heat flow occur?

10. Suppose 1.00 mol of oxygen is heated at constant pressure of 1.00 atm from 10.0°C to 25.0°C. (a) How much heat is absorbed by the gas? (b) Using the ideal gas law, calculate the change of volume of the gas in this process. (c) What is the work done by the gas during this expansion? (d) From the first law, calculate the change of internal energy of the gas in this process.

15.5 Heat Engines; 15.6 Refrigerators and Heat Pumps

11. (a) How much heat does an engine with an efficiency of 33.3% absorb in order to deliver 1.00 kJ of work? (b) How much heat is exhausted by the engine?

12. The efficiency of an engine is 0.21. For every 1.00 kJ of heat absorbed by the engine, how much (a) net work is done by it and (b) heat is released by it?

13. The United States generates about 5.0×10^{16} J of electric energy a day. This energy is equivalent to work, since it can be converted into work with almost 100% efficiency by an electric motor. (a) If this energy is generated by power plants with an average efficiency of 0.30, how much heat is dumped into the environment each day? (b) How much water would be required to absorb this heat if the water temperature is not to increase more than 2.0°C?

14. The intensity (power per unit area) of the sunlight incident on the Earth's surface, averaged over a 24-h period, is about 0.20 kW/m². If a solar power plant is to be built with an output capacity of 1.0×10^9 W, how big must the area of the solar energy collectors be for photocells operating at 20.0% efficiency?

15. An engine releases 0.450 kJ of heat for every 0.100 kJ of work it does. What is the efficiency of the engine?

16. An engine works at 30.0% efficiency. The engine raises a 5.00-kg crate from rest to a vertical height of 10.0 m, at which point the crate has a speed of 4.00 m/s. How much heat input is required for this engine?

17. How much heat does a heat pump with an inefficiency of 33.3% deliver when supplied with 1.00 kJ of electricity?

18. An air conditioner whose coefficient of performance is 2.00 removes 1.73×10^8 J of heat from a room per day. How much does it cost to run the air conditioning unit per day if electricity costs \$0.10 per kilowatt-hour? (Note that 1 kilowatt-hour = 3.6×10^6 J.)

15.7 Reversible Engines and Heat Pumps; 15.8 Carnot Cycle

19. In a certain steam engine, the boiler temperature is 127°C and the cold reservoir temperature is 27°C. While this engine does 2.0 kcal of work, what minimum amount of heat must be discharged into the cold reservoir?

20. Calculate the maximum possible efficiency of a heat engine that uses surface lake water at 18.0°C as a source of heat and rejects waste heat to the water 0.100 km below the surface where the temperature is 4.0°C.

21. An ideal refrigerator removes heat at a rate of 0.10 kW from its interior (+2.0°C) and exhausts heat at 40.0°C. How much electrical power is used?

22. A heat pump is used to heat a house with an interior temperature of 20.0°C. On a chilly day with an outdoor temperature of –10.0°C, what is the minimum work that the pump requires in order to deliver 1.0 kJ of heat to the house?

23. A coal-fired electrical generating station can use a higher T_H than a nuclear plant; for safety reasons the core of a nuclear reactor is not allowed to get as hot as coal. Suppose that $T_H = 727°C$ for a coal station but $T_H = 527°C$ for a nuclear station. Both power plants exhaust waste heat into a lake at $T_C = 27°C$. How much waste heat does each plant

exhaust into the lake to produce 1.00 MJ of electricity? Assume both operate as reversible engines.

24. (a) Calculate the efficiency of a reversible engine that operates between the temperatures 600.0°C and 300.0°C. (b) If the engine absorbs 100.0 kcal of heat from the hot reservoir, how much does it exhaust to the cold reservoir?

25. A reversible engine with a 30.0% efficiency has $T_C = 310.0$ K. (a) What is T_H? (b) How much heat is exhausted for every 0.100 kJ of work done?

26. Show that in a reversible engine the amount of heat Q_C is related to the work done W by

$$Q_C = -\frac{T_C}{T_H - T_C} W$$

27. An electrical power station generates steam at 500.0°C and condenses it with river water at 27°C. By how much would its theoretical maximum efficiency decrease if it had to switch to cooling towers that condense the steam at 47°C?

28. An ideal refrigerator keeps its contents at 0.0°C and exhausts heat into the kitchen at 40.0°C. For every 1.0 kJ of work done, (a) how much heat is exhausted? (b) How much heat is removed from the contents?

29. The outdoor temperature on a winter's day is –4°C. If you use 1.0 kJ of electrical energy to run a heat pump, how much heat does that put into your house at 21°C? Assume that the heat pump is reversible.

30. An oil-burning electric power plant uses steam at 773 K to drive a turbine, after which the steam is expelled at 373 K. The engine has an efficiency of 0.40. What is the theoretical maximum efficiency possible at those temperatures?

31. An inventor proposes a heat engine to propel a ship, using the temperature difference between the water at the surface and the water 10 m below the surface as the two reservoirs. If these temperatures are 15.0°C and 10.0°C, respectively, what is the maximum possible efficiency of the engine?

32. A heat engine uses the warm air at the ground as the hot reservoir and the cooler air at an altitude of several thousand meters as the cold reservoir. If the warm air is at 37°C and the cold air is at 25°C, what is the maximum possible efficiency for the engine?

33. A reversible refrigerator has an inefficiency of 0.25. How much work must be done to freeze 1.0 kg of liquid water initially at 0°C?

✦ 34. On a hot day, you are in a sealed, insulated room. The room contains a refrigerator, operated by an electric motor. The motor does work at the rate of 250 W when it is running. Assume the motor is ideal (no friction or electrical resistance) and that the refrigerator operates on a reversible cycle. In an effort to cool the room, you turn on the refrigerator and open its door. Let the temperature in the room be 320 K when this process starts, and the temperature in the cold compartment of the refrigerator be 256 K. At what net rate is heat added to (+) or subtracted from (–) the room and all of its contents?

✦✦ 35. Imagine that a car engine could be replaced by a Carnot engine with an ideal gas as the working substance. When the car is traveling at 65 miles per hour, the Carnot engine goes through its cycle 900.0 times per minute. The engine's hot reservoir is at 1000.0°C (the temperature of the exploding gas in a real car engine) and the cold reservoir is at 20.0°C (the outside temperature). During the isothermal expansion part of each cycle, the volume of the ideal gas increases by a factor of 10.0. The cylinders contain 0.223 mol of gas. What is the power output of the engine?

15.9 Entropy

36. List these in order of increasing entropy: (a) 0.5 kg of ice and 0.5 kg of (liquid) water at 0°C; (b) 1 kg of ice at 0°C; (c) 1 kg of (liquid) water at 0°C; (d) 1 kg of water at 20°C.

37. List these in order of increasing entropy: (a) 0.01 mol of N_2 gas in a 1-L container at 0°C; (b) 0.01 mol of N_2 gas in a 2-L container at 0°C; (c) 0.01 mol of liquid N_2.

38. An ice cube at 0.0°C is slowly melting. What is the change in the ice cube's entropy for each 1.00 g of ice that melts?

39. From Table 14.4 we know that approximately 540 kcal are needed to transform 1 kg of water at 100°C to steam at 100°C. What is the change in entropy of 1.0 kg of water evaporating at 100.0°C? (Specify whether the change in entropy is an increase, +, or a decrease, –.)

40. A large block of copper initially at 20.0°C is placed in a vat of hot water (80.0°C). For the first 1.0 J of heat that flows from the water into the block, find (a) the entropy change of the block, (b) the entropy change of the water, and (c) the entropy change of the universe. Note that the temperatures of the block and water are essentially unchanged by the flow of only 1.0 J of heat.

41. A large, cold (0.0°C) block of iron is immersed in a tub of hot (100.0°C) water. In the first 10.0 s, 10.0 kcal of heat are transferred, although the temperatures of the water and the iron do not change much in this time. Calculate the change in entropy of the system in this time.

42. On a cold winter day, the outside temperature is –15.0°C. Inside the house the temperature is +20.0°C. Heat flows out of the house through a window at a rate of 220.0 W. At what rate is the entropy of the universe changing due to this heat conduction through the window?

43. Within an insulated system, 100.0 kcal of heat is conducted through a copper rod from a hot reservoir at +200.0°C to a cold reservoir at +100.0°C. (The reservoirs are so big that this heat exchange does not change their temperatures appreciably.) What is the net change in entropy of the system, in kcal/K?

44. Plot the temperature versus entropy for the four stages of the Carnot engine discussed in Example 15.9. [*Hint*: First plot the constant temperature stages and then fill in the adiabatic stages.]

45. A student eats 2000 kcal per day. (a) Assuming that all of the food energy is released as heat, what is the rate of heat released (in watts)? (b) What is the rate of change of entropy of the surroundings if all of the heat is released into air at room temperature (20°C)?

15.10 Statistical Interpretation of Entropy

46. Suppose the macrostate of a system of 100 coins is specified by the number of heads. What is the entropy of the state with one head (in terms of Boltzmann's constant, k_B)?

47. For a system composed of two dice, let the macrostate be defined by the sum of the numbers showing on the top faces. What is the maximum entropy of this system in units of Boltzmann's constant, k_B?

48. Two dice are thrown. A macrostate is specified by the sum of the two numbers that come up on the dice. (a) How is a microstate specified for this system? (b) How many different microstates are there? (c) How many different macrostates are there? (d) What is the most probable macrostate? (e) What is the probability of getting this result? (f) What is the probability of rolling "snake eyes" (two 1s)?

49. Six coins are tossed simultaneously. The macrostate is specified by the number of "heads." (a) What is/are the most probable macrostate(s)? (b) What is/are the least probable macrostate(s)? (c) What is the probability of obtaining the most probable macrostate?

50. If 1.0-g of ice at 0.0°C melts into liquid water at 0.0°C, by what factor has the number of microstates increased?

51. If the number of microstates for a thermodynamic system doubles, how much has the system's entropy increased?

52. Four indistinguishable marbles must be placed into two distinguishable boxes. (a) How many microstates are there? (b) How many macrostates? (c) What is the most probable macrostate? (d) What is the entropy of that macrostate? (e) What is the least probable macrostate? (f) What is the entropy of the least probable macrostate?

COMPREHENSIVE PROBLEMS

53. The motor that drives a reversible refrigerator produces 148 W of useful power. The hot and cold temperatures of the heat reservoirs are 20.0°C and –5.0°C. What is the maximum amount of ice it can produce in 2.0 h from water that is initially at 8.0°C?

54. An engineer designs a ship that gets its power in the following way. The engine draws in warm water from the ocean, and after extracting some of the water's internal energy, returns the water to the ocean at a temperature 14.5°C lower than the ocean temperature. If the ocean is at a uniform temperature of 17°C, is this an efficient engine? Will the engineers design work?

55. A balloon contains 200.0 L of nitrogen gas at 20.0°C and at atmospheric pressure. How much energy must be added to raise the temperature of the nitrogen to 40.0°C while allowing the balloon to expand at atmospheric pressure?

56. An ideal gas is heated at a constant pressure of 2.0×10^5 Pa from a temperature of –73°C to a temperature of +27°C. The initial volume of the gas is 0.10 m³. The heat energy supplied to the gas in this process is 25 kJ. What is the increase in internal energy of the gas?

57. For a reversible engine, will you obtain a better efficiency by increasing the high-temperature reservoir by an amount ΔT or decreasing the low-temperature reservoir by the same amount ΔT?

58. A 0.50-kg block of iron [$c = 0.44$ kJ/(kg·K)] at 20.0°C is in contact with a 0.50-kg block of aluminum [$c = 0.900$ kJ/(kg·K)] at a temperature of 20.0°C. The system is completely isolated from the rest of the universe. Suppose heat flows from the iron into the aluminum until the temperature of the aluminum is 22.0°C. (a) From the first law, calculate the final temperature of the iron. (b) Estimate the entropy change of the system. (c) Explain how the result of part (b) shows that this process is impossible. [*Hint*: Since the system is isolated, $\Delta S_{System} = \Delta S_{Universe}$.]

59. List these in order of increasing entropy: (a) 1 mol of water at 20°C and 1 mol of ethanol at 20°C in separate containers; (b) a mixture of 1 mol of water at 20°C and 1 mol of ethanol at

20°C; (c) 0.5 mol of water at 20°C and 0.5 mol of ethanol at 20°C in separate containers; (d) a mixture of 1 mol of water at 30°C and 1 mol of ethanol at 30°C.

60. List these in order of increasing entropy: (a) 1000 He atoms moving at random velocities with an average speed of 400 m/s; (b) 1000 He atoms all moving at 400 m/s in the same direction; (c) 1000 He atoms all moving at 400 m/s in random directions.

61. If the pressure on a fish increases from 1.1 to 1.2 atm, its swim bladder decreases in volume from 8.16 mL to 7.48 mL while the temperature of the air inside remains constant. How much work is done on the air in the bladder?

62. A fish at a pressure of 1.1 atm has its swim bladder inflated to an initial volume of 8.16 mL. If the fish starts swimming horizontally, its temperature increases from 20.0°C to 22.0°C as a result of the exertion. (a) Since the fish is still at the same pressure, how much work is done by the air in the swim bladder? [*Hint*: First find the new volume from the temperature change.] (b) How much heat is gained by the air in the swim bladder? Assume air to be a diatomic ideal gas. (c) If this quantity of heat is lost by the fish, by how much will its temperature decrease? The fish has a mass of 5.00 g and its specific heat is about 3.5 J/(g·°C).

63. Suppose you mix 4.0 mol of a monatomic gas at 20.0°C and 3.0 moles of another monatomic gas at 30.0°C. If the mixture is allowed to reach equilibrium, what is the final temperature of the mixture? [*Hint*: Use energy conservation.]

64. A balloon contains 160 L of nitrogen gas at 25°C and 1.0 atm. How much energy must be added to raise the temperature of the nitrogen to 45°C while allowing the balloon to expand at atmospheric pressure?

65. The efficiency of a muscle during weight lifting is equal to the work done in lifting the weight divided by the total energy output of the muscle (work plus internal energy dissipated in the muscle). Let $g = 9.80$ m/s². Fill in this table to determine the efficiency of the muscle:

Table for Problem 65				
Mass Lifted (g)	**Height Raised (cm)**	**Internal Energy Dissipated (J)**	**Work Done (J)**	**Efficiency (%)**
16.4	0.577	1.39×10^{-3}		
8.09	0.745	1.64×10^{-3}		
1.05	0.880	1.85×10^{-3}		

66. (a) What is the entropy change of 1.00 mol of H_2O when it changes from ice to water at 0.0°C? (b) If the ice is in contact with an environment at a temperature of 10.0°C, what is the entropy change of the universe when the ice melts?

67. Estimate the entropy change of 850 g of water when it is heated from 20.0°C to 50.0°C. [*Hint*: Assume that the heat flows into the water at an average temperature.]

68. For a more realistic estimate of the maximum coefficient of performance of a heat pump, assume that a heat pump takes in heat from outdoors at *10°C below* the ambient outdoor temperature, to account for the temperature difference across its heat exchanger. Similarly, assume that the output must be *10°C hotter* than the house (which itself might be kept at 20°C), to make the heat flow into the house. Make a graph of the coefficient of performance of a reversible heat pump under these conditions as a function of outdoor temperature (from –15°C to +15°C in 5°C increments).

69. A 0.500-kg block of iron at 60.0°C is placed in contact with a 0.500-kg block of iron at 20.0°C. (a) The blocks soon come to a common temperature of 40.0°C. *Estimate* the entropy change of the universe when this occurs. [*Hint*: Assume that all the heat flow occurs at an average temperature for each block.] (b) Estimate the entropy change of the universe if, instead, the temperature of the hotter block increased to 80.0°C while the temperature of the colder block decreased to 0.0°C. [*Hint*: The answer is negative, indicating that the process is impossible.]

70. A container holding 1.20 kg of water at 20.0°C is placed in a freezer that is kept at –20.0°C. The water freezes and comes to thermal equilibrium with the interior of the freezer. What is the minimum amount of electrical energy required by the freezer to do this if it operates between reservoirs at temperatures of 20.0°C and –20.0°C?

71. A reversible heat engine has an efficiency of 33.3%, removing heat from a hot reservoir and rejecting heat to a cold reservoir at 0°C. If the engine now operates in reverse, how long would it take to freeze 1.0 kg of water at 0°C, if it operates on a power of 186 W?

✦ 72. Consider a heat engine that is *not* reversible. It has the *PV* cycle shown in Fig. 15.22. In the first stage (A) there is a constant temperature expansion while in contact with a warm reservoir at 373 K from $P_1 = 1.55 \times 10^5$ Pa and $V_1 = 2.00 \times 10^{-2}$ m^3 to $P_2 = 1.24 \times 10^5$ Pa and $V_2 = 2.50 \times 10^{-2}$ m^3. Then (B) a heat reservoir at the cooler temperature of 273 K is used to cool the gas at constant volume to 273 K from P_2 to P_3 $= 0.91 \times 10^5$ Pa. This is followed by (C) a constant temperature compression while still in contact with the cold reservoir at 273 K from P_3, V_2 to $P_4 = 1.01 \times 10^5$ Pa, V_1. The final stage (D)

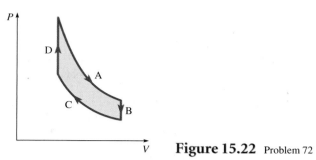

Figure 15.22 Problem 72

is heating the gas at constant volume from 273 K to 373 K by being in contact with the warm reservoir again, to return from P_4, V_1 to P_1, V_1. The engine uses 1.000 mol of a diatomic gas. (a) Fill out this chart, showing for each step in the cycle the work done, the heat input, and the change in internal energy.

Stage	W (Work Done)	ΔU (Change in Internal Energy)	Q (Net Heat Input)
A			
B			
C			
D			
Cycle ABCD			

(b) Find the efficiency of this engine. (c) Compare to the efficiency of a reversible engine.

73. (a) For the gas cycle in Problem 72 find the change in entropy of the cold reservoir in step B. Remember that the gas is always in contact with the cold reservoir. (b) What is the change in entropy of the hot reservoir in step D? (c) Using this information, find the change in entropy of the total system of gas plus reservoirs during the whole cycle.

✦ 74. A town is considering using its lake as a source of power. The average temperature difference from the top to the bottom is 15°C, and the average surface temperature is 22°C. (a) Assuming that the town can set up a reversible engine using the surface and bottom of the lake as heat reservoirs, what would be its efficiency? (b) If the town needs about 1.0×10^8 W of power to be supplied by the lake, how many m^3 of water does the heat engine use per second? (c) The surface area of the lake is 8.0×10^7 m^2 and the average incident intensity (over 24 h) of the sunlight is 200 W/m^2. Can the lake supply enough heat to meet the town's energy needs with this method?

ANSWERS TO PRACTICE PROBLEMS

15.1 Work was done on the gas; –28 kJ. **15.2** Heat flows into the gas; $Q = 3.8$ kJ **15.3** 359 kJ **15.4** The fire is irreversible: smoke and carbon dioxide and ash will not come together to form logs and twigs. The wood is in a highly ordered state, whereas the products of the fire are much less ordered. **15.5** 20% **15.6** 4 kW

15.7 1200 K **15.8** 178 MW **15.9** The adiabatic processes are needed to change the temperature of the working substance in the engine back and forth between T_C and T_H. **15.10** 0.03 J/K **15.11** +4.52k

APPENDIX A

Mathematics Review

A.1 ALGEBRA

There are two basic kinds of algebraic manipulations.

- The same operation can always be performed on both sides of an equation.
- Substitution is always permissible (if $a = b$, then any occurrence of a in any equation can be replaced with b).

   Products distribute over sums

$$a(b + c) = ab + ac \qquad (A\text{-}1)$$

The reverse—replacing $ab + ac$ with $a(b + c)$—is called *factoring*. Since dividing by c is the same as multiplying by $1/c$,

$$\frac{a + b}{c} = \frac{a}{c} + \frac{b}{c} \qquad (A\text{-}2)$$

   Equation (A-2) is the basis of the procedure for adding fractions. To add fractions, they must be expressed with a *common denominator*.

$$\frac{a}{b} + \frac{c}{d} = \frac{a}{b} \times \frac{d}{d} + \frac{c}{d} \times \frac{b}{b} = \frac{ad}{bd} + \frac{bc}{bd}$$

Now applying Eq. (A-2), we end up with

$$\frac{a}{b} + \frac{c}{d} = \frac{ad + bc}{bd} \qquad (A\text{-}3)$$

   To divide fractions, remember that dividing by c/d is the same as multiplying by d/c:

$$\frac{a}{b} \div \frac{c}{d} = \frac{\dfrac{a}{b}}{\dfrac{c}{d}} = \frac{a}{b} \times \frac{d}{c} = \frac{ad}{bc}$$

   A product in a square root can be separated:

$$\sqrt{ab} = \sqrt{a} \times \sqrt{b} \qquad (A\text{-}4)$$

Pitfalls to Avoid

These are some of the most common *incorrect* algebraic substitutions. Don't fall into any of these traps!

$$\sqrt{a + b} \neq \sqrt{a} + \sqrt{b}$$
$$\frac{a}{b + c} \neq \frac{a}{b} + \frac{a}{c}$$
$$\frac{a}{b} + \frac{c}{d} \neq \frac{a + c}{b + d}$$
$$(a + b)^2 \neq a^2 + b^2$$

In the last one, the cross term is missing: $(a + b)^2 = a^2 + 2ab + b^2$.

Graphs of Linear Functions

If the graph of y as a function of x is a straight line, then y is a *linear function* of x. The relationship can be written in the standard form

$$y = mx + b \qquad (A\text{-}5)$$

where m is the *slope* and b is the *y-intercept*. The slope measures how steep the line is. It tells how much y changes for a given change in x:

$$m = \frac{\Delta y}{\Delta x} = \frac{y_2 - y_1}{x_2 - x_1} \qquad (A\text{-}6)$$

The y-intercept is the value of y when $x = 0$. On the graph, the line crosses the y-axis at $y = b$.

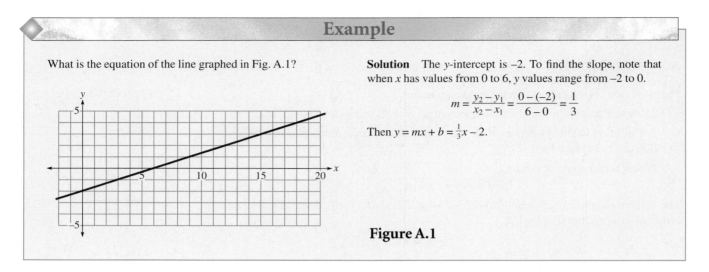

Example

What is the equation of the line graphed in Fig. A.1?

Solution   The y-intercept is -2. To find the slope, note that when x has values from 0 to 6, y values range from -2 to 0.

$$m = \frac{y_2 - y_1}{x_2 - x_1} = \frac{0 - (-2)}{6 - 0} = \frac{1}{3}$$

Then $y = mx + b = \frac{1}{3}x - 2$.

Figure A.1

A.2   SOLVING EQUATIONS

Solving an equation means using algebraic operations to isolate one variable. Many students tend to substitute numerical values into an equation as soon as possible. In many cases, that's a mistake. Although at first it may seem easier to manipulate numerical quantities than to manipulate algebraic symbols, there are several advantages to working with symbols:

- Symbolic algebra is much easier to follow than a series of numerical calculations. Plugging in numbers tends to obscure the logic behind your solution. If you need to trace back through your work (to find an error or review for an exam), it'll be much clearer if you have worked through the problem symbolically. It will also help your instructor when grading your homework papers or exams. When your work is clear, your instructor is better able to help you understand your mistakes. You may also get more partial credit on exams!

- Symbolic algebra lets you draw conclusions about how one quantity depends on another. For instance, working symbolically you might see that the horizontal range of a projectile is proportional to the *square* of the initial speed. If you had substituted the numerical value of the initial speed, you wouldn't notice that. In particular, when an algebraic symbol cancels out of the equation, you know that the answer doesn't depend on that quantity.

- On the most practical level, it's easy to make arithmetic or calculation errors. The later on in your solution that numbers are substituted, the fewer number of steps you have to check for such errors.

When solving equations that contain square roots, be careful not to assume that a square root is positive. The equation $x^2 = a$ has *two* solutions for x: $x = \pm\sqrt{a}$. (The symbol $\pm$ means *either + or −*.)

Solving Quadratic Equations

An equation is quadratic in x if it contains terms with no powers of x other than a squared term (x^2), a linear term (x^1), and a constant (x^0). Any quadratic equation can be put into the standard form:

$$ax^2 + bx + c = 0 \tag{A-7}$$

The quadratic formula gives the solutions to any quadratic equation written in standard form:

$$x = \frac{-b \pm \sqrt{b^2 - 4ac}}{2a} \qquad \text{(A-8)}$$

The symbol "±" (read "plus or minus") indicates that in general there are two solutions to a quadratic equation; that is, two values of x will satisfy the equation. One solution is found by taking the + sign and the other by taking the − sign in the quadratic formula. If $b^2 - 4ac = 0$, then there is only one solution (or, technically, the two solutions happen to be the same). If $b^2 - 4ac < 0$, then there is no solution to the equation (for x among the real numbers).

The quadratic formula still works if $b = 0$ and/or $c = 0$, although in such cases the equation can easily be solved without recourse to the quadratic formula.

Example

Solve the equation $5x(3 - x) = 6$.

Solution   First put the equation in standard quadratic form:

$$15x - 5x^2 = 6$$
$$-5x^2 + 15x - 6 = 0$$

We identify $a = -5$, $b = 15$, $c = -6$. Then

$$x = \frac{-b \pm \sqrt{b^2 - 4ac}}{2a}$$

$$= \frac{-15 \pm \sqrt{15^2 - 4 \times (-5) \times (-6)}}{-10}$$

$$\simeq \frac{-15 \pm 10.25}{-10} = 0.475 \text{ or } 2.525$$

Solving Simultaneous Equations

Simultaneous equations are a set of N equations with N unknown quantities. We wish to solve these equations *simultaneously* to find the values of all of the unknowns. We *must* have at least as many equations as unknowns. It pays to keep track of the number of unknown quantities and the number of equations in solving more challenging problems. If there are more unknowns than equations, then look for some other relationship between the quantities—perhaps some information given in the problem that has not been used.

One way to solve simultaneous equations is by *successive substitution*. Solve one of the equations for one unknown (in terms of the other unknowns). Substitute this expression into each of the other equations. That leaves $N - 1$ equations and $N - 1$ unknowns. Repeat until there is only one equation left with one unknown. Find the value of that unknown quantity, and then work backward to find all the others.

Example

Solve the equations $2x - 4y = 3$ and $x + 3y = -5$ for x and y.

Solution   First solve the second equation for x in terms of y:

$$x = -5 - 3y$$

Substitute $-5 - 3y$ for x in the first equation:

$$2 \times (-5 - 3y) - 4y = 3$$

This can be solved for y:

$$-10 - 10y = 3$$
$$-10y = 13$$
$$y = \frac{13}{-10} = -1.3$$

Now that y is known, use it to find x:

$$x = -5 - 3y = -5 - 3 \times (-1.3) = -1.1$$

It's a good idea to check the results by substituting into the original equations.

A.3   EXPONENTS AND LOGARITHMS

These identities show how to manipulate exponents.

$$a^{-x} = \frac{1}{a^x} \tag{A-9}$$

$$(a^x) \times (a^y) = a^{x+y} \tag{A-10}$$

$$\frac{a^x}{a^y} = (a^x) \times (a^{-y}) = a^{x-y} \tag{A-11}$$

$$(a^x) \times (b^x) = (ab)^x \tag{A-12}$$

$$(a^x)^y = a^{xy} \tag{A-13}$$

$$a^{1/n} = \sqrt[n]{a} \tag{A-14}$$

$$a^0 = 1 \quad \text{(for any } a \neq 0) \tag{A-15}$$

$$0^a = 0 \quad \text{(for any } a \neq 0) \tag{A-16}$$

A common mistake to avoid: $(a^x) \times (a^y) \neq a^{xy}$ [see Eq. (A-10)].

Logarithms

Taking a logarithm is the inverse of exponentiation:

$$x = \log_b y \quad \text{means that} \quad y = b^x \tag{A-17}$$

Thus, one undoes the other:

$$\log_b b^x = x \tag{A-18}$$

$$b^{\log_b x} = x \tag{A-19}$$

The two commonly used bases b are 10 (the *common* logarithm) and $e = 2.71828\ldots$ (the *natural* logarithm). The common logarithm is written "$\log_{10}$," or sometimes just "log" if base 10 is understood. The natural logarithm is usually written "ln" rather than "$\log_e$."

These identities are true for any base logarithm.

$$\log xy = \log x + \log y \tag{A-20}$$

$$\log \frac{x}{y} = \log x - \log y \tag{A-21}$$

$$\log x^a = a \log x \tag{A-22}$$

Here are some common mistakes to avoid:

$$\log (x + y) \neq \log x + \log y$$

$$\log (x + y) \neq \log x \times \log y$$

$$\log xy \neq \log x \times \log y$$

$$\log x^a \neq (\log x)^a$$

Semilog Graphs

A semilog graph uses a logarithmic scale on the vertical axis and a linear scale on the horizontal axis. Semilog graphs are useful when the data plotted is thought to be an exponential function. If

$$y = y_0 e^{ax}$$

then

$$\ln y = ax + \ln y_0$$

so a graph of ln y versus x will be a straight line with slope a and y-intercept ln y_0.

Rather than calculating ln y for each data point and plotting on regular graph paper, it is convenient to use special semilog paper. The vertical axis is marked so that the values of y can be plotted directly, but the markings are spaced proportional to the log of y. (If you are using a plotting calculator or a computer to make the graph, log scale should be chosen for one axis from the menu of options.) The slope a on a semilog graph is *not* $\Delta y/\Delta x$ since the logarithm is actually being plotted. The correct way to find the slope is

$$a = \frac{\Delta(\ln y)}{\Delta x} = \frac{\ln y_2 - \ln y_1}{x_2 - x_1}$$

Note that there cannot be a zero on a logarithmic scale.

The two graphs of Figs. A.2 and A.3 are linear and semilog plots, respectively, of the function $y = 3e^{-2x}$.

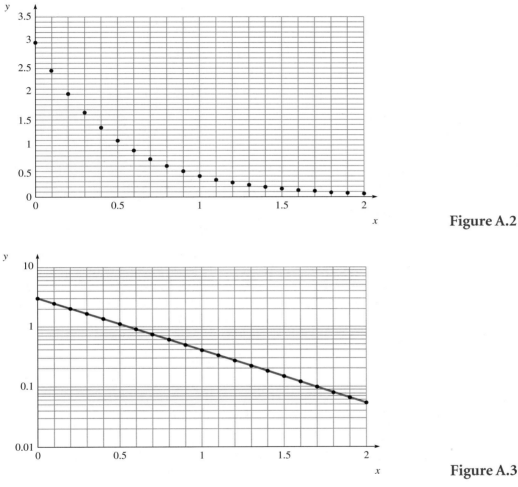

Figure A.2

Figure A.3

Log-Log Graphs

A log-log graph uses logarithmic scales for both axes. Log-log graphs are useful when the data plotted is thought to be a power function

$$y = Ax^n$$

For such a function,

$$\log y = n \log x + \log A$$

so a graph of $\log y$ versus $\log x$ will be a straight line with slope n and y-intercept $\log A$. The slope (n) on a log-log graph is found as

$$n = \frac{\Delta(\log y)}{\Delta(\log x)} = \frac{\log y_2 - \log y_1}{\log x_2 - \log x_1}$$

The graphs of Figs. A.4 and A.5 are linear and log-log plots, respectively, of the function $y = 130x^{3/2}$.

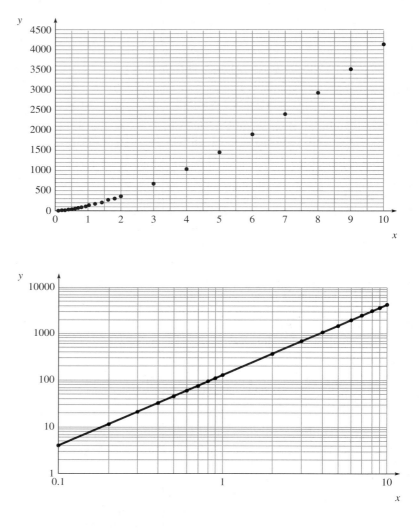

Figure A.4

Figure A.5

A.4   PROPORTIONS AND RATIOS

The notation

$$y \propto x$$

means that y is directly proportional to x. A proportionality can be written as an equation

$$y = kx$$

if the constant of proportionality k is written explicitly. Be careful: an equation can *look like* a proportionality without being one. For example, $V = IR$ means that $V \propto I$ *if and only if R is constant*. If R depends on I, then V is *not* proportional to I.

The notation

$$y \propto \frac{1}{x}$$

means that y is inversely proportional to x. The notation

$$y \propto x^n$$

means that y is proportional to the nth power of x.

Writing out proportions as ratios usually simplifies solutions when some common items in an equation are unknown but we do know the values of all but one of the proportional quantities. For example if $y \propto x^n$, we can write

$$\frac{y_1}{y_2} = \left(\frac{x_1}{x_2}\right)^n$$

Percentages

Percentages require careful attention. Look at these four examples:

"B is 30% of A" means $B = 0.30A$

"B is 30% larger than A" means $B = (1 + 0.30)A = 1.30A$

"B is 30% smaller than A" means $B = (1 - 0.30)A = 0.70A$

"A increases by 30%" means $\Delta A = +0.30A$

Example

If $P \propto T^4$, and T increases by 10.0%, by what percentage does P increase?

Solution

$$\Delta T = +0.100T_i$$
$$T_f = T_i + \Delta T = 1.100T_i$$
$$\frac{P_f}{P_i} = \left(\frac{T_f}{T_i}\right)^4 = 1.100^4 \approx 1.464$$

Therefore, P increases by about 46.4%.

A.5   APPROXIMATIONS

Binomial Approximations

A binomial is the sum of two terms. The general rule for the nth power of an algebraic sum is given by the binomial expansion:

$$(a + b)^n = a^n + na^{n-1}b + \frac{n(n-1)}{1 \times 2}\, a^{n-2}b^2 + \frac{n(n-1)(n-2)}{1 \times 2 \times 3}\, a^{n-3}b^3 + \cdots$$

The binomial approximations are used when a binomial in which one term is much smaller than the other is raised to a power n. Only the first two terms of the binomial expansion are of significant value; the other terms are dropped. A common case for physics problems is that in which $a = 1$, or can be made equal to one by factoring. The basic approximation forms are then given by

$$(1 + x)^n \approx 1 + nx \quad \text{when } x \ll 1 \tag{A-23}$$

$$(1 - x)^n \approx 1 - nx \quad \text{when } x \ll 1 \tag{A-24}$$

The power n can be any real number, including negative as well as positive numbers. It does not have to be an integer. An *estimate* of the error—the difference between the approximation and the exact expression—is given by

$$\text{error} \approx \tfrac{1}{2}n(n-1)x^2 \tag{A-25}$$

Of course, the larger term in a binomial is not necessarily 1, but the larger term can be factored out and then Eq. (A-23) or Eq. (A-24) applied. For instance, if $A \gg b$, then

$$(A + b)^n = \left[A \times \left(1 + \frac{b}{A}\right)\right]^n = A^n\left(1 + \frac{b}{A}\right)^n$$

Another common expansion is

$$e^x = 1 + x + \frac{x^2}{2!} + \frac{x^3}{3!} + \cdots$$

where, for any integer n, $n! = n \times (n-1) \times (n-2) \times \cdots \times [n - (n-1)]$; for example, $3! = 3 \times 2 \times 1 = 6$.

Small-Angle Approximations

Approximations for small angles appear in Section A.7 on trigonometry.

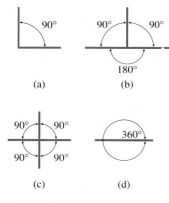

Figure A.6 (a) A right angle; (b) two adjacent right angles, or a straight line; (c) four adjacent right angles; (d) a full circle

A.6   GEOMETRY

Geometric Shapes

Table A.1 lists the geometric shapes that commonly appear in physics problems. It is often necessary to determine the area or volume of one of these simple shapes to complete the solution of a problem. The formulae for the properties associated with each geometric form are listed in the column to the right.

Angular Measure

An angle between two straight lines that meet at a single point is specified in degrees. If the two lines are perpendicular to each other, as shown in Fig. A.6a, the angular separation is said to be a right angle or 90°. In Fig. A.6b two such 90° angles are placed side by side; they add to 180°, so a straight line represents an angular separation of 180°. When four right angles are grouped as shown in Fig. A.6c, the angles add to 360° and a full circle contains 360° as shown in Fig. A.6d. An angle that is less than 90° is called an **acute** angle; one greater than 90° is called an **obtuse** angle.

Table A.1

Common Geometric Shapes Met in Physics Problems

Geometric Shape	Properties of Common Shapes	Geometric Shape	Properties of Common Shapes
Circle	Diameter $d = 2r$ Circumference $C = \pi d = 2\pi r$ Area $A = \pi r^2$	Sphere	Surface area $A = 4\pi r^2$ Volume $V = \frac{4}{3}\pi r^3$
Rectangle	Perimeter $P = 2b + 2h$ Area $A = bh$	Parallelepiped	Surface area $A = 2(ab + bc + ac)$ Volume $V = abc$
Right triangle	Perimeter $P = a + b + c$ Area $A = \frac{1}{2}\text{base} \times \text{height} = \frac{1}{2}ba$ Pythagorean theorem $c^2 = a^2 + b^2$ Hypotenuse $c = \sqrt{a^2 + b^2}$	Right circular cylinder	Surface area $A = 2\pi r^2 + 2\pi rh$ $= 2\pi r(r + h)$ Volume $V = \pi r^2 h$
Triangle	Area $A = \frac{1}{2}bh$		

When two lines meet, as shown in Fig. A.7, there are two possible angles that might be specified; one is the acute angle α and the other is the obtuse angle β. The symbol used to indicate an angle is $\angle$. When two angles placed adjacent to each other form a straight line, they are called supplementary angles; angles α and β in Fig. A.7 are supplementary angles. When two angles placed adjacent to each other form a right angle, they are called complementary angles.

Various triangles are shown in Fig. A.8. The sum of the interior angles of any triangle is 180°. An isosceles triangle has two sides of equal length; the angles opposite to the equal sides are equal angles. An equilateral triangle has all three sides of equal length; it is also equiangular. Right triangles have one right angle, 90°, and the sum of the other two angles is 90°, so those angles are acute angles. Commonly used right triangles have sides in the ratio of 3:4:5 and 5:12:13.

Triangles are similar when all three angles of one are equal to the three angles of the other. If two angles of one triangle are equal to two angles of the other, the third angles are necessarily equal and the triangles are similar. The ratio of corresponding sides of similar triangles are equal, as shown in Fig. A.9. Similar triangles of the same size are called congruent triangles.

Figure A.10 shows other useful relations among angles between intersecting lines. When two angles add to 180°, as $\angle\alpha + \angle\beta$ in three of the small figures, the angles are supplementary. Another small figure shows two angles, $\angle\alpha + \angle\beta$ adding to 90°, so in that case the angles are complementary.

For many physics problems it is convenient to use angles measured in radians rather than in degrees; the abbreviation for radians is rad. The arc length s measured along a circle is proportional to the angle between the two radii that define the arc, as shown in Fig. A.11. One radian is defined as the angle subtended when the arc length is equal to the length of the radius.

For θ measured in radians,

$$s = r\theta$$

When the angular displacement is all the way around the circle, 360°, the arc length is equal to the circumference of the circle.

$$s = 2\pi r = r\theta$$

The equivalent to 360° measured in radians is thus $\theta = 2\pi$ radians and the equivalence between radians and degrees is

$$1 \text{ rad} = \frac{360°}{2\pi} = 57.3° \quad \text{or} \quad 1° = \frac{2\pi}{360°} = 0.01745 \text{ rad}$$

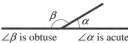

Figure A.7 Acute and obtuse angles

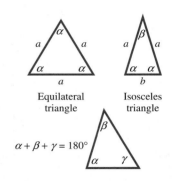

$$\alpha + \beta + \gamma = 180°$$

Figure A.8 Triangles

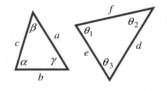

$$\alpha = \theta_1 \quad \beta = \theta_2 \quad \gamma = \theta_3$$
$$\frac{a}{d} = \frac{b}{e} = \frac{c}{f}$$

Figure A.9 Similar triangles

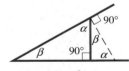
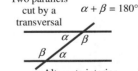

Figure A.10 Angles formed by intersecting lines

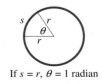

If $s = r$, $\theta = 1$ radian

Note that the radian has no physical dimensions; it is a ratio of two lengths so it is a pure number. We use the term *rad* to remind us of the angular units being used.

Figure A.11 Radian measure

A.7   TRIGONOMETRY

The basic trigonometric functions used in physics are shown in Fig. A.12. Note that in determining the function values, the units of length cancel, so the sine, cosine, and tangent functions are dimensionless.

Since the side opposite and the side adjacent to either of the angles in the right triangle are of lesser length than the hypotenuse according to the Pythagorean theorem, the values of the sine and cosine cannot exceed unity. The value of the tangent does exceed unity whenever the side opposite exceeds the side adjacent to the angle considered.

Figure A.13 shows the signs (positive or negative) associated with the trigonometric functions for an angle θ located in each of the four quadrants. The hypotenuse r is positive, so the sign for the sine or cosine is determined by the signs of x or y as measured along the positive or negative x- and y-axes. The sign of the tangent then depends on the signs of the sine and cosine. The angle θ is measured in a counterclockwise direction starting from the positive x-axis, which represents $0°$. Angles measured from the x-axis going in a clockwise direction (below the x-axis) are negative angles; an angle of $-60°$, which is located in the fourth quadrant, is the same as an angle of $+300°$. There are some useful trigonometric identities given in Table A.2.

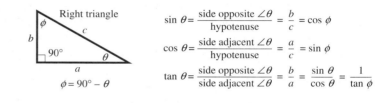

Figure A.12 Trigonometric functions used in physics problems; angles θ and ϕ are complementary angles.

$$\sin \theta = \frac{\text{side opposite } \angle\theta}{\text{hypotenuse}} = \frac{b}{c} = \cos \phi$$

$$\cos \theta = \frac{\text{side adjacent } \angle\theta}{\text{hypotenuse}} = \frac{a}{c} = \sin \phi$$

$$\tan \theta = \frac{\text{side opposite } \angle\theta}{\text{side adjacent } \angle\theta} = \frac{b}{a} = \frac{\sin \theta}{\cos \theta} = \frac{1}{\tan \phi}$$

$\phi = 90° - \theta$

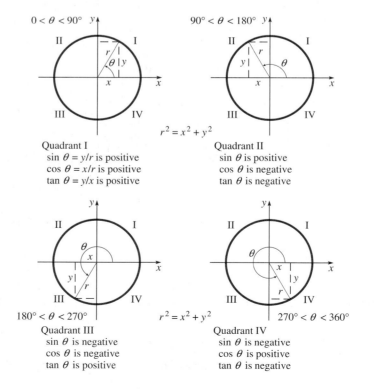

Figure A.13 Signs of trigonometric functions in various quadrants

Table A.2

Useful Trigonometric Identities

$\sin^2 \theta + \cos^2 \theta = 1$

$\sin 2\theta = 2 \sin \theta \cos \theta$

$\cos 2\theta = \cos^2 \theta - \sin^2 \theta = 2 \cos^2 \theta - 1 = 1 - 2 \sin^2 \theta$

$\tan 2\theta = \dfrac{2 \tan \theta}{1 - \tan^2 \theta}$

$\sin (\alpha \pm \beta) = \sin \alpha \cos \beta \pm \cos \alpha \sin \beta$

   For $\alpha = 90°$, $\sin (90° \pm \beta) = \cos \beta$

   For $\alpha = \beta$, $\sin 2\beta = 2 \sin \beta \cos \beta$

$\cos (\alpha \pm \beta) = \cos \alpha \cos \beta \mp \sin \alpha \sin \beta$

   For $\alpha = 90°$, $\cos (90° \pm \beta) = \mp \sin \beta$

   For $\alpha = \beta$, $\cos 2\beta = \cos^2 \beta - \sin^2 \beta = 1 - 2 \sin^2 \beta$

$\tan (\alpha \pm \beta) = \dfrac{\tan \alpha \pm \tan \beta}{1 \mp \tan \alpha \tan \beta}$

$\sin (-\theta) = -\sin \theta$

$\cos (-\theta) = \cos \theta$

$\tan (-\theta) = -\tan \theta$

$\sin \alpha + \sin \beta = 2 \sin \left[\frac{1}{2}(\alpha + \beta)\right] \cos \left[\frac{1}{2}(\alpha - \beta)\right]$

$\sin \alpha - \sin \beta = 2 \cos \left[\frac{1}{2}(\alpha + \beta)\right] \sin \left[\frac{1}{2}(\alpha - \beta)\right]$

Two laws that apply to any triangle labeled as shown in Fig. A.14:

Law of sines: $\dfrac{\sin \alpha}{a} = \dfrac{\sin \beta}{b} = \dfrac{\sin \gamma}{c}$

Law of cosines: $c^2 = a^2 + b^2 - 2ab \cos \gamma$ (where γ is the interior angle formed by the intersection of sides a and b)

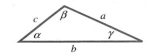

Figure A.14 A general triangle

Table A.3

Two Common Right Triangles

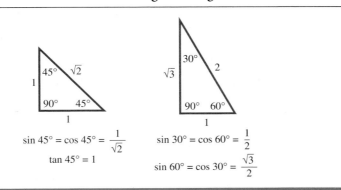

$\sin 45° = \cos 45° = \dfrac{1}{\sqrt{2}}$

$\tan 45° = 1$

$\sin 30° = \cos 60° = \dfrac{1}{2}$

$\sin 60° = \cos 30° = \dfrac{\sqrt{3}}{2}$

Small-Angle Approximations

These approximations are written for θ *in radians* and are valid when $\theta \ll 1$ rad.

$$\sin \theta \approx \theta \tag{A-26}$$

$$\cos \theta \approx 1 - \tfrac{1}{2}\theta^2 \tag{A-27}$$

$$\tan \theta \approx \theta \tag{A-28}$$

Figure A.15

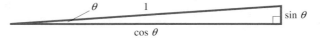

The sizes of the errors involved in using these approximations are roughly $\frac{1}{6}\theta^3$, $\frac{1}{24}\theta^4$, and $\frac{2}{3}\theta^3$ respectively. In *some* circumstances it may be all right to ignore the $\frac{1}{2}\theta^2$ term and write

$$\cos \theta \approx 1 \qquad\qquad (A\text{-}29)$$

The origin of these approximations can be illustrated using a right triangle of hypotenuse 1 with one very small angle θ (Fig. A.15). If θ is very small, then the adjacent side (cos θ) will be nearly the same length as the hypotenuse (1). Then we can think of those two sides as radii of a circle that subtend an angle θ. The relationship between the arc length s and the angle subtended is

$$s = \theta r$$

Since sin $\theta \approx s$ and $r = 1$, we have sin $\theta \approx \theta$. To find an approximate form for cos θ (but one more accurate than cos $\theta \approx 1$), we can use the Pythagorean theorem:

$$\sin^2 \theta + \cos^2 \theta = 1$$

$$\cos \theta = \sqrt{1 - \sin^2 \theta} \approx \sqrt{1 - \theta^2}$$

Now, using a binomial approximation,

$$\cos \theta \approx (1 - \theta^2)^{1/2} \approx 1 - \tfrac{1}{2}\theta^2$$

A.8   VECTORS

The distinction between vectors and scalars is discussed in Chapter 2. Scalars have magnitude while vectors have magnitude and direction. A vector is represented by an arrow of length proportional to the magnitude of the vector and aligned in a direction that corresponds to the vector direction.

In print, a vector is sometimes written in bold font, or in roman font with an arrow over it, or in bold font with an arrow over it (as done in this book). When writing by hand, a vector is designated by drawing an arrow over the symbol: $\vec{A}$. When we write just plain A, that stands for the *magnitude* of the vector. We also use absolute value bars to stand for the magnitude of a vector, so $A = |\vec{A}|$.

Addition and Subtraction of Vectors

When vectors are added or subtracted, the magnitudes and direction must be taken into account. When adding two collinear vectors $\vec{A}$ and $\vec{B}$, as shown in Fig.A.16a, the second vector ($\vec{B}$) is placed with its initial point, or origin, at the terminal point, or terminus, of the first vector ($\vec{A}$). The resultant vector ($\vec{C}$) is drawn from the origin of the first vector to the terminus of the second vector (tip to tail), as shown in Fig. A.16a. If collinear vectors are to be subtracted (Fig. A.16b), the procedure is the same, but the negative of one vector is used, which effectively reverses the direction of that vector. The resultant vector sum ($\vec{D}$) is again the magnitude of the vector formed by drawing an arrow from the tail of the first ($\vec{A}$) to the tip of the second ($\vec{B}$).

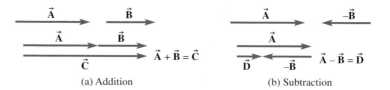

Figure A.16 Addition and subtraction of collinear vectors

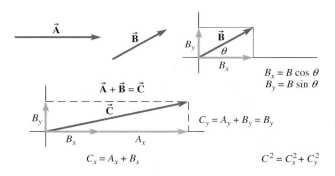

Figure A.17 Addition of two vectors by the component method

When the vectors are at an arbitrary angle with respect to each other, orthogonal components of each vector may be added or subtracted; the resultant is then found by applying the Pythagorean theorem to the newly formed orthogonal components.

In Fig. A.17, vector $\vec{A}$ is along the x-axis and vector $\vec{B}$ is at an angle θ with respect to the x-axis. Components of vector $\vec{B}$ are taken along the x- and y-axes. Then the x-components are added and separately the y-components are added. Since the y-component of vector $\vec{A}$ is zero, the y-component of the resultant vector $\vec{C}$ is the same as the y-component of vector $\vec{B}$. The sum of the x-components of vectors $\vec{A}$ and $\vec{B}$ is the x-component of the resultant vector $\vec{C}$.

Another method for adding vectors at an arbitrary angle involves placing the vectors tip to tail and then drawing from the tail of the first to the tip of the second, as shown in Fig. A.18. This is called the method of triangles, parallelogram method, or more simply, the geometric or graphical method.

Figure A.19 shows both methods of addition applied to two arbitrary vectors. One method requires graph paper and careful geometric construction of the vectors; the other method involves finding components on a set of common axes. The vectors specified in problems have their directions given with respect to a set of axes, so the component method is commonly used for finding the magnitude and direction of a resultant vector. The same result is found by either method. The direction of the resultant vector can be found from the resultant components.

$$\tan \phi = \frac{C_y}{C_x}$$

and

$$\phi = \tan^{-1} \frac{C_y}{C_x} = \arctan \frac{C_y}{C_x}$$

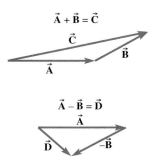

Figure A.18 Addition and subtraction of two vectors by the method of triangles

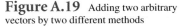

Figure A.19 Adding two arbitrary vectors by two different methods

Figure A.20 Multiplication of a vector by a scalar

Multiplication of Vectors: Product of a Vector and a Scalar

When a vector is multiplied by a scalar, the vector magnitude changes by the scalar factor, as shown in Fig. A.20. The direction of the vector does not change unless the scalar factor is negative, in which case the direction is reversed.

Cross Product of Two Vectors

One type of product between two vectors is the *cross product*, which is introduced in Chapter 19. It is denoted by

$$\vec{A} \times \vec{B} = \vec{C}$$

The product is a vector quantity; it has magnitude and direction. $\vec{A} \times \vec{B}$ is read as "$\vec{A}$ cross $\vec{B}$."

For two vectors, $\vec{A}$ and $\vec{B}$, separated by an angle θ (with θ chosen to be the *smaller* angle between the two), the magnitude of the cross product $\vec{C}$ is

$$|\vec{C}| = |\vec{A} \times \vec{B}| = AB \sin \theta$$

See Chapter 19 for the right-hand rule for finding the direction of a cross product.

The cross product depends on the order of the multiplication.

$$\vec{A} \times \vec{B} = -\vec{B} \times \vec{A}$$

The magnitude is $AB \sin \theta$ in both cases, but the direction of one cross product is opposite to the direction of the other.

APPENDIX B
Table of Selected Nuclides

Atomic Number Z	Element	Symbol	Mass Number A	Atomic Mass, *neutral atom* (u)	Percentage Abundance (or Decay Mode)	Half-life (if Unstable)
1	Hydrogen	H	1	1.007 825 0	99.989	
	(Deuterium)	(D)	2	2.014 101 8	0.0115	
	(Tritium)	(T)	3	3.016 049 3	β^-	12.33 yr
2	Helium	He	3	3.016 029 3	0.000 14	
			4	4.002 603 2	99.999 86	
3	Lithium	Li	6	6.015 122 3	7.6	
			7	7.016 004 0	92.4	
4	Beryllium	Be	7	7.016 929 2	EC	53.29 d
			8	8.005 305 1	2α	6.8×10^{-17} s
			9	9.012 182 1	100	
5	Boron	B	10	10.012 937 0	19.9	
			11	11.009 305 5	80.1	
6	Carbon	C	11	11.011 433 8	EC, β^+	20.39 min
			12	12.000 000 0	98.89	
			13	13.003 354 8	1.1	
			14	14.003 242 0	β^-	5730 yr
			15	15.010 599 3	β^-	2.449 s
7	Nitrogen	N	13	13.005 738 6	EC, β^+	9.965 min
			14	14.003 074 0	99.634	
			15	15.000 108 9	0.366	
8	Oxygen	O	15	15.003 065 4	EC, β^+	122.24 s
			16	15.994 914 6	99.757	
			17	16.999 131 5	0.038	
			18	17.999 160 4	0.205	
			19	19.003 578 7	β^-	26.91 s
9	Fluorine	F	19	18.998 403 2	100	
10	Neon	Ne	20	19.992 440 2	90.48	
			22	21.991 385 5	9.25	
11	Sodium	Na	22	21.994 436 8	EC, β^+	2.6019 yr
			23	22.989 769 7	100	
			24	23.990 963 3	β^-	14.9590 h
12	Magnesium	Mg	24	23.985 041 9	78.99	
15	Phosphorus	P	31	30.973 761 5	100	
			32	31.973 907 2	β^-	14.262 d
16	Sulfur	S	32	31.972 070 7	94.93	
19	Potassium	K	39	38.963 706 9	93.2581	
			40	39.963 998 7	0.0117; β^-	1.28×10^9 yr
20	Calcium	Ca	40	39.962 591 2	96.941	
27	Cobalt	Co	59	58.933 200 2	100	
			60	59.933 822 2	β^-	5.27 yr
36	Krypton	Kr	84	83.911 506 6	57.0	
			86	85.910 610 3	17.3	
			92	91.926 152 8	β^-	1.840 s

continued						

Atomic Number Z	Element	Symbol	Mass Number A	Atomic Mass, *neutral atom* (u)	Percentage Abundance (or Decay Mode)	Half-life (if Unstable)
37	Rubidium	Rb	85	84.911 789 3	72.165	
			93	92.922 032 8	β^-	5.84 s
38	Strontium	Sr	88	87.905 614 3	82.58	
			90	89.907 737 6	β^-	28.79 yr
39	Yttrium	Y	89	88.905 847 9	100	
			90	89.907 151 4	β^-	64.00 h
55	Cesium	Cs	133	132.905 446 9	100	
			141	140.920 044 0	β^-	24.94 s
56	Barium	Ba	138	137.905 241 3	71.70	
			141	140.914 406 4	β^-	18.27 min
82	Lead	Pb	206	205.974 449 0	24.1	
83	Bismuth	Bi	214	213.998 698 7	β^-	19.8 min
84	Polonium	Po	210	209.982 857 4	α	138.376 d
			218	218.008 965 8	α	3.10 min
86	Radon	Rn	222	222.017 570 5	α	3.8235 d
88	Radium	Ra	226	226.025 402 6	α	1600 yr
90	Thorium	Th	232	232.038 050 4	100; α	1.41×10^{10} yr
			234	234.043 595 5	β^-	24.10 d
92	Uranium	U	235	235.043 923 1	0.720; α	7.038×10^8 yr
			236	236.045 561 9	α	2.34×10^7 yr
			238	238.050 782 6	99.27; α	4.468×10^9 yr
			239	239.054 287 8	β^-	23.45 min

EC = electron capture.

Answers to Selected Questions and Problems

Chapter 1

Multiple Choice Questions
1. (b) 2. (a) 3. (b) 4. (c) 5. (d) 6. (b) 7. (d) 8. (b)
9. (d) 10. (c)

Problems
1. 7.7% 3. 4 5. 434 m/s 7. (a) 3; 5.74×10^{-3} kg (b) 1; 2 m
(c) 3; 4.50×10^{-3} m (d) 3; 4.50×10^{1} kg (e) 4; 1.009×10^{5} s
(f) 4; 9.500×10^{3} mL 9. (a) 1.29×10^{8} (b) 1.3×10^{8}
11. (a) -3.63×10^{7} (b) 2.268×10^{-178} 13. 0.278 m/s
15. (a) 220 markers (b) 221 markers 17. 0.14 W/cm^{2}
19. (a) 3.3×10^{-8} m (b) 3.3×10^{-2} μm (c) 1.3×10^{-6} in

21. 13.6 g/cm^{3} 23. 6.0×10^{-6} m^{3} 25. (a) $\sqrt{\dfrac{hG}{c^{5}}}$

(b) 1.3×10^{-43} s 27. (a) 8.8×10^{-2} m (b) 3.4 in

29. (a) $a = K\dfrac{v^{2}}{r}$, where K is a dimensionless constant.

(b) 21.0% 31. 0.46 s^{-1} 33. (a) 1.004 (1.00399)
(b) 1.02 (1.0198) (c) 1.2 (1.183) (d) 1.33 (1.288)
35. (a) 30° (b) 10° 37. 100 m 39. 10^{11} gal
41.

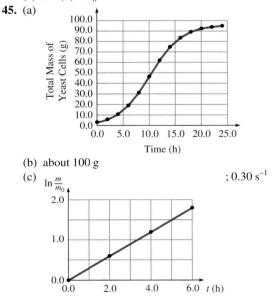

(a) 101.8°F (b) 0.9°F/h (c) no; the patient would die
before 12 hours passed and the temperature reached 113°F.
43. (a) a (b) $+v_0$
45. (a)

(b) about 100 g
(c) ; 0.30 s^{-1}

47. 56% 49. 42 km 51. $\dfrac{\text{kg} \cdot \text{m}}{\text{s}^{2}}$ 53. $90,000,000,000
55. (a) 4 (b) 21.6% 57. 104.5°F
59. (a) 2.4×10^{5} km/h (b) 10 min

Chapter 2

Multiple Choice Questions
1. (e) 2. (d) 3. (d) 4. (c) 5. (c) 6. (b) 7. (c) 8. (c)
9. (d) 10. (c)

Problems
1. 3200 N/cm 3. (a) 6.0×10^{10} N/m (b) 8.0 nm 5. 778 N
7. (a) the upward force on the Earth due to the skydiver; the
downward force on the parachute due to the skydiver; the
downward force on the air due to the skydiver
(b) the upward force on the skydiver due to the parachute; the
downward force on the air due to the parachute; the upward force
on the Earth due to the parachute (c) dynamic equilibrium
9. p = system of plant, soil, pot; c = cord; e = Earth; h = hook;
C = ceiling; s = system of plant, soil, pot, cord, hook

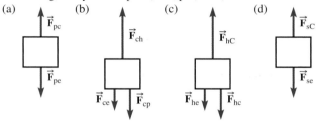

11. s = sailboat; e = Earth; w = wind;
l = lake; m = mooring line

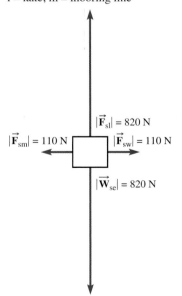

13. (a) 30 N to the right (b) 0 (c) 18 N downward

561

15. (a) 50.0 N upward (total for both feet)
(b) 650.0 N upward
(c) s = woman and chair system; e = Earth; f = floor

17. the upwardly directed drag due to the air; the downwardly directed force due to gravity
19. (a) 392 N   (b) 88.2 lb   **21.** (a) 1.9 N/cm   (b) 2.4 kg
23. (a) 1432 km   (b) 4664 km   **25.** (a) 5.6 N up   (b) 17.6 N
27. (a) 250 N   (b) 3100 N   (c) 250 N toward the satellite
29. (a) 1.98×10^{20} N   (b) the same
31. b = book; d = desk; e = Earth; h = hand

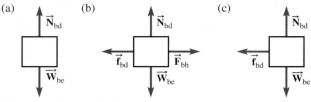

33. 12 N   **35.** (a) 0.41   (b) 40 N
37. (a) $\mu_s > 0.48$ (b) 0.60   (c) 0.48
39. t = table; e = Earth; 1 = block 1; 2 = block 2; 3 = block 3; 4 = block 4; h = horizontal force; 1234 = system of blocks (The blocks are numbered from left to right.)

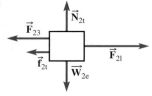

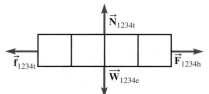

41. 1) the force of gravity; 2) the vertical force of the water opposing gravity and the forces of the water's currents (if any); 3) the force of the wind; 4) the force of the line tied to the mooring
43. Scale A reads 120 N; scale B reads 240 N.

45. Scale B reads 120 N; scale A reads 120 N.   **47.** (a) $\dfrac{k}{2}$ (b) $2k$

49. i = ice; e = Earth; s = stone; o = opponent's stone

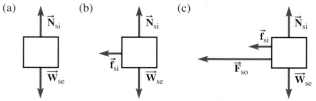

51. (a) 1) the gravitational forces between the magnet and the Earth; 2) the contact forces, normal and frictional, between the magnet and the photo; 3) the magnetic forces between the magnet and the refrigerator

(b) m = magnet; p = photo; e = Earth; r = refrigerator

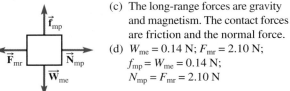

(c) The long-range forces are gravity and magnetism. The contact forces are friction and the normal force.
(d) $W_{me} = 0.14$ N; $F_{mr} = 2.10$ N;
$f_{mp} = W_{me} = 0.14$ N;
$N_{mp} = F_{mr} = 2.10$ N

53. (a) In both cases the two ropes pull on the scale with forces of 550 N in opposite directions, so the scales give the same reading.
(b) 550 N   **55.** (a) 6×10^{-6} N   (b) $F \propto r$ (c) The surface tension would only double, since the feet would only be twice as wide, yet they would have to support eight times the weight.
57. (a) 15 N   (b) 20 N
59.

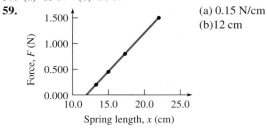

(a) 0.15 N/cm
(b) 12 cm

61. $\dfrac{k_1 k_2}{k_1 + k_2}$ **63.** 0.027 m/s^2 **65.** 98 N

Chapter 3

Multiple Choice Questions

1. (d)   **2.** (c)   **3.** (a)   **4.** (b)   **5.** (a)   **6.** (d)   **7.** (a)   **8.** (b)
9. (a)   **10.** (a)   **11.** (c)   **12.** (a)   **13.** (a)   **14.** (d)   **15.** (c)
16. (a)   **17.** (b)   **18.** (a)   **19.** (d)   **20.** (c)   **21.** (b)

Problems

1. 32 s   **3.** 8 m   **5.** 160 m   **7.** (a) 1.5 m/s   (b) 1.2 m/s
9. (a) 170 cm to the left   (b) 28 cm/s   (c) 9.4 cm/s to the left
11. 27 m/s west   **13.** 13 s   **15.** (a) -10 m/s^2 (b) 0   (c) 5.0 m
17. 2.5 m/s^2 **19.** 1.0×10^5 N in the direction of motion   **21.** 3.5 N
23. 17 kN   **25.** (a) 3.0 kN upward   (b) 3.3 m/s^2 downward
27. 23 N downward   **29.** (a) 86.4 m   (b) 14.4 m/s
31. (a) 4.0 m/s   (b) 14.0 m/s   **33.** 80 m   **35.** 1.5 m/s^2 northeast
39. No; it takes 236 m for the train to stop.   **41.** 85 m/s down
43. 5.0 m/s   **45.** 30.0 m/s   **47.** 13 m   **49.** (a) 5 m   (b) 1 km
51. (a) 72 N up   (b) 41 N up   (c) 0   (d) 4.3 m/s^2 down
53. (a) 567 N   (b) 629 N   **55.** 620 N   **57.** 4×10^{10} m
59. (a) 330 m/s up   (b) 16 m/s^2 up   **61.** $2v_0$ **63.** (a) 0.14 mi
(b) $2.2 \dfrac{\text{mi/h}}{\text{s}} = 0.98 \text{ m/s}^2$ in the direction of motion
65. (a) 46 m   (b) 30.0 m/s^2 up   **67.** 0.81 s   **69.** 1.8 m/s^2 down
71. (a) t_3 and t_4 (b) $t_0, t_2, t_5,$ and t_7 (c) t_1 and t_6
(d) $t_0, t_3,$ and t_7 (e) t_6 **73.** 68 s   **75.** (a) 25.0 km   (b) 150 s
(c) 76 km   (d) 1200 m/s downward   **77.** (a) 1.10mg
(b) 1.10mg **79.** $\Delta x = 160$ m   **81.** 3260 ft; 25.5 s
83. (a) 1 mm/s   (b) 20 ms   (c) 100 m/s

Chapter 4

Multiple Choice Questions

1. (c)   **2.** (a)   **3.** (d)   **4.** (d)   **5.** (a)   **6.** (a)   **7.** (a)   **8.** (c)
9. (d)   **10.** (e)

Problems

1. (a)

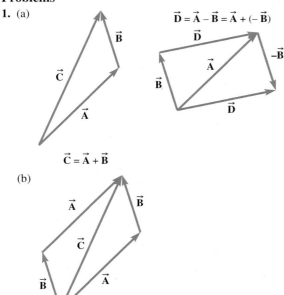

$$\vec{C} = \vec{A} + \vec{B}$$

(b)

$$\vec{C} = \vec{A} + \vec{B} = \vec{B} + \vec{A}$$

3. $\Delta\vec{r} = \vec{r}_n - \vec{r}_0$ **5.** x-comp = -17.3 m; y-comp = 10.0 m

7. (a) 2.0 units at 30° CCW from the $+y$-axis

(b) 2.0 units at 30° CW from the $+y$-axis

(c) x-comp = -1.0 unit; y-comp = $-\sqrt{3.0}$ units

9. 120 N **11.** (a) 9.4 cm, 32° CCW from the $+y$-axis

(b) 130 N, 27° CW from the $+x$-axis

(c) 16.3 m/s, 33° CCW from the $-x$-axis

(d) 2.3 m/s², 1.6° CCW from the $+x$-axis

13. 2.0 km, at 70° south of east **15.** 3.15 nm; 0.85 nm at 18° CCW from the horizontal **19.** 1810 N; 5 times the force with which Yoojin pulls; from the oak tree **21.** (a) 34 N (b) 39 N

23. $\dfrac{mg}{\cos\theta}$

25. (a) 75.2 N (b) 27.4 N (c) 0.364 (d) 80.0 N upward

27. (a) $W \tan 12°$ (b) $\dfrac{W}{\cos 12°}$ **29.** 23 km/h at 37° north of east

31. (a) 9.4 m/s at 45° north of east (b) 15 m/s² at 45° south of east

(c) Changing the direction of the velocity requires an acceleration.

33. (a)

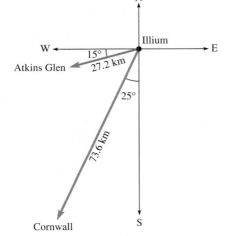

(b) 59.9 km at 85° north of east (c) 80 km/h at 85° north of east

35. (a)

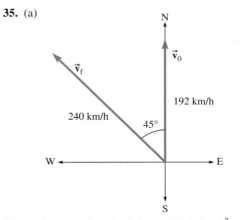

(b) 170 km/h at 7° south of west (c) 57 km/h² at 7° south of west

37. (a) $R = \dfrac{2v_0^2 \sin\theta \cos\theta}{g}$ (b) 221 m (c) 4 m **41.** (a) 2.0 s

(b) 2.0 s (c) 1.5 s **43.** (a) 3.50 s (b) 2.01 s (c) 80.8 m

45. (a)

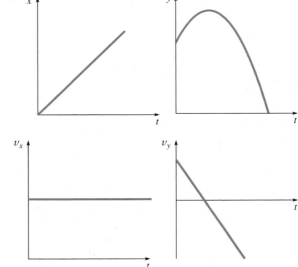

(b) 27.6 m/s at 25.0° above the horizontal (c) 37 m (d) 44 m above the ground **49.** 0.8 s **51.** (a) 69 N directed down the incline (b) 6.9 m/s² directed down the incline (c) 1.4 s

53. (a) $mg \tan\theta$ (b) $mg \tan\theta$ (c) $mg \tan\theta + \dfrac{ma}{\cos\theta}$

55. (a) 6.00×10^2 N directed along the 8.00×10^2-N vector

(b) 0.0414 m/s² in the same direction as the force **57.** (a) 88 N

(b) 2 s (c) 70 N (d) 10 kg **59.** 130 km/h north

61. 0.42 km/h **63.** 27° upstream **65.** (a) 76.37° N of E

(b) 2.717 h **67.** 39 s **69.** (a) 30.0° N of W (b) 9.1 min

71. 13 nautical miles at 23° south of east **73.** (a) 375.0 m

(b) 257.4 m at 10.5° N of W

75.   ; 400 N

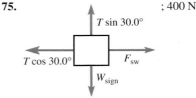

77. (a) 7.67 m (b) 13.9 m/s (c) 12.5 m/s **79.** 23 m

81. (a)

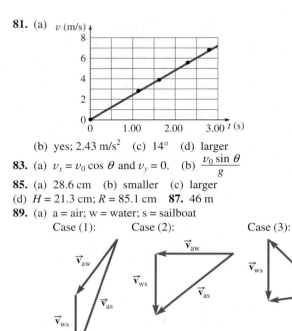

(b) yes; 2.43 m/s²   (c) 14°   (d) larger

83. (a) $v_x = v_0 \cos \theta$ and $v_y = 0$.   (b) $\dfrac{v_0 \sin \theta}{g}$

85. (a) 28.6 cm   (b) smaller   (c) larger
(d) $H = 21.3$ cm; $R = 85.1$ cm   **87.** 46 m

89. (a) a = air; w = water; s = sailboat

Case (1):   Case (2):   Case (3):

(b) 1 and 2   (c) all three

91. (a) 98 N   (b) 60° above the horizontal

93. (a) in front of   (b) 21.4 m   (c) 3.6 m   (d) earlier

Chapter 5

Multiple Choice Questions

1. (b)   **2.** (a)   **3.** (f)   **4.** (b)   **5.** (a)   **6.** (b)   **7.** (b)   **8.** (a)
9. (e)   **10.** (c)

Problems

1. 17 m   **3.** 0.105 rad/s   **5.** 26 rad/s   **7.** (a) 3.49 rad/s
(b) 0.45 m/s   **9.** 3800 ft   **11.** (a) 31 m/s   (b) 31 rad/s
13. 3.37 cm/s²   **15.** (a) 1.72×10^{-2} rad
(b) 514 m/s perpendicular to the average velocity
(c) 0.00595 m/s² perpendicular to the average velocity
(d) $\vec{a}_c = 0.00595$ m/s² perpendicular to the velocity, which is the
same as $\vec{a}_{av}$ within 3 significant figures.

17. (a) $\dfrac{mg}{\cos \phi}$ (b) $2\pi \sqrt{\dfrac{L \cos \phi}{g}}$ **19.** 7.9 m/s   **21.** 59°

23. (a) 2300 N   (b) 19 m/s   **25.** $\tan^{-1} \dfrac{v^2}{rg}$ **27.** 130 h

29. 3.6×10^{22} N   **31.** 16 h   **33.** (a) 13 N
(b) The bob has an upward acceleration, so the net F_y must be
upward and greater than the weight of the bob.   **35.** $g \sin \theta$
37. 4.0 rad/s²   **39.** (a) 1.7 rad/s²   (b) 0.56 rev
41. (a) 1.0×10^5 m/s   (b) 0.080 m/s²   (c) 5.0×10^{10} m/s²
43. 0.045 Hz   **45.** (a) 518.5 N   (b) 521.5 N   (c) 45 m
47. 0.0257 m/s²   **49.** 464 m/s   **51.** 1.80×10^6 degrees
53. (a) 7.3×10^{-5} rad/s   (b) 0.02 rad (arm-length ≈ 1 m and
finger-width ≈ 2 cm)   (c) 5 min   **55.** 200 km/s
57. smallest; 4.1 s   **59.** 0.40ω **61.** (a) 0.60 m/s
(b) The dolls do not stay on the record.   **63.** 1.0 rad/s

65. 110 μm/s   **67.** (a) 90°   (b) $T = \dfrac{2\pi m}{k}$ (c) $r = \dfrac{m}{k}v$

Chapter 6

Multiple Choice Questions

1. (e)   **2.** (b)   **3.** (b)   **4.** (a)   **5.** (c)   **6.** (c)   **7.** (c)   **8.** (c)
9. (b)   **10.** (b)   **11.** (f)

Problems

1. 75 J   **3.** No work is done.   **5.** 210 kJ   **7.** 720 kJ
9. (a) 0.70 J   (b) 0.37 m/s   **11.** 0   **13.** 5.8 MJ (meteor);
0.46 MJ (car); the meteor has more than 12 times the kinetic energy
of the car.   **15.** 1.6 J   **17.** 0   **19.** 5.2 J   **21.** (a) 0
(b) 3.4 kJ   (c) dissipated as heat   **23.** (a) 0   (b) 2.9 J
25. 2   **27.** 53 kJ   **29.** $v_1 = 25$ m/s; $v_2 = 18$ m/s; $v_3 = 21$ m/s
31. (a) −1.8 kJ   (b) 7.0 m/s   (c) 1100 N   (d) 5.7 m/s
33. 1.9 m   **35.** (a) $\sqrt{v^2 + 2gh}$ (b) The final speed is
independent of the angle.   **37.** 22.4 km/s   **39.** 0.33 m

41. (a) $\sqrt{5g(L-d)}$ (b) $\cos^{-1}\left(\dfrac{5d}{2L} - \dfrac{3}{2}\right)$ **43.** 13 m

45. (a) $d\sqrt{\dfrac{k}{m}}$ (b) d **47.** 2.5 kJ   **51.** 150 W   **53.** 60 kW

55. 6.2 g; the other 90% of the energy is lost as heat.
57. (a) 20 N   (b) 6.7 m/s   **59.** (a) 10 kW   (b) 5.8°
61. −52 kJ   **63.** (a) −500 J   (b) 3 GW
(c) 300,000 households   **65.** 60.0 km/s   **67.** 11.2 km/s
69. 43.5 km/s   **71.** (a) 2.62 kW   (b) 7.86 kW   **73.** 6.1 m
75. (a) 2200 kcal/day   (b) more than 0.51 lb   **77.** 27 N
79. 5×10^{24} J (equivalent to 50 million thermonuclear bombs)
81. 1.3 cm; 32 J   **83.** (a) 500 m³   (b) 600 kg   (c) 30 kJ
(d) 10 kW   (e) It decreases to $\frac{1}{8}$ of its initial value; the power
production of wind turbines is inconsistent, since modest changes
in wind speed produce large changes in power output.
85. approximately 3%–5%

Chapter 7

Multiple Choice Questions

1. (c)   **2.** (d)   **3.** (c)   **4.** (b)   **5.** (d)   **6.** (b)   **7.** (f)   **8.** (d)
9. (a)   **10.** (e)   **11.** (d)   **12.** (b)

Problems

1. 2.0 kg·m/s to the right   **3.** (a) 11 m/s   (b) 1300 N
5. 3 kg·m/s north   **7.** 20 kg·m/s in the −x-direction
9. 1.0×10^2 kg·m/s downward   **11.** 320 s   **13.** 6.0×10^3 N
opposite the car's direction of motion   **15.** (a) 750 kg·m/s upward
(b) 990 N·s downward   (c) 2500 N downward   **17.** 1.8 m/s
19. 2.6×10^5 m/s   **21.** 1500 kg   **23.** (4.2 cm, 0)
25. (1.9 m, 1.4 m)   **27.** (6 m/s, −4 m/s)   **31.** 3.0 m/s east
33. 0.20 kg   **35.** 43 m/s   **37.** 4.8 m/s   **39.** 0.066 m/s
45. (a) $\Delta p_{1x} = -1.00 m_1 v_0$; $\Delta p_{1y} = 0.751 m_1 v_0$
(b) $\Delta p_{2x} = m_1 v_0$; $\Delta p_{2y} = -0.751 m_1 v_0$; the momentum changes
for each mass are equal and opposite.   **47.** $1.73 v_{1f}$

49. $\left(\dfrac{v_0}{2}, -\dfrac{v_0}{2\sqrt{3}}\right)$ **51.** 5.0×10^9 kg·m/s   **53.** 34 N   **55.** 410 N

57. (a) 0.01 kg·km/h opposite the car's motion   (b) 0.01 kg·km/h
along the car's velocity   (c) 10^5 flies   **59.** 2.8 m/s
61. 0.83 m/s   **63.** (a) 2.5 m   (b) 4.0 m   **65.** 10^{-18} N

67. $\dfrac{1}{9}h$ **69.** 10 m/s   **71.** (a) $\dfrac{111}{2}$ (b) 1   (c) $\dfrac{111}{2}$

(c) 6.8 m/s^2 downward   **69.** (a) 10.0 cm   (b) 0.814   (c) 0.545
71. 0.4 mm/s   **73.** 270 m/s   **75.** 76 Pa   **77.** (a) 2.2 m/s up
(b) 21 kPa/s   **79.** (a) force is weight of column above the area,
definition of pressure   (b) 8.0 km   (c) lower limit
81. (a) 10 m   (b) The pump pushes the water up from below.

Chapter 10

Multiple Choice Questions

1. (c)   **2.** (b)   **3.** (b)   **4.** (a)   **5.** (a)   **6.** (a)   **7.** (c)   **8.** (b)
9. (c)   **10.** (a)   **11.** (f)   **12.** (e)   **13.** (e)   **14.** (f)   **15.** (f)
16. (e)   **17.** (c)   **18.** (c)   **19.** (j)   **20.** (k)

Problems

1. 0.1 mm   **3.** 2.2 cm   **5.** (a) 1.2×10^{-4} W   (b) 5×10^{-6} W; no
(c) 3.7×10^{-7} J   (d) 3.7×10^{-4} W; yes   **7.** 0.8 mm
9. tension: 1.5×10^{10} N/m^2; compression: 9.0×10^9 N/m^2
11. 8.7×10^{-5} m   **13.** 630 N   **15.** (a) 2.8×10^7 Pa
(b) 4.7×10^{-4} (c) 9.3×10^{-4} m   (d) 5.0×10^5 N
17. human: 3 cm^2; horse: 7 cm^2 **19.** volume: 7.7×10^{-4};
radius: 2.6×10^{-4} **21.** 7.5×10^5 N   **23.** 0.30 N   **25.** 0.63 m/s
27. 7.0 cm/s   **31.** 5.0 rad/s   **35.** 0.157 m/s; 24.7 m/s^2
37. (a) g (b) 0.78 m   **39.** −0.031 J   **41.** 0.250 Hz
43. (a)

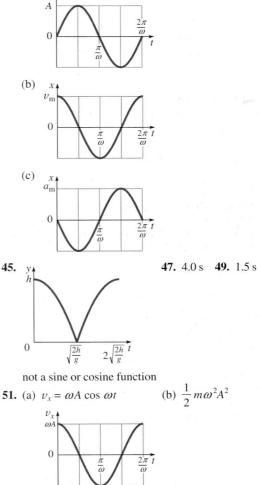

(b)

(c)

45.

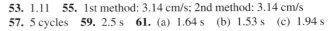

not a sine or cosine function

47. 4.0 s   **49.** 1.5 s

51. (a) $v_x = \omega A \cos \omega t$ (b) $\frac{1}{2} m\omega^2 A^2$

53. 1.11   **55.** 1st method: 3.14 cm/s; 2nd method: 3.14 cm/s
57. 5 cycles   **59.** 2.5 s   **61.** (a) 1.64 s   (b) 1.53 s   (c) 1.94 s

---

) **7.** (a)   **8.** (f)

) J   **11.** 4.5 N·m
. $(0.42s, 0.58s)$
25. (a) 540 N
: wall
or $\theta = 90°$, $T \rightarrow 0$.
much for a human,
0.0012 N·m
. 0.09 N·m

49. solid sphere: $K = \frac{7}{10}mv^2$; solid cylinder: $K = \frac{3}{4}mv^2$; hollow
cylinder: $K = mv^2$ **51.** (a) The drilled cylinder takes more time
because its rotational inertia is larger.   (b) 4% longer   **53.** $3r$
55. 0.0864 kg·m^2/s   **57.** 1.4×10^7 kg·m^2/s   **59.** 1.60 s
61. 1.5 rev/s   **63.** 3.15 rad/s   **65.** 2.10×10^6 N·m   **67.** 1.14
69. (a) 16 kg·m^2 (b) 8.0×10^7 J   (c) 320   (d) 120 km
73. 98 N·m   **75.** 5.4 rad/s   **77.** (a) 98 N   (b) This does not
help the person trying to lift the ladder, since the torque problem
is not alleviated by exerting a force at the point of rotation.

79. $\dfrac{m_2 g}{m_1 + m_2 + \dfrac{I}{R^2}}$ **81.** (a) 6.28 rad/s   (b) 0.955 kg·m^2/s

(c) friction   (d) 0.300 N   **83.** (b) $mr^2\omega$ (c) $\frac{1}{2}r^2\omega\Delta t$

85. (a) $\dfrac{I_0\omega_0}{I_0 + mR^2}$ (b) before: $K_{rot} = \frac{1}{2}I_0\omega_0^2$ and $L = I_0\omega_0$;

after: $K_{rot} = \frac{1}{2}\dfrac{I_0^2\omega_0^2}{I_0 + mR^2}$ and $L = I_0\omega_0$ **87.** (a) 735.0 N

(b) 0.88 m   (c) 0.55h **89.** 23 N   **91.** $T_1 = 67$ kN;

$T_2 = 250$ kN; $\vec{F}_p = 380$ kN at 51° with the horizontal
93. (a) 1.3 rev/s   (b) stays the same; no net torque (The falling
dumbbells are no longer part of the system.)
95. (a) 9.6 m/s   (b) 3.1 m/s   (c) 21 m/s

Chapter 9

Multiple Choice Questions

1. (b)   **2.** (b)   **3.** (d)   **4.** (a)   **5.** (a)   **6.** (b)   **7.** (a)   **8.** (a)
9. (d)   **10.** (c)

Problems

1. 50 atm   **3.** 22 kPa   **5.** 4.0 kN southward   **7.** (a) 625 N
(b) 6.25 mm   (c) 16.0   **9.** (a) 30 N   (b) 5.8 N·m   **11.** 2 atm
13. (a) 343 kPa   (b) 410 Pa   **15.** 2.9 N   **17.** 4.65×10^{-5} m^3;
$\dfrac{V_{Pt}}{V_{Al}} = 0.126$ **19.** (a) 2.2×10^5 Pa   (b) 1700 torr   (c) 2.2 atm
21. 114.0 cm Hg   **23.** (a) 5.6 cm   (b) 0.37 cm   **25.** 250 kg/m^3
27. (a) 8.8 N upward   (b) 9.6 N upward   **29.** 0.80g downward
31. (a) 0.910   (b) 1.28 cm   (c) 0.13 cm   **33.** 0.17 cm^3
35. 50 m/s   **37.** (a) 39.1 cm/s   (b) 78.5 cm^3/s   (c) 78.5 g/s
39. 1.9×10^5 N   **41.** (a) 78 W   (b) 390 kPa   (c) at the bottom
45. (a) 6850 Pa   (b) 0.685 N   **47.** 0.040 m^3/s   **49.** (a) 50 Pa
(b) 1100 Pa   (c) approximately 10 kPa   **51.** (a) 1.3×10^{-10} N
(b) 2.6×10^{-14} W   **53.** 0.4 Pa·s   **55.** 2.9 cm/s
57. Since m/v_t is constant, the drag force is primarily viscous.
59. 5 Pa   **61.** (a) $\gamma L\Delta s$ (b) $\Delta E = \gamma\Delta A$ **63.** (a) 1.1×10^8 Pa
(b) 1.1×10^8 N   **65.** 20%   **67.** (a) 1.4 N   (b) 0.43 N upward

63. 91 Hz   **65.** (a) $2\pi\sqrt{\dfrac{L\left(\dfrac{m_1}{3}+m_2\right)}{g\left(\dfrac{m_1}{2}+m_2\right)}}=2\pi\sqrt{\dfrac{2L(m_1+3m_2)}{3g(m_1+2m_2)}}$

(b) For $m_1 \gg m_2$, $T=2\pi\sqrt{\dfrac{2L}{3g}}$, and for $m_1 \ll m_2$, $T=2\pi\sqrt{\dfrac{L}{g}}$.

67. 2.1 m/s; 370 m/s^2 **69.** (b) $k=\dfrac{mg}{L}$

71. $y=(1.6\ \text{cm})\cos[(25\ \text{rad/s})t]$ **73.** 2.0°   **75.** $2\pi\sqrt{\dfrac{R}{g}}$

77. (a) 8×10^{-4} (b) 8.0 kN   (c) $5\times10^{-5}\ \text{m}^2$ (d) no

79. (a) 3.6 cm^2 (b) 7.8×10^6 Pa   (c) 3.3×10^{-4} m

81. (a) 5.1 N   (b) 7.7×10^{-2} J

Chapter 11

Multiple Choice Questions

1. (b)   **2.** (c)   **3.** (d)   **4.** (f)   **5.** (a)   **6.** (b)   **7.** (d)   **8.** (a)
9. (b)   **10.** (d)

Problems

1. 52 W/m^2 **3.** 1.7 MW   **5.** (a) 1.5 m/s   (b) 21 cm/s
7. 168 m/s   **9.** 16 ms   **11.** 0.375 m
13. (a) 340 Hz   (b) 3.0×10^8 Hz   **15.** 0.33 Hz
17. (a) 3.5 cm   (b) 6.0 cm   **19.** (a) 4.0 mm   (b) 1.0 m
(c) 0.010 s   (d) 100 m/s   (e) in the $+x$-direction (to the right)
21. (a)

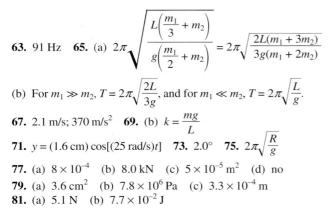

$y_{\max}=4.0$ cm; $\lambda=0.020$ cm; $v=1.2$ cm/s

(b)

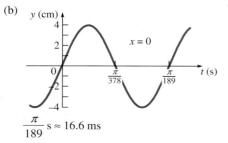

$\dfrac{\pi}{189}\ \text{s}\approx16.6\ \text{ms}$

23. $v_{\text{m}}=0.063$ m/s; $a_{\text{m}}=0.79$ m/s^2

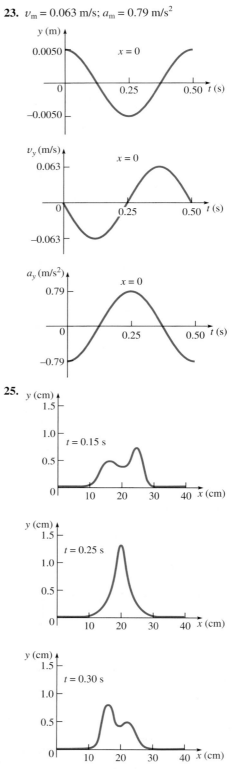

25.

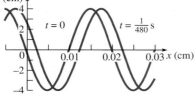

27. 96.0°

29. (a) 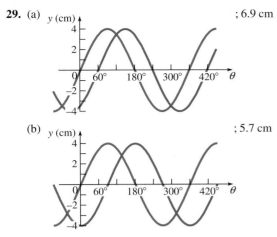 ; 6.9 cm

(b) ; 5.7 cm

31. 4.8 m/s **33.** 1.7 s **35.** (a) 0°; 9.0 cm (b) 180°; 3.0 cm
(c) 9:1 **37.** 80 μW/m² **39.** 7.8%
43. (a) 1350 m/s (b) 45.6 N (c) 0.76 m and 450.0 Hz
45. 4.5×10^{-4} kg/m **47.** 190 m **49.** 3 m
51. (a) $y(x, t) = (0.020\text{ m}) \sin[(1.6\text{ rad/s})\,t + (0.0016\text{ rad/m})x]$
(b) 0.031 m/s (c) 1.0 km/s **53.** $v \propto \sqrt{\lambda g}$ **55.** 80 km
57. 3.64 cm; 7.07 cm; 10.32 cm
59. 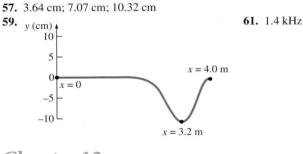   **61.** 1.4 kHz

Chapter 12

Multiple Choice Questions
1. (c) **2.** (a) **3.** (b) **4.** (c) **5.** (b) **6.** (c) **7.** (b) **8.** (c)
9. (b) **10.** (d)

Problems
1. 3.4 mm **3.** 173 ms **7.** 3.5 km/s **9.** (a) 338 m/s
(b) 2.8 km **11.** 0.099 W **13.** 26% **15.** (a) 125.0%
(b) 3.522 dB **17.** (a) 65.6 cm (b) 252.4 Hz **19.** 43.3 cm
21. (a) There is a displacement node (pressure antinode) at the
center of the rod and displacement antinodes (pressure nodes) at
the ends. (b) 5100 m/s (c) 13.1 cm (d) The ends move in
opposite directions, and thus, they are out of phase.
23. (a) 1.20 m (b) 90.0 cm (c) 338 m/s **25.** (a) 290.0 Hz
(b) 1.4% **27.** 2 Hz **29.** (a) 1.5 kHz (b) 500 Hz
33. 49 m/s **35.** (a) Mach 2.67 (b) 860 m/s **37.** 5.42 km
39. 34 cm **41.** (a) $\dfrac{2fv}{c - v}$ (b) $f_{\text{beat}} = \dfrac{2fv}{c}$
43. 17.9 Hz, 53.6 Hz, 89.3 Hz, 125 Hz
45. (a) 10 m (b) 2 ms (c) No, the bat will still be emitting
the first chirp. **47.** 3 kHz **49.** 15 m **51.** 0.0955 s
53. 534 m/s **55.** (a) 138 dB (b) 88.0 dB

Chapter 13

Multiple Choice Questions
1. (e) **2.** (d) **3.** (b) **4.** (c) **5.** (b) **6.** (d) **7.** (a) **8.** (c)
9. (c) **10.** (e)

Problems
1. (a) 29°C (b) 302 K **3.** (a) –6.0 K (b) –11°F
5. (a) 3.6 mm (b) 10.8 mm **7.** 3.8×10^{-4} mm²
9. (b) 2.4×10^{-3} **11.** (a) 44°C (b) –21°C **13.** 1.3 m
17. 150°C **21.** 7.31×10^{-26} kg **23.** 1.7×10^{27}
25. 10^{18} atoms **27.** 2.65×10^{25} atoms **29.** 8.9985 mol
31. 2.5×10^{19} molecules **35.** 0.0136 **37.** 1.55 **39.** 3.1 cm³
41. 3.3×10^{5} Pa **43.** (a) 90 ft³ (b) 5.3 h **45.** 4×10^{-17} Pa
47. 0.38 m³ **49.** 1550 K **51.** (a) 1.52×10^{5} J/m³
(b) 4.559×10^{7} J/m³ **53.** 50% **55.** $\dfrac{1}{\sqrt{2}}$
57. (a) 493 m/s (b) 461 m/s (c) 393 m/s **59.** yes
61. 1550 K **65.** 0.14°C **67.** (a) 100 nm (b) 200 nm
(c) 1 m **69.** 2.5×10^{4} s **71.** 4.8×10^{7} N/m² **73.** 3.05 mm
75. (b) 0.7 m **77.** (a) 52 cm (b) 12 m **81.** HNO₃
83. (a) 6.42×10^{-21} J (b) 0.25% **85.** 1.9×10^{14} molecules
87. 25 m/s

Chapter 14

Multiple Choice Questions
1. (a) **2.** (b) **3.** (d) **4.** (d) **5.** (c) **6.** (b) **7.** (c) **8.** (d)
9. (d) **10.** (b) **11.** (c) **12.** (c)

Problems
1. (a) 34 J (b) Yes; the increase in internal energy causes a
slight temperature increase. **3.** 4.90 kJ **5.** (a) 250 J
(b) all three **7.** 5.4 J **9.** 1.16×10^{-3} kWh **11.** 0.153 cal/K
13. 0.50 MJ **15.** 700 m **17.** (a) 581 kcal/K (b) 860 kcal/K
19. 177 kcal **21.** 0.030 kcal/(kg·K) **23.** 0.090 J
25. 27.5 kJ **27.** (a) 180°C (b) 20.9°C
29. 4.83×10^{21} molecules **31.** 36 g **33.** 539 cal/g
35. 730 kcal

37. (a)

Segment	Pressure	Temperature	Phase Change
AB	Decreases	Constant	Liquid to solid
BC	Constant	Increases	Solid to liquid
CD	Decreases	Constant	Liquid to vapor
DE	Constant	Decreases	Vapor to solid

(b) a is the critical point; b is the triple point.
39. 10 g **41.** 46.3 g **43.** 2 g **45.** 250 W
47. (a) 0.12 K/W (b) 2.7×10^{-4} K/W (c) 5.0×10^{-5} K/W
49. 6.67 W/m² **51.** (a) 0.32 W (b) 800 K/m (c) 0.16 W
(d) 0.64 W (e) 64°C **53.** –37°C
55. (a) the skier with the down jacket (b) The person with the
down jacket can stay outside 6.4 times longer. **57.** 105 W
59. 2.24 kW **61.** 4.8×10^{-5} m² **63.** (a) 8.9×10^{-3}°C/s
(b) 9.4 min **65.** 1.70°C/s **67.** (a) 103.81°C (b) 0.565 g/s
69. 0.58 L/h **71.** 0.189 kcal/(kg·K) **73.** 4.0 g
75. (a) 2.4 kcal (b) 350 g **77.** 6.6×10^{-3}°C
79. reduced to 75% of the original

81.

Animal	(a) BMR/kg	(b) BMR/m²
Mouse	210	1200
Dog	51	1000
Human	32	1000
Pig	18	1000
Horse	11	960

(a) It is true. (c) Radiative loss depends upon surface area.

83. 0.23 kg **85.** 7.86 g **87.** 140 m **89.** 22.6°C
91. (a) 7.00 times higher (b) 35.7°C; the dog is a much better regulator of temperature and, as a result, has more endurance.

Chapter 15

Multiple Choice Questions

1. (d) **2.** (d) **3.** (c) **4.** (c) **5.** (a) **6.** (c) **7.** (a) **8.** (c)
9. (c) **10.** (d) **11.** (e) **12.** (b) **13.** (d)

Problems

1. −290 J **3.** 95 J **5.** 101.3 J **7.** (a) 1372 J
(b) $\Delta U = 1216$ J; $Q = 2588$ J **9.** −5.00 kJ; the heat flows out of the gas and into the reservoir. **11.** (a) 3.00 kJ (b) 2.00 kJ
13. (a) 1.2×10^{17} J (b) 1.4×10^{13} kg **15.** 0.182
17. 3.00 kJ **19.** 6.0 kcal **21.** 14 W **23.** The coal-fired plant and the nuclear plant exhaust 0.43 MJ and 0.60 MJ of heat, respectively. **25.** (a) 433 K (b) 233 J **27.** 4.2%

29. 12 kJ **31.** 0.0174 **33.** 110 kJ **35.** 62.7 kW
37. (c), (a), (b) **39.** +1.4 kcal/K **41.** 9.81 cal/K
43. 56.6 cal/K **45.** (a) 100 W (b) 0.3 W/K
47. $1.79k_B$ **49.** (a) 3 heads and 3 tails (b) 6 heads or 6 tails
(c) $\dfrac{20}{64} = \dfrac{5}{16} = 0.3125$ **51.** $k \ln 2$ **53.** 31 kg **55.** 4.84 kJ
57. decreasing the low temperature reservoir
59. (c), (a), (b), (d) **61.** 0.079 J **63.** 24°C
65.

W (J)	e (%)
9.27×10^{-4}	40.0
5.91×10^{-4}	26.5
9.06×10^{-5}	4.67

67. 350 J/K

69. (a) 0.22 cal/K (b) −0.68 cal/K **71.** 15 min
73. (a) 7.61 J/K (b) −5.57 J/K (c) 2.04 J/K

CREDITS

Chapter 1
Opener: ©Lucasfilm, Ltd & ™ All Rights Reserved; p. 4: © Corbis R-F Website; 1.1(1): © Science, Visuals Unlimited; 1.1(2): © Hans Gelderblom/Getty Images; 1.1(3): Photo by Jennifer Merlis; 1.1(4): © Jonathan Blair/CORBIS; 1.1(5): © Vol. 34/PhotoDisc; 1.1(6): © Vol. 56/Corbis; 1.1(7): © Vol. 56/Corbis; 1.2: © Vol. 86/PhotoDisc; 1.4: © Yorgos Nikas/Getty Images.

Chapter 2
Opener: Bernard Photo Productions/Animals/Animals; p. 27 left: AP/Wide World Photos; 2.1: © Eric Meola/Getty Images; p. 29: © Doug Pensinger/Getty Images; 2.5: © David T. Roberts/Photo Researchers, Inc.; p. 33 top: © Vol. 85/Corbis; 2.8: © Steven Walker, Peter Arnold; p. 42: Photo by author R. C. Richardson; p. 56: © Mike Hewitt/Getty Images.

Chapter 3
Opener: Courtesy Al Macdonald; p. 64: © Cosmo Condina/Getty Images; 3.25: © T. Bannor/Custom Medical Stock Photo; p. 89: © Michael S. Yamashita /CORBIS.

Chapter 4
Opener: © John Bova/Photo Researchers; p. 104: © Vol. 35/Corbis; p. 117: © EyeWire R-F Website; 4.20: © Loren M. Winters/Visuals Unlimited; 4.49: Stouffer Productions/Animals/Animals.

Chapter 5
Opener: © Robert Landau/CORBIS; p. 149: © Vol. 29 /PhotoDisc; p. 153: © Vol. 7/Corbis; p. 159: © Chris Sattlberger/Getty Images; p. 165: © Robin Smith/Getty Images; p. 169: © Corbis R-F Website; 5.17: Supplied by Globe Photos; 5.21: © Richard T. Nowitz; 5.22: © MatthewStockman /Allsport; p. 176: © Angelo Hornak/CORBIS.

Chapter 6
Opener: © Mark Newman/PictureQuest; 6.1: © Galen Rowell/CORBIS; p. 189: Gerard.Sioen/FRANCEDIAS.COM; 6.8a: USA Archery; p. 195: NASA; 6.16: © Stephen Dalton/Animals/Animals; p. 199: NASA; 6.19: © Craig Jones/ Allsport; 6.25: © John Sohlden/Visuals Unlimited; 6.38: © James L. Amos/ CORBIS; 6.41: © Corbis R-F Website.

Chapter 7
Opener: Edward William Cooke (1811–80) Private Collection/Bridgeman Art Library; 7.3: © Erica Lansner/Globe Photos; 7.8: © Vol. 56/Corbis; 7.11b: © Loren M. Winters/Visuals Unlimited; 7.11c: © Michael Steele/Allsport; 7.17: © Loren M. Winters /Visuals Unlimited; p. 243 left: © Jeffrey L. Rotman/CORBIS; 7.22: © Doug Armand/Getty Images.

Chapter 8
Opener: © R.W. Jones/CORBIS; 8.5: AP/Wide World Photos; 8.20: © The Frank Lloyd Wright Foundation; 8.27: © Mike Powell/Allsport; 8.33: © Michael Newman/PhotoEdit; p. 276: © Kevin Fleming/CORBIS; 8.37a: Photography by Leah; 8.37b: Photography by Leah; 8.50: © Doug Pensinger/Allsport; 8.74a: © Tony Duffy/Allsport; 8.74b: © Jonathan Daniel/Allsport; 8.77: © Michel Hans, Agence Vandystadt, Paris–France/Allsport.

Chapter 9
Opener: © Corbis R-F Website; p. 309: © Tom Pantages; 9.10: © Susan Van Etten/PhotoEdit; 9.13: © PhotoDisc R-F Website; p. 316: © Corbis R-F Website; 9.14: © Takeshi Takahara/Photo Researchers, Inc.; 9.17: © David Wrobel/Visuals Unlimited; 9.27: © Courtesy NASA (adaptation); 9.28: © Larry Stepanowicz/Visuals Unlimited.

Chapter 10
Opener: © Ladybird Books; 10.1: © 1996 Amoz Eckerson/Visuals Unlimited; 10.5: Courtesy of Tim Zellner, Judd Wire Inc.; 10.6: © Natalie Forbes/CORBIS; p. 347: © John Springer/CORBIS; 10.9a: © Doug Pensinger/Getty Images; 10.9b: © SIU /Visuals Unlimited; 10.15: © Vol. 54/PhotoDisc; 10.28: © Bettmann/ CORBIS; 10.29: © Ray Malace Photo; 10.30: © Wolfgang Kaehler/CORBIS.

Chapter 11
Opener: AP/Wide World Photo; p. 376: © David Young-Wolff/PhotoEdit; 11.6a: © Loren M. Winters/Visuals Unlimited; 11.6b: © Loren M. Winters/Visuals Unlimited; p. 387: U.S. Navy photo by Photographer's Mate 1st Class Spike Call.

Chapter 12
Opener: Bernard Benoit//Science Photo Library/Photo Researchers; p. 406: Photo by Jennifer Merlis; 411: © Steve Kaufman/CORBIS; p. 413: U.S. Navy photo by Photographer's Mate Airman Lawrence Braxton; 12.4: Image courtesy of Berghaus Organ Company; 12.6: © The Picture Source/Terry Oakley; p. 429 top: © Loren M. Winters/Visuals Unlimited; p. 429 bottom: U.S. Navy photo by Ensign John Gay; p. 430: © Merlin D. Tuttle, Bat Conservation International; 12.18: AP/Wide World Photos.

Chapter 13
Opener: © John Eastcott/YVA Momatiuk; 13.6: © John Sohlden/Visuals Unlimited; p. 454: © Vol. 112/Corbis; 13.15a: © Joe McDonald/CORBIS; 13.15b: © Craig Lovell/CORBIS; 13.15c: © Ken Kaminesky/CORBIS.

Chapter 14
Opener: AP/Wide World Photos; 14.6: Photo by author R. C. Richardson; 14.18: © Montrose/Custom Medical Stock.

Chapter 15
Opener: © David Davis/Index Stock Imagery/PictureQuest.

Text Credits

Some examples and problems adapted from Alan H. Cromer, *Physics for the Life Sciences*. Copyright © 1977 by the McGraw-Hill Companies, Inc., New York. All rights reserved. Reprinted by permission of Alan H. Cromer.

Some examples and problems adapted from Alan H. Cromer, Study Guide for *Physics for the Life Sciences*. Copyright © 1977 by the McGraw-Hill Companies, Inc., New York. All rights reserved. Reprinted by permission of Alan H. Cromer.

Some examples and problems adapted from Alan H. Cromer, *Physics for the Life Sciences*, 2d. ed. Copyright © 1994 by the McGraw-Hill Companies, Inc. Primis Custom Publishing, Dubuque, Iowa. All rights reserved. Reprinted by permission of Alan H. Cromer.

Chapter 28: Quote by Richard Feynman from Richard Phillips Feynman, *The Character of Physical Law;* introduction by James Gleick. Copyright © 1994 by Modern Library, New York. All rights reserved. Reprinted by permission of MIT Press.

UNIT CONVERSIONS

Length
1 in = 2.540 cm
1 cm = 0.3937 in
1 ft = 30.48 cm
1 m = 39.37 in = 3.281 ft
1 mi = 5280 ft = 1.609 km
1 km = 0.6214 mi
1 light-year (ly) = 9.461×10^{15} m

Time
1 y = 365.24 d = 3.156×10^{7} s
1 d = 24 h = 1440 min = 8.64×10^{4} s

Speed
1 mi/h = 1.467 ft/s
$\quad$ = 1.609 km/h = 0.4470 m/s
1 km/h = 0.2778 m/s
$\quad$ = 0.6214 mi/h = 0.9113 ft/s
1 ft/s = 0.3048 m/s = 0.6818 mi/h
1 m/s = 3.281 ft/s = 3.600 km/h

Volume
1 liter (L) = 1000 cm^3 = 10^{-3} m^3
1 cm^3 = 0.0610 in^3 = 1 mL
1 m^3 = 1×10^{6} cm^3

Mass
1 kg = 1000 g
1 mass unit (u) = 1.6605×10^{-27} kg
$\quad$ [1 kg weighs 2.20 lb where g = 9.80 m/s^2]

Force
1 N = 0.2248 lb
1 lb = 4.448 N

Energy
1 J = 0.7376 ft·lb = 0.2389 cal
1 ft·lb = 1.36 J = 3.24×10^{-4} kcal
1 kcal = 4.186 kJ
1 Btu = 1055 J
1 kWh = 3.600 MJ = 860.0 kcal
1 eV = 1.602×10^{-19} J

Mass-Energy Equivalents
1 u = 931.5 MeV/c^2
$\quad$ = 1.492×10^{-10} J/c^2

Power
1 W = 1 J/s = 1.341×10^{-3} hp
1 hp = 550 ft·lb/s = 745.7 W

Pressure
1 Pa = 1 N/m^2 = 1.450×10^{-4} lb/in^2
1 atm = 0.1013 MPa = 14.7 lb/in^2
1 lb/in^2 = 6.90×10^{3} N/m^2
1 mm Hg = 1.333×10^{2} Pa
1 in Hg = 3.386×10^{3} Pa

Angle
1 rad = 57.30°
1° = 0.01745 rad
360° = 2π rad
1 rad/s = 9.55 rpm
1 rpm = 0.1047 rad/s

SI PREFIXES

Power	Prefix	Symbol
10^{12}	tera	T
10^{9}	giga	G
10^{6}	mega	M
10^{3}	kilo	k
10^{-1}	deci	d
10^{-2}	centi	c
10^{-3}	milli	m
10^{-6}	micro	μ
10^{-9}	nano	n
10^{-12}	pico	p
10^{-15}	femto	f

SI DERIVED UNITS

Quantity	Units		Equivalents	
Force	newton	N	J/m	$kg \cdot m/s^2$
Energy	joule	J	N·m	$kg \cdot m^2/s^2$
Power	watt	W	J/s	$kg \cdot m^2/s^3$
Pressure	pascal	Pa	N/m^2	$kg/(m \cdot s^2)$
Frequency	hertz	Hz	cycle/s	s^{-1}
Electric charge	coulomb	C		A·s
Electric potential	volt	V	J/C	$kg \cdot m^2/(A \cdot s^3)$
Electric resistance	ohm	Ω	V/A	$kg \cdot m^2/(A^2 \cdot s^3)$
Capacitance	farad	F	C/V	$A^2 \cdot s^4/(kg \cdot m^2)$
Magnetic field	tesla	T	N·s/(C·m)	$kg/(A \cdot s^2)$
Magnetic flux	weber	Wb	T·m^2	$kg \cdot m^2/(A \cdot s^2)$
Inductance	henry	H	V·s/A	$kg \cdot m^2/(A^2 \cdot s^2)$